Fuzzy Sets, Fuzzy Logic and Their Applications 2020

Fuzzy Sets, Fuzzy Logic and Their Applications 2020

Editor

Michael Gr. Voskoglou

MDPI • Basel • Beijing • Wuhan • Barcelona • Belgrade • Manchester • Tokyo • Cluj • Tianjin

Editor
Michael Gr. Voskoglou
University of Peloponnese
Greece

Editorial Office
MDPI
St. Alban-Anlage 66
4052 Basel, Switzerland

This is a reprint of articles from the Special Issue published online in the open access journal *Mathematics* (ISSN 2227-7390) (available at: https://www.mdpi.com/journal/mathematics/special_issues/Fuzzy_Sets2020).

For citation purposes, cite each article independently as indicated on the article page online and as indicated below:

LastName, A.A.; LastName, B.B.; LastName, C.C. Article Title. *Journal Name* **Year**, *Volume Number*, Page Range.

ISBN 978-3-0365-2006-3 (Hbk)
ISBN 978-3-0365-2007-0 (PDF)

Contents

About the Editor

Michael Gr. Voskoglou Professor Emeritus of Mathematical Sciences, School of Technological Applications, Graduate Technological Educational Institute of Western Greece. Visiting Researcher at the National Bulgarian Academy of Sciences in Sofia (1997–2000), Visiting Professor in postgraduate courses at the University of Warsaw (2009), at the University of Applied Sciences of Berlin (2010) and at the National Institute of Technology of Durgapur, India (2016).

He is the author of 16 books and of more than 500 papers published in reputed journals in around 30 countries (Scholar Google 2020: Citations 1524). He is the Editor-in-Chief of the *International Journal of Applications of Fuzzy Sets and Artificial Intelligence*, reviewer of the American Mathematical Society, and member of the Editorial Board of many reputed mathematical journals.

His research interests include algebra, fuzzy logic, Markov chains, artificial intelligence and mathematics education.

Preface to "Fuzzy Sets, Fuzzy Logic and Their Applications 2020"

A few decades ago, probability theory was the unique tool in the hands of experts for handling situations of the uncertainty appearing in problems of science, technology and everyday life. Probability, which is based on principles of the traditional, bivalent logic, is sufficient for tackling problems of uncertainty connected to randomness, but not those connected with imprecise or vague information. However, nowadays, with the development of fuzzy set theory introduced by Zadeh in 1965, and in extension of fuzzy logic, which is based on it, things have changed. In fact, these new mathematical tools have given the scientists the opportunity to model under conditions which are vague or not precisely defined, thus succeeding to solve mathematically problems whose statements are expressed in our natural language.

As a result, the spectre of their applications has been rapidly extended, covering all physical sciences, economics and management, expert systems, like financial planners, diagnostic, meteorological, information-retrieval, control systems, etc., industry, robotics, decision making, programming, medicine, biology, humanities, education, and almost all the other sectors of human activity, including human reasoning itself. It must be mentioned that fuzzy mathematics has also been significantly developed on a theoretical level, providing important insights even to branches of classical mathematics, like algebra, analysis, geometry, etc.

The present book contains the 24 articles accepted for publication among the 66 total manuscripts submitted to the Special Issue "Fuzzy Sets, Fuzzy Logic and Their Applications, 2020" of the MDPI *Mathematics* journal. This Special Issue is a continuity of last year's successful issue on the same subject, which has also been published in the form of a book containing 20 articles.

The 24 articles, which appear in the present book in the series that they have been published in Volumes 8 (2020) and 9 (2021) of the journal, cover a wide range of topics connected to the theory and applications of fuzzy sets and systems, fuzzy logic and their extensions/generalizations. These topics include, among others, elements from fuzzy graphs; fuzzy numbers; fuzzy equations; fuzzy linear spaces; intuitionistic fuzzy sets; soft sets; type-2 fuzzy sets, bipolar fuzzy sets, plithogenic sets, fuzzy decision making, fuzzy governance, fuzzy models in mathematics of finance, a philosophical treatise on the connection of the scientific reasoning with fuzzy logic, etc.

It is hoped that the book will become interesting and useful for those working in the area of fuzzy sets, fuzzy systems and fuzzy logic, as well as for those having the proper mathematical background and willing to become familiar with recent advances in fuzzy mathematics and fuzzy logic. As the Guest Editor of the Special Issue, I am grateful to the authors of the papers for their quality contributions, to the reviewers for their valuable comments towards the improvement of the submitted works, and to the administrative staff of the MDPI publications for the support to complete this project. Special thanks must be given to the Managing Editor of the Special Issue Ms. Grace Du for her excellent collaboration and valuable assistance.

Michael Gr. Voskoglou

Editor

mathematics

MDPI

Article

On Bipolar Fuzzy Gradation of Openness

Subhadip Roy [1], Jeong-Gon Lee [2,*], Syamal Kumar Samanta [3], Anita Pal [1] and Ganeshsree Selvachandran [4]

1 Department of Mathematics, National Institute of Technology Durgapur, Durgapur 713209, West Bengal, India; subhadip_123@yahoo.com (S.R.); anita.buie@gmail.com (A.P.)
2 Division of Applied Mathematics, Wonkwang University, Iksan 54538, Korea
3 Department of Mathematics, Visva Bharati, Santiniketan 731235, West Bengal, India; syamal_123@yahoo.co.in
4 Department of Actuarial Science and Applied Statistics, Faculty of Business and Information Science, UCSI University, Jalan Menara Gading, Cheras, Kuala Lumpur 56000, Malaysia; ganeshsree86@yahoo.com
* Correspondence: jukolee@wku.ac.kr

Received: 2 March 2020; Accepted: 25 March 2020; Published: 2 April 2020

Abstract: The concept of bipolar fuzziness is of relatively recent origin where in addition to the presence of a property, which is done in fuzzy theory, the presence of its counter-property is also taken into consideration. This seems to be much natural and realistic. In this paper, an attempt has been made to incorporate this bipolar fuzziness in topological perspective. This is done by introducing a notion of bipolar gradation of openness and to redefine the bipolar fuzzy topology. Furthermore, a notion of bipolar gradation preserving map is given. A concept of bipolar fuzzy closure operator is also introduced and its characteristic properties are studied. A decomposition theorem involving our bipolar gradation of openness and Chang type bipolar fuzzy topology is established. Finally, some categorical results of bipolar fuzzy topology (both Chang type and in our sense) are proved.

Keywords: bipolar gradation of openness; bipolar gradation of closedness; bipolar fuzzy topology; bipolar gradation preserving map

---

1. Introduction

From the very beginning of the invention of fuzzy sets by Zadeh [1], many authors have contributed towards fuzzifying the topological concept. Fuzzy topology was first introduced by Chang [2] in 1968. Since then, fuzzy topology had drawn the attention of many mathematicians and a foundation of systematic research began. Fuzzy topology, L-fuzzy topology, interval-valued fuzzy topology, and intuitionistic fuzzy topology ([3–6]) laid the foundation of new topological structures on some non-crisp sets. The lack of fuzziness in fuzzy topology was still a drawback to some extent. The Chang fuzzy topology is a crisp family of fuzzy subsets satisfying the properties of topology over some domain. However, a crisp collection never looked good for a proper justification for fuzzifying the topological concept. This absence of fuzziness in Chang fuzzy topology was pointed out by Sostak [7], Ying [8], Chattopadhyay et al. [9], Gregoroi [10], and Mondal [11]. Chattopadhyay et al. [12] introduced a notion of gradation, where every fuzzy set was associated with some grade of openness or closedness. With the concept of gradation of openness, they further studied fuzzy closure operator, gradation preserving maps, fuzzy compactness, and fuzzy connectedness ([9,12,13]). This concept of gradation has been used widely instead of direct fuzzification of some mathematical structures mainly in the field of topology by many researchers. Samanta [14] and Ghanim et al. [15] introduced gradation of uniformity and gradation of proximity, Thakur et al. [16] studied gradation of continuity, and Mondal et al. ([11,17–19]) introduced intuitionistic gradation and L-fuzzy gradation.

Bipolar fuzzy set ($\mathcal{BFS}$), a generalized concept of fuzzy set, has already found its way in the field of research as bipolarity in decisions often occurs in many practical problems. Unlike fuzzy set, the range of membership lies in [−1,1], where the range of membership (0,1] for some element is an indication of the satisfaction of the property, whereas the range of membership [−1,0) is an indication of the satisfaction of the counter-property. Some basic operations on bipolar fuzzy sets can be found in ([20,21]). Applications of bipolar fuzzy sets can be found in ([22–24]). Bipolar fuzzy topology ($\mathcal{BFT}$) studied by Azhagappan et al. [25] and Kim et al [26] are of Chang type. For a universal set X, $\mathcal{BF}(X)$ is the collection of all bipolar fuzzy sets of X and a bipolar fuzzy topology τ on X is a collection from $\mathcal{BF}(X)$ containing the null bipolar fuzzy set, absolute bipolar fuzzy set, finite intersection, and arbitrary union. Thus, for a bipolar fuzzy topological space ($\mathcal{BFTS}$) (X,τ), $\tau^+ = \{\mu^+ \in I^X;\ \mu \in \tau\}$ and $\tau^- = \{-\mu^- \in I^X;\ \mu \in \tau\}$ are fuzzy topologies of Chang type. In addition, (X,τ^+,τ^-) is a fuzzy bitopological space deduced from the bipolar fuzzy topology τ. Therefore, the study on bipolar fuzzy topology looks quite logical in the context of fuzzy topology as fuzzy topology can be considered as a special case of $\mathcal{BFT}$ and a $\mathcal{BFT}$ induces a special type of fuzzy bitopology. However the definition of a $\mathcal{BFTS}$ introduced in [25] looks similar to the definition of Chang fuzzy topological space where the bipolar fuzzy open sets are considered as a crisp collection over some universe. This looks to be a drawback in proper bifuzzification of the topological concept. Fuzzy set is a particular case of bipolar fuzzy set where the counter-property is absent i.e., counter-property takes the value 0 only—for example, "sweet and sour", "good and bad", "beauty and ugly", "matter and anti-matter", etc. By incorporating a bipolar gradation in the openness and closedness, we tried to rectify the previous drawbacks in bifuzzification of topological concept and thus introduce a modified definition of bipolar fuzzy topological space.

In this paper, we introduce a definition of bipolar gradation of openness of bipolar fuzzy subsets of X and give a new definition of bipolar fuzzy topological spaces. In our definition of bipolar fuzzy topology, each bipolar fuzzy subset is associated with a definite *bipolar gradation of openness* and non-openness. We have shown that the set of all bipolar fuzzy topologies in our sense form a complete lattice with an order relation defined in Definition 9. We also introduce *bipolar gradation preserving maps* and a decomposition theorem involving bipolar fuzzy topology in our sense and the same in Chang's sense is proved. Bipolar fuzzy closure operator is introduced and some of their characteristic properties are dealt with. Lastly, it is shown that the bipolar fuzzy topologies in our sense and the bipolar gradation preserving mapping is a topological category.

2. Preliminaries

Throughout the paper, the fuzzy topological space ($\mathcal{FTS}$) is considered in Chang's sense. Gradation of openness, gradation of closedness, and gradation preserving map will be called $\mathcal{GO}$, $\mathcal{GC}$, and $\mathcal{GP}$ map, respectively. Some straightforward proofs are omitted and some preliminary results related to this work are not discussed, which can be found in ([2,25–27]).

Definition 1 ([27]). *Let X be a non-empty set. Then, a pair $\mu = (\mu^-,\mu^+)$ is called a $\mathcal{BFS}$ in X, where $\mu^- : X \to [-1,0]$ and $\mu^+ : X \to [0,1]$ are two mappings. The positive membership function $\mu^+(x)$ denotes the satisfaction degree of an element x corresponding to the $\mathcal{BFS}$ μ and the negative membership function $\mu^-(x)$ denotes the satisfaction degree of an element x to the counter-property corresponding to the $\mathcal{BFS}$ μ. In particular, a $\mathcal{BFS}$ is said to be a null-$\mathcal{BFS}$ [25], denoted by $\tilde{0}$, where $\tilde{0} = (0^-,0^+)$ and $0^-(x) = 0$, $0^+(x) = 0$, for all $x \in X$. A $\mathcal{BFS}$ is said to be an absolute $\mathcal{BFS}$ [25], denoted by $\tilde{1}$, where $\tilde{1} = (1^-,1^+)$ and $1^-(x) = -1$, $1^+(x) = 1$, for all $x \in X$.*

Definition 2 ([27]). *Let X be a non-empty set and $\mu,\lambda \in \mathcal{BF}(X)$.*

(1) *μ is said to be a subset of λ, denoted by $\mu \subset \lambda$, if, for each $x \in X$, $\mu^+(x) \le \lambda^+(x)$ and $\mu^-(x) \ge \lambda^-(x)$.*
(2) *The complement of μ, denoted by $\mu^c = ((\mu^c)^-,(\mu)^c)^+)$, is a bipolar fuzzy set in X, defined as for each $x \in X$, $\mu^c(x) = (-1-\mu^-(x), 1-\mu^+(x))$.*

(3) *The intersection of* μ *and* λ*, denoted by* $\mu \cap \lambda$*, is a bipolar fuzzy set in* X*, defined as for each* $x \in X$, $(\mu \cap \lambda)(x) = (\mu^-(x) \vee \lambda^-(x), \mu^+(x) \wedge \lambda^+(x))$.
(4) *The union of* μ *and* λ*, denoted by* $\mu \cup \lambda$ *is a bipolar fuzzy set in* X*, defined for each* $x \in X$, $(\mu \cup \lambda)(x) = (\mu^-(x) \wedge \lambda^-(x), \mu^+(x) \vee \lambda^+(x))$.

Definition 3 ([25]). *Let* X *be a non-empty set. A collection of bipolar fuzzy subsets* τ *of* $\mathcal{BF}(X)$ *is said to be a* $\mathcal{BFT}$ *on* X*, if it satisfies the following conditions:*

(1) $\tilde{0}, \tilde{1} \in \tau$,
(2) *if* $\mu, \lambda \in \tau$*, then* $\mu \cap \lambda \in \tau$,
(3) *if* $\mu_i \in \tau$*, for each* $i \in \Delta$*, then* $\bigcup_{i\in\Delta} \mu_i \in \tau$.

Definition 4 ([26]). *Let* (X, τ_1) *and* (Y, τ_2) *be two bipolar fuzzy topological spaces. Then, a mapping* $f : (X, \tau_1) \to (Y, \tau_2)$ *is said to be continuous, if* $f^{-1}(V) \in \tau_1$ *for each* $V \in \tau_2$.

Definition 5 ([12]). *Let* X *be a non-empty set and* $\tau : I^X \to [0,1]$ *be a mapping. Then,* τ *is said to be a* $\mathcal{GO}$ *on* X*, if it satisfies the following conditions:*

(1) $\tau(\tilde{0}) = \tau(\tilde{1}) = 1$,
(2) $\tau(\mu_1 \cap \mu_2) \geq \tau(\mu_1) \wedge \tau(\mu_2)$,
(3) $\tau(\bigcup_{i\in\Delta} \mu_i) \geq \bigwedge_{i\in\Delta} \tau(\mu_i)$.

Definition 6 ([12]). *Let* X *be a non-empty set and* $\mathfrak{F} : I^X \to [0,1]$ *be a mapping. Then,* $\mathfrak{F}$ *is said to be a* $\mathcal{GC}$ *on* X*, if it satisfies the following conditions:*

(1) $\mathfrak{F}(\tilde{0}) = \mathfrak{F}(\tilde{1}) = 1$,
(2) $\mathfrak{F}(\mu_1 \cup \mu_2) \geq \mathfrak{F}(\mu_1) \wedge \mathfrak{F}(\mu_2)$,
(3) $\mathfrak{F}(\bigcap_{i\in\Delta} \mu_i) \geq \bigwedge_{i\in\Delta} \mathfrak{F}(\mu_i)$.

Remark 1 ([12]). The set of all $\mathcal{FTS}$ on X along with the order relation "$\leq$" forms a complete lattice.

Definition 7 ([12]). *Let* (X, τ) *and* (Y, τ') *be two* $\mathcal{FTS}$ *and* $f : X \to Y$ *be a mapping. Then,* f *is said to be a* $\mathcal{GP}$ *map if for each* $\mu \in I^Y$, $\tau'(\mu) \leq \tau(f^{-1}(\mu))$

Definition 8 ([13]). *Let* $(X, \mathfrak{F})$ *be a* $\mathcal{FTS}$ *with* $\mathfrak{F}$ *being a* $\mathcal{GC}$ *on* X*. For each* $r \in [0,1]$ *and for each* $\lambda \in I^X$*, the fuzzy closure of* λ *is defined as follows:*

$$cl(\lambda, r) = \cap\{\mu \in I^X : \mu \supseteq \lambda, \mathfrak{F}(\mu) \geq r\}.$$

3. Bipolar Gradation of Openness

In this section, we define bipolar gradation of openness (bipolar $\mathcal{GO}$), bipolar gradation of closedness (bipolar $\mathcal{GC}$) and prove some subsequent results.

Definition 9. *For any* $(r_1, s_1), (r_2, s_2) \in [-1, 0] \times [0, 1]$*, and for* $\{(r_i, s_i), i \in \Delta\}$*, define*

(1) $(r_1, s_1) \succeq (r_2, s_2)$ *if* $r_1 \leq r_2$ *and* $s_1 \geq s_2$,
(2) $(r_1, s_1) \succ (r_2, s_2)$ *if* $r_1 < r_2$ *and* $s_1 > s_2$,
(3) $(r_1, s_1) \prec (r_2, s_2)$ *if* $r_1 > r_2$ *and* $s_1 < s_2$,
(4) $\bigwedge_{i\in\Delta} (r_i, s_i) = (\bigvee_{i\in\Delta} r_i, \bigwedge_{i\in\Delta} s_i)$,
(5) $\bigvee_{i\in\Delta} (r_i, s_i) = (\bigwedge_{i\in\Delta} r_i, \bigvee_{i\in\Delta} s_i)$

Definition 10. *Let X be a non-empty set. Then, a mapping $\tau : \mathcal{BF}(X) \rightarrow [-1,0] \times [0,1]$ is said to be a bipolar $\mathcal{GO}$ on X, if it satisfies the following properties:*

(1) $\tau(\tilde{0}) = \tau(\tilde{1}) = (-1,1)$,
(2) $\tau(\mu_1 \cap \mu_2) \succeq \tau(\mu_1) \wedge \tau(\mu_2)$,
(3) $\tau(\bigcap_{i\in\Delta} \mu_i) \succeq \bigwedge_{i\in\Delta} \tau(\mu_i)$.

Example 1. *Let $X = \mathbb{R}$ be the set of all real numbers. Let T be the usual topology on $\mathbb{R}$ and T' be the topology generated by $\mathfrak{B} = \{(a,b] \ : \ a < b\}$. For $A \subseteq \mathbb{R}$ let χ_A denote the characteristic function of A. Define $\chi_A^* = (-\chi_A, \chi_A)$. Define a mapping $\tau : \mathcal{BF}(X) \rightarrow [-1,0] \times [0,1]$ by for each $\chi_A^* \in \mathcal{BF}(X)$,*

$$\tau(\chi_A^*) = \begin{cases} (-1,1) & \text{if } A \in T \\ (-\frac{1}{2}, \frac{1}{2}) & \text{if } A \in T' \backslash T \\ (0,0) & \text{otherwise.} \end{cases}$$

Then, τ is a bipolar $\mathcal{GO}$ on X.

Definition 11. *A mapping $\mathfrak{F} : \mathcal{BF}(X) \rightarrow [-1,0] \times [0,1]$ is said to be a bipolar $\mathcal{GC}$, if it satisfies the following properties:*

(1) $\mathfrak{F}(\tilde{0}) = \mathfrak{F}(\tilde{1}) = (-1,1)$,
(2) $\mathfrak{F}(\mu_1 \cup \mu_2) \succeq \mathfrak{F}(\mu_1) \wedge \mathfrak{F}(\mu_2)$,
(3) $\mathfrak{F}(\bigcap_{i\in\Delta} \mu_i) \succeq \bigwedge_{i\in\Delta} \mathfrak{F}(\mu_i)$.

Proposition 1. *Let τ be a bipolar $\mathcal{GO}$ on X. Then, a mapping $\mathfrak{F}_\tau : \mathcal{BF}(X) \rightarrow [-1,0] \times [0,1]$ defined by $\mathfrak{F}_\tau(\mu) = \tau(\mu^c)$, for all $\mu \in \mathcal{BF}(X)$, is a bipolar $\mathcal{GC}$ on X.*

Proof. We have $\mathfrak{F}_\tau(\tilde{0}) = \tau((\tilde{0})^c) = \tau(\tilde{1}) = (-1,1)$. Similarly, $\mathfrak{F}_\tau(\tilde{1}) = (-1,1)$.

$$\begin{aligned} \mathfrak{F}_\tau(\mu_1 \cup \mu_2) &= \tau((\mu_1 \cup \mu_2)^c) \\ &= \tau(\mu_1^c \cap \mu_2^c) \\ &\succeq \tau(\mu_1^c) \wedge \tau(\mu_2^c) \\ &= \mathfrak{F}_\tau(\mu_1) \wedge \mathfrak{F}_\tau(\mu_2), \end{aligned}$$

$$\begin{aligned} \mathfrak{F}_\tau(\bigcap_{i\in\Delta} \mu_i) &= \tau((\bigcap_{i\in\Delta} \mu_i)^c) \\ &= (\tau(\bigcup_{i\in\Delta} \mu_i^c)) \\ &\succeq \bigwedge_{i\in\Delta} \tau(\mu_i^c) \\ &= \bigwedge_{i\in\Delta} \mathfrak{F}_\tau(\mu_i). \end{aligned}$$

Consequently, the proof completes.

For a mapping $f : \mathcal{BF}(X) \rightarrow [-1,0] \times [0,1]$, let $f^- = \pi_1 \circ f$ and $f^+ = \pi_2 \circ f$. Then, f is a *bipolar* $\mathcal{GO}$, ($\mathcal{GC}$) iff $f^+, -f^-$ are $\mathcal{GO}$, ($\mathcal{GC}$) on X. □

Proposition 2. *Let $\mathfrak{F}$ be a bipolar $\mathcal{GC}$ on X. Then, a mapping $\tau_{\mathfrak{F}} : \mathcal{BF}(X) \rightarrow [-1,0] \times [0,1]$ defined by $\tau_{\mathfrak{F}}(\mu) = \mathfrak{F}(\mu^c)$, for all $\mu \in \mathcal{BF}(X)$, is a bipolar $\mathcal{GO}$ on X.*

Definition 12. *Let $\{\tau_k \ : \ k \in \Delta\}$ be a family of bipolar $\mathcal{GO}$ on X. Then, $\tau = \bigcap_{k\in\Delta} \tau_k$ is defined as, $\tau(\mu) = \bigwedge_{k\in\Delta} \tau_k(\mu)$.*

Proposition 3. *Arbitrary intersection of a family of bipolar $\mathcal{GO}$ is a bipolar $\mathcal{GO}$.*

Proof. Suppose that $\{\tau_k : k \in \Delta\}$ is a family of *bipolar* $\mathcal{GO}$ on X and $\tau = \underset{k\in\Delta}{\cap}\tau_k$. Clearly, we have $\tau(\tilde{0}) = \tau = (\tilde{1}) = (-1,1)$:

$$\begin{aligned}
\tau(\mu_1 \cap \mu_2) &= \underset{k\in\Delta}{\cap}\tau_k(\mu_1 \cap \mu_2)\\
&\succeq \underset{k\in\Delta}{\cap}(\tau_k(\mu_1) \wedge \tau_k(\mu_2))\\
&\succeq \underset{k\in\Delta}{\cap}\tau_k(\mu_1) \wedge \underset{k\in\Delta}{\cap}\tau_k(\mu_2)\\
&= \tau(\mu_1) \wedge \tau(\mu_2)
\end{aligned}$$

and

$$\begin{aligned}
\tau(\underset{i}{\cup}\mu_i) &= \underset{k}{\cap}\tau_k(\underset{i}{\cup}\mu_i)\\
&\succeq \underset{k}{\cap}\underset{i}{\wedge}\tau_k(\mu_i)\\
&= \underset{i}{\wedge}\underset{k}{\cap}\tau_k(\mu_i)\\
&= \underset{i}{\wedge}\tau(\mu_i).
\end{aligned}$$

Hence, τ is a *bipolar* $\mathcal{GO}$ on X. □

Remark 2. *Let X be a non-empty set. Define* $\tau_\circ, \tau_1 : \mathcal{BF}(X) \to [-1,0] \times [0,1]$ *by* $\tau_\circ(\tilde{0}) = \tau_\circ(\tilde{1}) = (-1,1)$, $\tau_\circ(\mu) = (0,0)$, *for all* $\mu \in \mathcal{BF}(X)\backslash\{\tilde{0},\tilde{1}\}$ *and* $\tau_1(\mu) = (-1,1)$, $\forall\mu \in \mathcal{BF}(X)$. *Then,* $\tau_\circ, \tau_1$ *are bipolar* $\mathcal{GO}$ *on X such that, for any bipolar* $\mathcal{GO}$ τ *on X,* $\tau_1 \succeq \tau \succeq \tau_\circ$ *i.e for any* $\mu \in \mathcal{BF}(X)$, $\tau_1(\mu) \succeq \tau(\mu) \succeq \tau_\circ(\mu)$.

Proposition 4. *Let* $\mathcal{M}_{\mathcal{BF}}(X)$ *denote the collection of all bipolar* $\mathcal{GO}$ *on X. Then,* $(\mathcal{M}_{\mathcal{BF}}(X), \succeq)$ *is a complete lattice.*

The proof follows from Proposition 3 and Remark 2.

Proposition 5. *Let* (X,τ) *be a* $\mathcal{BFTS}$, *where* τ *is a bipolar* $\mathcal{GO}$ *on X. Then, for each* $(r,s) \in [-1,0] \times [0,1]$, $\tau_{r,s} = \{\mu \in \mathcal{BF}(X) : \tau(\mu) \succeq (r,s)\}$ *is a is a Chang type* $\mathcal{BFT}$ *on X.*

Proof. We have $\tau(\tilde{0}) = \tau(\tilde{1}) = (-1,1) \succeq (r,s)$, for all $(r,s) \in [-1,0] \times [0,1]$. Therefore, we get $\tilde{0}, \tilde{1} \in \tau_{r,s}$. Let $\mu_1, \mu_2 \in \tau_{r,s}$. Then, we have

$$\begin{aligned}
&\tau(\mu_1) \succeq (r,s) \text{ and } \tau(\mu_2) \succeq (r,s)\\
&\tau(\mu_1 \cap \mu_2) \succeq \tau(\mu_1) \wedge \tau(\mu_2) \succeq (r,s) \wedge (r,s) = (r,s).
\end{aligned}$$

Hence, we obtain $\mu_1 \cap \mu_2 \in \tau_{r,s}$. Similarly, it can be shown that $\tau_{r,s}$ is closed under arbitrary union. Therefore, for each $(r,s) \in [-1,0] \times [0,1]$, $\tau_{r,s}$ is a Chang type $\mathcal{BFT}$ on X. □

Definition 13. *For each* $(r,s) \in [-1,0] \times [0,1]$, $\tau_{r,s}$ *is called the* $(r\text{-}s)$*-th level* $\mathcal{BFT}$ *on X with respect to the bipolar* $\mathcal{GO}$ τ.

Definition 14. *The family* $\{\tau_{r,s} : (r,s) \in [-1,0] \times [0,1]\}$ *is said to be a descending family if any* $(r_1,r_2) \succeq (s_1,s_2)$ *implies* $\tau_{r_1,r_2} \subset \tau_{s_1,s_2}$.

Proposition 6. *Let* (X,τ) *be a* $\mathcal{BFTS}$, *where* τ *is a bipolar* $\mathcal{GO}$ *on X and* $\{\tau_{r,s} : (r,s) \in [-1,0] \times [0,1]\}$ *be the family of all* $(r\text{-}s)$*-th level* $\mathcal{BFT}$ *on X with respect to the bipolar* $\mathcal{GO}$ τ. *Then, this family is descending family and and for each* $(r_1,r_2) \in [-1,0] \times [0,1]$,

$$\tau_{r_1,r_2} = \underset{(r_1,r_2)\succ(s_1,s_2)}{\cap} \tau_{s_1,s_2}.$$

Proof. Clearly, if $(r_1, r_2) \succeq (s_1, s_2)$, then $\tau_{r_1,r_2} \subset \tau_{s_1,s_2}$. Hence, $\{\tau_{r,s} : (r,s) \in [-1,0] \times [0,1]\}$ is a descending family of $\mathcal{BFT}$s on X.

Obviously, $\tau_{r_1,r_2} \subseteq \bigcap_{(r_1,r_2)\succ(s_1,s_2)} \tau_{s_1,s_2}$.

Next, let $\mu \in \cap\, \tau_{s_1,s_2}$, $\forall (r_1, r_2) \succ (s_1, s_2)$. Then, $\tau(\mu) \succeq (s_1, s_2)$, $\forall (r_1, r_2) \succ (s_1, s_2)$. Then, $\tau(\mu) \succeq \vee\{(s_1, s_2);\ (r_1, r_2) \succ (s_1, s_2)\} \Rightarrow \tau(\mu) \succeq (r_1, r_2) \Rightarrow \mu \in \tau_{r_1,r_2}$. Therefore, $\bigcap_{(r_1,r_2)\succ(s_1,s_2)} \tau_{s_1,s_2} \subseteq \tau_{r_1,r_2}$.

Hence, $\tau_{r_1,r_2} = \bigcap_{(r_1,r_2)\succ(s_1,s_2)} \tau_{s_1,s_2}$. □

Proposition 7. *Let* $\{T_{r,s} : (r,s) \in [-1,0] \times [0,1] \backslash \{(0,0)\}$ *be a non-empty descending family of Chang type* $\mathcal{BFT}$*s on X. Let* $\tau : \mathcal{BF}(X) \to [-1,0] \times [0,1]$ *be a mapping defined by* $\tau(\mu) = \vee\{(r,s) \in [-1,0] \times [0,1] \backslash \{(0,0)\};\ \mu \in T_{r,s}\}$. *Then,* τ *is a bipolar* $\mathcal{GO}$ *on X. Furthermore, if, for any* $(r_1, r_2) \in [-1,0] \times [0,1] \backslash \{(0,0)\}$

$$T_{r_1,r_2} = \bigcap_{(r_1,r_2)\succ(s_1,s_2)} T_{s_1,s_2}, \tag{1}$$

then $\tau_{r,s} = T_{r,s}$ *holds for all* $(r,s) \in [-1,0] \times [0,1] \backslash \{(0,0)\}$.

Proof. From the definition of τ, it is clear that $\tau(\tilde{0}) = \tau(\tilde{1}) = (-1, 1)$. Let $\mu_1, \mu_2 \in \mathcal{BF}(X)$ and let $\tau(\mu_i) = (l_i, k_i)$, $i = 1, 2$. If $(l_i, k_i) = (0, 0)$ for some i, then $\tau(\mu_1 \cap \mu_2) \succeq \tau(\mu_1) \wedge \tau(\mu_2)$. Without loss of generality, suppose $l_i < 0$ and $k_i > 0$. Let $l_i \leq s_1$ and $k_i \geq s_2$, $i = 1, 2$. Then, for any $\epsilon > 0$ with $l_i + \epsilon > 0$, there exist $r_1, r_2 \in [-1, 0)$ and $t_1, t_2 \in (0, 1]$ such that $\mu_i \in T_{r_i,t_i}$ and $l_i \leq r_i < l_i + \epsilon$ and $k_i - \epsilon < t_i \leq k_i$ and $k_i - \epsilon > 0$ for $i = 1, 2$. Now, let

$$\begin{aligned} r &= \max\{r_1, r_2\}, \quad l = \max\{l_1, l_2\}, \\ t &= \min\{t_1, t_2\}, \quad k = \min\{k_1, k_2\}. \end{aligned}$$

Then, $\mu_1 \cap \mu_2 \in T_{r,t}$ implies that $\tau(\mu_1 \cap \mu_2) \succeq (r, t) \succeq (l + \epsilon,\ k - \epsilon)$. Since $\epsilon > 0$ is arbitrary, it follows that $\tau(\mu_1, \mu_2) \succeq \tau(\mu_1) \wedge \tau(\mu_2)$.

Let $\mu_i \in \mathcal{BF}(X)$, for all $i \in \Delta$. Suppose that $\tau(\mu_i) = (l_i, k_i)$, for all $i \in \Delta$. Let $l = \bigvee_{i\in\Delta} l_i$, $k = \bigwedge_{i\in\Delta} k_i$. W.l.o.g, suppose $l < 0$ and $k > 0$. Let $\epsilon > 0$ be any number such that $k > \epsilon$ and $l + \epsilon < 0$. Then, $0 < k - \epsilon < k_i$ and $l + \epsilon > l_i$ for all $i \in \Delta$. Therefore, we have $\mu_i \in T_{l+\epsilon,\, k-\epsilon}$, for all $i \in \Delta$. Then, $\tau(\bigcup_{i\in\Delta} \mu_i) \succeq (l + \epsilon,\ k - \epsilon)$. Since $\epsilon > 0$ is arbitrary, it follows that $\tau(\bigcup_{i\in\Delta} \mu_i) \succeq (l,\ k)$. This implies that τ is a bipolar $\mathcal{GO}$ on X.

In order to show the next part, assume that $\{T_{r,s} : (r,s) \in [-1,0] \times [0,1] \backslash \{(0,0)\}$ satisfies the condition (1). Let $\mu \in T_{r_1,r_2}$. Then, $\tau(\mu) \succeq (r_1, r_2)$, so $\mu \in \tau_{r_1,r_2}$ and, consequently, $T_{r_1,r_2} \subset \tau_{r_1,r_2}$. Next, suppose that $\mu \in \tau_{r_1,r_2}$. Then, $\tau(\mu) \succeq (r_1, r_2)$. Let $\wedge\{l : \mu \in T_{l,k}\} = s_1 \leq r_1$ and $\vee\{k : \mu \in T_{l,k}\} = s_2 \geq r_2$. If $r_1 = 0$, $r_2 > 0$, then, for $\epsilon > 0$ with $r_2 - \epsilon > 0$, $\mu \in T_{r_1,r_2-\epsilon}$. Since $\epsilon > 0$ is arbitrary, $\mu \in \bigcap_{\epsilon>0} T_{r_1,r_2-\epsilon} = T_{r_1,r_2}$. Similarly, other cases can be dealt with. Thus, $\tau_{r,s} = T_{r,s}$ □

Remark 3. *The family* $\{\tau_{r,s} : (r,s) \in [-1,0] \times [0,1] \backslash \{(0,0)\}\}$ *of Proposition 7 is called the family of* $\mathcal{BFT}$*s associated with the bipolar* $\mathcal{GO}$, τ.

Remark 4. *Two bipolar* $\mathcal{GO}$ τ *and* τ' *on X is equal iff* $\tau_{r,s} = \tau'_{r,s}$, *for all* $(r,s) \in [-1,0] \times [0,1] \backslash \{(0,0)\}$.

Proposition 8. *Let* (X, T) *be a Chang type* $\mathcal{BFTS}$. *For each* $(r,s) \in [-1,0] \times [0,1] \backslash (0,0)$, *define a mapping* $T^{r,s} : \mathcal{BF}(X) \to [-1,0] \times [0,1]$ *by the rule*

$$T^{r,s}(\mu) = \begin{cases} (-1,1) & if\ \mu = \tilde{0}, \tilde{1} \\ (r,s) & if\ \mu \in T \backslash \{\tilde{0}, \tilde{1}\} \\ (0,0) & otherwise. \end{cases}$$

Then, $T^{r,s}$ is a bipolar $\mathcal{GO}$ on X such that $(T^{r,s})_{r,s} = T$

Definition 15. *Let T be a Chang type $\mathcal{BFT}$ on X; then, $T^{r,s}$ is called an (r-s)-th bipolar $\mathcal{GO}$ on X and $(X, T^{r,s})$ is called the (r-s)-th graded $\mathcal{BFTS}$.*

4. Bipolar Gradation Preserving Mapping

In a bipolar fuzzy setting, the continuity concept of a mapping is formulated in this section by introducing bipolar gradation preserving maps. Some of its properties are also studied.

Definition 16. *Let (X, τ) and (Y, τ') be two $\mathcal{BFTS}$s, where τ and τ' are bipolar $\mathcal{GO}$ on X and Y, respectively, and $f : X \to Y$ be a mapping. Then, f is called a bipolar gradation preserving map (bipolar $\mathcal{GP}$ map) if, for each $\mu \in \mathcal{BF}(Y)$, $\tau(f^{-1}(\mu)) \succeq \tau'(\mu)$.*

In the following Proposition, a relation between *bipolar gradation preserving* property with the continuity for a mapping over bipolar fuzzy topological spaces is established.

Proposition 9. *Let (X, τ) and (Y, τ') be two $\mathcal{BFTS}$s, where τ and τ' are bipolar $\mathcal{GO}$ on X and Y, respectively. Then, a mapping $f : X \to Y$ is a bipolar $\mathcal{GP}$ map iff $f : (X, \tau_{r,s}) \to (Y, \tau'_{r,s})$ is continuous for all $(r, s) \in [-1, 0] \times [0, 1] \backslash \{(0, 0)\}$.*

Proof. Suppose that f is a bipolar $\mathcal{GP}$ map and $\mu \in \tau'_{r,s}$. Then, $\tau'(\mu) \succeq (r, s)$. Since f is a bipolar $\mathcal{GP}$ map, it follows that $\tau(f^{-1}(\mu)) \preceq \tau'(\mu) \succeq (r, s)$. Hence, we get $f^{-1}(\mu) \in \tau_{r,s}$. Thus, $f : (X, \tau_{r,s}) \to (Y, \tau'_{r,s})$ is continuous for all $(r, s) \in [-1, 0] \times [0, 1] \backslash \{(0, 0)\}$.

Conversely, suppose that f is continuous for all $(r, s) \in [-1, 0] \times [0, 1] \backslash \{(0, 0)\}$. Let $\mu \in \mathcal{BF}(Y)$. If $\tau'(\mu) = (0, 0)$, then $\tau(f^{-1}(\mu)) \succeq \tau'(\mu)$. Let $\tau'(\mu) = (r, s)$, where $(r, s) \in [-1, 0] \times [0, 1] \backslash \{(0, 0)\}$. Then, $\mu \in \tau'_{r,s}$. Since f is continuous, it follows that $f^{-1}(\mu) \in \tau_{r,s}$. This implies that $\tau(f^{-1}(\mu)) \succeq (r, s) = \tau'(\mu)$. Consequently, f is a bipolar $\mathcal{GP}$ map. □

Proposition 10. *Let (X, T) and (Y, T') be two Chang type $\mathcal{BFTS}$s and $f : X \to Y$ be a mapping. Then, f is continuous iff $f : (X, T^{r,s}) \to (Y, (T')^{r,s})$ is a bipolar $\mathcal{GP}$ map for all $(r, s) \in [-1, 0] \times [0, 1] \backslash \{(0, 0)\}$.*

Proof. Suppose that $f : (X, T) \to (Y, T')$ is continuous. Take $\mu \in \mathcal{BF}(Y)$. Then, we have the following possibilities:

Case (1) If $\mu = \tilde{0}$ or $\tilde{1}$, then $f^{-1}(\tilde{0}) = \tilde{0}$ and $f^{-1}(\tilde{1}) = \tilde{1}$ and hence $(T^{r,s})(f^{-1}(\mu)) \succeq (T')^{r,s}(\mu)$.

Case (2) If $\mu \in T'$, then $(T')^{r,s}(\mu) = (r, s)$. By continuity of $f : (X, T) \to (Y, T')$, $f^{-1}(\mu) \in T$. Therefore, we get $(T^{r,s})(f^{-1}(\mu)) = (r, s)$. Thus, $(T^{r,s})(f^{-1}(\mu)) \succeq (T')^{r,s}(\mu)$.

Case (3) If $\mu \notin T'$, then $(T')^{r,s}(\mu) = (0, 0)$ and so $(T^{r,s})(f^{-1}(\mu)) \succeq (T')^{r,s}(\mu)$. Hence, $f : (X, T^{r,s}) \to (Y, (T')^{r,s})$ is a *bipolar $\mathcal{GP}$ map*.

The converse follows from Propositions 8 and 9. □

Proposition 11. *Let (X, τ), (Y, τ'), (Z, τ'') be three $\mathcal{BFTS}$s, where τ, τ', τ'' are bipolar $\mathcal{GO}$ on X, Y and Z respectively. If $f : (X, \tau) \to (Y, \tau')$ and $g : (Y, \tau') \to (Z, \tau'')$ are bipolar $\mathcal{GP}$ map, then $g \circ f : (X, \tau) \to (Z, \tau'')$ is a bipolar $\mathcal{GP}$ map.*

Proposition 12. *Let (X, τ) be a $\mathcal{BFTS}$ and $f : X \to Y$ be a mapping. Let $\{\tau'_{r,s} : (r, s) \in [-1, 0] \times [0, 1] \backslash \{(0, 0)\}\}$ be a descending family of Chang type $\mathcal{BFTS}$s on Y. Let τ' be the bipolar $\mathcal{GO}$ generated by this family. Suppose that, for each $(r, s) \in [-1, 0] \times [0, 1] \backslash \{(0, 0)\}$, $\mathfrak{B}_{r,s}$ be the base and $\xi_{r,s}$ be the subbase of $\tau'_{r,s}$. Then,*

(1) $f : (X, \tau) \to (Y, \tau')$ is a bipolar $\mathcal{GP}$ map iff $\tau(f^{-1}(\mu)) \succeq (r, s)$, for all $\mu \in \tau'_{r,s}$ and $(r, s) \in [-1, 0] \times [0, 1] \backslash \{(0, 0)\}$.

(2) $f : (X, \tau) \to (Y, \tau')$ is a bipolar $\mathcal{GP}$ map iff $\tau(f^{-1}(\mu)) \succeq (r,s)$, for all $\mu \in \mathfrak{B}_{r,s}$ and $(r,s) \in [-1,0] \times [0,1] \setminus \{(0,0)\}$.

(3) $f : (X, \tau) \to (Y, \tau')$ is a bipolar $\mathcal{GP}$ map iff $\tau(f^{-1}(\mu)) \succeq (r,s)$, for all $\mu \in \zeta_{r,s}$ and $(r,s) \in [-1,0] \times [0,1] \setminus \{(0,0)\}$.

5. Bipolar Fuzzy Closure Operator

A concept of bipolar fuzzy closure operator is introduced in this section and its characteristic properties are studied. As in the classical case of Kuratowski's closure operator, here it is shown that the bipolar fuzzy topology and the bipolar $\mathcal{GP}$ map are completely characterized by a bipolar fuzzy closure operator.

Let $(X, \mathfrak{F})$ be a $\mathcal{BFTS}$, where $\mathfrak{F}$ is a bipolar $\mathcal{GC}$ on X. For each $(r,s) \in [-1,0] \times [0,1] \setminus \{(0,0)\}$ and for $\lambda \in \mathcal{BF}(X)$, the $(r$-$s)$-th graded bipolar fuzzy closure ($\mathcal{BFC}$) of λ is defined by

$$Cl(\lambda, (r,s)) = \cap \{\mu \in \mathcal{BF}(X) \ : \ \mu \supseteq \lambda, \ \mathfrak{F}(\mu) \succeq (r,s)\}.$$

Proposition 13. *Let $(X, \mathfrak{F})$ be a $\mathcal{BFTS}$, where $\mathfrak{F}$ is a bipolar $\mathcal{GC}$ on X and let $Cl : \mathcal{BF}(X) \times [-1,0] \times [0,1] \setminus \{(0,0)\} \to \mathcal{BF}(X)$ be a $\mathcal{BFC}$ operator on $(X, \mathfrak{F})$. Then,*

(1) $Cl(\tilde{0}, (r,s)) = \tilde{0}$, $Cl(\tilde{1}, (r,s)) = \tilde{1}$, *for all* $(r,s) \in [-1,0] \times [0,1] \setminus \{(0,0)\}$.
(2) $\lambda \subseteq Cl(\lambda, (r,s))$, *for all* $\lambda \in \mathcal{BF}(X)$.
(3) $Cl(\lambda, (r_1, s_1)) \subseteq Cl(\lambda, (r_2, s_2))$ *if* $(r_2, s_2) \succeq (r_1, s_1)$.
(4) $Cl(\lambda_1 \cup \lambda_2, (r,s)) = Cl(\lambda_1, (r,s)) \cup Cl(\lambda_2, (r,s))$, *for all* $(r,s) \in [-1,0] \times [0,1] \setminus \{(0,0)\}$.
(5) $Cl(Cl(\lambda, (r,s)), (r,s)) = Cl(\lambda, (r,s))$, *for all* $(r,s) \in [-1,0] \times [0,1] \setminus \{(0,0)\}$.
(6) *If* $(r,s) = \vee_{i \in \Delta} \{(r_i, s_i);\ Cl(\lambda, (r_i, s_i)) = \lambda\}$, *then* $Cl(\lambda, (r,s)) = \lambda$.

Proposition 14. *Let $Cl : \mathcal{BF}(X) \times [-1,0] \times [0,1] \setminus \{(0,0)\} \to \mathcal{BF}(X)$ be a mapping satisfying $(1) - (4)$ of Proposition 13. Let $\mathfrak{F} : \mathcal{BF}(X) \to [-1,0] \times [0,1]$ be a mapping defined by $\mathfrak{F}(\lambda) = \vee\{(r,s);\ Cl(\lambda, (r,s)) = \lambda\}$ then $\mathfrak{F}$ is a bipolar $\mathcal{GC}$ on X. Again, $Cl = Cl_{\mathfrak{F}}$ iff the conditions (5) and (6) of Proposition 13 are satisfied by Cl.*

Proof. Clearly, $\mathfrak{F}(\tilde{0}) = \mathfrak{F}(\tilde{1}) = (-1, 1)$ by (1).

Let $\lambda_1, \lambda_2 \in \mathcal{BF}(X)$ and $\mathfrak{F}(\lambda_1) = (l_1, k_1)$, $\mathfrak{F}(\lambda_2) = (l_2, k_2)$. For $\epsilon > 0$, $\exists\ (r_i, s_i) \in [-1,0] \times [0,1] \setminus \{(0,0)\}$ such that $l_i \leq r_i < l_i + \epsilon$, $k_i - \epsilon < s_i \leq k_i$ and $Cl(\lambda_i, (r_i, s_i)) = \lambda_i$, $i = 1,2$. Let $r = r_1 \vee r_2$, $s = s_1 \wedge s_2$. Then, $(r,s) \preceq (r_i, s_i)$, $i = 1,2$ and hence $Cl(\lambda_1 \cup \lambda_2, (r,s)) = Cl(\lambda_1, (r,s)) \cup Cl(\lambda_2, (r,s)) = \lambda_1 \cup \lambda_2$ (By (iii)). Hence, $Cl(\lambda_1 \cup \lambda_2, (r,s)) = \lambda_1 \cup \lambda_2$. Thus, $\mathfrak{F}(\lambda_1 \cup \lambda_2) \succeq (r,s) \succeq (r_1, s_1) \wedge (r_2, s_2) \succeq (l_1 \vee l_2 + \epsilon, k_1 \wedge k_2 - \epsilon)$. Since $\epsilon > 0$ is arbitrary, $\mathfrak{F}(\lambda_1 \cup \lambda_2) \succeq (l_1 \vee l_2, k_1 \wedge k_2) = (l_1, k_1) \wedge (l_2, k_2) = \mathfrak{F}(\lambda_1) \wedge \mathfrak{F}(\lambda_2)$.

Let $\lambda_i \in \mathcal{BF}(X)$ and $\mathfrak{F}(\lambda_i) = (a_i, b_i)$, $\wedge_{i \in \Delta} \mathfrak{F}(\lambda_i) = (l,k)$ for all $i \in \Delta$ for all $i \in \Delta$. Without loss of generality, assume that $(l,k) \neq (0,0)$. For $\epsilon > 0$, $\exists\ (r_i, s_i) \in [-1,0] \times [0,1] \setminus \{(0,0)\}$ with $a_i \leq r_i < a_i + \epsilon$, $b_i - \epsilon < s_i \leq b_i$ such that $Cl(\lambda_i, (r_i, s_i)) = \lambda_i$, $\forall i \in \Delta$ and $(\vee_{i \in \Delta} r_i, \wedge_{i \in \Delta} s_i) \neq (0,0)$. Let $r = \vee_{i \in \Delta} r_i$, $s = \wedge_{i \in \Delta} s_i$. Then, $Cl(\lambda_i, (r,s)) = \lambda_i$, $\forall i \in \Delta$ (since $(r_i, s_i) \succeq (r,s)$, $i \in \Delta$). Thus, $Cl(\cap_{i \in \Delta} \lambda_i, (r,s)) \subset Cl(\lambda_i, (r,s)) = \lambda_i$, $\forall i \in \Delta$ (by (iv)) and hence $Cl(\cap_{i \in \Delta} \lambda_i, (r,s)) = \cap_{i \in \Delta} \lambda_i$. Thus, $\mathfrak{F}(\cap_{i \in \Delta} \lambda_i) \succeq (r,s) \succeq (l + \epsilon, k - \epsilon)$, since $\epsilon > 0$ is arbitrary $\mathfrak{F}(\cap_{i \in \Delta} \lambda_i) \succeq (l,k) \succeq \wedge_{i \in \Delta} \mathfrak{F}(\lambda_i)$.

In order to prove the next part, first suppose that Cl satisfies the conditions (1)–(6) of Proposition 13. Then,

$$\begin{aligned} & Cl_{\mathfrak{F}}(\lambda,(r,s)) \\ & = \cap\{\mu \supseteq \lambda : \mathfrak{F}(\mu) \succeq (r,s)\} \\ & = \cap\{\mu \supseteq \lambda : \bigvee_{i\in\Delta}\{(r_i,s_i);\ Cl(\mu,(r_i,s_i)) = \mu\} \succeq (r,s)\} \\ & = \cap\{\mu \supseteq \lambda : \forall\epsilon > 0,\ Cl(\mu,(r+\epsilon,s-\epsilon)) = \mu\} \\ & \subseteq Cl(\lambda,(r,s)). \end{aligned}$$

Again, by (2) $\lambda \subseteq Cl(\lambda,(r,s))$ and $Cl(\lambda,(r,s)) = Cl\Big(Cl(\lambda,(r,s)),(r+\epsilon,s-\epsilon)\Big)$ (by (2), (3), and (5)). Again, $Cl(\mu,(r+\epsilon,s-\epsilon)) = \mu \supseteq \lambda$, for all $\epsilon > 0$, implies, by (6), $\mu = Cl(\mu,(r,s)) \supseteq Cl(\lambda,(r,s))$. Thus,

$$Cl_{\mathfrak{F}}(\lambda,(r,s)) = \cap\{\mu \supseteq \lambda\ :\ \forall\epsilon > 0, Cl(\mu,(r+\epsilon,s-\epsilon)) = \mu\} \supseteq Cl(\lambda,(r,s)).$$

Therefore, we conclude that $Cl_{\mathfrak{F}}(\lambda,(r,s)) = Cl(\lambda,(r,s))$.

Next, suppose that $Cl_{\mathfrak{F}}(\lambda,(r,s)) = Cl(\lambda,(r,s))$ holds $\forall\lambda \in \mathcal{BF}(X)$. Since $Cl_{\mathfrak{F}}$ is the $\mathcal{BFC}$ operator generated by the bipolar $\mathcal{GC}$ $\mathfrak{F}$, it follows that $Cl_{\mathfrak{F}}$ satisfies conditions (1)–(6) of Proposition 13. Thus, by assumption, Cl also satisfies conditions (1)–(6) of Proposition 13. This completes the proof. □

Remark 5. *It can be easily verified that, if $Cl : \mathcal{BF}(X) \times [-1,0] \times [0,1]\backslash\{(0,0) \to \mathcal{BF}(X)$ is a $\mathcal{BFC}$ operator on X, then, for each $(r,s) \in [-1,0] \times [0,1]\backslash\{(0,0)\}$, $Cl_{r,s} : \mathcal{BF}(X) \to \mathcal{BF}(X)$ defined by $Cl_{r,s}(\lambda) = Cl(\lambda,(r,s))$ is a $\mathcal{BFC}$ operator of Chang type.*

Proposition 15. *Let (X,τ) be a Chang type $\mathcal{BFTS}$. Then, $Cl : \mathcal{BF}(X) \times [-1,0] \times [0,1]\backslash\{(0,0)\} \to \mathcal{BF}(X)$ is a $\mathcal{BFC}$ operator iff $Cl_{r,s} : \mathcal{BF}(X) \to \mathcal{BF}(X)$ is a Chang type $\mathcal{BFC}$ operator for the Chang type $\mathcal{BFTS}$ $(X,\tau_{r,s})$ for all $(r,s) \in [-1,0] \times [0,1]\backslash\{(0,0)\}$.*

Proof. Clearly, if Cl is a $\mathcal{BFC}$ operator for the $\mathcal{BFTS}$ (X,τ), then $Cl_{r,s}$ is a Chang type $\mathcal{BFC}$ operator for all $(r,s) \in [-1,0] \times [0,1]\backslash\{(0,0)\}$.

Conversely, suppose that $Cl_{r,s}$ is a Chang type $\mathcal{BFC}$ operator for the Chang type $\mathcal{BFTS}$ $(X,\tau_{r,s})$ for all $(r,s) \in [-1,0] \times [0,1]\backslash\{(0,0)\}$. Thus, the conditions (1), (2), (4), and (5) of Proposition 13 are satisfied. If $(r_1,s_1) \succeq (r_2,s_2)$, then, $\tau_{r_1,s_1} \subseteq \tau_{r_2,s_2}$. Therefore, condition (3) of Proposition 13 is satisfied. In order to prove condition (6), suppose that

$$(r,s) = \vee\{(u,v);\ Cl(\lambda,(u,v)) = \lambda.$$

Then, $\lambda^c \in \tau_{r+\epsilon,s-\epsilon}$ for all $\epsilon > 0$. Thus, we have $\lambda^c \in \bigcap_{\epsilon>0}\tau_{r+\epsilon,s-\epsilon}$, i.e., $\lambda^c \in \tau_{r,s}$. Therefore, we have $\lambda \in \mathfrak{F}_{r,s}$ and hence we conclude that $Cl(\lambda,(r,s)) = \lambda$. This completes the proof. □

Proposition 16. *Let $f : (X,\tau) \to (Y,\tau')$ be a mapping between two $\mathcal{BFTS}$s. Then, f is a bipolar $\mathcal{GP}$ map iff $f(Cl(\lambda,(r,s))) \subseteq Cl(f(\lambda),(r,s))$.*

Proof. By Proposition 9, f is a *bipolar $\mathcal{GP}$ map* iff $f : (X,\tau_{r,s}) \to (Y,\tau'_{r,s})$ is continuous for all $(r,s) \in [-1,0] \times [0,1]\backslash\{(0,0)\}$ iff $f(Cl(\lambda,(r,s))) \subseteq Cl(f(\lambda),(r,s))$. □

6. Category of Bipolar Fuzzy Topology

In this section, categorical behavior of bipolar fuzzy topological spaces is studied.

Let $\mathcal{C}_{\mathcal{BFT}}$ denote the category of all Chang type $\mathcal{BFTS}$s and continuous functions; $\mathcal{F}_{Top}$ denotes the category of all $\mathcal{BFTS}$s and bipolar $\mathcal{GP}$ maps in our sense; for each $(r,s) \in [-1,0] \times [0,1]\backslash\{(0,0)\}$, $\mathcal{F}^{r,s}_{Top}$ denotes the category of $(r\text{-}s)$-th graded $\mathcal{BFTS}$s and *bipolar $\mathcal{GP}$ maps*.

Proposition 17.

(1) $\mathcal{F}^{r,s}_{Top}$ *is a full subcategory of* $\mathcal{F}_{Top}$.
(2) *For each* $(r,s) \in [-1,0] \times [0,1] \backslash \{(0,0)\}$, $C_{\mathcal{BFT}}$ *and* $\mathcal{F}^{r,s}_{Top}$ *are isometric.*
(3) $\mathcal{F}^{r,s}_{Top}$ *is a bireflective full subcategory of* $\mathcal{F}_{Top}$.

Proof. The first two results follow from the facts: $(\tau_{r,s})^{r,s} = \tau$ if τ is a (r-s)-th *bipolar* $\mathcal{GO}$; $(T^{r,s})_{r,s} = T$ if T is a Chang type $\mathcal{BFT}$ and $f : (X, T) \to (Y, T')$ is continuous w.r.t the Chang type $\mathcal{BFT}$ iff $f : (X, T^{r,s}) \to (Y, (T')^{r,s})$ is a *bipolar* $\mathcal{GP}$ *map*, for all $(r,s) \in [-1,0] \times [0,1] \backslash \{(0,0)\}$. To prove (3), let us take a member (X, τ) of $\mathcal{F}_{Top}$. Then, for each $(r,s) \in [-1,0] \times [0,1] \backslash \{(0,0)\}$, $(X, (\tau_{r,s})^{r,s})$ is a $\mathcal{F}^{r,s}_{Top}$ member and also $I_X : (X, \tau) \to (X, (\tau_{r,s})^{r,s})$ is a *bipolar* $\mathcal{GP}$ *map*. Let (Y, τ') be a member of $\mathcal{F}^{r,s}_{Top}$ and $f : (X, \tau) \to (Y, \tau')$ be a *bipolar* $\mathcal{GP}$ *map*. Now, we only need to check whether $f : (X, (\tau_{r,s})^{r,s}) \to (Y, \tau')$ is a *bipolar* $\mathcal{GP}$ *map*. If $\mu = \tilde{0}$, then $\tau(f^{-1}(\tilde{0})) = \tau'(\tilde{0})$. Then, $(\tau_{r,s})^{r,s}(f^{-1}(\tilde{0})) = (\tau_{r,s})^{r,s}(\tilde{0}) \succeq \tau'(\tilde{0})$. Similarly, $(\tau_{r,s})^{r,s}(f^{-1}(\tilde{1})) \succeq \tau'(\tilde{1})$. If $\tau'(\mu) = (0,0)$, then, obviously $(\tau_{r,s})^{r,s}(f^{-1}(\mu)) \succeq \tau'(\mu)$. Let $\tau'(\mu) = (r,s)$. Then, $\tau(f^{-1}(\mu)) \succeq \tau'(\mu) \Rightarrow f^{-1}(\mu) \in \tau_{r,s}$. Then, $(\tau_{r,s})^{r,s}(f^{-1}(\mu)) \succeq (r,s) = \tau'(\mu)$. Thus, $f : (X, (\tau_{r,s})^{r,s}) \to (Y, \tau')$ is a *bipolar* $\mathcal{GP}$ *map*. □

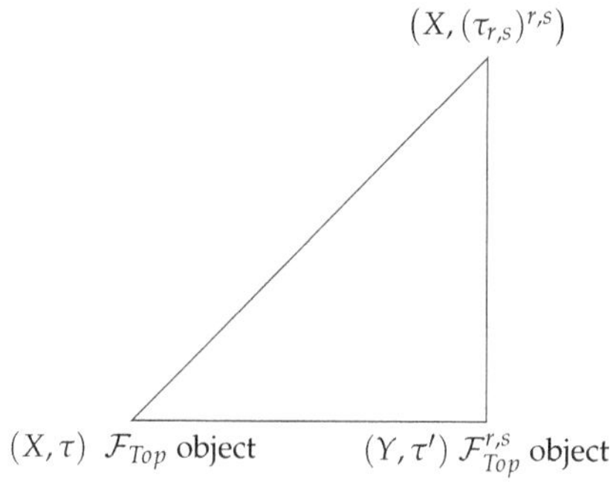

Remark 6. *From (2), (3) in Proposition 17* $C_{\mathcal{BFT}}$ *may be called a bireflective full subcategory of* $\mathcal{F}_{Top}$.

Proposition 18. *Let* $\{(X_i, \tau_i') : i \in \Delta\}$ *be a family of* $\mathcal{BFTS}$s *and* X *be a set such that* $f : X \to X_i$ *is a map for each* $i \in \Delta$. *Then, there exists a bipolar* $\mathcal{GO}$ τ *on* X *such that the following condition holds:*

(1) *for each* $i \in \Delta$, $f_i : (X, \tau) \to (X_i, \tau_i')$ *is a bipolar* $\mathcal{GP}$ *map.*
(2) *If* (Z, τ'') *is a* $\mathcal{BFTS}$, *then* $g : (Z, \tau'') \to (X, \tau)$ *is a bipolar* $\mathcal{GP}$ *map iff* $f_i \circ g$ *is a bipolar* $\mathcal{GP}$ *map for each* $i \in \Delta$.

Proof. (1) For each $(r,s) \in [-1,0] \times [0,1] \backslash \{(0,0)\}$ and for each $i \in \Delta$, we define

$$T_i^{r,s} = \{f_i^{-1}(\mu) \ : \ \mu \in (\tau_i')_{r,s}\},$$

where $(\tau_i')_{r,s} = \{\mu \in \mathcal{BF}(X_i) \ : \ \tau_i'(\mu) \succeq (r,s)\}$ is the (r-s)-th level $\mathcal{BFT}$ on X_i w.r.t τ_i'. It can be shown that $T_i^{r,s}$ is a $\mathcal{BFT}$ on X. Clearly, $\{T_i^{r,s} \ : \ (r,s) \in [-1,0] \times [0,1] \backslash \{(0,0)\}\}$ is a descending family. For each $(r,s) \in [-1,0] \times [0,1] \backslash \{(0,0)\}$, we define

$$\prod_{r,s} = \bigcup_{j \in \Delta} T_i^{r,s}.$$

Let $T_{r,s}$ be the $\mathcal{BFT}$ on X generated by $\prod_{r,s}$ as a subbase. Then, $\{T_{r,s} \ : \ (r,s) \in [-1,0] \times [0,1] \backslash \{(0,0)\}\}$ is a descending family. Then, there exists a *bipolar* $\mathcal{GO}$ τ on X associated with the family $\{T_{r,s} \ : \ (r,s) \in [-1,0] \times [0,1] \backslash \{(0,0)\}\}$, where $\tau(\mu) = \vee\{(r,s) \in [-1,0] \times [0,1] \backslash \{(0,0)\}; \ \mu \in T_{r,s}\}$.

First, we show that for each $i \in \Delta$, $f_i : (X, \tau) \to (X_i, \tau_i')$ is a *bipolar* $\mathcal{GP}$ *map*. Let $\mu \in \mathcal{BF}(X_i)$ and $\tau_i'(\mu) = (r,s)$, where $(r,s) \succ (0,0)$. Then, $f_i^{-1}(\mu) \in T_i^{r,s} \subset \prod_{r,s} \subset T_{r,s}$. Thus, $\tau(f_i^{-1}(\mu)) \succeq (r,s) = \tau_i'(\mu)$. Consequently, $f_i : (X, \tau) \to (X_i, \tau_i')$ is a *bipolar* $\mathcal{GP}$ *map*.

(2) If $g : (Z, \tau'')$ is a *bipolar* $\mathcal{GP}$ *map* and since, for each $i \in \Delta$, $f_i : (X, \tau) \to (X_i, \tau_i')$ is a *bipolar* $\mathcal{GP}$ *map*, by Proposition 11, the composition of two *bipolar* $\mathcal{GP}$ *map* $f_i \circ g$ is a *bipolar* $\mathcal{GP}$ *map* for each $i \in \Delta$.

Conversely, we have to show that $g : (Z, \tau'') \to (X, \tau)$ is a *bipolar* $\mathcal{GP}$ *map*. Let $(r,s) \in [-1,0] \times [0,1] \backslash \{(0,0)\}$ and $\mu \in \xi_{r,s}$. Then, $\mu \in T_i^{r,s}$ for some $i \in \Delta$. Then, there exists $\lambda \in (\tau_i')_{r,s}$ such that $f_i^{-1}(\lambda) = \mu$. Since $f_i \circ g$ is a *bipolar* $\mathcal{GP}$ *map* for each $i \in \Delta$, it follows that

$$\tau''((f_i \circ g)^{-1}(\lambda)) \succeq (r,s) \Rightarrow \tau''(g^{-1}(f_i^{-1}(\lambda)) \succeq (r,s) \Rightarrow \tau''(g^{-1}(\mu)) \succeq (r,s).$$

Hence, the result follows from Proposition 12. □

7. Conclusions

The notion of a bipolar fuzzy set is a generalization of a fuzzy set in the sense that a fuzzy set describes some property in a graded manner from its existence to its non existence by assigning values from 1 to 0, whereas a bipolar fuzzy set describes the same from the existence to the reverse existence through non-existence by taking values from 1 to –1 through 0. In this article, this idea of bipolarity is formalized in the topological sense by introducing a concept of *bipolar gradation of openness* to redefine bipolar fuzzy topology. Consequently, we introduce bipolar $\mathcal{GO}$ and bipolar $\mathcal{GC}$ and studied their properties. The relation between Chang type $\mathcal{BFT}$ and $\mathcal{BFT}$ in our sense is established successfully. The bipolar $\mathcal{GP}$ map and bipolar $\mathcal{FC}$ operator are studied. In addition, we have shown that the Chang type $\mathcal{BFT}$ and continuous function is a bireflective full subcategory of the topological category of $\mathcal{BFT}$ and bipolar $\mathcal{GP}$ maps in our sense. In the upcoming papers, we will study various topological properties including the compactness and connectedness in this setting.

Author Contributions: All authors have contributed equally to this paper in all aspects. All authors have read and agreed to the published version of the manuscript.

Funding: This research was supported by a Basic Science Research Program through the National Research Foundation of Korea (NRF) funded by the Ministry of Education (2018R1D1A1B07049321).

Conflicts of Interest: The authors declare no conflict of interest.

References

1. Zadeh, L.A. Fuzzy sets. *Inf. Control* **1965**, *8*, 338–353. [CrossRef]
2. Chang, C.L. Fuzzy topological spaces. *J. Math. Anal. Appl.* **1968**, *24*, 182–190. [CrossRef]
3. Hutton, B. Normality in fuzzy topological spaces. *J. Math. Anal. Appl.* **1975**, *50*, 74–79. [CrossRef]
4. Mondal, T.K.; Samanta, S.K. Connectedness in topology of interval-valued fuzzy sets. *Italian J. Pure Appl. Math.* **2005**, *18*, 33–50.
5. Mondal, T.K.; Samanta, S.K. Topology of interval-valued fuzzy sets. *Indian J. Pure Appl. Math.* **1999**, *30*, 23–38.
6. Zhang, J.; Shi, F.-G.; Zheng, C.-Y. On L-fuzzy topological spaces. *Fuzzy Sets Syst.* **2005**, *149*, 473–484. [CrossRef]
7. Sostak, A.P. On a fuzzy topological structure. In Proceedings of the 13th Winter School on Abstract Analysis, Section of Topology, Supp. Rend. Circ. Mathematical Palermo (Series II), Srni, Czech Republic, 5–12 January 1985; Volume 11, pp. 89–103.
8. Pao-Ming, P.; Ying-Ming, L. Fuzzy topology I, Neighbourhood structure of a fuzzy point and Moore-Smith convergence. *J. Math. Anal. Appl.* **1980**, *76*, 571–599. [CrossRef]
9. Hazra, R.N.; Samanta, S.K.; Chattopadhyay, K.C. Fuzzy topology redefined. *Fuzzy Sets Syst.* **1992**, *45*, 79–82. [CrossRef]
10. Gregori, V.; Vidal, A. Gradation of openness and Chang's fuzzy topologies. *Fuzzy Sets Syst.* **1996**, *109*, 233–244. [CrossRef]
11. Mondal, T.K.; Samanta, S.K. L-fuzzy gradation of openness and L-fuzzy gradation of proximity. *J. Korea Soc. Math. Educ. Ser. B Pure Appl. Math.* **2006**, *13*, 71–94.

12. Chattopadhyay, K.C.; Hazra, R.N.; Samanta, S.K. Gradation of openness: fuzzy topology. *Fuzzy Sets Syst.* **1992**, *49*, 237–242. [CrossRef]
13. Chattopadhyay, K.C.; Samanta, S.K. Fuzzy topology: Fuzzy closure operator, fuzzy compactness and fuzzy connectedness. *Fuzzy Sets Syst.* **1993**, *54*, 207–212. [CrossRef]
14. Samanta, S.K. Fuzzy proximities and fuzzy uniformities. *Fuzzy Sets Syst.* **1995**, *70*, 97–105. [CrossRef]
15. Ghanim, M.H.; Tantawy, O.A.; Selim, M.F. Gradation of uniformity and gradation of proximity. *Fuzzy Sets Syst.* **1996**, *79*, 373–382. [CrossRef]
16. Thakur, R.; Mondal, K.K.; Samanta, S.K. Gradation of continuity in fuzzy topological spaces. *Iranian J. Fuzzy Syst.* **2011**, *8* 143–159.
17. Mondal, T.K.; Samanta, S.K. On intuitionistic gradation of openness. *Fuzzy Sets Syst.* **2002**, *131*, 323–336. [CrossRef]
18. Mondal, T.K.; Samanta, S.K. Topology of interval-valued intuitionistic fuzzy sets. *Fuzzy Sets Syst.* **2001**, *119*, 483–494. [CrossRef]
19. Mondal, T.K.; Samanta, S.K. Intuitionistic gradation of openness: intuitionistic fuzzy topology. *Busefal* **1997**, *73*, 8–17.
20. Lee, K.M. Bipolar-valued fuzzy sets and their basic operations. In Proceedings of the International Conference on Intelligent Technologies, Bangkok, Thailand, 13–15 December 2000; pp. 307–312.
21. Zhang, W.-R. Bipolar fuzzy sets and relations: A computational framework forcognitive modeling and multiagent decision analysis. In Proceedings of the NAFIPS/IFIS/NASA'94 First, International Joint Conference of The North American Fuzzy Information Processing Society Biannual Conference, The Industrial Fuzzy Control and Intellige, San Antonio, TX, USA, 18–21 December 1994; pp. 305–309.
22. Jun, Y.B.; Park, C.H. Filters of BCH-algebras based on bipolar-valued fuzzy sets. *Int. Math. Forum.* **2009**, *4*, 631–643.
23. Lee, J.G.; Hur, K. Bipolar Fuzzy Relations. *Mathematics* **2019**, *7*, 1044. [CrossRef]
24. Akram, M. Bipolar fuzzy graphs. *Inform. Sci.* **2011**, *181*, 5548–5564. [CrossRef]
25. Azhagappan, A.; Kamaraj, M. Notes on bipolar valued fuzzy RW-closed and bipolar valued fuzzy RW-open sets in bipolar valued fuzzy topological spaces. *Int. J. Math. Arch.* **2016**, *7*, 30–36.
26. Kim, J.; Samanta, S.K.; Lim, P.K.; Lee, J.G.; Hur, K. Bipolar fuzzy topological spaces. *Ann. Fuzzy Math. Inform.* **2019**, *17*, 205–229. [CrossRef]
27. Lee, K.M. Comparison of interval-valued fuzzy sets, intuitionistic fuzzy sets and bipolar-valued fuzzy sets. *J. Fuzzy Logic Intell. Syst.* **2004**, *14*, 125–129.

mathematics

MDPI

Article

An Evolving Partial Consensus Fuzzy Collaborative Forecasting Approach

Tin-Chih Toly Chen [1], Yu-Cheng Wang [2,*] and Chin-Hau Huang [1]

1 Department of Industrial Engineering and Management, National Chiao Tung University, 1001, University Road, Hsinchu 300, Taiwan; tolychen@ms37.hinet.net (T.-C.T.C.); sasa76130@hotmail.com (C.-H.H.)

2 Department of Aeronautical Engineering, Chaoyang University of Technology, Taichung 41349, Taiwan

* Correspondence: tony.cobra@msa.hinet.net

Received: 7 March 2020; Accepted: 8 April 2020; Published: 10 April 2020

Abstract: Current fuzzy collaborative forecasting methods have rarely considered how to determine the appropriate number of experts to optimize forecasting performance. Therefore, this study proposes an evolving partial-consensus fuzzy collaborative forecasting approach to address this issue. In the proposed approach, experts apply various fuzzy forecasting methods to forecast the same target, and the partial consensus fuzzy intersection operator, rather than the prevalent fuzzy intersection operator, is applied to aggregate the fuzzy forecasts by experts. Meaningful information can be determined by observing partial consensus fuzzy intersection changes as the number of experts varies, including the appropriate number of experts. We applied the evolving partial-consensus fuzzy collaborative forecasting approach to forecasting dynamic random access memory product yield with real data. The proposed approach forecasting performance surpassed current fuzzy collaborative forecasting that considered overall consensus, and it increased forecasting accuracy 13% in terms of mean absolute percentage error.

Keywords: fuzzy collaborative forecasting; dynamic random access memory; partial consensus; fuzzy intersection

1. Introduction

Fuzzy collaborative forecasting combines fuzzy forecasting and collaborative intelligence [1]. Multiple experts apply fuzzy forecasting methods to forecast the same target and collaborate by consulting each other's forecast, subsequently modifying fuzzy forecasting method settings or forecasts [2]. In contrast with conventional forecasting methods that focus on maximizing forecasting accuracy, fuzzy collaborative forecasting methods attempt to optimize both forecasting precision and accuracy [3,4].

This paper proposes an evolving partial consensus fuzzy collaborative forecasting approach to enhance forecasting effectiveness for dynamic random access memory (DRAM) product yield. Most current fuzzy collaborative forecasting methods apply a fuzzy intersection (FI) to aggregate expert fuzzy forecasts [5]. Though this treatment effectively elevates forecasting precision in terms of the average range of fuzzy forecasts, it has a number of drawbacks as follows.

(1) The FI result usually covers a very narrow range. Though this improves forecasting precision for training data, the probability of missing test values increases [6].
(2) The FI result becomes the null set when there is no overall consensus among experts [7].

To overcome these drawbacks, a consensus among some experts, rather than all experts, can be sought instead. Chen [5] proposed the partial consensus FI (PCFI) operator, which can be non-null set if some experts can achieve a (partial) consensus. The PCFI also usually covers a wider range than the FI, reducing the possibility of missing a test value [8]. However, determining the appropriate number of experts to optimize forecasting performance remains an issue [9]. Therefore, this paper proposes an evolving partial consensus fuzzy collaborative forecasting approach, where multiple experts apply various fuzzy forecasting methods to forecast the same target, and the PCFI operator is employed to aggregate the forecasts. The appropriate number of experts can be determined by observing PCFI changes as the number of experts varies. Therefore, we propose the concept of an evolving PCFI (EPCFI) diagram. An important EPCFI diagram function is to determine the appropriate number of experts for a fuzzy collaborative forecasting task, which is critical for fuzzy group decision making [10–12]. The proposed evolving partial-consensus fuzzy collaborative forecasting approach is based on the EPCFI diagram.

Table 1 summarizes the differences between the proposed methodology and some current methods, and the specific contributions from the proposed methodology are as follows.

1. A systematic procedure is established to determine the appropriate number of experts for fuzzy collaborative forecasting.
2. The EPCFI diagram concept is introduced to analyze aggregation changes as the number of experts varies.
3. Experts are no longer forced to modify their fuzzy forecasts when an overall consensus cannot be achieved.

Table 1. Proposed and current collaborative fuzzy forecasting methods. FLR: fuzzy linear regression; PCFI: partial consensus fuzzy intersection; evolving PCFI; and ANN: artificial neural network.

Method	Forecast Source	Number of Sources	Forecasting Method	Aggregation Mechanism	Modification Mechanism
Chen [5]	Experts	Fixed	FLR	PCFI	Subjective modification
Zarandi et al. [13]	Agents	Fixed	Fuzzy inference rules	Weighted average	Genetic algorithm
Swaroop et al. [14]	Rules	Fixed	Fuzzy inference rules	Fuzzy union	ANN
Proposed	Experts	Dynamic	FLR	EPCFI	Not required

The remainder of this paper is organized as follows. Section 2 briefly reviews relevant previous studies, and Section 3 describes some models to fit fuzzy linear regressions (FLRs). Section 4 details the proposed evolving partial consensus fuzzy collaborative forecasting approach, and Section 5 presents experimental results applying the proposed approach to forecast the DRAM product yield. Section 6 summarizes and concludes the paper, and it discusses some topics for future investigation.

2. Literature Review

Cheikhrouhou et al. [15] built an autoregressive integrated moving average (ARIMA) model to forecast polyethylene bag demand. Experts subsequently judged unexpected future event effects on demand, and these became inputs to a Mamdani's fuzzy inference system (FIS) [16] to modify demand forecasts. However, FISs, including Mamdani's, Sugeno's [17], and adaptive network based FISs (ANFISs) [18] apply fuzzy rules that cannot guarantee test value inclusion in the corresponding fuzzy forecasts [2]. Consequently, a fuzzy union (S-norm) must be applied to aggregate the forecast results by using fuzzy inference rules to avoid missing a test value, which widens the fuzzy forecast range and sacrifices forecasting precision. Swaroop et al. [14] established an FIS to forecast the load on a power system, and this FIS became an input to an artificial neural network (ANN) to tune the load forecast.

Chen [19] proposed a fuzzy collaborative forecasting method in which each expert fitted an FLR to predict the effective cost per die for DRAM product. Fuzzy parameter values for the FLR were derived by solving various nonlinear programming problems. Thus, all test values were included in the corresponding fuzzy cost forecasts, at least for the training data. A fuzzy intersection, or the minimum T-norm, was then applied to aggregate expert fuzzy cost forecasts, which optimized forecasting precision in terms of an average fuzzy cost forecast range. A back propagation network (BPN) was subsequently constructed to defuzzify the aggregation result, optimizing forecasting accuracy measured as a root mean squared error (RMSE). Similar fuzzy collaborative forecasting methods have been subsequently proposed to forecast global CO_2 concentration [5], job cycle time [20], long-term load [21], and DRAM product unit cost [22].

Zhang et al. [23] and Ostrosi et al. [24] proposed an aggregation mechanism that minimized the sum of the squared differences between expert forecasts and the aggregation. However, the aggregation mechanism assumed that a consensus existed and directly derived the aggregation result. Gao et al. [25] aggregated expert fuzzy forecasts with a fuzzy weighted average, and then they measured the distance between each fuzzy forecast and the aggregation. Experts whose fuzzy forecast distance exceeded the same threshold were asked to modify their forecasts, and those that refused to do so were downweighted until the distance between each fuzzy forecast and the aggregation result was sufficiently small. However, fuzzy forecasts with heavier weights were usually closer to the aggregation result, even when they differed from other expert's forecasts, and these were not downweighted. Furthermore, calculating the fuzzy distance and deciding the appropriate threshold required additional subjective decisions, and sometimes consensus still could not be achieved after collaboration.

Herrera-Viedma et al. [26] argued that not only did the consensus represent the group's common perception of some values, it also represented the process to reach the consensus. After collaboration, experts directly modified fuzzy forecasting parameters or fuzzy forecasts to close the gap between each other. However, it still may not be possible to achieve a consensus in this process, i.e., expert fuzzy forecasts may not overlap. Chen [27] proposed a heterogeneous fuzzy collaborative forecasting method to predict semiconductor product yield, where experts fitted the yield learning process of the product with FLR by solving mathematical programming problems or training ANNs. Zarandi et al. [13] proposed a four-layer fuzzy multiagent system to forecast next-day stock prices based on collaboration among software agents. Chen and Wang [28] and Chen and Romanowski [29] proposed software agents, rather than real experts, for fuzzy collaborative forecasting to expedite collaboration. However, software agents usually follow pre-specified rules when fuzzy parameters need to be adjusted, which may result in unrealistic fuzzy forecasts.

3. Preliminary Models for Fitting a Fuzzy Linear Regression

Table 2 summarizes the abbreviations used throughout this paper; we use the following parameters and variables in the proposed methodology.

(1) (+): fuzzy addition.
(2) $\tilde{a}_i$: $i = 0 \sim m$: FLR coefficients.
(3) d: acceptable fuzzy forecast range.
(4) $\tilde{I}^{H/K}$: fuzzy intersection function.
(5) N: normalization function.
(6) $o \in R^+$: expert sensitivity to uncertainty in a fuzzy forecast (smaller values imply less sensitivity, and large values imply more sensitivity).
(7) $s \in [0, 1]$: required satisfaction level.
(8) s_j: satisfaction level at period j.
(9) $w \in R^+$: expert sensitivity to satisfaction level improvement (smaller values imply less sensitivity).
(10) x_{ji}; $i = 0 \sim m, j = 1 \sim n$: value for decision variable i forecasting y_j.
(11) y_j; $j = 1 \sim n$: test value at period j.
(12) $\tilde{y}_j$; $j = 1 \sim n$: fuzzy forecast at period j.
(13) $\mu_{\tilde{y}_j}$: membership function for $\tilde{y}_j$.

Table 2. Abbreviations used in this paper.

Acronym	Meaning
ANFIS	Adaptive network based fuzzy inference system
ANN	Artificial neural network
ARIMA	Autoregressive integrated moving average
BPN	Back propagation network
COG	Center-of-gravity
DRAM	Dynamic random access memory
EPCFI	Evolving partial-consensus fuzzy intersection
FI	Fuzzy intersection
FIS	Fuzzy inference system
FLR	Fuzzy linear regression
GA	Genetic algorithm
GD	Gradient descent
LP	Linear programming
MAE	Mean absolute error
MAPE	Mean absolute percentage error
NLP	Nonlinear programming
PCFI	Partial consensus fuzzy intersection
QP	Quadratic programming
RMSE	Root mean squared error
TFN	Triangular fuzzy number

Without a loss of generality, all fuzzy parameters and variables in the proposed methodology are given as or approximated with triangular fuzzy numbers (TFNs). We also assumed that all experts apply the following FLR to forecast the same target y based on decision variable values $\{x_i\}$ [19,30]:

$$\tilde{y}_j = \tilde{a}_0(+)\sum_{i=1}^{m} \tilde{a}_i x_{ji} \tag{1}$$

However, the proposed methodology could also be extended using other fuzzy forecasting methods.

Several mathematical models have been proposed to derive fuzzy parameter values in Equation (1). For example, Tanaka and Watada [31] proposed a linear programming method to minimize the fuzzy forecast sum of the ranges (or spreads), hence maximizing forecast precision. Taheri and Kelkinnama [32] solved another linear programming problem to minimize the sum of absolute errors. Peters [33] proposed a quadratic programming (QP) method to maximize average satisfaction level and, hence, improve forecast accuracy.

The simultaneous optimization of forecasting accuracy and precision has been pursued by many researchers, but it is somewhat challenging. Donoso et al. [34] proposed a compromise approach by minimizing the weighted sum of the sum of squared deviations between fuzzy forecast cores, actual values, and the sum of the squared ranges.

Chen and Lin [35] incorporated expert opinions into the Tanaka and Watada model and the Peters model, proposing two nonlinear programming (NLP) models, as follows.

NLP Model I:

$$\text{Min } Z_1 = \sum_{j=1}^{n} (y_{j3} - y_{j1})^o, \tag{2}$$

which is subject to

$$y_j \geq y_{j1} + s(y_{j2} - y_{j1}), \tag{3}$$

$$y_j \leq y_{j3} + s(y_{j2} - y_{j3}), \tag{4}$$

$$y_{j1} = a_{01} + \sum_{i=1}^{m} a_{i1} x_{ji}, \tag{5}$$

$$y_{j2} = a_{02} + \sum_{i=1}^{m} a_{i2}x_{ji}, \tag{6}$$

$$y_{j3} = a_{03} + \sum_{i=1}^{m} a_{i3}x_{ji}, \tag{7}$$

$$y_{j1} \leq y_{j2} \leq y_{j3}, \tag{8}$$

and

$$a_{i1} \leq a_{i2} \leq a_{i3}, \tag{9}$$

where $j = 1, 2, \ldots, n$ and $i = 0, 1, \ldots, m$. The objective function minimizes the high order sum of the fuzzy forecast ranges; Constraints (3) and (4) ensure that the membership of an actual value in the corresponding fuzzy forecast should be higher than s; Equations (5)–(7) are the decomposition of Equation (1); Constraints (8) and (9) define the sequence for the three TFN corners.

If o is large, it becomes difficult to optimize NLP Model I. Therefore, Chen and Wang [28] advised choosing $o \in [0, 4]$. When o is a positive integer, the model can be converted into an equivalent QP problem. Chen and Wang [36] also proposed a method to approximate the model with a QP problem. First, the y_j is normalized into [0, 1]:

$$y_j \to N(y_j) = \frac{y_j - \min_k y_k}{\max_k y_k - \min_k y_k}. \tag{10}$$

Hence:

$$N(y_{j3}) - N(y_{j1}) \in [0, 1], \tag{11}$$

since $y_{j3} \geq y_{j1}$. Chen and Wang method approximated the objective function with a quadratic, e.g., for $o = 1.5$:

$$\begin{aligned} &\sum_{j=1}^{n} (N(y_{j3}) - N(y_{j1}))^{1.5} \\ &\cong \sum_{j=1}^{n} (0.5027(N(y_{j3}) - N(y_{j1}))^2 + 0.5308(N(y_{j3}) - N(y_{j1})) - 0.0347) \end{aligned} \tag{12}$$

NLP Model II:

$$\text{Max } Z_2 = \sum_{j=1}^{n} s_j{}^{w}, \tag{13}$$

which is subject to

$$\sum_{j=1}^{n} (y_{j3} - y_{j1})^{o} \leq n \cdot d^{o}, \tag{14}$$

$$y_j \geq y_{j1} + s_j(y_{j2} - y_{j1}), \tag{15}$$

$$y_j \leq y_{j3} + s_j(y_{j2} - y_{j3}), \tag{16}$$

$$y_{j1} = a_{01} + \sum_{i=1}^{m} a_{i1}x_{ji}, \tag{17}$$

$$y_{j2} = a_{02} + \sum_{i=1}^{m} a_{i2}x_{ji}, \tag{18}$$

$$y_{j3} = a_{03} + \sum_{i=1}^{m} a_{i3}x_{ji}, \tag{19}$$

$$y_{j1} \leq y_{j2} \leq y_{j3}, \tag{20}$$

$$a_{i1} \leq a_{i2} \leq a_{i3}, \tag{21}$$

and

$$0 \leq s_j \leq 1. \tag{22}$$

The objective function maximizes the high order sum of satisfaction levels. Constraint (14) ensures that average fuzzy forecast range is narrower than d, Constraints (15) and (16) derive membership for an actual value in the corresponding fuzzy forecast, Equations (17)–(19) are the same decomposition for Equation (1) as used for NLP Model I, Constraints (20) and (21) define the sequence for the three TFN corners, and constraint (22) defines the range of the satisfaction level. When o and w are both positive integers, the model can be converted into an equivalent QP problem. Otherwise, Chen and Wang's method can also be applied to approximate the model with a QP problem, similarly to the case for NLP Model I.

Fuzzy forecasts generated by the NLP models can be diversified by varying o, s, d, and w, which lays the basis for collaboration.

4. Proposed Methodology

Figure 1 shows the proposed methodology system diagram. The proposed approach comprises four major steps as follows.

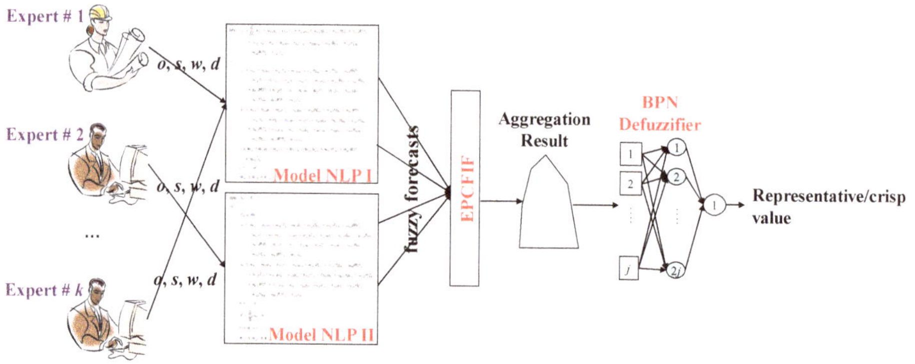

Figure 1. Proposed methodology procedure

Step 1. Experts apply either NLP model [35] to generate fuzzy forecasts with parameters o, s, w, and d, which are specified before formulating the models.

Step 2. An EPCFI is applied to aggregate expert fuzzy forecasts, where the aggregation is a polygonal fuzzy number, as shown in Figure 1.

Step 3. The appropriate number of experts is determined by observing aggregation result changes as the number of experts varies.

Step 4. A BPN is constructed to defuzzify the aggregation result, providing a representative/crisp value.

The proposed approach optimizes the NLP models to generate fuzzy forecasts and then constructs a BPN to defuzzify the aggregated result. In contrast, current methods use a BPN, an ANFIS, and other ANN types to directly generate forecasts [37–41].

4.1. EPCFI

Current fuzzy collaborative forecasting methods most commonly use an FI (i.e., minimum T-norm) to aggregate expert fuzzy forecasts:

$$\mu_{\tilde{y}_j}(x) = \min_k(\{\mu_{\tilde{y}_j(k)}(x)|k = 1 \sim K\}), \tag{23}$$

The prerequisite to apply an FI is that the expert fuzzy forecasts all include actual values, at least for the training (or learned) data. Otherwise, a fuzzy union (i.e., maximum T-conorm or S-norm) should be applied instead as the treatment taken in existing FISs.

A fuzzy intersection finds values common to expert fuzzy forecasts. Therefore, an FI can represent overall consensus among the experts. If each expert's fuzzy forecast is represented by a TFN, then the FI is a polygonal fuzzy number (see Figure 2) and its α cut can be expressed as:

$$\begin{aligned} \tilde{y}_j(\alpha) &= [\tilde{y}_j^L(\alpha), \tilde{y}_j^R(\alpha)] \\ &= [\max_k(\tilde{y}_j^L(k)(\alpha)), \min_k(\tilde{y}_j^R(k)(\alpha))] \end{aligned}, \tag{24}$$

where $[\tilde{y}_j^L(k)(\alpha), \tilde{y}_j^R(k)(\alpha)]$ is the α cut of $\tilde{y}_j(k)$.

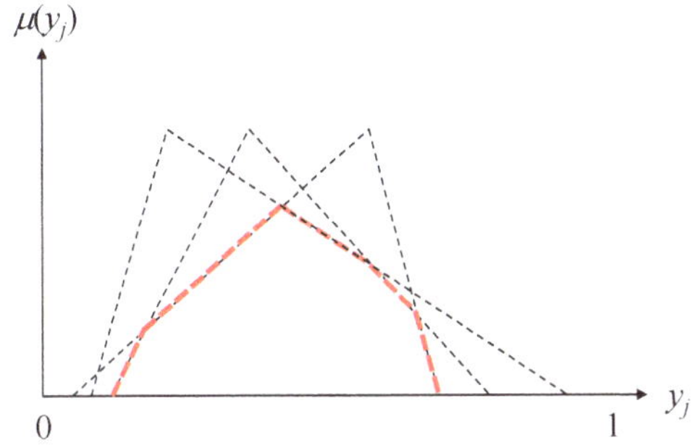

Figure 2. Fuzzy intersection (consensus).

When an overall consensus among experts cannot be achieved, the FI is the null set. In this situation, a consensus among some experts can be sought instead by using the PCFI operator [19].

Definition 1. *The H/K PCFI for fuzzy forecasts by K experts at period j, i.e.,* $\tilde{y}_j(1) \sim \tilde{y}_j(K)$, *is represented as* $\tilde{I}^{H/K}(\tilde{y}_j(1), \ldots, \tilde{y}_j(K))$, *such that:*

$$\mu_{\tilde{y}_j^{H/K}}(x) = \max_{\text{all}g}(\min(\mu_{\tilde{y}_j(g(1))}(x), \ldots, \mu_{\tilde{y}_j(g(H))}(x)))\forall x, \tag{25}$$

where $g() \in Z^+$; $1 \leq g() \leq K$; $g(p) \cap g(q) = \emptyset \ \forall \ p \neq q$; $H \geq 2$.

From Definition 1, each time a subset of size H is extracted from the set of K experts, membership for a value is determined by applying a minimum operator, representing a (partial) consensus among H experts. Since subsets do not overlap, the maximum operator is applied to aggregate memberships for a value.

For example, the 2/3 PCFI of $\tilde{y}_j(1) \sim \tilde{y}_j(K)$ can be expressed as:

$$\mu_{\tilde{y}_j^{2/3}}(x) = \max(\min(\mu_{\tilde{y}_j(1)}(x), \mu_{\tilde{y}_j(2)}(x)), \min(\mu_{\tilde{y}_j(1)}(x), \mu_{\tilde{y}_j(3)}(x)), \min(\mu_{\tilde{y}_j(2)}(x), \mu_{\tilde{y}_j(3)}(x)))\forall x, \tag{26}$$

as shown in Figure 3.

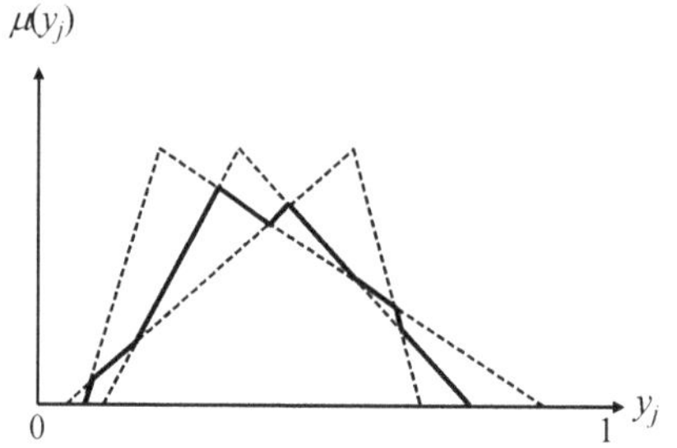

Figure 3. Partial consensus fuzzy intersection.

Property 1. *A PCFI is the hybrid of an FI and a fuzzy union for handling parts with and without consensus, respectively.*

An FI operator meets four requirements: boundary conditions, monotonicity, commutativity, and associativity. A fuzzy union operator also meets these requirements, but the boundary conditions for FI operators are contradictory to those for fuzzy union operators. Therefore, a PCFI operator meets three requirements: monotonicity, commutativity, and associativity. Thus, a PCFI eliminates the necessity to exclude fuzzy forecasts by radical experts or to force them to modify their fuzzy forecasts.

Theorem 1. *The PCFI result includes the FI result.*

Proof. The minimum of more items becomes smaller. Therefore:

$$\begin{aligned} \mu_{PCFI(\{\tilde{y}_j(k)\})}(x) &= \mu_{\tilde{y}_j^{H/K}}(x) \\ &= \max_{\text{all}g}(\min(\mu_{\tilde{y}_j(g(1))}(x),\ldots,\mu_{\tilde{y}_j(g(H))}(x))) \\ &\geq \max_{\text{all}g}(\min(\mu_{\tilde{y}_j(g(1))}(x),\ldots,\mu_{\tilde{y}_j(g(K))}(x))) \\ &= \mu_{\tilde{y}_j^{K/K}}(x) \\ &= \mu_{FI(\{\tilde{y}_j(k)\})}(x) \end{aligned} \tag{27}$$

since $H \leq K$. Thus,

$$PCFI(\{\tilde{y}_j(k)\}) \supseteq FI(\{\tilde{y}_j(k)\}).$$

□

Meaningful information can be determined by observing PCFI changes when the number of experts varies. To simplify this, we propose an EPCFI diagram, as shown in Figure 4 and defined as follows.

Definition 2. *An EPCFI diagram is a systematic representation of aggregation changes, i.e.,* $\tilde{y}_j^{H/K}$*, when the number of experts, H, varies.*

In the EPCFI diagram:

(1) $\tilde{y}_j^{2/4} \supseteq \tilde{y}_j^{3/4} \supseteq \tilde{y}_j^{4/4}$.

(2) If the consensus among all experts, i.e., $\tilde{y}_j^{4/4}$, is sought, the aggregation covers a very narrow range [4.99, 5.61]. A narrower range means higher forecasting precision, which is good for the training data but may increase the possibility of missing an actual value for test data.

(3) It is easier to reach a consensus among fewer experts, e.g., $\widetilde{y}_j^{3/4}$, and the aggregation covers a much wider range [2.89, 7.71].

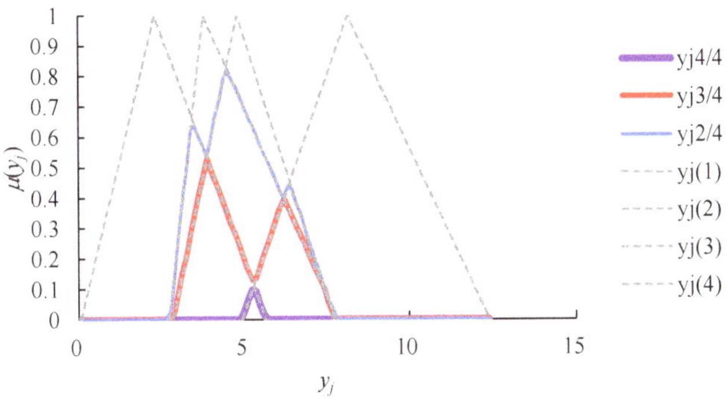

Figure 4. Typical EPCFI diagram.

A narrow PCFI can maximize the forecasting precision for training data, but the future situation may markedly differ from the past; hence, adopting a narrow PCFI (by considering consensus among all experts) is risky. Adopting a wider PCFI result by considering consensus among fewer experts provides a more robust outcome. Thus, the appropriate number of experts can be determined as follows.

(1) If $\widetilde{y}_j^{H+1/K}$ is much narrower than $\widetilde{y}_j^{H/K}$, then choosing $\widetilde{y}_j^{H/K}$ is less risky.
(2) If $\widetilde{y}_j^{H+1/K}$ is very close to $\widetilde{y}_j^{H/K}$, then $\widetilde{y}_j^{H+1/K}$ is preferable because the consensus among more experts should always be sought.

4.2. Back Propagating Network to Defuzzify the Aggregation

A BPN with the following configuration is constructed to defuzzify the aggregation.

(1) Input: BPN inputs include the value and membership for each EPCFI corner result. Deriving the representative value based on these corners is meaningful, because corner memberships are the same, which means that a consensus is achieved among experts. Consider the example shown in Figure 5. The EPCFI, in terms of $\widetilde{y}_j^{2/4}$, has seven corners; hence, there are 14 BPN inputs. However, the number of corners may differ from example to example. Therefore, the number of BPN inputs is determined by the maximum number of corners in all examples.
(2) Hidden layer: Many studies have shown that a single hidden layer is sufficient to fit complex nonlinear relationships [42]. The number of nodes in the hidden layer is twice the number of inputs [43,44].
(3) Output: o_j is compared with y_j.
(4) Learning rate: $\eta = 0.1$–1.0.
(5) Training algorithm: A gradient descent (GD) algorithm is used to prevent overfitting [45].
(6) Convergence criteria: Training terminates when the sum of squared error,

$$SSE = \sum_{j=1}^{n} (y_j - o_j)^2, \tag{28}$$

falls below a pre-specified threshold, or a maximal number of epochs have been run.

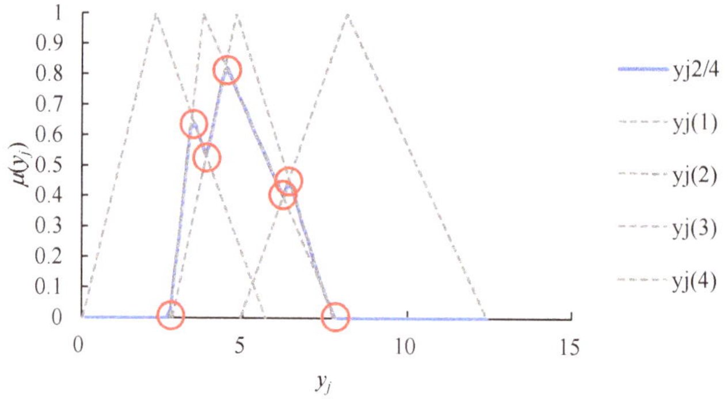

Figure 5. Determining inputs to the back propagation network (BPN).

Figure 6 shows how the BPN defuzzifier is incorporated in the proposed methodology.

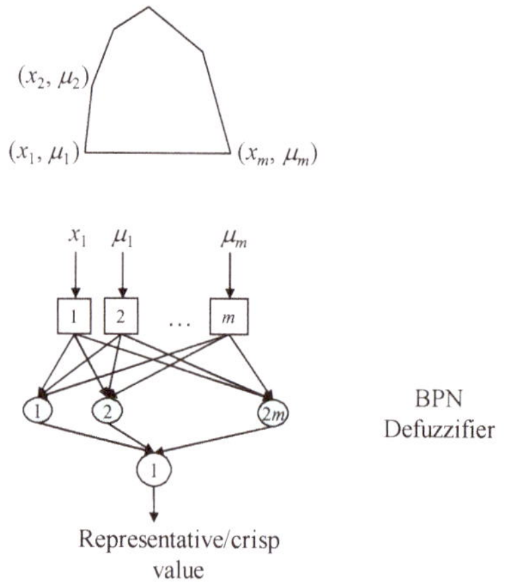

Figure 6. The role of the BPN defuzzifier.

5. Case Study: Forecasting DRAM Product Yield

5.1. Proposed Methodology Application

The proposed evolving partial-consensus fuzzy collaborative forecasting approach was applied to forecast DRAM (die) yield [36]. Product yield is the most critical performance measure for a DRAM factory [46,47], and accurately and precisely forecasting a future yield is essential to create a competitive production plan [48,49]. Improving DRAM yield can be modelled as a learning process that cannot be directly modelled with a conventional time series [50–52]:

$$\widetilde{Y}_t = \widetilde{Y}_0(\times)e^{-\frac{\widetilde{b}}{t}}, \tag{29}$$

After converting all terms on both sides to their logarithmic values:

$$\ln \widetilde{Y}_t = \ln \widetilde{Y}_0(-)\frac{\widetilde{b}}{t}, \tag{30}$$

which can be fitted as an FLR,

$$-\frac{1}{t} \to j, \tag{31}$$

$$\ln \widetilde{Y}_t \rightarrow \widetilde{y}_j, \quad (32)$$

$$\ln \widetilde{Y}_0 \rightarrow \widetilde{a}_0, \quad (33)$$

and

$$\widetilde{b} \rightarrow \widetilde{a}_1. \quad (34)$$

In the experiment, four experts applied various NLP methods to forecast DRAM yield with the following parameters:

Expert #1: NLP Model I ($o = 1$; $s = 0.5$; $w = 1$; $d = 0.4$).
Expert #2: NLP Model I ($o = 3$; $s = 0.35$; $w = 2$; $d = 0.55$).
Expert #3: NLP Model II ($o = 2$; $s = 0.4$; $w = 2$; $d = 0.35$).
Expert #4: NLP Model II ($o = 1$; $s = 0.25$; $w = 3$; $d = 0.7$).

Yield data were split into two parts, with the first six periods used to build the models and the remainder used for testing. NLP problems were solved using Lingo on a PC with a 3.6 GHz and 8 GB RAM i7-7700 CPU, which achieved an execution time of less than 3 s. Figure 7 shows the final expert forecasts.

Figure 8 shows the corresponding EPCFI diagram for the expert forecasts for Period 1. All expert fuzzy yield forecasts fell within a very narrow range, and the range shrank rapidly with each additional expert included. With only two experts, the aggregation range was very wide. Thus, a partial consensus among three experts, i.e., $\widetilde{y}_j^{3/4}$, seemed to be a reasonable choice.

Figure 9 shows the aggregation for three expert fuzzy yield forecasts, i.e., $\widetilde{y}_j^{3/4}$. The actual values at all periods except for those in Period 9 fell within the corresponding aggregation results.

Subsequently, the aggregation was defuzzified with a BPN. To this end, we first derived the aggregation corners for each period, as shown in Table 3. The maximum number of corners over all periods = 6, so the number of BPN inputs = 12 and number of nodes in the hidden layer = 24. The BPN was trained with the GD algorithm to prevent overfitting. Convergence criteria were established as:

(1) SSE < 10^{-6}; or
(2) 1000 epochs.

The BPN defuzzifier was implemented with the MATLAB® 2017 neural network toolbox on the same PC, with execution times of less than 1 s. Figure 10 shows defuzzification results.

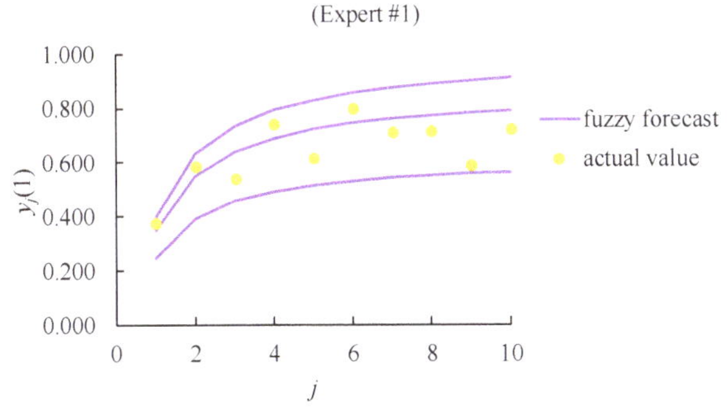

(**a**) Expert #1

Figure 7. *Cont.*

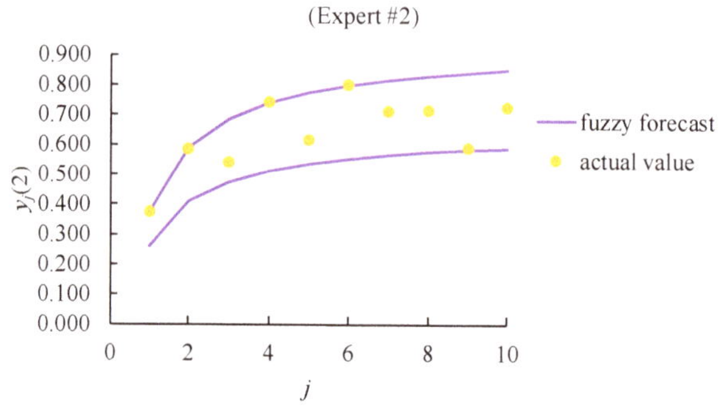

(**b**) Expert #2

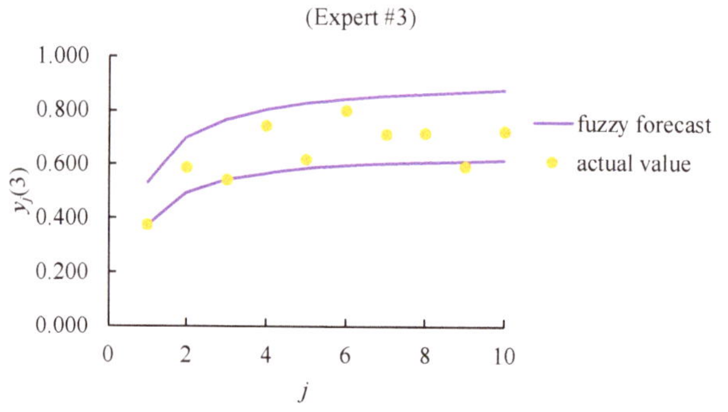

(**c**) Expert #3

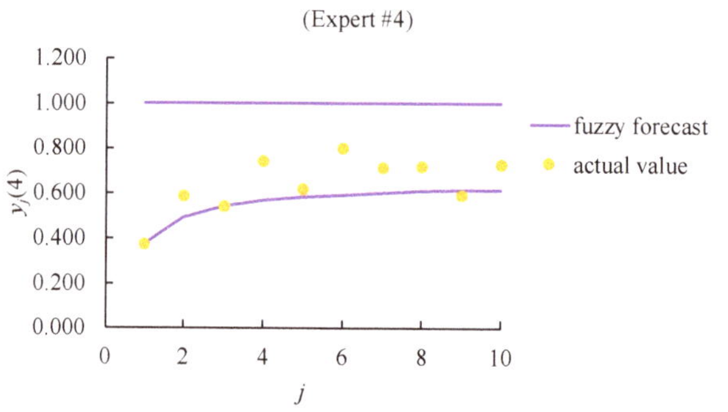

(**d**) Expert #4

Figure 7. Expert dynamic random access memory (DRAM) yield forecasts.

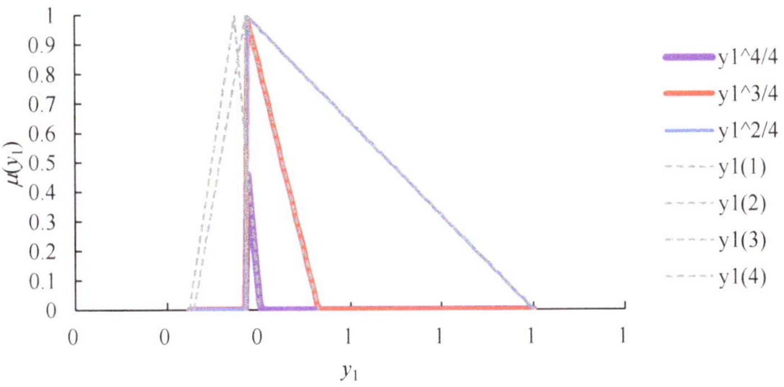

Figure 8. Example EPCFI diagram for Period 1.

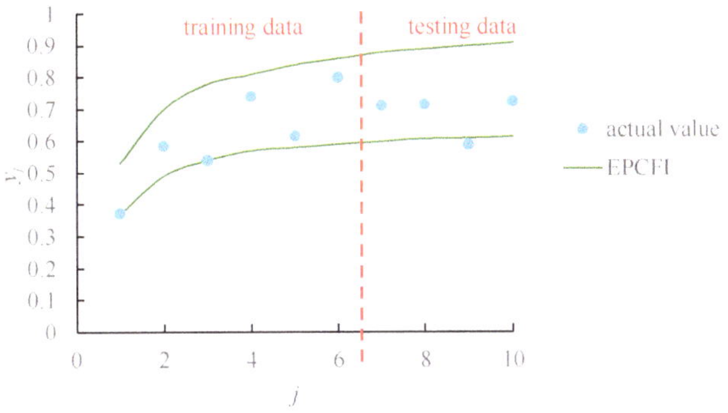

Figure 9. The aggregation results.

To validate the effectiveness of the proposed methodology, the optimized models and the trained BPN were applied to test (unlearned) data. Then, the forecasting accuracy was evaluated in terms of mean absolute error (MAE), mean absolute percentage error (MAPE), and RMSE as:

MAE = 0.04,
MAPE = 5.95%, and
RMSE = 0.06.

Table 3. Aggregation result corners

j	Corners
1	(0.37, 0.00), (0.38, 0.98), (0.53, 0.00)
2	(0.49, 0.00), (0.50, 0.66), (0.53, 0.84), (0.70, 0.00)
3	(0.54, 0.00), (0.54, 0.48), (0.60, 0.75), (0.69, 0.36), (0.69, 0.48), (0.78, 0.00)
4	(0.57, 0.00), (0.57, 0.39), (0.64, 0.71), (0.74, 0.29), (0.74, 0.51), (0.81, 0.00)
5	(0.58, 0.00), (0.58, 0.33), (0.66, 0.68), (0.77, 0.22), (0.79, 0.50), (0.84, 0.00)
6	(0.59, 0.00), (0.60, 0.31), (0.68, 0.66), (0.80, 0.18), (0.80, 0.48), (0.86, 0.00)
7	(0.60, 0.00), (0.60, 0.27), (0.69, 0.65), (0.81, 0.16), (0.82, 0.46), (0.88, 0.00)
8	(0.61, 0.00), (0.61, 0.26), (0.65, 0.70), (0.83, 0.13), (0.84, 0.42), (0.89, 0.00)
9	(0.61, 0.00), (0.61, 0.24), (0.71, 0.64), (0.84, 0.12), (0.84, 0.41), (0.90, 0.00)
10	(0.61, 0.00), (0.61, 0.23), (0.71, 0.63), (0.85, 0.10), (0.86, 0.38), (0.91, 0.00)

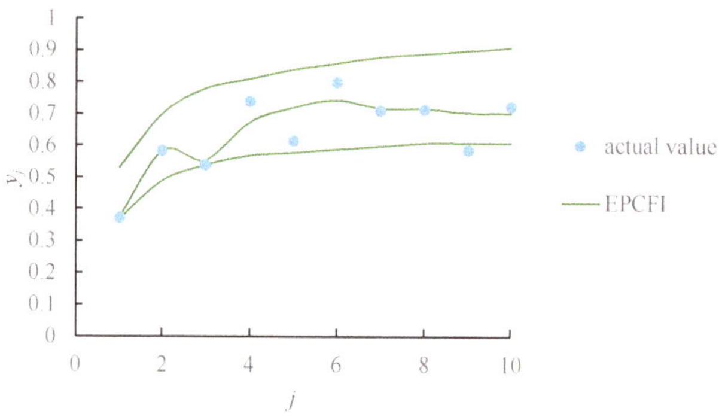

Figure 10. Defuzzification results.

5.2. Comparisons

Several current methods were applied to this case for comparison. First, we applied the Chen and Lin [35] fuzzy collaborative forecasting method based on overall consensus, employing an FI to aggregate expert fuzzy yield forecasts, as shown in Table 4. The number aggregation corners for each period was much smaller. Subsequently, we constructed a BPN to defuzzify the aggregation and provide a representative value. Figure 11 shows the final forecast results.

Table 4. Fuzzy intersection results for the Chen and Lin method.

j	Corners
1	(0.37, 0.00), (0.38, 0.46), (0.41, 0.00)
2	(0.59, 0.00), (0.59, 0.48), (0.64, 0.00)
3	(0.69, 0.00), (0.69, 0.34), (0.74, 0.00)
4	(0.74, 0.00), (0.74, 0.26), (0.80, 0.00)
5	(0.77, 0.00), (0.78, 0.20), (0.83, 0.00)
6	(0.80, 0.00), (0.80, 0.16), (0.85, 0.00)
7	(0.81, 0.00), (0.82, 0.14), (0.85, 0.00)
8	(0.83, 0.00), (0.84, 0.11), (0.87, 0.00)
9	(0.84, 0.00), (0.84, 0.11), (0.87, 0.00)
10	(0.85, 0.00), (0.85, 0.09), (0.88, 0.00)

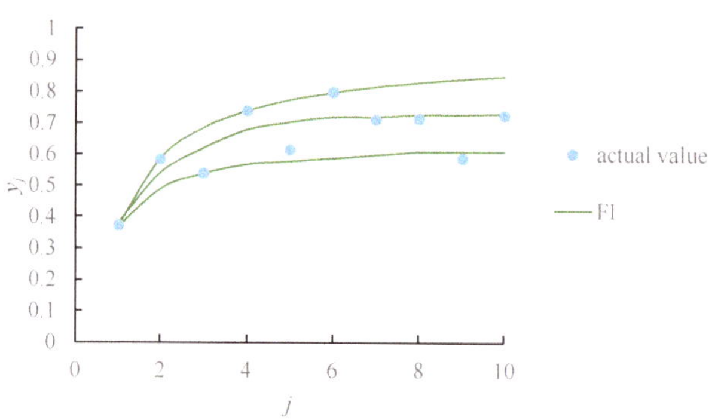

Figure 11. Chen and Lin method results based on overall consensus.

Forecast accuracy was evaluated in the same manner as for the proposed approach, with MAE = 0.04, MAPE = 6.84%, and RMSE = 0.07.

Subsequently, forecast performance was also evaluated for the case where the experts did collaborate. Expert fuzzy yield forecasts were defuzzified with the prevalent center-of-gravity (COG) method [53], as shown in Table 5.

Table 5. Forecasting performances for Chen and Lin method if experts did not collaborate.

Expert	MAE	MAPE	RMSE
1	0.06	9.26%	0.08
2	0.06	10.10%	0.09
3	0.04	7.08%	0.06
4	0.05	8.68%	0.08

Second, we applied the 6σ logistic regression method, which fitted collected yield data with the logistic regression model:

$$\log \hat{y}_j = -0.262 - \frac{0.697}{j}, \tag{35}$$

where $\sigma = 0.115$. This model achieved a coefficient of determination $R^2 = 0.87$, which was sufficiently high. Upper and lower yield forecast bounds were established by adding and subtracting 3σ to the yield forecast, respectively, as shown in in Figure 12. This method established very wide yield forecast ranges, so all actual values in the test data were included in the corresponding confidence intervals, elevating the hit rate to 100%. The forecasting accuracy when using the 6σ logistic regression method was evaluated as MAE = 0.04, MAPE = 6.36%, and RMSE = 0.06.

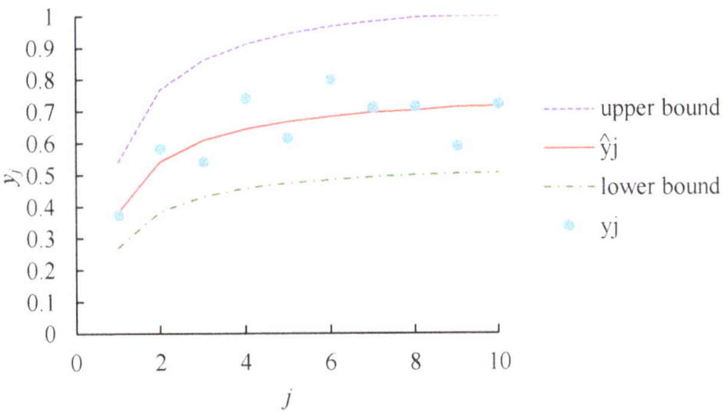

Figure 12. Forecast results found when the 6σ logistic regression method.

5.3. Discussion

Table 6 compares forecasting performances for the various considered methods. The proposed methodology achieved a superior forecast accuracy in terms of MAE, MAPE, and RMSE.

Table 6. Forecasting performance for the considered methods

Method	Hit Rate	Average Range	MAE	MAPE	RMSE
Expert #1	100%	0.34	0.06	9.26%	0.08
Expert #2	100%	0.26	0.06	10.10%	0.09
Expert #3	75%	0.26	0.04	7.08%	0.06
Expert #4	75%	0.40	0.05	8.68%	0.08
6σ logistic regression	100%	0.69	0.04	6.36%	0.06
Chen and Lin's fuzzy collaborative forecasting method [35]	75%	0.23	0.04	6.84%	0.07
The proposed methodology	75%	0.29	0.04	5.95%	0.06

Forecasting performance improved with expert collaboration. When experts achieved an overall consensus, forecasting precision improved by up to 43% in terms of the average range of the fuzzy yield forecasts. A comparable improvement was also achieved for partial consensus, with the forecasting accuracy in terms of MAPE improved by 31% after applying the proposed methodology.

The proposed evolving partial-consensus fuzzy collaborative forecasting approach surpassed Chen and Lin's fuzzy collaborative forecasting method for optimizing forecast accuracy for both fuzzy collaborative forecasting methods. This was most likely possible because $\tilde{y}_j^{3/4}$ had more corners than $\tilde{y}_j^{4/4}$, which gave the decision makers a higher degree of freedom in defuzzifying the aggregation. Hence, the possibility of finding actual values also improved.

Applying the common COG method to defuzzify the aggregation from the proposed methodology for each period achieved a forecast accuracy with MAE = 0.05, MAPE = 7.86%, and RMSE = 0.08, which was worse than that achieved when the BPN defuzzifier was applied, as shown in Figure 13. This confirmed the effectiveness of the BPN defuzzifier.

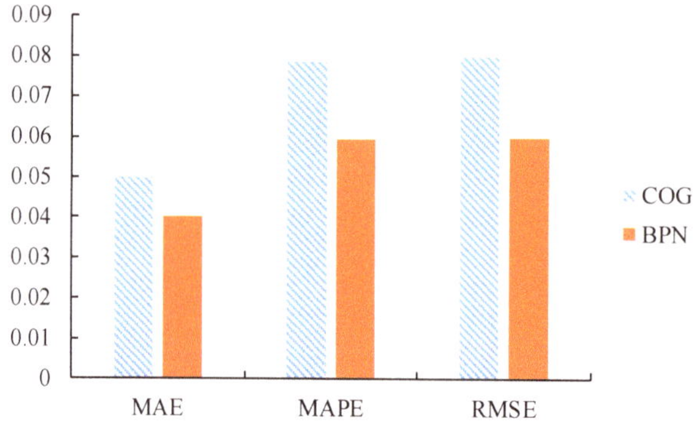

Figure 13. Forecasting performance for the proposed methodology with different defuzzifiers.

6. Conclusions

Most current fuzzy collaborative forecast methods apply an FI to aggregate expert fuzzy forecasts, a process that has several drawbacks. The PCFI operator [5] is useful to help overcome these drawbacks. However, how to determine the appropriate number of experts has not been considered—this can be assessed by observing PCFI changes when the number of experts varies. Therefore, this paper proposed EPCFI diagrams to simplify this comparison, as well as an evolving partial-consensus fuzzy collaborative forecasting approach based on the EPCFI diagrams.

The proposed evolving partial-consensus fuzzy collaborative forecasting approach was applied to forecast DRAM yield using real-world data, with the following conclusions.

(1) The proposed approach effectively improved forecast accuracy for test in terms of MAE, MAPE, and RMSE. The most significant advantage over current methods was up to 24% when MAPE was minimized.

(2) Compared with the Chen and Lin fuzzy collaborative forecasting method, the proposed methodology achieved a higher forecast accuracy at the expense of a slight increase in the average fuzzy yield forecast range due to increased degree of freedom in defuzzifying fuzzy the yield forecasts.

(3) Collaboration among experts was shown to be conducive to forecast performance. Forecast precision, in terms of average fuzzy yield forecast range, improved 43% after expert collaboration, and forecast accuracy (MAPE) also improved 31%

The proposed evolving partial-consensus fuzzy collaborative forecasting approach can be easily implemented with current data analysis software. The fuzzy forecasting method (Equation (1)) generalizes many fuzzy forecasting methods, such as the fuzzy moving average and the fuzzy ARIMA. Therefore, the proposed methodology can be easily applied to other problem types, such as fuzzy time series forecasts.

Future studies will investigate the application of the evolving partial-consensus fuzzy collaborative forecasting approach to more real cases to further investigate its effectiveness. Other mechanisms to measure partial expert consensus will also be explored. Experts often have unequal authority levels, and this situation should be incorporated when aggregating their fuzzy forecasts, such as through weighted aggregation operators like weighted FI, PCFI, and EPCFI.

Author Contributions: All authors equally contributed to the writing of this paper. All authors read and approved the final manuscript. Data curation, methodology and writing original draft: T.-C.T.C. and Y.-C.W.; writing—review and editing: T.-C.T.C., Y.-C.W., and C.-H.H. All authors have read and agreed to the published version of the manuscript.

Funding: This research received no external funding.

Conflicts of Interest: The authors declare no conflicts of interest.

References

1. Pedrycz, W. Collaborative architectures of fuzzy modeling. *Lect. Notes Comput. Sci.* **2008**, *5050*, 117–139.
2. Chen, T.C.T.; Honda, K. *Fuzzy Collaborative Forecasting and Clustering: Methodology, System Architecture, and Applications*; Springer: Cham, Switzerland, 2019.
3. Chen, T.C.T.; Honda, K. Introduction to fuzzy collaborative forecasting systems. In *Fuzzy Collaborative Forecasting and Clustering*; Springer: Berlin/Heidelberg, Germany, 2020; pp. 1–8.
4. Hernández, J.E.; Poler, R.; Mula, J.; Cordeiro, J.; Filipe, J. *Modelling collaborative forecasting in decentralized supply chain networks with a multiagent System. Balanced Automation Systems for Future Manufacturing Networks*; Springer: Berlin/Heidelberg, Germany, 2009; pp. 372–375.
5. Chen, T. A collaborative fuzzy-neural system for global CO2 concentration forecasting. *Int. J. Innov. Comput. Inf. Control.* **2012**, *8*, 7679–7696.
6. Chen, T. Forecasting the unit cost of a product with some linear fuzzy collaborative forecasting models. *Algorithms* **2012**, *5*, 449–468. [CrossRef]
7. Parreiras, R.O.; Ekel, P.Y.; Morais, D.C. Fuzzy set based consensus schemes for multicriteria group decision making applied to strategic planning. *Group Decis. Negot.* **2012**, *21*, 153–183. [CrossRef]
8. Lin, C.W.; Chen, T. 3D printing technologies for enhancing the sustainability of an aircraft manufacturing or MRO company—A multi-expert partial consensus-FAHP analysis. *Int. J. Adv. Manuf. Technol.* **2019**, *105*, 4171–4180. [CrossRef]
9. Richardson, M.; Domingos, P. Learning with knowledge from multiple experts. In Proceedings of the 20th International Conference on Machine Learning, Washington, DC, USA, 21–24 August 2003; pp. 624–631.
10. Chakraborty, C.; Chakraborty, D. A fuzzy clustering methodology for linguistic opinions in group decision making. *Appli. Soft Comput.* **2007**, *7*, 858–869. [CrossRef]
11. Kahraman, C.; Ruan, D.; Doğan, I. Fuzzy group decision-making for facility location selection. *Inf. Sci.* **2003**, *157*, 135–153. [CrossRef]
12. Boran, F.E.; Genç, S.; Kurt, M.; Akay, D. A multi-criteria intuitionistic fuzzy group decision making for supplier selection with TOPSIS method. *Expert Syst. Appl.* **2009**, *36*, 11363–11368. [CrossRef]
13. Zarandi, M.F.; Hadavandi, E.; Turksen, I.B. A hybrid fuzzy intelligent agent-based system for stock price prediction. *Int. J. Intell. Syst.* **2012**, *27*, 947–969. [CrossRef]
14. Swaroop, R.; Abdulqader, H.A.A. Load forecasting for power system planning using fuzzy-neural networks. In Proceedings of the World Congress on Engineering and Computer Science, San Francisco, CA, USA, 24–26 October 2012; pp. 24–26.
15. Cheikhrouhou, N.; Marmier, F.; Ayadi, O.; Wieser, P. A collaborative demand forecasting process with event-based fuzzy judgements. *Comput. Ind. Eng.* **2011**, *61*, 409–421. [CrossRef]
16. Amindoust, A.; Ahmed, S.; Saghafinia, A.; Bahreininejad, A. Sustainable supplier selection: A ranking model based on fuzzy inference system. *Appl. Soft Comput.* **2012**, *12*, 1668–1677. [CrossRef]

17. Kaur, A.; Kaur, A. Comparison of mamdani-type and sugeno-type fuzzy inference systems for air conditioning system. *Int. J. Soft Comput. Eng.* **2012**, *2*, 323–325.
18. Singh, R.; Kainthola, A.; Singh, T.N. Estimation of elastic constant of rocks using an ANFIS approach. *Appl. Soft Comput.* **2012**, *12*, 40–45. [CrossRef]
19. Chen, T. Applying the hybrid fuzzy c-means-back propagation network approach to forecast the effective cost per die of a semiconductor product. *Comput. Ind. Eng.* **2011**, *61*, 752–759. [CrossRef]
20. Chen, T. An effective fuzzy collaborative forecasting approach for predicting the job cycle time in wafer fabrication. *Comput. Ind. Eng.* **2013**, *66*, 834–848. [CrossRef]
21. Reagan, C.R.; Sari, S.R. Long term load forecasting in Tamil Nadu using fuzzy-neural technology. *Int. J. Eng. Innov. Technol.* **2014**, *3*, e8.
22. Chen, T.; Chiu, M.C. An improved fuzzy collaborative system for predicting the unit cost of a DRAM product. *Int. J. Intell. Syst.* **2015**, *30*, 707–730. [CrossRef]
23. Zhang, Z.; Xu, D.; Ostrosi, E.; Yu, L.; Fan, B. A systematic decision-making method for evaluating design alternatives of product service system based on variable precision rough set. *J. Intell. Manuf.* **2019**, *30*, 1895–1909. [CrossRef]
24. Ostrosi, E.; Bluntzer, J.B.; Zhang, Z.; Stjepandić, J. Car style-holon recognition in computer-aided design. *J. Comput. Des. Eng.* **2019**, *6*, 719–738. [CrossRef]
25. Gao, H.; Ju, Y.; Gonzalez, E.D.S.; Zhang, W. Green supplier selection in electronics manufacturing: An approach based on consensus decision making. *J. Clean. Prod.* **2020**, *245*, 118781. [CrossRef]
26. Herrera-Viedma, E.; Cabrerizo, F.J.; Kacprzyk, J.; Pedrycz, W. A review of soft consensus models in a fuzzy environment. *Inf. Fusion* **2014**, *17*, 4–13. [CrossRef]
27. Chen, T. A heterogeneous fuzzy collaborative intelligence approach for forecasting the product yield. *Appl. Soft Comput.* **2017**, *57*, 210–224. [CrossRef]
28. Chen, T.; Wang, Y.C. An agent-based fuzzy collaborative intelligence approach for precise and accurate semiconductor yield forecasting. *IEEE Trans. Fuzzy Syst.* **2014**, *22*, 201–211. [CrossRef]
29. Chen, T.; Romanowski, R. Forecasting the productivity of a virtual enterprise by agent-based fuzzy collaborative intelligence—with Facebook as an example. *Appl. Soft Comput.* **2014**, *24*, 511–521. [CrossRef]
30. Yoon, H.S.; Choi, S.H. The impact on life satisfaction of nursing students using the fuzzy regression model. *Int. J. Fuzzy Log. Intell. Syst.* **2019**, *19*, 59–66. [CrossRef]
31. Tanaka, H.; Watada, J. Possibilistic linear systems and their application to the linear regression model. *Fuzzy Sets Syst.* **1988**, *27*, 275–289. [CrossRef]
32. Taheri, S.M.; Kelkinnama, M. Fuzzy linear regression based on least absolutes deviations. *Iran. J. Fuzzy Syst.* **2012**, *9*, 121–140.
33. Peters, G. Fuzzy linear regression with fuzzy intervals. *Fuzzy Sets Syst.* **1994**, *63*, 45–55. [CrossRef]
34. Donoso, S.; Marín, N.; Vila, M.A. Quadratic programming models for fuzzy regression. In Proceedings of the International Conference on Mathematical and Statistical Modeling in Honor of Enrique Castillo, San Diego, CA, USA, 28–30 June 2006.
35. Chen, T.; Lin, Y.C. A fuzzy-neural system incorporating unequally important expert opinions for semiconductor yield forecasting. *Int. J. Uncertain. Fuzziness Knowl. Based Syst.* **2008**, *16*, 35–58. [CrossRef]
36. Chen, T.; Wang, Y.C. Semiconductor yield forecasting using quadratic-programming-based fuzzy collaborative intelligence approach. *Math. Probl. Eng.* **2013**, *2013*, 1–7. [CrossRef]
37. Feng, Y.; Zhang, W.; Sun, D.; Zhang, L. Ozone concentration forecast method based on genetic algorithm optimized back propagation neural networks and support vector machine data classification. *Atmos. Environ.* **2011**, *45*, 1979–1985. [CrossRef]
38. Talebizadeh, M.; Moridnejad, A. Uncertainty analysis for the forecast of lake level fluctuations using ensembles of ANN and ANFIS models. *Expert Syst. Appl.* **2011**, *38*, 4126–4135. [CrossRef]
39. Shrivastava, G.; Karmakar, S.; Kowar, M.K.; Guhathakurta, P. BPN model for long-range forecast of monsoon rainfall over a very small geographical region and its verification for 2012. *Geofizika* **2013**, *30*, 143–154.
40. Park, S.; Song, N.; Yu, W.; Kim, D. PSR: PSO-based signomial regression model. *Int. J. Fuzzy Logic Intell. Syst.* **2019**, *19*, 307–314. [CrossRef]
41. Chen, T. A fuzzy back propagation network for output time prediction in a wafer fab. *Appl. Soft Comput.* **2003**, *2*, 211–222. [CrossRef]

42. Hornik, K. Approximation capabilities of multilayer feedforward networks. *Neural Netw.* **1991**, *4*, 251–257. [CrossRef]
43. Wu, H.C.; Chen, T. CART–BPN approach for estimating cycle time in wafer fabrication. *J. Ambient Intell. Humaniz. Comput.* **2015**, *6*, 57–67. [CrossRef]
44. Lin, Y.C.; Wang, Y.C.; Chen, T.C.T.; Lin, H.F. Evaluating the suitability of a smart technology application for fall detection using a fuzzy collaborative intelligence approach. *Mathematics* **2019**, *7*, 1097. [CrossRef]
45. Wang, Y.C.; Chen, T. A fuzzy collaborative forecasting approach for forecasting the productivity of a factory. *Adv. Mech. Eng.* **2013**, *5*, 234571. [CrossRef]
46. Yunogami, T. *Technology management and competitiveness in the Japanese semiconductor industry. Recovering from Success, Innovation and Technology Management in Japan*; Oxford University Press: Oxford, UK, 2006; pp. 41–44.
47. Tirkel, I. Yield learning curve models in semiconductor manufacturing. *IEEE Trans. Semicond. Manuf.* **2013**, *26*, 564–571. [CrossRef]
48. Chang, J.H.; Tsai, M.H. Using simulation model to integrate production plan and dispatching system. In Proceedings of the IEEE International Symposium on Semiconductor Manufacturing, San Jose, CA, USA, 13–15 September 2005; pp. 95–98.
49. Wang, Y.C.; Chen, T.C.T. An FNLP approach for planning energy-efficient manufacturing: Wafer fabrication as an example. *Procedia Manuf.* **2019**, *38*, 439–446. [CrossRef]
50. Yavas, O.; Richter, E.; Kluthe, C.; Sickmoeller, M.; AG, Q. Wafer-edge yield engineering in leading-edge DRAM manufacturing. *Semicond. Fabtech* **2009**, *39*, 1–5.
51. Kim, H.G.; Han, Y.S.; Lee, J.H. Package yield enhancement using machine learning in semiconductor manufacturing. In Proceedings of the IEEE Advanced Information Technology, Electronic and Automation Control Conference, Chongqing, China, 19–20 December 2015; pp. 316–320.
52. Wang, Y.C.; Chen, T.C.T. A direct-solution fuzzy collaborative intelligence approach for yield forecasting in semiconductor manufacturing. *Procedia Manuf.* **2018**, *17*, 110–117. [CrossRef]
53. Chen, T.; Wang, M.J.J. Forecasting methods using fuzzy concepts. *Fuzzy Sets Syst.* **1999**, *105*, 339–352. [CrossRef]

mathematics

MDPI

Article

A Novel Technique to Solve the Fuzzy System of Equations

Nasser Mikaeilvand [1], Zahra Noeiaghdam [2], Samad Noeiaghdam [3,4,*] and Juan J. Nieto [5]

1 Department of Mathematics, Ardabil Branch, Islamic Azad University, Ardabil, Iran; Mikaeilvand@iauardabil.ac.ir
2 Department of Mathematics and Computer Science, Shahed University, Tehran, Iran; Z_Noeiaghdam@shahed.ac.ir
3 Department of Applied Mathematics and Programming, South Ural State University, Lenin Prospect 76, Chelyabinsk 454080, Russia
4 Baikal School of BRICS, Irkutsk National Research Technical University, Irkutsk, Russia
5 Departamento de Estatística, Análise Matemática e Optimización, Instituto de Matemáticas, Universidade de Santiago de Compostela, 15782 Santiago de Compostela, Spain; juanjose.nieto.roig@usc.es
* Correspondence: noiagdams@susu.ru or snoei@istu.edu

Received: 15 April 2020; Accepted: 21 May 2020; Published: 24 May 2020

Abstract: The aim of this research is to apply a novel technique based on the embedding method to solve the $n \times n$ fuzzy system of linear equations (FSLEs). By using this method, the strong fuzzy number solutions of FSLEs can be obtained in two steps. In the first step, if the created $n \times n$ crisp linear system has a non-negative solution, the fuzzy linear system will have a fuzzy number vector solution that will be found in the second step by solving another created $n \times n$ crisp linear system. Several theorems have been proved to show that the number of operations by the presented method are less than the number of operations by Friedman and Ezzati's methods. To show the advantages of this scheme, two applicable algorithms and flowcharts are presented and several numerical examples are solved by applying them. Furthermore, some graphs of the obtained results are demonstrated that show the solutions are fuzzy number vectors.

Keywords: fuzzy linear system; fuzzy number; fuzzy number vector; embedding method

1. Introduction

The fuzzy arithmetic has many applications in various sciences. For this reason, it is a noteworthy science in the scientific community that is constantly expanding [1–3]. In [4], Stanujkić et al. applied the fuzzy mathematics for solving decision-making problems, in [5], Stojić et al. presented a fuzzy model for determining the justifiability of investing in a road freight vehicle fleet, and in [6], Si et al. studied an approach to rank picture fuzzy numbers for decision-making problems. One of the main fields in fuzzy mathematics is solving fuzzy system of linear equations (FSLEs), which we refer to in this work. Fuzzy mathematics have real applications in heat transport, fluid flow, electromagnetism, and others. Based on numerical- and semi-analytical methods for solving FSLEs, we can study fuzzy bio-mathematical models. In [7], Abbasi et al. solved the fuzzy mathematical model of HIV infection and in [8], Mishra and Pandey applied the fuzzy system of equations to show the fuzzy mathematical model of computer viruses. Thus, studying the mathematical methods for solving the FSLE is important in theories and applications.

Solving and studying the FSLEs have been appeared in many research studies. In 1998, Friedman et al. [9] presented a model to solve the FSLEs by using the embedding approach. They replaced the original $n \times n$ fuzzy linear system by the $2n \times 2n$ crisp linear system. Additionally, they applied other methods for solving the FSLEs in [9,10]. Abbasbany et al. in [11–13], applied the decomposition method and steepest descent method for solving this problem. Allahviranloo et al. in [14–18], studied the fuzzy numbers and its application for solving fuzzy linear systems and a fuzzy matrix equation in the form $AXB = C$. In [19], some conditions of the existence of a fuzzy or interval solution of the $m \times n$ linear system were derived and in [20], the inner estimation of the solution set of a fuzzy linear system was found. Amirfakhrian in [21] considered the FSLEs by fuzzy distance, in [22] he presented the numerical solution of the FSLEs in the polynomial parametric form, and in [23] he studied the fuzzy matrix equations. Fariborzi Araghi et al. in [24] applied the inherited LU factorization, Mikaeilvand and Noeiaghdam in [25–27] discussed the fuzzy linear matrix equations, general solutions of $m \times n$ fuzzy linear systems, and least square solutions of inconsistent fuzzy linear matrix equations. Furthermore, Wang and Zheng illustrated the inconsistent and general fuzzy linear systems in [28,29]. Behera et al. [30] presented a new method for solving real and complex fuzzy systems of linear equations. Furthermore, in recent years the CESTAC method [31,32] based on the stochastic arithmetic has been applied to find the optimal iteration, the optimal approximation, and the optimal error of numerical methods to solve the FSLEs [33,34].

Ezzati in [35], changed the original $n \times n$ fuzzy linear system to two $n \times n$ crisp linear systems for solving the FSLEs. In comparison with Friedman's method, it is clear that solving two $n \times n$ linear systems is better than solving the $2n \times 2n$ linear system and hence, Ezzati's method is better than Friedman's method numerically. But, both of them have some faults:

1. They found both weak and strong solutions, and the weak solution was not the fuzzy linear system's solution;
2. Allahviranloo et al. [15] proved that the weak solution of the FSLEs, defined by Friedman et al. [9] and Ezzati [35] was not always a fuzzy number vector;
3. The kind of solution (strong or weak) can be determined only at the end of the solving process.

In this work, the embedding method is used to solve the $n \times n$ fuzzy linear system. For this aim, instead of solving the original $n \times n$ fuzzy linear system, we solve two $n \times n$ crisp linear systems in two steps. In the first step, if the $n \times n$ crisp linear system has a non-negative solution, the fuzzy linear system will have a fuzzy number vector solution, which will be found in the second step. In addition, because of some special properties of triangular fuzzy numbers, a special algorithm for solving fuzzy linear systems with triangular fuzzy numbers is proposed. Several theorems and lemmas are proved that show the number of operations in the new method is less than Ezzati's method. The solutions of the FSLEs can also be obtained by the fast and safe process, so it is better than Friedman's method. To show the abilities of the method, two algorithms and flowcharts are presented.

The organization of this paper is in the following form: Section 2 contains several definitions and theorems of fuzzy arithmetic and the FSLEs. Section 3 introduces a new idea to solve the FSLEs. In this section, Ezzati's method is modified and improved. Several theorems are also proved to show that the presented method is better than Friedman and Ezzati's methods. Furthermore, we will prove the number of operations in the presented method are less than their methods. Two applicable and efficient algorithms and flowcharts are also presented. In Section 4, several examples are solved by using the mentioned algorithms. Graphs of solutions are demonstrated to show that the obtained results are in the fuzzy form. Section 5 is the conclusion.

2. Preliminaries

In this paper, several definitions and details of fuzzy arithmetic are presented [4–6,9,10,36]. The methods of Friedman [9] and Ezzati [35] to solve the FSLEs are also considered. Ezzati's method has some problems in proving the theorems that we modify and represent them.

Definition 1 ([9,10]). *Let $\widetilde{p} = (\underline{p}(z), \overline{p}(z))$, $0 \le z \le 1$ be the arbitrary fuzzy number then the following criteria should be satisfied:*

(i) $\underline{p}(z)$ is a bounded monotonic increasing left continuous function;

(ii) $\overline{p}(z)$ *is a bounded monotonic decreasing left continuous function;*
(iii) $\underline{p}(z) \leq \overline{p}(z),\ 0 \leq z \leq 1.$

The set of all fuzzy numbers is denoted by $\mathbf{E}^1$. The crisp number k is called the singleton when $\overline{p}(z) = \underline{p}(z) = k,\ 0 \leq z \leq 1$.

Let $\widetilde{p} = (\underline{p}(z), \overline{p}(z)),\ \widetilde{q} = (\underline{q}(z), \overline{q}(z))$ be the arbitrary fuzzy functions and k be the scalar value. The operations between two fuzzy functions are defined as follows:

$$\underline{(p+q)}(z) = \underline{p}(z) + \underline{q}(z), \quad \overline{(p+q)}(z) = \overline{p}(z) + \overline{q}(z), \tag{1}$$

$$\underline{(p-q)}(z) = \underline{p}(z) - \overline{q}(z), \quad \overline{(p-q)}(z) = \overline{p}(z) - \underline{q}(z), \tag{2}$$

$$k\widetilde{p} = \begin{cases} (k\underline{p}(z), k\overline{p}(z)), & k \geq 0, \\ (k\overline{p}(z), k\underline{p}(z)), & k < 0. \end{cases} \tag{3}$$

Also $\widetilde{p} = \widetilde{q}$ if and only if $\underline{p}(z) = \underline{q}(z)$ and $\overline{p}(z) = \overline{q}(z)$.

Remark 1. *The triangular fuzzy number* $\tilde{p} = (c, \mu, \rho)$ *is defined as follows:*

$$\widetilde{p} = \begin{cases} \dfrac{x - c + \mu}{\mu}, & c - \mu \leq x \leq c, \\ \dfrac{c + \rho - x}{\rho}, & c \leq x \leq c + \rho, \\ 0, & o.w, \end{cases} \tag{4}$$

where $\mu, \rho > 0$. *It is clear that* $\underline{p}(z) = c - (1-z)\mu$, $\overline{p}(z) = c + (1-z)\rho$, *and* $\overline{p} - \underline{p} = (\mu + \rho)(1 - z)$. *The set of all triangular fuzzy numbers is denoted by* $\mathbf{TE}^1$.

Definition 2 ([9])**.** *Let* $(\tilde{v}_1, \tilde{v}_2, \ldots, \tilde{v}_j, \ldots, \tilde{v}_n)^T, \tilde{v}_j = (\underline{v}_j(z), \overline{v}_j(z)),\ 1 \leq j \leq n,\ 0 \leq z \leq 1$ *be the fuzzy number vector which is called the solution of FSLEs if and only if:*

$$\begin{aligned} \underline{\sum_{j=1}^{n} a_{ij} v_j(z)} &= \sum_{j=1}^{n} \underline{a_{ij} v_j}(z) = \underline{b_i}(z), \\ \overline{\sum_{j=1}^{n} a_{ij} v_j(z)} &= \sum_{j=1}^{n} \overline{a_{ij} v_j}(z) = \overline{b}_i(z), \quad i = 1, 2, \cdots, n. \end{aligned} \tag{5}$$

Finally, the methods of Friedman et al. [9] and Ezzati [35] to solve the FSLEs is reminded.

2.1. Friedman's Method

Friedman et al. [9] presented the FSLEs as:

$$Sv(z) = w(z), \tag{6}$$

where

$$\begin{aligned} v(z) &= \left(\underline{v}_1(z), \underline{v}_2(z), \cdots, \underline{v}_n(z), -\overline{v}_1(z), -\overline{v}_2(z), \cdots, -\overline{v}_n(z)\right)^T, \\ w(z) &= (\underline{w}_1(z), \underline{w}_2(z), \cdots, \underline{w}_n(z), -\overline{w}_1(z), -\overline{w}_2(z), \cdots, -\overline{w}_n(z))^T, \end{aligned} \tag{7}$$

and the elements of $S = (s_{ij})$, $1 \leq i, j \leq 2n$ are obtained based on the following conditions:

$$\begin{aligned} a_{ij} \geq 0 &\Rightarrow s_{ij} = s_{i+nj+n} = a_{ij}, \\ a_{ij} < 0 &\Rightarrow s_{ij+n} = s_{i+nj} = -a_{ij}. \end{aligned} \tag{8}$$

We note that for values s_{ij} which are determined by neglecting the criterion in Equation (8) we have $s_{ij} = 0$. The matrix S for $s_{ij} \geq 0, 1 \leq i, j \leq 2n$ can be formed as follows:

$$S = \begin{pmatrix} B & C \\ C & B \end{pmatrix}, \tag{9}$$

where B constructs by the positive entries of A and C constructs by the absolute values of the negative entries of A and $A = B - C$.

For the nonsingular matrix S, we have $v(z) = S^{-1}w(z)$. However, there is a high likelihood that the obtained solution does not have the proper fuzzy number vector. Therefore, the solution of the FSLEs can be defined in the following form:

Definition 3 ([9]). *Let Equation (6) have the unique solution in the form* $\tilde{v}(z) = \{(\underline{v}_i(z), -\overline{v}_i(z)), 1 \leq i \leq n\}$. *We define the fuzzy number vector* $\tilde{P} = \{(\underline{p}_i(z), \overline{p}_i(z)), 1 \leq i \leq n\}$ *as:*

$$\begin{aligned} \underline{p}_i(z) &= \min\{\underline{v}_i(z), \overline{v}_i(z), \underline{v}_i(1), \overline{v}_i(1)\}, \\ \overline{p}_i(z) &= \max\{\underline{v}_i(z), \overline{v}_i(z), \overline{v}_i(1), \underline{v}_i(1)\}, \end{aligned} \tag{10}$$

which is called the fuzzy solution of Equation (6). If $\underline{p}_i(z) = \underline{v}_i(z)$ *and* $\overline{p}_i(z) = \overline{v}_i(z)$, $1 \leq i \leq n$ *then* $\tilde{P}$ *is called a strong fuzzy solution. Otherwise,* $\tilde{P}$ *is called a weak fuzzy solution, which is not the solution of FSLE and is not always the fuzzy number vector.*

Allahviranloo et al. [15] showed that a weak solution of a FSLE is not always a fuzzy number vector and it is the main fault of Friedman's method.

2.2. Ezzati's Method

Consider the following FSLEs:

$$\begin{cases} a_{11}(\underline{v}_1(z) + \overline{v}_1(z)) + \ldots + a_{1n}(\underline{v}_n(z) + \overline{v}_n(z)) &= \underline{w}_1(z) + \overline{w}_1(z), \\ a_{21}(\underline{v}_1(z) + \overline{v}_1(z)) + \ldots + a_{2n}(\underline{v}_n(z) + \overline{v}_n(z)) &= \underline{w}_2(z) + \overline{w}_2(z), \\ \qquad \vdots & \qquad \vdots \\ a_{n1}(\underline{v}_1(z) + \overline{v}_1(z)) + \ldots + a_{nn}(\underline{v}_n(z) + \overline{v}_n(z)) &= \underline{w}_n(z) + \overline{w}_n(z), \end{cases} \tag{11}$$

where the solution of system in Equation (11) is in the following form:

$$\mathbf{g}(\mathbf{z}) = \begin{pmatrix} g_1(z) \\ g_2(z) \\ \vdots \\ g_n(z) \end{pmatrix} = \underline{\mathbf{v}}(\mathbf{z}) + \overline{\mathbf{v}}(\mathbf{z}) = \begin{pmatrix} \underline{v}_1(z) + \overline{v}_1(z) \\ \underline{v}_2(z) + \overline{v}_2(z) \\ \vdots \\ \underline{v}_n(z) + \overline{v}_n(z) \end{pmatrix}. \tag{12}$$

Since $(B+C)\underline{v}(z) = \underline{w}(z) + Cg(z)$ and $(B+C)\overline{v}(z) = \overline{w}(z) + Cg(z)$, hence $\underline{v}(z)$ or $\overline{v}(z)$ can be determined by solving the following system:

$$\begin{aligned} \underline{v}(z) &= (B+C)^{-1}(\underline{w}(z) + Cg(z)), \\ \overline{v}(z) &= (B+C)^{-1}(\overline{w}(z) + Cg(z)). \end{aligned} \tag{13}$$

Therefore, the solution of FSLEs in Equaiton (11) can be obtained by solving the system of Equation (13); that the vector of solution is unique. But it may still not be an appropriate fuzzy number vector.

Theorem 1 ([35]). *Let $\tilde{v}(z) = (\tilde{v}_1(z), \tilde{v}_2(z), \ldots, \tilde{v}_n(z))^T$ be the fuzzy solution of Equation (11) and the matrix $\mathbf{A}^{-1}$ exists. Then the solution of system:*

$$\boldsymbol{A}(\overline{\boldsymbol{v}}(z) + \underline{\boldsymbol{v}}(z)) = \overline{\boldsymbol{w}}(z) + \underline{\boldsymbol{w}}(z)\,, \tag{14}$$

for $\overline{w}(z) + \underline{w}(z) = (\overline{w}_1(z) + \underline{w}_1(z), \overline{w}_2(z) + \underline{w}_2(z), \ldots, \overline{w}_n(z) + \underline{w}_n(z))^T$ is in the following form $\overline{v}(z) + \underline{v}(z) = (\overline{v}_1(z) + \underline{v}_1(z), \overline{v}_2(z) + \underline{v}_2(z), \ldots, \overline{v}_n(z) + \underline{v}_n(z))^T$.

Since the number of operations for solving $n \times n$ system are less than $2n \times 2n$ system, thus Ezzati's method is better than Friedman's method. In Theorem 4 of Ezzati's method [35], the maximum number of multiplication operations (MNMOs) were obtained which has some failures. The modified form of this theorem is presented in the following form:

Theorem 2. *Let F_n be the MNMOs of Friedman's method for the solving system of Equation (6) with $v(z) = (\underline{v}_1(z), \underline{v}_2(z), \ldots, \underline{v}_n(z), -\overline{v}_1(z), -\overline{v}_2(z), \ldots, -\overline{v}_n(z))^T = S^{-1}w(z)$, and E_n is the MNMOs of Ezzati's method with $v(z) = (\underline{v}_1(z), \underline{v}_2(z), \ldots \underline{v}_n(z), \overline{v}_1(z), \overline{v}_2(z), \ldots, \overline{v}_n(z))^T$, then $F_n \geq E_n$ and $F_n - E_n = 2n^2$.*

Proof. Suppose $h_n(A)$ is the MNMOs of computing the matrix A^{-1}. Now, we can write:

$$\mathbf{S}^{-1} = \begin{pmatrix} D & E \\ E & D \end{pmatrix}, \tag{15}$$

where $D = \frac{1}{2}[(B+C)^{-1} + (B-C)^{-1}]$, and $E = \frac{1}{2}[(B+C)^{-1} - (B-C)^{-1}]$. Therefore, in order to determine S^{-1}, computing matrices $(B+C)^{-1}$ and $(B-C)^{-1}$ are required. It is clear that $h_n(S) = h_n(B+C) + h_n(B-C) = 2h_n(A)$. Since $\tilde{v}(z) \in E^1$; $\underline{v}(z)$ and $\overline{v}(z)$, in the simplest case are lines hence $F_n = 2h_n(A) + 8n^2$. For computing $\underline{v}(z) + \overline{v}(z) = (\underline{v}_1(z) + \overline{v}_1(z), \underline{v}_2(z) + \overline{v}_2(z), \ldots, \underline{v}_n(z) + \overline{v}_n(z))^T$ from Equation (11) and $\underline{v}(z) = (\underline{v}_1(z), \underline{v}_2(z), \ldots \underline{v}_n(z))^T$ from Equation (13) and according to Ezzati's method, the MNMOs are $h_n(A) + 2n^2$ and $h_n(B+C) + 4n^2$, respectively. Since $h_n(B+C) = h_n(A)$ thus $E_n = 2h_n(A) + 6n^2$, and finally $F_n - E_n = 2n^2$. □

Definition 4 ([35]). *Assume $\tilde{v}(z) = \{(\underline{v}_i(z), \overline{v}_i(z)),\ 1 \leq i \leq n\}$ is the unique solution of Equations (11) and (13). The fuzzy number vector $\tilde{P} = \{(\underline{p}_i(z), \overline{p}_i(z)),\ 1 \leq i \leq n\}$ is defined by:*

$$\begin{aligned} \underline{p}_i(z) &= \min\{\underline{v}_i(z), \overline{v}_i(z), \underline{v}_i(1)\}, \\ \overline{p}_i(z) &= \max\{\underline{v}_i(z), \overline{v}_i(z), \overline{v}_i(1)\}, \end{aligned} \tag{16}$$

which is called a fuzzy vector solution of Equations (11) and (13).

If $\underline{p}_i(z) = \underline{v}_i(z)$ and $\overline{p}_i(z) = \overline{v}_i(z)$, $1 \leq i \leq n$, then $\tilde{P}$ is called a strong fuzzy solution. Otherwise, $\tilde{P}$ is called a weak fuzzy solution which is not the fuzzy linear system's solution and is not always a fuzzy number vector.

Remark 2. *We know that Friedman et al. [9] and Ezzati [35] found two kinds of solutions, which are called the weak and strong solutions. The weak solution is not the system's solution and it is not always the fuzzy number vector [15]. Hence, finding the weak fuzzy solution is not an interesting problem. Furthermore, in these methods the kind of solutions (strong or weak) can be determined only in the end of solving process and is one of the important faults of these methods.*

In the next section, a novel method for solving $n \times n$ FSLEs is presented. Based on this method we can decrease the computing error because without carrying out a further computation, we can determine that the fuzzy linear system has no fuzzy number vector solution.

3. Proposed Method

In this section, a novel and applicable method to solve the FSLEs is presented. Several theorems and lemmas are illustrated to improve Ezzati's method [35]. By using these theorems we show the number of operations of our method are less than the methods of Ezzati [35] and Friedman [9].

Theorem 3. *Suppose that the inverse matrix of* $\boldsymbol{B} + \boldsymbol{C}$ *exists and* $\tilde{v}(z) = (\tilde{v}_1(z), \tilde{v}_2(z), \ldots, \tilde{v}_n(z))^T$ *is a fuzzy solution of Equation (11). Then* $\overline{v}(z) - \underline{v}(z) = (\overline{v}_1(z) - \underline{v}_1(z), \overline{v}_2(z) - \underline{v}_2(z), \ldots, \overline{v}_n(z) - \underline{v}_n(z))^T$ *is the solution of the following system:*

$$(\boldsymbol{B} + \boldsymbol{C})(\overline{v}(z) - \underline{v}(z)) = \overline{w}(z) - \underline{w}(z)\,, \tag{17}$$

where $\overline{w}(z) - \underline{w}(z) = (\overline{w}_1(z) - \underline{w}_1(z), \overline{w}_2(z) - \underline{w}_2(z), \ldots, \overline{w}_n(z) - \underline{w}_n(z))^T$.

Proof. Let $\tilde{v}_j(z) = (\underline{v}_j(z), \overline{v}_j(z)), 1 \leq j \leq n$ be the parametric form of $\tilde{v}_j$. For positive values a'_{ij} and a''_{ij} we have $a_{ij} = a'_{ij} - a''_{ij}$ and $a'_{ij} a''_{ij} = 0$ where a_{ij}, a'_{ij} and a''_{ij} are the coefficients of matrices A, B, and C respectively. By presenting Equation (11) to the parametric form, for $i = 1, 2, \ldots, n$ we get:

$$(a'_{i1} - a''_{i1})(\underline{v}_1(z), \overline{v}_1(z)) + \ldots + (a'_{in} - a''_{in})(\underline{v}_n(z), \overline{v}_n(z)) = (\underline{w}_i(z), \overline{w}_i(z)). \tag{18}$$

Hence,

$$a'_{i1}\underline{v}_1(z) - a''_{i1}\overline{v}_1(z) + a'_{i2}\underline{v}_2(z) - a''_{i2}\overline{v}_2(z) + \ldots + a'_{in}\underline{v}_n(z) - a''_{in}\overline{v}_n(z) = \underline{w}_i(z), \tag{19}$$

and

$$a'_{i1}\overline{v}_1(z) - a''_{i1}\underline{v}_1(z) + a'_{i2}\overline{v}_2(z) - a''_{i2}\underline{v}_2(z) + \ldots + a'_{in}\overline{v}_n(z) - a''_{in}\underline{v}_n(z) = \overline{w}_i(z). \tag{20}$$

Now, using Equation (19) from Equation (20) we can write:

$$(a'_{i1} + a''_{i2})(\overline{v}_1(z) - \underline{v}_1(z)) + (a'_{i2} + a''_{i2})(\overline{v}_2(z) - \underline{v}_2(z)) + \ldots + (a'_{in} + a''_{in})(\overline{v}_n(z) - \underline{v}_n(z)) = \overline{w}_i(z) - \underline{w}_i(z). \tag{21}$$

Therefore, $d(z) = \overline{v}(z) - \underline{v}(z) = (\overline{v}_1(z) - \underline{v}_1(z), \overline{v}_2(z) - \underline{v}_2(z), \ldots, \overline{v}_n(z) - \underline{v}_n(z))^T$ is the solution of $(B + C)(\overline{v}(z) - \underline{v}(z)) = \overline{w}(z) - \underline{w}(z)$. □

Theorem 4. *Suppose the inverse matrix of* $\boldsymbol{B} + \boldsymbol{C}$ *exists. Equation (11) does not have a fuzzy number vector solution, if the vector solution of the following system is not non-negative, i.e., at least one of the entries are negative:*

$$(\boldsymbol{B} + \boldsymbol{C})(\overline{v}(z) - \underline{v}(z)) = \overline{w}(z) - \underline{w}(z). \tag{22}$$

Proof. We know that, the vector solution of Equation (22) is $\overline{v}(z) - \underline{v}(z)$. Now, suppose that $\overline{v}(z) - \underline{v}(z)$ is not non-negative. So, according to Definition 1, the fuzzy number vector solution does not exist. It is clear that the matrix $(B + C)^{-1}$ is the non-positive matrix, i.e., at least one of the entries is positive because $(B + C)$ is the positive matrix. □

Triangular fuzzy numbers are simple and popular fuzzy numbers. Triangular fuzzy numbers also have a special property $\overline{w}(z) - \underline{w}(z) = (\rho' + \rho'')(1 - z)$. Hence when the right-hand side vector $\tilde{w}(z)$ is triangular, the parametric linear system of Equation (22) can be transformed into the crisp linear system.

Lemma 1. *Suppose that the inverse matrix of $(B + C)$ exists, and $\tilde{w}(z) \in TE^1$. Equation (11) does not have a fuzzy number vector solution, if the vector solution of the following system is not non-negative, i.e., at least one of the entries is negative:*

$$(B + C)(\mu' + \mu'') = (\rho' + \rho''), \tag{23}$$

where $(\mu' + \mu'')(1 - z) = \overline{v}(z) - \underline{v}(z)$, $(\rho' + \rho'')(1 - z) = \overline{w}(z) - \underline{w}(z)$.

Now, a new method to solve the FSLEs is presented. Assume that the inverse matrix of A in Equation (11) exists. For solving Equation (11), the following system:

$$(\mathbf{B} + \mathbf{C})(\overline{\mathbf{v}}(z) - \underline{\mathbf{v}}(z)) = \overline{\mathbf{w}}(z) - \underline{\mathbf{w}}(z), \tag{24}$$

should be solved where the matrices B and C were defined in Section 2.2. Let the solution of this system be in the following form:

$$\mathbf{d}(\mathbf{z}) = \begin{pmatrix} d_1(z) \\ d_2(z) \\ \vdots \\ d_n(z) \end{pmatrix} = \overline{\mathbf{v}}(\mathbf{z}) - \underline{\mathbf{v}}(\mathbf{z}) = \begin{pmatrix} \overline{v}_1(z) - \underline{v}_1(z) \\ \overline{v}_2(z) - \underline{v}_2(z) \\ \vdots \\ \overline{v}_n(z) - \underline{v}_n(z) \end{pmatrix}. \tag{25}$$

If $d = \overline{v} - \underline{v}$ is not non-negative, then we do not have the fuzzy number vector solution. Otherwise, in order to show the existence of fuzzy number vector solution for Equation (11), we continue our idea. At first, we should solve the following system:

$$\mathbf{A}(\overline{\mathbf{v}}(z) + \underline{\mathbf{v}}(z)) = \overline{\mathbf{w}}(z) + \underline{\mathbf{w}}(z). \tag{26}$$

According to Theorem 1, we know that this system has the solution in the following form:

$$\mathbf{g}(\mathbf{z}) = \begin{pmatrix} g_1(z) \\ g_2(z) \\ \vdots \\ g_n(z) \end{pmatrix} = \overline{\mathbf{v}}(\mathbf{z}) + \underline{\mathbf{v}}(\mathbf{z}) = \begin{pmatrix} \overline{v}_1(z) + \underline{v}_1(z) \\ \overline{v}_2(z) + \underline{v}_2(z) \\ \vdots \\ \overline{v}_n(z) + \underline{v}_n(z) \end{pmatrix}. \tag{27}$$

Finally, by solving systems of Equations (24) and (26) and finding $\mathbf{d}(\mathbf{z})$ and $\mathbf{g}(\mathbf{z})$ we have:

$$\begin{cases} \underline{v}(z) = \frac{\mathbf{g}(\mathbf{z}) - \mathbf{d}(\mathbf{z})}{2}, \\ \overline{v}(z) = \frac{\mathbf{g}(\mathbf{z}) + \mathbf{d}(\mathbf{z})}{2}. \end{cases} \tag{28}$$

If the conditions of Definition 1 are satisfied, then the solution of FSLEs for Equation (11) can be obtained by solving the crisp linear system of Equations (24) and (26) that the solution vector is the fuzzy number vector and unique. Otherwise, if at least one of the conditions is not true, the fuzzy linear system of Equation (11) does not have a fuzzy number vector solution.

Remark 3. *If $\tilde{w} \in TE^1$, then according to Lemma 1 the system of Equation (24) have the vector solution as $d' = \mu' + \mu''$ where $d(z) = d'(1-z)$. So, Equation (28) can be written in the following form:*

$$\begin{cases} \underline{v}(z) = \frac{\mathbf{g(z)-d'(1-z)}}{2}, \\ \overline{v}(z) = \frac{\mathbf{g(z)+d'(1-z)}}{2}. \end{cases} \tag{29}$$

Theorem 5. *Assume that n is any integer, $n \geq 2$, and denote by E_n and D_n the MNMOs that are required to calculate $v(z) = (\underline{v}_1(z), \underline{v}_2(z), \ldots \underline{v}_n(z), \overline{v}_1(z), \overline{v}_2(z), \ldots, \overline{v}_n(z))^T$, in Ezzati's method [35] and presented method then:*

$$\begin{cases} E_n - D_n = 2n^2, & \overline{v}(z) - \underline{v}(z) \geq 0, \\ E_n - D_n = h_n(A) + 4n^2, & o.w., \end{cases} \tag{30}$$

where $h_n(A)$ shows the MNMOs that are required to calculate A^{-1}.

Proof. According to Theorem 2, we have $E_n = 2h_n(A) + 6n^2$. Assume $d(z) = \overline{v}(z) - \underline{v}(z)$ is the non-negative matrix then in order to know that the fuzzy linear system of Equation (11) has the fuzzy number vector solution, we need to solve the system of Equation (26). So, for computing $\overline{v}(z) - \underline{v}(z) = (\overline{v}_1(z) - \underline{v}_1(z), \overline{v}_2(z) - \underline{v}_2(z), \ldots, \overline{v}_n(z) - \underline{v}_n(z))^T$, and $\overline{v}(z) + \underline{v}(z) = (\overline{v}_1(z) + \underline{v}_1(z), \overline{v}_2(z) + \underline{v}_2(z), \ldots, \overline{v}_n(z) + \underline{v}_n(z))^T$, from Equations (24) and (26), the maximum number of multiplication operations are $h_n(B+C) + 2n^2$ and $h_n(A) + 2n^2$, respectively. Clearly $h_n(B+C) = h_n(A)$. Hence $D_n = 2h_n(A) + 4n^2$ and $E_n - D_n = 2n^2$.

Otherwise, assume that $d(z) = \overline{v}(z) - \underline{v}(z)$ is not the non-negative matrix. According to Theorem 4 we do not have the fuzzy number vector solution for solving the FSLEs of Equation (11). If we do not have the fuzzy number vector solution, computing $\overline{v}(z) + \underline{v}(z)$ from the system of Equation (26) will be unnecessary. Thus, we need to compute $d(z) = \overline{v}(z) - \underline{v}(z)$. Therefore, in this case, the maximum number of multiplication operations are $h_n(B+c) + 2n^2$ or $h_n(A) + 2n^2$. Finally, we have $E_n \geq D_n$, $E_n - D_n = h_n(A) + 4n^2$. □

Lemma 2. *Let $\tilde{w}(z)$ be the triangular fuzzy number vector from Equation (11). Then $\tilde{v}(z)$ is the triangular fuzzy number vector solution from Equation (11).*

Lemma 3. *Suppose that in Theorem 5, $\tilde{w}(z)$ is the triangular fuzzy number vector from Equation (11), then $E_n \geq D_n$ and*

$$\begin{cases} E_n - D_n = 3n^2 - n, & \overline{v}(z) - \underline{v}(z) \geq 0, \\ E_n - D_n = h_n(A) + 5n^2, & o.w. \end{cases} \tag{31}$$

Proof. If $\tilde{w}(z)$ is the triangular fuzzy number vector from Equation (11) i.e., since $\tilde{w}(z) \in TE^1; \underline{w}(z)$ and $\overline{w}(z)$ in the simplest case is the line. So clearly, according to Theorem 2, we have $E_n = 2h_n(A) + 6n^2$, and according to Remark 1, we get $\overline{v}(z) = c + (1-z)\mu'$ and $\underline{v}(z) = c - (1-z)\mu''$. So, $\overline{v}(z) - \underline{v}(z) = (\mu' + \mu'')(1-z)$ and $\overline{w}(z) - \underline{w}(z) = (\rho' + \rho'')(1-z)$, and from the system of Equation (24), we have:

$$(B+C)(\mu' + \mu'')(1-z) = (\rho' + \rho'')(1-z). \tag{32}$$

If $r \neq 1$, the following relation can be obtained as:

$$(B+C)(\mu' + \mu'') = \rho' + \rho'', \tag{33}$$

where it is the crisp linear system. It is clear that for $r = 1$, the FSLE can be replaced by crisp linear system.

Now, we assume that $d' = (\mu' + \mu'')$ is non-negative. Then for understanding that whether the FSLEs of Equation (11) has the fuzzy number vector solution, we need to solve the system of Equation (26). So, for computing $(\mu' + \mu'') = (\mu_1' + \mu_1'', \mu_2' + \mu_2'', \dots, \mu_n' + \mu_n'')$ from Equation (33) and $\overline{v}(z) + \underline{v}(z) = (\overline{v}_1(z) + \underline{v}_1(z), \overline{v}_2(z) + \underline{v}_2(z), \dots, \overline{v}_n(z) + \underline{v}_n(z))^T$ from Equation (26) and $d(z) = (1-z)d'$ for the final solution in Equation (29) the maximum number of multiplication operations are $h_n(B+C) + n^2$, $h_n(A) + 2n^2$ and n respectively. Clearly $h_n(B+C) = h_n(A)$. So $D_n = 2h_n(A) + 3n^2 + n$, and

$$E_n \geq D_n, \quad E_n - D_n = 3n^2 - n. \tag{34}$$

Otherwise, assume that $d' = (\mu' + \mu'')$ is not non-negative. Then, according to Lemma 1 we do not have a fuzzy number vector solution for solving the fuzzy linear system of Equation (11). We know that if we do not have a fuzzy number vector solution, there is no necessity for computing $\underline{v}(z) + \overline{v}(z)$ from Equation (26). Thus, we need to compute $d' = (\mu' + \mu'')$. Therefore, in this case the maximum number of multiplication operations are $h_n(B+C) + n^2$ *or* $h_n(A) + n^2$. Then, we have $E_n \geq D_n, \quad E_n - D_n = h_n(A) + 5n^2$. □

Method's Algorithms

In Figures 1 and 2, the flowcharts of mentioned algorithms are presented.

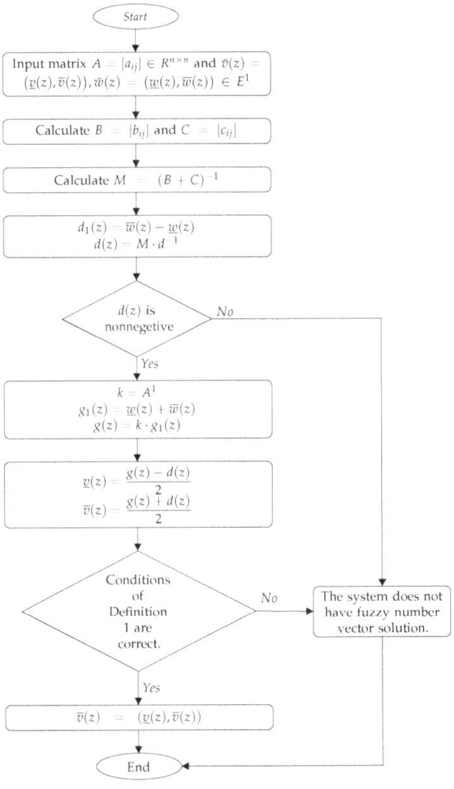

Figure 1. Flowchart of Algorithm 1.

Algorithm 1: Let **A** be the nonsingular matrix.

Step 1: Input matrix $A = [a_{ij}] \in R^{n\times n}$ and $\tilde{v}(z) = (\underline{v}(z), \overline{v}(z)), \tilde{w}(z) = (\underline{w}(z), \overline{w}(z)) \in E^1$.

Step 2: Calculate $B = [b_{ij}]$ and $C = [c_{ij}]$ as

$$\begin{cases} If\ a_{ij} > 0 \Rightarrow b_{ij} = a_{ij}; \quad else\ b_{ij} = 0, \\ If\ a_{ij} < 0 \Rightarrow c_{ij} = -a_{ij}; \quad else\ c_{ij} = 0. \end{cases}$$

Step 3: Calculate $M = (B + C)^{-1}$.

Step 4: Calculate $d_1(z) = \overline{w}(z) - \underline{w}(z)$ and $d(z) = M.d_1(z)$. If $d(z)$ is not nonnegative, go to Step 8.

Step 5: Calculate $k = A^{-1}, g_1(z) = \overline{w}(z) + \underline{w}(z)$ and $g(z) = k.g_1(z)$.

Step 6: Calculate $\underline{v}(z) = \frac{g(z)-d(z)}{2}$ and $\overline{v}(z) = \frac{g(z)+d(z)}{2}$.

Step 7: If conditions of Definition 1 are true then $\tilde{v}(z) = (\underline{v}(z), \overline{v}(z))$ and go to Step 9. Else go to Step 8.

Step 8: Show the message "The system does not have fuzzy number vector solution".

Step 9: End.

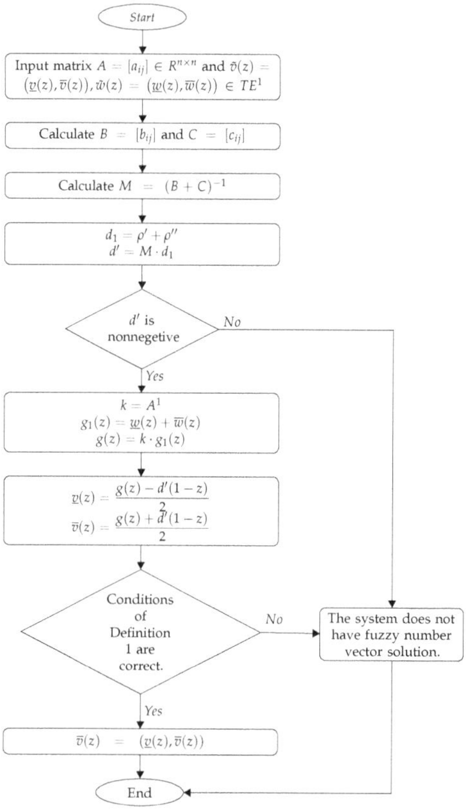

Figure 2. Flowchart of Algorithm 2.

Algorithm 2: The following algorithm is presented to triangular fuzzy linear system.

Step 1: Input matrix $A = [a_{ij}] \in R^{n\times n}$
and $\tilde{v}(z) = (\underline{v}(z), \overline{v}(z)), \tilde{w}(z) = (\underline{w}(z), \overline{w}(z)) \in TE^1$.
Step 2: Calculate $B = [b_{ij}]$ and $C = [c_{ij}]$ as

$$\begin{cases} If\ a_{ij} > 0 \Rightarrow b_{ij} = a_{ij}; \quad else\ b_{ij} = 0, \\ If\ a_{ij} < 0 \Rightarrow c_{ij} = -a_{ij}; \quad else\ c_{ij} = 0. \end{cases}$$

Step 3: Calculate $M = (B+C)^{-1}$.
Step 4: Calculate $d_1 = \rho' + \rho''$ and $d' = M.d_1$. If d' is not nonnegative, go to Step 8.
Step 5: Calculate $k = A^{-1}, g_1(z) = \overline{w}(z) + \underline{w}(z)$ and $g(z) = k.g_1(z)$.
Step 6: Calculate $\underline{v}(z) = \frac{g(z)-d'(1-z)}{2}$; and $\overline{v}(z) = \frac{g(z)+d'(1-z)}{2}$.
Step 7: If conditions of Definition 1 are true then $\tilde{v}(z) = (\underline{v}(z), \overline{v}(z))$ and go to Step 9. Else go to Step 8.
Step 8: Show the message "The system does not have fuzzy number vector solution".
Step 9: End.

4. Numerical Illustrations

In this section, some examples of the FSLEs are presented [9]. Furthermore, several graphs are demonstrated to show that the obtained solutions are fuzzy vector solutions.

Example 1 ([9]). *Consider the following* 2×2 *FSLEs:*

$$\begin{cases} \tilde{v}_1 - \tilde{v}_2 = (z, 2-z), \\ \tilde{v}_1 + 3\tilde{v}_2 = (4+z, 7-2z), \end{cases} \tag{35}$$

where $\tilde{w}$ *is a triangular vector of fuzzy numbers hence Algorithm 2 is applied. By using this algorithm we have:*
Step 1. *Input matrix:*

$$A = \begin{pmatrix} 1 & -1 \\ 1 & 3 \end{pmatrix}, \tag{36}$$

$$\tilde{v}(z) = (\underline{v}(z), \overline{v}(z)) \in TE^1, \quad \tilde{w}(z) = (\underline{w}(z), \overline{w}(z)) = \begin{pmatrix} (z, 2-z) \\ (4+z, 7-2z) \end{pmatrix}. \tag{37}$$

Step 2. *Calculate:*

$$B = [b_{ij}] = \begin{pmatrix} 1 & 0 \\ 1 & 3 \end{pmatrix}, \quad C = [c_{ij}] = \begin{pmatrix} 0 & 1 \\ 0 & 0 \end{pmatrix}. \tag{38}$$

Step 3. *Compute:*

$$M = (B+C)^{-1} = \begin{pmatrix} \frac{3}{2} & -\frac{1}{2} \\ -\frac{1}{2} & \frac{1}{2} \end{pmatrix}. \tag{39}$$

Step 4. *Calculate:*

$$d_1 = \rho' + \rho'' = \begin{pmatrix} \rho_1' + \rho_1'' \\ \rho_2' + \rho_2'' \end{pmatrix} = \begin{pmatrix} 2 \\ 3 \end{pmatrix}, \quad d' = M.d_1 = \begin{pmatrix} \frac{3}{2} \\ \frac{1}{2} \end{pmatrix}. \tag{40}$$

It is clear that $d' = \mu' + \mu''$ is the non-negative matrix, therefore go to the Step 5.

Step 5. *Calculate:*

$$k = A^{-1} = \begin{pmatrix} \frac{3}{4} & \frac{1}{4} \\ -\frac{1}{4} & \frac{1}{4} \end{pmatrix}, \quad g_1(z) = \overline{w}(z) + \underline{w}(z) = \begin{pmatrix} 2 \\ 11 - z \end{pmatrix}, \tag{41}$$

$$g(z) = k.g_1(z) = \begin{pmatrix} \frac{17 - z}{4} \\ \frac{9 - z}{4} \end{pmatrix}. \tag{42}$$

Step 6. *Compute:*

$$\underline{v}(z) = \frac{g(z) - d(z)}{2} = \begin{pmatrix} \underline{v}_1 \\ \underline{v}_2 \end{pmatrix} = \begin{pmatrix} 1.375 + 0.625z \\ 0.875 + 0.125z \end{pmatrix}, \tag{43}$$

$$\overline{v}(z) = \frac{g(z) + d(z)}{2} = \begin{pmatrix} \overline{v}_1 \\ \overline{v}_2 \end{pmatrix} = \begin{pmatrix} 2.875 - 0.875z \\ 1.375 - 0.375z \end{pmatrix}. \tag{44}$$

Since the conditions of Definition 1 are satisfied, the FSLEs has the fuzzy number vector solution. Now, we can show that the MNMOs of the presented method are less than the MNMOs of Friedman and Ezzati's methods. Suppose $h_n(A)$ is the MNMOs of A^{-1} and F_n, E_n and D_n are the MNMOs obtained from Friedman's method, Ezzati's method and the proposed method, respectively. So for $n = 2$ we have:

$$\begin{aligned} F_n &= 2h_n(A) + 8n^2 = 2h_n(A) + 32, \\ E_n &= 2h_n(A) + 6n^2 = 2h_n(A) + 24, \\ D_n &= 2h_n(A) + 3n^2 + n = 2h_n(A) + 14, \end{aligned}$$

and it is obvious that $D_n < E_n < F_n$. In Figure 3, the obtained solutions are presented for $z \in [0,1]$. Based on this figure we can clearly see that the obtained results are the fuzzy number vector solutions.

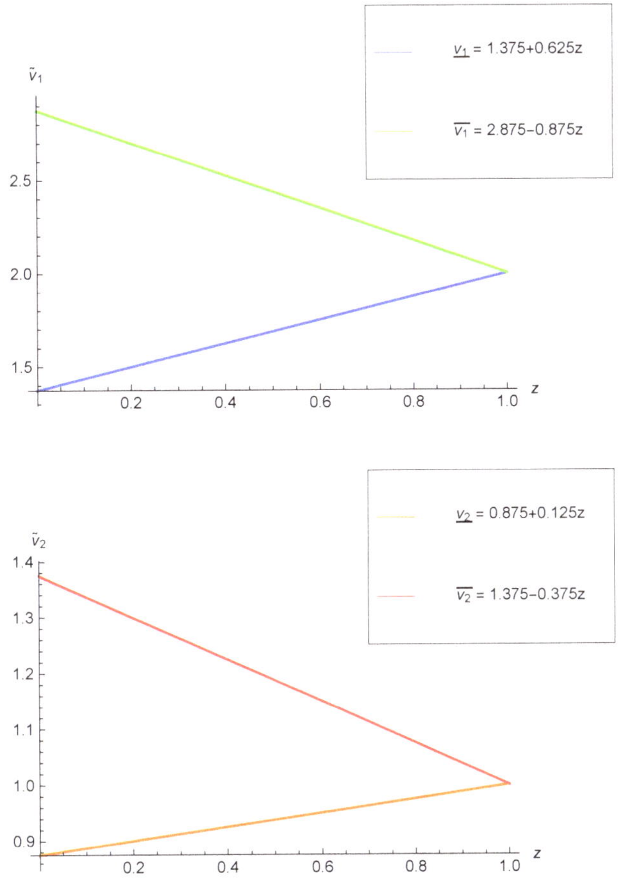

Figure 3. The solutions of example 1.

Example 2 ([9]). *Consider the* 3×3 *FSLEs:*

$$\begin{cases} \tilde{v}_1 + \tilde{v}_2 - \tilde{v}_3 = (z, 2-z), \\ \tilde{v}_1 - 2\tilde{v}_2 + \tilde{v}_3 = (2+z, 3), \\ 2\tilde{v}_1 + \tilde{v}_2 + 3\tilde{v}_3 = (-2, -1-z), \end{cases} \tag{45}$$

where $\tilde{Y}$ *is a triangular fuzzy number vector. By using Algorithm 2 we have:*
Step 1. *Input matrix:*

$$A = \begin{pmatrix} 1 & 1 & -1 \\ 1 & -2 & 1 \\ 2 & 1 & 3 \end{pmatrix}, \quad \tilde{w}(z) = (\underline{w}(z), \overline{w}(z)) = \begin{pmatrix} (z, 2-z) \\ (2+z, 3) \\ (-2, -1-z) \end{pmatrix}. \tag{46}$$

Step 2. *Compute:*

$$B = [b_{ij}] = \begin{pmatrix} 1 & 1 & 0 \\ 1 & 0 & 1 \\ 2 & 1 & 3 \end{pmatrix}, \quad C = [c_{ij}] = \begin{pmatrix} 0 & 0 & 1 \\ 0 & 2 & 0 \\ 0 & 0 & 0 \end{pmatrix}. \tag{47}$$

Step 3. *Calculate:*

$$M = (B + C)^{-1} = \begin{pmatrix} 5 & -1 & -3 \\ -2 & 1 & 1 \\ -1 & 0 & 1 \end{pmatrix}. \tag{48}$$

Step 4. *Compute:*

$$d_1 = \rho' + \rho'' = \begin{pmatrix} \rho'_1 + \rho''_1 \\ \rho'_2 + \rho''_2 \end{pmatrix} = \begin{pmatrix} 2 \\ 1 \\ 1 \end{pmatrix}, \quad d' = M.d_1 = \begin{pmatrix} 6 \\ -2 \\ -1 \end{pmatrix}. \tag{49}$$

Since $d' = \mu' + \mu''$ is not non-negative, therefore the system does not have a fuzzy number vector solution. In order to show the MNMOs of mentioned methods for $n = 3$ we get:

$$F_n = 2h_n(A) + 8n^2 = 2h_n(A) + 72,$$
$$E_n = 2h_n(A) + 6n^2 = 2h_n(A) + 54,$$
$$D_n = 2h_n(A) + n^2 = 2h_n(A) + 9.$$

In this case, we have $D_n < E_n < F_n$ and it shows that MNMOs of the presented method is less than two other methods.

Example 3. *Consider the following 2×2 FSLEs:*

$$\begin{cases} \tilde{v}_1 + \tilde{v}_2 = (4z, 6 - 2z), \\ \tilde{v}_1 + 2\tilde{v}_2 = (5z, 8 - 3z). \end{cases} \tag{50}$$

In this example, by applying Algorithm 1 we have:

Step 1. *Input matrix:*

$$A = \begin{pmatrix} 1 & 1 \\ 1 & 2 \end{pmatrix}, \quad \tilde{w}(z) = (\underline{w}(z), \overline{w}(z)) = \begin{pmatrix} (4z, 6 - 2z) \\ (5z, 8 - 3z) \end{pmatrix}. \tag{51}$$

Step 2. *Compute:*

$$B = [b_{ij}] = \begin{pmatrix} 1 & 1 \\ 1 & 2 \end{pmatrix}, \quad C = [c_{ij}] = \begin{pmatrix} 0 & 0 \\ 0 & 0 \end{pmatrix}. \tag{52}$$

Step 3. *Calculate:*

$$M = (B + C)^{-1} = \begin{pmatrix} 2 & -1 \\ -1 & 1 \end{pmatrix}. \tag{53}$$

Step 4. *Compute:*

$$d_1(z) = \overline{w}(z) - \underline{w}(z) = \begin{pmatrix} 6 - 6z \\ 8 - 8z \end{pmatrix}, \quad d(z) = M.d_1 = \begin{pmatrix} 4 - 4z \\ 2 - 2z \end{pmatrix}. \tag{54}$$

We know $d(z) = \overline{v}(z) - \underline{v}(z)$ is non-negative for $0 \le z \le 1$, therefore, in order to know about having the fuzzy number vector solution we need to go to Step 5.
Step 5. *Calculate:*

$$k = A^{-1} = \begin{pmatrix} 2 & -1 \\ -1 & 1 \end{pmatrix}, \quad g_1(z) = \overline{w}(z) + \underline{w}(z) = \begin{pmatrix} 6 + 2z \\ 8 + 2z \end{pmatrix}, \tag{55}$$

$$g(z) = k.g_1(z) = \begin{pmatrix} 4 + 2z \\ 2 \end{pmatrix}. \tag{56}$$

Step 6. *Compute:*

$$\underline{v}(z) = \frac{g(z) - d(z)}{2} = \begin{pmatrix} 3z \\ z \end{pmatrix}, \quad \overline{v}(z) = \frac{g(z) + d(z)}{2} = \begin{pmatrix} 4 - z \\ 2 - z \end{pmatrix}. \tag{57}$$

Since the conditions of Definition 1 are connected hence the vector solution is the fuzzy number vector solution. Thus, the FSLEs of Equation (50) has the fuzzy number vector solution. The MNMOs of this example for $n = 2$ can be obtained in the following form:

$$\begin{aligned} F_n &= 2h_n(A) + 8n^2 = 2h_n(A) + 32, \\ E_n &= 2h_n(A) + 6n^2 = 2h_n(A) + 24, \\ D_n &= 2h_n(A) + 4n^2 = 2h_n(A) + 16, \end{aligned}$$

and we have $D_n < E_n < F_n$. It is clear that the mentioned method is faster than two other methods. Figure 4 shows that based on the presented method and for $z \in [0, 1]$ the obtained solutions are the fuzzy number vector solutions.

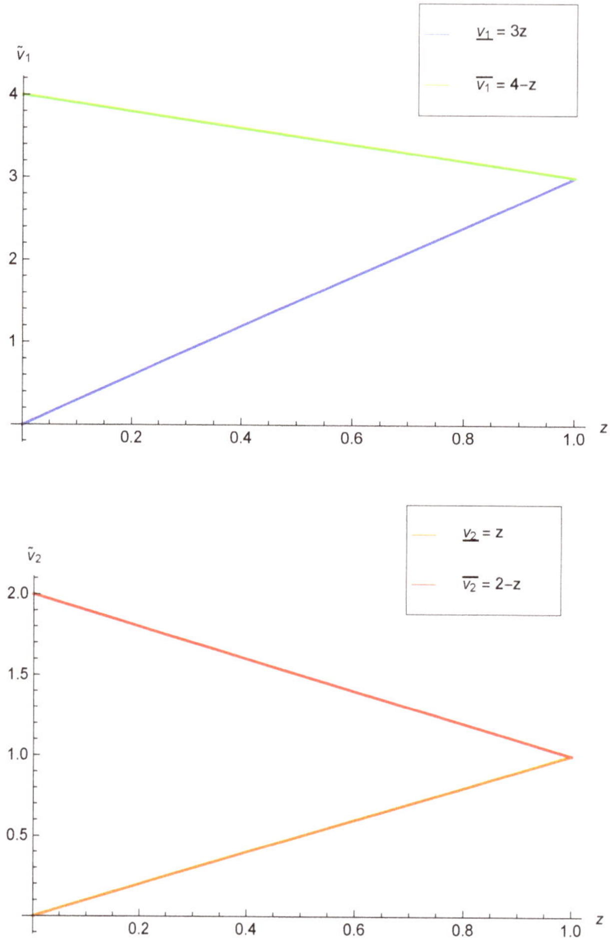

Figure 4. The solutions of example 2.

5. Conclusions

The aim of this work was to solve the fuzzy system of linear equations. In this study, we improved Friedman and Ezzati's methods [9,35] for solving fuzzy linear systems. Two kinds of solutions (weak and strong) were found and the weak solution was found to not be the system's solution and was not always the fuzzy number vector. Furthermore, in their methods the kind of solution could only be found at the end of the solving process. In addition, they had a high number of operations found in the mentioned solutions. In the proposed method, the initial fuzzy system was replaced by two $n \times n$ crisp linear systems in two steps. The presented algorithms and flowcharts showed the usefulness and contributions of the proposed approach. For all examples, after presenting the solving process, we compared the number of operations to show the abilities of the proposed method. It shows that the mentioned method decreased the MNMOs in comparison with Friedman and Ezzati's methods and it was one of the main novelties of this research. Solving the $m \times n$ system of fuzzy linear equations and fuzzy bio-mathematical models using the mentioned method will be the focus of our future research.

Author Contributions: All the authors designed the research and the techniques and methods, searched literature and wrote the paper. All the authors revised the manuscript. All the authors contributed equally to this work. All authors have read and agreed to the published version of the manuscript.

Funding: This research was funded by partially supported by Agencia Estatal de Investigación (AEI) of Spain under grant MTM2016-75140-P, co-financed by the European Community fund FEDER, and XUNTA de Galicia under grants ED431C 2019-02 and R2016-022.

Conflicts of Interest: The authors declare no conflict of interest.

References

1. Nieto, J.J.; Khastan, A.; Ivaz, K. Numerical solution of fuzzy differential equations under generalized differentiability. *Nonlinear Anal. Hybrid Syst.* **2009**, *3*, 700–707. [CrossRef]
2. Noeiaghdam, Z.; Allahviranloo, T.; Nieto, J.J. *q*-Fractional differential equations with uncertainty. *Soft Comput.* **2019**, *23*, 9507–9524. [CrossRef]
3. Noieaghdam, Z.; Noeiaghdam, S.; Nieto, J.J. A full fuzzy method for solving fuzzy fractional differential equations based on the generalized Taylor expansion. *arXiv* **2019**, arXiv:1912.04048.
4. Stanujkić, D.; Karabašević, D. An extension of the WASPAS method for decision-making problems with intuitionistic fuzzy numbers: A case of website evaluation. *Oper. Res. Eng. Sci. Theory Appl.* **2018**, *1*, 29–39. [CrossRef]
5. Stojić, G.; Sremac, S.; Vasiljković, I. A fuzzy model for determining the justifiability of investing in a road freight vehicle fleet. *Oper. Res. Eng. Sci. Theory Appl.* **2018**, *1*, 62–75. [CrossRef]
6. Si, A.; Das, S.; Kar, S. An approach to rank picture fuzzy numbers for decision making problems. *Appl. Manag. Eng.* **2019**, *2*, 54–64. [CrossRef]
7. Abbasi, R.; Hamidi Beheshti, M.T.; Mohraz, M. Modeling andstability analysis of HIV-1 as a time delay fuzzy TS system via LMIs. *Appl. Math. Model.* **2015**, *39*, 7134–7154. [CrossRef]
8. Mishra, B.K.; Pandey, S.K. Fuzzy epidemic model for the transmission of worms in computer network. *Nonlinear Anal. Real World Appl.* **2010**, *11*, 4335–4341. [CrossRef]
9. Friedman, M.; Ming, M.; Kandel, A. Fuzzy linear systems. *Fuzzy Sets Syst.* **1998**, *96*, 201–209. [CrossRef]
10. Ma, M.; Friedman, M.; Kandel, A. A new fuzzy arithmetic. *Fuzzy Sets Syst.* **1999**, *108*, 83–90. [CrossRef]
11. Abbasbandy, S.; Ezzati, R.; Jafarian, A. LU decomposition method for solving fuzzy system of linear equations. *Appl. Math. Comput.* **2006**, *172*, 633–643. [CrossRef]
12. Abbasbandy, S.; Jafarian, A. Steepest descent method for system of fuzzy linear equations. *Appl. Math. Comput.* **2006**, *175*, 823–833. [CrossRef]
13. Asady, B.; Abasbandy, S.; Alavi, M. Fuzzy general linear systems. *Appl. Math. Comput.* **2005**, *169*, 34–40. [CrossRef]
14. Allahviranloo, T.; Afshar Kermani, M. Solution of a fuzzy system of linear equation. *Appl. Math. Comput.* **2006**, *175*, 519–531. [CrossRef]
15. Allahviranloo, T.; Ghanbari, M.; Hosseinzadeh, A.A.; Haghi, E.; Nuraei, R. A note on Fuzzy linear systems. *Fuzzy Sets Syst.* **2011**, *177*, 87–92. [CrossRef]
16. Allahviranloo, T.; Hosseinzadeh Lotfi, F.; Khorasani Kiasari, M.; Khezerloo, M. On the fuzzy solution of LR fuzzy linear systems. *Appl. Math. Model.* **2013**, *37*, 1170–1176. [CrossRef]
17. Allahviranloo, T.; Nuraei, R.; Ghanbari, M.; Haghi, E.; Hosseinzadeh, A.A. A new metric for L-R fuzzy numbers and its application in fuzzy linear systems. *Soft Comput.* **2012**, *16*, 1743–1754. [CrossRef]
18. Allahviranloo, T.; Mikaeilvand, N.; Barkhordary, M. Fuzzy linear matrix equation. *Springer Sci.* **2009**, *8*, 165–177. [CrossRef]
19. Kargar, R.; Allahviranloo, T.; Rostami-Malkhalifeh, M.; Jahanshaloo, G.R. A Proposed Method for Solving Fuzzy System of Linear Equations. *Sci. World J.* **2014**, *2014*, 782093. [CrossRef]
20. Nuraei, R.; Allahviranloo, T.; Ghanbari, M. Finding an inner estimation of the solution set of a fuzzy linear system, Applied Mathematical Modelling. *Appl. Math. Comput.* **2013**, *37*, 5148–5161.
21. Amirfakhrian, M. Analyzing the solution of a system of fuzzy linear equations by a fuzzy distance. *Soft Comput.* **2012**, *16*, 1035–1041. [CrossRef]
22. Amirfakhrian, M. Numerical solution of a fuzzy system of linear equations with polynomial parametric form. *Int. J. Comput. Math.* **2007**, *84*, 1089–1097. [CrossRef]

23. Amirfakhrian, M.; Fallah, M.; Rodríguez-López, R. A method for solving fuzzy matrix equations. *Soft Comput.* **2018**, *22*, 2095–2103. [CrossRef]
24. Fariborzi Araghi, M.A.; Fallahzadeh, A. Inherited LU factorization for solving fuzzy system of linear equations. *Soft Comput.* **2013**, *17*, 159–163. [CrossRef]
25. Mikaeilvand, N.; Noeiaghdam, Z. The General Solutions of Fuzzy Linear Matrix Equations. *J. Math. Ext.* **2015**, *9*, 1–13.
26. Mikaeilvand, N.; Noeiaghdam, Z. The General Solutions of $m \times n$ Fuzzy Linear Systems. *Middle-East J. Sci. Res.* **2012**, *11*, 128–133.
27. Noieaghdam, Z.; Mikaelvand, N. Least squares solutions of inconsistent fuzzy linear matrix equations. *Int. J. Ind. Math.* **2012**, *4*, 365–374.
28. Wang, K.; Zheng, B. Inconsistent fuzzy linear systems. *Appl. Math. Comput.* **2006**, *181*, 973–981. [CrossRef]
29. Zheng, B.; Wang, K. General fuzzy linear systems. *Appl. Math. Comput.* **2006**, *181*, 1276–1286. [CrossRef]
30. Behera, D.; Chakraverty, S. A new method for solving real and complex fuzzy systems of linear equations. *Comput. Math. Model.* **2012**, *23*, 507–518. [CrossRef]
31. Fariborzi Araghi, M.A.; Noeighdam, S. Dynamical control of computations using the Gauss-Laguerre integration rule by applying the CADNA library. *Adv. Appl. Math. Sci.* **2016**, *16*, 1–18.
32. Fariborzi Araghi, M.A.; Noeiaghdam, S. A Valid Scheme to Evaluate Fuzzy Definite Integrals by Applying the CADNA Library. *Int. J. Fuzzy Syst. Appl.* **2017**, *6*, 1–20. [CrossRef]
33. Fariborzi Araghi, M.A.; Zarei, E. Dynamical control of computations using the iterative methods to solve fully fuzzy linear systems. In *Advances in Fuzzy Logic and Technology*; Springer: Cham, Switzerland, 2017; pp. 55–68.
34. Fariborzi Araghi, M.A.; Fattahi, H. Solving fuzzy linear systems in the stochastic arithmetic by applying CADNA library. In *International Conference on Evolutionary Computation Theory and Applications*; SciTePress: Paris, France, 2011; pp. 446–450.
35. Ezzati, R. Solving fuzzy linear systems. *Soft Comput.* **2011**, *15*, 193–197. [CrossRef]
36. Dubois, D.; Prade, H. Fuzzy sets and systems: Theory and applications. In *Mathematics in Science and Engineering*; Bellman, R., Eds.; Publishing House, Academic Press Inc.: New York, NY, USA, 1980.

mathematics

Review

A Philosophical Treatise on the Connection of Scientific Reasoning with Fuzzy Logic

Evangelos Athanassopoulos [1] and Michael Gr. Voskoglou [2,*]

[1] Independent Researcher, Giannakopoulou 39, 27300 Gastouni, Greece; evatha@gmail.com
[2] Department of Applied Mathematics, Graduate Technological Educational Institute of Western Greece, 22334 Patras, Greece
* Correspondence: voskoglou@teiwest.gr

Received: 4 May 2020; Accepted: 19 May 2020; Published:1 June 2020

Abstract: The present article studies the connection of scientific reasoning with fuzzy logic. Induction and deduction are the two main types of human reasoning. Although deduction is the basis of the scientific method, almost all the scientific progress (with pure mathematics being probably the unique exception) has its roots to inductive reasoning. Fuzzy logic gives to the disdainful by the classical/bivalent logic induction its proper place and importance as a fundamental component of the scientific reasoning. The error of induction is transferred to deductive reasoning through its premises. Consequently, although deduction is always a valid process, it is not an infallible method. Thus, there is a need of quantifying the degree of truth not only of the inductive, but also of the deductive arguments. In the former case, probability and statistics and of course fuzzy logic in cases of imprecision are the tools available for this purpose. In the latter case, the Bayesian probabilities play a dominant role. As many specialists argue nowadays, the whole science could be viewed as a Bayesian process. A timely example, concerning the validity of the viruses' tests, is presented, illustrating the importance of the Bayesian processes for scientific reasoning.

Keywords: inductive and deductive reasoning; fuzzy logic (FL); scientific method; probability and statistics; Bayesian probabilities

---

"Doubt is not a pleasant situation, but the excessive certainty is an unreasonable situation"
Voltaire (1694–1778)

1. Introduction

It is hard to deny that the rapid and impressive advances of the last 100–150 years in science and technology have been greatly based on the principles of the bivalent logic of Aristotle, which has played (and plays) a dominant role for more than 23 centuries for the progress of the Western civilization. On the contrary, ideas of multi-valued logics were traditionally connected to the culture and customs of the Eastern countries of Asia, being inherent in the philosophy of Buddha Siddhartha Gautama, who lived in India during the fifth century BC. Nevertheless, the formalization of those logics was performed in the West with the Zadeh's Fuzzy Logic (FL), an infinite-valued logic based on the mathematical theory of fuzzy set (FS) [1]. The instantaneous switch from truth (1) to falsity (0) can easily distinguish propositions of classical logic from those in FL with values lying in the interval [0, 1]. As it usually happens in science with such radical ideas, FL, when first introduced during the 1970s, was confronted with distrust and reserve by most mathematicians and other positive scientists. However, it has been eventually proved to be an effective generalization and complement of the classical logic and has found many and important applications to almost all sectors of human activity, e.g., [2]: Chapters 4–8, [3], etc.

Human reasoning is characterized by inaccuracies and uncertainties, which stem from the nature of humans and the world. In fact, none of our senses and observation instruments allows us to reach an absolute precision in a world which is based on the principle of continuity, as opposed to discrete values. Consequently, FL, introducing the concept of membership degree that allows a condition to be in a state other than true or false, provides a flexibility to formalize human reasoning. Another advantage of FL is that rules are set in natural language with the help of linguistic, and therefore fuzzy, variables. In this way, the potential of FL becomes strong enough to enable developing a model of human reasoning proceeding in natural language. Furthermore, since at the basis of reasoning in expert systems are human notions and concepts, the success of these systems depends upon the correspondence between human reasoning and their formalization. Thus, FL appears as a powerful theoretical framework for studying not only human reasoning, but also the structure of expert systems.

However, only a limited number of reports appear in the literature connecting FL to human reasoning; e.g., see [4–6]. This was our main motivation for performing the present study by continuing our previous analogous efforts. In fact, in a recent book written in the Greek language, the first author of this article studies the inductive reasoning under the light of FL and criticizes the excessive accusations of the Philosophy of Science against it [7]. Additionally, in an earlier work [8], the second author has introduced a model for analyzing human reasoning by representing its steps as fuzzy sets on a set of linguistic labels characterizing the individual's performance in each of those steps.

The target of the article at hand is to analyze the scientific method of reasoning under the light of FL. The rest of the article is organized as follows: Section 2 examines the inductive and deductive reasoning from the scope of FL. Section 3 explains how these two fundamental components of human reasoning act together for creating, improving and expanding the scientific knowledge and presents a graphical representation of the scientific progress through the centuries. Section 4 studies the unreasonable effectiveness of mathematics in the natural sciences (Winger's enigma), while Section 5 quantifies the degree of truth of the inductive and deductive arguments. The article closes with the general conclusions presented in Section 6. A schematic diagram of the proposed study is presented in Figure 1.

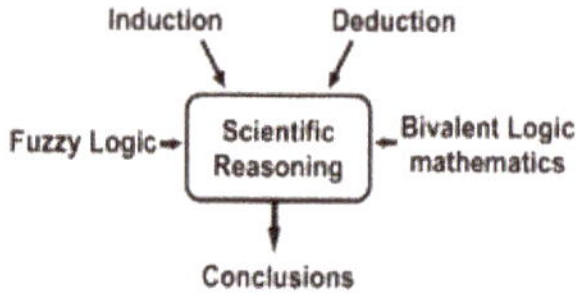

Figure 1. Schematic diagram of the proposed study.

The novelty of the present article is that it brings together philosophy, mathematics and fuzzy logic, thus becoming a challenge for further scientific discussion on the subject. This is the added value of this work.

2. Human Reasoning and Fuzzy Logic

Induction and deduction are the two fundamental mechanisms of human reasoning. Roughly speaking, induction is the process of going from the specific to general, whereas deduction is exactly the opposite process of going from the general to the specific. A third form of reasoning that does not fit to induction and deduction is the abduction. It starts with a series of incomplete observations and proceeds to the most possible conclusion. For example, abduction is often used by jurors, who make decisions based on the evidence presented to them, by doctors, who make a diagnosis based on test results, etc.

A typical form of a deductive argument is the following:

- If "Every A is B" and "C is an A", then "C is B".

For example, "All dogs are animals" and "Snoopy is a dog", therefore, "Snoopy is an animal". That is, deductive reasoning starts from a first premise and based on a second premise reaches a logical inference.

A deductive argument is always valid/consistent in the sense that, if its premises are true, then its inference is also true. However, if at least one of its premises is false, then the inference may also be false. For example, "All positive integers are even numbers", "5 is a positive integer", therefore "5 is an even number". In that case, the inference is false due to the fact that the first premise is false. However, the deductive argument remains a valid logical procedure. This is better understood, if one restates the above argument in the form: "If all positive integers were even numbers, then 5 would be an even number". In conclusion, the typical logic is interested only for the validity and not for the truth of an argument. Therefore, a deductive argument is always compatible to the principles of the bivalent logic, regardless whether or not its inference is true.

A characteristic example of the deductive mechanism can be found in the function of computers. A computer is unable to judge if the input data inserted to it is correct and, therefore, if the result obtained is correct and useful for the user. The only thing that it guarantees is that, if the input is correct, then the output will be correct too. Thus, the credo that was popular in the early days of computing, "Garbage in, garbage out" (GIGO), or "Rubbish in, rubbish out" (RIRO), is still valid.

An inductive argument makes generalizations from specific observations. It starts with a series of observations on the specific cases of a phenomenon and reaches a conclusion by making a broad/imperfect generalization. A typical form of an inductive argument is the following:

- The element x_1 of the set A has the property B.
- The element x_2 of the set A has the property B.
- ....
- ....
- The element x_n of the set A has the property B.
- Therefore, all the elements of the set A have the property B.

The inference of an inductive argument could be either true or false. However, an inductive argument is never valid/consistent, even if its inference is true, because it is not a procedure compatible to the principles of the typical logic. In fact, if all the premises of an inductive argument are true, there is always the possibility for its inference to be false.

Let us now restate the previous inference as follows:

- Therefore, all the elements of the set A possibly have the property B.

This is not an acceptable by the bivalent logic statement, because it does not satisfy the principle of the excluded middle. However, it is compatible to FL, which does not adopt the above principle. It is evident that the greater the value of n in the last premise of the above inductive argument (i.e., the greater the number of observations performed), the greater the degree of truth of its inference. In other words, an inductive argument, although not a logical process according to the standards of the bivalent logic, is acceptable when its inference is "translated" in the language of FL.

The validity of the deductive reasoning is very important for scientific research, because it enables the construction of extended arguments. An extended argument is understood to be a chain of simple deductive arguments A_1, A_2, ... , An, which has been built by putting as first premise of the next argument the inference of the previous one. In this way the validity of the argument A_1 is transferred to the final argument A_n.

In everyday life, however, people always want to know the truth in order to organize better, or even to protect their lives. Consequently, in such cases the significance of an argument has greater importance than its validity/precision. In Figure 2, retrieved from https://complementarytraining.net/strength-training-categorization/figure-6-precision-vs-significance, the extra precision on the left actually makes things worse for the poor man in danger, who has to spend too much time trying to understand the data and misses the opportunity to take the much needed action of getting out of the way. On the contrary, the rough/fuzzy warning on the right could save his life.

Figure 2. Validity/precision and significance in real world.

Figure 2 illustrates very successfully the importance of FL for real-life situations. Real-world knowledge generally has a different structure and requires different formalization than existing formal systems. FL, which according to Zadeh is "a precise logic of imprecision and approximate reasoning", serves as a link between classical logic and human reasoning/experience, which are two incommensurable approaches. Having a much higher generality than bivalent logic, FL is capable of generalizing any bivalent logic-based theory. Linguistic variables and fuzzy if-then rules are in effect a powerful modelling tool used widely in FL applications. In addition, FL is the basis for computing with words, i.e., computation with information described in the natural language. This is particularly useful when dealing with second order uncertainty, i.e., uncertainty about uncertainty [9].

Some important details about the history and evolution of FL, like Plato's ideas about the existence of a third area beyond "true" and "false", the Lukasiewicz's [10] and Tarski's multi-valued logics, etc., can be found in Section 2 of [11]. In addition, it is useful to notice here the connection of FL to the Reiter's logic for default reasoning [12]. An example of default reasoning is the following: "Whenever x is a bird, then in the absence of any information about the contrary (e.g., that x is a penguin or an ostrich or a Maltese falcon, etc.) one may assume that x flies".

Fuzzy mathematics is trying to cover an inherent weakness of crisp mathematics, which was very successfully stated by Einstein during his speech in the Prussian Academy of Sciences in 1921 as follows: "So far as lows of mathematics refer to reality, they are not certain. And so far as they are certain, they do not refer to reality".

FL, however, should not be viewed as the final solution for representing human knowledge about the world. Simply, it has offered a model that could easily be grasped by scientists and researchers alike as a step toward formalizing human reasoning. Because of this, Zadeh's basic notion of FS stimulated enormous research activity that has generated various extensions and generalizations (type-2 FS, interval-valued FS, intuitionistic FS, hesitant FS, Pythagorean FS, complex FS, neutrosophic sets, etc.), as well as several alternative theories for dealing with the several forms of the existing in the real world uncertainty (grey systems, rough sets, soft sets, etc.) [13]. Although none of those generalizations/theories look to be perfect for tackling all the types of uncertainty existing in the real word, the combination of all of them provides a powerful framework for this purpose.

It is of worth noting here that, before the development of FL by Zadeh, the unique tool to deal with the uncertainty appearing in everyday life and science used to be probability theory. Nevertheless, probability, which is based on the principles of the bivalent logic, is only capable to tackle the cases of uncertainty which are due to randomness. On the contrary, FL, its generalizations and all the alternative theories mentioned above, can also manage the several types of uncertainty that are due to imprecision. Probabilities and membership degrees, although they both function on the same interval

[0, 1], they are completely different notions ([14], Remark 3, p. 22). An additional example illustrating this difference is presented in Section 5 of the present article.

3. The Scientific Method of Thinking and the Curve of the Scientific Error

As we have previously seen, the truth of a deductive argument depends upon the truth or not of its premises. However, the premises must be true not only in a particular moment, but forever. The following example illustrates the significance of this remark:

- All swans are white.
- The bird in the lake is a swan.

- The bird in the lake is white.

This used to be a true deductive argument before the discovery of Australia, where a kind of black swan was traced in 1697 (Figure 3). Since then the argument turned to be false.

Figure 3. A black swan (cygnus aratus) in Australia (retrieved from Wikipedia.org).

Unfortunately, only a few arguments are known that could be used as premises, which remain true forever. Even the premise that "all humans are mortal" could be put under dispute. In fact, there is a theoretical possibility that the continuous progress of science and technology could enable in the remote future humans to live forever. Then, we would have a fact analogous to the case of the swans! In Physics, the classical law of "conservation of the energy" has been transformed now to "conservation of energy and mass". This seems, however, as a premise that could remain true forever.

Some premises of such kind can also be found in mathematics, which is considered as the most "solid" among the sciences; e.g., the arithmetic of the natural numbers. However, problems exist even in the area of mathematics. For example, the target of the leader of formalism David Hilbert (1862–1943) anticipated a complete and consistent (i.e., not permitting the existence of paradoxes) axiomatic development of all branches of mathematics [15]. Gödel's (1906–1978) two incompleteness theorems [16], however, put a definite end to Hilbert's ambitious plans. In fact, as a consequence of the second theorem, there is no formal system with a finite set of axioms and rules that can prove the consistency of another system, since it must prove its own consistency first, which is impossible. Thus, for a given system the best to hope is that, although by the first theorem is not complete (i.e., there exist propositions which cannot either proved or refuted inside it), it is consistent. Of course, this happens due to the inherent deficiencies of the formal systems and does not imply that the human ability to understand is restricted and therefore some truths will never be known.

Evolutionary anthropologists have estimated that the modern human (homo sapiens, which means wise man) occurred on the earth roughly to 200 thousand years ago. However, the existing witnesses lead to the conclusion that the first smart human behavior, connected to the stone-made tools and the isolation of fire, appeared much earlier, in the time of the homo erectus (upright man). The earliest occurrence of the homo erectus, who is the archaic ancestor of homo sapiens, is estimated about two million years ago. Homo erectus appears to be much more similar to modern humans than to his ancestors, the australopithecines, with a more humanlike gait, body proportions, height, and brain capacity. He was capable of starting speech, hunting and gathering in coordinated groups, caring for injured or sick group members, and possibly making art. However, his IQ remained at a constantly low level, being slightly higher than that of his ancestors.

Curiosity was a dominant characteristic of humans from the time of their occurrence on the earth. They made continuous observations trying to understand the world around them and they searched for explanations about the various natural phenomena. Inductive reasoning helped them on that step to create such explanations by making hypotheses, which were eventually transformed to empirical theories and applications for improving their lives and security; e.g., constructing stone and wood-made weapons and tools for hunting and cultivating the earth, using the fire for heating and cooking, etc.

Deductive arguments started to appear in human reasoning since the time of occurrence of the homo sapiens. Deductive reasoning became eventually the dominant component of human reasoning that led through the centuries to the development of the human civilization. In the first place, humans tried to verify by deduction the empirical theories created on the basis of earlier observations and inductive arguments. For this, they used as premises intuitively profound truths, which much later have been called axioms. On the basis of the axioms, whose number should be as small as possible, they proved, using deductive arguments, all the other inferences (laws, propositions, etc.) of the empirical theory. They also created new deductive inferences on the basis of the corresponding theory, usually leading to various useful applications.

This process is graphically represented on the left side of Figure 4, where $\alpha_1, \alpha_2, \alpha_3, \alpha_4, \ldots$ are observations of the real world that have been transformed, by induction, to the theory T_1. Theory T_1 was verified by deduction and additional deductive inferences $K_1, K_2, \ldots, K_s, \ldots, K_n$ were obtained. Next, a new series of observations $b_1, b_2, b_3, \ldots$, follow. If some of those observations are not compatible to the laws of theory T_1, a new theory T_2 is formed by induction (intuitively) to replace/extend T_1. The deductive verification of T_2 is based on axioms partially or even completely different to those of theory T_1 and new deductive inferences $L_1, L_2, \ldots, L_s, \ldots, L_m$ follow. The same process could be repeated (observations $c_1, c_2, c_3, \ldots$, theory T_3, inferences $M_1, M_2, \ldots, M_s, \ldots, M_k$, etc.) one or more times. In each case the new theory extends or rejects the previous one approaching more and more the absolute truth.

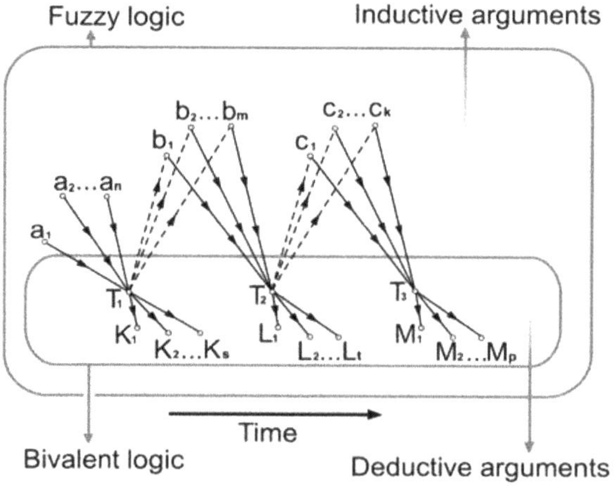

Figure 4. A graphical representation of the scientific method.

This procedure is known as the scientific method. The term was introduced in the 19th century, when significant terminologies appeared establishing clear boundaries between science and non-science.

However, the scientific method characterizes the development of science since at least the 17th century. Aristotle (384–322 BC) is recognized as the inventor of the scientific method due to his refined analysis of the logical implications contained in demonstrative discourse.

James Ladyman, Professor of Philosophy at the University of Bristol, UK, notes that the relationship between observations and theory is much more composite than it looks at first glance [17]. Our illustration of the scientific method in Figure 4 reveals that there exists a continuous "ping pong" between induction and deduction that characterizes this relationship.

The first book in the history of human civilization written on the basis of the principles of the scientific method is, according to the existing witnesses, the "Elements" of Euclid (365–300 BC) addressing the axiomatic foundation of Geometry. However, although the Euclidean Geometry remains still valid for small distances on the earth, the non-Euclidean Geometries of Lobachevsky (1792–1856) and Riemann (1826–1866), which were developed by changing the fifth Euclid's axiom of the parallel lines (for more details see Section 4), have replaced it for the great distances on the earth's surface and of the Universe ([18]: Section 3).

The non-Euclidean geometries helped Einstein to prove on a theoretical basis (deductively) his general relativity theory (alternatively termed as the new theory of gravity), which corrected the Newton's law of gravity in case of very strong gravitational forces (for more details see Section 4). Einstein's theory was experimentally verified by the irregularity of the Hermes' orbit around the sun, although at that time many scientists argued that this irregularity was due to the existence of an unknown planet near Hermes or a satellite of Hermes or a group of asteroids near the planet. However, the magnitude of the divergence of the light's journey, which was calculated during the eclipse of the sun on 29 May 1919, was the definite evidence for the soundness of Einstein's theory. In fact, the eclipse let some stars, which normally should be behind the sun, to appear beside it on the sky [19].

In many cases, a new theory knocks completely down a previously existing one. This happened, for example, with the geocentric theory (Almagest) of Ptolemy of Alexandria (100–170). That theory, being able to satisfactorily predict the movements of the planets and the moon, was considered to be true for centuries. However, it was finally proved to be wrong and has been replaced by the heliocentric theory of Copernicus (1473–1543). The Copernicus' theory was supported and enhanced a hundred years later by the observations/studies of Kepler and Galileo. But, although the idea of the earth and the other planets rotating around the sun has its roots at least in the time of the ancient Greek astronomer Aristarchus of Samos (310–230 BC), the heliocentric theory faced many obstacles for a long period of time, especially from the part of the church, before its final justification [20].

It becomes evident that the scientific method is highly based on the "Trial and Error" procedure. According to W. H. Thrope ([21]: p. 26), this term was devised by C. Lloyd Morgan (1852–1936). This procedure is characterized by repeated attempts, which are continued until success or until the subject stops trying. Sir Karl Raimund Popper (1902–1994), one of the 20th century's most influential philosophers of science, suggested the principle of falsification according to which a proposition can be characterized as scientific, only if it includes all the necessary criteria for its control, and therefore only if it could be falsified [22]. This principle gives actual emphasis to the second component of the "Trial and Error" procedure. Critiques on the ideas of Popper report that he used the principle of falsification to decrease the importance of induction for the scientific method [7,17].

All those discussed in this Section are depicted in Figure 5, representing the curve of the scientific error. This curve, which is connected to the scientific method illustrated in Figure 4, represents the scientific progress through the centuries (the smaller the error, the greater the progress). The horizontal axis in Figure 5 corresponds to the time (t) and the vertical one to the value of the scientific error (E). The starting point M of the curve, giving the maximal value of the scientific error, corresponds to the time t_0 of occurrence of the homo erectus on the earth. The curve is decreasing asymptotically to the axis of time approaching continuously the absolute truth, where the value of the scientific error is zero. The part MT_1 of the curve has been designed with dots, because no evidence of deductive reasoning existed before, which means that only empirical but not scientific progress was realized during that period. The point T_1 of the curve corresponds to the time t_1 of occurrence of the homo sapiens, while the value of -13.7×10^9 years is approximately the time of creation of the Universe (Big Bang) [19].

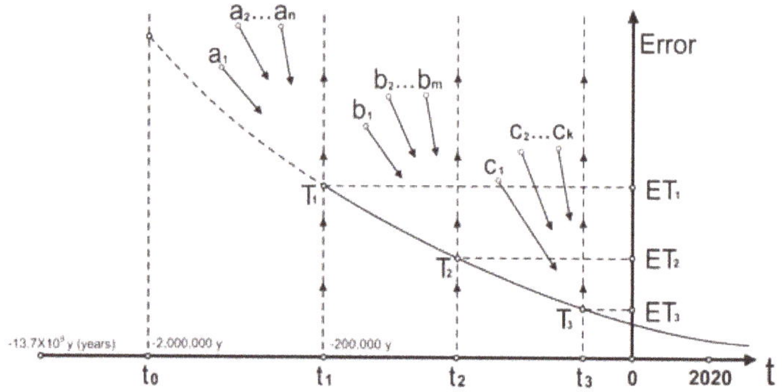

Figure 5. The curve of scientific error.

In conclusion, although deduction is the basis of the scientific method, almost all the scientific progress (with pure mathematics being probably the unique exception; see Section 4) has its roots to inductive reasoning. To put it on a different basis, quantifying the scientific reasoning within the interval [0, 1], only the values 0 and 1 correspond to deductive reasoning, which is connected to the bivalent logic. All the other, values correspond to inductive reasoning, which is supported by the FL.

It is worth emphasizing here that the error of induction is transferred to deductive reasoning through its premises. Therefore, the scientific error in its final form is actually a deductive and not an inductive error. However, many philosophers consider deduction as being an infallible method. This consideration is due to the following two reasons:

- Deduction is always a consistent method
- The existing theories are considered as being always true, which leads frequently to surprises.

The Scottish philosopher David Hume (1711–1776) argued that, with deductive reasoning only, humans would be starving. Although this could not happen nowadays due to our past knowledge obtained with the help of induction, without inductive reasoning no further scientific progress could be achieved. Therefore, a proper balance must exist between induction and deduction, the two fundamental forms of human reasoning.

4. The Unreasonable Effectiveness of Mathematics in the Natural Sciences

Mathematics is the fundamental tool for the development of science and technology. In fact, a single equation (or another mathematical representation) frequently has the potential to replace hundreds of written pages in explaining a phenomenon or situation. For example, the famous Einstein's equation $E = mC^2$ is enough to explain the relationship between mass (m) and energy (E) with the help of the light's speed (C).

The success of mathematics in the natural sciences appears in two forms, the energetic and the pathetic one. In the former case, the laws of nature are expressed mathematically by developing the corresponding mathematical theories so that they fit the existing observations. In the latter case, however, completely abstract mathematical theories, without any visible applications at the time of their creation, are utilized in unsuspicious time for the construction of physical models!

The Nobelist E. P. Winger, in his famous Richard Courant lecture at the New York University on 11 May 1959, characterized the success of mathematics in describing the architecture of the Universe as the "unreasonable effectiveness of mathematics in the natural sciences" [23]. Since then, this is usually referred as the Winger's enigma.

The Winger's enigma is the main argument of the philosophical school of mathematical realism in supporting the idea that mathematics exists independently of human reasoning. Consequently, the supporters of this idea believe that mathematics is actually discovered and not invented by humans. Although mathematical realism is nowadays under dispute by those arguing, under the light of various indications from the history of mathematics and the data of experiments performed by cognitive scientists and psychologists, that mathematics is a human invention or at least a mixture of inventions and discoveries ([18]: Section 4), it still has many supporters. Among those are the Platonists, arguing the existence of an eternal and unchanged "universe" of mathematical forms, the supporters of the MIT's cosmologist Max Tegmark theory that Universe is not simply described by mathematics, but IT IS mathematics, etc. ([18]: Section 2).

As it has already mentioned in Section 3, Euclid created the theoretical foundation of the traditional Geometry on the basis of 10 axioms, which were used to prove all the other known on that time geometric propositions and theorems. It is recalled that a fifth of those axioms, stated in its present form by Proclus (412–485), says that from a point outside a given straight line only one parallel can be drawn to this line. However, this axiom does not have the plainness of the rest of the Euclid's axioms. This gave during the centuries to many mathematicians the impulsion to try to prove the fifth axiom with the help of the other Euclid's axioms.

One of the latest among those mathematicians was the Russian Lobachevsky, who, when he failed to do so, decided to investigate what happens if the fifth axiom does not hold. Thus, replacing (on a theoretical basis) that axiom by the statement that AT LEAST TWO parallels can be drawn from the given point to the given line, he created the Hyperbolic Geometry, which is developed on a hyperbolic paraboloid's (saddle's) surface. The Riemann's Elliptic Geometry on the surface of a sphere followed, which is based on the assumption that NO PARALLEL can be drawn from the given point to the given straight line. Several other types of non-Euclidean Geometries can be also developed in an analogous way.

Approximately 50 years later, Einstein expressed his strong belief that the Newton's calculation of the gravitational force F between two masses m_1 and m_2 by the formula

$$F = G\frac{m_1 m_2}{r^2} \quad (1)$$

where G stands for the gravitational constant, was not correct for the existing in the Universe (outside the earth) strong gravitational forces (general relativity theory; see Section 3). This new approach was based on the fact that, according to Einstein's special theory of relativity (1905), the distance (r) and the time (t) are changing in a different way with respect to a motionless and to a moving observer.

To support his argument, Einstein introduced the concept of the four-dimensional time–space and after a series of intensive efforts (1908–1915) he finally managed to prove that the geometry of this space is non-Euclidean! For example, the non-Euclidean divergence of the radius r of a sphere of total surface S and mass m is, according to Einstein's theory, equal to $r - \sqrt{\frac{S}{4\pi}} = \frac{Gm}{3C^2}$. The theoretical foundation of the general relativity theory made Einstein to state with relief and surprise: "How is it possible for mathematics to fit so eminently to the natural reality?"

The non-Euclidian form of time–space is physically explained by its distortion created by the presence of mass or of an equivalent amount of energy. The level projection of this distortion is represented in Figure 6, retrieved from http://physics4u.gr/blog/2017/04/21. This appears analogous to the distortion created by a ball of bowling on the level of a trampoline.

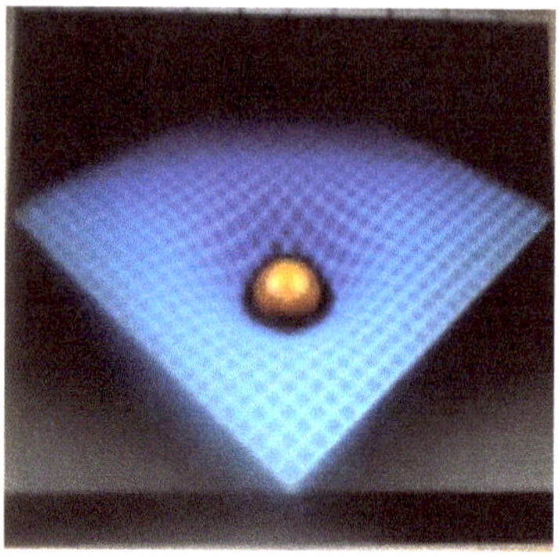

Figure 6. Level projection of the Universe's distortion created by the presence of mass.

Another characteristic example of the interaction between the energetic and the pathetic form of mathematics is the Knot Theory, initiated through an unsuccessful effort to represent the atom's structure. However, the theoretical research on the structure and properties of knots led eventually to the understanding of the mechanisms of the DNA!

It is probably useful to add here a personal experience about the Winger's enigma. The Ph.D. research of the second author of this article during the early 1980's was focused on "Iterated Skew Polynomial Rings" [24], a purely theoretical topic of abstract algebra. To the great surprise of the specialists on the subject, however, this topic has found recently two very important applications to the theory of Quantum Groups (a basic tool of Theoretical Physics) [25] and to Cryptography for analyzing the structure of certain codes [26].

Today the distinction between Pure and Applied Mathematics tends to lose its meaning, because almost all the known mathematical topics have already found applications to science, technology and/or to the everyday life situations. However, the research process that the pathetic approach of mathematics follows, could be considered as an exception (possibly the unique) of the general process followed by the scientific method. In fact, although the axioms introduced for the study of a mathematical topic via the pathetic approach are sometimes based on intuitional criteria or beliefs, the method that this approach follows is purely deductive. The results obtained are considered as correct, if the mathematical manipulation is proved to be correct and regardless whether or not they have a physical meaning. The amazing thing is, however, as said before, that those results are more or less finding unexpected practical applications in the near or remote future. This made the famous astrophysicist and bestselling author M. Livio wonder: "Is God a mathematician?" [27].

Thus, although inductive reasoning is actually the "mother" of the scientific and technological progress and mathematics is the main tool promoting this progress, the pathetic approach of mathematics is a "pure product" of deduction and the connected to it bivalent logic. On the other hand, FL and the related to it generalizations/theories give to the disdainful by the bivalent logic inductive reasoning its proper place and importance as a fundamental component of scientific reasoning.

5. Quantification of the Inductive and Deductive Inferences

As we have previously seen, according to bivalent logic, a given proposition is either true or false. This is really useful in case of a deductive inference, but it says almost nothing in case of an inductive argument, which is based on a series of observations. In fact, as we have seen in Section 2, from the scope of the bivalent logic an inductive argument is never valid, even if it is true. However, what really counts in practice is not the validity, but the degree of truth of the inductive argument. There is, therefore, a need to quantify the inductive arguments, i.e., to calculate their degree of truth. The main tools being available for making those calculations are Probability and Statistics, as well as FL and the related to it generalizations/theories.

Probability and Statistics are related areas of mathematics which, however, have fundamental differences. Probability is a theoretical branch of mathematics which deals with predicting the likelihood of future events. On the contrary, Statistics is an applied branch of mathematics, which tries to make sense of observations in the real world by analyzing the frequency of past events. The distinction between Probability and Statistics could be clarified better by tracing the thoughts of a gambler mathematician during a game with dice. If the gambler is a probabilist, he will think that each face of the dice comes up with probability $\frac{1}{6}$. If instead he is a statistician, he will think: "How I know that the dice are not loaded? I keep track how often each number comes up and once I am confident that the dice are fair I'll decide how to play". In other words, Probability enables one to predict the consequences of an ideal world, whereas Statistics enables him/her to measure the extent to which our world is ideal.

Consider, for example, the following argument: "The sun will always rise in the morning". From the statistical point of view this is a true argument, because the sun rises every morning from the time that humans occurred on earth and have started observing it. Therefore, the number of positive cases of this "experiment" is equal to the number of the total cases, which means that the corresponding frequency is equal to 1. Things, however, are not exactly the same for the probabilistic approach. In fact, it has been estimated by the cosmologists that the sun will "extinguish" in about 7 billion years from now, becoming a small white dwarf star, after being transformed to a red giant first. Therefore, setting $t = 7 \times 10^9 \times 365$ days, the probability that the given statement is true is equal to $p = \frac{t-1}{t} = 1 - \frac{1}{t}$. Since t is a very large number, $\frac{1}{t}$ is very close to 0, therefore, p is very close, but not equal, to 1. Thus, the previous argument is not absolutely true. The difference between the statistical and the probabilistic inference in this case is due to the fact that an experiment cannot be repeated an infinite number of times, which means that the value 1 of the corresponding frequency is not exact.

The A. F. Chalmer's generalization, however, that no scientific knowledge can be considered as being absolutely true, cannot be justified in the way that he attempts to do in [28]: Chapter 2. In fact, his argument that the corresponding probability is equal to the quotient of a finite number (positive observations) by an infinite number (total observations) is not correct, for the very simple reason that an experiment could not be repeated an infinite number of times.

Edwin T. Jaynes (1922–1998), Professor of Physics at the University of Washington, was one of the first suggesting in the middle of 1990s that probability theory is a generalization of the bivalent logic reducing to it in the special case that our hypotheses are considered to be either absolutely true or absolutely false [29]. Many eminent scientists have been inspired by the ideas of Jaynes. Among those is the Fields medalist David Mumford, who believes that Probability theory and statistical inference are emerging now as a better foundation of scientific models and even as essential ingredients for the foundation of mathematics [30]. Mumford reveals in [30] that he takes back his advice given to a graduate student during his Algebraic Geometry days in the 1970s: "Good grief, don't waste your time studying Statistics, it's all cookbook nonsense"!

But what happens in the everyday life? As we have already seen in Section 2, Probability, being a "child" of the bivalent logic, is effective in tackling only the cases of uncertainty which are due to randomness, but not those due to imprecision. For example, consider the expression "John is a good football player". According to the statistical approach, one should attend John playing in a series of games in order to decide if this argument is true. However, different observers could have different opinions about his playing skills. The probabilistic approach is also not suitable. The statement, for example, that "The probability for John to be a good player is 80%" means that John, being or not, according to the law of the excluded middle, a good player, is probably a good one, perhaps because his coach believes so. In such cases, FL comes to bridge the gap through the use of a suitable membership function. Saying for example, that "The membership degree of John in the fuzzy set of the good players is 0.8", one understands that John is a rather good player.

Nevertheless, the problem with FL is that the definition of the membership function, although it should always be compatible with the common logic, is not unique, depending on the observer's subjective criteria. The generalizations/extensions of the concept of fuzzy set and the related theories (see Section 2) have been developed on the purpose of managing better this weakness of FL, but each one of them succeeds to do so in certain cases only. However, the combination of FL with all those generalizations/theories provides a framework that treats satisfactorily all the existing in the real-world types of uncertainty.

Let us now turn our attention to the deductive arguments. As we have seen in Section 3, the error of the inductive reasoning is transferred to deduction through its premises. Therefore, the quantification of the deductive arguments becomes also a necessity. But, as it has already been discussed, the inference of a deductive argument is true under the CONDITION that its premises are true. In other words, if H denotes the hypothesis imposed by its premises and I denotes its inference, then the conditional probability P(I/H) expresses the degree of truth of the deductive argument. Then, by the Bayes's (1701–1761) formula, one finds that

$$P(I/H) = \frac{P(H/I)P(H)}{P(I)} \quad (2)$$

Equation (2), connects P(I/H) with the conditional probability P(H/I) of the inverse process. The final outcome depends on the value of the independent probability P(H) about the truth of the existing hypothesis, usually referred as the prior probability. The Bayes formula, although at first glance looks as a salient generalization of the original concept of probability, has been proved to be very important for scientific reasoning. Many philosophers assert that science as a whole could be viewed as a Bayesian process! (e.g., see [31,32], etc.) The New York Times reports that "Bayesian statistics are rippling through everything from physics to cancer research, ecology to psychology"; see also Figure 7, retrieved from [32]. Artificial Intelligence specialists utilize Bayesian software for helping machines to recognize patterns and make decisions. In a conference held at the New York University on November 2014 with title "Are Brains Bayesian?", specialists from all around the world had the opportunity to discuss how the human mind employs Bayesian algorithms to pre-check and decide.

Figure 7. Bayes's theorem guest appearance on the hit CBS show "Bing Bang Theory".

The following example, inspired from [32] and properly adapted here, illustrates the importance of the Bayes's theorem in real life situations. This is a timely example due to the current COVID-19 pandemic, because it concerns the creditability of medical tests.

EXAMPLE: In a town of 10,000 inhabitants, Mr. X makes a test for checking whether or not he is infected by a dangerous virus. It has been statistically estimated that 1% of the town's population has been infected by the virus, while the test has a 96% statistical success to diagnose both positive and negative cases of infections. The result of the test was positive and Mr. X concluded that the probability to be a carrier of the virus is 96%. Is that a correct?

The answer is no! For explaining this, let us consider the following two events:

- H: The test is positive, and
- I: The subject is a carrier of the virus.

From the given data it turns out that P(I) = 0.1 and P(H/I) = 0.96. Further, from the 10,000 inhabitants of the town, 100 are carriers of the virus and 9900 are not. Assume (on a theoretical basis) that all the inhabitants make the test. Then we should have $9900 \times 4\% = 396$ positive results from the non-carriers and 96 positive results from the carriers, i.e., 492 in total positive results. Therefore P(H) 0.492. Replacing the values of P(I), P(H/I) and P(H) to formula (1) one finds that P(I/H) $\approx$ 0.1951. Therefore, the probability for Mr. X to be a carrier of the virus is only 19.51%!

On the contrary, assuming that "H: The test is negative", we should have $9900 \times 96\% = 9504$ negative results from the non-carriers and four negative results from the carriers of the virus, i.e., 9508 in total negative results. Therefore, P(H) 0.958 and in this case Equation (2) gives that P(I/H) $\approx$ 0.1. Therefore, if the result of the test of Mr. X is negative, the probability to be a carrier is only 1%.

Assume now that Mr. X has some suspicious symptoms related to the virus and that for people having those symptoms the probability to have been infected is 80%. In that case P(I) = 0.8 and P(H/I) = 0.96. Further, 800 out of 1000 people having those symptoms are carriers of the virus and 200 are not. Therefore, we should have $800 \times 96\% = 768$ positive results from the carriers and $200 \times 4\% = 8$ positive results from the non-carriers, i.e., 776 in total positive results. Therefore, P(H) = 0.776 and Equation (2) gives now that P(I/H) $\approx$ 98.97%. Consequently, if the result of Mr. X's test is positive, then the probability to be a carrier exceeds the accuracy of the test!

The three alternative cases of the previous example and in particular the first one support the view of many specialists currently insisting that the tests performed without any special reason by those presenting no symptoms of COVID-19 are not effective and they simply burden the healthcare systems of their countries without purpose.

Further, the first and the third example enable one to study the sensitivity of the solution of the corresponding problem, which is expressed by the value of the conditional probability P(I/H). It becomes evident that the variable which affects dramatically the solution is the value of the independent probability P(I) of a subject to be a carrier of the virus. Figure 8 represents graphically the changes of the problem's solution P(I/H) with respect to the changes of the value of P(I).

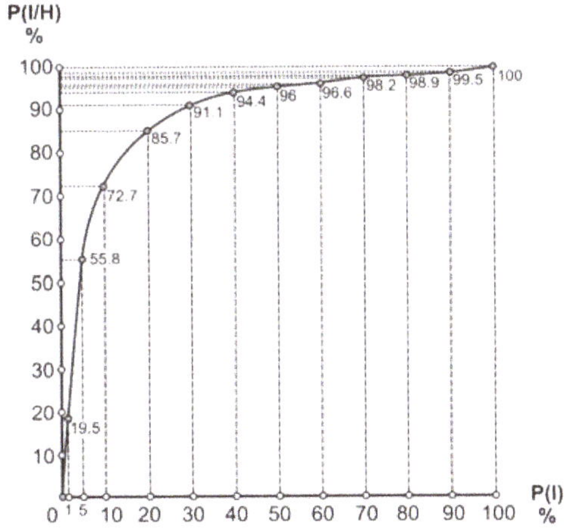

Figure 8. Changes of the problem's solution with respect to the values of P(I).

The Bayesian process appears frequently in the scientific method (Section 3). Consider for example the events:

- H: Newton's theory of gravity, and
- I: Velocity of Hermes around the sun.

In this case the value of the conditional probability P(H/I) will increase by replacing H with the general relativity theory.

In conclusion, the Bayesian process can distinguish science from pseudo-science more precisely than the Popper's falsification principle [22], by assigning a degree of truth to each possible case. On the contrary, the Popper's principle, based purely on the laws of the bivalent logic, characterizes each case only as either true or false.

6. Conclusions

The conclusions obtained by the discussion performed in this work could be summarized in the following statements:

- Induction and deduction are the two main types of human reasoning. However, despite the fact that the scientific method is based on deduction, almost all the scientific progress (with pure mathematics being possibly the unique exception) has its roots to inductive reasoning.
- FL restores the place and importance of the disdainful by the classical/bivalent logic induction as a fundamental component of the scientific reasoning.
- The success of mathematics in the natural sciences has been characterized by eminent scientists as being unreasonable (Winger's enigma). In fact, completely theoretical topics of mathematics find frequently, in unsuspicious time, important applications to science, technology and everyday life situations.
- The scientific method is characterized by the fact that the error of inductive reasoning is transferred to deduction through its premises. Therefore, there is a need of calculating the degree of truth of both inductive and deductive arguments. In the former case, the tools available for this quantification are probability and statistics, and of course FL and its generalizations for situations characterized by imprecision. In the latter case, it has recently turned out that the basic tool is the Bayesian probabilities. All these tools can help one to make a decision about a particular study and this is actually the managerial added value of the present study.
- Eminent scientists argue nowadays that the whole of science could be characterized as a Bayesian process, that distinguishes more precisely than the Popper's falsification principle the science from pseudo-science. A timely example, related to the validity of the viruses' tests, was presented here illustrating the importance of the Bayesian probabilities.

It must not be considered, however, that the wide and composite subject of this treatise is integrated here. We do hope, however, that the discussion performed in the present article could be promote a wider dialogue on this philosophical topic leading to further thoughts and research on the subject.

Author Contributions: E.A.: Methodology; Formal Analysis; Resources; Visualization. M.G.V.: Writing—Original Draft Preparation; Conceptualization; Resources; Data Curation; Visualization. All authors have read and agreed to the published version of the manuscript.

Funding: This research received no external funding.

Conflicts of Interest: The authors declare no conflict of interest.

References

1. Zadeh, L.A. Fuzzy Sets. *Inf. Control* **1965**, *8*, 338–353. [CrossRef]
2. Voskoglou, M.G. *Finite Markov Chain and Fuzzy Logic Assessment Models: Emerging Research and Opportunities*; Create space Independent Publishing Platform, Amazon: Columbia, SC, USA, 2017.
3. Voskoglou, M.G. (Ed.) *An Essential Guide to Fuzzy Systems*; Nova Science Publishers: Hauppauge, NY, USA, 2019.
4. Freksa, S. Fuzzy Logic: An Interface Between Logic and Human Reasoning. *IEEE Expert* **1994**, *9*, 20–24.

5. Dernoncourt, F. *Fuzzy Logic: Between Human Reasoning and Artificial Intelligence*; Report; Ecole Normale Supperieure: Paris, France, 2011.
6. Novák, V. Fuzzy Natural Logic: Towards Mathematical Logic of Human Reasoning. In *Towards the Future of Fuzzy Logic*; Seising, R., Trillas, E., Kacprzyk, J., Eds.; Studies in Fuzziness and Soft Computing; Springer: New York, NY, USA, 2015; Volume 325.
7. Athanassopoulos, E. Inductive Error and Fuzzy Logic—Criticism to the Philosophy of Science. In *Greek Language*; Self-Edition: Patras, Greece, 2019.
8. Voskoglou, M.G. Fuzzy Logic in Human Reasoning. *Bull. Electr. Eng. Inform.* **2012**, *2*, 158–168. [CrossRef]
9. Zadeh, L.A. Is there a need for Fuzzy Logic? *Inf. Sci.* **2008**, *178*, 1751–1779. [CrossRef]
10. Borkowski, L. On Three-valued Logic. In *Selected Works by Jan Łukasiewicz*; North–Holland: Amsterdam, The Netherlands, 1970; pp. 87–88.
11. Voskoglou, M.G. Methods for Assessing Human-Machine Performance under Fuzzy Conditions. *Mathematics* **2019**, *7*, 230. [CrossRef]
12. Reiter, R. A logic for default reasoning. *Artif. Intell.* **1980**, *13*, 81–132. [CrossRef]
13. Voskoglou, M.G. Generalizations of Fuzzy Sets and Related Theories. In *An Essential Guide to Fuzzy Systems*; Voskoglou, M., Ed.; Nova Science Publishers: Hauppauge, NY, USA, 2019; pp. 345–352.
14. Voskoglou, M.G. Fuzzy Sets, Grey System Theory and Computational Thinking. In *An Essential Guide to Fuzzy Systems*; Voskoglou, M., Ed.; Nova Science Publishers: Hauppauge, NY, USA, 2019; pp. 1–53.
15. Shapiro, S. *Thinking about Mathematics*; Oxford University Press: Oxford, UK, 2000.
16. Franzen, T. *Gödel's Theorem: An Incomplete Guide to its Use and Abuse*; A.K. Peters: Wellesley, MA, USA, 2005.
17. Ladyman, J. *Understanding the Philosophy of Science*; Routledge: Oxon, UK, 2002.
18. Voskoglou, M.G. Studying the Winger's "Enigma" about the Unreasonable Effectiveness of Mathematics in the Natural Sciences. *Am. J. Appl. Math. Stat.* **2017**, *5*, 95–100.
19. Singh, S. *Bing Bang—The Origin of the Universe*; Harper Perennian Publishers: New York, NY, USA, 2005.
20. Gingerich, O. *The Eye of the Heaven—Ptolemy, Copernicus, Kepler*; American Institute of Physics: New York, NY, USA, 1993.
21. Thrope, W.H. *The Origins and Rise of Ethology: The Science of the Natural Behavior of Animals*; Praeger: London, UK; New York, NY, USA, 1979.
22. Popper, K.R. *The Logic of Scientific Discovery*; Routlege: Oxon, UK, 2002; (Edition in German, 1934, translated in English, 1959).
23. Winger, E.P. The unreasonable effectiveness of mathematics in the natural sciences. *Commun. Pure Appl. Math.* **1960**, *13*, 1–14.
24. Voskoglou, M.G. A Contribution to the Study of Rings. Ph.D. Thesis, Department of Mathematics, University of Patras, Patras, Greece, 1982. (In Greek Language)
25. Majid, S. What is a Quantum Group? *Not. Am. Math. Soc.* **2006**, *53*, 30–31.
26. Lopez-Permouth, S. Matrix Representations of Skew Polynomial Rings with Semisimple Coefficient Rings. *Contemp. Math.* **2009**, *480*, 289–295.
27. Livio, M. *Is God a Mathematician?* Simon & Schuster: London, UK, 2009.
28. Chalmers, A.F. *What Is This Thing Called Science? An Assessment of the Nature and Status of Science and Its Methods*, 2nd ed.; University of Queensland Press: Brisbane, Lucia, 1992.
29. Janes, E.T. *Probability Theory: The Logic of Science, 8th Printing*; Cambridge University Press: Cambridge, UK, 2011.
30. Mumford, D. The Dawing of the Age of Stochasticity. In *Mathematics: Frontiers and Perspectives*; Amoid, V., Atiyah, M., Laxand, P., Mazur, B., Eds.; AMS: Providence, RI, USA, 2000; pp. 197–218.
31. Bertsch McGrayne, S. *The Theory that Would Not Die*; Yale University Press: New Haven, CT, USA; London, UK, 2012.
32. Hoghan, J. Bayes's Theorem: What is the Big Deal? January 2015. Available online: http://blogs.scientificamerican.com/cross-check/bayes-s-theorem-what-s-the-big-deal (accessed on 15 April 2020).

mathematics

MDPI

Article

A Method of Generating Fuzzy Implications from n Increasing Functions and n + 1 Negations

Maria N. Rapti and Basil K. Papadopoulos *

Section of Mathematics and Informatics, Department of Civil Engineering, School of Engineering, Democritus University of Thrace, 67100 Kimeria, Greece; marapti@civil.duth.gr

* Correspondence: papadob@civil.duth.gr

Received: 14 April 2020; Accepted: 22 May 2020; Published: 1 June 2020

Abstract: In this paper, we introduce a new construction method of a fuzzy implication from n increasing functions $g_i : [0,1] \to [0,\infty)$, $(g(0) = 0)$ $(i = 1,2,\ldots,n,\ n \in \mathbb{N})$ and $n+1$ fuzzy negations N_i $(i = 1,2,\ldots,n+1,\ n \in \mathbb{N})$. Imagine that there are plenty of combinations between n increasing functions g_i and $n+1$ fuzzy negations N_i in order to produce new fuzzy implications. This method allows us to use at least two fuzzy negations N_i and one increasing function g in order to generate a new fuzzy implication. Choosing the appropriate negations, we can prove that some basic properties such as the exchange principle (EP), the ordering property (OP), and the law of contraposition with respect to N are satisfied. The worth of generating new implications is valuable in the sciences such as artificial intelligence and robotics. In this paper, we have found a novel method of generating families of implications. Therefore, we would like to believe that we have added to the literature one more source from which we could choose the most appropriate implication concerning a specific application. It should be emphasized that this production is based on a generalization of an important form of Yager's implications.

Keywords: fuzzy implication; ordering property; least fuzzy negation; t-conditionality

1. Introduction

Fuzzy implications are the generalization of the classical (Boolean) implication in the interval of [0,1]. They play an important role in the area of fuzzy logic, decision theory, and fuzzy control. We can generate fuzzy implications from aggregation functions and fuzzy negations [1–5]. Other ways of generating fuzzy implications can be achieved by additive generating functions or by some initials implications [6–11]. Fuzzy implications are used for the application of the 'if-then' rule in fuzzy systems and inference processes, through Modus Ponens and Modus Tollens [12].

This paper is inspired by Yager's f-generated implications where $f : [0,1] \to [0,\infty]$ is a strictly decreasing and continuous function and $f(1) = 0$. In addition, a fuzzy implication $I : [0,1]^2 \to [0,1]$ is defined by: $I(x,y) = f^{-1}(xf(y))$, $x, y \in [0,1]$ with the understanding $0 \cdot \infty = 0$ (see [1] Definition 3.1.1). In this paper, we use functions $g_i : [0,1] \to [0,\infty)$, which are increasing and continuous, and also $g_i(0) = 0$. We present a new production machine of fuzzy implications. Such a type of generating fuzzy implications can be found in the literature [1,2,5–8], for example, $I_{RC} = 1 - x + xy$. The production of new fuzzy implications is accomplished with the help of any fuzzy negations and increasing functions. These generated fuzzy implications fulfill the necessary properties required to be fuzzy implications (see [1] Definition 1.1.1.). Moreover, if the negations are selected with certain properties, then the generated implications may also fulfill additional properties like the neutrality property (NP), exchange principle (EP), identity principle (IP), and some others. The worth of this production of implications could be estimated at artificial intelligence, robotics science, etc. [13–15]. This method of producing implications gives us the possibility, in a fuzzy environment, to find a large number of implications, which could help any researcher choose the most appropriate one.

The paper is organized as follows. In Section 2, we recall the basic concepts and definitions used in the paper. In Section 3, we study the new constructed method of fuzzy implications. Firstly, we present a constructed method using one increasing function g and two negations N_1, N_2, then a second method using two increasing functions g_1, g_2 and three negations N_1, N_2, N_3. Finally, we generalize our constructed method using n functions $g_1, g_2, \ldots, g_n$ and $n+1$ negations $N_1, N_2, \ldots, N_n, N_{n+1}$.

2. Preliminaries

In order to help the reader get familiar with the theory, we recall here some of the concepts and results employed in the rest of the paper.

Definition 1. (see [1] Definition 1.1.1). *A function $I : [0, 1]^2 \rightarrow [0, 1]$ is called a fuzzy implication if it satisfies, for all $x, x_1, x_2, y, y_1, y_2 \in [0, 1]$, the following conditions:*

$$x_1 \le x_2 \text{ then } I(x_1, y) \ge I(x_2, y), \quad \text{i.e., I is is decreasing in the first variable} \tag{1}$$

$$y_1 \le y_2 \text{ then } I(x, y_1) \le I(x, y_2), \quad \text{i.e., I is increasing in the second variable.} \tag{2}$$

$$I(0,0) = 1 \tag{3}$$

$$I(1,1) = 1 \tag{4}$$

$$I(1,0) = 0 \tag{5}$$

The set of all fuzzy implications will be denoted by FI.

A fuzzy negation N is a generalization of the classical complement or negation ¬, whose truth table consists of the two conditions: ¬0 = 1 and ¬1 = 0.

Definition 2. (see [1] Definition 1.4.1). *A function $N : [0,1] \rightarrow [0,1]$ is called a fuzzy negation if it satisfies the following conditions:*

$$N(0) = 1, \ N(1) = 0 \tag{6}$$

$$N \text{ is decreasing} \tag{7}$$

Definition 3. (see [1] Definition 1.4.2 (i)). *A fuzzy negation N is called strict if, in addition,*

$$N \text{ is strictly decreasing} \tag{8}$$

$$N \text{ is continuous} \tag{9}$$

Definition 4. (see [1] Definition 1.4.2 (ii)). *A fuzzy negation N is called strong if it is an involution, i.e.,*

$$N(N(x)) = x, \ x \in [0,1] \tag{10}$$

Definition 5. (see [1] Definition 1.4.2 (ii)). *A fuzzy negationN is said to be non-vanishing if*

$$N(x) = 0 \Leftrightarrow x = 1 \tag{11}$$

Definition 6. (see [1] Definition 1.4.2 (ii)). *A fuzzy negation N* is said to be non-filling if

$$N(x) = 1 \Leftrightarrow x = 0 \tag{12}$$

Definition 7. (see [1] Definition 1.4.15 (ii)). *Let $I \in FI$. The function $N_I : [0,1] \longrightarrow [0,1]$ defined by*

$$N_I(x) := I(x,0),\ x \in [0,1] \tag{13}$$

is called the natural negation of I.

Example 1. (see [1] example 1.4.4, [2] Section 2.1 Example 1). *Important negations that will be used throughout this paper are the standard negation$N_C = 1 - x$, the least or Godel, and the greatest or dual Godel fuzzy negations given respectively by*

$$N_{D_1}(x) = \begin{cases} 1 & \text{if } x = 0 \\ 0 & \text{if } x \in (0,1] \end{cases} \qquad N_{D_2}(x) = \begin{cases} 1 & \text{if } x \in [0,1) \\ 0 & \text{if } x = 1 \end{cases}$$

Definition 8. (see [1] Definitions 1.3.1, 1.5.1). *A Fuzzy ImplicationI is said to satisfy*

i. *The left neutrality property if:*

$$I(1,y) = y,\ y \in [0,1] \tag{14}$$

ii. *The exchange principle if:*

$$I(x,I(y,z)) = I(y,I(x,z)),\ x,y,z \in [0,1] \tag{15}$$

iii. *The identity principle if:*

$$I(x,x) = 1,\ x \in [0,1] \tag{16}$$

iv. *The ordering property if:*

$$I(x,y) = 1 \Leftrightarrow x \le y,\ x,y \in [0,1] \tag{17}$$

v. *The law of contraposition with respect toN if:*

$$I(x,y) = I(N(y),N(x)), \qquad x,y \in [0,1] \tag{18}$$

vi. *The law of left contraposition with respect toN if:*

$$I(N(x),y) = I(N(y),x),\ x,y \in [0,1] \tag{19}$$

vii. *The law of right contraposition with respect to N if:*

$$I(x,N(y)) = I(y,N(x)),\ x,y \in [0,1] \tag{20}$$

Definition 9. (see [1] Notations and Some Preliminaries). *We say that functions $f,\ g : [0,1]^n \to [0,1]$ are Φ-conjugate if there exists a $\varphi \in \Phi$ such that $g = f_\varphi$, where*

$$f_\varphi(x_1,x,\ldots,x_n) = \varphi^{-1}(f(\varphi(x_1),\varphi(x_2),\ldots\varphi(x_n))),\ x_1,x_2,\ldots,x_n \in [0,1]. \tag{21}$$

Definition 10. (see [1] Definition 2.2.1). *A function$S : [0,1]^2 \longrightarrow [0,1]$ is called a triangular conorm (t-conorm) if it satisfies, for all $x,y,z \in [0,1]$, the following conditions:*

$$S(x,y) = S(y,x) \tag{22}$$

$$S(x,S(y,z)) = S(S(x,y),z) \tag{23}$$

$$\text{If } y \le z, \text{ then } S(x,y) \le S(x,z), \text{ i.e., } S(x,\cdot) \text{ is increasing} \tag{24}$$

$$S(x,0) = x \tag{25}$$

Definition 11. (see [1] Definition 2.1.1). *A function* $S : [0,1]^2 \longrightarrow [0,1]$ *is called a triangular norm (t-norm) if it satisfies, for all* $x, y, z \in [0,1]$, *the following conditions:*

$$T(x,y) = T(y,x) \tag{26}$$

$$T(x,T(y,z)) = T(T(x,y),z) \tag{27}$$

$$\text{If } y \le z, \text{ then } T(x,y) \le T(x,z)\text{ , i.e., } T(x,\cdot) \text{ is increasing} \tag{28}$$

$$T(x,1) = x \tag{29}$$

Remark 1. (see [1] Propositions 1.1.8, 1.4.8 and Remarks 2.1.4 (vii), 2.2.5 (vii)). *It is proved that, if* $\varphi \in \Phi$, *T is a continuous t-norm, S is a continuous t-conorm, N is a fuzzy (strict, strong) negation, and I is a fuzzy implication, then* T_φ *is a t-norm,* S_φ *is a t-conorm,* N_φ *is a fuzzy (strict, strong) negation, and* I_φ *is a fuzzy implication.*

Definition 12. (see [1] Definition 2.4.1). *The equation* $p \rightarrow q \equiv \neg p \vee q$ *creates a new class of fuzzy implications.*

A function $I : [0,1]^2 \longrightarrow [0,1]$ *is called an (S, N)-Implication if there exist a t-conorm S and a fuzzy negation N such that:*

$$I(x,y) = S(N(x),y)\,, \qquad x,\ y \in [0,1] \tag{30}$$

Definition 13. (see [1] Subsection 7.3). *The equation* $(p \wedge q) \rightarrow r \equiv (p \rightarrow (q \rightarrow r))$ *is known as the law of importation and is a tautology in classical logic. The general form of the above equivalence is given by*

$$I(T(x,y),z) = I(x,I(y,z)), \quad x,y,z \in [0,1] \tag{31}$$

Definition 14. (see [1] Definition 7.4.1). *An implication I and a t-norm T satisfy the T-conditionality property if and only if*

$$T(x,I(x,y)) \le y, \quad x,y \in [0,1] \tag{32}$$

Proposition 1. (see [1] Definition 7.4.2). *If* $I \in FI$ *is such that there exist* $x, y \in (0, 1)$ *such that* $x > y$ *and* $I(x, y) = 1$, *then I does not satisfy (32) with any t-norm T.*

Proposition 2. (see [1] Definition 7.4.3). *Let* $I \in FI$, *a t-norm T satisfy (32),* N_I *is the natural negation of I and* N_T *is the natural negation of T, then* $N_I \le N_T$, *the natural negation of T.*

Definition 15. (see [1] Definition 1.6.1). *Let N be a fuzzy negation and I be a fuzzy implication. A function* $I_N : [0,1]^2 \rightarrow [0,1]$ *defined by*

$$I_N(x,y) = I(N(y),N(x)),\ x,y \in [0,1] \tag{33}$$

is called the N-reciprocal of I.

When N is the classical negation N_C, *then* I_N *is called the reciprocal of I and is denoted by* I'.

3. The Main Results

In this section, we give definitions of new generated implications and prove some useful properties of them.

3.1. Fuzzy Implications Generated by One Increasing Function g and Two fuzzy negations N_1, N_2

Theorem 1. *If N_1, N_2 are two fuzzy negations and $g : [0,1] \to [0,\infty)$ is an increasing and continuous function with $g(0) = 0$, then the function $I : [0,1]^2 \to [0,1]$ defined by*

$$I(x,y) = N_2\left(\frac{g(x)}{g(1)} \cdot N_1(y)\right), \quad x,y \in [0,1] \tag{34}$$

is a fuzzy implication.

Proof. Let $g : [0,1] \to [0,\infty)$ be an increasing and continuous function with $g(0) = 0$ and $x_1, x_2, y \in [0,1]$.

If $x_1 \le x_2$ then

$g(x_1) \le g(x_2) \Rightarrow \frac{g(x_1)}{g(1)} \cdot N_1(y) \le \frac{g(x_2)}{g(1)} \cdot N_1(y) \Rightarrow N_2\left(\frac{g(x_1)}{g(1)} \cdot N_1(y)\right) \ge N_2\left(\frac{g(x_2)}{g(1)} \cdot N_1(y)\right) I(x_1, y) \ge I(x_2, y)$, i.e., $I(\cdot, y))$ is decreasing, i.e., I satisfies (1)

Let $y_1, y_2, x \in [0,1]$. If $y_1 \le y_2$, then $N_1(y_1) \ge N_1(y_2) \Rightarrow \frac{g(x)}{g(1)} \cdot N_1(y_1) \ge \frac{g(x)}{g(1)} \cdot N_1(y_2) \Rightarrow N_2\left(\frac{g(x)}{g(1)} \cdot N_1(y_1)\right) \le N_2\left(\frac{g(x)}{g(1)} \cdot N_1(y_2)\right)$

$\Rightarrow I(x, y_1) \le I(x, y_2)$, i.e., $I(x, \cdot)$ is increasing, i.e., I satisfies (2)

$I(0,0) = N_2\left(\frac{g(0)}{g(1)} \cdot N_1(0)\right) = N_2(0) = 1$, i.e., I satisfies (3)

$I(1,1) = N_2\left(\frac{g(1)}{g(1)} \cdot N_1(1)\right) = N_2(0) = 1$, i.e., I satisfies (4)

$I(1,0) = N_2\left(\frac{g(1)}{g(1)} \cdot N_1(0)\right) = N_2(1) = 0$, i.e., I satisfies (5)

Therefore, $I \in FI$. □

Proposition 3. *Let I be the fuzzy implication of Theorem 1, then the fuzzy implication N-reciprocal of I is*

$$I_N(x,y) = I(N(y), N(x)) = N_2\left(\frac{g(N(y))}{g(1)} \cdot N_1(N(x))\right) \tag{35}$$

Proposition 4. *If $N_1 = N_2 = N$ are strong negations, then the fuzzy implication of Theorem 1 satisfies additionally the left neutrality property (14) and the exchange principle (15).*

Proof. $I(1,y) = N\left(\frac{g(1)}{g(1)} \cdot N(y)\right) = N(N(y)) = y$, $y \in [0,1]$, i.e., I satisfies (14)

$$I(a, I(b,x)) = N\left(\frac{g(a)}{g(1)} \cdot N(I(b,x))\right) = N\left(\frac{g(a)}{g(1)} \cdot N\left(N\left(\frac{g(b)}{g(1)} \cdot N(x)\right)\right)\right) \overset{\text{N: strong negation}}{=} N\left(\frac{g(a)g(b)}{g(1)^2} N(x)\right)$$
$$I(b, I(a,x)) = N\left(\frac{g(b)}{g(1)} \cdot N(I(a,x))\right) = N\left(\frac{g(b)}{g(1)} \cdot N\left(N\left(\frac{g(a)}{g(1)} \cdot N(x)\right)\right)\right) \overset{\text{N: strong negation}}{=} N\left(\frac{g(a)g(b)}{g(1)^2} N(x)\right)$$

Thus, we have $I(a, I(b,x)) = I(b, I(a,x))$, i.e., I satisfies (15). □

Theorem 2. *If $\varphi \in \Phi$ and I is the fuzzy implication of Theorem 1, then I_φ is a fuzzy implication.*

Proof. According to Remark 1, I_φ is a fuzzy implication. □

Proposition 5. *If* $N_1(x) = N_{D_1}(x) = \begin{cases} 1, & if\ x = 0 \\ 0, & if\ x \in (0,1] \end{cases}$ *(the least fuzzy negation), then the fuzzy implication of Theorem 1 satisfies the Identity Principle (16).*

Proof.

$$I(x,x) = N_2\left(\frac{g(x)}{g(1)} \cdot N_1(x)\right) \overset{x=0}{=} N_2\left(\frac{g(0)}{g(1)} \cdot N_1(0)\right) = N_1(0) = 1$$

$$I(x,x) = N_2\left(\frac{g(x)}{g(1)} \cdot N_1(x)\right) \overset{x\in(0,1]}{=} N_2\left(\frac{g(0)}{g(1)} \cdot N_1(0)\right) = N_1(0) = 1$$

i.e., I satisfies (16). □

Proposition 6. *If the fuzzy implication of Theorem 1 satisfies the Identity Principle (16), then it satisfies the Ordering Property (17).*

Proof. Let x,y ∈ [0,1] and x ≤ y, then I(x,y) ≥ I(y, y) = 1. Thus, I(x, y) = 1.

If $I(x,y) = 1 \Leftrightarrow$
$I(x,y) = N_2\left(\frac{g(x)}{g(1)} \cdot N_1(y)\right) = 1 \Leftrightarrow$
$\frac{g(x)}{g(1)} \cdot N_1(y) = 0 \Leftrightarrow$
$g(x) = 0$ or $N_1(y) = 0 \Leftrightarrow$
$x = 0 \leq y$ or $y = 1 \geq x$.
Thus, we have x ≤ y. □

Proposition 7. *The natural negation* N_I *of the fuzzy implication of Theorem 1 is*

$$N_I(x) = N_2\left(\frac{g(x)}{g(1)}\right)$$

Proof.

$$N_I(x) = I(x,0) = N_2\left(\frac{g(x)}{g(1)} \cdot N_1(0)\right) = N_2\left(\frac{g(x)}{g(1)}\right).$$

□

Proposition 8. *When* $N_1(x) = N_2(x) = N(x)$ *are strong negations, then the fuzzy implication of Theorem 1 is an (S, N)–implication.*

Proof. When $N_1(x) = N_2(x) = N(x)$ are strong negations, according to Theorem 1 and Proposition 4, the fuzzy implication $I(x,y) = N_2\left(\frac{g(x)}{g(1)} \cdot N_1(y)\right)$ satisfies (I1) and (EP). Moreover, if $N_1(x), N_2(x)$ are continuous negations, then N_I is also a continuous fuzzy negation. We deduce that I is an (S, N) – implication (see [1] Theorem 2.4.10). □

Example 2. *Let* $g(x) = x$, $N_2(x) = 1 - x$, $N_1(x) = 1 - x^2$.

Then, $I(x,y) = N_2\left(x\left(1 - y^2\right)\right) = 1 - x\left(1 - y^2\right) = 1 - x + xy^2$
The graph of the above surface is plotted in Figure 1.

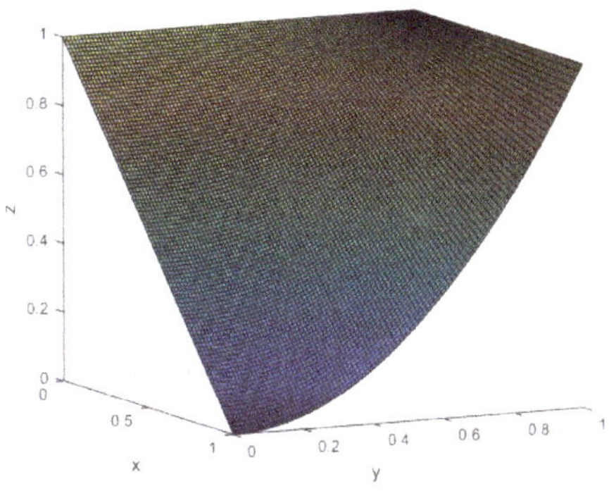

Figure 1. I implication of example 2.

Example 3. *Let* $g(x) = x$, $N_1(x) = N_2(x) = 1 - x$. *Thus,* $I(x, y) = N_2(x(1 - y)) = 1 - x(1 - y) = 1 - x + xy$ *Then, it is the Reinchenbach Implication.*

The graph of the above surface is plotted in Figure 2.

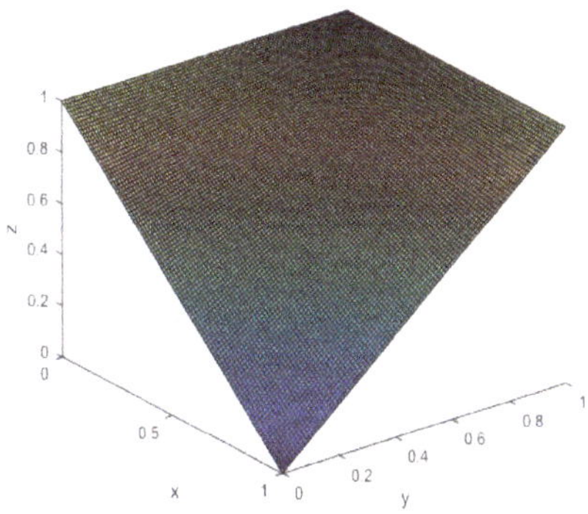

Figure 2. I implication of example 3.

Example 4. *Let* $g(x) = x$, $N_2(x) = 1 - x$, $N_1(x) = \frac{1-x}{1+x}$. *Then,* $I(x, y) = N_2\left(x \cdot \frac{1-y}{1+y}\right) = \frac{1+y-x+xy}{1+y}$.

The graph of the above surface is plotted in Figure 3.

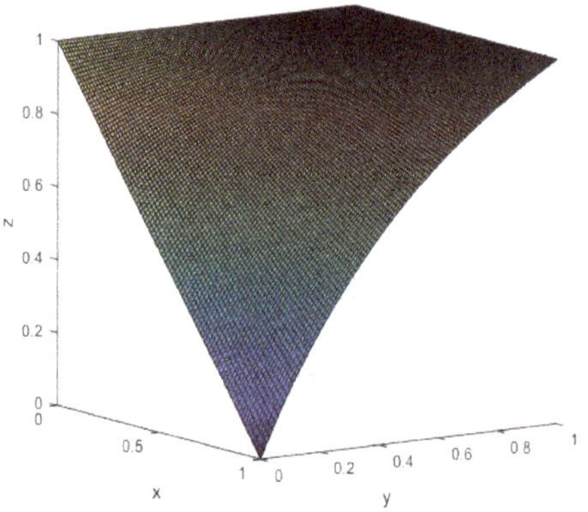

Figure 3. I implication of example 4.

Proposition 9. *Let N_1, N_2, N_3 be three fuzzy negations. Let us suppose also that N_3 is a strong fuzzy negation. If g: [0,1] → [0,∞) is an increasing and continuous function with g(0) = 0 and $g(x) = N_1(N_3(x))$, then the fuzzy implication $I(x,y) = N_2\left(\frac{g(x)}{g(1)} \cdot N_1(y)\right)$ defined in Theorem 1 satisfies the law of contraposition (18) with respect to N_3.*

Proof. g: [0,1] → [0,∞) is an increasing and continuous function

$$I(N_3(y), N_3(x)) = N_2\left(\frac{g(N_3(y))}{g(1)} \cdot N_1(N_3(x))\right) = N_2\left(\frac{N_1(N_3(N_3(y)))}{g(1)} \cdot N_1(N_3(x))\right) \overset{N_3\text{:strong negation}}{=}$$

$$N_2\left(\frac{N_1(y)}{g(1)} \cdot N_1(N_3(x))\right) = N_2\left(\frac{N_1(y)}{g(1)} \cdot g(x)\right) = N_2\left(\frac{g(x)}{g(1)} \cdot N_1(y)\right) = I(x,y), \text{ i.e., I satisfies (18).}$$ □

Lemma 1. (see [1] Proposition 1.5.3). *Let N_1, N_2, N_3 be three fuzzy negations with the properties N_1, N_3 being strict ones and N_3 additionally being a strong negation. If g: [0,1] → [0,∞) is an increasing and continuous function with g(0) = 0 and $g(x) = N_1(N_3(x))$, then the fuzzy implication $I(x,y) = N_2\left(\frac{g(x)}{g(1)} \cdot N_1(y)\right)$, $x, y \in [0,1]$ defined by Theorem 1 satisfies the left (L-CP) and the right (R-CP) law of the contraposition.*

Proof. According to Proposition 1.5.3 [1], I satisfies the left (19) and the right (20) law of the contraposition. □

Using Definition 14 and Proposition 2 of Section 2, we prove the following:

Proposition 10. *Let I be the fuzzy implication defined by Theorem 1, $I(x,y) = N_2\left(\frac{g(x)}{g(1)} \cdot N_1(y)\right)$, $x, y \in [0,1]$. Let us suppose that T is a t-norm and T satisfies (32), then:*

i. $N_2(x) \le N_1(x)$, *if* N_2 *is a strong negation.*
ii. $g(x) \ge g(1) \cdot N_1(N_T(x,y))$, *if* N_1 is a strong negation.

Proof. As I and T satisfy (TC), then $T(x, I(x,y)) \le y$ for all $x, y \in [0,1]$.

i. Let $x = 1$, we have $I(1,y) \le y$, $y \in [0,1]$ then $N_2\left(\frac{g(1)}{g(1)} \cdot N_1(y)\right) \le y \Rightarrow N_2(N_1(y)) \le y \Rightarrow$

$$N_2(N_2(N_1(y))) \ge N_2(y) \xrightarrow{N_2\,:\,\text{strong negation}} N_2(y) \le N_1(y).$$

ii. From Proposition 2, we have $N_I \leq N_T \Rightarrow N_2\left(\frac{g(x)}{g(1)}\right) \leq N_T(x,y) \xrightarrow{N_2\,:\,\text{strong negation}} N_2\left(N_2\left(\frac{g(x)}{g(1)}\right)\right) \geq N_2(N_T(x,y)) \Rightarrow \frac{g(x)}{g(1)} \geq N_2(N_T(x,y)) \Rightarrow g(x) \geq g(1)\cdot N_2(N_T(x,y))$

□

3.2. Fuzzy Implications Generated by Two Increasing Functions g_1, g_2 and Three Fuzzy Negations N_1, N_2, N_3

Theorem 3. *If $g_1(x),\ g_2(x) : [0,1] \to [0,\infty)$ are increasing and continuous functions with $g_1(0) = g_2(0) = 0$ and N_1, N_2 N_3 are fuzzy negations, then the function $I : [0,1]^2 \to [0,1]$ defined by*

$$I(x,y) = N_3\left(\frac{\frac{g_1(x)}{g_1(1)}N_1(y) + \frac{g_2(x)}{g_2(1)}N_2(y)}{2}\right), x, y \in [0,1] \tag{36}$$

is a fuzzy implication.

Proof.

Let $x_1, x_2, y \in [0,1]$.

If

$$x_1 \leq x_2 \Rightarrow \frac{g_1(x_1)}{g_1(1)} \leq \frac{g_1(x_2)}{g_1(1)} \text{ and } \frac{g_2(x_1)}{g_2(1)} \leq \frac{g_2(x_2)}{g_2(1)},$$
$$\text{then, } N_1(y)\frac{g_1(x_1)}{g_1(1)} \leq N_1(y)\frac{g_1(x_2)}{g_1(1)} \text{ and } N_2(y)\frac{g_2(x_1)}{g_2(1)} \leq N_2(y)\frac{g_2(x_2)}{g_2(1)}$$
$$\Longrightarrow \frac{N_1(y)\frac{g_1(x_1)}{g_1(1)}{}_1 + N_2(y)\frac{g_2(x_1)}{g_2(1)}}{2} \leq \frac{N_1(y)\frac{g_1(x_2)}{g_1(1)}{}_1 + N_2(y)\frac{g_2(x_2)}{g_2(1)}}{2} \Longrightarrow$$
$$N_3\left(\frac{N_1(y)\frac{g_1(x_1)}{g_1(1)}{}_1 + N_2(y)\frac{g_2(x_1)}{g_2(1)}}{2}\right) \leq N_3\left(\frac{N_1(y)\frac{g_1(x_2)}{g_1(1)}{}_1 + N_2(y)\frac{g_2(x_2)}{g_2(1)}}{2}\right)$$

$I(x_1, y) \geq I(x_2, y)$, i.e., $I(\cdot,y)$ is decreasing, i.e., I satisfies (1).

Let $y_1,\ y_2, x \in [0, 1]$

If $y_1 \leq y_2$, then $N_1(y_1) \geq N_2(y_2) \Rightarrow \frac{g_1(x)}{g(1)}N_1(y_1) \geq \frac{g_1(x)}{g(1)}N_1(y_2)$ and $\frac{g_2(x)}{g(1)}N_2(y_1) \geq \frac{g_2(x)}{g(1)}N_2(y_2)$

$$\Rightarrow \frac{\frac{g_1(x)}{g(1)}N_1(y_1) + \frac{g_1(x)}{g(1)}N_2(y_1)}{2} \geq \frac{\frac{g_2(x)}{g(1)}N_1(y_2) + \frac{g_2(x)}{g(1)}N_2(y_2)}{2} \Rightarrow N_3\left(\frac{\frac{g_1(x)}{g(1)}N_1(y_1) + \frac{g_1(x)}{g(1)}N_2(y_1)}{2}\right)$$
$$\leq N_3\left(\frac{\frac{g_2(x)}{g(1)}N_1(y_2) + \frac{g_2(x)}{g(1)}N_2(y_2)}{2}\right)$$

$\Rightarrow I(x, y_1) \leq I(x, y_2)$, i.e., $I(x, \cdot)$ is increasing, i.e., I satisfies (2).

$$I(0,0) = N(0) = 1 \text{ i.e., I satisfies } (3)$$
$$I(1,1) = N(0) = 1 \text{ i.e., I satisfies } (4)$$
$$I(1,0) = N(1) = 0 \text{ i.e., I satisfies } (5).$$

□

Proposition 11. *Let Ibe the fuzzy implication of Theorem 3. If* $N_1(x) = N_2(x) = N_3(x) = N(x)$ *are strong negations, then the neutrality property (14) and the exchange principle (15) are satisfied.*

Proof. $I(1,x) = N\left(\frac{N(x)+N(x)}{2}\right) = N\left(\frac{2N(x)}{2}\right) = N(N(x)) = x$, i.e., I satisfies (14)

$$I(a,I(b,x)) = N\left(\frac{\frac{g_1(a)}{g_1(1)}N(I(b,x))+\frac{g_2(a)}{g_2(1)}N(I(b,x))}{2}\right) =$$

$$N\left(\frac{\frac{g_1(a)}{g_1(1)}N\left(N\left(\frac{\frac{g_1(b)}{g_1(1)}N(x)+\frac{g_2(b)}{g_2(1)}N(x)}{2}\right)\right)+\frac{g_2(a)}{g_2(1)}N\left(N\left(\frac{\frac{g_1(b)}{g_1(1)}N(x)+\frac{g_2(b)}{g_2(1)}N(x)}{2}\right)\right)}{2}\right) \overset{N:\text{strong negation}}{=}$$

$$= N\left(\frac{\frac{g_1(a)}{g_1(1)}\cdot\frac{\frac{g_1(b)}{g_1(1)}N(x)+\frac{g_2(b)}{g_2(1)}N(x)}{2}+\frac{g_2(a)}{g_2(1)}\cdot\frac{\frac{g_1(b)}{g_1(1)}N(x)+\frac{g_2(b)}{g_2(1)}N(x)}{2}}{2}\right) =$$

$$= N\left(\frac{\frac{g_1(a)}{g_1(1)}\cdot\frac{g_1(b)}{g_1(1)}N(x)+\frac{g_1(a)}{g_1(1)}\cdot\frac{g_2(b)}{g_2(1)}N(x)+\frac{g_2(a)}{g_2(1)}\cdot\frac{g_1(b)}{g_1(1)}N(x)+\frac{g_2(a)}{g_2(1)}\cdot\frac{g_2(b)}{g_2(1)}N(x)}{4}\right).$$

$$I(b,I(a,x)) = N\left(\frac{\frac{g_1(b)}{g_1(1)}N(I(a,x))+\frac{g_2(b)}{g_2(1)}N(I(a,x))}{2}\right) =$$

$$N\left(\frac{\frac{g_1(b)}{g_1(1)}N\left(N\left(\frac{\frac{g_1(a)}{g_1(1)}N(x)+\frac{g_2(a)}{g_2(1)}N(x)}{2}\right)\right)+\frac{g_2(b)}{g_2(1)}N\left(N\left(\frac{\frac{g_1(a)}{g_1(1)}N(x)+\frac{g_2(a)}{g_2(1)}N(x)}{2}\right)\right)}{2}\right) \overset{N:\text{strong negation}}{=}$$

$$= N\left(\frac{\frac{g_1(b)}{g_1(1)}\cdot\frac{\frac{g_1(a)}{g_1(1)}N(x)+\frac{g_2(a)}{g_2(1)}N(x)}{2}+\frac{g_2(b)}{g_2(1)}\cdot\frac{\frac{g_1(a)}{g_1(1)}N(x)+\frac{g_2(a)}{g_2(1)}N(x)}{2}}{2}\right)$$

$$= N\left(\frac{\frac{g_1(a)}{g_1(1)}\cdot\frac{g_1(b)}{g_1(1)}N(x)+\frac{g_1(a)}{g_1(1)}\cdot\frac{g_2(b)}{g_2(1)}N(x)+\frac{g_2(a)}{g_2(1)}\cdot\frac{g_1(b)}{g_1(1)}N(x)+\frac{g_2(a)}{g_2(1)}\cdot\frac{g_2(b)}{g_2(1)}N(x)}{4}\right)$$

We conclude that (15) is satisfied. □

Proposition 12. *Let I be the fuzzy implication of Theorem 3 and* N_1, N_2, N_3 *be fuzzy negations. If* $N_1(x) = N_2(x) = N_{D_1}(x) = \begin{cases} 1, & if\ x = 0 \\ 0, & if\ x \in (0,1] \end{cases}$ *(the least fuzzy negation), then the identity principle (16) is satisfied.*

Proof.

$$I(x,y) = N_3\left(\frac{\frac{g_1(x)}{g_1(1)}N_1(y)+\frac{g_2(x)}{g_2(1)}N_2(y)}{2}\right) \overset{x=0}{=} N_3(0) = 1$$

$$I(x,y) = N_3\left(\frac{\frac{g_1(x)}{g_1(1)}N_1(y)+\frac{g_2(x)}{g_2(1)}N_2(y)}{2}\right) \overset{x\in(0,1]}{=} N_3(0) = 1$$

Thus, I satisfies (16). □

Proposition 13. *If the fuzzy implication of Theorem 3 satisfies the identity principle (16), then it satisfies the ordering property (17).*

Proof. Let x,y ∈ [0,1] and x ≤ y, then I(x,y) ≥ I(y, y) = 1. Thus, I(x, y) = 1.

If $I(x,y) = 1 \Leftrightarrow$

$$I(x,y) = N_3\left(\frac{\frac{g_1(x)}{g_1(1)}N_1(y) + \frac{g_2(x)}{g_2(1)}N_2(y)}{2}\right) = 1 \Longleftrightarrow \frac{\frac{g_1(x)}{g_1(1)}N_1(y) + \frac{g_2(x)}{g_2(1)}N_2(y)}{2} = 0 \Longleftrightarrow$$

$g_1(x)N_1(x) = 0$ and $g_2(x)N_2(x) = 0$ $(g_1, g_2, N_1, N_2 \geq 0)$
$g_1(x) = 0$ or $N_1(y) = 0$ and $g_2(x) = 0$ or $N_2(x) = 0$
$x = 0 \leq y$ or $y = 1 \geq x$.
Thus, we have $x \leq y$. □

Proposition 14. *Let N_1, N_2, N_3, N_4 be four fuzzy negations. Let us suppose also that N_4 is a strong fuzzy negation. If g_1, g_2: [0,1] → [0,∞) are increasing and continuous functions with $g_1(0) = g_2(0) = 0$ and $g_1(x) = N_1(N_4(x))$, and $g_2(x) = N_2(N_4(x))$, then the fuzzy implication defined in Theorem 3 satisfies the law of contraposition (16) with respect to N_4.*

Proof.

$$I(N_4(y), N_4(x)) = N_3\left(\frac{\frac{g_1(N_4(y))}{g_1(1)}N_1(N_4(x)) + \frac{g_2(N_4(y))}{g_2(1)}N_2(N_4(x))}{2}\right) =$$

$$= N_3\left(\frac{\frac{N_1(N_4(N_4(y)))}{g_1(1)}N_1(N_4(x)) + \frac{N_2(N_4(N_4(y)))}{g_2(1)}N_2(N_4(x))}{2}\right)$$

$$\overset{N_4\text{:strong negation}}{=} N_3\left(\frac{\frac{N_1(y)}{g_1(1)}N_1(N_4(x)) + \frac{N_2(y)}{g_2(1)}N_2(N_4(x))}{2}\right)$$

$$= N_3\left(\frac{\frac{g_1(x)}{g_1(1)}N_1(y) + \frac{g_2(x)}{g_2(1)}N_2(y)}{2}\right) = I(x,\ y)$$

Thus, I satisfies (16). □

Proposition 15. *The natural negation N_I of the fuzzy implication of Theorem 3 is*

$$N_I(x) = N_3\left(\frac{\frac{g_1(x)}{g_1(1)} + \frac{g_2(x)}{g_2(1)}}{2}\right)$$

Proof.

$$N_I(x) = I(x, 0) = N_3\left(\frac{\frac{g_1(x)}{g_1(1)}N_1(0) + \frac{g_2(x)}{g_2(1)}N_2(0)}{2}\right) = N_3\left(\frac{\frac{g_1(x)}{g_1(1)} + \frac{g_2(x)}{g_2(1)}}{2}\right).$$

□

Example 5. *Let $g_1(x) = g_2(x) = x$, $N_3(x) = 1 - x = N_1(x)$, $N_2(x) = 1 - x^2$*

Then, $I(x,y) = N_3\left(\frac{x(1-y)+x(1-y^2)}{2}\right) == N_3\left(\frac{2x-xy-xy^2}{2}\right) = \frac{2-2x+xy+xy^2}{2}$

The graph of the above surface is plotted in Figure 4.

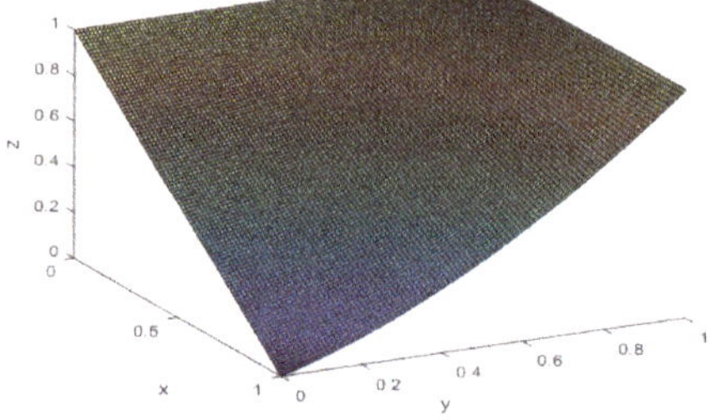

Figure 4. I implication of example 5.

Example 6. *Let* $g_1(x) = x^2$, $g_2(x) = x$, $N_3(x) = 1 - \sqrt{x}$, $N_1(x) = 1 - x$, $N_2(x) = 1 - x^2$.

Then, $I(x, y) = N_3\left(\frac{x^2(1-y)+x(1-y^2)}{2}\right) = 1 - \sqrt{\frac{x^2-x^2y+x-xy^2}{2}}$.

The graph of the above surface is plotted in Figure 5.

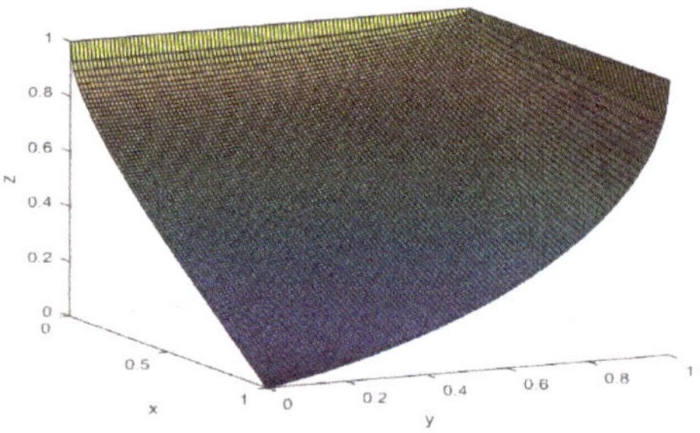

Figure 5. I implication of example 6.

3.3. Fuzzy Implications Generated by n Increasing Function $g_i(i = 1, 2, \ldots n, n \in \mathbb{N})$ and n + 1 Fuzzy Negations N_i , $(i = 1, 2, \ldots n + 1 , n \in \mathbb{N})$

Theorem 4. *If* $g_i(x)$*: [0,1]* $\rightarrow$ *[0,∞) are increasing and continuous functions, where* $g_i(0) = 0$ *(i = 1, 2, … n,* $n \in \mathbb{N}$*) and* N_i *are fuzzy negations (i = 1,2, … n + 1 ,* $n \in \mathbb{N}$*), then the function* $I : [0, 1]^2 \rightarrow [0, 1]$ *defined by*

$$I(x, y) = N_{n+1}\left(\sum_{i=1}^{n}\left(\frac{\frac{g_i(x)}{g_i(1)} \cdot N_i(y)}{n}\right)\right) x, y \in [0, 1], \tag{37}$$

is a fuzzy implication.

Proof. Let $x_1, x_2, y \in [0, 1]$.

If $x_1 \le x_2 \Rightarrow g_1(x_1) \le g_1(x_2)$, $g_2(x_1) \le g_2(x_2), \ldots, g_n(x_1) \le g_n(x_2) \Rightarrow \sum_{i=1}^{n} \frac{g_i(x_1)}{g_i(1)} \le \sum_{i=1}^{n} \frac{g_i(x_2)}{g_i(1)}$, then $N_1(y)g_1(x_1) \le N_1(y)g_1(x_2)$, $N_2(y)g_2(x_1) \le N_2(y)g_2(x_2), \ldots, N_n(y)\, g_n(x_1) \le N_n(y)g_n(x_2)$ $\sum_{i=1}^{n} \frac{1}{n}N_i(y)\frac{g_i(x_1)}{g_i(1)} \le \sum_{i=1}^{n} \frac{1}{n}N_i(y)\frac{g_i(x_2)}{g_i(1)} \Rightarrow N_{n+1}\left(\sum_{i=1}^{n}\left(\frac{1}{n}N_i(y)\frac{g_i(x_1)}{g_i(1)}\right)\right) \ge N_{n+1}\left(\sum_{i=1}^{n}\left(\frac{1}{n}N_i(y)\frac{g_i(x_2)}{g_i(1)}\right)\right)$

$\Rightarrow I(x_1, y) \ge I(x_2, y)$, i.e., I(,y) is decreasing, i.e., I satisfies (1)

Let $y_1, y_2, x \in [0, 1]$. If $y_1 \le y_2$, then $N_i(y_1) \ge N_i(y_2) \Rightarrow \frac{g_i(x)}{g_i(1)}N_i(y_1) \ge \frac{g_i(x)}{g_i(1)}N_i(y_2) \Rightarrow$ $\sum_{i=1}^{n} \frac{1}{n}\frac{g_i(x)}{g_i(1)}N_i(y_1) \ge \sum_{i=1}^{n} \frac{1}{n}\frac{g_i(x)}{g_i(1)}N_i(y_2) \Rightarrow$

$$N_{n+1}\left(\sum_{i=1}^{n}\left(\frac{1}{n}N\left(y_1\right)\frac{g_i(x)}{g_i(1)}\right)\right) \le N_{n+1}\left(\sum_{i=1}^{n}\left(\frac{1}{n}N\left(y_2\right)\frac{g_i(x)}{g_i(1)}\right)\right)$$

$\Rightarrow I\left(x, y_1\right) \le I\left(x, y_2\right)$, i.e., $I(x,\cdot)$ is increasing, i.e., I satisfies (2)

$I(0,0) = N(0) = 1$, i.e., I satisfies (3) $I(1,1) = N(0) = 1$, i.e., I satisfies (4)

$I(1,0) = N\left(\frac{1+1+\ldots+1}{n}\right) = N\left(\frac{n}{n}\right) = N(1) = 0$, i.e., I satisfies (5). □

Proposition 16. *Let I be the fuzzy implication of Theorem 4. If $N_{n+1}(x) = N_1(x) = \ldots = N_n(x) = N(x)$ are strong negations, then the left neutrality property (14) and the exchange principle (15) are satisfied.*

Proof. $I(1,y) = N\left(\sum_{i=1}^{n}\left(\frac{\frac{g_i(1)}{g_i(1)}\cdot N(y)}{n}\right)\right) = N(N(y)) = y,\ y \in [0,1]$, i.e., I satisfies (14)

$$I(a, I(b,x)) = N\left(\sum_{i=1}^{n}\left(\frac{\frac{g_i(a)}{g_i(1)}\cdot N(I(b,x))}{n}\right)\right)$$

$$= N\left(\sum_{i=1}^{n}\left(\frac{\frac{g_i(a)}{g_i(1)}\cdot N\left(N\left(\sum_{i=1}^{n}\left(\frac{\frac{g_i(b)}{g_i(1)}\cdot N(x)}{n}\right)\right)\right)}{n}\right)\right)$$

$$\overset{\text{N: strong negation}}{=} N\left(\sum_{i=1}^{n}\left(\frac{\frac{g_i(a)}{g_i(1)}\cdot\sum_{i=1}^{n}\left(\frac{\frac{g_i(b)}{g_i(1)}\cdot N(x)}{n}\right)}{n}\right)\right)$$

$$I(b, I(a,x)) = N\left(\sum_{i=1}^{n}\left(\frac{\frac{g_i(b)}{g_i(1)}\cdot N(I(a,x))}{n}\right)\right) =$$

$$= N\left(\sum_{i=1}^{n}\left(\frac{\frac{g_i(b)}{g_i(1)}\cdot N\left(N\left(\sum_{i=1}^{n}\left(\frac{\frac{g_i(a)}{g_i(1)}\cdot N(x)}{n}\right)\right)\right)}{n}\right)\right)$$

$$\overset{\text{N:strong negation}}{=} N\left(\sum_{i=1}^{n}\left(\frac{\frac{g_i(b)}{g_i(1)}\cdot\sum_{i=1}^{n}\left(\frac{\frac{g_i(a)}{g_i(1)}\cdot N(x)}{n}\right)}{n}\right)\right)$$

We conclude that (15) is satisfied, $I(a, I(b,x)) = I(b, I(a,x))$. □

Proposition 17. *Let* $I(x,y) = N_{n+1}\left(\sum_{i=1}^{n}\left(\frac{\frac{g_i(x)}{g_i(1)}\cdot N_i(y)}{n}\right)\right)$, $x, y \in [0,1]$, *the fuzzy implication defined by Theorem 4. Then, if* $N_1 = N_2 = \ldots N_n = N_{D_1} = \begin{cases} 1\ ,\ if\ x = 0 \\ 0\ ,\ if\ x \in (0,1] \end{cases}$ *(the least fuzzy negation), the identity principle (16) is satisfied.*

Proof.

$$I(x,x) = N_{n+1}\left(\sum_{i=1}^{n}\left(\frac{\frac{g_i(x)}{g_i(1)}\cdot N_i(x)}{n}\right)\right) \overset{x=0}{=} N_{n+1}(0) = 1$$

$$I(x,x) = N_{n+1}\left(\sum_{i=1}^{n}\left(\frac{\frac{g_i(x)}{g_i(1)}\cdot N_i(x)}{n}\right)\right) \overset{x\in(0,1]}{=} N_{n+1}(0) = 1$$

Thus, I satisfies (16). □

Proposition 18. *Let $N_1,\ldots,N_{n+1},N_{n+2}$ n + 2 be fuzzy negations. Let us suppose also that N_{n+2} is a strong fuzzy negation. If g_i: [0,1] → [0,∞) (i = 1, … , n , n ∈ ℕ) are increasing and continuous functions with $g_i(0) = \ldots = g_n(0) = 0$ and $g_i(x) = N_i(N_{n+2}(x))$, then the function $I : [0,1]^2 \to [0,1]$ defined by Theorem 4, $I(x,y) = N_{n+1}\left(\sum_{i=1}^{n}\left(\frac{\frac{g_i(x)}{g_i(1)}\cdot N_i(y)}{n}\right)\right)$ $x, y \in [0,1]$, satisfies the law of contraposition (18) with respect to N_{n+2}.*

Proof.

$$I(N_{n+2}(y),\ N_{n+2}(x)) = N_{n+1}\left(\sum_{i=1}^{n}\left(\frac{\frac{g_i(N_{n+2}(y))}{g_i(1)}\cdot N_i(N_{n+2}(x))}{n}\right)\right) =$$

$$= N_{n+1}\left(\sum_{i=1}^{n}\left(\frac{\frac{N_i(N_{n+2}(N_{n+2}(y)))}{g_i(1)}\cdot N_i(N_{n+2}(x))}{n}\right)\right)$$

$$\overset{N_{n+2}:\ strong\ negation}{=} N_{n+1}\left(\sum_{i=1}^{n}\left(\frac{\frac{N_i(y)}{g_i(1)}\cdot N_i(N_{n+2}(x))}{n}\right)\right) = I(x,y).$$

Thus, I satisfies (16). □

Proposition 19. *The natural negation N_I of the fuzzy negation of Theorem 4 is*

$$N_I(x) = N_{n+1}\left(\sum_{i=1}^{n}\left(\frac{\frac{g_i(x)}{g_i(1)}}{n}\right)\right)$$

Proof.

$$N_I(x) = I(x,0) = N_{n+1}\left(\sum_{i=1}^{n}\left(\frac{\frac{g_i(x)}{g_i(1)}\cdot N_i(0)}{n}\right)\right) = N_{n+1}\left(\sum_{i=1}^{n}\left(\frac{\frac{g_i(x)}{g_i(1)}}{n}\right)\right).$$

□

4. Conclusions

In this paper, a new production machine of fuzzy implications from n continuous increasing functions and n + 1 negation are introduced. We studied certain properties of these new fuzzy implications, as the left neutrality property (14), exchange principle (15), identity principle (16), ordering property (17), law of contraposition (18), and T-Conditionality (32), where some results are obtained if the fuzzy negations are strong or the least fuzzy negations. The advance of this method relies on the fact that we can combine a lot of fuzzy negations N_i and increasing functions g_i in order to generate fuzzy implications.

Finally, we believe that this production machine needs to be investigated further. It has been observed that in order to be satisfied, certain desirable properties by the implications generated by this method must use strong fuzzy negations or the least fuzzy negation. A question that arises is the following one:

Are there non-strong fuzzy negations that satisfy the left neutrality property (14) or the exchange principle (15)? In addition, in a future paper, we will study the behavior of non-continuous functions in terms of the validity of certain basic properties.

Author Contributions: Supervision, B.K.P.; Investigation, M.N.R. All authors have read and agreed to the published version of the manuscript.

Funding: This research received no external funding.

Conflicts of Interest: The authors declare no conflict of interest.

References

1. Baczynski, M.; Jayaram, B. *Fuzzy Implications*; Springer: Berlin/Heidelberg, Germany, 2008. [CrossRef]
2. Baczynski, M.; Jayaram, B. (S, N)-and R-implications; a state-of-the-art survey. *Fuzzy Sets Syst.* **2008**, *159*, 1836–1859. [CrossRef]
3. Baczynski, M.; Jayaram, B. QL-implications: Some properties and intersections. *Fuzzy Sets Syst.* **2010**, *161*, 158–188. [CrossRef]
4. Baczynski, M.; Jayaram, B. (U, N)-implications and their characterizations. *Fuzzy Sets Syst.* **2009**, *160*, 2049–2062. [CrossRef]
5. Durante, F.; Klement, E.P.; Meriar, R.; Sempi, C. Conjunctors and their residual implicators: Characterizations and construction methods. *Mediterr. J. Math.* **2007**, *4*, 343–356. [CrossRef]
6. Massanet, S.; Torrens, J. An overview of construction methods of fuzzy implications. In *Advances in Fuzzy Implication Functions. Studies in Fuzziness and Soft Computing*; Springer: Berlin/Heidelberg, 2013; Volume 300, pp. 1–30. [CrossRef]
7. Baczynski, M.; Jayaram, B.; Massanet, S.; Torrens, J. Fuzzy implications: Past, present, and future. In *Springer Handbook of Computational Intelligence. Springer Handbooks*; Springer: Berlin/Heidelberg, Germany, 2015; pp. 183–202. [CrossRef]
8. Sainio, E.; Turunen, E.; Mesiar, R. A characterization of fuzzy implications generated by generalized quantifiers. *Fuzzy Sets Syst.* **2008**, *159*, 491–499. [CrossRef]
9. Baczynski, M.; Jayaram, B. On the characterization of (S, N)-implications. *Fuzzy Sets Syst.* **2007**, *158*, 1713–1727. [CrossRef]
10. Massanet, S.; Torrens, J. Threshold generation method of construction of a new implication from two given ones. *Fuzzy Sets Syst.* **2012**, *205*, 50–75. [CrossRef]
11. Balasubramanian, J. Yager's new class of implications J_f and some classical tautologies. *Inf. Sci.* **2007**, *177*, 930–946. [CrossRef]
12. Zadeh, L.A. Outline of a new approach to the analysis of complex systems and decision processes. *IEEE Trans. Syst. Man Cybern.* **1973**, *SMC-3*, 28–44. [CrossRef]
13. Bogiatzis, A.C.; Papadopoulos, B.K. Producing fuzzy inclusion and entropy measures and their application on277 global image thresholding. *Evol. Syst.* **2018**, *9*, 331–353. [CrossRef]
14. Bogiatzis, A.C.; Papadopoulos, B.K. Local Thresholding of degraded or unevenly illuminated documents using fuzzy inclusion and entropy measures. *Evol. Syst.* **2019**, *10*, 593–619. [CrossRef]
15. Bogiatzis, A.C.; Papadopoulos, B. Global Image Thresholding Adaptive Neuro- Fuzzy Inference System Trained with Fuzzy Inclusion and Entropy Measures. *Symmetry* **2019**, *11*, 286. [CrossRef]

mathematics

MDPI

Article

Entropy Measures for Plithogenic Sets and Applications in Multi-Attribute Decision Making

Shio Gai Quek [1], Ganeshsree Selvachandran [1], Florentin Smarandache [2], J. Vimala [3], Son Hoang Le [4,*], Quang-Thinh Bui [5,6] and Vassilis C. Gerogiannis [7]

1 Department of Actuarial Science and Applied Statistics, Faculty of Business & Information Science, UCSI University, Jalan Menara Gading, Cheras, Kuala Lumpur 56000, Malaysia; queksg@ucsiuniversity.edu.my (S.G.Q.); Ganeshsree@ucsiuniversity.edu.my (G.S.)
2 Department of Mathematics, University of New Mexico, 705 Gurley Avenue, Gallup, NM 87301, USA; fsmarandache@gmail.com
3 Department of Mathematics, Alagappa University, Karaikudi, Tamil Nadu 630003, India; vimaljey@alagappauniversity.ac.in
4 Faculty of Information Technology, Ho Chi Minh City University of Technology (HUTECH), Ho Chi Minh City 700000, Vietnam
5 Institute of Research and Development, Duy Tan University, Da Nang 550000, Vietnam; qthinhbui@gmail.com
6 Faculty of Electrical Engineering and Computer Science, VŠB-Technical University of Ostrava, 70800 Ostrava-Poruba, Czech Republic
7 Department of Digital Systems, University of Thessaly, GR 41500 Larissa, Greece; vgerogian@uth.gr
* Correspondence: lh.son84@hutech.edu.vn

Received: 13 May 2020; Accepted: 10 June 2020; Published: 12 June 2020

Abstract: Plithogenic set is an extension of the crisp set, fuzzy set, intuitionistic fuzzy set, and neutrosophic sets, whose elements are characterized by one or more attributes, and each attribute can assume many values. Each attribute has a corresponding degree of appurtenance of the element to the set with respect to the given criteria. In order to obtain a better accuracy and for a more exact exclusion (partial order), a contradiction or dissimilarity degree is defined between each attribute value and the dominant attribute value. In this paper, entropy measures for plithogenic sets have been introduced. The requirements for any function to be an entropy measure of plithogenic sets are outlined in the axiomatic definition of the plithogenic entropy using the axiomatic requirements of neutrosophic entropy. Several new formulae for the entropy measure of plithogenic sets are also introduced. The newly introduced entropy measures are then applied to a multi-attribute decision making problem related to the selection of locations.

Keywords: neutrosophic set; plithogenic set; fuzzy set; entropy; similarity measure; information measure

1. Introduction

In recent years, there has been numerous authors who gave characterizations of entropy measures on fuzzy sets and their generalizations. Most notably, the majority of them had worked on developing entropy measures on intuitionistic fuzzy sets (IFS). Alongside with their introduction of new ways of entropy measures on IFS, these authors have also given some straightforward examples to show how their entropy measures can be applied to various applications including multi-attribute decision making (MADM) problems [1,2].

In 2016, Zhu and Li [3] gave a new definition for entropy measures on IFS. The new definition was subsequently compared against many other previous definitions of entropy measures on IFS. Montes et al. [4] proposed another new definition for entropy measures on intuitionistic fuzzy sets based on divergence. Both of these research groups [3,4] subsequently demonstrated the applications of their definition of IFS onto MADM problems, and both of them deployed examples of IFS, whose data values were not derived from real-life datasets but were predetermined by the authors to justify their new concepts. On the other hand, Farnoosh et al. [5] also gave their new definition for entropy measures on IFS, but they focused only on discussing its potential application in fault elimination of digital images rather than MADM. Ansari et al. [6] also gave a new definition of entropy measures on IFS in edge detection of digital images. Both research groups [5,6] did not provide examples on how their new definitions for entropy measures on IFS may be applied on MADM.

Some of the definitions of entropy measures defined for IFS were parametric in nature. Gupta et al. [7] defined an entropy measures on IFS, characterized by a parameter α. Meanwhile, Joshi and Kumar [8] independently (with respect to [7]) defined a new entropy measures on IFS, also characterized by a parameter α. An example on MADM was also discussed by Joshi and Kumar [8], once again involving a small, conceptual IFS like those encountered in the work of Zhu and Li [3] as well as Montes et al. [4]. The works by Joshi and Kumar [8] were subsequently followed by Garg et al. [9] who defined an entropy measure on IFS characterized by two parameters: (α, β). Like the previous authors, Garg et al. [9] discussed the application of their proposed entropy measure on MADM using a similar manner. In particular, they compared the effect of different parameters α, β on the results of such decision-making process. Besides, they had also compared the results yielded by the entropy measure on IFS from some other authors. Joshi and Kumar [10] also defined another entropy measure on IFS, following their own previous work on the classical fuzzy sets in [11] and also the work by Garg et al. in [9].

For various generalizations derived from IFS, such as inter-valued intuitionistic fuzzy sets (IVIFS) or generalized intuitionistic fuzzy soft sets (GIFSS), there were also some studies to establish entropy measures on some generalizations, followed by a demonstration on how such entropy measures can be applied to certain MADM problems. Recently, Garg [12] defined an entropy measure for inter-valued intuitionistic fuzzy sets and discussed the application of such entropy measures on solving MADM problems with unknown attribute weights. In 2018, Rashid et al. [13] defined another distance-based entropy measure on the inter-valued intuitionistic fuzzy sets. Again, following the conventions of the previous authors, they clarified the applications of their work on MADM problem using a simple, conceptual small dataset. Selvachandran et al. [14] defined a distance induced intuitionistic entropy for generalized intuitionistic fuzzy soft sets, for which they also clarified the applications of their work on MADM problems using a dataset of the same kind.

As for the Pythagorean fuzzy set (PFS) and its generalizations, an entropy measure was defined by Yang and Hussein in [15]. Thao and Smarandache [16] proposed a new entropy measure for Pythagorean fuzzy sets in 2019. Such new definitions of entropy in [16] discarded the use of natural logarithm as in [15], which is computationally intensive. Such work was subsequently followed by Athira et.al. [17,18], where an entropy measure was given for Pythagorean fuzzy soft sets—a further generalization of Pythagorean fuzzy sets. As for vague set and its generalizations, Feng and Wang [19] defined an entropy measure considering the hesitancy degree. Later, Selvachandran et al. [20] defined an entropy measure on complex vague soft sets. In the ever-going effort of establishing entropy measures for other generalizations of fuzzy sets, Thao and Smarandache [16] and Selvachandran et al. [20] were among the research groups who justified the applicability of their entropy measures using examples on MADM. Likewise, each of those works involved one or several (if more than one example provided in a work) small and conceptual datasets created by the authors themselves.

Besides IFS, PFS, vague sets and all their derivatives, there were also definitions of entropy established on some other generalizations of fuzzy sets in recent years, some came alongside with examples on MADM involving conceptual datasets as well [21]. Wei [22] defined an asymmetrical cross entropy measure for two fuzzy sets, called the fuzzy cross-entropy. Such cross entropy for interval neutrosophic sets was also studied by Sahin in [23]. Ye and Du [21] gave four different new ways entropy measures on interval-valued neutrosophic sets. Sulaiman et al. [24,25] defined entropy measures for interval-valued fuzzy soft sets and multi-aspect fuzzy soft sets. Hu et al. [26] gave an entropy measure for hesitant fuzzy sets. Al-Qudah and Hassan [27] gave an entropy measure for complex multi-fuzzy soft sets. Barukab et al. [28] gave an entropy measure for spherical fuzzy sets. Piasecki [29] gave some remarks and characterizations of entropy measures among fuzzy sets. In 2019, Dass and Tomar [30] further examined the legitimacy of some exponential entropy measures on IFS, such as those defined by Verna and Sharma in [31], Zhang and Jiang in [32], and Mishra in [33]. On the other hand, Kang and Deng [34] outlined the general patterns for which the formula for entropy measures could be formed, thus also applicable for entropy measures to various generalizations of fuzzy sets. Santos et al. [35] and Cao and Lin [36] derived their entropy formulas for data processing based on those for fuzzy entropy with applications in image thresholding and electroencephalogram.

With many entropy measures being defined for various generalizations of fuzzy sets, it calls upon a need to standardize which kind of functions are eligible to be used as entropy measures and which are not. One of the most notable and recent works in this field is accomplished by Majumdar [1], who established an axiomatic definition on the entropy measure on a single-valued neutrosophic set (SVNS). Such an axiomatic definition of entropy measure, once defined for a particular generalization of fuzzy set, serves as an invaluable tool when choosing a new entropy measure for a particular purpose. Moreover, with the establishment of such an axiomatic definition of entropy measure, it motivates researchers to work on deriving a collection of functions, which all qualify themselves to be used as entropy measures, rather than inventing a single standalone function as an entropy measure for a particular scenario.

In 2017, Smarandache [2] firstly established a concept of plithogenic sets, intended to serve as a profound and conclusive generalization from most (if not all) of the previous generalizations from fuzzy sets. This obviously includes the IFS, where most works had been done to establish its great variety of entropy measures. However, Smarandache [2] did not give any definitions on entropy measures for plithogenic sets.

Our work on this paper shall be presented as follows: Firstly, in Section 2, we mention all the prerequisite definitions needed for the establishment of entropy measures for plithogenic sets. We also derive some generalizations of those previous definitions. Such generalizations are necessary to further widen the scope of our investigation on the set of functions that qualifies as entropy measures for plithogenic sets. Then, in Section 3, we first propose new entropy measures for plithogenic sets in which requirements for any function to be an entropy measure of plithogenic sets are outlined in the axiomatic definition. Later in Section 3, several new formulae for the entropy measure of plithogenic sets are also introduced. In Section 4, we will apply a particular example of our entropy measure onto a MADM problem related to the selection of locations.

Due to the complexity and the novelty of plithogetic sets, as well as the scope constraints of this paper, the plithogenic set involved in the demonstration of MADM will be of a small caliber within 150 data values in total. Those data values contained in the plithogenic set example will be also conceptual in nature (only two to three digits per value). Such presentation, although it may be perceived as simple, is in alliance with the common practice of most renowned works done by the previous authors discussed before, whenever a novel way of entropy measure is invented and first applied on a MADM problem. Hence, such a start-up with a small and conceptual dataset does not hinder the justification on the practicability of the proposed notions. Quite the contrary, it enables even the most unfamiliar readers to focus on the procedure of such novel methods of dealing with MADM problems, rather than being overwhelmed by the immense caliber of computation encountered in dealing with up-to-date real-life datasets.

2. Preliminary

Throughout all the following of this article, let U be the universal set.

Definition 1 [1]. *A single valued neutrosophic sets (SVNS) on U is defined to be the collection*

$$A = \{(x, T_A(x), I_A(x), F_A(x)) : x \in U\}$$

where $T_A, I_A, F_A : U \to [0,1]$ *and* $0 \le T_A(x) + I_A(x) + F_A(x) \le 3$.

We denote $SVNS(U)$ *to be the collection of all SVNS on U.*

Majumdar [1] have established the following axiomatic definition for an entropy measure on SVNS.

Definition 2 [1]. *An entropy measure on SVNS is a function* $E_N : SVNS(U) \to [0,\ 1]$ *that satisfies the following axioms for all* $A \in SVNS(U)$:

I. $E_N(A) = 0$ *if A is a crisp set i.e.,* $(T_A(x), I_A(x), F_A(x)) \in \{(1,0,0), (0,0,1)\}$ *for all* $x \in U$.
II. $E_N(A) = 1$ *if* $(T_A(x), I_A(x), F_A(x)) = \left(\frac{1}{2}, \frac{1}{2}, \frac{1}{2}\right)$ *for all* $x \in U$.
III. $E_N(A) \ge E_N(B)$ *if A is contained in B i.e.,* $T_A(x) \le T_B(x),\ I_A(x) \ge I_B(x),\ F_A(x) \ge F_B(x)$ *for all* $x \in U$.
IV. $E_N(A) = E_N(A^c)$ *for all* $A \in SVNS(U)$.

In the study of fuzzy entropy, a fuzzy set with membership degree of 0.5 is a very special fuzzy set as it is the fuzzy set with the highest degree of fuzziness. Similarly, in the study of entropy for SVNSs, a SVNS with all its membership degree of 0.5 for all the three membership components is very special as it is the SVNS with the highest degree of uncertainty. Hence, we denote $A_{[\frac{1}{2}]} \in SVNS(U)$ as the SVNS with $(T_A(x), I_A(x), F_A(x)) = \left(\frac{1}{2}, \frac{1}{2}, \frac{1}{2}\right)$ for all $x \in A$. Such axiomatic descriptions in Definition 2 of this paper, defined by Majumdar [1], serve as the cornerstone for establishing similar axiomatic descriptions for the entropy measures on other generalizations of fuzzy sets, which shall certainly include that for plithogenic sets by Smarandache [2].

We however, disagree with (iii) of Definition 2. As an illustrative example, let A be empty, then A has zero entropy because it is of *absolute certainty* that A "does not contain any element". Whereas a superset of A, say $B \in SVNS(U)$, may have higher entropy because it may not be crisp. Thus, we believe that upon establishing (iii) of Definition 2, the authors in [1] concerned only the case where A and B are very close to the entire U. Thus, on the establishment of entropy measures on plithogenic sets in this article, only axioms (i), (ii) and (iv) of Definition 2 will be considered.

Together with axioms (i), (ii), and (iv) of Definition 2, the following two well-established generalizations of functions serve as **our motives of defining the entropies**, which allows different users to customize to their respective needs.

Definition 3 [1]. *Let* $\mathrm{T} : [0,1]^2 \to [0,1]$, *be a function satisfying the following for all* $p, q, r, s \in [0,\ 1]$.

1. $\mathrm{T}(p,q) = \mathrm{T}(q,p)$ *(commutativity)*
2. $\mathrm{T}(p,q) \le \mathrm{T}(r,s)$, *if* $p \le r$ *and* $q \le s$
3. $\mathrm{T}(p, \mathrm{T}(q,r)) = \mathrm{T}(\mathrm{T}(p,q), r)$ *(associativity)*
4. $\mathrm{T}(p,0) = 0$
5. $\mathrm{T}(p,1) = p$

Then T *is said to be a T-norm function.*

Example 1. *"minimum" is a T-norm function.*

Definition 4 [1]. *Let* $\mathrm{S} : [0,1]^2 \to [0,1]$, *be a function satisfying the following for all* $p, q, r, s \in [0,1]$.

1. $\mathrm{S}(p,q) = \mathrm{S}(q,p)$ *(commutativity)*
2. $\mathrm{S}(p,q) \le \mathrm{S}(r,s)$, *if* $p \le r$ *and* $q \le s$

3. $\mathrm{S}(p, \mathrm{S}(q, r)) = \mathrm{S}(\mathrm{S}(p, q), r)$ *(associativity)*
4. $\mathrm{S}(p, 0) = p$
5. $\mathrm{S}(p, 1) = 1$

   Then, S is said to be an S-norm (or a T-conorm) function.

Example 2. *"maximum" is an S-norm function.*

In the study of fuzzy logic, we also find ourselves seeking functions that measure the central tendencies, as well as a given position of the dataset, besides maximum and minimum. Such measurement often involves more than two entities and those entities can be ordered of otherwise. This is the reason we introduce the concept of *M-type* function and the *S-type* function, defined on all finite (not ordered) sets with entries in $[0, 1]$. Due to the commutativity of S-norm function, S-type function is thus a further generalization of S-norm function as it allows more than two entities. In all the following of this article, let us denote $\Phi_{[0,1]}$ as the collection of all finite sets with entries in $[0, 1]$. To avoid ending up with too many brackets in an expression, it is convenient to denote the image of $\{a_1, a_2, \cdots, a_n\}$ under f as $f(a_1, a_2, \cdots, a_n)$.

Definition 5 [ ]. *Let* $f : \Phi_{[0,1]} \to [0, 1]$, *be a function satisfying the following:*

(i) $f(0, 0, \cdots, 0) = 0$
(ii) $f(1, 1, \cdots, 1) = 1$

   Then f is said to be an M-type function.

Remark 1. *"maximum", "minimum", "mean", "interpolated inclusive median", "interpolated exclusive median", "inclusive first quartile", "exclusive 55th percentile",* $1 - \prod_{k \in K}(1 - k)$ *and* $1 - \sqrt[|K|]{\prod_{k \in K}(1 - k)}$ *are some particular examples of M-type functions.*

Definition 6 [ ]. *Let* $f : \Phi_{[0,1]} \to [0, 1]$, *be a function satisfying the following:*

(i) $f(0, 0, \cdots, 0) = 0.$
(ii) *If* $K \in \Phi_{[0,1]}$ *contains at least an element* $k > 0$, *then* $f(K) > 0$.
(iii) *If* $K \in \Phi_{[0,1]}$ *contains at least an element* $k = 1$, *then* $f(K) = 1$.
(iv) *For every two sets from* $\Phi_{[0,1]}$ *with the same cardinality:* $K = \{k_1, k_2, \cdots, k_n\}$ *and* $R = \{r_1, r_2, \cdots, r_n\}$. *If* $k_j \geq r_j$ *for all* j, *then* $f(K) \geq f(R)$.

   Then f is said to be an S-type function.

Remark 2. *"maximum",* $1 - \prod_{k \in K}(1 - k)$ *and* $1 - \sqrt[|K|]{\prod_{k \in K}(1 - k)}$ *are some particular examples of S-type functions.*

Lemma 1. *If f is an S-type function, then it is also an M-type function.*

Proof. As $\{1, 1, \cdots, 1\}$ contains one element which equals to 1, $f(1, 1, \cdots, 1) = 1$, thus the lemma follows. □

Remark 3. *The converse of this lemma is not true however, as it is obvious that "mean" is an M-type function but not an S-type function.*

All of these definitions and lemmas suffice for the establishment of our entropy measure for plithogenic sets.

3. Proposed Entropy Measure for Plithogenic Sets

In [2], Smarandache introduced the concept of plithogenic set. Such a concept is as given in the following definition.

Definition 7 [2]. *Let U be a universal set. Let $P \subseteq U$. Let A be a set of attributes. For each attribute $a \in A$: Let S_a be the set of all its corresponding attribute values. Take $V_a \subseteq S_a$. Define a function $d_a : P \times V_a \to [0,1]$, called the attribute value appurtenance degree function. Define a function $c_a : V_a \times V_a \to [0,1]$, called the attribute value contradiction (dissimilarity) degree function, which further satisfies:*

(i) $c_a(v,v) = 0$, *for all* $v \in V_a$.
(ii) $c_a(v_1,v_2) = c_a(v_2,v_1)$, *for all* $v_1, v_2 \in V_a$.

Then:

(a) $\mathbf{R} = \langle P, A, V, d, c\rangle$ *is said to form a plithogenic set on U.*
(b) d_a *is said to be the attribute value appurtenance fuzzy degree function (abbr. AFD-function) for a in* $\mathbf{R}$, *and $d_a(x,v)$ is called the appurtenance fuzzy degree of x in v.*
(c) c_a *is said to be the attribute value contradiction (dissimilarity) fuzzy degree function (abbr. CFD-function) for a in* $\mathbf{R}$, *and $c_a(v_1,v_2)$ is called the contradiction fuzzy degree between v_1 and v_2.*

Remark 4. *If $P = U$, $A = \{a_0\}$, $V_{a_0} = \{v_1, v_2, v_3\}$, $c_{a_0}(v_1,v_2) = c_{a_0}(v_2,v_3) = 0.5$ and $c_{a_0}(v_1,v_3) = 1$, then* $\mathbf{R}$ *is reduced to a single valued neutrosophic set (SVNS) on U.*

Remark 5. *If $P = U$, $V_a = \{v_1, v_2\}$ for all $a \in A$, $d_a : P \times V_a \to [0,1]$ is such that $0 \le d_a(x,v_1) + d_a(x,v_2) \le 1$ for all $x \in P$ and for all $a \in A$, $c_a(v_1,v_2) = c_a(v_2,v_1) = 1$ for all $a \in A$, then* $\mathbf{R}$ *is reduced to a generalized intuitionistic fuzzy soft set (GIFSS) on U.*

Remark 6. *If $P = U$, $A = \{a_0\}$, $V_{a_0} = \{u_1, v_1, u_2, v_2\}$, $d_{a_0} : P \times V_{a_0} \to [0,1]$ is such that $0 \le d_{a_0}(x,v_1) + d_{a_0}(x,v_2) \le 1$, $0 \le d_{a_0}(x,u_1) \le d_{a_0}(x,v_1)$ and $0 \le d_{a_0}(x,u_2) \le d_{a_0}(x,v_2)$ all satisfied for all $x \in P$, and $c_{a_0}(u_1,u_2) = c_{a_0}(v_1,v_2) = 1$, then* $\mathbf{R}$ *is reduced to an inter-valued intuitionistic fuzzy set (IVIFS) on U.*

Remark 7. *If $P = U$, $A = \{a_0\}$, $V_{a_0} = \{v_1, v_2\}$, $d_{a_0} : P \times V_{a_0} \to [0,1]$ is such that $0 \le d_{a_0}(x,v_1) + d_{a_0}(x,v_2) \le 1$ for all $x \in P$, and $c_{a_0}(v_1,v_2) = 1$, then* $\mathbf{R}$ *is reduced to an intuitionistic fuzzy set (IFS) on U.*

Remark 8. *If $P = U$, $A = \{a_0\}$, $V_{a_0} = \{v_1, v_2\}$, $d_{a_0} : P \times V_{a_0} \to [0,1]$ is such that $0 \le d_{a_0}(x,v_1)^2 + d_{a_0}(x,v_2)^2 \le 1$ for all $x \in P$, and $c_{a_0}(v_1,v_2) = 1$, then* $\mathbf{R}$ *is reduced to a Pythagorean fuzzy set (PFS) on U.*

Remark 9. *If $P = U$, $A = \{a_0\}$ and $V_{a_0} = \{v_0\}$, then* $\mathbf{R}$ *is reduced to a fuzzy set on U.*

Remark 10. *If $P = U$, $A = \{a_0\}$, $V_{a_0} = \{v_0\}$, and $d_{a_0} : P \times V_{a_0} \to \{0,1\} \subset_{\neq} [0,1]$, then* $\mathbf{R}$ *is reduced to a classical crisp set on U.*

In all the following, the collection of all the plithogenic sets on U shall be denoted as $\mathrm{PLFT}(U)$.

Definition 8. *Let $= \langle P, A, V, d, c\rangle \in \mathrm{PLFT}(U)$. The compliment for* $\mathbf{R}$, *is defined as*

$$\overline{\mathbf{R}} = \left\langle P, A, V, \bar{d}, c\right\rangle,$$

where $\bar{d}_a = 1 - d_a$ for all $a \in A$.

Remark 11. *This definition of compliment follows from page 42 of* [2].

Remark 12. *It is clear that* $\overline{\mathbf{R}} \in \mathrm{PLFT}(U)$ *as well.*

With the all these definitions established, we now proceed to define a way of measurement of entropy for plithogenic sets. In the establishment of such entropy measures, we must let all the AFD-functions $\{d_a : a \in A\}$ and all the CFD-functions $\{c_a : a \in A\}$ to participate in contributing to the overall entropy measures of $= \langle P, A, V, d, c\rangle \in \mathrm{PLFT}(U)$.

We now discuss some common traits of how each element from $\{d_a : a \in A\}$ and $\{c_a : a \in A\}$ shall contribute to the overall entropy measures of **R**, all of which are firmly rooted in our conventional understanding of entropy as a quantitative measurement for the amount of disorder.

Firstly, on the elements of $\{d_a : a \in A\}$: In accordance with Definition 7, each $d_a(x, v)$ is the appurtenance fuzzy degree of $x \in P$, over the attribute value $v \in V_a$ (V_a in turn belongs to the attribute $a \in A$). Note that $d_a(x, v) = 1$ indicates *absolute certainty* of membership of x in v; whereas $d_a(x, v) = 0$ indicates *absolute certainty* of non-membership of x in v. Hence, any $d_a(x, v)$ satisfying $d_a(x, v) \in \{0, 1\}$ must be regarded as contributing *zero* magnitude to the overall entropy measure of **R**, as absolute certainty implies zero amount of disorder. On the other hand, $d_a(x, v) = 0.5$ indicates *total uncertainty* of the membership of x in v, as 0.5 is in the middle of 0 and 1. Hence, any $d_a(x, v)$ satisfying $d_a(x, v) = 0.5$ must be regarded as contributing *the greatest* magnitude to the overall entropy measure of **R**, as total uncertainty implies the highest possible amount of disorder.

Secondly, on the elements of $\{c_a : a \in A\}$: For each attribute $a \in A$, $c_a(v_1, v_2) = 0$ indicates that the attribute values v_1, v_2 are of identical meaning (synonyms) with each other (e.g., "big" and "large"), whereas $c_a(v_1, v_2) = 1$ indicates that the attribute values v_1, v_2 are of opposite meaning to each other (e.g., "big" and "small"). Therefore, in the case of $c_a(v_1, v_2) = 0$ and $\{d_a(x, v_1), d_a(x, v_2)\} = \{0, 1\}$, it implies that x is absolutely certain to be inside one v_i among $\{v_1, v_2\}$, while outside of the other, even though v_1 and v_2 carry identical meaning to each other. Such collection of $\{c_a(v_1, v_2), d_a(x, v_1), d_a(x, v_2)\}$ is therefore of the highest possible amount of disorder, because their combined meaning implies an analogy to the statement of "x is very large and not big" or "x is not large and very big". As a result, such collection of $\{c_a(v_1, v_2), d_a(x, v_1), d_a(x, v_2)\}$ aforementioned must be regarded as contributing *the greatest* magnitude to the overall entropy measure of **R**. Furthermore, in the case of $c_a(v_1, v_2) = 1$ and $\{d_a(x, v_1), d_a(x, v_2)\} \subset_{\neq} \{0, 1\}$, it implies that x is absolutely certain to be inside both v_1 and v_2 (or outside both v_1 and v_2), even though v_1 and v_2 carry opposite meaning with each other. Likewise, such collection of $\{c_a(v_1, v_2), d_a(x, v_1), d_a(x, v_2)\}$ is of the highest possible amount of disorder, because their combined meaning implies an analogy to the statement of "x something is very big and very small" or "x something is not big and not small". As a result, such a collection of $\{c_a(v_1, v_2), d_a(x, v_1), d_a(x, v_2)\}$ aforementioned must be regarded as contributing *the greatest* magnitude to the overall entropy measure of **R** as well.

We now define the three axioms of entropy on plithogenic sets, analogous to the axioms (i), (ii), and (iv) in Definition 2 respectively.

Definition 9. *An entropy measure on plithogenic sets, is a function* $E : \mathrm{PLFT}(U) \to [0, 1]$ *satisfying the following three axioms*

(i) *(analogy to (i) in Definition 2): Let* $\mathbf{R} = \langle P, A, V, d, c\rangle \in \mathrm{PLFT}(U)$ *satisfying the following conditions for all* $(x, v_1, v_2) \in P \times V_a \times V_a$:

    (a) $d_a : P \times V_a \to \{0, 1\}$ *for all* $a \in A$.
    (b) $\{d_a(x, v_1), d_a(x, v_2)\} = \{0, 1\}$ *whenever* $c_a(v_1, v_2) \geq 0.5$.
    (c) $\{d_a(x, v_1), d_a(x, v_2)\} \subset_{\neq} \{0, 1\}$ *whenever* $c_a(v_1, v_2) < 0.5$.

    Then $E(\mathbf{R}) = 0$.

(ii) *(analogy to (ii) in Definition 2). Let* $\mathbf{R} = \langle P, A, V, d, c\rangle \in \mathrm{PLFT}(U)$ *satisfying* $d_a : P \times V_a \to \{0.5\}$ *for all* $a \in A$. *Then* $E(\mathbf{R}) = 1$.

(iii) *(analogy to (iv) in Definition 2). For all* $\mathbf{R} = \langle P, A, V, d, c\rangle \in \mathrm{PLFT}(U)$, $E(\overline{\mathbf{R}}) = E(\mathbf{R})$ *holds.*

The three axioms in Definition 9 thus serve as general rules for which any functions must fulfill to be used as entropy measures on plithogenic sets. However, the existence of such functions satisfying these three axioms needs to be ascertained. To ensure that we do have an abundance of functions satisfying these axioms, we must therefore propose and give characterization to such functions with explicit examples and go to the extent of proving that each one among our proposed examples satisfy all these axioms. Such a procedure of proving the existence of many different entropy functions is indispensable. This is because in practical use, the choices of an entropy measure will fully depend on the type of scenario examined, as well as the amount of computing power available to perform such computations, without jeopardizing the axioms of entropy measures as mentioned. It is only by doing so that users are guaranteed to have plenty of room to customize an entropy measure of plithogenic sets suited for their particular needs. In light of this motivation, a theorem showing a collection of functions satisfying those axioms is presented in this paper.

Theorem 1. *Let* m_1, m_2, m_3 *be any M-type functions. Let* s_1, s_2 *be any S-type functions. Let* Δ *be any function satisfying the following conditions:*

(i) $\Delta(1-c) = \Delta(c), \Delta(0) = \Delta(1) = 0, \Delta(0.5) = 1.$
(ii) $\Delta(c)$ *is increasing within [0, 0.5]. In other words,* $\Delta(c_1) \le \Delta(c_2)$ *whenever* $0 \le c_1 < c_2 \le 0.5$.

Let ω *be any function satisfying the following conditions:*

(i) $\omega(x) = 0$ *for all* $x \in [0, 0.5]$, $\omega(1) = 1$.
(ii) $\omega(c)$ *is increasing within [0.5, 1]. In other words,* $\omega(c_1) \le \omega(c_2)$ *whenever* $0 \le c_1 < c_2 \le 1$.

Define $\varepsilon_{\Delta,a} : P \times V_a \to [0,1]$, *where* $\varepsilon_{\Delta,a}(x,v) = \Delta(d_a(x,v))$ *for all* $(x,v) \in P \times V_a$. *Define* $\varphi_{\omega,a} : P \times V_a \times V_a \to [0,1]$, *where:*

$$\varphi_{\omega,a}(x,v_1,v_2) = \omega(1 - c_a(v_1,v_2)) \cdot \left|d_a(x,v_1) - d_a(x,v_2)\right| + \omega(c_a(v_1,v_2)) \cdot \left|d_a(x,v_1) + d_a(x,v_2) - 1\right|$$

for all $(x, v_1, v_2) \in P \times V_a \times V_a$.

Then, any function $E : \mathrm{PLFT}(U) \to [0, 1]$, *in the form of*

$$E(\mathbf{R}) = m_3\Big\{m_2\Big\{ m_1\Big\{s_2\Big\{\varepsilon_{\Delta,a}(x,v),\ s_1\{\varphi_{\omega,a}(x,v,u) : u \in V_a\}\Big\}\Big\} : v \in V_a\Big\} : a \in A\Big\} : x \in P\Big\}$$

for all $\mathbf{R} = \langle P, A, V, d, c\rangle \in \mathrm{PLFT}(U)$, *are all entropy measures on plithogenic sets.*

Proof. + Axiom (i): Taking any arbitrary $u, v \in V_a$, $a \in A$ and $x \in P$.

a. As $d_a(x,v) \in \{0,1\}$, $\varepsilon_{\Delta,a}(x,v) = \Delta(d_a(x,v)) = 0$.
b. Whenever $c_a(v_1,v_2) \ge 0.5$, it follows that $1 - c_a(v_1,v_2) \le 0.5$, which implies $\omega(1 - c_a(v_1,v_2)) = 0$.

Thus, $\varphi_{\omega,a}(x,v_1,v_2) = \omega(c_a(v_1,v_2)) \cdot \left|d_a(x,v_1) + d_a(x,v_2) - 1\right|$.

Since $\{d_a(x,v_1), d_a(x,v_2)\} = \{0,1\}$, $d_a(x,v_1) + d_a(x,v_2) - 1 = 0$ follows, which further implies that $\varphi_{\omega,a}(x,v_1,v_2) = 0$.

c. whenever $c_a(v_1,v_2) < 0.5$, it implies $\omega(c_a(v_1,v_2)) = 0$.

Thus, $\varphi_{\omega,a}(x,v_1,v_2) = \omega(1 - c_a(v_1,v_2)) \cdot \left|d_a(x,v_1) - d_a(x,v_2)\right|$.

Since $\{d_a(x,v_1), d_a(x,v_2)\} \subset_{\neq} \{0,1\}$, $d_a(x,v_1) - d_a(x,v_2) = 0$ follows, which further implies that $\varphi_{\omega,a}(x,v_1,v_2) = 0$.

Hence, $\varphi_{\omega,a}(x,v,u) = \varepsilon_{\Delta,a}(x,v) = 0$ follows for all u, v, a, x.

As a result,

$$\begin{aligned} E(\mathbf{R}) &= m_3\Big\{m_2\Big\{\, m_1\Big\{s_2\Big\{\varepsilon_{\Delta,a}(x,v),\ s_1\{\varphi_{\omega,a}(x,v,u) : u \in V_a\}\Big\}\Big\} : v \in V_a\Big\} : a \in A\Big\} : x \in P\Big\} \\ &= m_3\{m_2\{\, m_1\{s_2\{0,\ s_1\{0 : u \in V_a\}\ \} : v \in V_a\} : a \in A\} : x \in P\,\} \\ &= m_3\{m_2\{\, m_1\{s_2\{0,0\} : v \in V_a\} : a \in A\} : x \in P\,\} \\ &= m_3\{m_2\{\, m_1\{0 : v \in V_a\} : a \in A\} : x \in P\,\} = 0. \end{aligned}$$

+ Axiom (ii): Taking any arbitrary $v \in V_a, a \in A$ and $x \in P$.
As $d_a : P \times V_a \to \{0.5\}$ for all $a \in A$, we have
$d_a(x,v) = 0.5$ for all v, a, x. This further implies that
$\varepsilon_{\Delta,a}(x,v) = \Delta(d_a(x,v)) = 1$ for all v, a, x.
As a result,

$$\begin{aligned} E(\mathbf{R}) &= m_3\Big\{m_2\Big\{\, m_1\Big\{s_2\Big\{\varepsilon_{\Delta,a}(x,v),\ s_1\{\varphi_{\omega,a}(x,v,u) : u \in V_a\}\Big\}\Big\} : v \in V_a\Big\} : a \in A\Big\} : x \in P\Big\} \\ &= m_3\{m_2\{\, m_1\{s_2\{1,\ s_1\{\varphi_{\omega,a}(x,v,u) : u \in V_a\}\} : v \in V_a\} : a \in A\} : x \in P\} \\ &= m_3\{m_2\{\, m_1\{1 : v \in V_a\} : a \in A\} : x \in P\} = 1. \end{aligned}$$

+ Axiom (iii): $\overline{d}_a = 1 - d_a$ follows by Definition 8. This will imply the following

(a) $\Delta\big(\overline{d}_a(x,v)\big) = \Delta(1 - d_a(x,v)) = \Delta(d_a(x,v)) = \varepsilon_{\Delta,a}(x,v)$.

(b) First, we have

$$\begin{aligned} \Big|\overline{d}_a(x,v_1) - \overline{d}_a(x,v_2)\Big| &= \big|(1 - d_a(x,v_1)) - (1 - d_a(x,v_2))\big| \\ &= \big|-d_a(x,v_1) + d_a(x,v_2)\big| \\ &= \big|d_a(x,v_1) - d_a(x,v_2)\big| \end{aligned}$$

and

$$\begin{aligned} \Big|\overline{d}_a(x,v_1) + \overline{d}_a(x,v_2) - 1\Big| &= \big|(1 - d_a(x,v_1)) + (1 - d_a(x,v_2)) - 1\big| \\ &= \big|1 - d_a(x,v_1) + 1 - d_a(x,v_2) - 1\big| \\ &= \big|1 - d_a(x,v_1) - d_a(x,v_2)\big| \\ &= \big|d_a(x,v_1) + d_a(x,v_2) - 1\big|. \end{aligned}$$

Therefore, it follows that

$$\begin{aligned} &\omega(1 - c_a(v_1,v_2))\cdot\Big|\overline{d}_a(x,v_1) - \overline{d}_a(x,v_2)\Big| + \omega(c_a(v_1,v_2))\cdot\Big|\overline{d}_a(x,v_1) + \overline{d}_a(x,v_2) - 1\Big| \\ &= \omega(1 - c_a(v_1,v_2))\cdot\big|d_a(x,v_1) - d_a(x,v_2)\big| + \omega(c_a(v_1,v_2))\cdot\big|d_a(x,v_1) + d_a(x,v_2) - 1\big| \\ &= \varphi_{\omega,a}(x,v_1,v_2) \end{aligned}$$

Since

$$E(\mathbf{R}) = m_3\Big\{m_2\Big\{m_1\Big\{s_2\Big\{\varepsilon_{\Delta,a}(x,v), s_1\{\varphi_{\omega,a}(x,v,u) : u \in V_a\}\Big\}\Big\} :\ v \in V_a\Big\} : a \in A\Big\} :\ x \in P\Big\}$$

$E(\mathbf{R}) = E(\overline{\mathbf{R}})$ now follows. □

Remark 13. *As* $\varepsilon_{\Delta,a}(x,v) = \Delta(d_a(x,v))$ *and*

$$\varphi_{\omega,a}(x,v,u) = \omega(1 - c_a(v,u))\cdot\big|d_a(x,v) - d_a(x,u)\big| + \omega(c_a(v,u))\cdot\big|d_a(x,v) + d_a(x,u) - 1\big|$$

It follows that

$$E(\mathbf{R}) = m_3\left\{ m_2\left\{ m_1\left\{ s_2\left\{ \Delta(d_a(x,v)),\ s_1\left\{ \begin{array}{c} \omega(1-c_a(v,u))\cdot \\ \left|d_a(x,v)-d_a(x,u)\right|+ \\ \omega(c_a(v,u))\cdot \\ \left|d_a(x,v)+d_a(x,u)-1\right| \end{array} : u \in V_a \right\} \right\} : v \in V_a \right\} : a \in A \right\} : x \in P \right\}$$

Such a version of the formula serves as an even more explicit representation of $E(\mathbf{R})$.

Remark 14. *For instance, the following is one of the many theoretical ways of choosing* $\{m_1, m_2, m_3, s_1, s_2, \Delta, \omega\}$ *to form a particular entropy measure on plithogenic sets.*

(a) $\omega(c) = \begin{cases} 0,\ 0 \le c < \frac{1}{2} \\ 2\left(c-\frac{1}{2}\right),\ \frac{1}{2} \le c \le 1 \end{cases}$, *for all* $c \in [0,1]$.

(b) $\Delta(c) = \begin{cases} 2c,\ 0 \le c < \frac{1}{2} \\ 2(1-c),\ \frac{1}{2} \le c \le 1 \end{cases}$, *for all* $c \in [0,1]$.

(c) $s_1(K) = \text{maximum}(K)$, *for all* $K \in \Phi_{[0,1]}$.

(d) $s_2(K) = 1 - \prod_{k \in K}(1-k)$, *for all* $K \in \Phi_{[0,1]}$.

(e) $m_1(K) = \text{mean}(K)$, *for all* $K \in \Phi_{[0,1]}$.

(f) $m_2(K) = \text{median}(K)$, *for all* $K \in \Phi_{[0,1]}$.

(g) $m_3(K) = \text{mode}(K)$, *for all* $K \in \Phi_{[0,1]}$.

In practical applications, however, the choices of $\{m_1, m_2, m_3, s_1, s_2, \Delta, \omega\}$ will depend on the type of scenario examined, as well as the amount of computing power available to perform such computations. Such abundance of choices is a huge advantage, because it allows each user plenty of room of customization suited for their own needs, without jeopardizing the principles of entropy functions.

4. Numerical Example of Plithogenic Sets

In this section, we demonstrate the utility of the proposed entropy functions for plithogenic sets using an illustrative example of a MADM problem involving a property buyer making a decision whether to live in Town P or Town B.

4.1. Attributes and Attributes Values

Three different addresses within Town P are selected: $P = \{p, q, r\}$. Another four different addresses within Town B are selected as well: $B = \{\alpha, \beta, \gamma, \delta\}$. All the seven addresses are investigated by that person based on 3 attributes as follows:

$$A = \left\{ \begin{array}{c} \text{Services near the address } (j),\ \text{Security near the address } (s), \\ \text{Public transport near the address } (t) \end{array} \right\}$$

For each of the 3 attributes, the following attribute values are considered:

$$V_j = \{\text{School}(u_1),\ \text{Bank}(u_2),\ \text{Factory}(u_3),\ \text{Construction Site}(u_4),\ \text{Clinic}(u_5)\}$$
$$V_s = \{\text{Police on Patrol}(v_1),\ \text{Police Station}(v_2),\ \text{CCTV Coverage}(v_3),\ \text{Premise Guards}(v_4)\}$$
$$V_t = \{\text{Bus}(w_1),\ \text{Train}(w_2),\ \text{Taxi}(w_3),\ \text{Grab services}(w_4)\}$$

4.2. Attribute Value Appurtenance Degree Functions

In light of the limitation of one person doing the investigation, there could possibly be some characteristics of Town P left unknown or unsure of. As a result, our example involved in this paper, though small in caliber, shall provide a realistic illustration of such phenomena.

Thus, in our example: Let the attribute value appurtenance degree functions for Town P be given in Tables 1–3 (as deduced by the property buyer).

Table 1. Attribute value appurtenance fuzzy degree function for $j \in A$ on Town P (d_j).

V_j / Addresses in Town P	u_1	u_2	u_3	u_4	u_5
p	1.0	1.0	0.0	0.0	1.0
q	0.0	0.0	1.0	0.8	0.0
r	1.0	0.9	0.0	0.3	1.0

Table 2. Attribute value appurtenance fuzzy degree function for $s \in A$ on Town P (d_s).

V_s / Addresses in Town P	v_1	v_2	v_3	v_4
p	0.1	1.0	0.9	0.8
q	0.9	0.0	0.8	0.9
r	0.1	1.0	0.8	0.7

Table 3. Attribute value appurtenance fuzzy degree function for $t \in A$ on Town P (d_t).

V_t / Addresses in Town P	w_1	w_2	w_3	w_4
p	0.9	0.9	0.9	0.1
q	0.8	0.8	0.1	0.9
r	0.9	1.0	0.1	0.8

For example:
$d_j(p, u_1) = 1.0$ indicates that schools exist near address p in town P.
$d_t(q, w_4) = 0.9$ indicates that Grab services are very likely to exist near address q in town P.
$d_s(r, v_2) = 1.0$ indicates that police stations exist near address r in town P.
Similarly, let the attribute value appurtenance degree functions for Town B be given in Tables 4–6 (as deduced by the property buyer):

Table 4. Attribute value appurtenance fuzzy degree function for $j \in A$ on Town B (h_j).

V_j / Addresses in Town B	u_1	u_2	u_3	u_4	u_5
α	0.0	1.0	1.0	0.0	1.0
β	1.0	0.0	1.0	0.8	0.0
γ	0.4	0.5	0.6	0.4	0.6
δ	0.0	0.1	0.1	0.2	0.9

Table 5. Attribute value appurtenance fuzzy degree function for $s \in A$ on Town B (h_s).

V_s / Addresses in Town B	v_1	v_2	v_3	v_4
α	0.9	0.8	0.9	0.8
β	0.2	0.1	0.5	0.4
γ	0.8	0.9	0.8	0.5
δ	0.1	0.2	0.6	0.5

Table 6. Attribute value appurtenance fuzzy degree function for $t \in A$ on Town B (h_t).

V_t / Addresses in Town B	w_1	w_2	w_3	w_4
α	0.5	0.5	0.3	0.4
β	0.0	0.9	0.9	0.9
γ	0.8	0.0	0.1	0.1
δ	0.9	0.1	0.8	0.9

4.3. Attribute Value Contradiction Degree Functions

Moreover, each of the attributes of a town may be dependent on one another. For example, in a place where schools are built, clinics should be built near to the schools, whereas factories should be built far from the schools. Moreover, the police force should spread their manpower patrolling across the town away from a police station. As a result, our example involved in this paper, though small in caliber, shall provide a realistic illustration of such phenomena as well.

Thus, as an example, let the attribute value contradiction degree functions for the attributes $j, s, t \in A$ be given in Tables 7–9: (as deduced by the property buyer), to be used for both towns.

Table 7. Attribute value contradiction degree functions for $j \in A$ (c_j).

V_j	u_1	u_2	u_3	u_4	u_5
u_1	0.0	0.2	1.0	0.7	0.0
u_2	0.2	0.0	0.9	0.5	0.1
u_3	1.0	0.9	0.0	0.2	0.9
u_4	0.7	0.5	0.2	0.0	0.5
u_5	0.0	0.1	0.9	0.5	0.0

Table 8. Attribute value contradiction degree functions for $s \in A$ (c_s).

V_s	v_1	v_2	v_3	v_4
v_1	0.0	1.0	0.5	0.5
v_2	1.0	0.0	0.5	0.5
v_3	0.5	0.5	0.0	0.1
v_4	0.5	0.5	0.1	0.0

Table 9. Attribute value contradiction degree functions for $t \in A$ (c_t).

V_t	w_1	w_2	w_3	w_4
w_1	0.0	0.3	0.1	0.1
w_2	0.3	0.0	0.0	0.1
w_3	0.1	0.0	0.0	0.9
w_4	0.1	0.1	0.9	0.0

In particular,

$c_j(u_1, u_3) = 1.0$ indicates that schools and factories should not be in the same place, because it is not healthy to the students.

$c_j(u_1, u_5) = 0.0$ indicates that schools and clinics should be available together, so that any student who falls ill can visit the clinic.

$c_s(v_1, v_2) = 1.0$, because it is very inefficient for police to patrol only nearby a police station itself, instead of places of a significant distance to a police station. This also ensures that police force will be present in all places, as either a station or a patrol unit will be present.

$c_t(w_1, w_2) = 0.0$, because all train stations must have buses going to/from it. On the other hand, one must also be able to reach a train station from riding a bus.

$c_t(w_3, w_4) = 0.9$ due to the conflicting nature of the two businesses.

4.4. Two Plithogenic Sets Representing Two Towns

From all attributes of the two towns given, we thus form two plithogenic sets representing each of them

(a) $\mathbf{R} = \langle P, A, V, d, c \rangle$, which describes Town P
(b) $\mathbf{T} = \langle B, A, V, h, c \rangle$, which describes Town B

Intuitively, it is therefore evident that the property buyer should choose Town P over Town B as his living place. One of the many reasons being, in Town P schools and factories are unlikely to appear near one address within the town, whereas in Town B there exist addresses where schools and factories are both nearby (so both schools and factories are near to each other). Moreover, in Town P, the police force is more efficient as they spread their manpower across the town, rather than merely patrolling near their stations and even leaving some addresses unguarded as in Town B. On top of this, there exists places in town where taxi and grab services are near to each other, which can cause conflict or possibly vandalism to each other's property. Town P is thus deemed less "chaotic", whereas Town B is deemed more "chaotic".

As a result, our entropy measure must be able to give Town P as having lower entropy than Town B, under certain choices of $\{m_1, m_2, m_3, s_1, s_2, \Delta, \omega\}$ which are customized for the particular use of the property buyer.

4.5. An Example of Entropy Measure on Two Towns

Choose the following to form $E : \text{PLFT}(U) \to [0, 1]$ in accordance with Theorem 1:

(a) $\omega(c) = \begin{cases} 0, \ 0 \leq c < \frac{1}{2} \\ 2\left(c - \frac{1}{2}\right), \ \frac{1}{2} \leq c \leq 1 \end{cases}$, for all $c \in [0, 1]$.

(b) $\Delta(c) = \begin{cases} 2c, \ 0 \leq c < \frac{1}{2} \\ 2(1 - c), \ \frac{1}{2} \leq c \leq 1 \end{cases}$, for all $c \in [0, 1]$.

(c) $s_1(K) = s_2(K) = 1 - \sqrt[|K|]{\prod_{k \in K}(1 - k)}$, for all $K \in \Phi_{[0,1]}$.

(d) $m_1(K) = m_2(K) = m_3(K) = \text{mean}(K)$, for all $K \in \Phi_{[0,1]}$.

Then, by the calculation in accordance with Theorem 1 which is subsequently highlighted in Figure 1.

We have $E(\mathbf{R}) = \frac{0.05541+0.14126+0.25710}{3} = 0.15126$, and $E(\mathbf{T}) = \frac{0.54868+0.43571+0.39926}{3} = 0.46122$.

Town P is concluded to have lower entropy, and, therefore, is less "chaotic", compared to Town B.

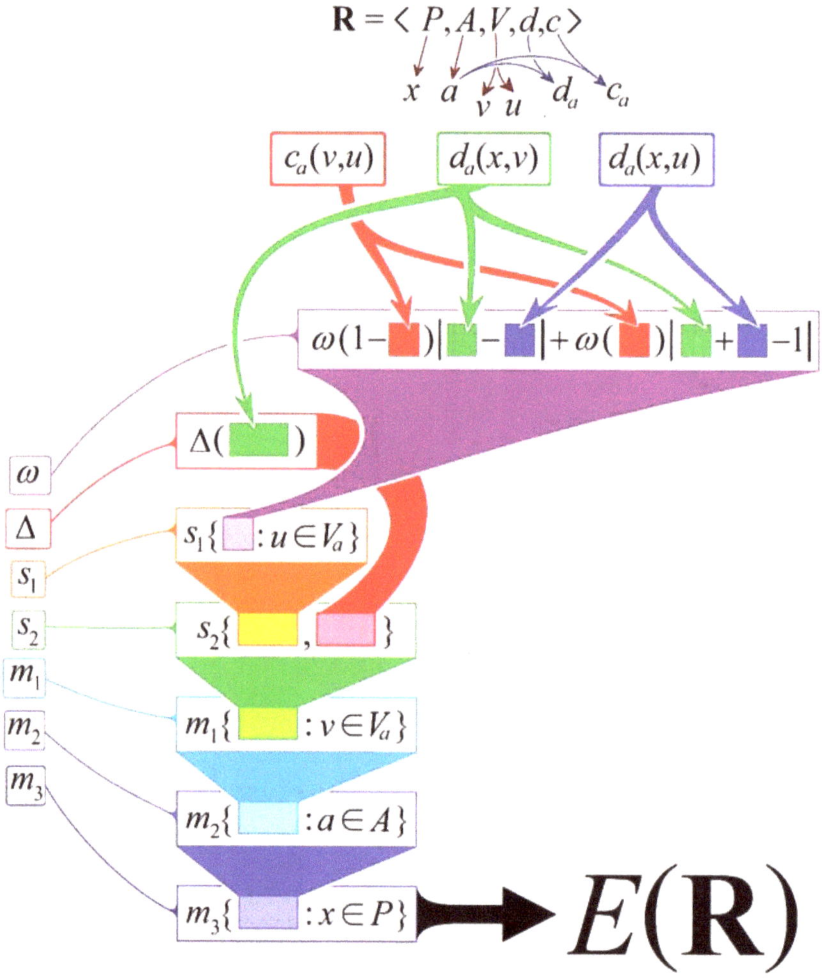

Figure 1. The entire workflow of determining the entropy measure **R** of a plithogenic set.

5. Conclusions

The plithogenic set $\mathbf{R} = \langle P, A, V, d, c \rangle$ is an improvement to the neutrosophic model whereby each attribute is characterized by a degree of appurtenance d that describes belongingness to the given criteria, and every pair attribute is characterized by a degree of contradiction c that describes the amount of similarity or opposition between two attributes. In Section 3 of this paper, we have introduced new entropy measures for plithogenic sets $E(\mathbf{R})$. The axiomatic definition of the plithogenic entropy was defined using some of the axiomatic requirements of neutrosophic entropy and some additional conditions. Some formulae for the entropy measure of plithogenic sets have been introduced in Theorem 1 and these formulas have been developed further to satisfy characteristics of plithogenic sets such as satisfying exact exclusion (partial order) and containing a contradiction or dissimilarity degree between each attribute value and the dominant attribute value. The practical application of the proposed plithogenic entropy measures was demonstrated by applying it to a multi-attribute decision making problem related to the selection of locations.

Future works related to the plithogenic entropy include studying more examples of entropy measures for plithogenic sets with structures different from the one mentioned in Theorem 1, and to apply the different types of entropy measure for plithogenic sets onto real life datasets. We are also working on developing entropy measures for other types of plithogenic sets such as plithogenic intuitionistic fuzzy sets and plithogenic neutrosophic sets, and the study of the application of these measures in solving real world problems using real life datasets [36–43].

Author Contributions: Concept: S.G.Q. and G.S.; methodology: S.G.Q., G.S., F.S. and J.V.; flowchart: S.G.Q.; software: S.H.L. and Q.-T.B.; validation: S.H.L. and V.C.G.; data curation: Q.-T.B. and S.G.Q.; writing—original draft preparation: S.G.Q. and G.S.; writing—review and editing: F.S., J.V., S.H.L., Q.-T.B., V.C.G. All authors have read and agreed to the published version of the manuscript.

Funding: This research received no external funding.

Conflicts of Interest: The authors declare no conflict of interest.

References

1. Majumdar, P. On new measures of uncertainty for neutrosophic sets. *Neutrosophic Sets Syst.* **2017**, *17*, 50–57.
2. Smarandache, F. *Plithogeny, Plithogenic Set, Logic, Probability, and Statistics, Brussels, Belgium*; Pons: Brussels, Belgium, 2017.
3. Zhu, Y.-J.; Li, D. A new definition and formula of entropy for intuitionistic fuzzy sets. *J. Intell. Fuzzy Syst.* **2016**, *30*, 3057–3066. [CrossRef]
4. Montes, S.; Pal, N.R.; Montes, S. Entropy measures for Atanassov intuitionistic fuzzy sets based on divergence. *Soft Comput.* **2018**, *22*, 5051–5071. [CrossRef]
5. Farnoosh, R.; Rahimi, M.; Kumar, P. Removing noise in a digital image using a new entropy method based on intuitionistic fuzzy sets. In Proceedings of the 2016 IEEE International Conference on Fuzzy Systems (FUZZ-IEEE), Vancouver, BC, Canada, 23–27 August 2016; pp. 1328–1332.
6. Ansari, M.D.; Mishra, A.R. New Divergence and Entropy Measures for Intuitionistic Fuzzy Sets on Edge Detection. *Int. J. Fuzzy Syst.* **2017**, *20*, 474–487. [CrossRef]
7. Gupta, P.; Arora, H.D.; Tiwari, P. Generalized entropy for intuitionistic fuzzy sets. *Malays. J. Math. Sci.* **2016**, *10*, 209–220.
8. Joshi, R.; Kumar, S. A new intuitionistic fuzzy entropy of order-α with applications in multiple attribute decision making. In *Advances in Intelligent Systems and Computing*; Deep, K., Bansal, J.C., Das, K.N., Lal, A.K., Garg, H., Nagar, A.K., Pant, M., Eds.; Springer: Singapore, 2017.
9. Garg, H.; Agarwal, N.; Tripathi, A. Generalized Intuitionistic Fuzzy Entropy Measure of Order α and Degree β and Its Applications to Multi-Criteria Decision Making Problem. *Int. J. Fuzzy Syst. Appl.* **2017**, *6*, 86–107. [CrossRef]
10. Joshi, R.; Kumar, S. A New Parametric Intuitionistic Fuzzy Entropy and its Applications in Multiple Attribute Decision Making. *Int. J. Appl. Comput. Math.* **2018**, *4*, 52. [CrossRef]
11. Joshi, R.; Kumar, S. A new exponential fuzzy entropy of order-(α, β) and its application in multiple attribute decision-making problems. *Commun. Math. Stat.* **2017**, *5*, 213–229. [CrossRef]
12. Garg, H. Generalized Intuitionistic Fuzzy Entropy-Based Approach for Solving Multi-attribute Decision-Making Problems with Unknown Attribute Weights. *Proc. Natl. Acad. Sci. India Sect. A Phys. Sci.* **2017**, *89*, 129–139. [CrossRef]
13. Rashid, T.; Faizi, S.; Zafar, S. Distance Based Entropy Measure of Interval-Valued Intuitionistic Fuzzy Sets and Its Application in Multicriteria Decision Making. *Adv. Fuzzy Syst.* **2018**, *2018*, 1–10. [CrossRef]
14. Selvachandran, G.; Maji, P.K.; Faisal, R.Q.; Salleh, A.R. Distance and distance induced intuitionistic entropy of generalized intuitionistic fuzzy soft sets. *Appl. Intell.* **2017**, *47*, 132–147. [CrossRef]
15. Yang, M.-S.; Hussain, Z. Fuzzy Entropy for Pythagorean Fuzzy Sets with Application to Multicriterion Decision Making. *Complexity* **2018**, *2018*, 1–14. [CrossRef]
16. Thao, N.X.; Smarandache, F. A new fuzzy entropy on Pythagorean fuzzy sets. *J. Intell. Fuzzy Syst.* **2019**, *37*, 1065–1074. [CrossRef]
17. Athira, T.M.; John, S.J.; Garg, H. A novel entropy measure of Pythagorean fuzzy soft sets. *AIMS Math.* **2020**, *5*, 1050–1061. [CrossRef]

18. Athira, T.; John, S.J.; Garg, H. Entropy and distance measures of Pythagorean fuzzy soft sets and their applications. *J. Intell. Fuzzy Syst.* **2019**, *37*, 4071–4084. [CrossRef]
19. Feng, W.; Wang, H. An improved fuzzy entropy of Vague Sets considering hesitancy degree. In *Proceedings of the 2018 Joint International Advanced Engineering and Technology Research Conference (JIAET 2018)*; Atlantis Press: Beijing, China, 2018.
20. Selvachandran, G.; Garg, H.; Quek, S.G. Vague Entropy Measure for Complex Vague Soft Sets. *Entropy* **2018**, *20*, 403. [CrossRef]
21. Ye, J.; Du, S. Some distances, similarity and entropy measures for interval-valued neutrosophic sets and their relationship. *Int. J. Mach. Learn. Cybern.* **2017**, *10*, 347–355. [CrossRef]
22. Wei, G. Picture fuzzy cross-entropy for multiple attribute decision making problems. *J. Bus. Econ. Manag.* **2016**, *17*, 491–502. [CrossRef]
23. Şahin, R. Cross-entropy measure on interval neutrosophic sets and its applications in multicriteria decision making. *Neural Comput. Appl.* **2015**, *28*, 1177–1187. [CrossRef]
24. Sulaiman, N.H.; Kamal, N.L.A.M. A subsethood-based entropy for weight determination in interval-valued fuzzy soft set group decision making. In *Proceedings of the 25th National Symposium on Mathematical Sciences: Mathematical Sciences as the Core of Intellectual Excellence*; AIP Publishing: Melville, NY, USA, 2017; p. 020062.
25. Sulaiman, N.H.; Mohamad, D. Entropy for Multiaspect Fuzzy Soft Sets and Its Application in Group Decision Making. In *Proceedings of the Lecture Notes in Electrical Engineering*; Springer Science and Business Media LLC: Berlin, Germany, 2019; pp. 289–296.
26. Hu, J.; Yang, Y.; Zhang, X.; Chen, X. Similarity and entropy measures for hesitant fuzzy sets. *Int. Trans. Oper. Res.* **2017**, *25*, 857–886. [CrossRef]
27. Al-Qudah, Y.; Hassan, N. Complex Multi-Fuzzy Soft Set: Its Entropy and Similarity Measure. *IEEE Access* **2018**, *6*, 65002–65017. [CrossRef]
28. Barkub, O.; Abdullah, S.; Ashraf, S.; Arif, M.; Khan, S.A. A New Approach to Fuzzy TOPSIS Method Based on Entropy Measure under Spherical Fuzzy Information. *Entropy* **2019**, *21*, 1231. [CrossRef]
29. Piasecki, K. Some remarks on axiomatic definition of entropy measure. *J. Intell. Fuzzy Syst.* **2017**, *33*, 1945–1952. [CrossRef]
30. Dass, B.; Tomar, V.P. Exponential entropy measure defined on intuitionistic fuzzy set. *Our Herit.* **2019**, *67*, 313–320.
31. Verma, R.; Sharma, B.D. Exponential entropy on intuitionistic fuzzy sets. *Kybernetika* **2013**, *49*, 114–127.
32. Zhang, Q.-S.; Jiang, S.-Y. A note on information entropy measures for vague sets and its applications. *Inf. Sci.* **2008**, *178*, 4184–4191. [CrossRef]
33. RajMishra, A.; Hooda, D.S.; Jain, D. On Exponential Fuzzy Measures of Information and Discrimination. *Int. J. Comput. Appl.* **2015**, *119*, 1–7. [CrossRef]
34. Kang, B.; Deng, Y. The Maximum Deng Entropy. *IEEE Access* **2019**, *7*, 120758–120765. [CrossRef]
35. Santos, H.; Couso, I.; Bedregal, B.; Takáč, Z.; Minárová, M.; Asiaın, A.; Barrenechea, E.; Bustince, H. Similarity measures, penalty functions, and fuzzy entropy from new fuzzy subsethood measures. *Int. J. Intell. Syst.* **2019**, *34*, 1281–1302. [CrossRef]
36. Cao, Z.; Lin, C.-T. Inherent Fuzzy Entropy for the Improvement of EEG Complexity Evaluation. *IEEE Trans. Fuzzy Syst.* **2017**, *26*, 1032–1035. [CrossRef]
37. Bui, Q.-T.; Vo, B.; Do, H.-A.N.; Hung, N.Q.V.; Snášel, V. F-Mapper: A Fuzzy Mapper clustering algorithm. *Knowl.-Based Syst.* **2020**, *189*, 105107. [CrossRef]
38. Witarsyah, D.; Fudzee, M.F.M.; Salamat, M.A.; Yanto, I.T.R.; Abawajy, J. Soft Set Theory Based Decision Support System for Mining Electronic Government Dataset. *Int. J. Data Warehous. Min.* **2020**, *16*, 39–62. [CrossRef]
39. Le, T.; Vo, M.; Kieu, T.; Hwang, E.; Rho, S.; Baik, S. Multiple Electric Energy Consumption Forecasting Using a Cluster-Based Strategy for Transfer Learning in Smart Building. *Sensors* **2020**, *20*, 2668. [CrossRef]
40. Fan, T.; Xu, J. Image Classification of Crop Diseases and Pests Based on Deep Learning and Fuzzy System. *Int. J. Data Warehous. Min.* **2020**, *16*, 34–47. [CrossRef]

41. Selvachandran, G.; Quek, S.G.; Lan, L.T.H.; Son, L.H.; Giang, N.L.; Ding, W.; Abdel-Basset, M.; Albuquerque, V.H.C. A New Design of Mamdani Complex Fuzzy Inference System for Multi-attribute Decision Making Problems. *IEEE Trans. Fuzzy Syst.* **2020**, 1. [CrossRef]
42. Son, L.H.; Ngan, R.T.; Ali, M.; Fujita, H.; Abdel-Basset, M.; Giang, N.L.; Manogaran, G.; Priyan, M.K. A New Representation of Intuitionistic Fuzzy Systems and Their Applications in Critical Decision Making. *IEEE Intell. Syst.* **2020**, *35*, 6–17. [CrossRef]
43. Thong, N.; Lan, L.; Chou, S.-Y.; Son, L.; Dong, D.D.; Ngan, T.T. An Extended TOPSIS Method with Unknown Weight Information in Dynamic Neutrosophic Environment. *Mathematics* **2020**, *8*, 401. [CrossRef]

mathematics

MDPI

Article

Applying Fixed Point Techniques to Stability Problems in Intuitionistic Fuzzy Banach Spaces

P. Saha [1], T. K. Samanta [2], Pratap Mondal [3], B. S. Choudhury [1] and Manuel De La Sen [4,*]

[1] Department of Mathematics, Indian Institute of Engineering Science and Technology, Shibpur, Howrah, West Bengal 711103, India; parbati_saha@yahoo.co.in (P.S.); binayak12@yahoo.co.in (B.S.C.)

[2] Department of Mathematics, Uluberia College, Uluberia, Howrah, West Bengal 711315, India; mumpu_tapas5@yahoo.co.in

[3] Department of Mathematics, Bijoy Krishna Girls' College, Howrah, Howrah, West Bengal 711101, India; pratapmondal111@gmail.com

[4] Institute of Research and Development of Processes, Faculty of Science and Technology, University of the Basque Country, Campus of Leioa, Bizkaia, 48940 Leioa, Spain

* Correspondence: manuel.delasen@ehu.eus

Received: 17 February 2020; Accepted: 10 June 2020; Published: 15 June 2020

Abstract: In this paper we investigate Hyers-Ulam-Rassias stability of certain nonlinear functional equations. Considerations of such stabilities in different branches of mathematics have been very extensive. Again the fuzzy concepts along with their several extensions have appeared in almost all branches of mathematics. Here we work on intuitionistic fuzzy real Banach spaces, which is obtained by combining together the concepts of fuzzy Banach spaces with intuitionistic fuzzy sets. We establish that pexiderized quadratic functional equations defined on such spaces are stable in the sense of Hyers-Ulam-Rassias stability. We adopt a fixed point approach to the problem. Precisely, we use a generxalized contraction mapping principle. The result is illustrated with an example.

Keywords: Hyers-Ulam stability; pexider type functional equation; intuitionistic fuzzy normed spaces; alternative fixed point theorem

1. Introduction

In this paper, we derive Hyers-Ulam-Rassias stability results for certain functional equations in the context of intuitionistic fuzzy Banach spaces (IFBS). The problem of stability that we study here was for the first time mathematically formulated by Ulam [1]. It was partly solved and further generalized by Hyers [2] and Rassias [3]. Today we know such stability problems as the problems of the Hyers-Ulam-Rassias (H-U-R) stability. It has many extended forms and has been studied in several domains of mathematics including differential equations [4], functional equations [5], isometries [6], etc. Our interest is in the study of such stabilities for certain functional equations. H-U-R stability for functional equations on linear spaces has been discussed in quite a large number of papers, some of which are noted in [7–14].

The fuzzy concept was mathematically introduced by Zadeh [15] in 1965. Over the following years it was adopted in almost all the domains of mathematics including linear algebra and functional analysis. The idea of a fuzzy set has many extensions of itself. One such extension is the concept of intuitionistic fuzzy set introduced by Atanassov [16]. Here we have an additional degree of membership, which is sometimes referred to as the degree of non-belongingness.

In this paper we consider the intuitionistic fuzzy linear spaces as defined by S. Shakeri [17]. It is a generalization of the definition of fuzzy normed linear space given by Mirmostafaee [18]. Stability of functional equations on the above-mentioned space have been considered in works like [19–21]. Precisely in this paper we consider the H-U-R stability problem for pexiderized quadratic functional equations. These equations are generalized quadratic functional equations and appeared in the literature in works like [22–24]. Amongst several approaches to H-U-R stability problems we adopt the fixed point approach where the stability is established through an application of a fixed point theorem obtained in complete generalized metric spaces [25].

2. Mathematical Background

The following is the definition of a pexiderized quadratic functional equation.
A mapping $f: R \to R$ is said to be a quadratic form if $f(x) = cx^2$ for all $x, c \in R$.

Let X and Y be a real vector space and a Banach space, respectively, and corresponding to a mapping $f: X \to Y$, consider the functional equation

$$f(x+y) + f(x-y) = 2f(x) + 2f(y) \tag{1}$$

Any solution of Equation (1) is termed as quadratic mapping. Particularly, if $X = Y = R$, the quadratic form $f(x) = cx^2$ is a solution of (1). The form

$$f(x+y) + f(x-y) = 2g(x) + 2h(y) \tag{2}$$

is known as pexiderized quadratic functional equation [26,27], which is an extension of the above definition of quadratic functional equation.

Definition 1 ([28,29]). *Consider the set L^* and the order relation $\leq_{L^*}$ defined by*

$$Ł^* = \{ (\alpha_1, \alpha_2) : (\alpha_1, \alpha_2) \in [0,1]^2 \text{ and } \alpha_1 + \alpha_2 \leq 1 \},$$

$$(\alpha_1, \alpha_2) \leq_{L^*} (\beta_1, \beta_2) \Leftrightarrow \alpha_1 \leq \beta_1, \alpha_2 \geq \beta_2, \forall (\alpha_1, \alpha_2), (\beta_1, \beta_2) \in L^*.$$

Then $(L^*, \leq_{L^*})$ is a complete lattice.
The elements $0_{L^*} = (0,1)$ and $1_{L^*} = (1,0)$ are its units.

Definition 2 ([16]). *An intuitionistic fuzzy set A of E where E is a nonempty set, is $A = \{ (x, \mu_A(x), \nu_A(x)) : x \in E \}$, in which case the functions $\mu_A : E \to [0,1]$ and $\nu_A : E \to [0,1]$ are the degree of membership and the degree of non-membership respectively for every $x \in E$ satisfying $0 \leq \mu_A(x) + \nu_A(x) \leq 1$.*
For our notational purposes we denote an intuitionistic fuzzy set on X by any function $A_{\mu,\nu} = X \to L^*$ given by $A_{\mu,\nu}(x) = (\mu_A(x), \nu_A(x))$ with $\mu_A, \nu_A : X \to [0,1]$ satisfying $0 \leq \mu_A(x) + \nu_A(x) \leq 1$.

Definition 3 ([30]). *A triangular norm (t-norm) on L^* is a mapping $\Gamma : (L^*)^2 \to L^*$ satisfying the following conditions:*

(a) $(\forall \alpha \in L^*)(\Gamma(\alpha, 1_{L^*}) = \alpha)$ *(boundary condition),*
(b) $(\forall (\alpha, \beta) \in (L^*)^2)(\Gamma(\alpha, \beta) = \Gamma(\beta, \alpha))$ *(commutativity),*
(c) $(\forall (\alpha, \beta, \gamma) \in (L^*)^3)(\Gamma(\alpha, \Gamma(\beta, \gamma)) = \Gamma(\Gamma(\alpha, \beta), \gamma))$ *(associativity),*
(d) $(\forall (\alpha, \alpha', \beta, \beta') \in (L^*)^4)(\alpha \leq_{L^*} \alpha'$ *and* $\beta \leq_{L^*} \beta' \Rightarrow \Gamma(\alpha, \beta) \leq_{L^*} \Gamma(\alpha', \beta'))$ *(monotonicity).*

If Γ is continuous then Γ is called a continuous t-norm.

Definition 4 ([30]). *A triangular conorm (t-conorm) on L^* is a mapping $S : (L^*)^2 \to L^*$ satisfying the following conditions:*

(a) $(\forall \alpha \in L^*)(S(\alpha, 0_{L^*}) = \alpha)$ *(boundary condition),*
(b) $(\forall (\alpha, \beta) \in (L^*)^2)(S(\alpha, \beta) = S(\beta, \alpha))$ *(commutativity),*

(c) $(\forall(\alpha,\beta,\gamma)\in(L^*)^3)\,(S(\alpha,S(\beta,\gamma))=S(S(\alpha,\beta),\gamma))$ *(associativity)*,
(d) $(\forall(\alpha,\alpha',\beta,\beta')\in(L^*)^4)\,(\alpha\leq_{L^*}\alpha'$ *and* $\beta\leq_{L^*}\beta'\Rightarrow S(\alpha,\beta)\leq_{L^*}S(\alpha',\beta'))$ *(monotonicity)*.

Example 1. *Let*

$$M(\alpha,\beta)=(\min\{\alpha_1,\beta_1\},\max\{\alpha_2,\beta_2\})$$

for all $\alpha=(\alpha_1,\alpha_2),\ \beta=(\beta_1,\beta_2)\in L^*$.
Then $M(\alpha,\beta)$ *is a continuous t-norm.*

Definition 5 ([30]). *A continuous t-norm* Γ *on* L^* *is said to be continuous t-representable if we can find a continuous t-norm* $*$ *and a continuous t-conorm* $\diamond$ *on* $[0,1]$ *such that for all* $x=(\alpha_1,\alpha_2),\ y=(\beta_1,\beta_2)\in L^*,\ \Gamma(x,y)=(\alpha_1*\beta_1,\alpha_2\diamond\beta_2)$
We now define the iterated sequence Γ^n *recursively by* $\Gamma^1=\Gamma$ *and*

$$\Gamma^n(x^{(1)},x^{(2)},\cdots,x^{(n+1)})=\Gamma(\Gamma^{(n-1)}(x^{(1)},x^{(2)}\cdots,x^{(n)}),x^{(n+1)}),$$

$\forall\, n\geq 2,\ x^{(i)}\in L^*.$

Intuitionistic fuzzy normed linear space was defined by Saadati [31]. Shakeri [17] has stated this definition in a more compact form. We state the definition in the form used by Shakeri [17]

Definition 6 ([17]). *We call the triple* $(X,P_{\mu,\nu},\Gamma)$ *an intuitionistic fuzzy normed space (briefly IFN-space) if X is a vector space,* Γ *is a continuous t-norm and* $P_{\mu,\nu}$ *is a mapping* $X\times(0,\infty)\to L^*$ *which is an intuitionistic fuzzy set satisfying the following conditions:*
for all $x,y\in X$ *and* $t,s>0$,

(i) $P_{\mu,\nu}(x,0)=0_{L^*}$;
(ii) $P_{\mu,\nu}(x,t)=1_{L^*}$ *if and only if* $x=0$;
(iii) $P_{\mu,\nu}(\alpha x,t)=P_{\mu,\nu}\left(x,\frac{t}{|\alpha|}\right)$ *for all* $\alpha\neq 0$;
(iv) $P_{\mu,\nu}(x+y,t+s)\geq_{L^*}\Gamma(P_{\mu,\nu}(x,t),P_{\mu,\nu}(y,s))$.

It can be noted that $P_{\mu,\nu}$ *has the form* $P_{\mu,\nu}(x,t)=(\mu_x(t),\nu_x(t))=(\mu(x,t),\nu(x,t))$ *such that* $0\leq\mu_x(t)+\nu_x(t)\leq 1$ *for all* $x\in X$ *and* $t>0$*. Then with* μ *and* ν *the above definition reduces to the more explicit form used in* [31].

Definition 7 ([17]). (1) *The sequence* $\{x_n\}$ *is said to be convergent to a point* $x\in X$ *if*

$$P_{\mu,\nu}(x_n-x,s)\to 1_{L^*}\ \text{as}\ n\to\infty\ \ for\ every\ \ s>0.$$

(2) *A sequence* $\{x_n\}$ *in an IFN-space* $(X,P_{\mu,\nu},M)$ *is said to be a Cauchy sequence if given any* $\varepsilon>0$ *and* $s>0$*, we can find* $n_0\in N$ *such that*

$$P_{\mu,\nu}(x_n-x_m,s)>_{L^*}(1-\varepsilon,\varepsilon),\ \forall n,m\geq n_0$$

(3) *An IFN-space* $(X,P_{\mu,\nu},M)$ *is said to be complete if every Cauchy sequence in* $(X,P_{\mu,\nu},M)$ *is convergent in* $(X,P_{\mu,\nu},M)$*. A complete intuitionistic fuzzy normed space is called an intuitionistic fuzzy Banach space.*

We require the following fixed point result to establish our result of stability in this paper.

Definition 8 ([25]). *Let* X *be a nonempty set. A function* $d:X\times X\to[0,\infty]$ *is called a generalized metric on* X *if* d *satisfies*

(i) $d\,(p\,,q) = 0$ *if and only if* $p = q$;
(ii) $d\,(p\,,q) = d\,(q\,,p)$ *for all* $p\,,q \in X$;
(iii) $d\,(p\,,q) \leq d\,(p\,,r) + d\,(r\,,q)$ *for all* $p\,,q\,,r \in X$. *Then* (X,d) *is called a generalized metric space.*

Theorem 1 ([12,23,32]). *Let* $(X\,,d)$ *be a complete generalized metric space and let* $J : X \to X$ *be a strictly contractive mapping with Lipschitz constant* $0 < L < 1$, *that is,*

$$d(J\,p\,,J\,q) \leq L\,d\,(p\,,q),$$

for all $p\,,q \in X$. *Then for each* $p \in X$, *either*

$$d\,(J^k\,p\,,J^{\,k+1}\,p) = \infty,\ \forall\, k \geq 0$$

or,

$$d\,(J^k\,p\,,J^{\,k+1}\,p) < \infty\quad \forall\, k \geq k_o$$

for some non-negative integers k_0. *Moreover, if the second alternative holds then*

(1) *the sequence* $\{J^k\,p\}$ *converges to a fixed point* $q^\star$ *of* J;
(2) $q^\star$ *is the unique fixed point of* J *in the set*

$$Y = \{q \in X : d\,(J^{k_0}\,p\,,q) < \infty\};$$

(3) $d(q\,,q^\star) \leq (\frac{1}{1-L})\,d\,(q\,,J\,q)$ *for all* $q \in Y$.

3. The Hyers-Ulam-Rassias Stability Result

Throughout the result of the paper, X is considered to be a normed linear space, $(Y\,,P_{\mu,\nu}\,,M)$ an IF-real Banach space, $(Z\,,P\,'_{\mu,\nu}\,,M)$ an IFN-space and M is continuous $t-$ norm defined in Example 2, also consider

$$\begin{aligned} &M_1\,(x\,,t) \\ &= M^2\left\{P\,'_{\mu,\nu}\left(\phi\,(x\,,x)\,,\frac{t}{3}\right),\,P\,'_{\mu,\nu}\left(\phi\,(x\,,0)\,,\frac{t}{3}\right),\right. \\ &\qquad\qquad \left. P\,'_{\mu,\nu}\left(\phi\,(0\,,x)\,,\frac{t}{3}\right)\right\} \end{aligned} \tag{3}$$

where $\phi : X \times X \to Z$.

Lemma 1. *Let* $(Z\,,P'_{\mu,\nu},\,M)$ *be an IFN-space. Let* $\phi : X \times X \to Z$ *be a mapping and further let* $E = \{g \,|\, g : X \to Y\}$. *Let* $d : E \times E \to [0,\infty]$ *be defined by*

$$\begin{aligned} &d\,(g\,,h) \\ &:= \inf\{k \in R^+ : P_{\mu,\nu}(g(x) - h(x)\,,kt) \geq_{L^*} M_1\,(x\,,t)\ \textit{for all}\ x \in X\,,t > 0\} \end{aligned}$$

and $g\,,h \in E$.
Then $(E\,,d)$ *is a complete generalized metric space.*

Proof. Let $f\,,g\,,h \in E$ and $d\,(f\,,g) = k_1 < \infty, d\,(g\,,h) = k_2 < \infty$.
Then $P_{\mu,\nu}(f(x) - g(x)\,,k_1\,t) \geq_{L^*} M_1\,(x\,,t)$ and $P_{\mu,\nu}(g(x) - h(x)\,,k_2\,t) \geq_{L^*} M_1\,(x\,,t)$
Therefore $P_{\mu,\nu}(f(x) - h(x)\,,(k_1 + k_2)\,t)$
$\geq_{L^*} M\,(P_{\mu,\nu}(f(x) - g(x)\,,k_1\,t)\,,P_{\mu,\nu}(g(x) - h(x)\,,k_2\,t)\,)$ (by property iv of Definition 6)

$\geq_{L^*}$ $M\,(M_1\,(x\,,\,t)\,,\,M_1\,(x\,,\,t))$ (by the monotonicity property) $=$ $_{L^*}$ $M_1\,(x\,,\,t)$ (Idempotent property)s for all $x\,\in\,X, t\,>\,0$.

Hence $d\,(f\,,\,h)\,\leq\,k_1\,+\,k_2$ so that $d\,(f,\,h)\,\leq\,d\,(f\,,\,g)\,+\,d\,(g\,,\,h)$ which is the triangle inequality. The other axioms are obvious, and hence, $(E\,,\,d)$ is a generalized metric space. Now we prove that $(E\,,\,d)$ is complete.

Let $\{\,g_n\}$ be a Cauchy sequence in $(E\,,\,d)$. Now for each fixed $x\in\,X$ and for every $t>0$ and $\epsilon\,>\,0$ there exists $\lambda\,>\,0$ such that $M_1\,(x\,,\,\frac{t}{\lambda})\,>\,1-\varepsilon$. Since $\{\,g_n\}$ is a Cauchy sequence in $(E,\,d)$ corresponding to $\lambda\,>\,0$, there exists $n_0\,\in\,N$ such that $d(g_n,\,g_m)\,<\,\lambda$ for all $m\,,\,n\geq\,n_0$.

Since $g_n,\,g_m\,\in E$ so we find,

$$d\,(g_n\,,\,g_m)\;=\;\inf\{\,k\,\in\,R^+\,:\,P_{\mu,\nu}\,(g_n\,(x)\,-\,g_m\,(x)\,,\,kt)\,\geq\,M_1\,(x\,,\,t)\,\}$$

That is,

$$d\,(g_n\,,\,g_m)\;=\;\inf\{\,k\,\in\,R^+\,:\,P_{\mu,\nu}\,(g_n\,(x)\,-\,g_m\,(x)\,,\,t)\,\geq\,M_1\,(x\,,\,\frac{t}{k})\,\}$$

then there exists $k_3\,\in\,[0\,,\,\infty)$ such that $d\,(g_n\,,\,g_m)\,\leq\,k_3\,<\,\lambda$ for all $m\,,\,n\geq\,n_0$ and $P_{\mu,\nu}\,(g_n\,(x)\,-$ $g_m\,(x)\,,\,t)\,\geq\,M_1\,(x\,,\,\frac{t}{k_3})\,\geq\,M_1\,(x\,,\,\dfrac{t}{\lambda})\,>\,1-\varepsilon$, as $P_{\mu,\nu}\,(x\,,\,t)$ is non-decreasing with respect to t for all $m\,,\,n\geq\,n_0$.

Thus, for fixed $x\,\in\,X$, $\{\,g_n\,(x)\,\}$ is a Cauchy sequence in Y. Again since Y is Banach space, every $x\in\,X$, there exists $g(x)\,\in\,Y$ such that $g_n\,(x)\,\to g\,(x)$ as $n\to\infty$. Then the mapping $g\,:\,X\,\to\,Y$ is such that $g_n\,(x)\,\to g\,(x)$ as $n\to\infty$ for all $x\,\in\,X$.

Again, $\{\,g_n\,\}$ is a Cauchy sequence in $(E,\,d)$ therefore for $\epsilon\,>\,0, t\,>\,0$ there exists $n_0\,\in\,N$ such that $d\,(g_n,\,g_m)\,<\,\epsilon\quad\forall\,m,\,n\,\geq\,n_0$ and hence there exists $k'\,\in\,[0,\,\infty)$ such that $d\,(g_n,\,g_m)\,\leq\,k'\,<\,\epsilon\;\forall\,m,\,n\,\geq\,n_0$

$$P_{\mu,\nu}\,(g_n\,(x)\,-\,g_m\,(x)\,,\,t)\,\geq\,M_1\,(x\,,\,\frac{t}{k'})\geq\,M_1\,(x\,,\,\frac{t}{\epsilon}).$$

That is

$$P_{\mu,\nu}\,(g_m\,(x)\,-\,g_n\,(x)\,,\,\epsilon\,t)\,\geq\,M_1\,(x\,,\,t),\;\forall\,n\,,\,m\,\geq\,n_0$$

Now let $\epsilon\,,\,\delta\,>\,0$ be given and $m,\,n\,>\,n_0\,,\,t\,>\,0$, then

$$\begin{aligned}
&P_{\mu,\nu}\,(g_n\,(x)\,-\,g\,(x)\,,\,(\epsilon\,+\,\delta)\,t)\\
\geq_{L^*}\;&M\,\{P_{\mu,\nu}\,(g_n\,(x)\,-\,g_m\,(x)\,,\,\epsilon\,t)\,,\,P_{\mu,\nu}\,(g_m\,(x)\,-\,g\,(x)\,,\,\delta\,t)\}\\
\geq_{L^*}\;&M\,\{\,M_1\,(x\,,\,t)\,,\,P_{\mu,\nu}\,(g_m\,(x)\,-\,g\,(x)\,,\,\delta\,t)\}\\
\geq_{L^*}\;&M\,\{\,M_1\,(x\,,\,t)\,,\,1_{L^*}\}\,[\,bytakinglimitas\,m\,\to\,\infty]\,=_{L^*}\,M_1\,(x\,,\,t))
\end{aligned}$$

that is, $d\,(g_n,\,g)\,\leq\,\epsilon\,+\,\delta$ for all $x\,\in\,X$ and $m,\,n\,\geq\,n_0$.

Taking $\delta\to 0$ we have a mapping $g\,:\,X\,\to\,Y$ such that

$$g\,(x)\;=\;P_{\mu,\nu}\;-\;\lim_{n\to\infty}\,g_n\,(x)\,\in\,E\,.$$

Therefore, $(E\,,\,d)$ is a complete generalized metric space. □

For our purpose, we denote

$$Let\;\;D\,f\,(x\,,\,y)\;=\;f\,(x\,+\,y)\,+\,f\,(x\,-\,y)\,-\,2\,g\,(x)\,-\,2\,h\,(y)\tag{4}$$

Theorem 2. *Let X be a linear space, $(Z, P'_{\mu,\nu}, M)$ be an IFN-space, $\phi : X \times X \to Z$ be such that*

$$P'_{\mu,\nu}(\phi(2x, 2x), t) \geq_{L^*} P'_{\mu,\nu}(\alpha\phi(x, x), t) \tag{5}$$

for some real α with $0 < \alpha < 2$, $(\forall x \in X, t > 0)$ and

$$\lim_{n\to\infty} P'_{\mu,\nu}(\phi(2^n x, 2^n x), 2^n t) = 1_{L^*}$$

where $x \in X$ and $t > 0$. Further let $(Y, P_{\mu,\nu}, M)$ be a complete IFN-space. If $f, g, h : X \to Y$ are odd mappings such that

$$P_{\mu,\nu}(Df(x, y), t) \geq_{L^*} P'_{\mu,\nu}(\phi(x, y), t) \tag{6}$$

$(\forall x \in X, t > 0)$, where $Df(x,y)$ is given by Equation (4). Then there exists a unique additive mapping $A : X \to Y$ define by $A(x) := \lim_{n\to\infty} \left(\frac{f(2^n x)}{2^n}\right)$ for all $x \in X$ satisfying

$$P_{\mu,\nu}(f(x) - A(x), t) \geq_{L^*} M_1(x, t(2-\alpha)) \tag{7}$$

and

$$P_{\mu,\nu}(A(x) - g(x) - h(x), t) \geq_{L^*} M_1\left(x, \frac{t \times 3(2-\alpha)}{5-\alpha}\right). \tag{8}$$

Proof. Interchanging the role of x and y in Equation (6) we get

$$P_{\mu,\nu}(f(x+y) - f(x-y) - 2g(y) - 2h(x), t)$$
$$\geq_{L^*} P'_{\mu,\nu}(\phi(y, x), t) \tag{9}$$

Also from Equation (6) and using Equation (9) we get

$$P_{\mu,\nu}(2f(x+y) - 2g(x) - 2h(y) - 2g(y) - 2h(x), 2t)$$
$$\geq_{L^*} M\left\{P'_{\mu,\nu}(\phi(x, y), t), P'_{\mu,\nu}(\phi(y, x), t)\right\}$$

that is,

$$P_{\mu,\nu}(f(x+y) - g(x) - h(y) - g(y) - h(x), t)$$
$$\geq_{L^*} M\left\{P'_{\mu,\nu}(\phi(x, y), t), P'_{\mu,\nu}(\phi(y, x), t)\right\} \tag{10}$$

Now putting $y = 0$ in Equation (10) we have

$$P_{\mu,\nu}(f(x) - g(x) - h(x), t)$$
$$\geq_{L^*} M\left\{P'_{\mu,\nu}(\phi(x, 0), t), P'_{\mu,\nu}(\phi(0, x), t)\right\} \tag{11}$$

Replacing y by x in Equation (11) we get

$$P_{\mu,\nu}(f(y) - g(y) - h(y), t)$$
$$\geq_{L^*} M\left\{P'_{\mu,\nu}(\phi(y, 0), t), P'_{\mu,\nu}(\phi(0, y), t)\right\} \tag{12}$$

Hence using Equations (10)–(12) we get

$$P_{\mu,\nu}(f(x+y) - f(x) - f(y), 3t)$$
$$\geq_{L^*} M^5\left\{P'_{\mu,\nu}(\phi(x, y), t), P'_{\mu,\nu}(\phi(y, x), t)\right.$$

$$P'_{\mu,\nu}(\phi(x,0),t,)\, P'_{\mu,\nu}(\phi(0,x),t)$$
$$P'_{\mu,\nu}(\phi(y,0),t),P'_{\mu,\nu}(\phi(0,y),t)$$

Therefore

$$P_{\mu,\nu}(f(x+y)-f(x)-f(y),t)$$
$$\geq_{L^*} M^5\left\{P'_{\mu,\nu}\left(\phi(x,y),\frac{t}{3}\right),P'_{\mu,\nu}\left(\phi(y,x),\frac{t}{3}\right),\right.$$
$$P'_{\mu,\nu}\left(\phi(x,0),\frac{t}{3}\right),P'_{\mu,\nu}\left(\phi(0,x),\frac{t}{3}\right),$$
$$\left.P'_{\mu,\nu}\left(\phi(y,0),\frac{t}{3}\right),P'_{\mu,\nu}\left(\phi(0,y),\frac{t}{3}\right)\right\} \tag{13}$$

Also we put $y=x$ in Equation (13)

$$P_{\mu,\nu}(f(2x)-2f(x),t)$$
$$\geq_{L^*} M^5\left\{P'_{\mu,\nu}\left(\phi(x,x),\frac{t}{3}\right),P'_{\mu,\nu}\left(\phi(x,x),\frac{t}{3}\right),\right.$$
$$P'_{\mu,\nu}\left(\phi(x,0),\frac{t}{3}\right),P'_{\mu,\nu}\left(\phi(0,x),\frac{t}{3}\right),$$
$$\left.P'_{\mu,\nu}\left(\phi(x,0),\frac{t}{3}\right),P'_{\mu,\nu}\left(\phi(0,x),\frac{t}{3}\right)\right\}$$
$$=M^2\left\{P'_{\mu,\nu}\left(\phi(x,x),\frac{t}{3}\right),P'_{\mu,\nu}\left(\phi(x,0),\frac{t}{3}\right),\right.$$
$$\left.P'_{\mu,\nu}\left(\phi(0,x),\frac{t}{3}\right)\right\}$$
$$=M_1(x,t)$$

that is,

$$P_{\mu,\nu}(f(2x)-2f(x),t)\geq_{L^*} M_1(x,t) \tag{14}$$

Now define a mapping $J: E\to E$ by $Jg(x)=\frac{1}{2}g(2x)$ for all $g\in E$ and $x\in X$, where (E,d) is a complete generalized metric space as in Lemma 1. We now prove that J is a strictly contractive mapping of E with the Lipschitz constant $\frac{\alpha}{2}$.

Let $g,h\in E$ and $\epsilon>0$. Then there exists $k'\in R^+$ satisfying $P_{\mu,\nu}(g(x)-h(x),k't)\geq_{L^*} M_1(x,t)$ such that $d(g,h)\leq k'<d(g,h)+\epsilon$ for any $\epsilon>0$.

Then $\inf\{k\in R^+: P_{\mu,\nu}(g(x)-h(x),kt)\geq_{L^*} M_1(x,t)\}\leq k'<d(g,h)+\epsilon$ that is, $\inf\left\{k\in R^+: P_{\mu,\nu}(\frac{g(2x)}{2}-\frac{h(2x)}{2},\frac{kt}{2})\geq_{L^*} M_1(2x,t)\right\}<d(g,h)+\epsilon$ that is, $\inf\left\{k\in R^+: P_{\mu,\nu}(Jg(x)-Jh(x),\frac{kt}{2})\geq_{L^*} M_1(2x,t)\right\}<d(g,h)+\epsilon$ that is, $\inf\left\{k\in R^+: P_{\mu,\nu}(Jg(x)-Jh(x),\frac{k\alpha t}{2})\geq_{L^*} M_1(x,t)\right\}<d(g,h)+\epsilon$ as $M_1(2^n x,t)=M_1(x,\frac{t}{\alpha^n})$ or, $d\{\frac{2}{\alpha}(Jg,Jh)\}<d(g,h)+\epsilon$ or, $d\{(Jg,Jh)\}<\frac{\alpha}{2}\{d(g,h)+\epsilon\}$. Taking $\epsilon\to 0$ we get $d\{(Jg,Jh)\}<\frac{\alpha}{2}\{d(g,h)\}$. Therefore, J is strictly contractive mapping with Lipschitz constant $\frac{\alpha}{2}$.

Also from Equation (14) $d(f,Jf)\leq\frac{1}{2}$ and $d(Jf,J^2f)\leq\frac{\alpha}{2}d(f,Jf)<\infty$. Again, replacing x by $2^n x$ in Equation (14) we get $P_{\mu,\nu}(f(2^{n+1}x)-2f(2^n x),t)\geq_{L^*} M_1(2^n x,t)$ or, $P_{\mu,\nu}(\frac{f(2^{n+1}x)}{2^{n+1}}-\frac{f(2^n x)}{2^n},\frac{t}{2^{n+1}})\geq_{L^*} M_1(2^n x,t)$

$$\geq_{L^*} M_1(x,\frac{t}{\alpha^n})$$

or, $P_{\mu,\nu}\left(J^{n+1}f(x) - J^{n}f(x), t\frac{(\frac{\alpha}{2})^{n}}{2}\right) \geq_{L^*} M_1(x,t)$

Hence, $d(J^{n+1}f, J^{n}f) \leq \frac{1}{2}(\frac{\alpha}{2})^{n} < \infty$ has Lipschitz constant $\frac{\alpha}{2} < 1$ for $n \geq n_0 = 1$.

Therefore, by Theorem 1 there exists a mapping $A : X \to Y$ such that the following holds:

1. A is a fixed point of J for which $A(2x) = 2A(x)$ for all $x \in X$.

Further, A is a unique fixed point of J in the set $E_1 = \{g \in E : d(J^{n_0}f, g) = d(Jf, g) < \infty\}$. Therefore, $d(Jf, A) < \infty$.

Also from Equation (14) $d(Jf, f) \leq \frac{1}{2} < \infty$. Thus $f \in E_1$. Now, $d(f, A) \leq d(f, Jf) + d(Jf, A) < \infty$. Thus, there exists $k \in (0, \infty)$ satisfying

$$P_{\mu,\nu}(f(x) - A(x), kt) \geq_{L^*} M_1(x,t)$$

for all $x \in X, t > 0$;

2. $d(J^{n}f, A)$
$= \inf\{k \in R^{+} : P_{\mu,\nu}(J^{n}f(x) - A(x), kt) \geq_{L^*} M_1(x,t)\}$
$= \inf\{k \in R^{+} : P_{\mu,\nu}(f(2^{n}x) - A(2^{n}x), 2^{n}kt) \geq_{L^*} M_1(x, (\frac{2}{\alpha})^{n}t)\}$

Therefore, $d(J^{n}f, A) \leq (\frac{\alpha}{2})^{n} \to 0$ as $n \to \infty$. This implies the equality

$$A(x) = \lim_{n\to\infty} J^{n}f(x) = \lim_{n\to\infty}\frac{f(2^{n}x)}{2^{n}} \tag{15}$$

for all $x \in X$.

3. $d(f, A) \leqslant \frac{1}{1-L}d(f, Jf)$ with $f \in E_1$ which implies the inequality

$$d(f, A) \leq \frac{1}{1-\frac{\alpha}{2}} \times \frac{1}{2} = \frac{1}{2-\alpha}$$

then it follows that

$$P_{\mu,\nu}(A(x) - f(x), \frac{1}{2-\alpha}t) \geq_{L^*} M_1(x,t)$$

It implies that

$$P_{\mu,\nu}(A(x) - f(x), t) \geq_{L^*} M_1(x, (2-\alpha)t) \tag{16}$$

for all $x \in X; t > 0$.

Replacing x and y by $2^{n}x$ and $2^{n}y$ in Equation (13) we have

$$\begin{aligned} &P_{\mu,\nu}\left(\frac{f(2^{n}(x+y))}{2^{n}} - \frac{f(2^{n}x)}{2^{n}} - \frac{f(2^{n}y)}{2^{n}}, t\right) \\ &\geq_{L^*} M^{5}\left\{P'_{\mu,\nu}\left(\phi(2^{n}x, 2^{n}y), \frac{2^{n}t}{3}\right), P'_{\mu,\nu}\left(\phi(2^{n}y, 2^{n}x), \frac{2^{n}t}{3}\right),\right. \\ &\quad P'_{\mu,\nu}\left(\phi(2^{n}x, 0), \frac{2^{n}t}{3}\right), P'_{\mu,\nu}\left(\phi(0, 2^{n}x), \frac{2^{n}t}{3}\right), \\ &\quad\left. P'_{\mu,\nu}\left(\phi(2^{n}y, 0), \frac{2^{n}t}{3}\right), P'_{\mu,\nu}\left(\phi(0, 2^{n}y), \frac{2^{n}t}{3}\right)\right\} \end{aligned} \tag{17}$$

Taking the limit as $n \to \infty$ in Equation (17) and using

$$\lim_{n\to\infty} P'_{\mu,\nu}(\phi(2^{n}x, 2^{n}y), 2^{n}t) = 1_{L^*}$$

we have

$$P_{\mu,\nu}(A(x+y) - A(x) - A(y), t) = 1_{L^*}$$

$$A(x+y) = A(x) + A(y) \tag{18}$$

that is, A is additive.

Also from Equation (11) we have

$$\begin{aligned}
&P_{\mu,\nu}(A(x) - g(x) - h(x), t\frac{5-\alpha}{3}) \\
&= P_{\mu,\nu}(A(x) - f(x) + f(x) - g(x) - h(x), t + \frac{2-\alpha}{3}t) \\
&\geq_{L^*} M\left(P_{\mu,\nu}(A(x) - f(x), t), P_{\mu,\nu}\left(f(x) - g(x) - h(x), \frac{2-\alpha}{3}t\right)\right) \\
&\geq_{L^*} M\left(M_1(x, (2-\alpha)t), M\left(P'_{\mu,\nu}\left(\phi(x,0), \frac{2-\alpha}{3}t\right),\right.\right. \\
&\qquad \left.\left. P'_{\mu,\nu}\left(\phi(0,x), \frac{2-\alpha}{3}t\right)\right)\right) \\
&\geq_{L^*} M(M_1(x, (2-\alpha)t), M_1(x, (2-\alpha)t)) \\
&\geq_{L^*} M_1(x, (2-\alpha)t)
\end{aligned}$$

Therefore,

$$P_{\mu,\nu}(A(x) - g(x) - h(x), t) \geq_{L^*} M_1\left(x, \frac{t \times 3(2-\alpha)}{5-\alpha}\right).$$

Again, A is the unique fixed point of J with the following property that there exists $u \in (0, \infty)$ such that

$$P_{\mu,\nu}(f(x) - A(x), ut) \geq_{L^*} M_1(x, t)$$

for all $x \in X$ and $t > 0$ [23]. This establishes the uniqueness of A. This completes the proof of the theorem. □

Theorem 3 ([23]). *Let X be a linear space and $(Z, P'_{\mu,\nu}, M)$ be an IFN-space. Let $\phi : X \times X \to Z$ be such that*

$$P'_{\mu,\nu}(\phi(2x, 2x), t) \geq_{L^*} P'_{\mu,\nu}(\alpha\phi(x,x), t) \tag{19}$$

for some real α with $0 < \alpha < 4$, $(\forall x \in X, t > 0)$ and

$$\lim_{n\to\infty} P'_{\mu,\nu}(\phi(2^n x, 2^n x), 4^n t) = 1_{L^*}$$

for all $x, y \in X$ and $t > 0$. Let $(Y, P_{\mu,\nu}, M)$ be a complete IFN-space. If $f, g, h : X \to Y$ are even mappings with $f(0) = g(0) = h(0) = 0$ such that

$$P_{\mu,\nu}(Df(x,y), t) \geq_{L^*} P'_{\mu,\nu}(\phi(x,y), t) \tag{20}$$

($\forall x \in X, t > 0$), where D is given by Equation (4). Then there exists a unique quadratic mapping $Q : X \to Y$ defined by $Q(x) := \lim_{n\to\infty}\left(\frac{f(2^n x)}{4^n}\right)$ for all $x \in X$ satisfying

$$P_{\mu,\nu}(f(x) - Q(x), t) \geq_{L^*} M_1(x, t(4-\alpha)) \tag{21}$$

and

$$P_{\mu,\nu}(Q(x) - g(x), t) \geq_{L^*} M_1\left(x, \frac{t \times 6(4-\alpha)}{10-\alpha}\right). \tag{22}$$

also

$$P_{\mu,\nu}(Q(x) - h(x), t) \geq_{L^*} M_1\left(x, \frac{t \times 6(4-\alpha)}{10-\alpha}\right).$$

Proof. Putting $y = x$ in Equation (20)

$$P_{\mu,\nu}(f(2x) - 2g(x) - 2h(x), t) \geq_{L^*} P'_{\mu,\nu}(\phi(x,x), t) \tag{23}$$

Also putting $x = 0$ in Equation (20)

$$P_{\mu,\nu}(2f(y) - 2h(y), t) \geq_{L^*} P'_{\mu,\nu}(\phi(0,y), t) \tag{24}$$

Again putting $y = 0$ in Equation (20)

$$P_{\mu,\nu}(2f(x) - 2g(x), t) \geq_{L^*} P'_{\mu,\nu}(\phi(x,0), t) \tag{25}$$

Now using Equations (20), (24), (25)

$$P_{\mu,\nu}\{f(x+y) + f(x-y) - 2f(x) - 2f(y), 3t\}$$

$$\begin{aligned} &= P_{\mu,\nu}\{f(x+y) + f(x-y) - 2g(x) - 2h(y) - \\ &\qquad \{2f(y) - 2h(y)\} - \{2f(x) - 2g(x)\}, 3t\} \\ &\geq_{L^*} M^2\{P'_{\mu,\nu}(\phi(x,y),t), P'_{\mu,\nu}(\phi(0,y),t), P'_{\mu,\nu}(\phi(x,0),t)\} \end{aligned}$$

Therefore

$$P_{\mu,\nu}\{f(x+y) + f(x-y) - 2f(x) - 2f(y), t\}$$

$$\begin{aligned} &\geq_{L^*} M^2\left\{P'_{\mu,\nu}\left(\phi(x,y), \frac{t}{3}\right), P'_{\mu,\nu}\left(\phi(0,y), \frac{t}{3}\right),\right. \\ &\qquad \left. P'_{\mu,\nu}\left(\phi(x,0), \frac{t}{3}\right)\right\} \end{aligned} \tag{26}$$

Now putting $y = x$ in Equation (26) we get

$$P_{\mu,\nu}(f(2x) - 4f(x), t)$$

$$\begin{aligned} &\geq_{L^*} M^2\left(P'_{\mu,\nu}\left(\phi(x,x), \frac{t}{3}\right), P'_{\mu,\nu}\left(\phi(0,x), \frac{t}{3}\right),\right. \\ &\qquad \left. P'_{\mu,\nu}\left(\phi(x,0), \frac{t}{3}\right)\right) \\ &= M_1(x,t) \end{aligned}$$

Thus,

$$P_{\mu,\nu}(f(2x) - 4f(x), t) \geq_{L^*} M_1(x, t)$$

Similar to before [23], we consider the set $E := \{g : X \to Y\}$ and introduce a complete generalized metric on E. Again, define a mapping $J : E \to E$ by $Jg(x) = \frac{1}{4}g(2x)$ for all $g \in E$ and $x \in X$. And in a similar way as before we prove that J is strictly contractive mapping with Lipschitz constant $\frac{\alpha}{4}$ and $d(f, Jf) \leq \frac{1}{4}$.

Therefore by Theorem 1 there exists a mapping $Q : X \to Y$ such that the followings hold:

1. Q is a fixed point of J, that is, $Q(2x) = 4Q(x)$ for all $x \in X$.

The mapping Q is a unique fixed point of J in the set $E_1 = \{g \in E : d(J^{n_0}f, g) = d(Jf, g) < \infty\}$ and there exists $k \in (0, \infty)$ satisfying

$$P_{\mu,\nu}(f(x) - Q(x), kt) \geq_{L^*} M_1(x, t)$$

for all $x \in X, t > 0$;

2. $d(J^n f, Q) \leq (\frac{\alpha}{4})^n \to 0$ as $n \to \infty$. This implies the equality

$$Q(x) = \lim_{n\to\infty} J^n f(x) = \lim_{n\to\infty} \frac{f(2^n x)}{4^n}$$

3. $d(f, Q) \leqslant \frac{1}{1-L} d(f, Jf)$ with $f \in E_1$, which implies the inequality

$$d(f, Q) \leq \frac{1}{1-\frac{\alpha}{4}} \times \frac{1}{4} = \frac{1}{4-\alpha}$$

then it follows that

$$P_{\mu,\nu}(Q(x) - f(x), \frac{1}{4-\alpha}t) \geq_{L^*} M_1(x, t)$$

It implies that

$$P_{\mu,\nu}(Q(x) - f(x), t) \geq_{L^*} M_1(x, (4-\alpha)t)$$

for all $x \in X; t > 0$.

Replacing x and y by $2^n x$ and $2^n y$ in Equation(26) we have

$$P_{\mu,\nu}\left\{\frac{f(2^n(x+y))}{4^n} + \frac{f(2^n(x-y))}{4^n} - \frac{2f(2^n x)}{4^n} - \frac{2f(2^n y)}{4^n}, t\right\}$$
$$\geq_{L^*} M^2\left\{P'_{\mu,\nu}\left(\phi(2^n x, 2^n y), \frac{4^n t}{3}\right), P'_{\mu,\nu}\left(\phi(0, 2^n y), \frac{4^n t}{3}\right),\right.$$
$$\left.P'_{\mu,\nu}\left(\phi(2^n x, 0), \frac{4^n t}{3}\right)\right\}$$

Taking limit as $n \to \infty$ we get

$$P_{\mu,\nu}(Q(x+y) + Q(x-y) - 2Q(x) - 2Q(y), t) = 1_{L^*}$$

that is, $Q(x+y) + Q(x-y) = 2Q(x) + 2Q(y)$ that is, Q is quadratic.

Also from Equation (25) we have

$$P_{\mu,\nu}(Q(x) - g(x), \frac{10-\alpha}{6}t)$$

$$= P_{\mu,\nu}(Q(x) - f(x) + f(x) - g(x), t + \frac{(4-\alpha)}{6}t)$$

$$\geq_{L^*} M\left(P_{\mu,\nu}(Q(x) - f(x), t), P_{\mu,\nu}\left(f(x) - g(x), \frac{(4-\alpha)}{2.3}t\right)\right)$$

$$\geq_{L^*} M\left(M_1(x, (4-\alpha)t), M\left(P'_{\mu,\nu}\left(\phi(x, 0), \frac{(4-\alpha)}{3}t\right)\right)\right)$$

$$\geq_{L^*} M\left(M_1(x, (4-\alpha)t), M_1(x, (4-\alpha)t)\right)$$

$$\geq_{L^*} M_1(x, (4-\alpha)t)$$

Therefore,

$$P_{\mu,\nu}(Q(x) - g(x), t) \geq_{L^*} M_1\left(x, \frac{t \times 6(4-\alpha)}{10-\alpha}\right).$$

Similarly,

$$P_{\mu,\nu}(Q(x) - h(x), t) \geq_{L^*} M_1\left(x, \frac{t \times 6(4-\alpha)}{10-\alpha}\right).$$

□

Corollary 1. *Let* $p < 1$ *be a non-negative real number and* X *be norm linear space with norm* $\|.\|$, $(Z, P'_{\mu,\nu}, M)$ *be an IFN-space,* $(Y, P_{\mu,\nu}, M)$ *be a complete IFN-space and* $z_0 \in Z$. *If* $f, g, h : X \to Y$ *are odd mappings such that*

$$P_{\mu,\nu}(f(x+y) + f(x-y) - 2g(x) - 2h(y), t)$$
$$\geq_{L*} P'_{\mu,\nu}(z_0(\|x\|^p + \|y\|^p), t)$$
$$(x, y \in X, t > 0, z_0 \in Z)$$

then there exists a unique additive mapping $A : X \to Y$ *such that*

$$P_{\mu,\nu}(f(x) - A(x), t) \geq_{L^*} P'_{\mu,\nu}\left(z_0\|x\|^p, \frac{t}{6}(2-2^p)\right)$$

and $P_{\mu,\nu}(A(x) - g(x) - h(x), t) \geq_{L^*} P'_{\mu,\nu}\left(z_0\|x\|^p, \frac{(2-2^p)}{10-2^{p+1}}t\right)$
for all $x \in X$ *and* $t > 0, z_0 \in Z$.

Proof. Define $\phi(x, y) = z_0(\|x\|^p + \|y\|^p)$, then the corollary is proved exactly as Theorem 2 with $\alpha = 2^p$. □

Corollary 2. *Let* $p < 2$ *be a non-negative real number and* X *be norm linear space with norm* $\|.\|$, $(Z, P'_{\mu,\nu}, M)$ *be an IFN-space,* $(Y, P_{\mu,\nu}, M)$ *be a complete IFN-space and* $z_0 \in Z$. *If* $f, g, h : X \to Y$ *are even mappings such that*

$$P_{\mu,\nu}(f(x+y) + f(x-y) - 2g(x) - 2h(y), t)$$
$$\geq_{L*} P'_{\mu,\nu}(z_0(\|x\|^p + \|y\|^p), t)$$
$$(x, y \in X, t > 0, z_0 \in Z)$$

then there exists a unique quadratic mapping $Q : X \to Y$ *such that*

$$P_{\mu,\nu}(f(x) - Q(x), t) \geq_{L^*} P'_{\mu,\nu}\left(z_0\|x\|^p, \frac{t}{6}(4-2^p)\right)$$

and $P_{\mu,\nu}(Q(x) - g(x), t) \geq_{L^*} P'_{\mu,\nu}\left(z_0\|x\|^p, \frac{(4-2^p)}{10-2^p}t\right)$ *for all* $x \in X$ *and* $t > 0, z_0 \in Z$.

Proof. Define $\phi(x, y) = z_0 (\|x\|^p + \|y\|^p)$. Then the corollary is proved exactly as Theorem 3 with $\alpha = 2^p$ □

Example 2. *Let* $(X, \|.\|)$ *be a Banach algebra and let Z be a normed linear space, M a continuous t-norm as defined in Example 1. Then* $(X, P_{\mu,\nu}, M)$ *is a complete IFN-space in which* $P_{\mu,\nu}(x,t) = (\mu_x(t), \nu_x(t))$.
Define $f, g, h : X \to X$, *by* $f(x) = x^2 + A\|x\|x_0$, $g(x) = x^2 + B\|x\|x_0$, $h(y) = y^2 + C\|y\|x_0$, $\|x_0\| = 1$ *in X and* A, B, C *are positive real numbers.*
Then $\|f(x+y) + f(x-y) - 2g(x) - 2h(y)\| \leq 2(A+B)\|x\| + 2(A+C)\|y\|$
for all $x, y \in X$.
Let $\phi : X \times X \to Z$ *be defined as*
$\phi(x, y) = 2(A+B)\|x\| z_0 + 2(A+C)\|y\| z_0$ *for all* $x, y \in X$ *and* z_0 *be a unit vector in Z.*
Thus, $P_{\mu,\nu}(f(x+y) + f(x-y) - 2g(x) - 2h(y), t)$

$$\geq_{L^*} P'_{\mu,\nu}(2(A+B)\|x\| z_0 + 2(A+C)\|y\| z_0, t)$$

$$=_{L^*} P'_{\mu,\nu}(\phi(x,y), t)$$

for all $x, y \in X$ *and* $t > 0$.
Then $P_{\mu,\nu}(\phi(2x, 2y), t) \geq P'_{\mu,\nu}(2\phi(x, y), t)$ *for all* $x, y \in X$ *and* $t > 0$. *Hence, all the conditions of Theorem 3 are valid for* $\alpha = 2 < 4$.
Therefore, f can be approximated by a mapping $Q : X \to X$ *such that*

$$P_{\mu,\nu}(f(x) - Q(x), t)$$

$$\geq_{L^*} M_1(x, 2t)$$

$$= P'_{\mu,\nu}\left(\|x\| z_0, \frac{t}{6\min\{(A+B), (2A+B+C), (A+C)\}}\right)$$

for all $x, y \in X$ *and* $t > 0$.

4. Conclusions

Our consideration in this paper is a pexiderized quadratic functional equation, which is an extension of the quadratic functional equation. It may be possible to extend the cubic and higher order functional equations on similar lines. In our proof of the main theorem, we have made extensive use of the characteristics of intuitionistic fuzzy Banach spaces. As a future problem, we can think of the problem of Hyers-Ulam-Rassias stability for more general forms of functional equations in intuitionistic fuzzy linear spaces.

5. Data Availability

The data used to support the findings of this study are available from the corresponding author upon request.

Author Contributions: Conceptualization, P.S., T.K.S. and P.M.; methodology, T.K.S., P.M. and B.S.C.; validation, P.S., P.M., B.S.C. and M.D.L.S.; formal analysis, T.K.S. and P.M.; writing—original draft preparation, P.S. and B.S.C.; writing—review and editing, P.S., T.K.S., B.S.C. and M.D.L.S.; supervision, P.S., B.S.C. and M.D.L.S.; funding acquisition, M.D.L.S. All authors have read and agreed to the published version of the manuscript.

Funding: This work was supported by the Basque Government under the Grant IT 1207-19.

Acknowledgments: The fifth author thanks the Basque Government for Grant IT 1207-19. The suggestions of the referees are acknowledged.

Conflicts of Interest: The authors declare that there are no conflicts of interest regarding the publication of this paper.

References

1. Ulam, S.M. *Problems in Modern Mathematics*; Science Editions; Wiley: New York, NY, USA, 1964; Chapter VI.
2. Hyers, D.H. On the stability of the linear functional equation. *Proc. Nat. Acad. Sci. USA* **1941**, *27*, 222–224. [CrossRef] [PubMed]
3. Rassias, T.M. On the stability of the linear mapping in Banach spaces. *Proc. Am. Math. Soc.* **1978**, *72*, 297–300. [CrossRef]
4. Jung, S.M. Hyers-Ulam stability of linear differential equations of first order, II. *App. Math. Lett.* **2006**, *19*, 854–858. [CrossRef]
5. Grabiec, A. The generalized Hyers-Ulam stability of a class of functional equations. *Publ. Math. Debrecen* **1996**, *48*, 217–235.
6. Dong, Y. On approximate isometries and application to stability of a function. *J. Math. Anal. Appl.* **2015**, *426*, 125–137. [CrossRef]
7. Aoki, T. On the stability of the linear transformation in Banach spaces. *Math. Soc. Jpn.* **1950**, *2*, 64–66. [CrossRef]
8. Cholewa, P.W. Remarks on the stability of functional equations. *Aequ. Math.* **1984**, *27*, 76–86. [CrossRef]
9. Czerwik, S.On the stability of the quadratic mappings in normed spaces. *Abh. Math. Sem. Univ. Hamburg* **1992**, *62*, 59–64. [CrossRef]
10. Gavruta, P.A generalization of the Hyers-Ulam-Rassias Stability of approximately additive mappings. *J. Math. Anal. Appl.* **1994**, *184*, 431–436. [CrossRef]
11. Isac, G.; Rassias, T.H. Stability of ψ -additive mapping: Applications to nonlinear analysis. *Int. J. Math. Math. Sci.* **1996**, *19*, 219–228. [CrossRef]
12. Mihet, D. The fixed point method for fuzzy stability of the Jensen functional equation. *Fuzzy Sets Syst.* **2009**, *160*, 1663–1667. [CrossRef]
13. Samanta, T.K.; Kayal, N.C.; Mondal, P. The stability of a general quadratic functional equation in fuzzy Banach spaces. *J. Hyperstruct.* **2012**, *1*, 71–87.
14. Samanta, T.K.; Mondal, P.; Kayal, N.C. The generalized Hyers-Ulam-Rassias stability of a quadratic functional equation in fuzzy Banach spaces. *Ann. Fuzzy Math. Inform.* **2013**, *6*, 59–68.
15. Zadeh, L.A. Fuzzy sets. *Inf. Control* **1965**, *8*, 338–353. [CrossRef]
16. Atanassov, K.T. Intuitionistic fuzzy sets. *Fuzzy Sets Syst.* **1986**, *20*, 87–96. [CrossRef]
17. Shakeri, S. Intutionistic fuzzy stability of Jenson type mapping. *J. Nonlinear Sci. Appl.* **2009**, *2*, 105–112. [CrossRef]
18. Mirmostafee, A.K.; Moslehian, M.S. Fuzzy versions of Hyers-Ulam-Rassias theorem. *Fuzzy Sets Syst.* **2008**, *159*, 720–729. [CrossRef]
19. Kayal, N.C.; Samanta, T.K.; Saha, P.; Choudhury, B.S. A Hyers-Ulam-Rassias stability result for functional equations in intuitionistic fuzzy Banach spaces. *Iran. J. Fuzzy Syst.* **2016**, *13*, 87–96.
20. Mondal, P.; Kayal, N.C.; Samanta, T.K. Stability of a quadratic functional equation in intuitionistic fuzzy banach spaces. *J. New Results. Sci.* **2016**, *10*, 52–59.
21. Wang, Z.; Rassias, T.M.; Saadati, R. Intuitionistic fuzzy stability of Jensen-type quadratic functional equations. *Filomat* **2014**, *28*, 663–676. [CrossRef]
22. Mohiuddine, S.A.; Sevli, H. Stability of Pexiderized quadratic functional equation in intuitionistic fuzzy normed space. *J. Comp. Appl. Math.* **2011**, *235*, 2137–2146. [CrossRef]
23. Mondal, P.; Kayal, N.C.; Samanta, T.K. The stability of Pexider type functional equation in intuitionistic fuzzy banach spaces via fixed point technique. *J. Hyperstruct.* **2015**, *4*, 37–49.
24. Xu, T.Z.; Rassias, M.J.; Xu, W.X.; Rassias, J.M. A fixed point approach to the intuitionistic fuzzy stability of quintic and sextic functional equations. *Iran. J. Fuzzy Syst.* **2012**, *9*, 21–40.
25. Diaz, J.B.; Margolisi, B. A fixed point theorem of the alternative for contractions on a generalized complete metric space. *Bull. Am. Math. Soc.* **1968**, *74*, 305–309. [CrossRef]
26. Jung, S.M. On the Hyers-Ulam stability of functional equations that have the quadratic property. *J. Math. Anal. Appl.* **1998**, *222*, 126–137. [CrossRef]
27. Jung, S.M. Quadratic functional equations of Pexider type. *J. Math. Math. Sci.* **2000**, *24*, 351–359. [CrossRef]
28. Atanassov, K.T. Geometrical interpretation of the elements of the intuitionistic fuzzy objects. *Int. J. Bioautomat.* **2016**, *20*, S27–S42.

29. Deschrijiver, G.; Kerre, E.E. On the relationship between some extensions of fuzzy set theory. *Fuzzy Sets Syst.* **2003**, *23*, 227–235. [CrossRef]
30. Deschrijver, G.; Cornelis, C.; Kerre, E.E. On the representation of intuitionistic fuzzy t-norms and t-conorms. *IEEE Trans. Fuzzy Sust.* **2004**, *12*, 45–61. [CrossRef]
31. Saadati, R.; Park, J.H. On Intuitionistic fuzzy topological spaces. *Chaos Solitons Fractals* **2006**, *27*, 331–344. [CrossRef]
32. Cadariu, L.; Radu, V. Fixed points and stability for functional equations in probabilistic metric and random normed spaces. *Fixed Point Theory Appl.* **2009**, *2009*, 589143. [CrossRef]

mathematics

MDPI

Article

An Extension of Fuzzy Competition Graph and Its Uses in Manufacturing Industries

Tarasankar Pramanik [1,†], G. Muhiuddin [2,*,†], Abdulaziz M. Alanazi [2,†] and Madhumangal Pal [3,†]

1 Department of Mathematics, Khanpur Gangche High School, Paschim Medinipur 721201, India; tarasankar.math07@gmail.com
2 Department of Mathematics, University of Tabuk, Tabuk 71491, Saudi Arabia; am.alenezi@ut.edu.sa
3 Department of Applied Mathematics with Oceanology and Computer Programming, Vidyasagar University, Midnapore 721102, Inida; mmpalvu@mail.vidyasagar.ac.in
* Correspondence: chishtygm@gmail.com
† These authors contributed equally to this work.

Received: 11 May 2020; Accepted: 10 June 2020; Published: 19 June 2020

Abstract: Competition graph is a graph which constitutes from a directed graph (digraph) with an edge between two vertices if they have some common preys in the digraph. Moreover, Fuzzy competition graph (briefly, FCG) is the higher extension of the crisp competition graph by assigning fuzzy value to each vertex and edge. Also, Interval-valued FCG (briefly, IVFCG) is another higher extension of fuzzy competition graph by taking each fuzzy value as a sub-interval of the interval $[0, 1]$. This graph arises in many real world systems; one of them is discussed as follows: Each and every species in nature basically needs ecological balance to survive. The existing species depends on one another for food. If there happens any extinction of any species, there must be a crisis of food among those species which depend on that extinct species. The height of food crisis among those species varies according to their ecological status, environment and encompassing atmosphere. So, the prey to prey relationship among the species cannot be assessed exactly. Therefore, the assessment of competition of species is vague or shadowy. Motivated from this idea, in this paper IVFCG is introduced and several properties of IVFCG and its two variants interval-valued fuzzy k-competition graphs (briefly, IVFKCG) and interval-valued fuzzy m-step competition graphs (briefly, IVFMCG) are presented. The work is helpful to assess the strength of competition among competitors in the field of competitive network system. Furthermore, homomorphic and isomorphic properties of IVFCG are also discussed. Finally, an appropriate application of IVFCG in the competition among the production companies in market is presented to highlight the relevance of IVFCG.

Keywords: interval-valued fuzzy competition graph; interval-valued fuzzy p competition graph; interval-valued fuzzy neighbourhood graph; interval-valued m-step fuzzy competition graph; homomorphism of graph products

1. Introduction

Cohen [1] first developed the concept of competition graph (CG) to solve the problem of the food web in ecology. The problem of a food web is to describe the predator-prey relationship among species in the community. Food web is a relationship network framed to describe the relationships among food habits of species. It is a fact that there is a predator-prey relation in ecosystem among the species. The plants are the main source of energy for all the living entity. Species are classified into few levels depending on the predator-prey relationship; for example, primary producer (plants), primary consumer (herbivorous), secondary consumer (carnivorous) and omnivorous. In ecosphere, the plants those are the primary producers can produce through photosynthesis. Herbivorous eats only plants for energy, carnivorous takes herbivorous as their food. There is no unique choice of food to omnivorous. From primary producers to secondary consumers there is a food chain among themselves. But the food web is not same at all as a food chain. An example of a food web is shown in Figure 1. In this figure, grasses are the main source of food and grasshoppers eat them, frogs eat grasshoppers, snakes eat frogs, peacocks have snakes but, eagle depends on snakes as well as grasshoppers. If some species say grasshopper, abolished in this food web, other species (here, eagle) who depend on the abolished species may either exterminate or may have to make every effort for existence adapting another food habit depending on ecological nook, habitat and surrounding atmosphere. Same species may have different food habits in different places depending on ecosystem, habitat and surrounding atmosphere. In this example, shown in Figure 1, it is considered that in a certain ecosystem, 70–80% eagle depends on the grasshopper and 30–40% on snake for his food need. These can be transferred to its similar correspondence to interval-valued fuzzy number as [0.7, 0.8] and [0.3, 0.4] respectively. Peacock has 100% dependence on snake, the snake has 100% dependence on frog and frog has that on a grasshopper. Grasshopper depends only on grass. If any two species depends on the same species, there must be a competition between those two species. Being motivated by this idea, we can model up this natural phenomenon as an IVFCG. In addition to ecology, this graph model has many uses in circuit designing, economical model and coding as well as energy systems, etc.

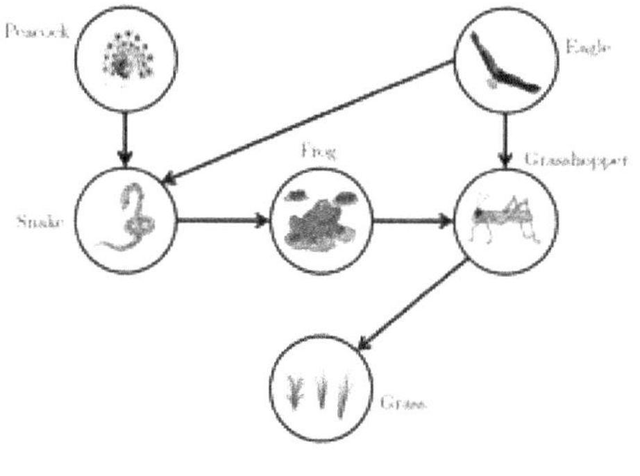

Figure 1. An example of a food web.

We have generalized the model to its more realistic cases as an IVFCG. In IVFCG, the vertices and edges may be considered as an interval of numbers instead of precise numbers.

Graph theory has an extensive sector of applications in the real world. In 1975, Rosenfeld [2] generalizes the Euler's graph theory model to fuzzy graph (briefly, FG) theory. Before generalizing graph theory, he has studied the fuzzy relation (briefly, FR) of fuzzy sets and he also introduced several types of FGs. The scope of FG theory is widening fast for its demand in society. The FG theory is being extensively used to solve the problems on the system where there is a network which is either physical, biological or artificial such as, the neuron in the human brain, rail routing system, transportation problem, traffic signaling system, scheduling problem, etc. In the fuzzy field of mathematics, there are various types of FGs which are classified as follows:

1. the set of all vertices is crisp and the set of all edges is fuzzy
2. the set of all vertices is fuzzy and the set of all edges is crisp

3. the set of all vertices is fuzzy and the set of all edges is fuzzy
4. the sets of all vertices and edges are crisp with fuzzy connectivity.

Among these, the most studied type of FGs is the third one, as this is the more general case of FGs. Fuzzy systems are applied to the problems where approximate reasoning is involved.

A *fuzzy set* (FS) δ is a pair $\delta = (S, \delta)$ the membership function (MF) $\delta : S \rightarrow [0, 1]$ where S is a vertex set. There are different types of FSs which are extended further such as BFS, interval-valued FS (IVFS), intuitionistic FS, etc.

Zadeh [3] developed the concept of IVFSs as a generalization of FSs in which the membership values lies in $[0, 1]$ instead of a precise number. Since the IVFS is an interval number, it is more strong enough to consider real-world problems than the traditional FSs. Therefore, it has more area of applications such as medical diagnosis, multivalued logic, fuzzy control, approximate reasoning, intelligent control, etc.

1.1. Motivation and Main Contribution of the Proposed Work

As we have seen, there is competition in most of our real-world problem, especially in industries, ecology, or wherever the economy is involved. This competition depends on certain parameters. These parameters can be anything like time, money, demand, etc. In the case of competitive real-world problems, the contestant has to accurately determine who his competitors are and how strong they are. In a system where many competitors are related to each other in different ways, it is possible to make this diagnosis accurately with the help of a mathematical model. But one thing to note about these parameters is that their values are never specified in the case of real-world problems. Time is an important parameter when marketing a product in such a market industry. But in this case, it is very true that no one can say in advance exactly when a product can be marketed. It can be said that the product can be marketed at any time interval like 1 to 2 months or 30 days to 45 days, etc. Due to this kind of vagueness in the quality of parameters, we will use fuzzy mathematical model instead of using any crisp mathematical model. However, the simple fuzzy set system is multi-valued but cannot express the idea of the 'interval' properly. So in this paper, we have proposed to extend our existing fuzzy system to an interval-valued fuzzy system in competition graph model.

Main contribution of the proposed work is to find the strength of competition among competitors exists in a network so that the competitors can decide their strong competitors and take positive steps to achieve its profit. IVFCG is useful rather than other methods because:

1. most of real-world problems are those networks whose nodes have vague parameters and this method deal with such type of networks well.
2. if the parameters associated with the nodes of the networks are of interval then the method is very much useful in dealing such.
3. an efficient algorithmic approach.

Authors' contribution towards the development of interval-valued fuzzy competition graph and making use of it in market competition is listed in Table 1.

Table 1. Comparison of the work to the existing research work.

Author	Year	Contributions	Remarks
Cohen [1]	1968	Use of interval graphs in food webs	Deals only with crisp graph
Kim et al. [4]	1995	p-Competition graph of a digraph	Further variation of crisp competition graph
Brigham et al. [5]	1995	Tolerance competition graph	Deals with the competition graphs where tolerances matter
Cho et al. [6]	2000	m-Step competition graph of a digraph	Another variation of a competition graph
Sonnatag and Teichert [7]	2004	Competition hypergraphs	Competition is studied in hypergraphs
Samanta and Pal [8]	2013	Fuzzy k-Competition graphs and p-Competition graphs	Fuzziness is considered in the earlier two types of crisp graphs
Pramanik et al. [9]	2017	Fuzzy ϕ-tolerance competition graphs	Fuzziness is considered in more general version of tolerance competition graphs
Pramanik et al. [10]	2016	Interval-valued fuzzy ϕ-tolerance competition graphs	More general fuzzy system is considered in fuzzy ϕ-tolerance competition graphs
Pramanik et al. (This paper)	—	In this paper, fuzzy values of all the network problems related to competition are also taken as intervals. As a result much more generalizations have been made	More generalized concept than all previous existing research works.

1.2. Review of Previous Works

To represent any network in the mathematical model we use graphs. The graph is dealt with several physical, biological, social, economic relationships very well. For example, friendship is a social relationship network which is modelled as a graph for several community sites such as Facebook, Twitter, LinkedIn, etc. in many forms and they have several problems to solve related to this network. In the cases where the impreciseness in relations comes, the corresponding relationship network can be modelled as an FG model. In 2003, Bhutani and Battou [11] consider the operations on FGs where the m-strong property is reserved. The necessity of finding strongness in FGs demands the contribution of Bhutani and Rosenfeld [12] to find strong arcs in FGs. The reader may look for more characterization of FGs in [13,14].

There are a lot of variations in CGs described in Cohen's work [1]. Several researchers have found various derivations of competition graphs. Such as Cho et al. [6] developed the m-step CG of a digraph. The p-CG of a digraph has been defined by Kim et al. [4,15]. The tolerance CG is defined by Brigham et al. [5]. The competition hypergraphs have been found in Sonnatag et al. [7]. Recent work on FKCG and p-competition FGs is available in [8]. Nayeem and Pal [16] have worked to find the shortest path in a network where the relationship between the nodes is imprecise. A detailed survey of the works on CG can be found in [17]. Recently, the fuzzy tolerance graph [18] is further extended to fuzzy ϕ-tolerance CG by Pramanik et al. [9]. To emphasize real-world problem Samanta and Pal [19,20] have studied fuzzy planar graph. Pramanik et al. [21] have generalized the fuzzy planar graph by introducing the IVFSs instead of traditional FSs. Rashmanlou and Pal [22] have studied several properties on highly irregular interval-valued FGs (IVFG). To find the shortest path in a complex network is very emerging work in this modern edge. There are various techniques to find shortest paths in a network. The bipolar fuzzy hypergraph is an extension of fuzzy hypergraph by introducing bipolar fuzzy vertex sets (or simply, bipolar FS (BFS)) and bipolar FR instead of traditional FSs. The bipolar FG (BFG) is introduced by Samanta and Pal [23] which has emerging importance in a complex networking system. Colouring problem is also a challenging task in the research field nowadays. Samanta et al. [24] have introduced a new approach to colour an FG in a vague sense. Rashmanlou et al. [25] have worked on bipolar fuzzy graphs which is an extension of fuzzy graphs. In 2014, Rashmanlou and Pal [26] have studied the properties of isometry on interval-valued fuzzy graphs. Balanced interval-valued fuzzy graphs [27] and Antipodal interval-valued fuzzy graphs [28] are another two types of fuzzy graphs which are introduced by Rashmanlou and Pal. For further studies on FGs and its variations the works of literature [29–31] may be very helpful.

There may occur challenging situations in a system's operation characterized by a degree of vagueness and/or uncertainty. Voskoglou [32] uses principles of fuzzy logic to develop a general model representing such kind of situations. He also introduced a stochastic method for the description of a finite Markov chain as the main steps of mathematical modelling process in [33]. In 2012, a fuzzy model [34] has been developed by him to describe the process of Analogical Reasoning. Gil et al. [35] have determined the travel and delay times in a road ending in a traffic light under different traffic flows and traffic light cycles using a microscopic traffic simulator. To find the approximate measure of the behavior of the plant Hedrea et al. [36] uses TP-based model transformation method in order to obtain a Tensor Product-based model of magnetic levitation systems. Deveci et al. [37] developed a quantitative assessment framework for public bus operators to translate the passenger demands into service quality specifications. Recently, Deveci et al. [38] have developed a multi-criteria decision-making model considering technical, economic, environmental and social criteria to assess Ireland's most promising offshore wind sites. In airlines, crew scheduling problem is a challenging problem. Deveci and Demirel have proposed a solution and made a survey on this in [39]. Canitez and Deveci [40] have presented a model framework so that public transport system and multi-stakeholder can better manage car sharing applications. In 2015, Deveci et al. [41] studied fuzzy-based multi-criteria decision making methods to solve the carbon di-oxide geological storage location selection problem.

In this paper, IVFCG is defined and investigated several properties on this graph. Also, several variations of this graph class such as interval-valued m-step FCG, IVFKCG, etc. are introduced. The homomorphism and isomorphism properties of several IVFCG products have also been studied. An application on the competition of producers for their products is discussed. This application and the application on ecosystem discussed earlier shows the importance of IVFCG.

The arrangements for the paper are as follows:

After a short inception in Section 1, previous works have been reviewed in Section 1.2. In Section 2, the needful preliminaries that have been surveyed are placed. The main work of IVFCG is introduced in Section 3. Introducing Definition of IVFCG, many results have been studied there. Section 4 describes an interesting idea to apply in the real field. Homomorphism properties of IVFG products have been studied in Section 5. Next, the conclusion has been drawn in Section 6.

2. Preliminaries

A *FS* δ on a set S is a function $\delta : S \to [0,1]$, known as the MF. The *support* of δ is $\text{supp}(\delta) = \{d \in S | \delta(d) \neq 0\}$ and the *core* of δ is $\text{core}(\delta) = \{d \in S | \delta(d) = 1\}$. The *support length* is $s(\delta) = |\text{supp}(\delta)|$ and the *core length* is $c(\delta) = |\text{core}(\delta)|$. The height of δ is $h(\delta) = \max\{\delta(d) | d \in S\}$. The FS δ is said to be *normal* if $h(\delta) = 1$.

A *FG* is defined on a non-empty finite set S equipped with FS δ defined by a MF $\delta : S \to [0,1]$ and a FR θ on the FS δ such that $\theta(p,q) \leq \delta(p) \wedge \delta(q)$ for all $p, q \in S$, where $\wedge$ represents minimum. A fuzzy edge (p,q), $p,q \in S$ is said to be independent strong [31] if $\theta(p,q) \geq \frac{1}{2}\min\{\delta(p),\delta(q)\}$ and is called weak, otherwise. The *degree* of a vertex d of a FG $G=(S,\delta,\theta)$ is $deg(d) = \sum_{c \in S-\{d\}} \theta(d,c)$. The *order* of a FG G is $O(G) = \sum_{c\in S} \delta(c)$. The *size* [22] of a FG G is $S(G) = \sum \theta(c,d)$.

A *directed FG or, fuzzy digraph (FDG)* [42] $\overrightarrow{F} = (S,\delta,\nu)$ defined on a non-empty set S equipped with a fuzzy MF $\delta : S \to [0,1]$ and a FR $\nu : S \times S \to [0,1]$ such that for all $c,d \in S$, $\nu(\overrightarrow{c,d}) \leq \delta(c) \wedge \delta(d)$.

As ν need not be symmetric, an FDG may consists of two directed edges between two vertices with opposite directions. These edges are called parallel edges. There exists a loop at a vertex $c \in S$, if $\nu(\overrightarrow{c,c}) \neq 0$.

Every FG corresponds to an undirected FG $F=(S,\delta,\theta)$ where $\theta(c,d) = \max\{\nu(\overrightarrow{c,d}), \nu(\overrightarrow{d,c})\}\ \forall\, c,d \in S$ and this undirected FG is called the *underlying FG* [31].

A complete FDG is an FDG $\overrightarrow{F} = (S,\delta,\nu)$ in which the relation $\nu(\overrightarrow{c,d}) = \nu(\overrightarrow{d,c}) = \delta(c) \wedge \delta(d)$ for all $c,d \in S$ holds.

To introduce the CG, Cohen defined a digraph $\overrightarrow{F} = (S, \overrightarrow{E})$ which nicely represents an ecological problem of food web. In food web, species are represented as vertex p in $S(\overrightarrow{F})$ and an arc $\overrightarrow{(p,s)}$ in $\overrightarrow{E}(\overrightarrow{F})$ means that p preys on species s. A vertex $p \in S(\overrightarrow{F})$ represents a species in the food web and arc $\overrightarrow{(p,s)} \in \overrightarrow{E}(\overrightarrow{F})$ means that p is dependent on the species s. If a prey s is dependent on two different species then it is said that the two species compete for the prey s. Therefore, each species in the food web are interdependent and this interdependence is designed by Cohen as competition graph model. An undirected graph $G=(S,E)$ of a digraph $\overrightarrow{F} = (S, \overrightarrow{E})$ with same vertex set S is said to be CG if between any two vertices p,q there is an edge in E, such that the arcs $\overrightarrow{(p,s)}, \overrightarrow{(q,s)}$ are in $\overrightarrow{E(\overrightarrow{F})}$. Several fields like channel assignment, energy systems, modeling of complex economic, coding, etc. uses the study of CG.

In an FDG $\overrightarrow{F} = (S,\delta,\nu)$, the fuzzy *out-neighbourhood* [31] of a vertex $d \in S$ is a FS $\Delta^+(d) = (S_v^+, m_v^+)$, where $S_v^+ = \{c | \nu(\overrightarrow{d,c}) > 0\}$ and $m_v^+ : S_v^+ \to [0,1]$ is defined by $m_v^+ = \nu(\overrightarrow{d,c})$.

In an FDG $\overrightarrow{F} = (S,\delta,\nu)$, the fuzzy *in-neighbourhood* [31] of a vertex $d \in S$ is a FS $\Delta^-(d) = (S_v^-, m_v^-)$, where $S_v^- = \{c | \nu(\overrightarrow{c,d}) > 0\}$ and $m_v^- : S_v^- \to [0,1]$ is defined by $m_v^- = \nu(\overrightarrow{c,d})$.

In a FG $F=(S,\delta,\theta)$, the fuzzy *neighbourhood* [43] of a vertex $d \in S$ is the FS $\Delta(d) = (S_v, m_v)$, where $S_v = \{c | \theta(c,d) > 0\}$ and $m_v : S_v \to [0,1]$ is defined by $m_v = \theta(c,d)$.

A FS $\Delta_m^+(d) = (S_v^+, m_v^+)$, where $S_v^+ = \{c | \theta_m(\overrightarrow{d,c}) = \min\{\nu(\overrightarrow{d,c_1}), \nu(\overrightarrow{c_1,c_2}), \ldots, \nu(\overrightarrow{c_m,c})\} > 0, vc_1c_2\ldots c_mc$ is a path from d to $c\}$ and $m_v^+ : X_v^+ \to [0,1]$ is said to be the *m-step fuzzy out-neighbourhood* [31] of a vertex $d \in S$ of a directed FG $\overrightarrow{F} = (S,\delta,\nu)$.

The *FCG* [31] of an FDG $\overrightarrow{F} = (S,\delta,\nu)$ is an undirected graph $\mathcal{C}(\overrightarrow{F}) = (S,\delta,\theta)$ which has the same fuzzy vertex set as in $\overrightarrow{F}$ and has a fuzzy edge between two vertices $c,d \in S$ in $\mathcal{C}(\overrightarrow{F})$ if and only if $\Delta^+(c) \cap \Delta^+(d)$ is non-empty FS in $\overrightarrow{F}$. The membership value of the edge (c,d) in $\mathcal{C}(\overrightarrow{F})$ is $\theta(c,d) = (\delta(c) \wedge \delta(d))h(\Delta^+(c) \cap \Delta^+(d))$.

The *m-step FCG* [31] of an FDG $\overrightarrow{F} = (S,\delta,\nu)$ is denoted by $\mathcal{C}_m(\overrightarrow{F})$ and is defined by $\mathcal{C}_m(\overrightarrow{F}) = (S,\delta,\theta)$ where $\theta(c,d) = (\delta(c) \wedge \delta(d))h(\Delta_m^+(c) \cap \Delta_m^+(d))$ for all $c,d \in S$.

An *interval number* [44] L is an interval $[l^-, l^+]$ with $0 \leq l^- \leq l^+ \leq 1$. For any two interval numbers $L_1 = [l_1^-, l_1^+]$ and $L_2 = [l_2^-, l_2^+]$ the followings are defined:

1. $L_1 + L_2 = [l_1^-, l_1^+] + [l_2^-, l_2^+] = [l_1^- + l_2^- - l_1^- \cdot l_2^-, l_1^+ + l_2^+ - l_1^+ \cdot l_2^+]$,
2. $\min\{L_1, L_2\} = [\min\{l_1^-, l_2^-\}, \min\{l_1^+, l_2^+\}]$,
3. $\max\{L_1, L_2\} = [\max\{l_1^-, l_2^-\}, \max\{l_1^+, l_2^+\}]$,
4. $L_1 \leq L_2 \Leftrightarrow l_1^- \leq l_2^-$ and $l_1^+ \leq l_2^+$,

5. $L_1 = L_2 \Leftrightarrow l_1^- = l_2^-$ and $l_1^+ = l_2^+$,
6. $L_1 < L_2 \Leftrightarrow L_1 \leq L_2$ but $L_1 \neq L_2$,
7. $kL_1 = [kl_1^-, kl_2^+]$, where $0 \leq k \leq 1$.

2.1. Some Terminology of FGs

The *fuzzy subgraph* [45] of a FG $F = (S, \delta, \theta)$ is a FG $F' = (S, \tau, \nu)$ with $\tau(c) \leq \delta(c)$ for all $c \in S$ and $\nu(c,d) \leq \theta(c,d)$ for all $c, d \in S$.

Definition 1. *A FG $F = (S, \delta, \theta)$ is said to be complete if $\theta(c,d) = \min\{\delta(c), \delta(d)\}$ for all $c, d \in S$.*

Strong edge in a FG is defined in many ways in various literature. Among them the definition stated in [46] is more suitable for our purpose. We use this definition in our work too.

Definition 2. *A FG $F = (S, \delta, \theta)$ is called the bipartite FG if there are two non-empty vertex sets S_1 and S_2 such that $\theta(d_1, d_2) = 0$ if $d_1, d_2 \in S_1$ or $d_1, d_2 \in S_2$. Further, if $\theta(d_1, d_2) = \min\{\delta(d_1), \delta(d_2)\}$ for all $d_1 \in S_1$ and $d_2 \in S_2$, then F is called a complete bipartite FG.*

An effective edge [47] in a FG $F = (S, \delta, \theta)$ is an edge (c, d) such that the condition $\theta(c,d) = \min\{\delta(c), \delta(d)\}$ holds. The end vertices of the effective edge are called effective adjacent vertices. The number of effective incident edges on a vertex d of a FG is the effective incident degree of the FG. A FG is a complete FG if its all the edges are effective incident. The effective incident degree of a pendent vertex in a FG is defined as 1. If one end vertex of a fuzzy edge of a FG is fuzzy pendent vertex then the fuzzy edge is call *fuzzy pendent edge* [8]. The membership value of the fuzzy pendent edge is the minimum among the membership values of the fuzzy end vertices.

If the degree of a vertex d of a FG $F = (S, \delta, \theta)$ is a fixed positive real number, say, k for all $d \in S$ then the FG F is said to be *regular* [48]. The FG F is called *totally regular FG* [48] if each vertex of F has same total degree k. If in a FG F there are at least two vertices which are adjacent with distinct degrees, the FG is said to be *irregular* [49]. If every two adjacent vertices of the FG have different degrees then the FG is said to be *neighbourly irregular* [49]. If there are at least two adjacent vertices which have distinct total degrees, is said to be totally irregular. The FG is said to be *neighbourly total irregular* [49] if every two adjacent vertices have distinct total degrees. A FG is said to be highly irregular [49] if every vertex of G is adjacent to vertices with distinct degrees.

Definition 3. *The crisp graph $F^* = (S, \delta^*, \theta^*)$ corresponding to a FG $F = (S, \delta, \theta)$ with same vertex set and $\delta^* = \{c \in S | \delta(c) > 0\}$ and $\theta^* = \{(c,d) \in S \times S | \theta(c,d) > 0\}$ is called the underlying crisp graph of the FG F.*

The *complement* [45] of FG $F = (S, \delta, \theta)$ is the FG $F' = (S, \delta', \theta')$ where $\delta'(c) = \delta(c)$ for all $c \in S$ and

$$\theta'(c,d) = \begin{cases} 0, & \text{if } \theta(c,d) > 0, \\ \delta(c) \wedge \delta(d), & \text{otherwise.} \end{cases}$$

Definition 4 ([50])**.** *Let δ be a FS defined by $\delta : S \to [0,1]$ and θ is a FR where $\overrightarrow{\theta} : S \times S \to [0,1]$ such that for all $c, d \in S$, $\overrightarrow{\theta}(c,d) \leq \delta(c) \wedge \delta(d)$. Then $\overrightarrow{F} = (S, \delta, \overrightarrow{\theta})$ is said to be an FDG.*

Since $\overrightarrow{\theta}$ is well defined, an FDG does not have more than two directed edges with opposite directions between any two vertices. The membership value of a directed edge $\overrightarrow{(c,d)}$ is denoted by $\overrightarrow{\theta}(c,d)$. The loop at a vertex c is mathematically expressed as $\overrightarrow{\theta}(c,c) \neq 0$. Since, in an FDG $\overrightarrow{\theta}(c,d)$ and $\overrightarrow{\theta}(d,c)$ may have different values, $\overrightarrow{\theta}$. The *underlying crisp graph of FDG* is the graph similarly obtained except the directed arcs are replaced by undirected edges.

2.2. Fuzzy Hypergraphs

Goetschel [51] introduced fuzzy hypergraphs. The Definition of fuzzy hypergraph is given below

Definition 5. *Let S be a non-empty finite set and let $\mathcal{E}$ be a finite family of nontrivial FSs on S (or subsets of S) such that $S = \bigcup\{\text{supp}(A)|A \in \mathcal{E}\}$. Then the pair $\mathcal{H}= (S, \mathcal{E})$ is a fuzzy hypergraph on S.*

S and $\mathcal{E}$ are respectively vertex set and fuzzy edge set of $\mathcal{H}$. The height of $\mathcal{H}$, $h(\mathcal{H})$, is defined by $h(\mathcal{H}) = max\{h(A)|A \in \mathcal{E}\}$. A fuzzy hypergraph is *simple* if $\mathcal{E}$ has no repeated fuzzy edges and whenever $A, B \in \mathcal{E}$ and $A \subseteq B$, then $A = B$. A fuzzy hypergraph $\mathcal{H}= (S, \mathcal{E})$ is *support simple* if whenever $A, B \in \mathcal{E}$, $A \subseteq B$ and $supp(A) = supp(B)$, then $A = B$. Suppose $A = (X_1, \theta) \in F$, $X_1 \subseteq S$ and $c \in (0, 1]$. The $c-$cut of A, A^c, is defined by $A^c = \{c \in S|\theta(c) \geq c\}$. If $\mathcal{E}^c = \{A^c| \in \mathcal{E}/\{\phi\}\}$ and $S^c = \bigcup\{A^c|A \in \mathcal{E}\}$. If $\mathcal{E}^c \neq \phi$, then the (crisp) hypergraph $H^c = (S^c, \mathcal{E}^c)$ is the *$c-$ level hypergraph* of $\mathcal{H}$.

Suppose $\mathcal{H}_1 = (S, \mathcal{E}_1)$ and $\mathcal{H}_2 = (S, \mathcal{E}_2)$ are fuzzy hypergraphs. Then $\mathcal{H}_1$ is partial hypergraph of $\mathcal{H}_2$ if $\mathcal{E}_1 \subseteq \mathcal{E}_2$. A FS $A = (S, \theta)$ with $\theta : S \to [0, 1]$ is an *elementary FS* if θ is constant function or θ has range $\{0, a\}, 0 \neq a$. An *elementary fuzzy hypergraph* is a fuzzy hypergraph in which all fuzzy edges are elementary.

A fuzzy hypergraph $\mathcal{H}= (S, \mathcal{E})$ is a *m tempered fuzzy hypergraph* of a crisp hypergraph $H^* = (S, E)$ if there exists a FS $A = (S, m)$ such that $m : S \to (0, 1]$ and $\mathcal{E}= \{\theta_{E_i}|E_i \in E\}$ where

$$\theta_{E_i}(c) = \begin{cases} min\{m(e)|e \in E_i\} & \text{if } c \in E_i \\ 0, & otherwise \end{cases}$$

A *fuzzy transversal* $\mathcal{T}= (S, \tau)$ of $\mathcal{H}$ is a FS defined on S with the property that $\tau_{h(A)} \cap \theta_{h(A)} \neq \phi$ for each $A \in \mathcal{E}$ (recall that $h(A)$ is the height of A). A *minimal fuzzy transversal* $\mathcal{T}$ for $\mathcal{H}$ is a transversal of $\mathcal{H}$ with the property that if $T_1 < T$, then T_1 is not a fuzzy transversal of $\mathcal{H}$.

2.3. Fuzzy Intersection Graphs

McAllister [52] introduced fuzzy intersection graphs. The Definition of fuzzy intersection graph is now given.

Definition 6. *$\mathcal{F}= \{A_1 = (S, m_1), A_2 = (S, m_2) \ldots, A_n = (S, m_n)\}$ be a finite family of FSs defined on a set S and consider $\mathcal{F}$ as crisp vertex set $S = \{d_1, d_2 \ldots, v_n\}$. The fuzzy intersection graph of $\mathcal{F}$ is the FG $Int(\mathcal{F})= (S, \delta, \theta)$ where $\delta : S \to [0, 1]$ is defined by $\delta(v_i) = h(A_i)$ and $\theta : S \times S \to [0, 1]$ is defined by*

$$\theta(v_i, v_j) = \begin{cases} h(A_i \cap A_j), & if\, l \neq j \\ 0, & if\, l = j. \end{cases}$$

2.4. Bipolar FGs

There are several real relationship network system, where each nodes or relation between them simultaneously have some properties and as well as have opposite properties. For example, in almost every social networking system a member may have two or more properties among them there are two properties are very opposite to each other. Any member of the system may 'like' some other member or he may 'dislike' the member. This concept introduces a new generalised FS which is called BFS system. The elements of the set have some positive membership values and some negative membership values.

Zhang [53], first introduced the concept of BFS as a generalisation of FS. For example, set of all foods constitutes a set with the property 'sweetness of food', then this set must be a FS. This property indicates there must have another property 'bitterness of food' which also should be traced out. Positive membership values and negative membership values are set by defining grade of sweetness and grade of bitterness of food respectively. Other tastes like salty, sour, pungent (e.g. chili), etc. are irrelevant to the corresponding property. So membership values of tastes of these foods are taken as zero.

The Definition of BFS is given as follows. Let S be a nonempty set. A BFS T on S is an object having the form $T = \{(c, m^+(c), m^-(c)) | \ c \in S\}$, where $m^+ : S \to [0,1]$ and $m^- : S \to [-1,0]$ are mappings. If $m^+(c) \neq 0$ and $m^-(c) = 0$, then we say that c has only the positive satisfaction for T. Similarly, if $m^+(c) = 0$ and $m^-(c) \neq 0$, it is to be said that the vertex c somewhat satisfies the counter property of T. There may have possibility that a vertex c with $m^+(c) \neq 0$ and $m^-(c) \neq 0$ may satisfy MF so that some its properties overlaps that of its counter property over some portion of S. For the BFS $T = \{(c, m^+(c), m^-(c)) | c \in S\}$, we simply write $T = (m^+, m^-)$.

For every two BFSs $L = (m_L^+, m_L^-)$ and $T = (m_J^+, m_J^-)$ on S,

$(L \cap T)(c) = (min(m_L^+(c), m_J^+(c)), max(m_L^-(c), m_J^-(c)))$.

$(L \cup T)(c) = (max(m_L^+(c), m_J^+(c)), min(m_L^-(c), m_J^-(c)))$.

Akram [44,54] introduced BFGs and investigated some properties of it. The formal Definition is given as follows.

Definition 7. *A BFG on a set S is the pair $B = (L, T)$ where $L = (m_L^+, m_L^-)$ is a BFS on S and $T = (m_J^+, m_J^-)$ is a BFS on $E \subseteq S \times S$ such that $m_J^+(c,q) \leq \min\{m_L^+(c), m_L^+(q)\}$ and $m_J^-(c,q) \geq \max\{m_L^-(c), m_L^-(q)\}$ for all $(c,q) \in E$. Here L is called bipolar fuzzy vertex set of S, T is the bipolar fuzzy edge set of E. Thus $B = (L, T)$ is a BFG.*

A BFG $B = (L, T)$ is said to be strong if
$m_J^+(c,q) = min(m_L^+(c), m_L^+(q))$ and $m^-(c,q) = max(m_L^-(c), m_L^-(q))$.

The Definition of strong BFG is given below.

Definition 8. *The complement [44] of a strong BFG B is $\bar{B} = (\bar{L}, \bar{T})$ where $\bar{L} = (\bar{m}_L^+, \bar{m}_L^-)$ is a BFS on $\bar{S}$ and $\bar{T} = (\bar{m}_J^+, \bar{m}_J^-)$ is a BFS on $\bar{E} \subseteq \bar{S} \times \bar{S}$ such that*

(1) $\bar{S} = S$,

(2) $\bar{m}_L^+(c) = m_L^+(c)$ and $\bar{m}_L^-(c) = m_L^-(c)$ for all $c \in S$,

(3)

$$\bar{m}_J^+(c,q) = \begin{cases} 0, & \text{if } m_J^+(c,q) > 0, \\ m_L^+(c) \wedge m_L^+(q), & \text{otherwise.} \end{cases}$$

$$\bar{m}_J^-(c,q) = \begin{cases} 0, & \text{if } m_J^-(c,q) < 0, \\ m_L^-(c) \vee m_L^-(q), & \text{otherwise.} \end{cases}$$

Definition 9 ([44]). *Let $F = (L, T)$ be a BFG where $L = (m_1^+, m_1^-)$ and $T = (m_2^+, m_2^-)$ be two BFSs on a non-empty finite set S and $E \subseteq S \times S$ respectively. The graph F is called complete BFG if $m_2^+(c,d) = \min\{m_1^+(c), m_1^+(d)\}$ and $m_2^-(c,d) = \max\{m_1^-(c), m_1^-(d)\}$ for all $c, d \in S$.*

Regular BFGs are also important subclass of BFGs.

Definition 10 ([55]). *Let $F = (L, T)$ be a BFG where $L = (m_1^+, m_1^-)$ and $T = (m_2^+, m_2^-)$ be two BFSs on a non-empty finite set S and $E \subseteq S \times S$ respectively. If $d^+(c) = k_1, d^-(c) = k_2$ for all $c \in S$, k_1, k_2 are two real numbers, then the graph is called (k_1, k_2)-regular BFG.*

Definition 11 ([55]). *Let $F = (L, T)$ be a BFG where $L = (m_1^+, m_1^-)$ and $T = (m_2^+, m_2^-)$ be two BFSs on a non-empty finite set S and $E \subseteq S \times S$ respectively. $td(c) = (td^+(c), td^-(c))$ is the total degree of a vertex $c \in S$ where $td^+(c) = \sum_{(c,d) \in E} m_2^+(c,d) + m_1^+(c), td^-(c) = \sum_{(c,d) \in E} m_2^-(c,d) + m_1^-(c)$. If all the vertices of a BFG are of total degree, then the graph is called totally regular BFG.*

An IVFS L on a set S is a mapping $\theta_L : S \to [0,1] \times [0,1]$, called the MF, i.e. $\theta_L(c) = [\theta_L^-(c), \theta_L^+(c)]$. The support of L is $\text{supp}(L) = \{c \in S | \theta_L^-(c) \neq 0\}$ and the core of L is $\text{core}(L) = \{c \in S | \theta_L^-(c) = 1\}$. The support length is $s(L) = |\text{supp}(L)|$ and the core length is $c(L) = |\text{core}(L)|$. The height of L is $h(L) = \max\{\theta_L(c) | c \in S\} = [h^-(L), h^+(L)] = [\max\{\theta_L^-(c)\}, \max\{\theta_L^+(c)\}], \forall c \in S$.

Let $F = \{L_1, L_2, \cdots, L_n\}$ be a finite family of interval-valued fuzzy subsets on a set S. The fuzzy intersection of two IVFSs (IVFSs) L_1 and L_2 is an IVFS defined by

$$\begin{aligned} L_1 \cap L_2 &= \Big\{ \Big(c, \Big[\min\{\theta_{L_1}^-(c), \theta_{L_2}^-(c)\}, \\ &\qquad \min\{\theta_{L_1}^+(c), \theta_{L_2}^+(c)\}\Big]\Big) : c \in S \Big\} \end{aligned}$$

The fuzzy union of two IVFSs L_1 and L_2 is a IVFS defined by

$$\begin{aligned} L_1 \cup L_2 &= \Big\{ \Big(c, \Big[\max\{\theta_{L_1}^-(c), \theta_{L_2}^-(c)\}, \\ &\qquad \max\{\theta_{L_1}^+(c), \theta_{L_2}^+(c)\}\Big]\Big) : c \in S \Big\} \end{aligned}$$

Fuzzy out-neighbourhood of a vertex $d \in S$ of an interval-valued fuzzy directed graph (IVFDG) $\overrightarrow{F} = (S, L, \overrightarrow{T})$ is the IVFS $\Delta^+(d) = (X_v^+, m_v^+)$ where $X_v^+ = \{c : \theta_T(\overrightarrow{d,c}) > 0\}$ and $m_v^+ : X_v^+ \to [0,1] \times [0,1]$ defined by $m_v^+ = \theta_T(\overrightarrow{d,c}) = [\theta_T^-(\overrightarrow{d,c}), \theta_T^+(\overrightarrow{d,c})]$

Here T is an interval-valued FR on a set S, is denoted by $\theta_T : S \times S \to [0,1] \times [0,1]$ such that

$$\theta_T^-(c,q) \leq \min\{\theta_L^-(c), \theta_L^-(q)\}$$
$$\theta_T^+(c,q) \leq \min\{\theta_L^+(c), \theta_L^+(q)\}$$

Consider $L = [\theta_L^-, \theta_L^+]$ is an IVFS on S and $T = [\theta_T^-, \theta_T^+]$ is an IVFS on $S \times S$ then the triplet $F = (S, L, T)$ is said to be an IVFG. An edge (c,d), $c, d \in S$ in an IVFG is said to be independent strong if $\theta_T^-(c,d) \geq \frac{1}{2}\min\{\theta_L^-(c), \theta_L^-(d)\}$. An interval-valued FDG (IVFDG) $\overrightarrow{F} = (S, L, \overrightarrow{T})$ is an IVFG where the FR $\overrightarrow{T}$ is antisymmetric.

An IVFG $Z = (S, L, T)$ is said to be *complete IVFG* if $\theta^-(c,d) = \min\{\delta^-(c), \delta^-(d)\}$ and $\theta^+(c,d) = \min\{\delta^+(c), \delta^+(d)\}$, $\forall c, d \in S$. An IVFG is said to be *bipartite* if there are two vertex sets S_1 and S_2 such that $S_1 \cup S_2 = S$ and $S_1 \cap S_2 = \phi$ where $\theta^+(d_1, d_2) = 0$ if $d_1, d_2 \in S_1$ or $d_1, d_2 \in S_2$ and $\theta^+(d_1, d_2) > 0$ if $d_1 \in S_1$ (or S_2) and $d_2 \in S_2$ (or S_1).

The *Cartesian product* [44] $Z_1 \times Z_2$ of two IVFGs $Z_1 = (S_1, L_1, T_1)$ and $Z_2 = (S_2, L_2, T_2)$ is defined as a pair $(S_1 \times S_2, L_1 \times L_2, T_1 \times T_2)$ such that

1. $\begin{cases} \theta_{L_1 \times L_2}^-(p_1, p_2) = \min\{\theta_{L_1}^-(p_1), \theta_{L_2}^-(p_2)\} \\ \theta_{L_1 \times L_2}^+(p_1, p_2) = \min\{\theta_{L_1}^+(p_1), \theta_{L_2}^+(p_2)\}, \end{cases}$

for all $p_1 \in S_1, p_2 \in S_2$,

2. $$\begin{cases} \theta^-_{T_1\times T_2}((c,p_2),(c,q_2)) = \min\{\theta^-_{L_1}(c), \theta^-_{T_2}(p_2,q_2)\} \\ \theta^+_{T_1\times T_2}((c,p_2),(c,q_2)) = \min\{\theta^+_{L_1}(c), \theta^+_{T_2}(p_2,q_2)\}, \end{cases}$$

for all $c \in S_1$ and $(p_2,q_2) \in E_2$,

3. $$\begin{cases} \theta^-_{T_1\times T_2}((p_1,q),(q_1,q)) = \min\{\theta^-_{T_1}(p_1,q_1), \theta^-_{L_2}(q)\} \\ \theta^+_{T_1\times T_2}((p_1,q),(q_1,q)) = \min\{\theta^+_{T_1}(p_1,q_1), \theta^+_{L_2}(q)\}, \end{cases}$$

for all $(p_1,q_1) \in E_1$ and $q \in S_2$

The composition $Z_1[Z_2] = (S_1 \circ S_2, L_1 \circ L_2, T_1 \circ T_2)$ of two IVFGs Z_1 and Z_2 of the graphs Z_1^* and Z_2^* is defined as follows:

1. $$\begin{cases} \theta^-_{L_1\circ L_2}(p_1,p_2) = \min\{\theta^-_{L_1}(p_1), \theta^-_{L_2}(p_2)\} \\ \theta^+_{L_1\circ L_2}(p_1,p_2) = \min\{\theta^+_{L_1}(p_1), \theta^+_{L_2}(p_2)\}, \end{cases}$$

for all $p_1 \in S_1, p_2 \in S_2$,

2. $$\begin{cases} \theta^-_{T_1\circ T_2}((c,p_2),(c,q_2)) = \min\{\theta^-_{L_1}(c), \theta^-_{T_2}(p_2,q_2)\} \\ \theta^+_{T_1\circ T_2}((c,p_2),(c,q_2)) = \min\{\theta^+_{L_1}(c), \theta^+_{T_2}(p_2,q_2)\}, \end{cases}$$

for all $c \in S_1$ and $(p_2,q_2) \in E_2$,

3. $$\begin{cases} \theta^-_{T_1\circ T_2}((p_1,q),(q_1,q)) = \min\{\theta^-_{T_1}(p_1,q_1), \theta^-_{L_2}(q)\} \\ \theta^+_{T_1\circ T_2}((p_1,q),(q_1,q)) = \min\{\theta^+_{T_1}(p_1,q_1), \theta^+_{L_2}(q)\}, \end{cases}$$

for all $(p_1,q_1) \in E_1$ and $q \in S_2$

4. $$\begin{cases} \theta^-_{T_1\circ T_2}((p_1,p_2),(q_1,q_2)) = \min\{\theta^-_{L_2}(p_2), \theta^-_{L_2}(q_2), \theta^-_{T_1}(p_1,q_1)\} \\ \theta^+_{T_1\circ T_2}((p_1,p_2),(q_1,q_2)) = \min\{\theta^+_{L_2}(p_2), \theta^+_{L_2}(q_2), \theta_{T_1}(p_1,q_1)\}, \end{cases}$$

otherwise.

The *union* $Z_1 \cup Z_2 = (S_1 \cup S_2, L_1 \cup L_2, T_1 \cup T_2)$ of two IVFGs Z_1 and Z_2 of the graphs Z_1^* and Z_2^* is defined as follows:

1. $$\begin{cases} \theta^-_{L_1\cup L_2}(c) = \theta^-_{L_1}(c) \text{ if } c \in S_1 \text{ and } c \notin S_2 \\ \theta^-_{L_1\cup L_2}(c) = \theta^-_{L_2}(c) \text{ if } c \in S_2 \text{ and } c \notin S_1 \\ \theta^-_{L_1\cup L_2}(c) = \max\{\theta^-_{L_1}(c), \theta^-_{L_2}(c)\} \text{ if } c \in S_1 \cap S_2. \end{cases}$$

2. $$\begin{cases} \theta^+_{L_1\cup L_2}(c) = \theta^+_{L_1}(c) \text{ if } c \in S_1 \text{ and } c \notin S_2 \\ \theta^+_{L_1\cup L_2}(c) = \theta^+_{L_2}(c) \text{ if } c \in S_2 \text{ and } c \notin S_1 \\ \theta^+_{L_1\cup L_2}(c) = \max\{\theta^+_{L_1}(c), \theta^+_{L_2}(c)\} \text{ if } c \in S_1 \cap S_2. \end{cases}$$

3. $$\begin{cases} \theta^-_{T_1\times T_2}(c,q) = \theta^-_{T_1}(c,q) \text{ if } (c,q) \in E_1 \text{ and } (c,q) \notin E_2 \\ \theta^-_{T_1\times T_2}(c,q) = \theta^-_{T_2}(c,q) \text{ if } (c,q) \in E_2 \text{ and } (c,q) \notin E_1 \\ \theta^-_{T_1\times T_2}(c,q) = \max\{\theta^-_{T_1}(c,q), \theta^-_{T_2}(c,q)\} \text{ if } (c,q) \in E_1 \cap E_2. \end{cases}$$

4. $$\begin{cases} \theta^+_{T_1\times T_2}(c,q) = \theta^+_{T_1}(c,q) \text{ if } (c,q) \in E_1 \text{ and } (c,q) \notin E_2 \\ \theta^+_{T_1\times T_2}(c,q) = \theta^+_{T_2}(c,q) \text{ if } (c,q) \in E_2 \text{ and } (c,q) \notin E_1 \\ \theta^+_{T_1\times T_2}(c,q) = \max\{\theta^+_{T_1}(c,q), \theta^+_{T_2}(c,q)\} \text{ if } (c,q) \in E_1 \cap E_2. \end{cases}$$

The *join* $Z_1 + Z_2 = (S_1 + S_2, L_1 + L_2, T_1 + T_2)$ of two IVFGs Z_1 and Z_2 of the graphs Z_1^* and Z_2^* is defined as follows:

1. $$\begin{cases} \theta^-_{L_1+L_2}(c) = (\theta^-_{L_1} \cup \theta^-_{L_2})(c) \\ \theta^+_{L_1+L_2}(c) = (\theta^+_{L_1} \cup \theta^+_{L_2})(c) \end{cases}$$

if $c \in S_1 \cup S_2$,

2. $\begin{cases} \theta^-_{T_1+T_2}(c,q) = (\theta^-_{T_1} \cup \theta^-_{T_2})(c,q) \\ \theta^+_{T_1+T_2}(c,q) = (\theta^+_{T_1} \cup \theta^+_{T_2})(c,q) \end{cases}$

if $(c,q) \in E_1 \cap E_2$,

3. $\begin{cases} \theta^-_{T_1+T_2}(c,q) = \min\{\theta^-_{L_1}(c), \theta^-_{L_2}(q)\} \\ \theta^+_{T_1+T_2}(c,q) = \min\{\theta^+_{L_1}(c), \theta^+_{L_2}(q)\} \end{cases}$

for all $(c,q) \in E'$, where E' is the set of all edges joining the nodes of S_1 and S_2.

A *homomorphism* [48] between two FGs $F_1 = (S, \delta_1, \theta_1)$ and $F_2 = (S, \delta_2, \theta_2)$ is a map $f : S_1 \to S_2$ which satisfies $\delta_1(c) \leq \delta_2(f(c))$ for all $c \in S_1$ and $\theta_1(c,q) \leq \theta_2(f(c), f(q))$ for all $c, q \in S_1$ where S_1 is the set of vertices of F_1 and S_2 is that of F_2. A FG F_1 is said to be *homomorphic* to F_2 if there exist a homomorphism between F_1 and F_2.

An *isomorphism* [48] between two FGs $F_1 = (S, \delta_1, \theta_1)$ and $F_2 = (S, \delta_2, \theta_2)$ is a bijective homomorphism $f : S_1 \to S_2$ which satisfies $\delta_1(c) = \delta_2(f(c))$ for all $c \in S_1$ and $\theta_1(c,q) \leq \theta_2(f(c), f(q))$ for all $c, q \in S_1$ where S_1 is the set of vertices of F_1 and S_2 is that of F_2. A FG F_1 is said to be *isomorphic* to F_2 if there exist an isomorphism between F_1 and F_2.

3. Interval-Valued FCG

In this section, IVFCG is defined and investigated some properties.

Definition 12 (Interval-valued FCG). *Interval-valued FCG (IVFCG) of an IVFDG $\overrightarrow{Z} = (S, L, \overrightarrow{T})$ is an undirected graph $IVFC(\overrightarrow{Z}) = (S, L, T')$ whose vertex membership value is same as that of IVFDG and membership value of the edge (c,d) is an interval number $\theta_{T'}(c,d) = [\theta^-_{T'}(c,d), \theta^+_{T'}(c,d)]$ where,*

$$\theta^-_{T'}(c,d) = (\theta^-_L(c) \wedge \theta^-_L(d))\, h^-(\Delta^+(c) \cap \Delta^+(d))$$
$$\theta^+_{T'}(c,d) = (\theta^+_L(c) \wedge \theta^+_L(d))\, h^+(\Delta^+(c) \cap \Delta^+(d))$$

for all $c, d \in S$.

Example 1. *Let us consider an IVFDG shown in Figure 2. All the membership values of vertices and edges are arbitrarily taken and depicted in Figure 2.*

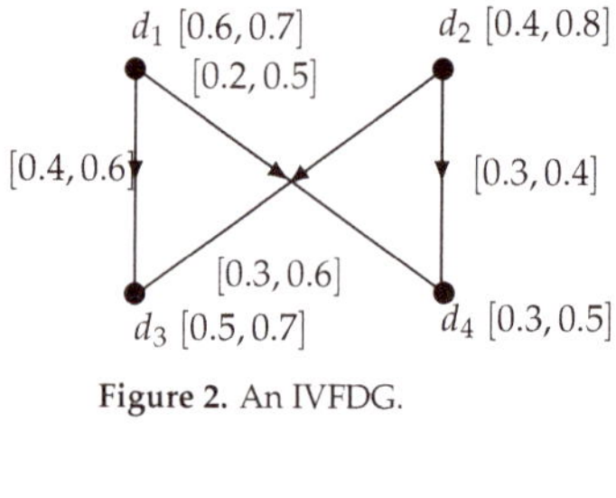

Figure 2. An IVFDG.

All the obvious computations are done as follows:

$$\Delta^+(d_1) = \{(d_3, [0.4, 0.6]),\ (d_4, [0.2, 0.5])\},$$
$$\Delta^+(d_2) = \{(d_3, [0.3, 0.6]),\ (d_4, [0.3, 0.4])\}$$
$$\Delta^+(d_3) = \phi,\ \Delta^+(d_4) = \phi,$$
$$\therefore \Delta^+(d_1) \cap \Delta^+(d_2) = \{(d_3, [0.4, 0.6]),\ (d_4, [0.2, 0.5])\}.$$

Then, $h^-\ (\Delta^+(d_1) \cap \Delta^+(d_2))$
$= 0.3$ *and* $h^+\ (\Delta^+(d_1) \cap \Delta^+(d_2)) = 0.6$.

Hence, the IVFCG of the IVFDG is obtained and shown in Figure 3.

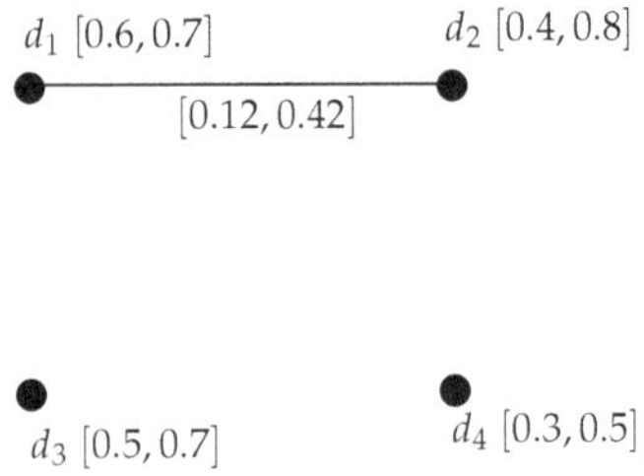

Figure 3. IVFCG of the IVFDG shown in Figure 2.

Theorem 1. *Let* $\overrightarrow{Z} = (S, L, \overrightarrow{T})$ *be an IVFDG. An edge* (c, d) *of* $IVFC(\overrightarrow{Z})$ *is independent strong if and only if* $h^-(\Delta^+(c) \cap \Delta^+(d)) > 0.5$ *provided that* $\Delta^+(c) \cap \Delta^+(d)$ *has one and only one element.*

Proof. Since $\Delta^+(c) \cap \Delta^+(d)$ has one and only one element let, $\Delta^+(c) \cap \Delta^+(d) = \{(w, [m^-, m^+])\}$, where $[m^-, m^+]$ is interval-valued fuzzy membership value of the vertex w. Then $h^-(\Delta^+(c) \cap \Delta^+(d)) = m^-$. So, $\theta^-_{T'}(c, d) = (\theta^-_L(c) \wedge \theta^-_L(d))h^-(\Delta^+(c) \cap \Delta^+(d)) = m^- \times (\theta^-_L(c) \wedge \theta^-_L(d)) > \frac{1}{2}(\theta^-_L(c) \wedge \theta^-_L(d))$ if and only if $m^- = h^-(\Delta^+(c) \cap \Delta^+(d)) > 0.5$. Hence the theorem follows. □

It is evident that, if all the edges of an IVFDG are independent strong then, the corresponding IVFCG may or may not have an independent strong edge. For this, an example is shown in Figure 4.

(**a**) IVFDG whose all edges are strong

(**b**) IVFCG whose all edges are not strong

Figure 4. An example that an IVFCG have no independent strong edge although all the edges are independent strong in IVFDG.

But in the next theorem, a result is obtained for the case when all the edges of a IVFDG are independent strong.

Theorem 2. *Let all the edges of an IVFDG $\overrightarrow{Z} = (S, L, \overrightarrow{T})$ be independent strong. Then $\frac{\theta^-_{T'}(c,d)}{(\theta^-_L(c) \wedge \theta^-_L(d))^2} > 0.5$ for all $c, d \in S$ in $IVFC(\overrightarrow{Z})$, provided $\theta^-_L(c) \wedge \theta^-_L(d) \neq 0$.*

Proof. Since all the edges of $\overrightarrow{Z} = (S, L, \overrightarrow{T})$ is independent strong then $\theta^-_T(\overrightarrow{c,d}) > \frac{1}{2}(\theta^-_L(c) \wedge \theta^-_L(d))$ i.e., $\frac{\theta^-_T(\overrightarrow{c,d})}{\theta^-_L(c) \wedge \theta^-_L(d)} > 0.5$. For all $c, d \in S$ such that $\theta^-_{T'}(c,d) \neq 0$ let $\Delta^+(c) \cap \Delta^+(d)$ has at least one element. Let $\Delta^+(c) \cap \Delta^+(d) = \{(w_1, [m^-_1, m^+_1]), (w_2, [m^-_2, m^+_2]), \ldots, (w_k, [m^-_k, m^+_k])\}$, where $[m^-_i, m^+_i]$ are membership values of w_i, $l = 1, 2, \ldots, k$. This shows that $[m^-_i, m^+_i] = [\min\{\theta^-_T(\overrightarrow{c,w_i}), \theta^-_T(\overrightarrow{d,w_i})\}, \min\{\theta^+_T(\overrightarrow{c,w_i}), \theta^+_T(\overrightarrow{d,w_i})\}]$. Therefore, $h^-(\Delta^+(c) \cap \Delta^+(d)) = \max\{m^-_1, m^-_2, \ldots, m^-_k\} = m^-_{\max}$ (say). Obviously, $m^-_{\max} > \theta^-_T(\overrightarrow{c,d})$ shows that

$$\frac{m^-_{\max}}{\theta^-_L(c) \wedge \theta^-_L(d)} > \frac{\theta^-_T(\overrightarrow{c,d})}{\theta^-_L(c) \wedge \theta^-_L(d)} > 0.5.$$

Therefore,

$$\theta^-_{T'}(c,d) = (\theta^-_L(c) \wedge \theta^-_L(d)) h^-(\Delta^+(c) \cap \Delta^+(d))$$

$$\text{or,} \quad \frac{\theta^-_{T'}(c,d)}{\theta^-_L(c) \wedge \theta^-_L(d)} = m^-_{\max}$$

$$\text{or,} \quad \frac{\theta^-_{T'}(c,d)}{(\theta^-_L(c) \wedge \theta^-_L(d))^2} = \frac{m^-_{\max}}{\theta^-_L(c) \wedge \theta^-_L(d)} > 0.5.$$

□

Definition 13. *An IVFG $F_1 = (S_1, L_1, T_1)$ is said to be homomorphic to an IVFG $F_2 = (S_2, L_2, T_2)$ if there exist a homomorphism $f : S_1 \to S_2$ such that $\theta^-_{L_1}(c) \leq \theta^-_{L_2}(f(c))$, $\theta^+_{L_1} \leq \theta^+_{L_2}(f(c))$ for all $c \in S_1$ and $\theta^-_{T_1}(c,d) \leq \theta^-_{T_2}(f(c), f(d))$, $\theta^+_{T_1}(c,d) \leq \theta^+_{T_2}(f(c), f(d))$ for all $c, d \in S_1$.*

If this homomorphism is bijective then the IVFG is said to be isomorphic.

Definition 14. *An IVFG $Z_1 = (S_1, L_1, T_1)$ is said to be isomorphic to an IVFG $Z_2 = (S_2, L_2, T_2)$ if there exist a bijective homomorphism $f : S_1 \to S_2$ such that $\theta^-_{L_1}(c) = \theta^-_{L_2}(f(c))$, $\theta^+_{L_1} = \theta^+_{L_2}(f(c))$ for all $c \in S_1$ and $\theta^-_{T_1}(c,d) = \theta^-_{T_2}(f(c), f(d))$, $\theta^+_{T_1}(c,d) = \theta^+_{T_2}(f(c), f(d))$ for all $c, d \in S_1$.*

Next theorem shows that, if an IVFDG is complete then its underlying competition graph and undirected graph are homomorphic to each other.

Theorem 3. *An IVFCG of a complete IVFDG $\overrightarrow{Z} = (S, L, \overrightarrow{T})$ is homomorphic to underlying undirected graph of $\overrightarrow{Z}$.*

Proof. An IVFCG has same vertex set as that of IVFDG $\overrightarrow{Z}$ with their respective fuzzy membership values. So, there exist at least one homomorphism $f : S(IVFC(\overrightarrow{Z})) \to S(\overrightarrow{Z})$ such that, $\theta^-_L(c) = \theta^-_L(f(c))$, $\theta^+_L = \theta^+_L(f(c))$ for all $c \in S$. Since $\overrightarrow{Z}$ is complete, $\theta^-_T(f(c), f(d)) = \theta^-_L(f(c)) \wedge \theta^-_L(f(d))$ and $\theta^+_T(f(c), f(d)) = \theta^+_L(f(c)) \wedge \theta^+_L(f(d))$. As $f^-(\Delta^+(f(c)) \cap \Delta^+(f(d))) \leq 1$, $\theta^-_{T'}(c,d) = (\theta^-_L(f(c)) \wedge \theta^-_L(f(d))) f^-(\Delta^+(f(c)) \cap$

$\Delta^+(f(d))) \leq \theta_L^-(f(c)) \wedge \theta_L^-(f(d)) = \theta_T^-(f(c), f(d))$. Similarly, $\theta_{T'}^+(c,d) \leq \theta_T^+(f(c), f(d))$. Hence, the result follows. □

Remark 1. *Although an IVFCG is homomorphic to an underlying undirected graph of a complete IVFDG, there does not exist any isomorphism between them. As, for every triangular orientation of three vertices c, d, w (a complete graph of three or more vertices must consists of it) there exists at most one edge say, c, d between them. Hence, $\theta_{T'}(c,w) = [0,0] \neq \theta_T(f(c), f(w))$.*

Interval-Valued FKCG and m-Step Competition Graphs

Here, we introduce two particular types of competition graphs called IVF k-competition graphs and m-step competiton graphs.

Definition 15. *Let k be a non-negative integer. The IVFKCG $IVFC_k(\overrightarrow{Z})$ of an IVFDG $\overrightarrow{Z} = (S, L, \overrightarrow{T})$ is an undirected FG $IVFC_k(\overrightarrow{Z}) = (S, L, T')$ which has the same fuzzy vertex set as that of $\overrightarrow{Z}$ and has a fuzzy edge between two vertices $c, d \in S$ in $IVFC_k(\overrightarrow{Z})$ if and only if $s(\Delta^+(c) \cap \Delta^+(d)) > k$. The edge membership value between c and d in $IVFC_k(\overrightarrow{Z})$ is $\theta_{T'}^-(c,d) = \frac{k'-k}{k'}(\theta_L^-(c) \wedge \theta_L^-(d))h^-(\Delta^+(c) \cap \Delta^+(d))$ and $\theta_{T'}^+(c,d) = \frac{k'-k}{k'}(\theta_L^+(c) \wedge \theta_L^+(d))h^+(\Delta^+(c) \cap \Delta^+(d))$ where, $k' = s(\Delta^+(c) \cap \Delta^+(d))$.*

Theorem 4. *Let $\overrightarrow{Z} = (S, L, \overrightarrow{T})$ be an IVFDG. If $s(\Delta^+(c) \cap \Delta^+(d)) = 2k$, then the edge (c,d) is independent strong in $IVFC_k(\overrightarrow{Z})$.*

Proof. By the Definition of IVFKCG the edge membership value of an edge (c,d) in $IVFC_k(\overrightarrow{Z})$ is $\theta_T^-(c,d) = \frac{k'-k}{k'}(\theta_L^-(c) \wedge \theta_L^-(d))h^-(\Delta^+(c) \cap \Delta^+(d))$ and $\theta_T^+(c,d) = \frac{k'-k}{k'}(\theta_L^+(c) \wedge \theta_L^+(d))h^+(\Delta^+(c) \cap \Delta^+(d))$ where, $k' = s(\Delta^+(c) \cap \Delta^+(d))$. Then $\theta_T^-(c,d) = \frac{k'-k}{k'}(\theta_L^-(c) \wedge \theta_L^-(d))h^-(\Delta^+(c) \cap \Delta^+(d)) > \frac{k'-k}{k'}(\theta_L^-(c) \wedge \theta_L^-(d)) = 0.5(\theta_L^-(c) \wedge \theta_L^-(d))$ as $h^-(\Delta^+(c) \cap \Delta^+(d)) > 0$. Therefore, $\frac{\theta_T^-(c,d)}{\theta_L^-(c) \wedge \theta_L^-(d)} > 0.5$. Hence, (c,d) is an independent strong edge. □

Definition 16. *The IVFMCG of an IVFDG $\overrightarrow{Z} = (S, L, \overrightarrow{T})$ is denoted by $IVFC_m(\overrightarrow{Z})$ and is defined by $IVFC_m(\overrightarrow{Z}) = (S, L, T')$ where the membership value of the edge (c,d) is $\theta_{T'}(c,d) = [\theta_{T'}^-(c,d), \theta_{T'}^+(c,d)]$, where $\theta_{T'}^-(c,d) = (\theta_L^-(c) \wedge \theta_L^-(d))h^-(\Delta_m^+(c) \cap \Delta_m^+(d))$ and $\theta_{T'}^+(c,d) = (\theta_L^+(c) \wedge \theta_L^+(d))h^+(\Delta_m^+(c) \cap \Delta_m^+(d))$.*

Example 2. *An example of interval-valued fuzzy 2-step CG of the IVFDG of Figure 5a is shown in Figure 5b.*

In Figure 5a, the vertices c and q have 2-step common neighbourhood c and therefore, the vertices c and q has an edge in interval-valued fuzzy 2-step CG as shown in Figure 5b.

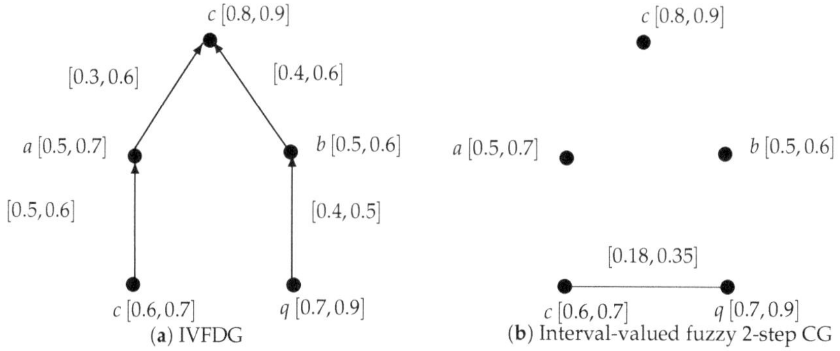

Figure 5. An example of interval-valued fuzzy 2-step CG.

Definition 17. *Let $\overrightarrow{Z} = (S, L, \overrightarrow{T})$ be an IVFDG. Let d be a common vertex of m-step fuzzy out-neighbourhoods of vertices $c_1, c_2, \cdots, c_n$, n being any positive integer. The m-step vertex $d \in S$ is said to be independent strong vertex if $\theta_m^-(\overrightarrow{c_i, d}) > 0.5$ for all $l = 1, 2, \ldots, n$. The strength of the vertex d is denoted by $s_m(d)$ and is defined by $s_m(d) = [s_m^-(d), s_m^+(d)]$ where $s_m^- = \frac{\sum_{l=1}^n \theta_m^-(\overrightarrow{c_i, d})}{n}$ and $s_m^+ = \frac{\sum_{l=1}^n \theta_m^+(\overrightarrow{c_i, d})}{n}$.*

Theorem 5. *If a vertex (prey) d of $\overrightarrow{Z}$ is independent strong, then $s_m^-(d) > 0.5$, but the converse is not necessarily true.*

Proof. Let $\overrightarrow{Z} = (S, L, \overrightarrow{T})$ be an IVFDG. Let d be a common vertex of m-step fuzzy out-neighbourhoods of the vertices $c_1, c_2, \cdots, c_n$, n being any positive integer. As the vertex d is independent strong then $\theta_m^-(\overrightarrow{c_i, d}) > 0.5$ for all $l = 1, 2, \ldots, n$. Therefore, $s_m^-(d) = \frac{\theta_m^-(\overrightarrow{c_1, d}) + \theta_m^-(\overrightarrow{c_2, d}) + \cdots + \theta_m^-(\overrightarrow{c_n, d})}{n} > \frac{0.5 + 0.5 + \cdots + 0.5}{n} = 0.5$.

Conversely let, $s_m^-(d) > 0.5$. Now, $s_m^-(d)$ is the average of n real numbers which is greater than 0.5 does not always mean that each n number is greater than 0.5. □

Theorem 6. *If all vertices (preys) of $\overrightarrow{Z}$ are independent strong, then all the edges of $IVFC_m(\overrightarrow{Z})$ are independent strong.*

Proof. Let all the vertices of $\overrightarrow{Z} = (S, L, \overrightarrow{T})$ are independent strong. Let $IVFC_m(\overrightarrow{Z}) = (S, L, T')$ where $\theta_{T'}(c, d) = [\theta_{T'}^-(c, d), \theta_{T'}^+(c, d)] = [(\theta_L^-(c) \wedge \theta_L^-(d))h^-(\Delta_m^+(c) \cap \Delta_m^+(d)), (\theta_L^+(c) \wedge \theta_L^+(d))h^+(\Delta_m^+(c) \cap \Delta_m^+(d))]$ be an IVFMCG of IVFDG $\overrightarrow{Z}$. If $\Delta_m^+(c) \cap \Delta_m^+(d)$ be empty set then there does not exists any edge between c and d in $IVFC_m(\overrightarrow{Z})$. If $\Delta_m^+(c) \cap \Delta_m^+(d)$ be non-empty then obviously $h^-(\Delta_m^+(c) \cap \Delta_m^+(d)) > 0.5$ as all the edges of $\overrightarrow{Z}$ are independent strong and hence $(\theta_L^-(c) \wedge \theta_L^-(d))h^-(\Delta_m^+(c) \cap \Delta_m^+(d)) > 0.5$ which implies that all the edges of $IVFC_m(\overrightarrow{Z})$ are independent strong. □

Theorem 7. *The $IVFC_m(\overrightarrow{Z})$ of $\overrightarrow{Z} = (S, L, \overrightarrow{T})$ has no edge if $m > |S|$.*

Proof. If $m > |S|$, the number of vertices in $\overrightarrow{Z}$ then it is obvious that there can not exist any fuzzy directed path of length m between any two vertices c, d of S. Then $\Delta_m^+(c) \cap \Delta_m^+(d)$ is a null set. Hence membership value of each pair of vertices is zero which means there can not have any edge in $IVFC_m(\overrightarrow{Z})$. □

4. An Application of IVFCG in Manufacturing Industries

Every manufacturing industry has several production company and markets to sell the product. Any production company produces their products as per market demands. They are also liable to transport the products to the market so that the end user can use their product within a reasonable time. They wish to deliver with minimum cost as much as they can. Market has the time-bound factor to get the production from company within a reasonable cost. Market has various opportunities to choose the company as well as company can choose market for their sake. So, there is fair competition between companies. The problem is to find out which companies are in competition and the strengths of their competition to achieve markets that they serve, considering all the cases of production, demands and the time that they can spare. This problem can be modeled as an IVFCG by considering the following correspondences:

- Companies and markets are treated as vertices.
- The membership values of vertices that are taken as companies is a sub-interval of $[0,1]$. The significance of this interval number is that every company has a minimum and maximum capability to produce the product. We have assigned a grade to each power of capabilities within the min-max range. So, the interval becomes a fuzzy interval number.
- Similarly, assigning grade for demands that the market has, each vertex associated to a fuzzy interval number.
- The company and market are connected, that is, they have an edge if they both have the same time tenure to transport or take the product. A grade is assigned to each time within the tenure. This membership grade is also a fuzzy interval number.

Assuming the company and market have higher membership values than that of their shared time, i.e., membership value of each edge is less than the minimum of membership values of all the vertices, the problem is well-defined for an IVFCG model.

To find the strength of competitions among companies in manufacturing industries, the calculation flow is shown as a flowchart in the Figure 6. To explain the problem, in particular, let us consider the following example.

Three companies namely, C_1, C_2 and C_3 produces certain product. Each company has a capability to produce 20–70%, 87–98% and 90–100% of demands respectively. Each of these shadowiness in capability of production can be corresponded to interval-valued fuzzy numbers as $[0.2, 0.7]$, $[0.87, 0.98]$ and $[0.9, 1]$ respectively, in fuzzy sense. There are two markets M_1 and M_2. They have also 90–100% and 85–95% demands in market respectively. Amount of demands are also shadowy. These can be corresponded to interval-valued fuzzy numbers as $[0.9, 1]$ and $[0.85, 0.95]$ respectively. Similarly, the interval-valued fuzzy numbers for transportation time corresponding to the edges (C_1, M_1), (C_1, M_2), (C_2, M_1), (C_2, M_2), (C_3, M_1) and (C_3, M_2) can be taken as $[0.1, 0.4]$, $[0.2, 0.6]$, $[0.85, 0.9]$, $[0.75, 0.95]$, $[0.8, 0.95]$ and $[0.8, 0.9]$ respectively. The relationship is shown in Figure 7. Note that this is an interval-valued fuzzy complete bipartite graphs.

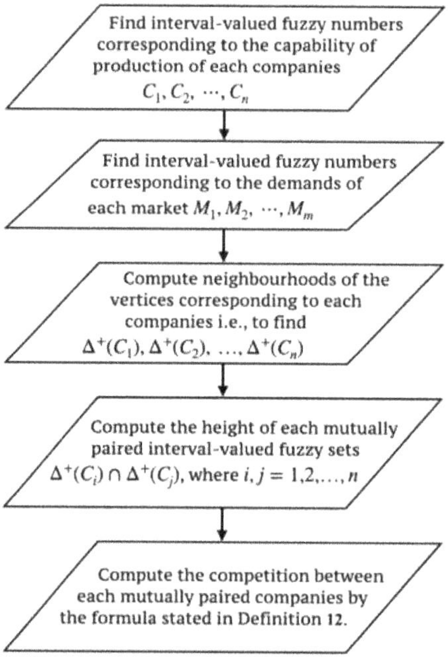

Figure 6. Flowchart of the work flow to compute the strength of competitions among companies in manufacturing industries.

Now,
$$\Delta^+(C_1) = \{M_1[0.1, 0.4], M_2[0.2, 0.6]\}$$
$$\Delta^+(C_2) = \{M_1[0.85, 0.9], M_2[0.75, 0.95]\}$$
$$\Delta^+(C_3) = \{M_1[0.8, 0.95], [0.8, 0.9]\}$$

Then,
$$h(\Delta^+(C_1) \cap \Delta^+(C_2))$$
$$= h(\{M_1[0.1, 0.4], M_2[0.2, 0.6]\}) = [0.2, 0.6]$$
$$h(\Delta^+(C_1) \cap \Delta^+(C_3))$$
$$= h(\{M_1[0.1, 0.4], M_2[0.2, 0.6]\}) = [0.2, 0.6]$$
$$h(\Delta^+(C_2) \cap \Delta^+(C_3))$$
$$= h(\{M_1[0.8, 0.9], M_2[0.75, 0.9]\}) = [0.8, 0.9]$$

Therefore,

$$\theta_{T'}(C_1, C_2) = [0.2 \times 0.2, 0.7 \times 0.6]$$
$$= [0.04, 0.42]$$
$$\theta_{T'}(C_1, C_3) = [0.2 \times 0.2, 0.7 \times 0.6]$$
$$= [0.04, 0.42]$$
$$\theta_{T'}(C_2, C_3) = [0.87 \times 0.8, 0.98 \times 0.9]$$
$$= [0.696, 0.882] \simeq [0.70, 0.88]$$

The corresponding IVFCG of Figure 7 is shown in Figure 8. The membership value (degree) of competition among the companies is shown in Table 2.

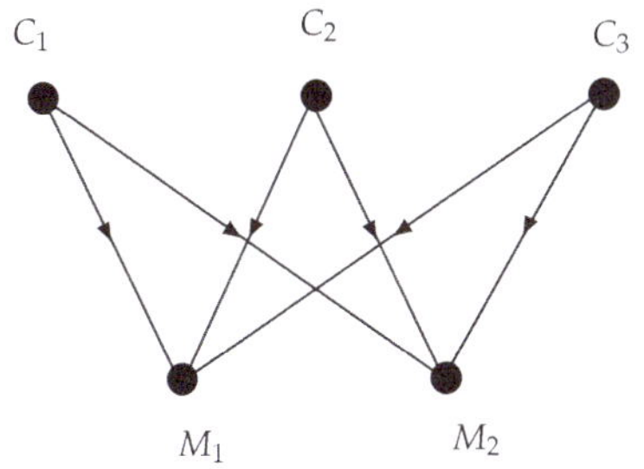

Figure 7. The relationship between companies and markets.

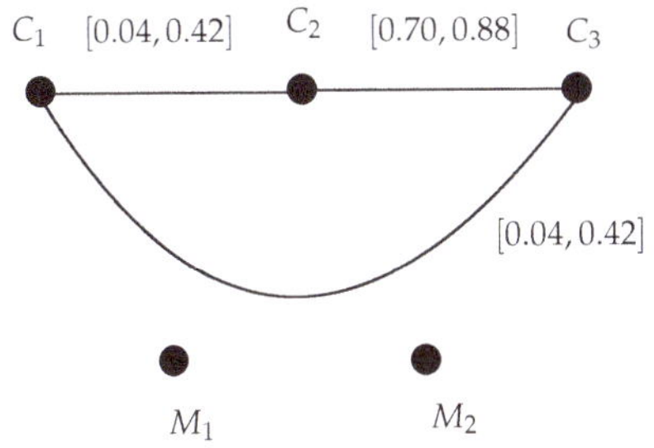

Figure 8. IVFCG of Figure 7.

Table 2. Degree of Competition among the Companies.

Companies	Degree of Competition	Competition in %
C_1, C_2	[0.04, 0.42]	[4, 42]
C_2, C_3	[0.70, 0.88]	[70, 88]
C_3, C_1	[0.04, 0.42]	[4, 42]

A complete analysis of the result is shown in the Table 3.

Table 3. Analysis of the result obtained in the problem of manufacturing industries.

Description of the Result	Result Obtained	Analysis of the Result
Highest degree of competition among companies	$[0.70, 0.88]$	This result shows that the companies have at least 70% and at most 88% competitions in the market (Computations made using the formula stated in Definition 1)
Independent strength of competition between the companies C_2 and C_3	$\left[\frac{0.70}{\min\{0.87,0.9\}}, \frac{0.88}{\min\{0.98,1\}}\right] = [0.80, 0.90] > [0.5, 0.5]$	The height of interval-valued fuzzy set $\Delta^+(C_2) \cap \Delta^+(C_3)$ is $[0.8,0.9]$ which is greater than $[0.5, 0.5]$. So there is a strong competition between the two companies C_2 and C_3 (Refer Theorem 2)

The diagrammatic representation is shown in Figure 9.

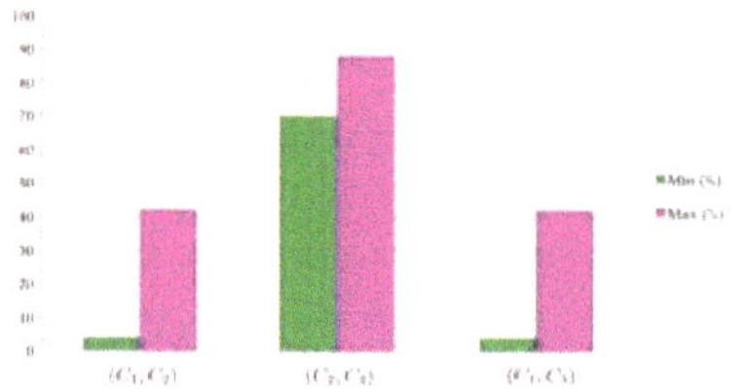

Figure 9. Competition among three companies.

5. Implications

In the case of any kind of competitive interconnected system, each competitor verifies the ability and capability of his opponent. The observations we present are useful in determining the capabilities and capabilities of all competitors present in such systems. The strength and intensity of competition between any two competitors can be determined within an interval. As a result, although the strength of competition is correct, it is within an interval, so the scope of application of the method is wide. Theoretically, it has been shown the cases when and where the strength of a competitor becomes higher.

6. Conclusions

There are many works have been done on fuzzy competition graphs and its extensions. After the work of FCG, we feel the importance of IVFCG as many real problems like time-bound network-based technology, neurology, ecology, market demand, etc. demands the uses of such type of modelling introduced in this paper. There is a great deal to handle with homomorphism and isomorphism of IVFCG products that have done by proving them in this paper. The proposed method of IVFCG is much more useful for the analysis of any network related to competition. This method is very useful for solving real-world problems. Here interval-valued fuzzy set is used instead of a general fuzzy set. One of the biggest problems in the world of this civilization is the constant competition of the manufacturing industries. Here, the competitive strength of the manufacturing industries determined and described the position of a company in the market. But, the problem of manufacturing industries is even bigger. There is a need to solve various problems starting from economic problems to business communication, business relations etc. However, many real problems can occur where a relationship is bipolar, for example, let's say two companies produce two types of products in a market where there is no competition but great cooperation. For example, if one company produces petrol-powered cars and the other company produces petrol, there should be no competition between them. In all these cases the problem can be solved by using bipolar fuzzy set in the case of Competition graphs. There are also opportunities to solve various real problems using intuitionistic fuzzy sets.

Author Contributions: Conceptualization, M.P.; Methodology, T.P.; Validation, G.M. and A.M.A.; Writing—original draft, G.M.; Writing—review and editing, A.M.A. All authors have read and agreed to the published version of the manuscript.

Funding: This research received no external funding.

Acknowledgments: The authors are grateful to the Editor-in-Chief and Honorable reviewers of the journal "Mathematics" for their suggestions to improve the quality and presentation of the paper.

Conflicts of Interest: The authors declare no conflict of interest.

References

1. Cohen, J.E. *Interval Graphs and Food Webs: A Finding and a Problem*; RAND Corporation: Santa Monica, CA, USA, 1968.
2. Rosenfeld, A. *Fuzzy Graphs*; Academic Press: New York, NY, USA, 1975; pp. 77–95.
3. Zadeh, L.A. Similarity relations and fuzzy orderings. *Inf. Sci.* **1971**, *3*, 177–200. [CrossRef]
4. Kim, S.R.; McKee, T.A.; McMorris, F.; Roberts, F.S. p-Competition graphs. *Linear Algebra Appl.* **1995**, *217*, 167–178. [CrossRef]
5. Brigham, R.C.; McMorris, F.R.; Vitray, R.P. Tolerance competition graphs. *Linear Algebra Appl.* **1995**, *217*, 41–52. [CrossRef]
6. Cho, H.H.; Kim, S.R.; Nam, Y. The m-step competition graph of a digraph. *Discret. Appl. Math.* **2000**, *105*, 115–127. [CrossRef]
7. Sonnatag, M.; Teichert, H.M. Competition hypergraphs. *Discret. Appl. Math.* **2004**, *143*, 324–329. [CrossRef]
8. Samanta, S.; Pal, M. Fuzzy k-competition graphs and p-competition fuzzy graphs. *Fuzzy Inf. Eng.* **2013**, *5*, 191–204. [CrossRef]
9. Pramanik, T.; Samanta, S.; Sarkar, B.; Pal, M. Fuzzy ϕ-tolerance competition graphs. *Soft Comput.* **2017**, *21*, 3723–3734. [CrossRef]
10. Pramanik, T.; Samanta, S.; Pal, M.; Mondal, S.; Sarkar, B. Interval-valued fuzzy ϕ-tolerance competition graphs. *SpringerPlus* **2016**, *5*, 1981. [CrossRef]
11. Bhutani, K.R.; Battou, A. On M-strong fuzzy graphs. *Inf. Sci.* **2003**, *155*, 103–109. [CrossRef]
12. Bhutani, K.R.; Rosenfeld, A. Strong arcs in fuzzy graphs. *Inf. Sci.* **2003**, *152*, 319–322. [CrossRef]
13. Craine, W.L. Characterization of fuzzy interval graphs. *Fuzzy Sets Syst.* **1994**, *68*, 181–193. [CrossRef]
14. Ghosh, P.; Kundu, K.; Sarkar, D. Fuzzy graph representation of a fuzzy concept lattice. *Fuzzy Sets Syst.* **2010**, *161*, 1669–1675. [CrossRef]
15. Isaak, G.; Kim, S.R.; McKee, T.A.; McMorris, F.R.; Roberts, F.S. 2-competition graphs. *SIAM J. Disc. Math.* **1992**, *5*, 524–538. [CrossRef]
16. Nayeem, S.M.A.; Pal, M. Shortest path problem on a network with imprecise edge weight. *Fuzzy Optim. Decis. Mak.* **2005**, *4*, 293–312. [CrossRef]
17. Cable, C.; Jones, K.F.; Lundgren, J.R.; Seager, S. Niche graphs. *Discret. Appl. Math.* **1989**, *23*, 231–241. [CrossRef]
18. Samanta, S.; Pal, M. Fuzzy tolerance graphs. *Int. J. Latest Trends Math.* **2011**, *1*, 57–67.
19. Samanta, S.; Pal, M.; Pal, A. New concepts of fuzzy planar graphs. *Int. J. Adv. Res. Artif. Intell.* **2014**, *3*, 52–59. [CrossRef]
20. Samanta, S.; Pal, M. Fuzzy planar graph. *IEEE Trans. Fuzzy Syst.* **2015**, *23*, 1936–1942. [CrossRef]
21. Pramanik, T.; Samanta, S.; Pal, M. Interval-valued fuzzy planar graphs. *Int. J. Mach. Learn. Cybern.* **2016**, *7*, 653–664. [CrossRef]

22. Rashmanlou, H.; Pal, M. Some properties of highly irregular interval-valued fuzzy graphs. *World Appl. Sci. J.* **2013**, *27*, 1756–1773.
23. Samanta, S.; Pal, M. Irregular bipolar fuzzy graphs. *Int. J. Appl. Fuzzy Sets* **2012**, *2*, 91–102.
24. Samanta, S.; Pramanik, T.; Pal, M. Fuzzy colouring of fuzzy graphs. *Afr. Mat.* **2016**, *27*, 37–50. [CrossRef]
25. Rashmanlou, H.; Samanta, S.; Pal, M.; Borzooei, R.A. A study on bipolar fuzzy graphs. *J. Intell. Fuzzy Syst.* **2015**, *28*, 571–580. [CrossRef]
26. Rashmanlou, H.; Pal, M. Isometry on interval-valued fuzzy graphs. *Int. J. Fuzzy Math. Arch.* **2014**, *3*, 28–35.
27. Rashmanlou, H.; Pal, M. Balanced interval-valued fuzzy graphs. *J. Phys. Sci.* **2013**, *17*, 43–57.
28. Rashmanlou, H.; Pal, M. Antipodal interval-valued fuzzy graphs. *Int. J. Appl. Fuzzy Sets Artif. Intell.* **2013**, *3*, 107–130.
29. Samanta, S.; Pal, M. Fuzzy threshold graphs. *CiiT Int. J. Fuzzy Syst.* **2011**, *3*, 1–9.
30. Samanta, S.; Pal, M. Some more results on bipolar fuzzy sets and bipolar fuzzy intersection graphs. *J. Fuzzy Math.* **2014**, *22*, 253–262.
31. Samanta, S.; Akram, M.; Pal, M. *m*-step fuzzy competition graphs. *J. Appl. Math. Comput.* **2015**, *47*, 461–472. [CrossRef]
32. Voskoglou, M. Applications of Fuzzy Logic to Systems' Modelling. *Int. J. Fuzzy Syst. Appl.* **2015**, *3*, 1–15. [CrossRef]
33. Voskoglou, M. 3.7—A Stochastic Model for the Modelling Process. In *Mathematical Modelling*; Haines, C., Galbraith, P., Blum, W., Khan, S., Eds.; Woodhead Publishing: Cambridge, UK, 2007; pp. 149–157. [CrossRef]
34. Voskoglou, M.G. A Fuzzy Model For Analogical Problem Solving. *Int. J. Fuzzy Log. Syst.* **2012**, *2*, 10–11. [CrossRef]
35. Gil, R.A.; Johanyák, Z.C.; Kovács, T. Surrogate model based optimization of traffic lights cycles and green period ratios using microscopic simulation and fuzzy rule interpolation. *Int. J. Artif. Intell.* **2018**, *16*, 20–40.
36. Hedrea, E.L.; Precup, R.E.; Bojan-Dragos, C.A. Results on Tensor Product-based Model Transformation of Magnetic Levitation Systems. *Acta Polytech. Hung.* **2019**, *16*.
37. Deveci, M.; Öner, S.C.; Canıtez, F.; Öner, M. Evaluation of service quality in public bus transportation using interval-valued intuitionistic fuzzy QFD methodology. *Res. Transp. Bus. Manag.* **2019**, 33, 100387. [CrossRef]
38. Deveci, M.; Cali, U.; Kucuksari, S.; Erdogan, N. Interval type-2 fuzzy sets based multi-criteria decision-making model for offshore wind farm development in Ireland. *Energy* **2020**, 117317. [CrossRef]
39. Deveci, M.; Demirel, N.C. A survey of the literature on airline crew scheduling. *Eng. Appl. Artif. Intell.* **2018**, *74*, 54–69. [CrossRef]
40. Canitez, F.; Deveci, M. An integration model for car sharing and public transport: Case of Istanbul. In Proceedings of the Accessible sur ResearchGate, Istanbul, Turkey, 2–3 November 2017.
41. Deveci, M.; Demirel, N.Ç.; John, R.; Özcan, E. Fuzzy multi-criteria decision making for carbon dioxide geological storage in Turkey. *J. Nat. Gas Sci. Eng.* **2015**, *27*, 692–705. [CrossRef]
42. Mordeson, J.N.; Nair, P.S. Cycles and cocycles of fuzzy graphs. *Inf. Sci.* **1996**, *90*, 39–49. [CrossRef]
43. Samanta, S.; Pal, M.; Pal, A. Some more results on fuzzy *k*-competition graphs. *Int. J. Adv. Res. Artif. Intell.* **2014**, *3*, 60–67. [CrossRef]
44. Akram, M.; Dudek, W.A. Interval-valued fuzzy graphs. *Comput. Math. Appl.* **2011**, *61*, 289–299. [CrossRef]
45. Mordeson, J.N.; Nair, P.S. *Fuzzy Graphs and Hypergraphs*, 1st ed.; Physica: Heidelberg, Germany, 2000. [CrossRef]
46. Eslahchi, C.; Onaghe, B.N. Vertex strength of fuzzy graphs. *Int. J. Math. Math. Sci.* **2006**, *2006*, 1–9. [CrossRef]
47. Nagoorgani, A.; Hussain, R.J. Fuzzy effective distance *k*-dominating sets and their applications. *Int. J. Algorith. Comput. Math.* **2009**, *2*, 25–36.
48. Nagoorgani, A.; Radha, K. On regular fuzzy graphs. *J. Phys. Sci.* **2008**, *12*, 33–40.
49. Nagoorgani, A.; Latha, A. On irregular fuzzy graphs. *Appl. Math. Sci.* **2012**, *6*, 517–523.
50. Mordeson, J.N.; Nair, P.S. Successor and source of (fuzzy) finite state machines and (fuzzy) directed graphs. *Inf. Sci.* **1996**, *95*, 113–124. [CrossRef]
51. Goetschel, R.H. Introduction to fuzzy hypergraphs and Hebbian structures. *Fuzzy Sets Syst.* **1995**, *76*, 113–130. [CrossRef]

52. McAllister, M.L.N. Fuzzy intersection graphs. *Comput. Math. Appl.* **1988**, *15*, 871–886. [CrossRef]
53. Zhang, W.R. Bipolar fuzzy sets and relations: A computational framework for cognitive modeling and multiagent decision analysis. *Proc. IEEE Conf.* **1994**, *309*, 305–309.
54. Akram, M. Bipolar fuzzy graphs with applications. *Knowl. Based Syst.* **2013**, *39*, 1–8. [CrossRef]
55. Akram, M. Interval-valued fuzzy line graphs. *Neural Comput. Appl.* **2012**, *21*, 145–150. [CrossRef]

mathematics

MDPI

Article

Solvability of a Bounded Parametric System in Max-Łukasiewicz Algebra

Martin Gavalec* **and Zuzana Němcová**

Faculty of Informatics and Management, University of Hradec Králové, 50003 Hradec Králové, Czech Republic; zuzana.nemcova@uhk.cz
* Correspondence: martin.gavalec@uhk.cz; Tel.: +420-493-332-248

Received: 26 April 2020; Accepted: 19 June 2020; Published: 23 June 2020

Abstract: The max-Łukasiewicz algebra describes fuzzy systems working in discrete time which are based on two binary operations: the maximum and the Łukasiewicz triangular norm. The behavior of such a system in time depends on the solvability of the corresponding bounded parametric max-linear system. The aim of this study is to describe an algorithm recognizing for which values of the parameter the given bounded parametric max-linear system has a solution—represented by an appropriate state of the fuzzy system in consideration. Necessary and sufficient conditions of the solvability have been found and a polynomial recognition algorithm has been described. The correctness of the algorithm has been verified. The presented polynomial algorithm consists of three parts depending on the entries of the transition matrix and the required state vector. The results are illustrated by numerical examples. The presented results can be also applied in the study of the max-Łukasiewicz systems with interval coefficients. Furthermore, Łukasiewicz arithmetical conjunction can be used in various types of models, for example, in cash-flow system.

Keywords: max-min algebra; fuzzy max-T algebra; Łukasiewicz triangular norm; max-Łukasiewicz algebra; parametric solvability

MSC: 90C15

---

1. Introduction

The max-Łukasiewicz algebra (max-Łuk algebra, for short), is one of the so-called max-T fuzzy algebras, which are defined for various triangular norms T.

A max-T fuzzy algebra works with variables in the unit interval $\mathcal{I} = \langle 0, 1\rangle$ and uses the binary operations of maximum and a t-norm, T, instead of the conventional operations of addition and multiplication. Formally, a max-T fuzzy algebra is a triplet $(\mathcal{I}, \oplus, \otimes_T)$, where $\mathcal{I} = \langle 0, 1\rangle$ and $\oplus =$ max, $\otimes_T = T$ are binary operations on $\mathcal{I}$. By $\mathcal{I}(m, n)$, $\mathcal{I}(n)$, we denote the set of all matrices, vectors, of the given dimensions over $\mathcal{I}$. The operations $\oplus$, $\otimes_T$ are extended to matrices and vectors in the standard manner. Similarly, partial orderings on $\mathcal{I}(m, n)$ and $\mathcal{I}(n)$ are induced by the linear ordering on $\mathcal{I}$.

The triangular norms (t-norms, for short) were introduced in [1], in connection with probabilistic metric spaces. The t-norms interpretations are mainly the conjunction in fuzzy logics and intersection of fuzzy sets. Therefore, they find applications in many domains, for example in decision making processes, game theory and statistics, information and data processing or risk management. The t-norms and t-conorms belong to basic notions in the theory of fuzzy sets. The following four main t-norms: Łukasiewicz, Gödel, product and drastic (and many others) can be found in [2].

The Łukasiewicz norm is often characterized as a logic of absolute or metric comparison.

The Łukasiewicz conjunction is defined by formula

$$x \otimes_L y = \max\{x + y - 1, 0\}. \tag{1}$$

The Gödel norm is defined as the minimum of the entries (the truth degrees of the constituents). Gödel logic is the simplest norm; it is often characterized as a logic of relative comparison

$$x \otimes_G y = \min(x, y). \tag{2}$$

The product norm is defined by the formula

$$x \otimes_p y = x \cdot y. \tag{3}$$

The drastic triangular norm (the "weakest norm") is a basic example of a non-divisible t-norm on any partially ordered set. This t-norm is defined by the formula

$$x \otimes_{\mathrm{d}} y = \begin{cases} \min(x, y) & \text{if} \max(x, y) = 1, \\ 0 & \text{if} \max(x, y) < 1. \end{cases} \tag{4}$$

The max-T algebras with the above mentioned t-norms have various applications and their steady states and optimization methods have been intensively studied, see, for example, [3–7]. The algebras with interval entries have been studied in [8–10].

In the particular case when T is the Gödel t-norm, we get an important max-min algebra which is useful in solving various problems in fuzzy scheduling and optimization. Max-min algebra belongs to the so-called tropical mathematics, which has many applications and brings a great number of contributions to mathematical theory. Interesting monographs [11–14] and collections of papers [15–19] come from tropical mathematics and its applications.

Tropical algebras are often used for describing and studying systems working in discrete time stages. The state of the system in stage k is described by the state vector, $x(k)$. Then the transition matrix, A, determines the transition of the system to the next stage. In more detail, the next state of the system, $x(k+1)$, is obtained by multiplication $A \otimes x(t) = x(t+1)$. During the work of the system, it can happen that, after some time, the system reaches a steady state. In algebraic notation, the state vectors of steady states are eigenvectors of the transition matrix with some eigenvalue $\lambda \in \mathcal{I}$: $A \otimes x = \lambda \otimes x$.

The eigenproblem in max-min algebra has been frequently investigated, and many interesting results have been found. The structure of the eigenspace has been described and algorithms for computing the largest eigenvector have been suggested, see for example [20,21]. The eigenvectors in a max-T algebra, for various triangular norms T, have applications in fuzzy set theory. Such eigenvectors have been studied in [5,7,22]. The eigenvalues and eigenvectors are important characteristics of the system described by the fuzzy algebra. For the case of the drastic and product t-norms, the structure of the eigenspace has been studied in [5,7]. Finally, [22] describes the case of a Łukasiewicz fuzzy algebra.

Łukasiewicz arithmetical conjunction has applications in many model situations. The operation subtracts 1 from the sum of the components and takes the maximum with zero.This leads to the idea that the result of the operation is a remainder that is over the unit. Thus, the Łukasiewicz conjunction can be used, for example, in describing backup of data on a computer, the maximal capacity of an oil tank or lump payment in finances.

Such applications often lead to systems of max-Łuk linear equations. There is no inverse operation to $\oplus$ in max-Łuk algebra, therefore the transfer of variables from one side of equation to the other side is not possible. As a consequence, solving the one-sided linear systems (with variables, say, on the left-hand side of the equations) requires an approach different from solving the two-sided systems (with variables on both sides).

The aim of this paper is to present an algorithm for recognizing solvability of a given one-sided max-Łuk linear system with bounded variables, in dependence of a linear parameter factor on the right side, see (9) and (10) for an exact formulation.

This problem has not yet been studied in the parametrized version. The main contribution of this paper is description of the recognition algorithm, which has crucial role in the investigation of interval eigenvectors. The algorithm for recognizing the solvability of a given one-sided max-Łuk linear system can be shortly summarized in the following steps:

1. permute the equations in the system so that the right-hand side will be decreasing, that is
$$0 \le 1 - b_1 \le 1 - b_2 \le \cdots \le 1 - b_m \le 1, \tag{5}$$
2. recognize the solvability for some λ with $1 - b_m < \lambda \le 1$, according to Theorem 3 (case A), by verifying $C \otimes_L y^\star(\lambda^m_{\max}) = \lambda^m_{\max} \otimes_L b$,
3. recognize the solvability for some λ with $0 \le \cdots \le 1 - b_h < \lambda \le 1 - b_{h+1} \le \ldots 1$, according to Theorem 4 (case B), by verifying $C \otimes_L y^\star(\lambda^h_{\max}) = \lambda^h_{\max} \otimes_L b$. This step may be repeated, if necessary, with different indices $h \le m$,
4. recognize the solvability for some λ with $0 \le \lambda \le 1 - b_1$, according to Theorem 5 (case C), by verifying $\underline{y}_j \le \bigwedge_{i \in M}(1 - c_{ij})$, for every $j \in N$.
5. the system is solvable if the answer is positive at least once in steps 2, 3 or 4. Otherwise, the system is insolvable for any value of λ.

The structure of this paper is the following. Section 2 contains a case study based on an interactive cash-flow system, which shows motivation for solving linear systems in max-Łuk algebra. The problem is formulated in Section 3, where the preparatory results are also presented. The main results are described in Section 4. Illustrative numerical examples related to the case study from Section 2 are shown with details in Section 5. Discussion, comparison of the results with other papers, as well as future developments, are given in Conclusions.

2. Case Study: Interactive Cash-Flow System

Consider an interactive cash-flow system created by a network of n cooperating banks, $B_1, B_2, \ldots, B_n$. Assume that the cooperation is performed in stages. During the run of the system, variable interest rates of the banks mutually influence each other. In each stage, every bank B_i chooses a cash-flow cooperation with some other bank B_j (choice $i = j$ is also possible) in order to achieve the optimal profit, expressed by the value of the interest rate achieved for the next stage.

The system can be modeled as a discrete events system (DES). For any bank B_i, variable $x_i(k)$ shows the interest rate value in stages $k = 1, 2, \ldots,$ the vector $x(k) = \left(x_1(k), x_2(k), \ldots, x_n(k)\right)^T$ is called the state vector of DES in stage k. The change of the next state-vector values during the transition of the DES to the state vector $x(k+1)$ depends on the entries a_{ij} of the so-called transition matrix A.

The possible increase of the profit coming from the cooperation of B_i with bank B_j is equal to a_{ij} (including the lump payment). Thus, the efficient sum of a_{ij} and $x_i(k)$ is only the part exceeding 1 (that is, exceeding 100%), in the case when B_i chooses B_j for cooperation in the stage k.

Optimization of the variable interest rate in stage $k+1$ leads every B_i to such a choice of B_j, where the efficient increase of the profit is maximal. That is, $x_i(k+1) = \max\limits_{j \in N} \max\left(a_{ij} + x_j(k) - 1, 0\right)$. In max-Łuk notation the optimal choice can be written as

$$x_i(k+1) = \bigoplus_{j \in N} a_{ij} \otimes_L x_j(k), \text{ or} \tag{6}$$

$$x(k+1) = A \otimes_L x(k). \tag{7}$$

For simplicity we assume that the system is homogeneous (that is, A does not change from stage to stage).

In real life, the matrix and vector entries are not always exact values. For example, if (7) is applied for prediction, then the transition matrix is not exactly known, it is only an estimation, belonging to some interval $A \in \mathbf{A} = [\underline{A}, \overline{A}]$. Analogously, the state vector belongs to some interval $x \in \mathbf{X} = [\underline{x}, \overline{x}]$. We say that the DES is considered with interval coefficients.

For formulas with interval coefficients, it must be decided which values from the corresponding interval will be taken. One possibility is to take all values (using the universal quantifier). The other possibility is to use the existential quantifier and only require that there is some value from the interval, such that the formula is satisfied.

If there are more interval variables in the formula in consideration, then the quantifiers can be combined. For example, various types of quantified notions in max-min algebra are described in [23,24].

By recurrent application of (7), the sequence of state vectors (also called: orbit of the DES) x, $A \otimes x, \ldots, A^k \otimes x, \ldots$, where $A^k = A \otimes \ldots \otimes A$ (k times), can be created. The orbit represents a predicted evolution of interest rates. Two natural questions arise:

Q1. Can the orbit reach a fixed given state vector value?

Q2. Can the orbit reach a steady state (such a state which does not change from stage to stage)?

Q1 requires to recognize whether, in some stage k, there is a value $y = x(k)$ such that $b = x(k+1)$ for a given vector $b \in \mathcal{I}(n)$. If we consider the problem in the interval arithmetic, then we get the state vector variable $y \in [\underline{y}, \overline{y}]$. Moreover, we can generalize the problem by adding a parameter $\lambda \in \mathcal{I}$ to the given value b. Then the original question is beeing solved as a special subcase with $\lambda = 1$.

Therefore, question Q1 can be solved as one-sided bounded parametric problem studied in Sections 3 and 4. The main result is Theorem 6, which describes a necessary and sufficient condition for solvability of the system (9) and (10).

The computations answering to Q1 are illustrated by Example 1 (positive answer) and Example 2 (negative answer) in Section 5, with detailed interpretation.

Q2 is connected with the eigenproblem of the transition matrix. A steady state is characterized by the equation $x(k+1) = x(k)$ or, equivalently, by $A \otimes_L x = x$. That is, steady states are equivalent to max-Łuk eigenvectors of the transition matrix A. Usually the eigenvectors are considered in a more general form, with added the so-called eigenvalue $\lambda \in \mathcal{I}$. That is, $x \in \mathcal{I}(n)$ is an eigenvector of matrix $A \in \mathcal{I}(n,n)$ with eigenvalue $\lambda \in \mathcal{I}$ if $A \otimes_L x = \lambda \otimes_L x$. The eigenvectors in max-Łuk algebra have been studied in [3,6], and recently, in a more general context, in [25].

If we wish to answer Q2 in the interval arithmetics, then we have to consider $A \in \mathbf{A} = [\underline{A}, \overline{A}]$ and $x \in \mathbf{X} = [\underline{x}, \overline{x}]$. According to the choice of universal/existential quantifier for $A \in \mathbf{A}$, and for $x \in \mathbf{X}$, various types of interval eigenproblem have been studied by various authors over max/plus and max-min algebra.

For example, $\mathbf{X}$ is called a strongly tolerable eigenvector of $\mathbf{A}$ if

$$(\exists \lambda)(\exists A \in \mathbf{A})(\forall x \in \mathbf{X})\big[A \otimes x = \lambda \otimes x\big] \tag{8}$$

In words, we ask for the existence of λ and $A \in \mathbf{A}$ such that every $x \in \mathbf{X}$ is an eigenvector of A with eigenvalue λ (we shortly say that every $x \in \mathbf{X}$ is tolerated by A).

Analogous problem has been solved in max-min algebra in [23], where it has been shown that the problem can be reduced to the solvability of the system $\tilde{C} \otimes y = \lambda \otimes \tilde{b}$ using generators of the interval matrix $\mathbf{A}$. The main idea of the algorithm is to find a certificate matrix of the given instance of dimension $n \times n$, as a max-min linear combination of generators. The necessary coefficients of this linear combinations can be computed by solving an auxiliary one-sided max-min linear system of dimension $n^2 \times n^2$.

By analogy, this approach can easily be transferred from max-min to max-Łuk algebra, with a single exception of recognizing the solvability for the auxiliary one-sided linear system of dimension $n^2 \times n^2$. Namely, to recognize the parametric solvability of a one-sided linear system is substantially more complicated problem in max-Łuk algebra than it is in max-min algebra. In fact, it is this manuscript, where an efficient algorithm for the necessary solvability problem has been formulated.

Till now, the specific methods of max-Łuk algebra have only been presented at Conference EURO 2019 in Dublin. The extended version of this presentation is in preparation and will be submitted soon. The recognition method described in this manuscript, plays important role in the proofs of the following two theorems.

Theorem 1 ([23]). *Let an interval matrix* $\mathbf{A} = [\underline{A}, \overline{A}]$ *and an interval vector* $\mathbf{X} = [\underline{x}, \overline{x}]$ *be given. Then,* $\mathbf{X}$ *is a strongly tolerable eigenvector of* $\mathbf{A}$ *if and only if* $\tilde{C} \otimes_L y = \lambda \otimes_L \tilde{b}$ *is solvable for some* $\lambda \in \mathcal{I}$.

Theorem 2 ([23]). *The recognition problem of whether a given interval vector* $\mathbf{X}$ *is a strong tolerance eigenvector of a given interval matrix* $\mathbf{A}$ *in max-min algebra, is solvable in* $O(n^5)$ *time.*

3. Bounded Parametric Systems of Max-Łuk Linear Equations

In view of the motivation inspired by the case study in Section 2, the solvability problem for a bounded parametric linear system in max-Łuk algebra is studied in this paper.

We consider the system

$$C \otimes_L y = \lambda \otimes_L b, \tag{9}$$

$$\underline{y} \leq y \leq \overline{y}, \tag{10}$$

with fixed matrix $C \in \mathcal{I}(m, n)$ and the right-hand side vector $b \in \mathcal{I}(m)$. The basic question is whether the system is solvable for some value $0 < \lambda \in \mathcal{I}$ of the parameter. In other words, we are looking for a necessary and sufficient condition allowing the recognition of whether there is a $\lambda \in \mathcal{I} \setminus \{0\}$ such that (9) and (10) is solvable (the case $\lambda = 0$ is trivial).

In the sequel, we use the notation $M = \{1, 2, \ldots, m\}$ and $N = \{1, 2, \ldots, n\}$. The set of all solutions to (9) without any constraint is denoted by $S(C, \lambda \otimes_L b)$; the solution set with the upper bound is $S(C, \lambda \otimes_L b, \overline{y})$, and the solution set with both upper and lower bound is denoted by $S(C, \lambda \otimes_L b, \overline{y}, \underline{y})$. That is, we have to recognize whether $S(C, \lambda \otimes_L b, \overline{y}, \underline{y}) \neq \emptyset$, for some $\lambda \in \mathcal{I}$ or not.

Without any loss of generality, we assume till the end of the paper that the right-hand side vector $b \in \mathcal{I}(m)$ satisfies the monotonicity condition

$$1 \geq b_1 \geq b_2 \geq \cdots \geq b_m \geq 0. \qquad \text{Then} \tag{11}$$

$$0 \leq 1 - b_1 \leq 1 - b_2 \leq \cdots \leq 1 - b_m \leq 1. \tag{12}$$

System (9) is equivalent to

$$(\forall i \in M) \left[\bigoplus_{j \in N} (c_{ij} \otimes_L y_j) = \lambda \otimes_L b_i \right], \tag{13}$$

which is further equivalent to

$$(\forall i \in M)\, (\forall j \in N)\, \left[c_{ij} \otimes_L y_j \leq \lambda \otimes_L b_i \right], \tag{14}$$

$$(\forall k \in M)\, (\exists j \in N)\, \left[c_{kj} \otimes_L y_j = \lambda \otimes_L b_k \right]. \tag{15}$$

In view of the definition of $\otimes_L$, the inequality in (14) takes one of the following forms

$$0 < c_{ij} + y_j - 1 \le \lambda + b_i - 1, \tag{16}$$
$$0 \ge c_{ij} + y_j - 1, \qquad 0 < \lambda + b_i - 1, \tag{17}$$
$$0 \ge c_{ij} + y_j - 1, \qquad 0 \ge \lambda + b_i - 1. \tag{18}$$

We shall use the notation $H(\lambda) = \{i \in M; 0 < \lambda + b_i - 1\}$ (for short: H if λ is clear from the context). For $i \in M \setminus H$ we have $0 \ge \lambda + b_i - 1$. Therefore,

$$\lambda \otimes_L b_i = \lambda + b_i - 1 \qquad \text{for } i \in H, \tag{19}$$
$$\lambda \otimes_L b_i = 0 \qquad \text{for } i \in M \setminus H. \tag{20}$$

For brevity, we write $d_{ij} = b_i - c_{ij}$ for every $i \in M, j \in N$.

Lemma 1. *If $y \in S(C, \lambda \otimes_L b, \overline{y}, \underline{y})$, then*

1. $(\forall j \in N)\left[y_j \le (\lambda + \bigwedge_{i \in H} d_{ij})\right]$,
2. $(\forall j \in N)\left[y_j \le \bigwedge_{i \in M \setminus H} (1 - c_{ij})\right]$,
3. $(\forall j \in N)\left[\underline{y}_j \le y_j \le \overline{y}_j\right]$.

Proof. Let $j \in N$ be fixed.

(i) For every $i \in H$ we have $0 < \lambda + b_i - 1$, which implies $c_{ij} + y_j - 1 \le \lambda + b_i - 1$, in view of (16) and (17). That is, $y_j \le \lambda + b_i - c_{ij} = \lambda + d_{ij}$. As a consequence, $y_j \le \bigwedge_{i \in H} (\lambda + d_{ij}) = \lambda + \bigwedge_{i \in H} d_{ij}$.

(ii) For $i \in M \setminus H$ we have $0 \ge \lambda + b_i - 1$, which implies $0 \ge c_{ij} + y_j - 1$, by (18). Then $y_j \le 1 - c_{ij}$, that is, $y_j \le \bigwedge_{i \in M \setminus H} (1 - c_{ij})$.

(iii) The assertion follows directly from the definition. □

If the equality $c_{kj} \otimes_L y_j = \lambda \otimes_L b_k$ in (15) holds, then we say that y_j is active in row k. If so, we write $k \in A_j(\lambda)$ and $A_j = A_j(\lambda)$ is then called the activity set of the variable y_j.

There are two possible activity subcases:

$$y_j = \lambda + d_{kj} \qquad \text{for } k \in H, \tag{21}$$
$$0 \le y_j \le 1 - c_{kj} \qquad \text{for } k \in M \setminus H. \tag{22}$$

Namely, if $k \in H$, then $0 < \lambda + b_i - 1 = \lambda \otimes_L b_k$. Then also $c_{kj} \otimes_L y_j > 0$, which gives $c_{kj} + y_j - 1 = \lambda + b_i - 1$. As a consequence, $y_j = \lambda + b_k - c_{kj} = \lambda + d_{kj}$. On the other hand, if $k \in M \setminus H$, then $0 \ge \lambda + b_i - 1$. That is, $\lambda \otimes_L b_k = 0$. Then, also $c_{kj} \otimes_L y_j = 0$, which implies $c_{kj} + y_j - 1 \le 0$, and $y_j \le 1 - c_{kj}$.

In subcase (21) with $k \in H$, we have $y_j = \lambda + d_{kj} \le \lambda + \bigwedge_{i \in H} d_{ij}$. As a consequence,

$$d_{kj} = \bigwedge_{i \in H} d_{ij}. \tag{23}$$

In subcase (22) with $k \in M \setminus H$, we get, using Lemma 1(ii),

$$0 \le y_j \le \bigwedge_{i \in M \setminus H} (1 - c_{ij}) \le 1 - c_{kj}. \tag{24}$$

Lemma 2. *Assume $C \in \mathcal{I}(m,n)$, $b \in \mathcal{I}(m)$ with the monotonicity condition (11), $y, \overline{y}, \underline{y} \in \mathcal{I}(n)$, $\lambda \in \mathcal{I}$ and $h \in M$ with $1 - b_h < \lambda \leq 1 - b_{h+1}$. Then $H = \{1, 2, \ldots, h\}$ and the following statements are equivalent:*

1. $y \in S(C, \lambda \otimes_L b, \overline{y}, \underline{y})$,
2. $y \in S(C_H, \lambda \otimes_L b_H, \overline{y} \wedge \overline{y}^h, \underline{y})$,

where the submatrix C_H (subvector b_H) consists of the rows in C_i (in b_i) with $i \in H$. Analogously, the vector $\overline{y}^h \in \mathcal{I}(n)$ with $\overline{y}^h_j = \bigwedge_{i \in M \setminus H}(1 - c_{ij})$ for every $j \in N$ is constructed from the rows C_i of C, $i \in M \setminus H$.

Proof. (i) $\Rightarrow$ (ii).

Assume (9)–(11). Then (14) and (15) are satisfied. In particular, considering only the rows $i, k \in H \subseteq M$, we get

$$C_H \otimes_L y = \lambda \otimes_L b_H. \tag{25}$$

The inequalities $\underline{y} \leq y \leq \overline{y}$ follow by assumption (i). Moreover, (24) implies $y \leq \overline{y}^h = \bigwedge_{i \in M \setminus H}(1 - C_i)$. Summarizing, we have

$$y \in S(C_H, \lambda \otimes_L b_H, \overline{y} \wedge \overline{y}^h, \underline{y}). \tag{26}$$

(ii) $\Rightarrow$ (i). Conversely, assume (26). Then $\bigoplus_{j \in N}(c_{ij} \otimes_L y_j) = \lambda \otimes_L b_i$, for every $i \in H$.

Moreover, for $i \in M \setminus H$ and $j \in N$, we have $y_j \leq 1 - c_{ij}$, by assumption $y \leq \overline{y}_H$. Therefore, $c_{ij} + y_j - 1 \leq 0$, which implies $c_{ij} \otimes_L y_j = 0 = \lambda \otimes_L b_i$ since $i \in M \setminus H$, which gives $\lambda + b_i - 1 \leq 0$. The inequalities $\underline{y} \leq y \leq \overline{y}$ follow immediately. □

For $\lambda \in \mathcal{I}$ and for every $i \in M, j \in N$, we define

$$y^\star_{ij}(\lambda) = \begin{cases} \lambda + d_{ij} & \text{if } i \in H \\ 1 - c_{ij} & \text{if } i \in M \setminus H. \end{cases} \tag{27}$$

Furthermore, we define $y^\star(\lambda) \in \mathcal{I}(n)$ by putting, for $j \in N$,

$$y^\star_j(\lambda) = \overline{y}_j \wedge \bigwedge_{i \in H}(\lambda + d_{ij}) \wedge \bigwedge_{i \in M \setminus H}(1 - c_{ij}). \tag{28}$$

Lemma 3. *Let $C \in \mathcal{I}(m,n)$, $b \in \mathcal{I}(m)$, $\lambda \in \mathcal{I}$ and $\overline{y} \in \mathcal{I}(n)$. Then*

1. $(\forall i \in M)(\forall j \in N)\ c_{ij} \otimes_L y^\star_j(\lambda) \leq \lambda \otimes_L b_i$,
2. $y^\star(\lambda) \leq \overline{y}$,
3. *$y^\star(\lambda)$ is the maximal vector in $\mathcal{I}(n)$ fulfilling conditions (i) and (ii).*

Proof. It is easy to verify, using the definition of $\otimes_L$, that $c_{ij} \otimes_L y^\star_{ij}(\lambda) = \lambda \otimes_L b_i$ for every $i \in M, j \in N$. Then, assertions (i) and (ii) follow from the definition of $y^\star(\lambda)$.

(iii) Assume $y \in \mathcal{I}(n)$ satisfies conditions (i) and (ii) with $y^\star(\lambda)$ replaced by y. Let $i \in M, j \in N$. By (i) we have $c_{ij} \otimes_L y_j \leq \lambda \otimes_L b_i$. Suppose, by contradiction, that there is $j \in N$ such that $y_j > y^\star_j(\lambda)$. Then, in view of (28), there is $i \in M$ such that $y_j > y^\star_{ij}(\lambda)$. We consider two cases.

Case (a): $i \in H$. Then $y_j > \lambda + b_i - c_{ij}$, which implies $c_{ij} + y_j - 1 > \lambda + b_i - 1$. Thus, $c_{ij} \otimes_L y_j > \lambda \otimes_L b_i$, a contradiction.

Case (b): $i \in M \setminus H$. Then $y_j > 1 - c_{ij}$ implies $c_{ij} + y_j - 1 > 0$. Thus, $c_{ij} \otimes_L y_j > 0 = \lambda \otimes_L b_i$, a contradiction.

Given that i, j are arbitrary, $y \leq y^\star(\lambda)$ follows. □

Lemma 4. *Let $C \in \mathcal{I}(m,n)$, $b \in \mathcal{I}(m)$, $\lambda \in \mathcal{I}$ and $\overline{y} \in \mathcal{I}(n)$. Then the following statements are equivalent.*

1. $S(C, \lambda \otimes_L b, \overline{y}) \neq \emptyset$,
2. $y^\star(\lambda) \in S(C, \lambda \otimes_L b, \overline{y})$.

If $y \in S(C, \lambda \otimes_L b, \overline{y})$, *then* $y \leq y^\star(\lambda)$.

Proof. The assertion of the lemma follows directly from Lemma 3. □

Lemma 5. *Let* $C \in \mathcal{I}(m,n)$, $b \in \mathcal{I}(m)$, $\lambda \in \mathcal{I}$ *and let* $\underline{y}, \overline{y} \in \mathcal{I}(n)$ *with* $\underline{y} \leq \overline{y}$. *Then the following statements are equivalent.*

1. $S(C, \lambda \otimes_L b, \overline{y}, \underline{y}) \neq \emptyset$,
2. $y^\star(\lambda) \in S(C, \lambda \otimes_L b, \overline{y}, \underline{y})$.

If $y \in S(C, \lambda \otimes_L b, \overline{y}, \underline{y})$, *then* $y \leq y^\star(\lambda)$.

Proof. Assume that $y \in S(C, \lambda \otimes_L b, \overline{y}, \underline{y})$. Then, in particular, $y \in S(C, \lambda \otimes_L b, \overline{y})$, which implies $y^\star(\lambda) \in S(C, \lambda \otimes_L b, \overline{y})$ and $y \leq y^\star(\lambda)$, in view of Lemma 3. Furthermore, $\underline{y} \leq y \leq y^\star(\lambda)$ implies $y^\star(\lambda) \in S(C, \lambda \otimes_L b, \overline{y})$. The converse implication is trivial. □

Remark 1. *The assertions of Lemma 5 are often expressed by saying that for fixed* $\lambda \in \mathcal{I}$, $y^\star(\lambda)$ *is the maximal possible candidate for a solution of the system* (9) *and* (10).

Remark 2. *By a standard definition, the minimum of the empty subset of* $\mathcal{I}$ *is the maximal element in* $\mathcal{I}$. *Hence, if there is no* $i \in M$ *with* $\lambda \leq 1 - b_i$, *then* $\bigwedge_{i \in M \setminus H} (1 - c_{ij}) = I$, *and*

$$y_j^\star(\lambda) = \overline{y}_j \wedge \bigwedge_{i \in H} (\lambda + d_{ij}).$$

Similarly, if there is no $i \in M$ *with* $\lambda > 1 - b_i$, *then* $\bigwedge_{i \in H} (\lambda + d_{ij}) = I$, *and*

$$y_j^\star(\lambda) = \overline{y}_j \wedge \bigwedge_{i \in M \setminus H} (1 - c_{ij}).$$

4. Parametric Solvability Problem of Max-Łuk Linear Equations

The main result of this paper is description of a recognition algorithm for the parametric solvability problem. The problems (9) and (10) will be discussed according to

$$h(\lambda) = \max\{\, i \in M;\ \lambda > 1 - b_i \,\}. \tag{29}$$

Similarly to Remark 2, we put $h(\lambda) = \max \emptyset = 0$ if $\lambda \leq 1 - b_i$ for all $i \in M$. For $j \in N$ we also use the notation

$$y_j^{\star\star}(\lambda) = \bigwedge_{i \in H} (\lambda + d_{ij}) \wedge \bigwedge_{i \in M \setminus H} (1 - c_{ij}). \tag{30}$$

We consider three cases: (A) $h(\lambda) = m$, (B) $0 < h(\lambda) < m$ and (C) $h(\lambda) = 0$. The solvability in case A is described by the following theorem, with the notation

$$\lambda_{\max}^m = 1 \wedge \bigwedge_{j \in N} \bigvee_{i \in M} (\overline{y}_j - d_{ij}). \tag{31}$$

Theorem 3. *Case (A). Assume* $C \in \mathcal{I}(m,n)$ *and* $b \in \mathcal{I}(m)$ *with the monotonicity condition* (11). *Then the following statements are equivalent.*

1. $S(C, \lambda \otimes_L b, \overline{y}, \underline{y}) \neq \emptyset$ *for some* λ *with* $1 \geq \lambda > 1 - b_m$,
2. $S(C, \lambda^m_{\max} \otimes_L b, \overline{y}, \underline{y}) \neq \emptyset$.

Proof. (i) ⇒ (ii) Assume $\lambda \in \mathcal{I}$ is given with $1 \geq \lambda > 1 - b_m$ and $S(C, \lambda \otimes_L b, \overline{y}, \underline{y}) \neq \emptyset$.

We have $1 \geq \lambda > 1 - b_m \geq 1 - b_i$ for every $i \in M$. That is, $H = M$ and $M \setminus H = \emptyset$. In view of Remark 2, for every $j \in N$

$$y^\star_j(\lambda) = \overline{y}_j \wedge \bigwedge_{i \in M} (\lambda + d_{ij}). \tag{32}$$

The assumption $S(C, \lambda \otimes_L, \overline{y}, \underline{y}) \neq \emptyset$ implies $C \otimes_L y^\star(\lambda) = \lambda \otimes_L b$. That is, (10), (14) and (15) are satisfied with $y_j = y^\star_j(\lambda)$.

Now we consider $\kappa \in \mathcal{I}$ with $\lambda \leq \kappa \leq 1$. We look for a necessary and sufficient condition such that (10), (14) and (15) hold for $y_j = y^\star_j(\kappa)$.

Since $\lambda \leq \kappa$, we have $y^\star_j(\lambda) \leq y^\star_j(\kappa)$, for every $j \in N$. Therefore, $\underline{y}_j \leq y^\star_j(\lambda)$ implies $\underline{y}_j \leq y^\star_j(\kappa)$. That is, $\underline{y} \leq y^\star(\kappa)$ for every $\lambda \leq \kappa \leq 1$. The conditions for the upper bound inequality $y^\star(\kappa) \leq \overline{y}$ will be discussed later.

First, we verify conditions (14) and (15). In view of the assumption, we have $1 \geq \kappa \geq \lambda > 1 - b_m \geq 1 - b_i$ for every $i \in M$. That is, $H = M$ and $M \setminus H = \emptyset$. Then, for every $j \in N$

$$y^{\star\star}_j(\lambda) = \bigwedge_{i \in M} (\lambda + d_{ij}) = \lambda + \bigwedge_{i \in M} d_{ij}. \tag{33}$$

Similarly, for every $j \in N$

$$y^{\star\star}_j(\kappa) = \kappa + \bigwedge_{i \in M} d_{ij}. \tag{34}$$

It follows that the equalities

$$c_{ij} + y^{\star\star}_j(\lambda) - 1 = \lambda + b_i - 1, \tag{35}$$

$$c_{ij} + y^{\star\star}_j(\kappa) - 1 = \kappa + b_i - 1 \tag{36}$$

are equivalent. Similarly,

$$c_{ij} + y^{\star\star}_j(\lambda) - 1 \leq \lambda + b_i - 1, \tag{37}$$

$$c_{ij} + y^{\star\star}_j(\kappa) - 1 \leq \kappa + b_i - 1 \tag{38}$$

are equivalent.

Therefore, the assumption $y^{\star\star}(\lambda) \in S(C, \lambda \otimes_L b)$ is equivalent to $y^{\star\star}(\kappa) \in S(C, \kappa \otimes_L b)$, for every $\kappa \in \mathcal{I}$ with $\lambda \leq \kappa \leq 1$.

To achieve also the upper bound inequality $y^{\star\star}(\kappa) \leq \overline{y}$, further conditions must be imposed on κ. Namely, for every $j \in N$ the condition

$$y^{\star\star}_j(\kappa) = \kappa + \bigwedge_{i \in M} d_{ij} \leq \overline{y}_j \tag{39}$$

must be added. As a consequence, we get

$$\kappa \leq \bigwedge_{j \in N} \bigvee_{i \in M} (\overline{y}_j - d_{ij}). \tag{40}$$

Now, with the notation

$$\lambda_{\max}^{m} = 1 \wedge \bigwedge_{j \in N} \bigvee_{i \in M} (\overline{y}_j - d_{ij}) \tag{41}$$

we have

$$y^{\star\star}(\kappa) \le \overline{y} \quad \Leftrightarrow \quad \kappa \le \lambda_{\max}^{m}. \tag{42}$$

Therefore, $y^{\star\star}(\kappa) \in S(C, \lambda_{\max}^{m} \otimes_L b, \overline{y}, \underline{y})$. The converse implication (ii) $\Rightarrow$ (i) is trivial. □

In case B we have $0 < h(\lambda) < m$. That is,

$$0 \le \dots \le 1 - b_h < \lambda \le 1 - b_{h+1} \le \dots 1. \tag{43}$$

Write $\Lambda(h) = (1 - b_h, 1 - b_{h+1}\rangle$, for brevity. In view of (19) and (20), we have $H = \{1, 2, \dots, h\}$ (with $0 < \lambda + b_i - 1$ for $i \in H$) and $M \setminus H = \{h+1, h+2, \dots, m\}$ (with $0 \ge \lambda + b_i - 1$ for $i \in M \setminus H$). Moreover, we denote

$$N(h) = \left\{ j \in N;\ \left((\overline{y}_j \wedge \overline{y}_j^h) - \bigwedge\nolimits_{i \in H} d_{ij} \right) > 1 - b_h \right\}, \tag{44}$$

$$\lambda_{\max}^{h} = (1 - b_{h+1}) \wedge \bigwedge\nolimits_{j \in N(h)} \left((\overline{y}_j \wedge \overline{y}_j^h) - \bigwedge\nolimits_{i \in H} d_{ij} \right). \tag{45}$$

Theorem 4. *Case (B). Assume $C \in \mathcal{I}(m, n)$, $b \in \mathcal{I}(m)$ and $h \in M$. Further assume the monotonicity condition (11) holds. Then the following statements are equivalent.*

1. $S(C, \lambda \otimes_L b, \overline{y}, \underline{y}) \neq \emptyset$ *for some* $\lambda \in \Lambda(h)$,
2. $S(C_H, \lambda_{\max}^{h} \otimes_L b_H, \overline{y} \wedge \overline{y}^h, \underline{y}) \neq \emptyset$.

Proof. For fixed $\lambda \in \Lambda_h$, (i) is equivalent (in view of Lemma 5) to the statement

$$y^{\star}(\lambda) \in S(C, \lambda \otimes_L b, \overline{y}, \underline{y}), \tag{46}$$

which is further equivalent (in view of Lemma 2) to

$$y^{\star}(\lambda) \in S(C_H, \lambda \otimes_L b_H, \overline{y} \wedge \overline{y}^h, \underline{y}). \tag{47}$$

The proof will be completed by demonstrating that (47) is equivalent to (ii). We assume (47) for fixed $\lambda \in \Lambda(h)$, and we describe conditions under which (47) also holds for arbitrary $\kappa \in \Lambda(h)$ with $\lambda \le \kappa$.

We shall verify the restrictions $\underline{y} \le y^{\star}(\kappa) \le \overline{y} \wedge \overline{y}^h$ and the activity of the variables $y^{\star}(\kappa)$ in every row $k \in H$ of the matrix C_H with vector b_H.

The lower restriction follows by monotonicity $\underline{y}_j \le y_j^{\star}(\lambda) \le y_j^{\star}(\kappa)$, for every $j \in N$. The upper restriction

$$y_j^{\star}(\kappa) = \overline{y}_j \wedge \left(\kappa + \bigwedge\nolimits_{i \in H} d_{ij} \right) \wedge \bigwedge\nolimits_{i \in M \setminus H} (1 - c_{ij}) \le \overline{y}_j \wedge \overline{y}_j^h \tag{48}$$

follows by Lemma 2 directly from

$$y_j^{\star}(\kappa) = \overline{y}_j \wedge \overline{y}_j^h \wedge \left(\kappa + \bigwedge\nolimits_{i \in H} d_{ij} \right) \le \overline{y}_j \wedge \overline{y}_j^h. \tag{49}$$

As a consequence, the activity condition (22) is fulfilled in all rows $k \in M \setminus H$ for every variable $y_j^{\star}(\kappa)$ with $\in N$.

To verify also the second activity condition (21) for at least one variable $j \in N$ in every row $k \in H$, we denote by μ_j^h the break-point at which the (min/plus)-linear function

$$y_j^\star(\kappa) = \left(\overline{y}_j \wedge \overline{y}_j^h\right) \wedge \left(\kappa + \bigwedge_{i\in H} d_{ij}\right) \tag{50}$$

of the variable κ changes its direction. In other words,

$$y_j^\star(\kappa) = \begin{cases} \kappa + \bigwedge_{i\in H} d_{ij} & \text{if } \kappa \leq \mu_j^h, \\ \overline{y}_j \wedge \overline{y}_j^h & \text{if } \kappa \geq \mu_j^h. \end{cases} \tag{51}$$

By condition $1 - b_i < \kappa$, for $i \in H$, we get $0 < \kappa + b_i - 1 \leq \kappa + b_i - c_{ij} = \kappa + d_{ij}$.

At the break-point, both parts of the function (50) have the same value. That is,

$$\mu_j^h + \bigwedge_{i\in H} d_{ij} = \overline{y}_j \wedge \overline{y}_j^h, \tag{52}$$

or, equivalently,

$$\mu_j^h = \left(\overline{y}_j \wedge \overline{y}_j^h\right) - \bigwedge_{i\in H} d_{ij}. \tag{53}$$

□

Claim 1. *Assume $j \in N$, $\lambda \in \Lambda(h)$. If $y_j^\star(\lambda)$ is active in $k \in H$, then $1 - b_h < \lambda \leq \mu_j^h$. Moreover, for every $\kappa \in \Lambda(h)$ with $\kappa \leq \mu_j^h$, $y_j^\star(\kappa)$ is active in k.*

Proof of Claim 1. By assumption, $y_j^\star(\lambda) = \lambda + d_{kj} = \lambda + \bigwedge_{i\in H} d_{ij}$, in view of (21) and (23). Then, $\lambda \leq \mu_j^h$, and the activity of $y_j^\star(\lambda)$ in k is described by the formula

$$c_{kj} + \lambda + \bigwedge_{i\in H} d_{ij} - 1 = \lambda + b_k - 1, \tag{54}$$

while the activity of $y_j^\star(\kappa)$ under assumption $\kappa \leq \mu_j^h$ is described by

$$c_{kj} + \kappa + \bigwedge_{i\in H} d_{ij} - 1 = \kappa + b_k - 1. \tag{55}$$

As (54) and (55) are equivalent, the assertion of Claim 1 follows.

In view of (44), (45) and (53) we have

$$N(h) = \left\{j \in N;\ \mu_j^h > 1 - b_h\right\}, \tag{56}$$

$$\lambda_{\max}^h = (1 - b_{h+1}) \wedge \bigwedge_{j\in N(h)} \mu_j^h. \tag{57}$$

□

Claim 2. *Assume $y^\star(\lambda) \in S(C_H, \lambda \otimes_L b_H, \overline{y} \wedge \overline{y}^h, \underline{y})$. If $\lambda \leq \kappa \in \Lambda(h)$, with $\kappa \leq \mu_j^h$ for every $j \in N(h)$, then $y^\star(\kappa) \in S(C_H, \lambda_{\max}^h \otimes_L b_H, \overline{y} \wedge \overline{y}^h, \underline{y})$.*

Proof of Claim 2. By assumption, for every $k \in H$ there is a $j \in N$ such that $y^\star(\lambda)$ is active in k. Then by Claim 1, under assumption $\lambda \leq \kappa \leq \bigwedge_{j\in N} \mu_j^h$, for every $k \in H$ there is a $j \in N(h)$ such that $y^\star(\kappa)$ is active in k. That is, $y^\star(\kappa) \in S(C_H, \lambda_{\max}^h \otimes_L b_H, \overline{y} \wedge \overline{y}^h, \underline{y})$. □

In case C we have $H = \emptyset$ and $M \setminus H = M$. That is, $\lambda \leq 1 - b_i$ for all $i \in M$. The solvability in case C is described by the following theorem.

Theorem 5. *Case (C). Assume $C \in \mathcal{I}(m,n)$, $b \in \mathcal{I}(m)$ and $\underline{y} \leq \overline{y} \in \mathcal{I}(n)$, with the monotonicity condition (11). Then the following statements are equivalent*

1. $S(C, \lambda \otimes_L b, \overline{y}, \underline{y}) \neq \emptyset$, *for some λ with $0 \leq \lambda \leq 1 - b_1$,*
2. $\underline{y}_j \leq \bigwedge_{i \in M}(1 - c_{ij})$, *for every $j \in N$.*

Proof. Assume $0 \leq \lambda \leq 1 - b_1$ and $y \in S(C, \lambda \otimes_L b, \overline{y}, \underline{y})$. By Lemma 5(ii), this is equivalent to $y^\star(\lambda) \in S(C, \lambda \otimes_L b, \overline{y}, \underline{y})$. For every $j \in N$ we have, in view of Remark 2,

$$\underline{y}_j \leq y^\star_j(\lambda) = \overline{y}_j \wedge \bigwedge_{i \in M} (1 - c_{ij}). \tag{58}$$

The equivalence (i) $\Leftrightarrow$ (ii) follows immediately. □

Theorem 6. *Assume $C \in \mathcal{I}(m,n)$, $b \in \mathcal{I}(m)$ and $\underline{y} \leq \overline{y} \in \mathcal{I}(n)$, with the monotonicity condition (11). The bounded parametric system (9) and (10) is solvable for some $\lambda \in \mathcal{I}$ if and only if at least one of the following statements is fulfilled*

1. $h(\lambda) = m$, $S(C, \lambda^m_{\max} \otimes_L b, \overline{y}, \underline{y}) \neq \emptyset$,
2. $0 < h(\lambda) < m$, $\lambda \in \Lambda(h)$, $S(C_H, \lambda^h_{\max} \otimes_L b_H, \overline{y} \wedge \overline{y}^h, \underline{y}) \neq \emptyset$,
3. $h(\lambda) = 0$ *and* $\underline{y}_j \leq \bigwedge_{i \in M}(1 - c_{ij})$ *for every $j \in N$.*

Proof. For the convenience of the reader, we recall the previous definitions.

$$h(\lambda) = \max\{\, i \in M;\ \lambda > 1 - b_i \,\}, \tag{59}$$

$$\Lambda(h) = (1 - b_h, 1 - b_{h+1}\rangle, \tag{60}$$

$$N(h) = \left\{ j \in N;\ \left((\overline{y}_j \wedge \overline{y}^h_j) - \bigwedge\nolimits_{i \in H} d_{ij} \right) > 1 - b_h \right\}, \tag{61}$$

$$\lambda^m_{\max} = 1 \wedge \bigwedge_{j \in N} \bigvee_{i \in M} (\overline{y}_j - d_{ij}), \tag{62}$$

$$\lambda^h_{\max} = (1 - b_{h+1}) \wedge \bigwedge\nolimits_{j \in N(h)} \left((\overline{y}_j \wedge \overline{y}^h_j) - \bigwedge\nolimits_{i \in H} d_{ij} \right). \tag{63}$$

Assume that the system (9) and (10) is solvable for some $\lambda \in \mathcal{I}$. Clearly, one of these possibilities takes place: (a) $h(\lambda) = m$, (b) $0 < h(\lambda) < m$, $\lambda \in \Lambda(h)$, (c) $h(\lambda) = 0$. The assertion of the theorem then follows from Theorems 3–5. □

Theorem 7. *Suppose that $C \in \mathcal{I}(m,n)$, $b \in \mathcal{I}(m)$ and $\underline{y}, \overline{y} \in \mathcal{I}(n)$. The problem of recognizing the solvability of the bounded parametric max-Łuk linear system*

$$C \otimes_L y = \lambda \otimes_L b \tag{64}$$

with bounds $\underline{y} \leq y \leq \overline{y}$ for some value of the parameter $\lambda \in \mathcal{I}$ can be solved in $O(mn^2)$ time.

Proof. In view of Theorem 6, the solvability of the bounded max-Łuk linear system (64) for some $\lambda \in \mathcal{I}$ can be verified by verifying the solvability of (64) for the values $\lambda^1_{\max}, \lambda^2_{\max}, \dots \lambda^m_{\max}$ and verifying the condition $\underline{y}_j \leq \bigwedge_{i \in M}(1 - c_{ij})$, for every $j \in N$.

For every $h = 1, 2, \dots, m$, $\lambda^h_{\max}$ can be computed in $O(n^2)$ time and the computation of $y^\star(\lambda^h_{\max})$ requires $O(n)$ time. The verification of $y^\star(\lambda^h_{\max}) \in S(C_H, \lambda^h_{\max} \otimes_L b_H, \overline{y}, \underline{y})$ needs $O(n)$ time, while condition (ii) in Theorem 5 can be verified in $O(mn)$ time. Thus, the total computational complexity is $O(mn^2)$. □

5. Numerical Examples

Example 1 (A numerical illustration to Q1—solvable case). *Assume that transition matrix* C *and required state vector* b *are given. Our goal is to recognize whether there is* $0 < \lambda \in \mathcal{I}$ *and* $y \in \mathcal{I}(n)$ *with* $\underline{y} \leq y \leq \overline{y}$ *such that* $C \otimes_L y = \lambda \otimes_L b$*. In other words, we ask whether the system (9) and (10) with entries (65) is solvable for some* $0 < \lambda \in \mathcal{I}$*.*

$$C = \begin{pmatrix} 0.6 & 0.5 & 0.5 & 0.8 & 0.8 \\ 0.5 & 0.7 & 0.6 & 0.5 & 0.9 \\ 0.3 & 0.9 & 0.3 & 0.3 & 0.0 \\ 0.1 & 0.7 & 0.9 & 0.2 & 0.9 \\ 0.9 & 0.2 & 0.2 & 0.6 & 0.8 \end{pmatrix}, \ b = \begin{pmatrix} 0.7 \\ 0.6 \\ 0.3 \\ 0.1 \\ 0.1 \end{pmatrix}, \ \underline{y} = \begin{pmatrix} 0.1 \\ 0.0 \\ 0.1 \\ 0.3 \\ 0.1 \end{pmatrix}, \ \overline{y} = \begin{pmatrix} 0.8 \\ 0.6 \\ 0.6 \\ 0.9 \\ 0.5 \end{pmatrix}. \tag{65}$$

Applying Theorem 6, we get a positive answer. Namely, the system (9) and (10) is solvable for $\lambda \in (0.3, 0.4\rangle$ *with solution* $y^\star = (0.1, 0.1, 0.1, 0.3, 0.1)$ *and has no solution for* $\lambda \in (0, 0.3\rangle \cup (0.4, 1\rangle$*. Therefore, the orbit of the DES considered in case study in Section 2 can reach the state* $\lambda \otimes b$ *for every* $0.3 < \lambda \leq 0.4$*, if the starting state is* $y^\star$*. On the other hand, the DES cannot reach the state* $b = 1 \otimes_L b$ *nor can reach any state* $\lambda \otimes_L b$ *if* $\lambda \leq 0.3$ *or* $\lambda > 0.4$*.*

Details of the computation are shown below. We use the method described in Theorem 6. By Definition (59), we get five different values of $h(\lambda)$ *and distinguish the following cases: (a)* $h(\lambda) = 5$*, (b)* $h(\lambda) = 1, 2, 3$ *and (c)* $h(\lambda) = 0$*.*

Case (a). We have $H = \{1, 2, 3, 4, 5\}$*,* $M \setminus H = \emptyset$ *and* $\lambda \in (0.9, 1\rangle$*. By (62), we get* $\lambda^5_{\max} = 1$*. Using (32) we compute the maximal possible candidate for a solution:* $y^\star(\lambda^5_{\max}) = y^\star(1) = (0.2, 0.4, 0.2, 0.5, 0.2)^T$*. Clearly,* $\underline{y} \leq y^\star(1)$*. It remains to see whether* $y^\star(1)$ *is a solution to (9).*

$$C \otimes_L y^\star(1) = \begin{pmatrix} 0.6 & 0.5 & 0.5 & 0.8 & 0.8 \\ 0.5 & 0.7 & 0.6 & 0.5 & 0.9 \\ 0.3 & 0.9 & 0.3 & 0.3 & 0.0 \\ 0.1 & 0.7 & 0.9 & 0.2 & 0.9 \\ 0.9 & 0.2 & 0.2 & 0.6 & 0.8 \end{pmatrix} \otimes \begin{pmatrix} 0.2 \\ 0.4 \\ 0.2 \\ 0.5 \\ 0.2 \end{pmatrix} = \begin{pmatrix} 0.3 \\ 0.1 \\ 0.3 \\ 0.1 \\ 0.1 \end{pmatrix}$$
$$\neq \begin{pmatrix} 0.7 \\ 0.6 \\ 0.3 \\ 0.1 \\ 0.1 \end{pmatrix} = 1 \otimes \begin{pmatrix} 0.7 \\ 0.6 \\ 0.3 \\ 0.1 \\ 0.1 \end{pmatrix} = \lambda^5_{\max} \otimes_L b$$

In view of Lemma 5, $C \otimes_L y^\star(1) \neq \lambda^5_{\max} \otimes_L b$ *implies that the system has no solution when* $0.9 < \lambda \leq 1$*.*

Case (b). Three subcases are considered.

For $h(\lambda) = 3$*, we have* $H = \{1, 2, 3\}$*,* $M \setminus H = \{4, 5\}$ *and* $\Lambda(3) = (0.7, 0.9\rangle$*. Using (53) and (61), we compute* $\mu^3 = (0.1, 0.9, 0.1, 0.5, 0.4)^T$*,* $N(3) = \{2\}$*. Then* $\lambda^3_{\max} = 0.9 \wedge 0.9 = 0.9 \in \Lambda(3)$*, in view of (63). Applying this value we get* $y^\star(0.9) = (0.1, 0.3, 0.1, 0.4, 0.1)^T$*, in view of (28). The candidate* $y^\star(0.9)$ *fulfills* $\underline{y} \leq y^\star_j(0.9)$*, but it is not a solution to the system, because* $C \otimes_L y^\star(0.9) \neq \lambda^3_{\max} \otimes_L b$*. Therefore, the system has no solution when* $0.7 < \lambda \leq 0.9$*.*

For $h(\lambda) = 2$ *we have* $H = \{1, 2\}$*,* $M \setminus H = \{3, 4, 5\}$ *and* $\Lambda(2) = (0.4, 0.7\rangle$*. Similarly to the previous subcase, we can calculate* $\mu^2 = (0, 0.2, 0.1, 0.5, 0.4)^T$ *and* $N(2) = \{4\}$*. Then* $\lambda^2_{\max} = 0.7 \wedge 0.5 = 0.5 \in \Lambda(2)$*. The obtained result* $y^\star(0.5) = (0.1, 0.1, 0.1, 0.4, 0.1)^T$ *fulfills the condition* $\underline{y} \leq y^\star(0.5)$*, but again,* $y^\star(0.5)$ *is not a solution to the system, which means that there are no solutions when* $0.4 < \lambda \leq 0.7$*.*

For $h(\lambda) = 1$ *we have* $H = \{1\}$*,* $M \setminus H = \{2, 3, 4, 5\}$ *and* $\Lambda(1) = (0.3, 0.4\rangle$*. Similarly to the previous subcases,* $\mu^1 = (0, -0.1, -0.1, 0.5, 0.2)^T$ *and* $N(1) = \{4\}$*, and so* $\lambda^1_{\max} = 0.4 \wedge 0.5 = 0.4 \in \Lambda(1)$*. For* $\lambda^1_{\max} = 0.4$ *we compute* $y^\star(0.4) = (0.1, 0.1, 0.1, 0.3, 0.1)^T$*. This candidate fulfills* $\underline{y} \leq y^\star_j(0.4)$ *and also is* *a solution to the system, because of the equality $C \otimes_L y^\star(0.4) = 0.4 \otimes_L b$. It follows that the system (9) and (10) considered in this example is solvable when $\lambda = 0.4$.*

Case (c). In this case we have $h(\lambda) = 0$, i.e., $H = \emptyset$, $M \setminus H = M$ and $\lambda \in (0, 0.3\rangle$. The maximal candidate $y^\star(\lambda) = (0.1, 0.1, 0.1, 0.2, 0.1)^T$ satisfies $C \otimes_L y^\star(\lambda) = \lambda \otimes_L b$, but the requirement of $\underline{y} \le y^\star(\lambda)$ is not fulfilled. As a consequence, the considered system is not solvable when $0 < \lambda \le 0.3$.

Example 2 (A numerical illustration to Q1—insolvable case). *Similarly to Example 1, transition matrix C and required state vector b are given. Again, we wish to recognize whether there is $0 < \lambda \in \mathcal{I}$ and $y \in \mathcal{I}(n)$ with $\underline{y} \le y \le \overline{y}$ such that $C \otimes_L y = \lambda \otimes_L b$. In this example, the matrix C is the same, only the vectors b, $\underline{y}$ and $\overline{y}$ have different entries.*

$$C = \begin{pmatrix} 0.6 & 0.5 & 0.5 & 0.8 & 0.8 \\ 0.5 & 0.7 & 0.6 & 0.5 & 0.9 \\ 0.3 & 0.9 & 0.3 & 0.3 & 0.0 \\ 0.1 & 0.7 & 0.9 & 0.2 & 0.9 \\ 0.9 & 0.2 & 0.2 & 0.6 & 0.8 \end{pmatrix}, \; b = \begin{pmatrix} 0.8 \\ 0.8 \\ 0.8 \\ 0.5 \\ 0.5 \end{pmatrix}, \; \underline{y} = \begin{pmatrix} 0.1 \\ 0.2 \\ 0.3 \\ 0.8 \\ 0.5 \end{pmatrix}, \; \overline{y} = \begin{pmatrix} 0.7 \\ 0.4 \\ 0.7 \\ 0.9 \\ 0.8 \end{pmatrix}. \tag{66}$$

Applying the method described in Theorem 6, we get a negative result: the system has no solution for any $\lambda \in \mathcal{I}$. As a consequence, neither b, nor any of its multiples $\lambda \otimes_L b$ can be reached by the orbit of the DES.

The details of the computation are shown below. By Definition (59), we get three different values of $h(\lambda)$ and distinguish the following cases: (a) $h(\lambda) = 5$, (b) $h(\lambda) = 3$ and (c) $h(\lambda) = 0$.

Case (a). We have $H = \{1, 2, 3, 4, 5\}$, $M \setminus H = \emptyset$ and $\lambda \in (0.5, 1\rangle$. By (62), we get $\lambda^5_{\max} = 0.6$. From entries (66) we compute the maximal possible candidate for solution, $y^\star(\lambda^5_{\max}) = y^\star(0.6) = (0.2, 0.4, 0.2, 0.5, 0.2)^T$. Clearly, $\underline{y} \le y^\star(0.6)$. It remains to verify whether $y^\star(0.6)$ is a solution to (9).

In view of Lemma 5, $C \otimes_L y^\star(6) \ne \lambda^5_{\max} \otimes_L b$ implies that the system has no solution when $0.5 < \lambda \le 1$.

Case (b). In this case, only one subcase has to be considered.

For $h(\lambda) = 3$ we have $H = \{1, 2, 3\}$, $M \setminus H = \{4, 5\}$ and $\Lambda(3) = (0.2, 0.5\rangle$. Using (53) and (61), we compute $\mu^3 = (-0.1, 0.4, -0.1, 0.4, 0.2)^T$, $N(3) = \{2, 4\}$. Then $\lambda^3_{\max} = 0.5 \wedge 0.4 = 0.4 \in \Lambda(3)$, in view of (63). Applying this value we get $y^\star(0.4) = (0.1, 0.3, 0.1, 0.4, 0.1)^T$, in view of (28). The candidate $y^\star(0.4)$ does not fulfill $\underline{y} \le y^\star_j(0.4)$ and at the same time is not a solution to the system, because $C \otimes_L y^\star(0.4) \ne \lambda^3_{\max} \otimes_L b$. Therefore, the system has no solution when $0.2 < \lambda \le 0.5$.

Case (c). In this case we have $h(\lambda) = 0$, i.e., $H = \emptyset$, $M \setminus H = M$ and $\lambda \in (0, 0.2\rangle$. The maximal candidate $y^\star(\lambda) = (0.1, 0.1, 0.1, 0.2, 0.1)^T$ satisfies $C \otimes_L y^\star(\lambda) = \lambda \otimes_L b$, but the requirement of $\underline{y} \le y^\star(\lambda)$ is not fulfilled. As a consequence, the considered system is not solvable when $0 < \lambda \le 0.2$.

6. Conclusions

In this study, existence of a bounded solution to a one-sided linear system in max-Łuk algebra has been considered in dependence on a given linear parameter factor on the fixed side of the system. Equivalent solvability conditions have been found and a polynomial-time recognition algorithm has been suggested. The correctness of the algorithm has been exactly demonstrated. The work of the algorithm has been illustrated by numerical examples.

The results are new: although the solvability of a one-sided linear system in max-Łuk algebra in the non-parametric case can easily be verified, the method of recognizing the solvability of the parameterized system has not yet been known.

The presented results can be applied in the study of the max-Łukasiewicz systems with interval coefficients. Łukasiewicz arithmetical conjunction can also be used in various types of optimization problems, for example, in the study of interactive cash-flows. Furthermore, the suggested recognition algorithm plays an important role in the investigation of interval eigenvectors.

An advantage of the presented algorithm is, that not only the existence or non-existence of the solution is recognized; the solution values are computed as well, in the solvable case. A possible generalization of the results for other t-norms, different from the Łukasiewicz t-norm and minimum (the Gödel t-norm), remains open for future research.

Author Contributions: All authors contributed equally to this work. All authors have read and agreed to the published version of the manuscript.

Funding: This research was funded by the Czech Science Foundation (GAČR) #18-01246S.

Acknowledgments: The authors appreciate the valuable ideas and suggestions of J. Plavka (Technical University of Košice) expressed in personal discussions about this manuscript.

Conflicts of Interest: The authors declare no conflict of interest. The funders had no role in the design of the study; in the collection, analyses, or interpretation of data; in the writing of the manuscript, or in the decision to publish the results.

References

1. Schweizer, B.; Sklar, A. Statistical metric spaces. *Pac. J. Math.* **1960**, *10*, 313–334. [CrossRef]
2. Gottwald, S. *A Treatise on Many-Valued Logics*; Studies in Logic and Computation; Research Studies Press: Baldock, UK, 2001. [CrossRef]
3. Gavalec, M.; Němcová, Z. Steady states of max-Łukasiewicz fuzzy systems. *Fuzzy Sets Syst.* **2017**, *325*, 58–68. [CrossRef]
4. Cimler, R.; Gavalec, M.; Zimmermann, K. An optimization problem on the image set of a (max, min) fuzzy operator. *Fuzzy Sets Syst.* **2018**, *341*, 113–122. [CrossRef]
5. Gavalec, M.; Rashid, I.; Cimler, R. Eigenspace structure of a max-drast fuzzy matrix. *Fuzzy Sets Syst.* **2014**, *249*, 100–113. [CrossRef]
6. Rashid, I.; Gavalec, M.; Sergeev, S. Eigenspace of a three-dimensional max-Łukasiewicz fuzzy matrix. *Kybernetika* **2012**, *48*, 309–328.
7. Rashid, I.; Gavalec, M.; Cimler, R. Eigenspace structure of a max-prod fuzzy matrix. *Fuzzy Sets Syst.* **2016**, *303*, 136–148. [CrossRef]
8. Gavalec, M.; Plavka, J.; Tomášková, H. Interval eigenproblem in max-min algebra. *Liner Algebra Appl.* **2014**, *440*, 24–33. [CrossRef]
9. Gavalec, M.; Plavka, J.; Ponce, D. Tolerance types of interval eigenvectors in max-plus algebra. *Inf. Sci.* **2017**, *367*, 14–27. [CrossRef]
10. Gavalec, M.; Plavka, J.; Ponce, D. Tolerance and weak tolerance of interval eigenvectors in fuzzy algebra. *Fuzzy Sets Syst.* **2019**, *369*, 145–156. [CrossRef]
11. Gavalec, M. *Periodicity in Extremal Algebras*; Gaudeamus: Hradec Králpvé, Czech Republic, 2004.
12. Golan, J.S. *Semirings and Their Applications*; Springer: Berlin/Heidelberg, Germany, 2013.
13. Gondran, M.; Minoux, M. *Graphs, Dioids and Semirings: New Models and Algorithms*; Springer: Berlin/Heidelberg, Germany, 2008; Volume 41.
14. Kolokoltsov, V.; Maslov, V.P. *Idempotent Analysis and Its Applications*; Springer: Berlin/Heidelberg, Germany, 1997; Volume 401.
15. Mysková, H. Interval eigenvectors of circulant matrices in fuzzy algebra. *Acta Electrotech. Inform.* **2012**, *12*, 57. [CrossRef]
16. Mysková, H. Weak stability of interval orbits of circulant matrices in fuzzy algebra. *Acta Electrotech. Inform.* **2012**, *12*, 51. [CrossRef]
17. Plavka, J. On the weak robustness of fuzzy matrices. *Kybernetika* **2013**, *49*, 128–140.
18. Tan, Y.J. Eigenvalues and eigenvectors for matrices over distributive lattices. *Linear Algebra Appl.* **1998**, *283*, 257–272. [CrossRef]
19. Zimmermann, K. *Extremální Algebra*; Útvar vědeckých informací Ekonomického ústavu ČSAV: Praha, Czech Republic, 1976.
20. Cuninghame-Green, R.A. *Minimax Algebra*; Springer: Berlin/Heidelberg, Germany, 2012; Volume 166.
21. Gavalec, M. Monotone eigenspace structure in max-min algebra. *Linear Algebra Appl.* **2002**, *345*, 149–167. [CrossRef]

22. Gavalec, M.; Němcová, Z.; Sergeev, S. Tropical linear algebra with the Łukasiewicz T-norm. *Fuzzy Sets Syst.* **2015**, *276*, 131–148. [CrossRef]
23. Gavalec, M.; Ponce, D.; Plavka, J. Strong tolerance of interval eigenvectors in fuzzy algebra. *Fuzzy Sets Syst.* **2019**, *369*, 145–156. [CrossRef]
24. Gavalec, M.; Plavka, J.; Ponce, D. EA/AE-Eigenvectors of Interval Max-Min Matrices. *Mathematics* **2020**, *8*, 882. [CrossRef]
25. Wang, Q.; Qin, N.; Yang, Z.; Sun, L.; Peng, L.; Wang, Z. On Monotone Eigenvectors of a Max-T Fuzzy Matrix. *J. Appl. Math. Phys.* **2018**, *6*. [CrossRef]

mathematics

MDPI

Article

Similarity Measure of Lattice Ordered Multi-Fuzzy Soft Sets Based on Set Theoretic Approach and Its Application in Decision Making

Sabeena Begam S [1], Vimala J [1], Ganeshsree Selvachandran [2], Tran Thi Ngan [3,*] and Rohit Sharma [4]

1 Department of Mathematics, Alagappa University, Karaikudi 630004, India; sabeekisma23@gmail.com (S.B.S.); vimaljey@alagappauniversity.ac.in (V.J.)
2 Department of Actuarial Science and Applied Statistics, Faculty of Business and Information Science, UCSI University, Jalan Menara Gading, Cheras, Kuala Lumpur 56000, Malaysia; ganeshsree86@yahoo.com or Ganeshsree@ucsiuniversity.edu.my
3 Faculty of Computer Science and Engineering, Thuyloi University, 175 Tay Son, Dong Da, Hanoi 010000, Vietnam
4 Department of Electronics & Communication Engineering, SRM Institute of Science and Technology, Ghaziabad 201203, India; rohitapece@gmail.com
* Correspondence: ngantt@tlu.edu.vn; Tel.: +84-0989-040-450

Received: 26 June 2020; Accepted: 17 July 2020; Published: 31 July 2020

Abstract: Many effective tools in fuzzy soft set theory have been proposed to handle various complicated problems in different fields of our real life, especially in decision making. Molodtsov's soft set theory has been regarded as a newly emerging mathematical tool to deal with uncertainty and vagueness. Lattice ordered multi-fuzzy soft set ($\mathscr{LMFSS}$) has been applied in forecasting process. However, similarity measure is not used in this application. In our research, similarity measure of $\mathscr{LMFSS}$ is proposed to calculate the similarity between two $\mathscr{LMFSS}$s. Moreover, some of its properties are introduced and proved. Finally, an application of $\mathscr{LMFSS}$ in decision making using similarity measure is analysed.

Keywords: soft set; fuzzy soft set; multi-fuzzy set; multi-fuzzy soft set; $\mathscr{LMFSS}$; similarity measure of $\mathscr{LMFSS}$

MSC: 06D72

---

1. Introduction

Lattice theory plays an important role in many fields in everyday life. The notion of lattices was introduced by Richard Dedekind. Further, Garrett Birrkhoff [1] started the general development of lattice theory in the mid 1930s. George Gratzer [2] vitally developed the theory of lattices and discussed about the applications of lattice theory. In 1965, the notion of fuzzy set was introduced by Zadeh [3] to handle uncertainty in various fields of everyday life. Fuzzy set theory is the generalization of classical set theory, whose range values are within the integer 0 and 1 to the interval [0, 1]. Many researchers such as Xu et al., Roy et al., Majumdar and Samanta, Tripathy et al., are attracted by the concept of fuzzy sets and they have developed new notions of fuzzy sets and applied them in many fields of science and technology, economics, medical science. There are several types of fuzzy set extensions in fuzzy set theory, including intuitionistic fuzzy set, interval-valued fuzzy set, vague set, picture fuzzy set and complex fuzzy set.

In real world complicated problems in engineering, social science, economics, medical science, ... classical mathematics methods are always not successful because uncertainty is always present in these problems. A wide range of existing theories such as fuzzy set theory and it extensions in decision making [4], sensitivity analysis in MCDM problems [5], rough set theory in forecasting [6], probability theory, vague set theory and interval mathematics are well known. Apart from that, mathematical approaches are useful to model vagueness. Each of these concepts has its inherent difficulties as pointed out in Reference [7].

In 1999, Molodtsov initiated a new concept of soft set theory, which can be seen as a novel mathematical tool for dealing with uncertainty which is free from the limitations. This theory is useful in different fields such as decision making, data analysis and forecasting and so forth. The soft set model can be combined with other mathematical models. Maji et al. [8] presented the concept of generalized fuzzy soft set theory, which is based on a combination of the fuzzy set and soft set theory. In addition, this concept has proven to be very useful in many different fields. One of the most common applications of soft set is decision making support. Tran Thi Ngan et al. [9] proposed fuzzy aggregation operators and constructed a dental disease diagnosis from X-ray images using these proposed operators. Tran Manh Tuan et al. [10] introduced similarity measure extensions in fuzzy and neutrosophic sets. These measures were implemented in predicting linkage in a co-authorship network with higher performance. Complex fuzzy sets were used in multi-criteria decision making problems [11]. Complex fuzzy t-norms and t-conorms were introduced in this research and applied in solving decision making problem. The complex fuzzy measures on Mamdani complex fuzzy inference systems (Mamdani CFIS) were also presented in Reference [12]. Mamdani fuzzy inference system was extended on complex fuzzy sets. By experimenting on different data sets, Mamdani CFIS worked better comparing with Adaptive Neuro Complex Fuzzy Inference System(ANFIS) and Mamdani Fuzzy Inference System (FIS).

Sebastian and Ramakrishnan [13] proposed the concept of multi-fuzzy sets theory, which is a more general fuzzy set using ordinary fuzzy sets as building blocks and its membership function is an ordered sequence of ordinary fuzzy membership functions. In Reference [14], the notion of multi-fuzzy sets provides a new method to represent some problems which are difficult to explain in other extensions of fuzzy set theory, such as color of pixels. Yong et al. [15] introduced the concept of multi-fuzzy soft set by combining the multi-fuzzy set and soft set models and provided its application in decision making under an imprecise environment. Afterwards, Dey and Pal [16] generalized the notion of multi-fuzzy soft set and its application to decision making.

Similarity measure is an important topic for dealing with uncertain data. In recent years, many researchers have introduced different similarity measures between fuzzy sets, vague sets, soft sets, fuzzy soft sets, intuitionistic fuzzy sets and intuitionistic fuzzy soft sets. Similarity measures have been extensively studied from many aspects and applied in different fields such as decision making, pattern recognition, region extraction, coding theory, image processing, signal detection, security verification systems, medical diagnosis and so on. In 2008, Majumdar and Samanta [17] initiated the study of uncertainty measures of soft sets and also introduced some new similarity measures for fuzzy soft sets, which is based on distance, set theoretic approach and matching function in Reference [18]. Liu et al. [19] proposed similarity measures and entropy of fuzzy soft sets and its properties. Feng and Zheng [20] studied new similarity measures for fuzzy soft sets based on distance measures.

Many various similarity measures on fuzzy sets (FS) were proposed. Peng [21] proposed similarity and distance measure on Pythagorean FS (PFS). These new measures overcame the limitations of introduced similarity or distance measures. This was prove by examples and application in pattern recognition. Fei et al. [22] introduced a novel vector valued similarity measure between two intuitionistic fuzzy sets (IFS). This measure included the uncertainty and similarity for intuitionistic fuzzy sets. This measure was applied in solving classification problem on Iris dataset from UCI. The uncertainty supported the classification when the similarity between two classes was the same. Song et al. [23] reviewed available similarity measures and proposed a similarity measure on IFS based on the functions of IFS (membership, non-membership and hesitation functions). The application on medical data set showed the advantages of proposed measure comparing with other presented ones.

The similarity measure on complex multi-fuzzy soft set was presented in Reference [24]. This measure was used to evaluate the alternatives in order to make accuracy decision.

Not only measure the similarity among fuzzy sets, other similarity measures on different objects were also constructed. Chenlei Lv et al. [25] defined a measure to calculate nasal similarity among faces in 3D space. This measure was mainly based on the shape comparison, it was applied into facial classification and identification via a hierarchical structure on public facial datasets. To define users with the same behavior in recommender system based on collaborative filtering, a similarity measure was introduced by Gazdar and Hidri [26]. By experimental results on three UCI data sets, proposed similarity measure obtained higher performance in accuracy and ranking-oriented metrics. Tran Manh Tuan et al. [27] proposed complex fuzzy similarity measures and their weighted versions. These similarity measures were applied in a new rule reduction in order to construct an effective decision making support system.

In 2014, Zhang and Shu [28] extended the idea of multi-fuzzy soft set and introduced the notion of possibility multi-fuzzy soft set and applied it to a decision making problem and also discussed the similarity between two possibility multi-fuzzy soft set and its application to medical diagnosis. Yousef Al-Qudah and Nasruddin Hassan introduced axiomatic definitions of entropy and similarity measure for Complex multi-fuzzy soft set in Reference [24]. Selvachandran et al. proposed distance and distance induced intuitionistic entropy measures for the generalized intuitionistic fuzzy soft set model in Reference [29]. The information measures of distance and similarity for the complex vague soft set model was introduced by Selvachandran et al. in References [30,31] respectively. Vimala et al. initiated new theories such as fuzzy lattice ordered group [32], anti-lattice ordered fuzzy soft group [33], lattice ordered interval-valued hesitant fuzzy soft sets [34] and applied it to decision making problems [35] and also introduced the new concept of complex intuitionistic fuzzy soft lattice ordered group and its weighted distance measures in Reference [36]. Further, Sabeena begam et al. [37] derived the concept of lattice approach on multi-fuzzy soft set and also illustrated its application using forecasting process. The similarity measure between two $\mathscr{LMFSS}$s was not proposed. Later, the algebraic aspects of $\mathscr{LMFSS}$ such as new concept of modular and distributive $\mathscr{LMFSS}$ were presented in Reference [38]. Its properties were also established.

The main purpose of this paper is to introduce the concept of similarity between two lattice ordered multi-fuzzy soft sets. This is a new similarity measure between two $\mathscr{LMFSS}$s. To illustrate the proposed measure, two numerical examples are performed step by step. Our proposed fuzzy similarity measure has some main advantages. Firstly, this measure is simple and very efficient to evaluate. Secondly, this measure is introduced in multi-dimension using the lattice structure that makes it be easier for explanation in many problems. The properties of proposed measure and an application in decision making using this measure are also presented. Although the good properties of a similarity measure between two vectors were pointed out in Reference [39]. But in this paper, we are checking the similarity measure based on soft set theory. Thus, properties mentioned in Reference [39] were not suitable to evaluate our measure.

This paper is organized as follows—in Section 2, fundamentals of fuzzy set theory, soft set theory, fuzzy soft set theory, multi-fuzzy set theory, multi-fuzzy soft set theory, $\mathscr{LMFSS}$ and its operations which are useful for subsequent discussions are presented. Novel similarity measures between two $\mathscr{LMFSS}$s are discussed in Section 3. Section 4 discusses the application of similarity measures in two $\mathscr{LMFSS}$s. In Section 5, some conclusions and further works are provided.

2. Preliminaries

In this section, we summarize some of the important concepts related to this paper pertaining to soft set, multi-fuzzy set, multi-fuzzy soft set and lattice ordered multi-fuzzy soft set.

2.1. Fuzzy Sets and Fuzzy Soft Sets

Fuzzy set is a type of very important mathematical structure to represent a collection of objects whose boundary is vague.

Definition 1 ([3]). *Let U be a non-empty set. Let $A = \{(x, \mu(x)) : x \in U\}$ with $\mu : U \longrightarrow [0,1]$. Then A is said to be a fuzzy set over U, and μ is said to be the membership function of A.*

Denote the power set of U by $\wp(U)$. Molodtsov defined the concept of a soft set in the following way:

Definition 2 ([7]). *Let U be a non-empty set. Let E be a set of parameters. Let $\mathcal{G} : E \longrightarrow \wp(U)$. Then $(\mathcal{G}, E) = \{(\varepsilon, \mathcal{G}(\varepsilon)) : \varepsilon \in E\}$ is said to be a soft set on U. In particular, $\mathcal{G}(\varepsilon_0)$ is said to be the set of ε_0-approximate elements of $(\mathcal{G}, E)$.*

Denote the collection of all the fuzzy sets (as defined in Definition 1) over U by $\mathscr{F}(U)$. Maji described the concept of fuzzy soft set in the following manner:

Definition 3 ([8]). *Let U be a non-empty set. Let E be a set of parameters. Let $\mathcal{H} : E \longrightarrow \mathscr{F}(U)$. Then $(\mathcal{H}, E) = \{(\varepsilon, \mathcal{H}(\varepsilon)) : \varepsilon \in E\}$ is said to be a fuzzy soft set on U. In particular, $\mathcal{H}(\varepsilon_0)$ is said to be the fuzzy set of ε_0-approximate elements of $(\mathcal{H}, E)$.*

2.2. Multi-Fuzzy Sets and Multi-Fuzzy Soft Sets

In 2011, the concept of fuzzy sets was generalized to multi-fuzzy sets by Sebastian and Ramakrishnan. Multi-fuzzy set theory as a mathematical tool to deal with life problems that have multi-dimensional characterization properties.

Definition 4 ([13]). *Let U be a non-empty set. Let J be a set of indices. Let*

$$\widetilde{A} = \{\left(x, (\mu_j(x))_{j \in J}\right) : x \in U\}, \tag{1}$$

with $\mu_j : U \longrightarrow [0,1]$ for all $j \in J$. Then

i. *$\widetilde{A}$ is said to be a multi-fuzzy set in U.*
ii. *J is said to be the index of $\widetilde{A}$.*
iii. *$|J|$ is said to be the dimension of $\widetilde{A}$.*
iv. *$(\mu_j)_{j \in J}$ is said to be the multi-membership function of $\widetilde{A}$.*

Remark 1. *For the particular case of $J = \{1, 2, \ldots, p\} \subset \mathbb{N}$ in Definition 4:*

i. $(\mu_j(x))_{j \in J} = (\mu_1(x), \mu_2(x), \ldots, \mu_p(x))$.
ii. $(\mu_j)_{j \in J} = (\mu_1, \mu_2, \ldots, \mu_p)$.
iii. $|J| = p$.

The collection of all multi-fuzzy sets in U shall be denoted as $\mathscr{MFS}(U)$. In particular, of all multi-fuzzy sets in U with the index J shall be denoted as $\mathscr{MF}_J\mathscr{S}(U)$.

Definition 5 ([14]). *Let U be a non-empty set. Let $\widetilde{A}, \widetilde{B} \in \mathscr{MF}_J\mathscr{S}(U)$ which both have the same index J. Denote $(\mu_j)_{j \in J}$ and $(\nu_j)_{j \in J}$ to be the multi-membership functions of $\widetilde{A}$ and $\widetilde{B}$ respectively. Then, $\widetilde{A}$ is said to be a multi-fuzzy subset of $\widetilde{B}$, denoted as $\widetilde{A} \sqsubseteq \widetilde{B}$, if $\mu_j(x) \leqslant \nu_j(x)$ for all $j \in J$ and for all $x \in U$.*

In other words, the relationship of multi-fuzzy subset is only defined among multi-fuzzy sets sharing the same index. As a result, whenever given $\widetilde{P} \sqsubseteq \widetilde{Q}$, it is well understood that $\widetilde{P}$ and $\widetilde{Q}$ must have the same index.

Definition 6 ([14]). *Let U be a non-empty set. Let $\widetilde{A}, \widetilde{B} \in \mathscr{MFS}(U)$. $\widetilde{A}$ and $\widetilde{B}$ are said to be equal, denoted as $\widetilde{A} = \widetilde{B}$, if both $\widetilde{A} \sqsubseteq \widetilde{B}$ and $\widetilde{B} \sqsubseteq \widetilde{A}$ holds.*

Definition 7 ([14]). *Let U be a non-empty set. Let $\widetilde{A}, \widetilde{B} \in \mathscr{MFS}(U)$ which both have the same index J. Denote $(\mu_j)_{j\in J}$ and $(\nu_j)_{j\in J}$ to be the multi-membership functions of $\widetilde{A}$ and $\widetilde{B}$ respectively. The union and intersection of $\widetilde{A}$ and $\widetilde{B}$, denoted as $\widetilde{A} \sqcup \widetilde{B}$ and $\widetilde{A} \sqcap \widetilde{B}$ respectively, is defined to be the following multi-fuzzy sets in U:*

$$\widetilde{A} \sqcup \widetilde{B} = \{\left(x, (max\{\mu_j(x), \nu_j(x)\})_{j\in J}\right) : x \in U\} \tag{2}$$

$$\widetilde{A} \sqcap \widetilde{B} = \{\left(x, (min\{\mu_j(x), \nu_j(x)\})_{j\in J}\right) : x \in U\}. \tag{3}$$

Yang et al. initiated multi-fuzzy soft set, which can be seen as an extension of multi-fuzzy set and soft set model. It is defined as follows:

Definition 8 ([15]). *Let U be a non-empty set. Let E be a set of parameters. Let $\widetilde{\mathcal{F}} : E \longrightarrow \mathscr{MFS}(U)$. Then $(\widetilde{\mathcal{F}}, E) = \{(\varepsilon, \widetilde{\mathcal{F}}(\varepsilon)) : \varepsilon \in E\}$ is said to be a multi-fuzzy soft set on U.*

The collection of all multi-fuzzy soft sets on U shall be denoted as $\mathscr{MFSS}(U)$. In particular, the collection of all multi-fuzzy soft sets on U with index in J shall be denoted as $\mathscr{MF}_J\mathscr{SS}(U)$.

Remark 2. *Given any $\varepsilon \in E$ in Definition 8. As $\widetilde{\mathcal{F}}(\varepsilon) \in \mathscr{MFSS}(U)$, $\widetilde{\mathcal{F}}(\varepsilon)$ comes with its index J_ε by Definition 4. However, J_{ε_1} and J_{ε_2} need not be equal given any $\{\varepsilon_1, \varepsilon_2\} \subseteq E$.*

Definition 9 ([15]). *Let U be a non empty set. Let A, B be two sets of parameters. Let $(\widetilde{\mathcal{F}}, A), (\widetilde{\mathcal{G}}, B) \in \mathscr{MFSS}(U)$. $(\widetilde{\mathcal{F}}, A)$ is said to be a multi-fuzzy soft subset of $(\widetilde{\mathcal{G}}, B)$ if*

i. $A \subseteq B$.
ii. $\widetilde{\mathcal{F}}(\alpha) \sqsubseteq \widetilde{\mathcal{G}}(\alpha)$ *for all* $\alpha \in A$

In such a case, we write $(\widetilde{\mathcal{F}}, A) \subseteq (\widetilde{\mathcal{G}}, B)$.

For ease on notation, the notation of $\sqsubseteq$ used in Reference [14] was not carried forward to Reference [15].

Definition 10 ([15]). *Let U be a non empty sets. Let A, B be two sets of parameters. Let $(\widetilde{\mathcal{F}}, A), (\widetilde{\mathcal{G}}, B) \in \mathscr{MFSS}(U)$. $(\widetilde{\mathcal{F}}, A)$ and $(\widetilde{\mathcal{G}}, B)$ are said to be equal if both $(\widetilde{\mathcal{F}}, A) \subseteq (\widetilde{\mathcal{G}}, B)$ and $(\widetilde{\mathcal{G}}, B) \subseteq (\widetilde{\mathcal{F}}, A)$ holds. In such a case, we write $(\widetilde{\mathcal{F}}, A) = (\widetilde{\mathcal{G}}, B)$.*

Definition 11 ([15]). *The union of $(\widetilde{\mathcal{F}}, A), (\widetilde{\mathcal{G}}, B) \in \mathscr{MFSS}(U)$ is defined to be $(\widetilde{\mathcal{H}}, C) \in \mathscr{MFSS}(U)$ where $C = A \cup B$, and with* $\widetilde{\mathcal{H}}(\varepsilon) = \begin{cases} \widetilde{\mathcal{F}}(\varepsilon) & \text{if } \varepsilon \in A \setminus B \\ \widetilde{\mathcal{G}}(\varepsilon) & \text{if } \varepsilon \in B \setminus A \\ \widetilde{\mathcal{F}}(\varepsilon) \sqcup \widetilde{\mathcal{G}}(e) & \text{if } \varepsilon \in A \cap B \end{cases}$
for all $\varepsilon \in C$.

In such a case, we write $(\widetilde{\mathcal{H}}, C) = (\widetilde{\mathcal{F}}, A) \cup (\widetilde{\mathcal{G}}, B)$.

Definition 12 ([15]). *The intersection of $(\widetilde{\mathcal{F}}, A), (\widetilde{\mathcal{G}}, B) \in \mathscr{MFSS}(U)$ is defined to be $(\widetilde{\mathcal{K}}, D) \in \mathscr{MFSS}(U)$ where $D = A \cap B$, and with $\widetilde{\mathcal{K}}(\varepsilon) = \widetilde{\mathcal{F}}(\varepsilon) \sqcap \widetilde{\mathcal{G}}(\varepsilon)$ for all $\varepsilon \in D$.*

In such a case, we write $(\widetilde{\mathcal{K}}, D) = (\widetilde{\mathcal{F}}, A) \cap (\widetilde{\mathcal{G}}, B)$.

Definition 13 ([15]). *The conjunction of* $(\widetilde{\mathcal{F}}, A), (\widetilde{\mathcal{G}}, B) \in \mathscr{MFSS}(U)$ *is defined to be* $(\widetilde{\mathcal{L}}, P) \in \mathscr{MFSS}(U)$ *where* $P = A \times B$, *and with* $\widetilde{\mathcal{L}}(\alpha, \beta) = \widetilde{\mathcal{F}}(\alpha) \sqcap \widetilde{\mathcal{G}}(\beta)$ *for all* $(\alpha, \beta) \in P$.

In such a case, we write $(\widetilde{\mathcal{L}}, P) = (\widetilde{\mathcal{F}}, A) \wedge (\widetilde{\mathcal{G}}, B)$.

Definition 14 ([15]). *The disjunction of* $(\widetilde{\mathcal{F}}, A), (\widetilde{\mathcal{G}}, B) \in \mathscr{MFSS}(U)$ *is defined to be* $(\widetilde{\mathcal{T}}, P) \in \mathscr{MFSS}(U)$ *where* $P = A \times B$, *and with* $\widetilde{\mathcal{T}}(\alpha, \beta) = \widetilde{\mathcal{F}}(\alpha) \sqcup \widetilde{\mathcal{G}}(\beta)$ *for all* $(\alpha, \beta) \in P$.

In such a case, we write $(\widetilde{\mathcal{T}}, P) = (\widetilde{\mathcal{F}}, A) \vee (\widetilde{\mathcal{G}}, B)$.

2.3. Lattice Ordered Multi-Fuzzy Soft Set ($\mathscr{LMFSS}$)

Throughout this paper, $\widetilde{X}$ refers to the initial universe, $P(\widetilde{X})$ is the power set of $\widetilde{X}$, $\widetilde{E}$ is a set of parameters and $\widetilde{\mathcal{A}} \subseteq \widetilde{E}$.

In all the following passages, the concept of the lattice is referred to the one encountered in the literature of partially ordered sets (posets). We now combine the concepts of lattices and multi-fuzzy soft sets to obtain a new hyprid structure called lattice ordered multi-fuzzy soft set.

Definition 15 ([37]). *Let* U *be a non-empty set. Let* J *be a set of indices. Let* Λ *be a lattice of parameters. Let* $\widetilde{\mathcal{R}} : \Lambda \longrightarrow \mathscr{MF}_J\mathscr{S}(U)$ *be such that, for all* $\lambda_1, \lambda_2 \in \Lambda$: $\lambda_1 \leqslant \lambda_2$ *implies* $\widetilde{\mathcal{R}}(\lambda_1) \sqsubseteq \widetilde{\mathcal{R}}(\lambda_2)$. *Then* $(\widetilde{\mathcal{R}}, \Lambda) = \{(\varepsilon, \widetilde{\mathcal{R}}(\lambda)) : \lambda \in \Lambda\}$ *is said to be a lattice ordered multi-fuzzy soft set on* U *with index* J.

The collection of all lattice ordered multi-fuzzy soft sets on U with index in J shall be denoted as $\mathscr{LMF}_J\mathscr{SS}(U)$.

Furthermore, the collection of all lattice ordered multi-fuzzy soft sets on U shall be denoted as $\mathscr{LMFSS}(U)$. Note that a lattice ordered multi-fuzzy soft set can only be established upon a collection of multi-fuzzy soft sets all sharing a particular index in J. As a result, $\mathscr{LMFSS}(U) = \bigcup_{\text{all } J} \mathscr{LMF}_J\mathscr{SS}(U)$.

Example 1 ([37]). *There are three international companies in India who manufactures vehicles for the world market, namely x, y and z. Their most frequent buyer is known to be Japan. Besides Japan, they advertise their products mainly to countries of the G20 (due to their economic advancements), as well as countries in Asia (due to geographic closeness). Note that Japan belongs to both G20 and Asia, but none of G20 or Asia fully covers the other.*

Thus it is now desired to assess the sales performance of these three companies in four regions—Japan, G20, Asia, and worldwide. The assessment is purely based on the amount of revenue generated and is given as numbers from 0 (no revenue) to 1 (highest revenue). As a result, the wider the region of consideration, the higher the score as more sales take place in that wider region.

On the other hand, as each of these companies manufactures bikes, cars and trucks, the committee agreed to look at these three products separately.

In this example, let the results be summarized as the following tables for each of the 4 regions (Tables 1–4).

Table 1. Sales performances in Japan.

	Bikes	**Cars**	**Trucks**
Company x	0.2	0.3	0.6
Company y	0.5	0.3	0.4
Company z	0.6	0.5	0.3

Table 2. Sales performances among countries of the G20.

	Bikes	**Cars**	**Trucks**
Company x	0.5	0.4	0.8
Company y	0.5	0.5	0.5
Company z	0.7	0.6	0.5

Table 3. Sales performances in Asia.

	Bikes	Cars	Trucks
Company x	0.3	0.5	0.7
Company y	0.6	0.4	0.7
Company z	0.8	0.7	0.4

Table 4. Sales performances worldwide.

	Bikes	Cars	Trucks
Company x	0.8	0.6	0.8
Company y	0.7	0.6	0.9
Company z	0.9	0.8	0.6

With all of these inputs:

a. *Denote the 3 companies with the non-empty set* $U_0 = \{x, y, z\}$.

b. *Denote the 4 regions with the lattice of parameters* $\Lambda_0 = \{\alpha = Japan, \beta = G20, \gamma = Asia, \xi = worldwide\}$, *for which* $\alpha \leqslant \beta \leqslant \xi$ *and* $\alpha \leqslant \gamma \leqslant \xi$, *but no such relationship established between* β *and* γ.

c. *Denote the 3 categories of assessment with the set of indexes* $J_0 = \{b = bike, c = car, t = truck\}$. *And take* $(\mu_j)_{j\in J_0} = (\mu_b, \mu_c, \mu_t)$, *thus fixing the order of appearance of the elements of* J_0 *in the presentation of* $(\mu_j)_{j\in J_0}$.

We have thus formed a lattice ordered multi-fuzzy soft set as shown:

$$\begin{aligned}
&(\widetilde{\mathcal{R}}_0, \Lambda_0)\\
&= \{\widetilde{\mathcal{R}}_0(\alpha) = \{\tfrac{x}{(0.2,0.3,0.6)}, \tfrac{y}{(0.5,0.3,0.4)}, \tfrac{z}{(0.6,0.5,0.3)}\}\\
&\quad \widetilde{\mathcal{R}}_0(\beta) = \{\tfrac{x}{(0.5,0.4,0.8)}, \tfrac{y}{(0.5,0.5,0.5)}, \tfrac{z}{(0.7,0.6,0.5)}\}\\
&\quad \widetilde{\mathcal{R}}_0(\gamma) = \{\tfrac{x}{(0.3,0.5,0.7)}, \tfrac{y}{(0.6,0.4,0.7)}, \tfrac{z}{(0.8,0.7,0.4)}\}\\
&\quad \widetilde{\mathcal{R}}_0(\xi) = \{\tfrac{x}{(0.8,0.6,0.8)}, \tfrac{y}{(0.7,0.6,0.9)}, \tfrac{z}{(0.9,0.8,0.6)}\}\} \in \mathscr{LMF}_{J_0}\mathscr{SS}(U_0) \subseteq \mathscr{LMFSS}(U_0).
\end{aligned}$$

3. Similarity between Two $\mathscr{LMFSS}$s Based on Set Theoretic Approach

In this section, we introduce the concept of similarity measure of two $\mathscr{LMFSS}$s and further results on similarity measure of two $\mathscr{LMFSS}$s. In all the following context, denote $a \wedge b$ to be the minimum of a and b, denote $a \vee b$ to be the maximum of a and b.

Definition 16. *Let* $(\widetilde{\mathcal{R}}, \Gamma), (\widetilde{\mathcal{Q}}, \Gamma) \in \mathscr{LMF}_J\mathscr{SS}(U)$.

The similarity measure of $(\widetilde{\mathcal{R}}, \Gamma)$ *and* $(\widetilde{\mathcal{Q}}, \Gamma)$ *over* U *is defined as*

$$\mathbf{S}((\widetilde{\mathcal{R}}, \Gamma), (\widetilde{\mathcal{Q}}, \Gamma)) = \frac{\sum_{j\in J} S_j((\widetilde{\mathcal{R}}, \Gamma), (\widetilde{\mathcal{Q}}, \Gamma))}{|J|}$$

where

$$S_j((\widetilde{\mathcal{R}}, \Gamma), (\widetilde{\mathcal{Q}}, \Gamma)) = \frac{\sum_{\gamma\in\Gamma} \vee_{x\in U}\{\mu_{\widetilde{\mathcal{R}}(\gamma),j}(x) \wedge \mu_{\widetilde{\mathcal{Q}}(\gamma),j}(x)\}}{\sum_{\gamma\in\Gamma} \vee_{x\in U}\{\mu_{\widetilde{\mathcal{R}}(\gamma),j}(x) \vee \mu_{\widetilde{\mathcal{Q}}(\gamma),j}(x)\}}$$

Example 2. *Let* $U_0 = \{x_1, x_2, x_3, x_4\}$. *Let* $J_0 = \{1, 2, 3\}$.
Let $\Gamma_0 = \{\gamma_1, \gamma_2, \gamma_3, \gamma_4\}$, *for which* $\gamma_1 \leqslant \gamma_2 \leqslant \gamma_3 \leqslant \gamma_4$.
Let $(\widetilde{\mathcal{R}}_0, \Gamma_0), (\widetilde{\mathcal{Q}}_0, \Gamma_0) \in \mathscr{LMF}_{J_0}\mathscr{SS}(U_0)$ *which are defined as follows:*

$$(\widetilde{\mathcal{R}}_0, \Gamma_0) = \{\widetilde{\mathcal{R}}_0(\gamma_1) = \{\tfrac{x_1}{(0.2,0.1,0.3)}, \tfrac{x_2}{(0.1,0.2,0.4)}, \tfrac{x_3}{(0.18,0.23,0.45)}, \tfrac{x_4}{(0.3,0.6,0.2)}\},$$

$$\widetilde{\mathcal{R}}_0(\gamma_2) = \{\frac{x_1}{(0.28,0.49,0.5)}, \frac{x_2}{(0.33,0.42,0.63)}, \frac{x_3}{(0.29,0.33,0.52)}, \frac{x_4}{(0.4,0.68,0.3)}\},$$
$$\widetilde{\mathcal{R}}_0(\gamma_3) = \{\frac{x_1}{(0.6,0.72,0.69)}, \frac{x_2}{(0.47,0.73,0.8)}, \frac{x_3}{(0.5,0.62,0.8)}, \frac{x_4}{(0.5,0.7,0.4)}\},$$
$$\widetilde{\mathcal{R}}_0(\gamma_4) = \{\frac{x_1}{(0.8,0.85,0.93)}, \frac{x_2}{(0.52,0.8,0.9)}, \frac{x_3}{(0.6,0.7,0.9)}, \frac{x_4}{(0.7,0.9,1.0)}\}\}$$

and

$$(\widetilde{\mathcal{Q}}_0, \Gamma_0) = \{\widetilde{\mathcal{Q}}_0(\gamma_1) = \{\frac{x_1}{(0.1,0.4,0.2)}, \frac{x_2}{(0.3,0.1,0.2)}, \frac{x_3}{(0.2,0.1,0.3)}, \frac{x_4}{(0.1,0.4,0.3)}\},$$
$$\widetilde{\mathcal{Q}}_0(\gamma_2) = \{\frac{x_1}{(0.23,0.5,0.4)}, \frac{x_2}{(0.4,0.25,0.36)}, \frac{x_3}{(0.32,0.4,0.5)}, \frac{x_4}{(0.32,0.5,0.4)}\},$$
$$\widetilde{\mathcal{Q}}_0(\gamma_3) = \{\frac{x_1}{(0.43,0.8,0.7)}, \frac{x_2}{(0.5,0.6,0.7)}, \frac{x_3}{(0.4,0.71,0.6)}, \frac{x_4}{(0.4,0.62,0.5)}\},$$
$$\widetilde{\mathcal{Q}}_0(\gamma_4) = \{\frac{x_1}{(0.7,0.82,1.0)}, \frac{x_2}{(0.8,0.75,1.0)}, \frac{x_3}{(0.8,0.9,0.95)}, \frac{x_4}{(0.8,0.7,0.6)}\}\}.$$

where $(\mu_j)_{j \in J_0} = (\mu_{j_1}, \mu_{j_2}, \mu_{j_3})$.
By Definition 16, we obtain,

$$S_1((\widetilde{\mathcal{R}}, \Gamma), (\widetilde{\mathcal{Q}}, \Gamma)) =$$

$$= \frac{\sum_{\gamma \in \Gamma} \vee_{x \in U} \{\mu_{\widetilde{\mathcal{R}}(\gamma), j_1}(x) \wedge \mu_{\widetilde{\mathcal{Q}}(\gamma), j_1}(x)\}}{\sum_{\gamma \in \Gamma} \vee_{x \in U} \{\mu_{\widetilde{\mathcal{R}}(\gamma), j_1}(x) \vee \mu_{\widetilde{\mathcal{Q}}(\gamma), j_1}(x)\}}$$

$$= \frac{(0.1 \vee 0.1 \vee 0.18 \vee 0.1) + (0.23 \vee 0.33 \vee 0.29 \vee 0.32) + (0.43 \vee 0.47 \vee 0.4 \vee 0.4) + (0.7 \vee 0.52 \vee 0.6 \vee 0.7)}{0.2 \vee 0.3 \vee 0.2 \vee 0.3 + 0.28 \vee 0.4 \vee 0.32 \vee 0.4 + 0.6 \vee 0.5 \vee 0.5 \vee 0.5 + 0.8 \vee 0.8 \vee 0.8 \vee 0.8}$$

$$= \frac{0.8 + 0.33 + 0.47 + 0.7}{0.3 + 0.4 + 0.6 + 0.8}$$

$$= 0.8$$

$$S_2((\widetilde{\mathcal{R}}, \Gamma), (\widetilde{\mathcal{Q}}, \Gamma)) =$$

$$= \frac{\sum_{\gamma \in \Gamma} \vee_{x \in U} \{\mu_{\widetilde{\mathcal{R}}(\gamma), j_2}(x) \wedge \mu_{\widetilde{\mathcal{Q}}(\gamma), j_2}(x)\}}{\sum_{\gamma \in \Gamma} \vee_{x \in U} \{\mu_{\widetilde{\mathcal{R}}(\gamma), j_2}(x) \vee \mu_{\widetilde{\mathcal{Q}}(\gamma), j_2}(x)\}}$$

$$= \frac{(0.1 \vee 0.1 \vee 0.1 \vee 0.4) + (0.49 \vee 0.25 \vee 0.33 \vee 0.5) + (0.72 \vee 0.6 \vee 0.62 \vee 0.62) + (0.82 \vee 0.75 \vee 0.7 \vee 0.7)}{0.4 \vee 0.2 \vee 0.23 \vee 0.6 + 0.5 \vee 0.42 \vee 0.4 \vee 0.68 + 0.8 \vee 0.73 \vee 0.71 \vee 0.7 + 0.85 \vee 0.8 \vee 0.9 \vee 0.9}$$

$$= \frac{0.4 + 0.5 + 0.72 + 0.82}{0.6 + 0.68 + 0.8 + 0.9}$$

$$= 0.82$$

$$S_3((\widetilde{\mathcal{R}}, \Gamma), (\widetilde{\mathcal{Q}}, \Gamma)) =$$

$$= \frac{\sum_{\gamma \in \Gamma} \vee_{x \in U} \{\mu_{\widetilde{\mathcal{R}}(\gamma), j_3}(x) \wedge \mu_{\widetilde{\mathcal{Q}}(\gamma), j_3}(x)\}}{\sum_{\gamma \in \Gamma} \vee_{x \in U} \{\mu_{\widetilde{\mathcal{R}}(\gamma), j_3}(x) \vee \mu_{\widetilde{\mathcal{Q}}(\gamma), j_3}(x)\}}$$

$$= \frac{(0.2 \vee 0.2 \vee 0.3 \vee 0.2) + (0.4 \vee 0.36 \vee 0.5 \vee 0.3) + (0.69 \vee 0.7 \vee 0.6 \vee 0.4) + (0.93 \vee 0.9 \vee 0.9 \vee 0.6)}{0.3 \vee 0.4 \vee 0.45 \vee 0.3 + 0.5 \vee 0.63 \vee 0.52 \vee 0.4 + 0.7 \vee 0.8 \vee 0.8 \vee 0.5 + 1.0 \vee 1.0 \vee 0.95 \vee 1.0}$$

$$= \frac{0.3 + 0.5 + 0.7 + 0.93}{0.45 + 0.63 + 0.8 + 1.0}$$

$$= 0.84$$

Then the similarity between $(\widetilde{\mathcal{R}}_0, \Gamma_0)$ *and* $(\widetilde{\mathcal{Q}}_0, \Gamma_0)$ *over* U_0 *is thus*

$$\mathbf{S}((\widetilde{\mathcal{R}}_0, \Gamma_0), (\widetilde{\mathcal{Q}}_0, \Gamma_0)) = \frac{\sum_{j \in J_0} S_j((\widetilde{\mathcal{R}}_0, \Gamma_0), (\widetilde{\mathcal{Q}}_0, \Gamma_0))}{|J_0|} = \frac{0.80 + 0.82 + 0.84}{3} = 0.82.$$

Theorem 1. *Let* $(\widetilde{\mathcal{R}},\Gamma),(\widetilde{\mathcal{Q}},\Gamma),(\widetilde{\mathcal{P}},\Gamma)\in\mathscr{LMF}_J\mathscr{SS}(U)$. *Then the following holds:*

i. $\mathbf{S}((\widetilde{\mathcal{R}},\Gamma),(\widetilde{\mathcal{Q}},\Gamma))=\mathbf{S}((\widetilde{\mathcal{Q}},\Gamma),(\widetilde{\mathcal{R}},\Gamma))$.

ii. $0\leq\mathbf{S}((\widetilde{\mathcal{R}},\Gamma),(\widetilde{\mathcal{Q}},\Gamma))\leq 1$.

iii. $\mathbf{S}((\widetilde{\mathcal{R}},\Gamma),(\widetilde{\mathcal{Q}},\Gamma))=1\Rightarrow(\widetilde{\mathcal{R}},\Gamma)=(\widetilde{\mathcal{Q}},\Gamma)\neq\emptyset$

iv. $\mathbf{S}((\widetilde{\mathcal{R}},\Gamma),(\widetilde{\mathcal{Q}},\Gamma))=0\Rightarrow(\widetilde{\mathcal{R}},\Gamma)\cap(\widetilde{\mathcal{Q}},\Gamma)=\emptyset$.

v. $(\widetilde{\mathcal{R}},\Gamma)\subseteq(\widetilde{\mathcal{Q}},\Gamma)\subseteq(\widetilde{\mathcal{P}},\Gamma)\Rightarrow\mathbf{S}((\widetilde{\mathcal{R}},\Gamma),(\widetilde{\mathcal{P}},\Gamma))\leq\mathbf{S}((\widetilde{\mathcal{Q}},\Gamma),(\widetilde{\mathcal{P}},\Gamma))$.

Proof. (i) For

$$\begin{aligned}
S_j((\widetilde{\mathcal{R}},\Gamma),(\widetilde{\mathcal{Q}},\Gamma)) &= \frac{\sum_{\gamma\in\Gamma}\vee_{x\in U}\{\mu_{\widetilde{\mathcal{R}}(\gamma),j}(x)\wedge\mu_{\widetilde{\mathcal{Q}}(\gamma),j}(x)\}}{\sum_{\gamma\in\Gamma}\vee_{x\in U}\{\mu_{\widetilde{\mathcal{R}}(\gamma),j}(x)\vee\mu_{\widetilde{\mathcal{Q}}(\gamma),j}(x)\}}\\
&= \frac{\sum_{\gamma\in\Gamma}\vee_{x\in U}\{\mu_{\widetilde{\mathcal{Q}}(\gamma),j}(x)\wedge\mu_{\widetilde{\mathcal{R}}(\gamma),j}(x)\}}{\sum_{\gamma\in\Gamma}\vee_{x\in U}\{\mu_{\widetilde{\mathcal{Q}}(\gamma),j}(x)\vee\mu_{\widetilde{\mathcal{R}}(\gamma),j}(x)\}}\\
&= S_j((\widetilde{\mathcal{Q}},\Gamma),(\widetilde{\mathcal{R}},\Gamma))
\end{aligned}$$

we have,

$$\begin{aligned}
\mathbf{S}((\widetilde{\mathcal{R}},\Gamma),(\widetilde{\mathcal{Q}},\Gamma)) &= \frac{\sum_{j\in J}S_j((\widetilde{\mathcal{R}},\Gamma),(\widetilde{\mathcal{Q}},\Gamma))}{|J|}\\
&= \frac{\sum_{j\in J}S_j((\widetilde{\mathcal{Q}},\Gamma),(\widetilde{\mathcal{R}},\Gamma))}{|J|}\\
&= \mathbf{S}((\widetilde{\mathcal{Q}},\Gamma),(\widetilde{\mathcal{R}},\Gamma))
\end{aligned}$$

(ii) Proof of this condition is trivally followed from the Definition 16.
(iii)

$$\begin{aligned}
\mathbf{S}((\widetilde{\mathcal{R}},\Gamma),(\widetilde{\mathcal{Q}},\Gamma))=1 &\Rightarrow \frac{\sum_{j\in J}S_j((\widetilde{\mathcal{R}},\Gamma),(\widetilde{\mathcal{Q}},\Gamma))}{|J|}=1\\
&\Rightarrow S_j((\widetilde{\mathcal{R}},\Gamma),(\widetilde{\mathcal{Q}},\Gamma))=1\\
&\Rightarrow \frac{\sum_{\gamma\in\Gamma}\vee_{x\in U}\{\mu_{\widetilde{\mathcal{R}}(\gamma),j}(x)\wedge\mu_{\widetilde{\mathcal{Q}}(\gamma),j}(x)\}}{\sum_{\gamma\in\Gamma}\vee_{x\in U}\{\mu_{\widetilde{\mathcal{R}}(\gamma),j}(x)\vee\mu_{\widetilde{\mathcal{Q}}(\gamma),j}(x)\}}=1\\
&\Rightarrow \sum_{\gamma\in\Gamma}\vee_{x\in U}\{\mu_{\widetilde{\mathcal{R}}(\gamma),j}(x)\wedge\mu_{\widetilde{\mathcal{Q}}(\gamma),j}(x)\}=\sum_{\gamma\in\Gamma}\vee_{x\in U}\{\mu_{\widetilde{\mathcal{R}}(\gamma),j}(x)\vee\mu_{\widetilde{\mathcal{Q}}(\gamma),j}(x)\}\\
&\Rightarrow \mu_{\widetilde{\mathcal{R}}(\gamma),j}(x)=\mu_{\widetilde{\mathcal{Q}}(\gamma),j}(x)\neq\emptyset\\
&\Rightarrow (\widetilde{\mathcal{R}},\Gamma)=(\widetilde{\mathcal{Q}},\Gamma)\neq\emptyset
\end{aligned}$$

(iv)

$$\begin{aligned}
\mathbf{S}((\widetilde{\mathcal{R}},\Gamma),(\widetilde{\mathcal{Q}},\Gamma))=0 &\Rightarrow \frac{\sum_{j\in J}S_j((\widetilde{\mathcal{R}},\Gamma),(\widetilde{\mathcal{Q}},\Gamma))}{|J|}=0\\
&\Rightarrow S_j((\widetilde{\mathcal{R}},\Gamma),(\widetilde{\mathcal{Q}},\Gamma))=0\\
&\Rightarrow \frac{\sum_{\gamma\in\Gamma}\vee_{x\in U}\{\mu_{\widetilde{\mathcal{R}}(\gamma),j}(x)\wedge\mu_{\widetilde{\mathcal{Q}}(\gamma),j}(x)\}}{\sum_{\gamma\in\Gamma}\vee_{x\in U}\{\mu_{\widetilde{\mathcal{R}}(\gamma),j}(x)\vee\mu_{\widetilde{\mathcal{Q}}(\gamma),j}(x)\}}=0\\
&\Rightarrow \sum_{\gamma\in\Gamma}\vee_{x\in U}\{\mu_{\widetilde{\mathcal{R}}(\gamma),j}(x)\wedge\mu_{\widetilde{\mathcal{Q}}(\gamma),j}(x)\}=0\\
&\Rightarrow \mu_{\widetilde{\mathcal{R}}(\gamma),j}(x)\cap\mu_{\widetilde{\mathcal{Q}}(\gamma),j}(x)=\emptyset\\
&\Rightarrow (\widetilde{\mathcal{R}},\Gamma)\cap(\widetilde{\mathcal{Q}},\Gamma)=\emptyset
\end{aligned}$$

(v) Since $(\widetilde{\mathcal{R}},\Gamma) \subseteq (\widetilde{\mathcal{Q}},\Gamma) \subseteq (\widetilde{\mathcal{P}},\Gamma)$,

$$\begin{aligned}
&\Rightarrow \quad \mu_{\widetilde{\mathcal{R}}(\gamma),j}(x) \le \mu_{\widetilde{\mathcal{Q}}(\gamma),j}(x) \le \mu_{\widetilde{\mathcal{P}}(\gamma),j}(x)\\
&\Rightarrow \quad \frac{\sum_{\gamma\in\Gamma}\vee_{x\in U}\{\mu_{\widetilde{\mathcal{R}}(\gamma),j}(x)\wedge\mu_{\widetilde{\mathcal{P}}(\gamma),j}(x)\}}{\sum_{\gamma\in\Gamma}\vee_{x\in U}\{\mu_{\widetilde{\mathcal{R}}(\gamma),j}(x)\vee\mu_{\widetilde{\mathcal{P}}(\gamma),j}(x)\}} \le \frac{\sum_{\gamma\in\Gamma}\vee_{x\in U}\{\mu_{\widetilde{\mathcal{Q}}(\gamma),j}(x)\wedge\mu_{\widetilde{\mathcal{P}}(\gamma),j}(x)\}}{\sum_{\gamma\in\Gamma}\vee_{x\in U}\{\mu_{\widetilde{\mathcal{Q}}(\gamma),j}(x)\vee\mu_{\widetilde{\mathcal{P}}(\gamma),j}(x)\}}\\
&\Rightarrow \quad S_j((\widetilde{\mathcal{R}},\Gamma),(\widetilde{\mathcal{P}},\Gamma)) \le S_j((\widetilde{\mathcal{Q}},\Gamma),(\widetilde{\mathcal{P}},\Gamma))\\
&\Rightarrow \quad \frac{\sum_{j\in J} S_j((\widetilde{\mathcal{R}},\Gamma),(\widetilde{\mathcal{P}},\Gamma))}{|J|} \le \frac{\sum_{j\in J} S_j((\widetilde{\mathcal{Q}},\Gamma),(\widetilde{\mathcal{P}},\Gamma))}{|J|}\\
&\Rightarrow \quad \mathbf{S}((\widetilde{\mathcal{R}},\Gamma),(\widetilde{\mathcal{P}},\Gamma)) \le \mathbf{S}((\widetilde{\mathcal{Q}},\Gamma),(\widetilde{\mathcal{P}},\Gamma)).
\end{aligned}$$

□

Next, we discuss about the application of similarity measure.

4. Application of $\mathscr{LMFSS}$ Using Similarity Measure in Decision Making

In this section, an application for the decision making by using the similarity measure of two $\mathscr{LMFSS}$s to analyse the rainfall in 2016 and 2017 with expected rainfall.

Let $U_0 = \{x_1, x_2, x_3, x_4\}$ be the universal set, where x_1, x_2, x_3, x_4 stands for the set of four cities in India. Let $\Gamma_0 = \{\gamma_1, \gamma_2, \gamma_3, \gamma_4\}$ as the parameters which is consider as the set of rainfalls in rainy season, where γ_1 stands for "Hour-wise rainfall" which includes 1 h, 2 h, and 3 h, γ_2 stands for "Day-wise rainfall" which includes 2 day, 3 day and 4 day, γ_3 stands for "Week-wise rainfall" which includes 2 week, 3 week, 4 week and γ_4 stands for "Month-wise rainfall" which includes 3 month, 4 month, 5 month respectively.

In this example, suppose $(\widetilde{\mathcal{R}}_0, \Gamma_0) \in \mathscr{LMFSS}(U_0)$ represents the expected rainfall in India, defined as follows:

$$\begin{aligned}
(\widetilde{\mathcal{R}}_0,\Gamma_0) = \{\mathcal{R}_0(\gamma_1) &= \{\tfrac{x_1}{(0,0,0.12)}, \tfrac{x_2}{(0.1,0,0)}, \tfrac{x_3}{(0,0.13,0)}, \tfrac{x_4}{(0,0,0)}\},\\
\mathcal{R}_0(\gamma_2) &= \{\tfrac{x_1}{(0.1,0.12,0.13)}, \tfrac{x_2}{(0.15,0.1,0.13)}, \tfrac{x_3}{(0.16,0.17,0.1)}, \tfrac{x_4}{(0.1,0.13,0.1)}\},\\
\mathcal{R}_0(\gamma_3) &= \{\tfrac{x_1}{(0.15,0.14,0.16)}, \tfrac{x_2}{(0.16,0.12,0.14)}, \tfrac{x_3}{(0.18,0.19,0.2)}, \tfrac{x_4}{(0.13,0.14,0.11)}\}\\
\mathcal{R}_0(\gamma_4) &= \{\tfrac{x_1}{(0.18,0.2,1)}, \tfrac{x_2}{(0.17,0.15,0.19)}, \tfrac{x_3}{(0.19,0.2,0.21)}, \tfrac{x_4}{(0.15,0.16,0.12)}\}\}.
\end{aligned}$$

Tabulation of $(\widetilde{\mathcal{R}}_0, \Gamma_0)$ is given in Table 5 and Figure 1.

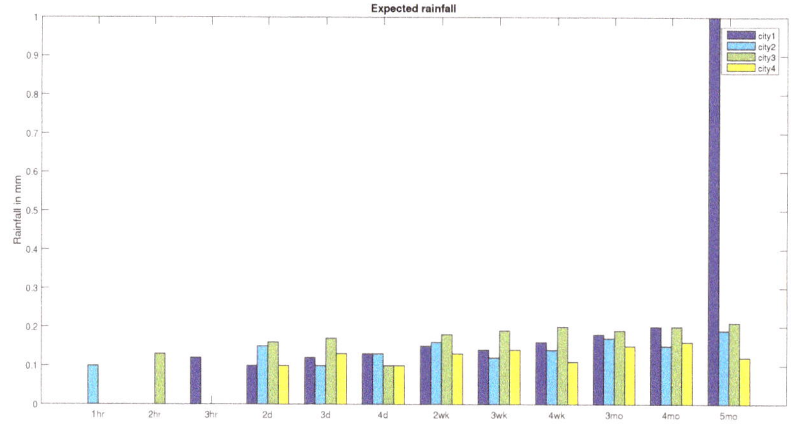

Figure 1. The expected rainfall in 2016 and 2017.

Table 5. Tabulation of $(\widetilde{\mathcal{R}}_0, \Gamma_0)$ Representing the Expected Rainfall.

$(\widetilde{\mathcal{R}}_0, \Gamma_0)$	x_1	x_2	x_3	x_4
γ_1	(0,0,0.12)	(0.1,0,0)	(0,0.13,0)	(0,0,0)
γ_2	(0.1,0.12,0.13)	(0.15,0.1,0.13)	(0.16,0.17,0.1)	(0.1,0.13,0.1)
γ_3	(0.15,0.14,0.16)	(0.16,0.12,0.14)	(0.18,0.19,0.2)	(0.13,0.14,0.11)
γ_4	(0.18,0.2,1)	(0.17,0.15,0.19)	(0.19,0.2,0.21)	(0.15,0.16,0.12)

Now let $(\widetilde{\mathcal{H}}_1, \Gamma_0), (\widetilde{\mathcal{H}}_2, \Gamma_0) \in \mathscr{LMFSS}(U_0)$ represents the recorded rainfall in India for the year 2016 and 2017 respectively, defined as follows:

$$\begin{aligned}
(\widetilde{\mathcal{H}}_1, \Gamma_0) = \{\mathcal{H}_1(\gamma_1) &= \{\tfrac{x_1}{(0,0,0)}, \tfrac{x_2}{(0,0,0.1)}, \tfrac{x_3}{(0,0,0)}, \tfrac{x_4}{(0.11,0,0)}\},\\
\mathcal{H}_1(\gamma_2) &= \{\tfrac{x_1}{(0,0.13,0)}, \tfrac{x_2}{(0.1,0,0.14)}, \tfrac{x_3}{(0,0,0.2)}, \tfrac{x_4}{(0.15,0,0.2)}\},\\
\mathcal{H}_1(\gamma_3) &= \{\tfrac{x_1}{(0.1,0.15,0)}, \tfrac{x_2}{(0.15,0.1,0.17)}, \tfrac{x_3}{(0.1,0,0.21)}, \tfrac{x_4}{(0.2,0.1,0.21)}\}\\
\mathcal{H}_1(\gamma_4) &= \{\tfrac{x_1}{(0.3,0.18,0.1)}, \tfrac{x_2}{(0.17,0.12,0.2)}, \tfrac{x_3}{(0.17,0.21,0.23}, \tfrac{x_4}{(0.22,0.13,0.23)}\}\},\\
(\widetilde{\mathcal{H}}_2, \Gamma_0) = \{\mathcal{H}_2(\gamma_1) &= \{\tfrac{x_1}{(0.1,0,0.13)}, \tfrac{x_2}{(0,0.12,0.1)}, \tfrac{x_3}{(0.1,0.14,0.1)}, \tfrac{x_4}{(0.1,0.11,0.13)}\},\\
\mathcal{H}_2(\gamma_2) &= \{\tfrac{x_1}{(0.13,0.1,0.15)}, \tfrac{x_2}{(0.1,0.14,0.12)}, \tfrac{x_3}{(0.13,0.16,0.13)}, \tfrac{x_4}{(0.2,0.18,0.18)}\},\\
\mathcal{H}_2(\gamma_3) &= \{\tfrac{x_1}{(0.14,0.13,0.17)}, \tfrac{x_2}{(0.15,0.17,0.18)}, \tfrac{x_3}{(0.17,0.18,0.16)}, \tfrac{x_4}{(0.25,0.31,0.26)}\}\\
\mathcal{H}_2(\gamma_4) &= \{\tfrac{x_1}{(0.16,0.18,0.3)}, \tfrac{x_2}{(0.19,0.18,0.2)}, \tfrac{x_3}{(0.2,0.23,0.3)}, \tfrac{x_4}{(0.28,0.34,0.28)}\}\}.
\end{aligned}$$

Tabulations of $(\widetilde{\mathcal{H}}_1, \Gamma_0)$ and $(\widetilde{\mathcal{H}}_2, \Gamma_0)$ are presented as in Tables 6 and 7 and Figures 2 and 3 respectively.

Table 6. Tabulation of $(\widetilde{\mathcal{H}}_1, \Gamma_0)$ Representing the Rainfall 2016.

$(\widetilde{\mathcal{H}}_1, \Gamma_0)$	x_1	x_2	x_3	x_4
γ_1	(0,0,0)	(0,0,0.1)	(0,0,0)	(0.11,0,0)
γ_2	(0,0.13,0)	(0.1,0,0.14)	(0,0,0.2)	(0.15,0,0.2)
γ_3	(0.1,0.15,0)	(0.15,0.1,0.17)	(0.1,0,0.21)	(0.2,0.1,0.21)
γ_4	(0.3,0.18,0.1)	(0.17,0.12,0.2)	(0.17,0.21,0.23)	(0.22,0.13,0.23)

Table 7. Tabulation of $(\widetilde{\mathcal{H}}_2, \Gamma_0)$ Representing the Rainfall in 2017.

$(\widetilde{\mathcal{H}}_2, \Gamma_0)$	x_1	x_2	x_3	x_4
γ_1	(0.1,0,0.13)	(0,0.12,0.1)	(0.1,0.14,0.1)	(0.1,0.11,0.13)
γ_2	(0.13,0.1,0.15)	(0.1,0.14,0.12)	(0.13,0.16,0.13)	(0.2,0.18,0.18)
γ_3	(0.14,0.13,0.17)	(0.15,0.17,0.18)	(0.17,0.18,0.16)	(0.25,0.31,0.26)
γ_4	(0.16,0.18,0.3)	(0.19,0.18,0.2)	(0.2,0.23,0.3)	(0.28,0.34,0.28)

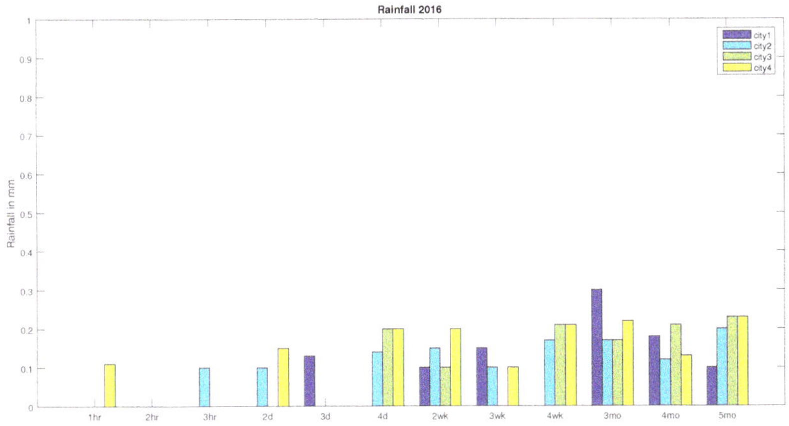

Figure 2. The recorded rainfall in India for the year 2016.

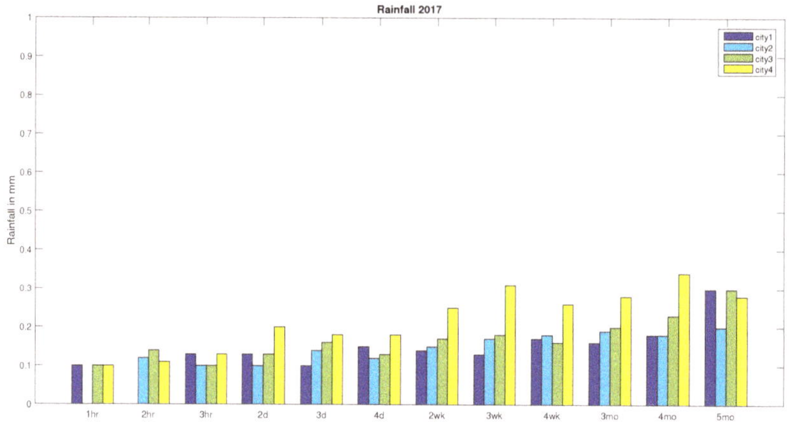

Figure 3. The recorded rainfall in India for the year 2017.

In order to make the decision of whether the rainfall in 2016 or the rainfall in 2017 is the expected rainfall in India, we use similarity measure on $\mathscr{LMFSS}s$ to calculate the similarity between the expected rainfall and the rainfall in 2016 ($\mathbf{S}((\widetilde{\mathcal{R}}_0, \Gamma_0), (\widetilde{\mathcal{H}}_1, \Gamma_0))$); the similarity between the expected rainfall and the rainfall in 2017 ($\mathbf{S}((\widetilde{\mathcal{R}}_0, \Gamma_0), (\widetilde{\mathcal{H}}_2, \Gamma_0))$). Comparing the obtained results, the higher similarity means the closer to expected rainfall.

First we calculate the similarity measure between $(\widetilde{\mathcal{R}}_0, \Gamma_0)$ and $(\widetilde{\mathcal{H}}_1, \Gamma_0)$:

$$
\begin{aligned}
& S_1((\widetilde{\mathcal{R}}_0, \Gamma_0), (\widetilde{\mathcal{H}}_1, \Gamma_0)) \\
= & \frac{(0 \vee 0 \vee 0 \vee 0) + (0 \vee 0.1 \vee 0 \vee 0.1) + (0.1 \vee 0.15 \vee 0.1 \vee 0.13) + (0.18 \vee 0.17 \vee 0.17 \vee 0.15)}{0 \vee 0.1 \vee 0 \vee 0.11 + 0.1 \vee 0.15 \vee 0.16 \vee 0.15 + 0.15 \vee 0.16 \vee 0.18 \vee 0.2 + 0.3 \vee 0.17 \vee 0.19 \vee 0.22} \\
= & \frac{0 + 0.1 + 0.15 + 0.18}{0.1 + 0.18 + 0.2 + 0.18} = 0.39
\end{aligned}
$$

$$
\begin{aligned}
& S_2((\widetilde{\mathcal{R}}_0, \Gamma_0), (\widetilde{\mathcal{H}}_1, \Gamma_0)) \\
= & \frac{(0 \vee 0 \vee 0 \vee 0) + (0.12 \vee 0 \vee 0 \vee 0) + (0.14 \vee 0.1 \vee 0 \vee 0.1) + (0.18 \vee 0.12 \vee 0.2 \vee 0.13)}{0 \vee 0 \vee 0 \vee 0 + 0.13 \vee 0.1 \vee 0.17 \vee 0.13 + 0.15 \vee 0.12 \vee 0.19 \vee 0.14 + 0.2 \vee 1 \vee 0.21 \vee 0.16} \\
= & \frac{0 + 0.12 + 0.14 + 0.2}{0 + 0.17 + 0.19 + 1} = 0.33
\end{aligned}
$$

$$
\begin{aligned}
& S_3((\widetilde{\mathcal{R}}_0, \Gamma_0), (\widetilde{\mathcal{H}}_1, \Gamma_0)) \\
= & \frac{(0 \vee 0 \vee 0 \vee 0) + (0 \vee 0.13 \vee 0.1 \vee 0.1) + (0 \vee 0.14 \vee 0.2 \vee 0.11) + (0.1 \vee 0.19 \vee 0.21 \vee 0.12)}{0.12 \vee 0.1 \vee 0 \vee 0 + 0.13 \vee 0.14 \vee 0.2 \vee 0.2 + 0.16 \vee 0.17 \vee 0.21 \vee 0.21 + 1 \vee 0.2 \vee 0.23 \vee 0.23} \\
= & \frac{0 + 0.13 + 0.2 + 0.21}{0.12 + 0.2 + 0.21 + 1.0} = 0.35
\end{aligned}
$$

Hence $\mathbf{S}((\widetilde{\mathcal{R}}_0, \Gamma_0), (\widetilde{\mathcal{H}}_1, \Gamma_0)) = \frac{0.39 + 0.33 + 0.35}{3} = 0.36$.

Next we calculate the similarity measure between $(\widetilde{\mathcal{R}}_0, \Gamma_0)$ and $(\widetilde{\mathcal{H}}_2, \Gamma_0)$:

$$
\begin{aligned}
& S_1((\widetilde{\mathcal{R}}_0, \Gamma_0), (\widetilde{\mathcal{H}}_2, \Gamma_0)) \\
= & \frac{(0 \vee 0 \vee 0 \vee 0) + (0.1 \vee 0.1 \vee 0.13 \vee 0.1) + (0.14 \vee 0.15 \vee 0.17 \vee 0.13) + (0.16 \vee 0.17 \vee 0.19 \vee 0.15)}{0.1 \vee 0.1 \vee 0.1 \vee 0.11 + 0.13 \vee 0.15 \vee 0.16 \vee 0.2 + 0.15 \vee 0.16 \vee 0.18 \vee 0.25 + 0.18 \vee 0.19 \vee 0.2 \vee 0.28} \\
= & \frac{0 + 0.13 + 0.17 + 0.19}{0.1 + 0.2 + 0.25 + 0.28} = 0.59
\end{aligned}
$$

$$
\begin{aligned}
& S_2((\widetilde{\mathcal{R}}_0,\Gamma_0),(\widetilde{\mathcal{H}}_2,\Gamma_0)) \\
= & \frac{(0 \vee 0 \vee 0.13 \vee 0) + (0.1 \vee 0.1 \vee 0.16 \vee 0.13) + (0.13 \vee 0.12 \vee 0.18 \vee 0.14) + (0.18 \vee 0.15 \vee 0.2 \vee 0.16)}{0 \vee 0.12 \vee 0.14 \vee 0.11 + 0.12 \vee 0.14 \vee 0.17 \vee 0.18 + 0.14 \vee 0.17 \vee 0.19 \vee 0.31 + 0.2 \vee 0.18 \vee 0.23 \vee 0.34} \\
= & \frac{0.13 + 0.16 + 0.18 + 0.2}{0.14 + 0.18 + 0.31 + 0.34} = 0.691
\end{aligned}
$$

$$
\begin{aligned}
& S_3((\widetilde{\mathcal{R}}_0,\Gamma_0),(\widetilde{\mathcal{H}}_2,\Gamma_0)) \\
= & \frac{(0.12 \vee 0 \vee 0 \vee 0) + (0.13 \vee 0.12 \vee 0.1 \vee 0.1) + (0.16 \vee 0.14 \vee 0.16 \vee 0.11) + (0.3 \vee 0.19 \vee 0.21 \vee 0.12)}{0.13 \vee 0.1 \vee 0.1 \vee 0.13 + 0.15 \vee 0.13 \vee 0.13 \vee 0.18 + 0.17 \vee 0.18 \vee 0.2 \vee 0.26 + 1.0 \vee 0.2 \vee 0.3 \vee 0.28} \\
= & \frac{0.12 + 0.13 + 0.16 + 0.21}{0.13 + 0.18 + 0.26 + 1.0} = 0.395
\end{aligned}
$$

Hence $\mathbf{S}((\widetilde{\mathcal{R}}_0,\Gamma_0),(\widetilde{\mathcal{H}}_2,\Gamma_0)) = \frac{0.59+0.691+0.395}{3} = 0.559$.

It is clear from the above results, that $(\widetilde{\mathcal{H}}_2,\Gamma_0)$ has significantly greater similarity to $(\widetilde{\mathcal{R}}_0,\Gamma_0)$, as compared with $(\widetilde{\mathcal{H}}_1,\Gamma_0)$ to $(\widetilde{\mathcal{R}}_0,\Gamma_0)$. So we conclude that the rainfall in 2016 is not an expected rainfall and the rainfall in 2017 is an expected rainfall in India.

5. Discussion

In this paper, our motivation to introduce the concept of similarity between two $\mathscr{LMFSS}$ is achieved. This similarity measure satisfies the good properties of similarity measures. Advantages of similarity measure on lattice ordered multi-fuzzy soft set include:

- It is simple and very efficient to evaluate.
- Many authors defined the fuzzy similarity measure in one dimension but some problems are difficult to explain in one dimension. To avoid the difficulties, we introduce the similarity measure in multi-dimension using the lattice structure.

The disadvantage of the proposed similarity measure is that it is only applicable to lattice ordered structures and does not work for other fuzzy structures.

Some properties of proposed measure are stated and proved by a theorem. Apart from that, an application for the decision making by using the similarity measure of two $\mathscr{LMFSS}$ to analyse the rainfall is obtained in this research. This application shows that our proposed measure is worth to use.

6. Conclusions

Multi-fuzzy soft set and its extensions are used in many different applications in decision making. The similarity measure on complex multi-fuzzy soft set has been proposed. $\mathscr{LMFSS}$ was applied in solving forecast problems, but the similarity on $\mathscr{LMFSS}$ was not introduced. In this paper, the concept of similarity measure of $\mathscr{LMFSS}$ is introduced. The numerical examples are presented in detail to illustrate the proposed similarity measure. We also define some properties of similarity measure on two $\mathscr{LMFSS}$s. These properties are proved by Theorem 3.3. Finally, an application of this similarity measure in decision making is presented.

In further works, we are going to extend the operations and properties of $\mathscr{LMFSS}$ using similarity measure. Besides, the using of this similarity measure in solving other real life problems will be studied.

Author Contributions: Methodology: V.J. and T.T.N., writing—original draft preparation: S.B.S. and R.S.; software: V.J. and S.B.S.; validation: T.T.N.; writing—review and editing: T.T.N. and Ganeshsree Selvachandran. All authors have read and agreed to the published version of the manuscript.

Funding: The article has been written with the joint financial support of RUSA-Phase 2.0 grant sanctioned vide letter No.F 24-51/2014-U, Policy (TN Multi-Gen), Dept. of Edn. Govt. of India, Dt. 09.10.2018, UGC-SAP (DRS-I) vide letter No.F.510/8/DRS-I/2016(SAP-I) Dt. 23.08.2016, DST-PURSE 2nd Phase programme vide letter No. SR/PURSE Phase 2/38 (G) Dt. 21.02.2017 and DST (FST - level I) 657876570 vide letter No.SR/FIST/MS-I/2018/17 Dt. 20.12.2018.

Acknowledgments: The authors would like to thank the Editor-in-Chief and the anonymous reviewers for their valuable comments and suggestions.

Conflicts of Interest: The authors declare no conflict of interest.

References

1. Birkhoff, G. Abstract linear dependence and lattices. *Am. J. Math.* **1935**, *57*, 800–804. [CrossRef]
2. Gratzer, G. *Lattice Theory: Foundation*; Springer Basel AG: Berlin/Heidelberg, Germany, 2011.
3. Zadeh, L.A. Fuzzy sets. *Inf. Control* **1965**, *8*, 338–353. [CrossRef]
4. Si, A.; Das, S.; Kar, S. An approach to rank picture fuzzy numbers for decision making problems. *Decis. Mak. Appl. Manag. Eng.* **2019**, *2*, 54–64. [CrossRef]
5. Mukhametzyanov, I.; Pamucar, D. A sensitivity analysis in MCDM problems: A statistical approach. *Decis. Mak. Appl. Manag. Eng.* **2018**, *1*, 51–80. [CrossRef]
6. Sharma, H.K.; Kumari, K.; Kar, S. A rough set approach for forecasting models. *Decis. Mak. Appl. Manag. Eng.* **2020**, *3*, 1–21. [CrossRef]
7. Molodtsov, D.A. Soft set theory-first results. *Comput. Math. Appl.* **1999**, *37*, 19–31. [CrossRef]
8. Maji, P.K.; Roy, A.R. Fuzzy soft sets. *J. Fuzzy Math.* **2001**, *3*, 589–602.
9. Ngan, T.T.; Tuan, T.M.; Minh, N.H.; Dey, N. Decision making based on fuzzy aggregation operators for medical diagnosis from dental X-ray images. *J. Med. Syst.* **2016**, *40*, 280. [CrossRef]
10. Tuan, T.M.; Chuan, P.M.; Ali, M.; Ngan, T.T.; Mittal, M. Fuzzy and neutrosophic modeling for link prediction in social networks. *Evol. Syst.* **2019**, *10*, 629–634. [CrossRef]
11. Ngan, T.T.; Lan, L.T.H.; Ali, M.; Tamir, D.; Son, L.H.; Tuan, T.M.; Naphtali, R.; Kandel, A. Logic connectives of complex fuzzy sets. *Rom. J. Inf. Sci. Technol.* **2018**, *21*, 344–358.
12. Selvachandran, G.; Quek, S.G.; Lan, L.T.H.; Giang, N.L.; Ding, W.; Abdel-Basset, M.; Albuquerque, V.H.C. A New Design of Mamdani Complex Fuzzy Inference System for Multi-attribute Decision Making Problems. *IEEE Trans. Fuzzy Syst.* **2019**. [CrossRef]
13. Sebastian, S.; Ramakrishnan, T.V. Multi-fuzzy sets: An extension of fuzzy sets. *Fuzzy Inform. Eng.* **2011**, *3*, 35–43. [CrossRef]
14. Sebastian, S.; Ramakrishnan, T.V. A Study on Multi-Fuzziness. Unpublished Ph.D. Thesis, Kannur University, Kannur, India, 2011.
15. Yong, Y.; Xia, T.; Congcong, M. The Multi-fuzzy soft set and its Application in Decision Making. *Appl. Math. Model.* **2013**, *37*, 4915–4923. [CrossRef]
16. Dey, A.; Pal, M. Generalised Multi-fuzzy soft set and its application in decision making. *Pac. Sci. Rev. A Nat. Sci. Eng.* **2017**, *17*, 23–28. [CrossRef]
17. Majumdar, P.; Samanta, S.K. Similarity measure of soft sets. *New Math. Nat. Comput.* **2008**, *4*, 1–12. [CrossRef]
18. Majumdar, P.; Samanta, S.K. On Similarity measure of fuzzy soft sets. *Int. J. Adv. Soft Comput. Its Appl.* **2011**, *3*, 1–8.
19. Liu, Z.; Qin, K.; Pei, Z. Similarity measure and entropy of fuzzy soft sets. *Sci. World J.* **2014**. [CrossRef]
20. Feng, Q.; Zheng, W. New similarity measures of fuzzy soft sets based on distance measures. *Ann. Fuzzy Math. Inform.* **2014**, *7*, 669–686
21. Peng, X. New similarity measure and distance measure for Pythagorean fuzzy set. *Complex Intell. Syst.* **2019**, *5*, 101–111. [CrossRef]
22. Fei, L.; Wang, H.; Chen, L.; Deng, Y. A new vector valued similarity measure for intuitionistic fuzzy sets based on OWA operators. *Iran. J. Fuzzy Syst.* **2019**, *16*, 113–126.
23. Song, Y.; Wang, X.; Quan, W.; Huang, W. A new approach to construct similarity measure for intuitionistic fuzzy sets. *Soft Comput.* **2019**, *23*, 1985–1998. [CrossRef]
24. Al-Qudah, Y.; Hassan, N. Complex multi-fuzzy soft set: Its entropy and similarity measure. *IEEE Access* **2018**, *6*, 65002–65017. [CrossRef]
25. Lv, C.; Wu, Z.; Wang, X.; Zhou, M.; Toh, K.A. Nasal similarity measure of 3D faces based on curve shape space. *Pattern Recognit.* **2019**, *88*, 458–469. [CrossRef]
26. Gazdar, A.; Hidri, L. A new similarity measure for collaborative filtering based recommender systems. *Knowl.-Based Syst.* **2020**, *188*, 105058. [CrossRef]
27. Tuan, T.M.; Lan, L.T.H.; Chou, S.Y.; Ngan, T.T.; Son, L.H.; Giang, N.L.; Ali, M. M-CFIS-R: Mamdani Complex Fuzzy Inference System with Rule Reduction Using Complex Fuzzy Measures in Granular Computing. *Mathematics* **2020**, *8*, 707. [CrossRef]

28. Zhang, H.D.; Shu, L. Possibility multi-fuzzy soft set and its decision making. *J. Intell. Fuzzy Syst.* **2014**, *27*, 2115–2125. [CrossRef]
29. Selvachandran, G.; Maji, P.K.; Faisal, R.Q.; Salleh, A.R. Distance and distance induced intuitionistic entropy of generalized intuitionistic fuzzy soft sets. *Appl. Intell.* **2017**, *47*, 132–147. [CrossRef]
30. Selvachandran, G.; Maji, P.K.; Abed, I.E.; Salleh, A.R. Complex vague soft sets and its distance measures. *J. Intell. Fuzzy Syst.* **2016**, *31*, 55–68. [CrossRef]
31. Selvachandran, G.; Garg, H.; Alaroud, M.H.S.; Salleh, A.R. Similarity measure of complex vague soft sets and its application to pattern recognition. *Int. J. Fuzzy Syst.* **2018**, *20*, 1901–1914. [CrossRef]
32. Vimala, J.; Arockia Reeta, J.; Anusuya Ilamathi, V.S. A Study On Fuzzy Soft Cardinality In Lattice Ordered Fuzzy Soft Group And Its Application In Decision Making Problems. *J. Intell. Fuzzy Syst.* **2018**, *34*, 1535–1542. [CrossRef]
33. Reeta, J.A.; Vimala, J. Implementation of Anti-Lattice Ordered Fuzzy Soft Groups and its Matrix Operation in Deciding Process. *J. Intell. Fuzzy Syst.* **2018**, *35*, 4857–4864. [CrossRef]
34. Pandipriya, A.R.; Vimala, J.; Sabeena Begam, S. Lattice ordered interval-valued hesitant fuzzy soft sets in decision making problem. *Int. J. Eng. Technol.* **2018**, *7*, 52–55. [CrossRef]
35. Pandipriya, A.R.; Vimala, J.; Peng, X.D.; Sabeena Begam, S. A Decision Making Approach on L-IVHFSS Setting. *Adv. Intell. Syst. Comput.* **2019**, *910*, 219–225.
36. Rajareega, S.; Vimala, J.; Preethi, D. Complex Intuitionistic Fuzzy Soft Lattice Ordered Group and Its Weighted Distance Measures. *Mathematics* **2020**, *8*, 705. [CrossRef]
37. Sabeena Begam, S.; Vimala, J. Application of lattice ordered multi-fuzzy soft set in forecasting process. *J. Intell. Fuzzy Syst.* **2019**, *36*, 2323–2331. [CrossRef]
38. Sabeena Begam, S.; Vimala, J.; Preethi, D. A novel study on the algebraic applications of special class of lattice ordered multi-fuzzy soft sets. *J. Discret. Math. Sci. Cryptogr.* **2019**, *22*, 883–899. [CrossRef]
39. Egghe, L. Good properties of similarity measures and their complementarity. *J. Am. Soc. Inf. Sci. Technol.* **2010**, *61*, 2151–2160. [CrossRef]

mathematics

MDPI

Article

AHP-TOPSIS Inspired Shopping Mall Site Selection Problem with Fuzzy Data

Neha Ghorui [1], Arijit Ghosh [2], Ebrahem A. Algehyne [3], Sankar Prasad Mondal [4,*] and Apu Kumar Saha [5]

1 Department of Mathematics, Prasanta Chandra Mahalanobis Mahavidyalaya, Kolkata 700108, India; neha.mundhra@thebges.edu.in
2 Department of Mathematics, St. Xavier's College (Autonomous), Kolkata 700016, India; arijitghosh@sxccal.edu
3 Department of Mathematics, Faculty of Sciences, University of Tabuk, Tabuk-71491, Saudi Arabia; e.algehyne@ut.edu.sa
4 Department of Applied Science, Maulana Abul Kalam Azad University of Technology, West Bengal, Haringhata 741249, India
5 Department of Mathematics, National Institute of Technology, Agartala, Jirania 799046, India; apu.math@nita.ac.in
* Correspondence: sankarprasad.mondal@makautwb.ac.in; Tel.: +91-9635578078

Received: 28 July 2020; Accepted: 11 August 2020; Published: 17 August 2020

Abstract: In the consumerist world, there is an ever-increasing demand for consumption in urban life. Thus, the demand for shopping malls is growing. For a developer, site selection is an important issue as the optimal selection involves several complex factors and sub-factors for a successful investment venture. Thus, these tangible and intangible factors can be best solved by the Multi Criteria Decision Making (MCDM) models. In this study, optimal site selection has been done out of multiple alternative locations in and around the city of Kolkata, West Bengal, India. The Fuzzy Analytic Hierarchy Process (FAHP) and Fuzzy Technique for Order of Preference by Similarity to Ideal Solution (FTOPSIS) has been applied for shopping mall site selection. The AHP is used to obtain the crispified weight of factors. Imprecise linguistic terms used by the decision-maker are converted to Triangular Fuzzy Numbers (TFNs). This research used integrated sub-factors fuzzy weights using FAHP to FTOPSIS for ranking of the alternatives. Hardly any research is done with the use of sub-factors. In this study, seven factors and seventeen sub-factors are considered, the authors collected data from different locations with the help of municipal authorities and architects. This work further provides useful guidelines for shopping mall selection in different states and countries.

Keywords: site selection; shopping mall site selection; linguistic terms for fuzzy variable; fuzzy AHP; fuzzy TOPSIS

1. Introduction

Shopping malls can be considered as one of the most important growth points for Business strategies. There exist multiple factors which are responsible for a decision-maker (DM) to select the optimal site for shopping mall construction. Detailed study has been conducted to identify all the factors and sub-factors related to site selection. The weights are assigned to each of the factors and sub-factors with the help of an expert decision-maker (DM). The needs for shopping malls are increasing throughout the country. As several factors influence the selection of the best site, it can be considered as an application of Multi Criteria Decision Making (MCDM). MCDM is considered as the most significant branch of Operation Research, as it incorporates complex decisions of people's lives. There exists multiple MCDM models. The researcher uses MCDM techniques depending on the problem of the decision making. The MCDM model Analytic Hierarchy Process (AHP) introduced by the author of Reference [1] is one of the powerful techniques to obtain the factors' and sub-factors' weight. The model is widely used in numerous fields of engineering, economics, and operations management. The factors' and sub-factors' importance are calculated by the pair-wise comparison matrix. The pure AHP model lacks the ability to capture uncertainty, thus several researchers have integrated Fuzzy with AHP to capture the impreciseness in decision making. Thus, in the present research, we used FAHP to determine the factors' and sub-factors' fuzzy weight.

The Technique of Order Preference by a Similarity to Ideal Solution (TOPSIS) developed by the authors of Reference [2] is a logistic approach to select the best alternatives in real life problems, when several conflicting qualitative and quantitative criteria exist for the evaluation. The idea of this technique is that the best alternative is closest to the positive ideal solution (PIS) and farthest from the negative ideal solution (NIS). Decision making problems with uncertainty nowadays play an important role [3–8]. The FTOPSIS, an extension of classical TOPSIS to fuzzy domain, was introduced by the authors of Reference [9] and uses fuzzy numbers instead of crisp values.

The selection of the parameters for the site selection of a shopping mall have originality. Moreover, the discussion of the universal design criteria affecting the criteria and sub-criteria has not been done in any paper before this. The study compares and evaluates the proposed sites for the shopping mall using the MCDM model. During the course of research, it was analyzed by experts and the opinion was taken from the people so as to bring the real customer needs and wants into the picture. Moreover, the study also explains the direct and indirect relation between the factors and sub-factors.

The fuzzy domain is a suitable approach in handling real life composite applications. The present paper focuses on the minute specifications of the sub-factors and applies Triangular Fuzzy Numbers (TFNs) to give proper weightage to factors and sub-factors using Fuzzy AHP. Finally, the best site is assessed using the Fuzzy TOPSIS approach. The use of fuzzy set theory with the MCDM technique enables the decision-making problem to deal with vagueness and uncertainty.

1.1. Background of the Study

In this modern era, the way we perceive shopping mall structures has changed. The shinier the structure, the more global the brand stores situated in it will be. Buying in a shopping mall makes the experience special since there are a lot of choices under a single roof, which adds to the appeal of the shopping experience [10]. Standalone brand stores have a higher possibility of offering more discounts as compared to the stores in the malls since they have less taxes and other overhead charges to pay. It is also seen that people often associate price with quality, and this leads them to have negative views about products in single stores. Since there are a large number of shopping malls with the same products, choosing which one to purchase from has become a trick [11]. This is because the same products might be available via online stores at a discounted price and it might be preferred since it is available sitting in one's house [12]. Hence, to create a niche for themselves, they must stand out. One of the ways to do so is understanding the target customer group [13,14]. The malls are also being frequently visited by pregnant women, new mothers, children, and people with disabilities, and so the design should be such that it is comfortable for all. If the building is designed in such a way so as to take all these factors into account, then it becomes a part of the universal design of architecture [15]. Universal design is such a design which is accessible by people of all ages [16]. The literature for this is limited since studies have not been done on these specific parts of the society [17,18]. This is because the architects are more often than not people who do not need special help and therefore they usually do not consider this section of the people [19], hence for them to show empathy and design a structure is similar to telling a child to paint a bird when he has never seen it. Therefore, a good understanding of the special needs must be incorporated into building the universal design so that everyone feels at home [20]. Earlier important features like the parking capacity and area size have now been subdued by environmental criterion, design aesthetics, and ergonomics, with studies proving that the latter criteria have a more potent effect on the customer satisfaction [21]. Ergonomic factors and aesthetics such as background and interior design offer more entertainment and comfort to the customers [22]. Environmental criteria like accessing facilities, noise pollution, air pollution, and traffic jams also play a key role in making a mall attractive to its prospective customers. Location selection is an important factor since it has long-term risks and costs associated with the motivation of the work [23,24]. It involves identification, evaluation, and selection among the options available. The process is influenced by qualitative, quantitative criteria such as related and supporting industries, firm strategy and rivalry, proximity to raw materials and markets, infrastructure conditions, market size and demand conditions, investment costs, natural conditions and human resources to name a few [25,26]. Hence, selecting a site for a shopping mall seems to be a major factor which decides the success of the business.

1.2. Background of Fuzzy AHP and Fuzzy TOPSIS

The authors of [27] applied the Fuzzy TOPSIS approach for transshipment site selection in Istanbul, Turkey. The factors' weights were determined by the AHP methodology. The authors of [28] used hybrid Fuzzy AHP (FAHP) to calculate the weights of varying locations in Vietnam and then TOPSIS was applied for ranking of the alternatives.

The authors of [29] applied an integrated approach of Hesitant Fuzzy Sets (HFSs) to TOPSIS to select the best hospital site in Istanbul. The customers outweigh the cost of travelling with the alternative shopping opportunities and this helps to understand the inter-metropolitan trade areas of shopping centers [30].

The authors of [31] opined that evaluation and determination of negative and positive characteristics of one location to another is a tough job. The authors of [32] used multinomial logit model and principal component analysis to take parking facilities, quality, value, satisfaction, and variety into consideration. The authors of [33] analyzed socialization, functionality, convenience, and recreation using Structural Equation Modeling. The authors of [34] conclude that the selection of a shopping mall site is not an easy task and the closer to reality the solution comes, the better the model. Multi Criteria Decision Making (MCDM) takes this into account by considering all the criteria and determining an optimal result. The authors of Reference [19] did a MCDM study on selection of safety equipment using universal design principles. The authors of Reference [35] solved a landfill selection problem in Canada by using a fuzzy MCDM approach. The authors of Reference [36] used MCDM on hesitant fuzzy linguistic terms for site selection of wind turbines. MCDM was used by the author of [37] in site selection of a GIS-based solar farm. The author of [38] referred to MCDM in the field of information technology and IT management. The authors of [39] used MCDM for the TOPSIS method in the field of operations research. The authors of [40] used MCDM for formulating energy policies. The authors of [41] used MCDM for formulating a technique on renewable energy. The authors of [42] used it for logistic suppliers under supply chain management, the authors of [43] have used it for knowledge management, the authors of [44] have adopted MCDM in the field of tourism management, and the authors of [45] used it for construction management.

MCDM makes the decision on the basis of multiple criteria. The maker of the decision is supposed to choose between the non-quantifiable multiple criteria and quantifiable multiple criteria. The preference of the decision-makers (DMs) are of the foremost importance since the options are very close and their personal preferences plays an important role in finding out the optimal solution. In order to tackle the uncertainties of the data, fuzzy methods can be integrated with the MCDM methods.

1.3. Motivation of the Study

Earlier, several researchers have used fuzzy numbers coupled with AHP and TOPSIS in an uncertain environment of decision making. Basically, for the selection of an optimal site, the researchers have taken factors' weight to rank the best. Hardly any research has been done with the use of sub-factors for the evaluation of the potential site under fuzzy environment. Here, we integrated two fuzzy MCDM methods to select the best site. At first, fuzzy AHP is applied to figure out the importance of 7 factors. Finally, the sub-factors' fuzzy weights are calculated for further evaluation. FTOPSIS is used for selection of an optimal site considering the seventeen sub-factors.

1.4. Novelties of the Work

The present study advances in the following way:

- The important factors and sub-factors are studied in a detailed way. These sub-factors capture minute detail considered by developers for new shopping mall site selection. Hardly any other papers have explored all these dimensions.
- Questionnaires were made. The municipal authorities and architects were interviewed to get a clear idea about the uncertain factors such as land cost, population density, and population growth rate.
- The TFN FAHP and TFN FTOPSIS were employed to get the proper weightage and optimal site.
- The linguistic terms assigned by the decision experts were converted to TFN in an effective and efficient process.
- Sensitivity analysis was conducted, and a comparative analysis was done with the proposed problem to illustrate how the ranking fluctuates with the change or elimination of certain factors and sub-factors.

1.5. Structure of the Paper

The remainder of the paper is constructed in the following way: Section 2 reflects the design and methodology of the proposed research. The definitions and algebraic operations of respective Fuzzy sets and Fuzzy numbers are illustrated. The AHP, FAHP, and FTOPSIS methodology are also discussed. Section 3 discusses the application, data source, and numerical problem of the study. A brief discussion of the sub-factors is executed and the final ranking is evaluated. Section 4 briefly describes the importance of shopping malls in the economic growth and discusses the result obtained. Section 5 discusses the sensitivity analysis and numerical simulation. Finally, the conclusion is presented in Section 6.

2. Design of the Proposed Model

Step 1: Identification of the research problem and defining the goal. Determination of factors and sub-factors by interviewing the experts and following literature reviews.

Step 2: Application of FAHP methodology to evaluate the importance of factors' and sub-factors' weight. The seven prospective locations in and around the city of Kolkata are chosen.

Step 3: The FTOPSIS model is applied for ranking of the best location for shopping mall site selection. The FTOPSIS approach is one of the best MCDM techniques when the problem has several imprecise qualitative and quantitative attributes. The model efficiently addresses the complex decision-making problem of optimal site selection. In Figure 1 we mentioned the steps for the proposed methodology.

2.1. Preliminaries Concepts

2.1.1. Fuzzy Sets

The fuzzy set concept was introduced by the author of Reference [46] to handle the uncertainties prevailing in the real-life problem. A fuzzy set is defined as:

$\widetilde{S} = \{(x, \mu_{\widetilde{S}}(x))\} | x \in X\}$, where X is a subset of the real numbers, R. Here, $\mu_{\widetilde{S}}(x)$ denotes the membership function. For each x there exists a membership which ranges from zero to one. The fuzzy set concept is widely used in real-life problems, where the information is imprecise and vague. There exist different forms of fuzzy numbers such as: Triangular Fuzzy Number (TFN), Trapezoidal Fuzzy Number (TrFN), Intuionistic Fuzzy Number (IFN), etc. The different forms of fuzzy numbers are used depending on the nature of the problem. TFN is widely used for its efficiency to deal with applications where vagueness and uncertainty surrounds the decision making. The hesitancy or fuzziness of the DM's can be expressed in TFN while constructing a comparison matrix and decision matrix to rank the best alternatives.

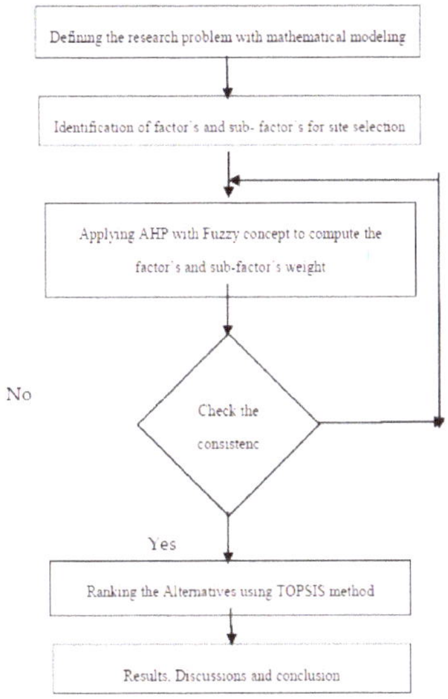

Figure 1. Flow chart for the proposed study.

A TFN, denoted as $\widetilde{S} = (p, r, t)$, consists of the following membership function:

$$\mu_{\widetilde{S}}(x) = \begin{cases} \frac{x-p}{r-p} for\ p \le x \le r \\ \frac{t-x}{t-r} for\ r \le x \le t \\ 0\ otherwise \end{cases} \quad (1)$$

The Figure 2 represents the pictorial representation of triangular fuzzy number. The variable "r" represents the maximum membership value of 1 and is considered to be the most promising value. The variables "p" and "t" denote the smallest viable value and the largest viable value, respectively. The triplet (p, r, t) which describes the fuzziness of a particular event enables the field of practicable evaluation.

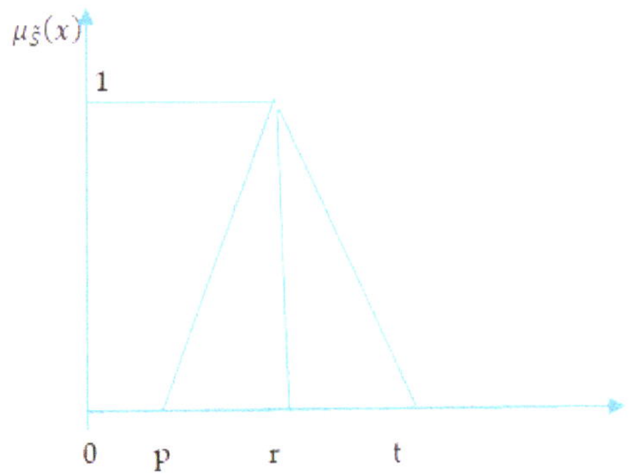

Figure 2. Diagram representing Triangular Fuzzy Number (TFN), $\widetilde{S} = (p, r, t)$.

Note 1:

1. When $p = r = t$, the TFN becomes a crisp value.
2. TFN can be visualized as a special case of TrFN, as when the two most achievable values become equal, then it is a TFN.

Definition 1. *Let $\widetilde{S_1} = (p_1, r_1, t_1)$ and $\widetilde{S_2} = (p_2, r_2, t_2)$ be two TFN's, then the distance between the two can be determined by the vertex method as:*

$$d\left(\widetilde{S_1}, \widetilde{S_2}\right) = \sqrt{1/3\left[(p_1 - p_2)^2 + (r_1 - r_2)^2 + (t_1 - t_2)^2\right]} \quad (?)$$

2.1.2. Arithmetic Operations of Fuzzy Numbers

Assume two TFN's: $\widetilde{S_1} = (p_1, r_1, t_1)$ and $\widetilde{S_2} = (p_2, r_2, t_2)$. The arithmetic operation properties [28] for two TFN: $\widetilde{S_1}$ and $\widetilde{S_2}$, are:

$$(p_1, r_1, t_1) + (p_2, r_2, t_2) = (p_1 + p_2,\ r_1 + r_2,\ t_1 + t_2) \quad (3)$$

$$(p_1, r_1, t_1) - (p_2, r_2, t_2) = (p_1 - t_2,\ r_1 - r_2,\ t_1 - p_2) \quad (4)$$

$$(p_1, r_1, t_1) \times (p_2, r_2, t_2) = (p_1 \times p_2,\ r_1 \times r_2,\ t_1 \times t_2) \quad (5)$$

$$\frac{(p_1, r_1, t_1)}{(p_2, r_2, t_2)} = \left(\frac{p_1}{t_2}, \frac{r_1}{r_2}, \frac{t_1}{p_2}\right), \text{ where } p_i > 0, r_i > 0,\ t_i > 0,\ i = 1,2 \quad (6)$$

$$\widetilde{S_1^-} = \left(\frac{1}{t_1}, \frac{1}{r_1}, \frac{1}{p_1}\right), \text{ where } p_1 > 0,\ r_1 > 0,\ t_1 > 0 \quad (7)$$

2.1.3. Fuzzy Analytic Hierarchy Process

The Analytic Hierarchy Process (AHP) was first developed by the author of Reference [1], and is a widely used scientific method in Multi Criteria Decision Making (MCDM). AHP helps the decision-makers to solve the complex decisions with heuristic methods. Evaluation of factors' and sub-factors' weights are important for ranking the optimal site selection. AHP creates a framework of the problem hierarchy with the construction of comparison matrices to give subjective judgments about the factor's which are considered highly responsible in ranking the best. In this paper, FAHP is used rather than AHP as the fuzzy environment takes into consideration the uncertainty and impreciseness of the decision experts. The fuzzy AHP methodology is based on the pairwise comparison approach, applied to understand the relative importance of factors and sub-factors. The AHP coupled with fuzzy logic enables the decision-makers a suitable approach in obtaining more realistic results in decision-making problems. The steps of FAHP are described below.

Step 1: Construction of a comparison matrix in terms of TFN by a decision expert or a group of decision experts.

Assume a group of 'M' decision-makers involved in the pairwise comparison of factors and sub-factors. Thus, 'm' set of matrices are obtained, $S_m = \{s_{klm}\}$: where $s_{klm} = (p_{klm}, r_{klm}, t_{klm})$ expresses the relative preference of k factor to l factor as decided by the 'm' decision expert.

$$\left.\begin{aligned} p_{kl} &= \min_{m=1,2,\ldots,M}(p_{klm}) \\ r_{kl} &= \sqrt[M]{\prod_{m=1}^{M} r_{klm}} \\ t_{kl} &= \max_{m=1,2,\ldots,M}(t_{klm}) \end{aligned}\right\} \quad (8)$$

Step 2: Defuzzification of TFN.

A TFN $s_{kl} = (p_{kl}, r_{kl}, t_{kl})$ can be defuzzified to a crisp value using the method proposed by the authors of Reference [47]:

$$(s_{kl}^{\alpha})^{\beta} = \left[\beta . p_{kl}^{\alpha} + (1-\beta).\ t_{kl}^{\alpha}\right],\ 0 \leq \beta \leq 1,\ 0 \leq \alpha \leq 1, \tag{9}$$

where α signifies the preference display of the evaluator and β signifies the risk factor of the uncertain conditions. The method explicitly express fuzzy logic owing to the uncertainty of the decision makers. The uncertainty is maximum when $\alpha = 0$ and stability increases in decision making with increasing α. Moreover, β can be defined as the risk factor of the decision maker. $\beta = 1$ implies highly pessimistic whereas $\beta = 0$ implies highly optimistic.

$p_{kl}^{\alpha} = (r_{kl} - p_{kl}) \times \alpha + p_{kl}$ denotes the lower bound of α-cut for s_{kl}, and $t_{kl}^{\alpha} = t_{kl} - \alpha.(t_{kl} - r_{kl})$ denotes the upper bound of α-cut for s_{kl}.

Step 3: Construction of a comparison matrix in terms of crisp values.

Generalized representation of defuzzified comparison matrix:

$$\left((s_{kl}^{\alpha})^{\beta}\right) = \begin{pmatrix} 1 & (s_{12}^{\alpha})^{\beta} & . & . & (s_{1n}^{\alpha})^{\beta} \\ (s_{21}^{\alpha})^{\beta} & 1 & . & . & (s_{2n}^{\alpha})^{\beta} \\ . & . & 1 & . & . \\ . & . & . & 1 & . \\ (s_{n1}^{\alpha})^{\beta} & (s_{n2}^{\alpha})^{\beta} & . & . & 1 \end{pmatrix}$$

Step 4: Normalization of the defuzzified matrix:

$$N_{kl} = \frac{s_{kl}}{\sum_{k=1}^{m} s_{kl}},\ \text{where} k = 1, 2, \ldots, m; l = 1, 2, \ldots, n; \tag{10}$$

Step 5: Estimation of factors' and sub-factors' weights:

$$E = \frac{N^{th} rootvalue}{\sum N^{th} root} \tag{11}$$

Step 6: Checking the Consistence Index $(C.I)$ of the matrix:

$$(C.I) = \frac{\gamma_{max} - n}{n-1} \tag{12}$$

where n is the size of the matrix.

Step 7: Calculation of Consistence Ratio $(C.R)$:

$$C.R = \frac{C.I}{R.I} \tag{13}$$

where $R.I$ denotes the Random Index, whose value differs with the size of the matrix "n".

The value of $C.R \leq 0.1$ indicates the matrix to be consistent, which implies that no further evaluation is required.

2.1.4. Technique for Order Preference by Similarity to Ideal Solution (TOPSIS) and (FTOPSIS)

The TOPSIS approach is one of the commonly used MCDM techniques, which was first developed by the authors of Reference [2]. The basic formulation of TOPSIS is to rank the alternatives, thus giving an idea as to which alternative is most preferred. The TOPSIS method isclassified as a distance measure method in which the optimal alternative obtained is farthest from the negative ideal solution (NIS) and closest to the positive ideal solution (PIS). The linguistic human decisions can be reflected better with Fuzzy TOPSIS (FTOPSIS). The approach is useful in handling the complexity of the situation involving several factors and their sub-factors. In this research, for the selection of the best site to construct a shopping mall, it is dependent on multiple conflicting factors and sub-factors, thus the MCDM method FTOPSIS introduced by the authors of Reference [9] is one of the most suitable and reliable methods. The fuzzy logic extends our goal to obtain more sensitive results in this regard. The steps of FTOPSIS are described below.

Step 1: Construction of the decision matrix

Step 2: To evaluate the normalized fuzzy decision matrix:

$$\left.\begin{aligned} \check{N} &= [n_{cd}]_{ij}, c = 1,2,\ldots,i; d = 1,2,\ldots..,j \\ n_{cd} &= \left(\frac{p_{cd}}{t^*_d}, \frac{r_{cd}}{t^*_d}, \frac{t_{cd}}{t^*_d}\right) d \in B,\ t^*_d = \max t_{cd} \\ n_{cd} &= \left(\frac{p^*_j}{t_{cd}}, \frac{p^*_j}{r_{cd}}, \frac{p^*_j}{p_{cd}}\right) d \in N.B,\ p^*_j = \min p_{cd} \end{aligned}\right\} \tag{14}$$

where B and N.B stand for Benefit factors and Non-Benefit factors, respectively.

Step 3: To obtain the weighted fuzzy normalized matrix by multiplying the sub-factors' fuzzy weights with the normalized fuzzy value:

$$H = [h_{cd}]_{ij}\ c = 1,2,\ldots,i;\ d = 1,2,\ldots.,j \tag{15}$$

where

$$h_{cd} = n_{cd} \times \hat{W}_d,\ c = 1,\ 2,\ldots.,i; d = 1,\ 2,\ldots,j \tag{16}$$

Step 4: Calculate the Fuzzy Positive Ideal Solution (FPIS) (T^+) and Fuzzy Negative Ideal Solution (FNIS) (T^-), where h_c^+ denotes the maximum value of h_{cd} and h_c^- denotes the minimum value of h_{cd}:

$$\left.\begin{aligned} T^+ &= \{h_1^+, h_2^+, \ldots., h_j^+\} = \left\{(\max h_{cd}|d \in K_a), (\min h_{cd}|d \in K_b)\right\}, \\ T^- &= \{h_1^-, h_2^-, \ldots., h_j^-\} = \left\{(\min h_{cd}|d \in K_a), (\max h_{cd}|d \in K_b)\right\} \end{aligned}\right\} \tag{17}$$

where K_a relates to the benefit factors and K_b relates to the non-benefit factors.

Step 5: Determine the distance of the alternatives from the PIS and NIS. The two Euclidean distances for individual alternatives can be calculated as follows:

$$\left.\begin{aligned} U_c^+ &= \textstyle\sum_{d=1}^{j} d(h_{cd}, h_c^+),\ c = 1,\ 2,\ \ldots..,\ i \\ U_c^- &= \textstyle\sum_{d=1}^{j} d(h_{cd}, h_c^-),\ c = 1,\ 2,\ \ldots..,\ i \end{aligned}\right\} \tag{18}$$

where d (.,.) denotes the distance between two fuzzy numbers.

Step 6: Determine the relative closeness to the ideal alternatives:

$$R_c = \frac{U_c^-}{U_c^- + U_c^+},\ c = 1,\ 2,\ \ldots.,\ i \tag{19}$$

Step 7: Rank the alternatives:

The alternatives are ranked based on the score obtained by R_c, and the larger value of R_c signifies the better alternatives.

Now, we fix the Linguistic terms and the corresponding TFN in Table 1 as follows:

Table 1. Linguistic terms and the corresponding TFN.

Scale Definition	TFN Scale	TFN Reciprocal Scale
Just Equal	(1,1,1)	(1,1,1)
Equally Important	(1/2,1,3/2)	(2/3,1,2)
Fairly Important	(3/2,2,5/2)	(2/5,1/2,2/3)
Strongly Important	(5/2,3,7/2)	(2/7,1/3,2/5)
Strongly more Important	(7/2,4,9/2)	(2/9,1/4,2/7)
Absolutely Important	(5,5,5)	(1/5,1/5,1/5)

3. Empirical Study of Shopping Mall Site Selection

Enhanced modern lifestyle has changed the shopping style of people from small independent retail shops to shopping malls. The key importance lies in the fact that shopping malls provide many variations which the population demands. The customer finds comfort while shopping in malls, as under one roof they get exclusive clothing brands, department stores, food courts or restaurants, game section for children, multiplex, family salon and spa, etc. People get attracted towards shopping malls as it benefits the customer with window shopping too. Numerous conflicting factors and sub-factors impact the evaluation of the best site.

The main aim of the proposed research is to find the best location for the construction of a shopping mall from a given set of alternatives. The emerging choice of people over shopping malls builds great interest in investors to invest in shopping mall businesses. Choosing the best site from a set of different locations is tough, as individual locations have the corresponding robustness and flaws. Maximum return is the main aim of an investor, thus selection of the best site requires scientific as well as mathematical modeling in an uncertain and imprecise environment.

Accessibility is one of the major factors of site selection. One of the factors of selecting a shopping mall site is based on how visible it is from the highway and how much road connectivity it has. In case there is already huge traffic on the site, it will become difficult to construct anything new since traffic congestion is not liked by shoppers. In case the shopping mall site is faraway, then in order to increase footfall, a shuttle service, promotions, and reward programs will go a long way. A destination shopping mall is also a good concept since the land price can be optimized. To make it a success, the characteristics of the future customers such as age, income, and brand preferences need to be studied in detail since only then can the shopping mall be made to cater to the retail and leisure offerings alike. Configuration and size are also important factors for site selection since proper use of the catchment area will lead to maximization of resources. Mixed use developments are one of the most liked structures by the developers since the various segments complement each other and a synergy is achieved which helps in appropriate circulation of all asset classes. In the long run, the scope for further expansion must also be kept in mind. The shopping mall should have such an infrastructure which enables maximum utilization of space. Current and future competition, performance of the retail spaces, and the consumption patterns are some of the other factors which influence the site selection.

Considering the above discussed factors and sub-factors, to summing up more, discussion related to the factors and their sub-factors are done in a more elaborative way, as follows.

a. Population Density

   (i) Population growth rate—The population of a particular place is an important factor [48–50] when considering the place as a building site for a shopping mall. The primary reason is that the greater the possibility of consumption in an area, the greater the predicted sales from the shopping mall, which in turn will increase the growth of the area and eventually lead to an increase in the happiness index. In case the area consists of people from the young and middle generations, then there are greater chances of good population growth, which in turn will lead to higher numbers of customers for the mall, thereby leading to a boom in the business.

b. Transportation [48,50]

   (i) Proximity to Metro—Having access to the metro will lead to an increase in the footfall in the mall. The reason is that greater connectivity of the mall acts as a catalyst for people who have the money and are looking for places to purchase from. The metro is also one of the most favored forms of transport by the younger generation since it is cost-friendly, has AC, reduces hassle of flagging down buses, and is available in greater numbers. In case the metro is not available, it creates a negative mindset in the customer's mind as to the mall being in a remote place.

(ii) Proximity to Railway—Railways are still one of the most widely used modes of transportation in India. If the shopping mall is situated near a Railway station, it will help people who have to do emergency buying or last-minute shopping. It also acts as a place of leisure for those who arrive earlier than their scheduled time and wish to roam nearby. This might lead to them suddenly chancing upon a product they would not have bought otherwise and doing an impulse buy. A large number of people doing an impulse buying will lead to a substantial income for the mall.

(iii) Proximity to Highway—Highways are visualized as roads with long roads and trees on both sides. The opening up of a shopping mall on the highway will attract customers who go out for short journeys and wish to shop and relax themselves. It will also act as a boom for those who might suddenly need few things in the midst of their long drive. Further, the inter-district buses that have a drop and pick-up of customers from such places will also provide an influx of customers who might chance upon something fancy before resuming their journey.

c. Regional Growth [49]

(i) Administrative offices—Administrative offices are places where people are usually stuck at a 9–5 job. They do not wish to go to different standalone stores at the end of the day in order to purchase their necessities and prefer a shopping mall which has stores of different kinds under its roof. This leads to an increase in their sales since once a preference is formed by the employees of the office, it leads to them being loyal customers for at least as long as they work at that job.

(ii) Business Hub—Shopping malls which are large in size and which have too many kinds of stores in it make the mall a business hub. For example, if the mall has a Café Coffee Day store, then it attracts people who wish to do a business meet or recruitment while sipping coffee. In case it so happens that the mall boasts of having apparels, stationery, grocery, food court, play pen, to name a few, then it leads the businesses to become interdependent, which leads the customer to get all his needs satisfied under one roof, which eventually boosts the business.

(iii) School and college—Having schools and colleges nearby helps the mall receive youngsters in large numbers. Students these days have after school or after college classes which gives them some time to spend outside before going for their class. This time spent in the mall equals to small purchases done by the students which increases the revenue of the mall. Parents often wait outside the schools for their children's classes to finish. During this time, the prospect of a shopping mall looks extremely appealing to them. Educators also frequently visit malls situated near their institutions since it provides a break-time respite from their schedule.

d. Cost [48–50]

(i) Land—Cost of the land needs to be such that the revenue earned from the mall can cover it many times over in a few years after its inception. Cost of the land varies in various places since land in an upcoming place will be more in valuation as compared to a land in a suburban place. Similarly, land around a disputed area might be lesser in cost as compared to a land in a non-disputed area. Moreover, it should be remembered that cost of land is a fixed cost and hence it should be recovered accordingly.

(ii) Construction—The construction cost of the land should also be considered as a fixed cost since it is not every day that the mall will be constructed. Renovations, expansion, and extra floors might be added, but it will not be the same as a new construction, it will be considered as add-on's. The materials chosen for construction should be of high quality so that it can withstand the perils of nature for a long time. Compromising with the construction aspect will bring a heavy cost to the mall later on.

e. Attractive design

(i) External design—The external design of the mall should be such that it is attractive, pleasing, and eye-catching at the first glance. The first thing the prospective customer looks at is the external design, which lays down the fact of whether it is a premium mall or a common one. In case the first thought is negative, it leads the customer to form a bad opinion mentally and he may undermine the mall saying that the products he is looking for are surely not available inside.

(ii) Internal design—The internal design should be spacious as well as well-optimized. It means that in case a store needs a larger area and has the scope for an increase in sales, it should be provided with it if space is available, while simultaneously being smart enough to close down the stores which are performing poorly. Proper display of the products should be done so that the customer gets the desire to purchase whether he had a previous intent or not. Proper labeling of the floors and the stores should be done for easy accessibility of the customers.

f. Provision for nearby Parking—Proper parking space availability serves as a major determinant for many buyers. In cases where adequate parking space is unavailable, it causes the customer to go and probably never come back. Moreover, the parking lot fee should also be reasonable and in line with the nearby areas so that people are not discouraged from parking in the mall and find their parking space elsewhere, since if such a situation arises, it will lead the customer to spend his money elsewhere.

g. Environmental health [48,50]

(i) Noise Pollution—Noise dims the brain cells and makes us move away from the source producing it. If an area has continuous noise pollution, then the people living there have a bad quality of life. In case the shopping mall is situated in such an area, it forms a bad impression on the prospective customers and they might not come for a revisit, which will lead to a loss of customers and revenue.

(ii) Air Pollution—Air is the basic ingredient of human life. From the first breath that a baby takes which marks its arrival in the world to the last breath a person takes which marks his departure, the air quality is something which might make or break a person. Increased exposure to air pollution decreases the life span of an individual and hence, if the shopping mall is built in such a place which has air pollution, it will lead to difficulty in breathing for the customers and employees and will not be a hit for sure.

Now, in Table 2 we represent the factors and sub-factors of the associated problem as follows:

Table 2. Representing the factors and sub-factors.

1.	Population $\left(\widetilde{f}^{1}\right)$	a.	Current population density $\left(\widetilde{f}^{11}\right)$
		b.	Socioeconomic state $\left(\widetilde{f}^{12}\right)$
		c.	Population growth rate $\left(\widetilde{f}^{13}\right)$
2.	Transportation $\left(\widetilde{f}^{2}\right)$	a.	Accessibility $\left(\widetilde{f}^{21}\right)$
		b.	Distance to transportation vehicles $\left(\widetilde{f}^{22}\right)$
3.	Regional Growth $\left(\widetilde{f}^{3}\right)$	a.	Administrative offices $\left(\widetilde{f}^{31}\right)$
		b.	Business hub $\left(\widetilde{f}^{32}\right)$
		c.	School and college$\left(\widetilde{f}^{33}\right)$

Table 2. *Cont.*

4.	Cost $(\widetilde{f}^4)$	a.	Land $(\widetilde{f}^{41})$
		b.	Construction $(\widetilde{f}^{42})$
5.	Building structure (Design) $(\widetilde{f}^5)$	a.	External design(Architecture) $(\widetilde{f}^{51})$
		b.	Internal design(Infrastructure) $(\widetilde{f}^{52})$
		c.	Parking $(\widetilde{f}^{53})$
6.	Provision of nearby Parking $(\widetilde{f}^6)$	a.	Capacity of the parking space $(\widetilde{f}^{61})$
		b.	Parking rate per hour $(\widetilde{f}^{62})$
7.	Environmental health $(\widetilde{f}^7)$	a.	Noise Pollution $(\widetilde{f}^{71})$
		b.	Air Pollution $(\widetilde{f}^{72})$

From Figure 3 we observe the relation between alternatives and factors or sub factors.

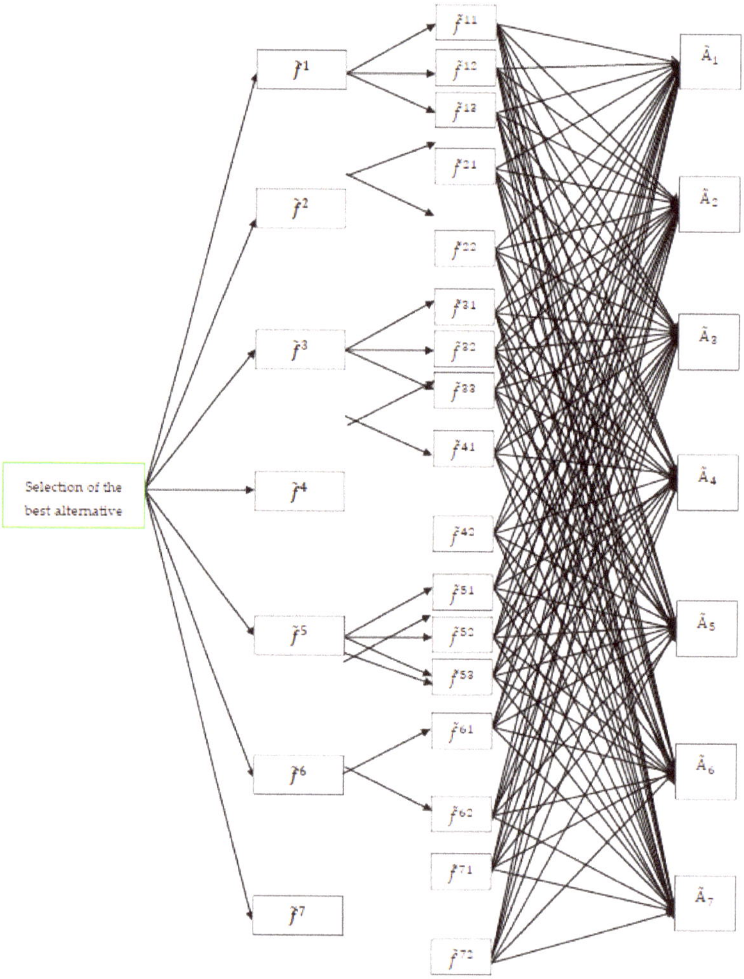

Figure 3. Hierarchical Structure representing the alternatives with factors and sub-factors.

In the present research, different locations in and around Kolkata(Calcutta) are taken as alternatives. The alternatives selected are (the location is mentioned in Google map, presented in Figure 4):

➢ Howrah (A_1)
➢ Chinsurah (A_2)
➢ Uttarpara (A_3)
➢ Dunlop (A_4)
➢ Ballygunge (A_5)
➢ Behala (A_6)
➢ New Town (A_7)

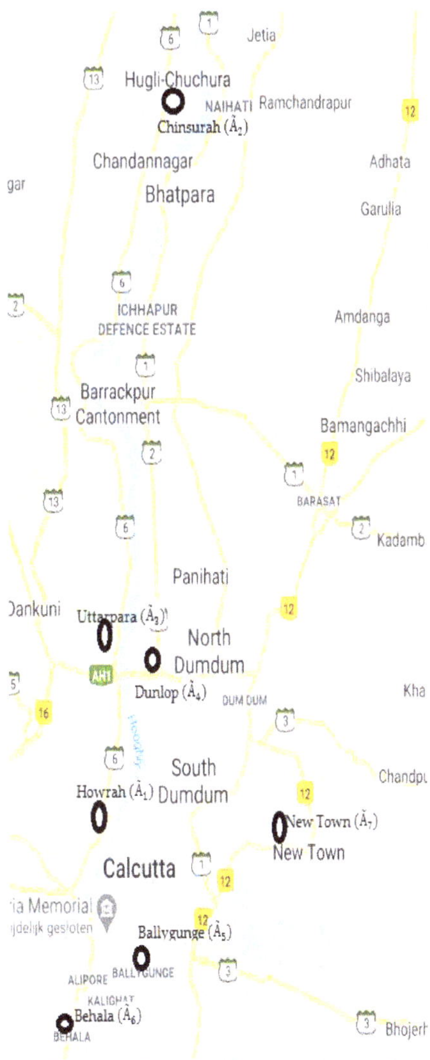

Figure 4. The symbol ○ in Google maps represents different site locations considered in and around the city of Kolkata(Calcutta) for this study.

3.1. Data Source for the Study

The data was collected from the various municipal authorities in the related locations of West Bengal. The municipal representatives were interviewed about the questions related to the factors and sub-factors of the shopping mall site selection problem. They provided information regarding population density and several important factors relevant for this research. The registrar of land, revenue, and expert architects were asked about the prevailing land price as it plays a vital role in site selection.

3.2. NumericalStudy

Step 1: We are now constructing Table 3 based on the comparison matrix with respect to factors using Table 1.

Table 3. Construction of the comparison matrix with respect to factors using Table 1.

$S_{77} = ( )$

	$\widetilde{f^1}$	$\widetilde{f^2}$	$\widetilde{f^3}$	$\widetilde{f^4}$	$\widetilde{f^5}$	$\widetilde{f^6}$	$\widetilde{f^7}$
$\widetilde{f^1}$	1	$\left(\frac{2}{5},\frac{1}{2},\frac{2}{3}\right)$	$\left(\frac{2}{7},\frac{1}{3},\frac{2}{5}\right)$	$\left(\frac{7}{2},4,\frac{9}{2}\right)$	$\left(\frac{3}{2},2,\frac{5}{2}\right)$	$\left(\frac{2}{3},1,2\right)$	$\left(\frac{1}{2},1,\frac{3}{2}\right)$
$\widetilde{f^2}$	$\left(\frac{3}{2},2,\frac{5}{2}\right)$	1	$\left(\frac{1}{2},1,\frac{3}{2}\right)$	$\left(\frac{2}{5},\frac{1}{2},\frac{2}{3}\right)$	$\left(\frac{3}{2},2,\frac{5}{2}\right)$	$\left(\frac{3}{2},2,\frac{5}{2}\right)$	$\left(\frac{2}{5},\frac{1}{2},\frac{2}{3}\right)$
$\widetilde{f^3}$	$\left(\frac{5}{2},3,\frac{7}{2}\right)$	$\left(\frac{2}{3},1,2\right)$	1	$(5,5,5)$	$\left(\frac{5}{2},3,\frac{7}{2}\right)$	$\left(\frac{3}{2},2,\frac{5}{2}\right)$	$\left(\frac{2}{3},1,2\right)$
$\widetilde{f^4}$	$\left(\frac{2}{9},\frac{1}{4},\frac{2}{7}\right)$	$\left(\frac{3}{2},2,\frac{5}{2}\right)$	$\left(\frac{1}{5},\frac{1}{5},\frac{1}{5}\right)$	1	$\left(\frac{3}{2},2,\frac{5}{2}\right)$	$\left(\frac{1}{2},1,\frac{3}{2}\right)$	$\left(\frac{3}{2},2,\frac{5}{2}\right)$
$\widetilde{f^5}$	$\left(\frac{2}{5},\frac{1}{2},\frac{2}{3}\right)$	$\left(\frac{2}{7},\frac{1}{3},\frac{2}{5}\right)$	$\left(\frac{2}{9},\frac{1}{4},\frac{2}{7}\right)$	$\left(\frac{2}{5},\frac{1}{2},\frac{2}{3}\right)$	1	$\left(\frac{2}{9},\frac{1}{4},\frac{2}{7}\right)$	$\left(\frac{1}{2},1,\frac{3}{2}\right)$
$\widetilde{f^6}$	$\left(\frac{1}{2},1,\frac{3}{2}\right)$	$\left(\frac{2}{5},\frac{1}{2},\frac{2}{3}\right)$	$\left(\frac{2}{5},\frac{1}{2},\frac{2}{3}\right)$	$\left(\frac{2}{3},1,2\right)$	$\left(\frac{7}{2},4,\frac{9}{2}\right)$	1	$\left(\frac{3}{2},2,\frac{5}{2}\right)$
$\widetilde{f^7}$	$\left(\frac{2}{3},1,2\right)$	$\left(\frac{3}{2},2,\frac{5}{2}\right)$	$\left(\frac{1}{2},1,\frac{3}{2}\right)$	$\left(\frac{2}{5},\frac{1}{2},\frac{2}{3}\right)$	$\left(\frac{2}{3},1,2\right)$	$\left(\frac{2}{5},\frac{1}{2},\frac{2}{3}\right)$	1

Step 2: Defuzzification of TFN

Here, $\alpha = 0.5$ signifies the preference display of the evaluator and $\beta = 0.5$ signifies the risk factor of the uncertain conditions. The defuzzification step is performed as follows:

$$p_{31}^{\alpha} = (3-2.5)\times 0.5 + 2.5t_{31}^{\alpha} = 3.5-(3.5-3)\times 0.5\ \left(s_{31}^{0.5}\right)^{0.5} = 0.5\times 2.75 + (1-0.5)\times 3.25 = 3$$

Applying the above-mentioned process, the calculation is conducted for all other elements and the defuzzified matrix is represented in Table 4.

Table 4. Defuzzified comparison matrix.

$S_{88} =$

	$\widetilde{f^1}$	$\widetilde{f^2}$	$\widetilde{f^3}$	$\widetilde{f^4}$	$\widetilde{f^5}$	$\widetilde{f^6}$	$\widetilde{f^7}$
$\widetilde{f^1}$	1	0.52	0.34	4	2	1.17	1
$\widetilde{f^2}$	2	1	1	0.52	2	2	0.52
$\widetilde{f^3}$	3	1.17	1	5	3	2	1.17
$\widetilde{f^4}$	0.25	2	0.2	1	2	1	2
$\widetilde{f^5}$	0.52	0.34	0.25	0.52	1	0.25	1
$\widetilde{f^6}$	1	0.52	0.52	1.17	4	1	2
$\widetilde{f^7}$	1.17	2	1	0.52	1.17	0.52	1

The matrix was normalized using Equation (10) and then the priority weights for the factors were calculated using Equation (11). Finally, the Consistency Ratio(CR) values were obtained to determine the consistency of the matrix. The Table 5 depicts weights of the factors.

Table 5. The factors' obtained weights.

Factors	$\tilde{f}^1$	$\tilde{f}^2$	$\tilde{f}^3$	$\tilde{f}^4$	$\tilde{f}^5$	$\tilde{f}^6$	$\tilde{f}^7$
Priority Weight	0.141	0.15	0.26	0.115	0.063	0.15	0.125

The factor "Regional Growth" scored the maximum weight of 0.26, whereas the factor "Provisions of nearby parking" scored the lowest weight.

In the similar order, the sub-factors' matrices were obtained and their respective fuzzy weights were calculated in the exact process as those above. After obtaining the sub-factor fuzzy weights, global weight was computed by the product of individual sub-factor weight with the respective factor fuzzy weight. The global fuzzy weights obtained are ultimately required for ranking the best alternative using the FTOPSIS method. The Table 6 represents Global Fuzzy weights.

Table 6. Representing Global Fuzzy weights.

Factors' Fuzzy Weight	Sub-Factors' Fuzzy Weight	Global Fuzzy Weight
$\widetilde{W}^1 = (0.08, 0.14, 0.24)$	$\widetilde{W}^{11} = (0.1, 0.12, 0.16)$	$\hat{W}^{11} = (0.01, 0.02, 0.04)$
	$\widetilde{W}^{12} = (0.28, 0.36, 0.47)$	$\hat{W}^{12} = (0.02, 0.05, 0.11)$
	$\widetilde{W}^{13} = (0.39, 0.52, 0.67)$	$\hat{W}^{13} = (0.03, 0.07, 0.16)$
$\widetilde{W}^2 = (0.086, 0.15, 0.24)$	$\widetilde{W}^{21} = (0.31, 0.5, 0.92)$	$\hat{W}^{21} = (0.03, 0.07, 0.22)$
	$\widetilde{W}^{22} = (0.27, 0.5, 0.80)$	$\hat{W}^{22} = (0.02, 0.07, 0.19)$
$\widetilde{W}^3 = (0.16, 0.26, 0.44)$	$\widetilde{W}^{31} = (0.12, 0.15, 0.20)$	$\hat{W}^{31} = (0.02, 0.04, 0.09)$
	$\widetilde{W}^{32} = (0.29, 0.38, 0.5)$	$\hat{W}^{32} = (0.05, 0.10, 0.22)$
	$\widetilde{W}^{33} = (0.34, 0.47, 0.64)$	$\hat{W}^{33} = (0.05, 0.12, 0.28)$
$\widetilde{W}^4 = (0.07, 0.12, 0.18)$	$\widetilde{W}^{41} = (0.18, 0.2, 0.23)$	$\hat{W}^{41} = (0.01, 0.024, 0.04)$
	$\widetilde{W}^{42} = (0.70, 0.8, 0.91)$	$\hat{W}^{42} = (0.049, 0.096, 0.164)$
$\widetilde{W}^5 = (0.04, 0.06, 0.10)$	$\widetilde{W}^{51} = (0.1, 0.12, 0.15)$	$\hat{W}^{51} = (0.004, 0.007, 0.015)$
	$\widetilde{W}^{52} = (0.30, 0.39, 0.51)$	$\hat{W}^{52} = (0.012, 0.02, 0.051)$
	$\widetilde{W}^{53} = (0.35, 0.49, 0.66)$	$\hat{W}^{53} = (0.014, 0.029, 0.066)$
$\widetilde{W}^6 = (0.08, 0.15, 0.26)$	$\widetilde{W}^{61} = (0.18, 0.2, 0.22)$	$\hat{W}^{61} = (0.014, 0.03, 0.057)$
	$\widetilde{W}^{62} = (0.72, 0.8, 0.9)$	$\hat{W}^{62} = (0.058, 0.12, 0.234)$
$\widetilde{W}^7 = (0.07, 0.12, 0.23)$	$\widetilde{W}^{71} = (0.31, 0.5, 0.80)$	$\hat{W}^{71} = (0.022, 0.06, 0.212)$
	$\widetilde{W}^{72} = (0.27, 0.5, 0.80)$	$\hat{W}^{72} = (0.019, 0.06, 0.184)$

The global fuzzy sub-factors obtained will be used in ranking of the alternatives using TOPSIS. In this article, the factor $\left(\widetilde{f}^4\right)$ "Cost", $\left(\widetilde{f}^7\right)$ "Environmental health" and the sub-factor $\left(\widetilde{f}^{62}\right)$"Parking rate per hour" are non-beneficial attributes, whereas the other factors and sub-factors are beneficial attributes. The steps of numerical FTOPSIS has been executed below.

Step 1: Preference of the alternatives with respect to the sub-factors in linguistic terms using Table 7.

Table 7. Linguistic preference in terms of TFN.

Linguistic Variable	TFN
Excellent (E)	(7,9,9)
Good (G)	(5,7,9)
Fair (F)	(3,5,7)
Poor (P)	(1,3,5)
Very Poor (VP)	(1,1,3)

The table constructed using Linguistic preference in terms of TFN of Table 7 is represented in Table 8.

Table 8. Construction of decision matrix in terms of linguistic terms using Table 7.

	$\widetilde{f}^{11}$	$\widetilde{f}^{12}$	$\widetilde{f}^{13}$	$\widetilde{f}^{21}$	$\widetilde{f}^{22}$	$\widetilde{f}^{31}$	$\widetilde{f}^{32}$	$\widetilde{f}^{33}$	$\widetilde{f}^{41}$	$\widetilde{f}^{42}$	$\widetilde{f}^{51}$	$\widetilde{f}^{52}$	$\widetilde{f}^{53}$	$\widetilde{f}^{61}$	$\widetilde{f}^{62}$	$\widetilde{f}^{71}$	$\widetilde{f}^{72}$
A_1	E	F	E	E	E	E	G	G	G	G	F	G	P	VP	F	E	E
A_2	F	F	F	F	G	E	F	F	F	G	G	F	P	VP	P	P	VP
A_3	G	E	F	F	G	P	VP	G	P	F	G	F	F	P	F	P	F
A_4	E	G	G	E	G	P	F	G	G	F	G	F	G	G	P	E	E
A_5	E	E	F	E	E	F	F	G	E	E	E	E	G	E	E	G	G
A_6	E	G	E	G	G	F	F	G	G	G	G	G	F	P	G	F	G
A_7	P	G	F	F	G	P	F	P	F	F	E	E	E	G	F	VP	VP

Step 2: Normalizing of the matrix using Equation (14)

Step 3: Computation of weighted normalized matrix using Equation (15)

Keeping in mind the page layout, the Weighted normalized matrix of all the sub-factors has been divided into three tables (Tables 9–11).

Table 9. Weighted normalized matrix for factor 1 and 2

	f^{11}	f^{12}	f^{13}	f^{21}	f^{22}
A_1	0.007,0.02,0.04	0.006,0.03,0.08	0.023,0.07,0.16	0.023,0.07,0.22	0.015,0.07,0.19
A_2	0.003,0.01,0.03	0.006,0.03,0.08	0.009,0.04,0.12	0.009,0.04,0.17	0.01,0.05,0.19
A_3	0.005,0.015,0.04	0.015,0.05,0.11	0.009,0.04,0.12	0.009,0.04,0.17	0.01,0.05,0.19
A_4	0.007,0.02,0.04	0.01,0.04,0.11	0.016,0.05,0.12	0.023,0.07,0.22	0.01,0.05,0.19
A_5	0.007,0.02,0.04	0.015,0.05,0.11	0.009,0.04,0.12	0.023,0.07,0.22	0.015,0.07,0.19
A_6	0.007,0.02,0.04	0.01,0.04,0.11	0.023,0.07,0.16	0.016,0.05,0.22	0.01,0.05,0.19
A_7	0.001,0.006,0.022	0.01,0.04,0.11	0.009,0.04,0.12	0.009,0.04,0.17	0.01,0.05,0.19

Table 10. Weighted normalized matrix for factor 3, 4 and 5

	f^{31}	f^{32}	f^{33}	f^{41}	f^{42}	f^{51}
A_1	0.01, 0.04, 0.09	0.03, 0.07, 0.22	0.03, 0.09, 0.28	0.001, 0.003, 0.008	0.02, 0.04, 0.10	0.001, 0.004, 0.12
A_2	0.01, 0.04, 0.09	0.02, 0.05, 0.17	0.02, 0.06, 0.22	0.001, 0.005, 0.01	0.02, 0.04, 0.10	0.002, 0.005, 0.015
A_3	0.002, 0.013, 0.05	0.005, 0.01, 0.07	0.03, 0.09, 0.28	0.002, 0.008, 0.01	0.02, 0.06, 0.164	0.002, 0.005, 0.015
A_4	0.002, 0.013, 0.05	0.02, 0.05, 0.17	0.03, 0.09, 0.28	0.001, 0.003, 0.008	0.02, 0.06, 0.164	0.002, 0.005, 0.015
A_5	0.006, 0.02, 0.07	0.02, 0.05, 0.17	0.03, 0.09, 0.28	0.001, 0.001, 0.003	0.02, 0.02, 0.02	0.003, 0.007, 0.015
A_6	0.006, 0.02, 0.07	0.02, 0.05, 0.17	0.03, 0.09, 0.28	0.001, 0.003, 0.008	0.02, 0.04, 0.10	0.002, 0.005, 0.015
A_7	0.002, 0.013, 0.05	0.02, 0.05, 0.17	0.005, 0.04, 0.154	0.001, 0.005, 0.01	0.02, 0.06, 0.164	0.001, 0.004, 0.12

Table 11. Weighted normalized matrix for factor 5, 6 and 7.

	f^{52}	f^{53}	f^{61}	f^{62}	f^{71}	f^{72}
A_1	0.007, 0.01, 0.051	0.001, 0.009, 0.04	0.001, 0.003, 0.02	0.008, 0.024, 0.08	0.002, 0.007, 0.03	0.002, 0.007, 0.02
A_2	0.004, 0.011, 0.04	0.001, 0.009, 0.04	0.001, 0.003, 0.02	0.01, 0.04, 0.234	0.004, 0.02, 0.212	0.006, 0.06, 0.184
A_3	0.004, 0.011, 0.04	0.005, 0.016, 0.05	0.001, 0.01, 0.03	0.008, 0.024, 0.08	0.004, 0.02, 0.212	0.003, 0.01, 0.06
A_4	0.004, 0.011, 0.04	0.007, 0.02, 0.066	0.008, 0.02, 0.057	0.01, 0.04, 0.234	0.002, 0.007, 0.03	0.002, 0.007, 0.02
A_5	0.009, 0.02, 0.051	0.007, 0.02, 0.066	0.01, 0.03, 0.057	0.006, 0.01, 0.03	0.002, 0.008, 0.04	0.002, 0.008, 0.04
A_6	0.007, 0.01, 0.051	0.005, 0.016, 0.05	0.001, 0.01, 0.03	0.006, 0.02, 0.05	0.003, 0.012, 0.07	0.002, 0.008, 0.04
A_7	0.009, 0.02, 0.051	0.01, 0.029, 0.066	0.01, 0.03, 0.057	0.008, 0.024, 0.08	0.007, 0.06, 0.212	0.006, 0.06, 0.184

Step 4: Determination of the Fuzzy Positive Ideal Solution (FPIS) (T^+) and Fuzzy Negative Ideal Solution (FNIS) (T^-), where h_c^+ denotes the maximum value of h_{cd} and h_c^- denotes the minimum value of h_{cd}.

Due to page layout, the Fuzzy Positive Ideal Solution (FPIS) and Fuzzy Negative Ideal Solution (FNIS) has been divided into three tables (Tables 12–14).

Table 12. First part of Fuzzy Positive Ideal Solution (FPIS) and Fuzzy Negative Ideal Solution (FNIS).

T^+	0.007, 0.02, 0.04	0.015, 0.05, 0.11	0.023, 0.07, 0.16	0.023, 0.07, 0.22	0.015, 0.07, 0.19	0.01, 0.04, 0.09
T^-	0.001, 0.006,0.022	0.006, 0.03, 0.08	0.009, 0.04, 0.12	0.009, 0.04, 0.17	0.01, 0.05, 0.19	0.002, 0.013, 0.05

Table 13. Second part of FPIS and FNIS.

T^+	0.03, 0.07, 0.22	0.03, 0.09, 0.28	0.001, 0.001, 0.003	0.02, 0.02, 0.02	0.003, 0.007, 0.12	0.009, 0.02, 0.051
T^-	0.005, 0.01, 0.07	0.005, 0.04, 0.154	0.002, 0.008, 0.01	0.02, 0.06, 0.164	0.001, 0.004, 0.015	0.004, 0.01, 0.04

Table 14. Third part of FPIS and FNIS.

T^+	0.01, 0.029, 0.066	0.01, 0.03, 0.057	0.006, 0.01, 0.03	0.002, 0.007, 0.03	0.002, 0.007, 0.02
T^-	0.001, 0.009, 0.04	0.001, 0.003, 0.02	0.01, 0.04, 0.234	0.007, 0.06, 0.212	0.006, 0.06, 0.184

Table 15 describes the distance measure from FPIS, FNIS and relative closeness coefficient, as well as the final ranking.

Table 15. Calculation of Relative Closeness Coefficient (Final Ranking).

Alternatives	$U_c^+(UcP)$	$U_c^-(UcN)$	$R_c=U_c^-/(U_c^-+U_c^+)$	Rank
A_1	0.157	0.65	0.805	2
A_2	0.66	0.168	0.203	7
A_3	0.547	0.38	0.410	5
A_4	0.39	0.425	0.521	4
A_5	0.153	0.675	0.815	1
A_6	0.275	0.553	0.668	3
A_7	0.610	0.251	0.291	6

The Relative Closeness (R_c) depicts the ranking. From the Figure 5 we can see that the larger value of R_c indicates the most preferred alternatives. In this study, the alternative $\widetilde{A}_5$ scores the maximum R_c value of 0.815, whereas, the alternative $\widetilde{A}_2$ scores the lowest value of 0.203. The ranking is obtained on the basis of surveys conducted and data analysis performed. Further sensitivity analysis is conducted to assess the impact of modified weights for different factors and sub-factors.

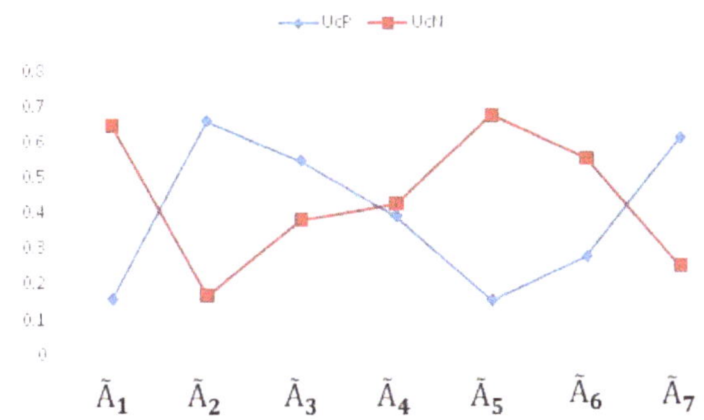

Figure 5. The geometric distance from FPIS and FNIS using Table 15.

4. Sensitivity Analysis and Numerical Simulation

The results of the final ranking are summarized in Table 16. The best shopping mall site selection is $\widetilde{A}_5$. To analyze the impact, due to interchange of sub-factors' weight or elimination of certain sub-factors, sensitivity analysis was conducted. The different rankings obtained are depicted in Table 16 and Figure 6

Table 16. Different Rankings obtained under different cases.

Alternatives	Case1	Case2	Case3
A_1	2	2	1
A_2	7	7	6
A_3	5	6	5
A_4	4	4	4
A_5	1	1	2
A_6	3	3	3
A_7	6	5	7

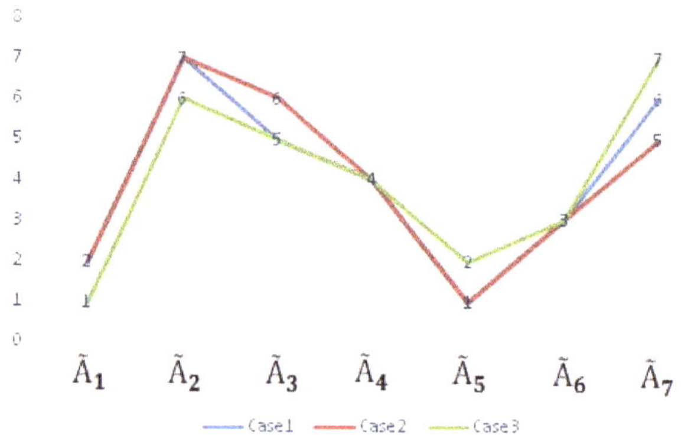

Figure 6. Ranking of alternate sites as per different cases of Sensitivity Analysis from Table 16.

In the analysis, three cases are taken:

Case 1: The sub-factors Current population density $(\widetilde{f^{11}})$ and Administrative offices $(\widetilde{f^{31}})$ fuzzy weights were interchanged; thus, a different ranking was obtained.

Case 2: The sub-factors Capacity of the parking space $\left(\widetilde{f^{61}}\right)$ and Parking rate per hour $\left(\widetilde{f^{62}}\right)$ were eliminated, as these factors are vital, elimination of these yields different ranking.

Case 3: The sub-factors Noise Pollution $(\widetilde{f^{71}})$ and Air Pollution $(\widetilde{f^{72}})$ were eliminated and thus a different ranking was obtained.

Sensitivity analysis shows that different weightages for sub-factors leads to changes in the rankings. These rankings obtained through sensitivity analysis clearly depict the sensitive nature of the optimal ranking based on market dynamics. Change in the environmental factors and socio-economic scenario in the proposed area under consideration leads to changes in the rankings.

5. Discussion

Generation of employment is one of the economic factors of a shopping mall. They create direct as well as indirect employment. Direct employment means the jobs which are directly employed within the establishment and indirect employment are the jobs which are created because of the establishment such as logistics, supplies from local stores, cleaners, and guards, to name a few. Additionally, the income received by these workers are spent on the local shops and services which acts as a multiplier for the economy since the malls lead to income generation for the workers as well as the local businesses, thus uplifting the economy of the area. According to our study, Ballygunge (A_5), located on the heart of the city of Kolkata, is the best site for shopping mall construction as it has the maximum relative closeness coefficient, followed by Howrah (A_1), Behala (A_6), Dunlop (A_4), Uttarpara (A_3), New Town (A_7), and Chinsurah (A_2).

6. Conclusions

Shopping mall site selection can be identified as a critical phenomenon as it can affect the environmental and social-economic factors. The optimal site selection is a complex complicated process as the decision experts must have a broader view and knowledge about the factors and sub-factors, some are quantitative, and few are imprecise qualitative. The desired site enables the investors to invest as it benefits them with the maximum return. Building a shopping mall involves a huge amount of investment. For a developer, selecting the right site is of paramount importance. Precise site selection will lead to profitability of the investor, otherwise it will be disastrous. Proper site selection will ensure larger footfall; thus, a large population will benefit and the economy of the region will get a boost. An improper selection of site will result in a loss to the investor which will have a cascading negative impact over the economy. A shopping mall built in an improper location will face difficulty in selling their floor areas and the purchasers of the floor areas will face losses due to inadequate tread, leading to an unsustainable existence. The application takes into account more of the sub-factors than the factors. A detailed study was carried out in respect of the attributes. A highly knowledgeable, skilled, experienced expert's opinion has been taken into consideration regarding factors and sub-factors for the preferential ranking of the alternatives. The MCDM techniques FAHP and FTOPSIS were used to select the best location. The study used sub-factors for ranking of the alternatives. Seven locations around the city of Kolkata (Calcutta), West Bengal, were considered in the study, where conflicting preferences regarding criteria and sub-criteria exist. Thus, these MCDM techniques can be considered as a practical approach to rank the best site, when multiple complex imprecise constraints exist. Sensitivity analysis gives a clear idea regarding influence of market dynamics in the ranking process. The result of FTOPSIS and sensitivity analysis obtained showed that the 'Ballygunge' region of Kolkata is the superior alternative for shopping mall site selection. The possible explanation for the best ranking is that 'Ballygunge' is one of the most developed regions in Kolkata and excels in a majority of the factors and sub-factors considered in this study. Thus, the rank of 'Ballygunge' region scored a spot of 1 with our ranking method and 1 or 2 under sensitivity analysis.

For future research, a group of decision-makers can be used for the evaluation. Depending on the locations, factors can be added or removed. Different MCDM techniques such as Preference Ranking Organization Method for Enrichment Evaluation (PROMETHEE), Vlse Kriterijumska Optimizacija I Kompromisno Resenje (VIKOR), etc., can be applied with fuzzy logic, interval number, hesitant fuzzy sets (HFS), etc. This research can be extended as a practical approach for ranking the alternatives considering various conflicting factors and sub-factors for real-life problems like super-specialty hospital site selection, sports academy site selection, etc.

Author Contributions: Conceptualization, N.G., E.A.A. and S.P.M.; Data curation, A.G.; Formal analysis, A.G. and S.P.M.; Investigation, N.G., E.A.A. and S.P.M.; Methodology, A.G., E.A.A. and S.P.M.; Project administration, S.P.M.; Writing—review & editing, A.K.S. All authors have read and agreed to the published version of the manuscript.

Funding: This research received no external funding.

Acknowledgments: The third author gratefully acknowledge the support of the University of Tabuk, Ministry of Education in Saudi Arabia.

Conflicts of Interest: The authors declare no conflict of interest.

References

1. Satty, T.L. *The Analytic Hierarchy Process*; McGraw-Hill: New York, NY, USA, 1980.
2. Hwang, C.L.; Yoon, K. Methods for multiple attribute decision making. In *Multiple Attribute Decision Making*; Springer: Berlin/Heidelberg, Germany, 1981; pp. 58–191.
3. Garg, H. A new generalized improved score function of interval-valued intuitionistic fuzzy sets and applications in expert systems. *Appl. Soft Comput.* **2016**, *38*, 988–999. [CrossRef]
4. Kumar, K.; Garg, H. TOPSIS method based on the connection number of set pair analysis under interval-valued intuitionistic fuzzy set environment. *Comput. Appl. Math.* **2018**, *37*, 1319–1329.

5. Sarkar, B. An inventory model with reliability in an imperfect production process. *Appl. Math. Comput.* **2012**, *218*, 4881–4891.
6. Sarkar, B.; Sana, S.S.; Chaudhuri, K. An imperfect production process for time varying demand with inflation and time value of money—An EMQ model. *Expert Syst. Appl.* **2011**, *38*, 13543–13548. [CrossRef]
7. Abdel-Basset, M.; Mohamed, M.; Smarandache, F. An extension of neutrosophic AHP–SWOT analysis for strategic planning and decision-making. *Symmetry* **2018**, *10*, 116. [CrossRef]
8. Selvachandran, G.; Quek, S.G.; Smarandache, F.; Broumi, S. An extended technique for order preference by similarity to an ideal solution (TOPSIS) with maximizing deviation method based on integrated weight measure for single-valued neutrosophic sets. *Symmetry* **2018**, *10*, 236. [CrossRef]
9. Sodhi, B.; Prabhakar, T.V. A simplified description of Fuzzy TOPSIS. *arXiv* **2012**, arXiv:1205.5098.
10. Cheng, E.W.; Li, H.; Yu, L. A GIS approach to shopping mall location selection. *Build. Environ.* **2007**, *42*, 884–892. [CrossRef]
11. Raajpoot, N.A.; Sharma, A.; Chebat, J.C. The role of gender and work status in shopping center patronage. *J. Bus. Res.* **2008**, *61*, 825–833. [CrossRef]
12. DeLisle, J.R. US shopping center classifications: Challenges and opportunities. *Res. Rev.* **2005**, *12*, 96–101.
13. Allard, T.; Babin, B.J.; Chebat, J.C. When income matters: Customers evaluation of shopping malls' hedonic and utilitarian orientations. *J. Retail. Consum. Serv.* **2009**, *16*, 40–49. [CrossRef]
14. Severin, V.; Louviere, J.J.; Finn, A. The stability of retail shopping choices over time and across countries. *J. Retail.* **2001**, *77*, 185–202. [CrossRef]
15. Connell, B.R. *The Principles of Universal Design, Version 2.0*; Trade Press Publishing Corporation: Milwaukee, WI, USA, 1997; Available online: http://www.design.Ncsu.edu/cud/univ.design/princoverviewhtml (accessed on 20 July 2020).
16. Story, M.F. Maximizing usability: The principles of universal design. *Assist. Technol.* **1998**, *10*, 4–12. [CrossRef] [PubMed]
17. Stephanidis, C.; Savidis, A. Universal access in the information society: Methods, tools, and interaction technologies. *Univers. Access Inf. Soc.* **2001**, *1*, 40–55. [CrossRef]
18. Zajicek, M.; Brewster, S. *Design Principles to Support Older Adults*; Springer: Berlin/Heidelberg, Germany, 2004.
19. Yılmaz Kaya, B.; Dağdeviren, M. Selecting occupational safety equipment by MCDM approach considering universal design principles. *Hum. Factors Ergon. Manuf. Serv. Ind.* **2016**, *26*, 224–242. [CrossRef]
20. Darses, F.; Wolff, M. How do designers represent to themselves the users' needs? *Appl. Ergon.* **2006**, *37*, 757–764. [CrossRef]
21. Vink, P. (Ed.) *Comfort and Design: Principles and Good Practice*; CRC Press: Boca Raton, FL, USA, 2004.
22. Ahmed, Z.U.; Ghingold, M.; Dahari, Z. Malaysian shopping mall behavior: An exploratory study. *Asia Pac. J. Mark. Logist.* **2007**, *19*, 331–348. [CrossRef]
23. Aćimović, S.; Mijušković, V. Key logistics location selection factors in retail business. *Ekon. Ideje Praksa* **2016**, *19*, 57–70.
24. Făgărăşan, M.; Cristea, C. Logistics center location: Selection using multicriteria decision making. *Ann. Univ. Oradea Fascicle Manag. Technol. Eng.* **2015**, *24*, 157–162. [CrossRef]
25. Durmuş, A.; Turk, S.S. Factors influencing location selection of warehouses at the intra-urban level: Istanbul case. *Eur. Plan. Stud.* **2014**, *22*, 268–292. [CrossRef]
26. Yang, J.; Lee, H. An AHP decision model for facility location selection. *Facilities* **1997**, *15*, 241–254. [CrossRef]
27. Önüt, S.; Soner, S. Transshipment site selection using the AHP and TOPSIS approaches under fuzzy environment. *Waste Manag.* **2008**, *28*, 1552–1559. [CrossRef] [PubMed]
28. Wang, C.N.; Huang, Y.F.; Chai, Y.C.; Nguyen, V.T. A multi-criteria decision making (MCDM) for renewable energy plants location selection in Vietnam under a fuzzy environment. *Appl. Sci.* **2018**, *8*, 2069. [CrossRef]
29. Senvar, O.; Otay, I.; Bolturk, E. Hospital site selection via hesitant fuzzy TOPSIS. *IFAC-PapersOnLine* **2016**, *49*, 1140–1145. [CrossRef]
30. Suárez-Vega, R.; Santos-Peñate, D.R.; Dorta-González, P. Location models and GIS tools for retail site location. *Appl. Geogr.* **2012**, *35*, 12–22. [CrossRef]
31. Nema, A.K.; Gupta, S.K. Optimization of regional hazardous waste management systems: An improved formulation. *Waste Manag.* **1999**, *19*, 441–451. [CrossRef]
32. Hauser, J.R.; Koppelman, F.S. Alternative perceptual mapping techniques: Relative accuracy and usefulness. *J. Mark. Res.* **1979**, *16*, 495–506. [CrossRef]

33. Terblanché, N.S. The perceived benefits derived from visits to a super regional shopping centre: An exploratory study. *S. Afr. J. Bus. Manag.* **1999**, *30*, 141–146. [CrossRef]
34. Farahani, R.Z.; SteadieSeifi, M.; Asgari, N. Multiple criteria facility location problems: A survey. *Appl. Math. Model.* **2010**, *34*, 1689–1709. [CrossRef]
35. Cheng, S.; Chan, C.W.; Huang, G.H. Using multiple criteria decision analysis for supporting decisions of solid waste management. *J. Environ. Sci. Health Part A* **2002**, *37*, 975–990. [CrossRef]
36. Aktas, A.; Kabak, M. A model proposal for locating wind turbines. *Procedia Comput. Sci.* **2016**, *102*, 426–433. [CrossRef]
37. Uyan, M. GIS-based solar farms site selection using analytic hierarchy process (AHP) in Karapinar region, Konya/Turkey. *Renew. Sustain. Energy Rev.* **2013**, *28*, 11–17. [CrossRef]
38. Oztaysi, B. A decision model for information technology selection using AHP integrated TOPSIS-Grey: The case of content management systems. *Knowl. Based Syst.* **2014**, *70*, 44–54. [CrossRef]
39. Liu, S.; Chan, F.T.; Ran, W. Multi-attribute group decision-making with multi-granularity linguistic assessment information: An improved approach based on deviation and TOPSIS. *Appl. Math. Model.* **2013**, *37*, 10129–10140. [CrossRef]
40. Abid, F.; Bahloul, S. Selected MENA countries'attractiveness to G7 investors. *Econ. Model.* **2011**, *28*, 2197–2207. [CrossRef]
41. Papadopoulos, A.; Karagiannidis, A. Application of the multi-criteria analysis method Electre III for the optimisation of decentralised energy systems. *Omega* **2008**, *36*, 766–776. [CrossRef]
42. Chen, C.T.; Pai, P.F.; Hung, W.Z. An integrated methodology using linguistic PROMETHEE and maximum deviation method for third-party logistics supplier selection. *Int. J. Comput. Intell. Syst.* **2010**, *3*, 438–451. [CrossRef]
43. Li, M.; Jin, L.; Wang, J. A new MCDM method combining QFD with TOPSIS for knowledge management system selection from the user's perspective in intuitionistic fuzzy environment. *Appl. Soft Comput.* **2014**, *21*, 28–37. [CrossRef]
44. Liu, C.H.; Tzeng, G.H.; Lee, M.H. Improving tourism policy implementation–The use of hybrid MCDM models. *Tour. Manag.* **2012**, *33*, 413–426. [CrossRef]
45. Kanapeckiene, L.; Kaklauskas, A.; Zavadskas, E.K.; Seniut, M. Integrated knowledge management model and system for construction projects. *Eng. Appl. Artif. Intell.* **2010**, *23*, 1200–1215. [CrossRef]
46. Zadeh, L.A. Fuzzy sets. *Inf. Control* **1965**, *8*, 338–353. [CrossRef]
47. Chang, C.W.; Wu, C.R.; Lin, H.L. Applying fuzzy hierarchy multiple attributes to construct an expert decision making process. *Expert Syst. Appl.* **2009**, *36*, 7363–7368. [CrossRef]
48. Cheng, E.W.; Li, H.; Yu, L. The analytic network process (ANP) approach to location selection: A shopping mall illustration. *Constr. Innov.* **2005**, *5*, 83–98. [CrossRef]
49. Erdin, C.; Akbaş, H.E. A comparative analysis of fuzzy topsis and geographic information systems (gis) for the location selection of shopping malls: A case study from turkey. *Sustainability* **2019**, *11*, 3837. [CrossRef]
50. Kazemi, A.; Amiri, M. Selecting Shopping Center Site Using MADM Techniques. In Proceedings of the International Conference on Education, E-Governance, Law and Business, Dubai, UAE, 1–2 January 2017.

mathematics

MDPI

Article

Octahedron Subgroups and Subrings

Jeong-Gon Lee [1,*], **Young Bae Jun** [2] **and Kul Hur** [1]

[1] Division of Applied Mathematics, Wonkwang University, 460, Iksan-daero, Iksan-Si, Jeonbuk 54538, Korea; kulhur@wku.ac.kr
[2] Department of Mathematics Education, Gyeongsang National University, Jinju 52828, Korea; skywine@gmail.com
* Correspondence: jukolee@wku.ac.kr

Received: 1 August 2020; Accepted: 25 August 2020; Published: 28 August 2020

Abstract: In this paper, we define the notions of *i*-octahedron groupoid and *i*-OLI [resp., *i*-ORI and *i*-OI], and study some of their properties and give some examples. Also we deal with some properties for the image and the preimage of *i*-octahedron groupoids [resp., *i*-OLI, *i*-ORI and *i*-OI] under a groupoid homomorphism. Next, we introduce the concepts of *i*-octahedron subgroup and normal subgroup of a group and investigate some of their properties. In particular, we obtain a characterization of an *i*-octahedron subgroup of a group. Finally, we define an *i*-octahedron subring [resp., *i*-OLI, *i*-ORI and *i*-OI] of a ring and find some of their properties. In particular, we obtain two characterizations of *i*-OLI [resp., *i*-ORI and *i*-OI] of a ring and a skew field, respectively.

Keywords: octahedron set; *i*-octahedron subgroupoid; *i*-octahedron ideal; *i*-sup-property, *i*-octahedron subgroup; *i*-octahedron subring

MSC: 20N25

1. Introduction

In 1965, Zadeh [1] proposed the concept of fuzzy sets as a generalization of crisp sets, in order express mathematically uncertainty problems. After then, Zadeh [2] and Atanassov [3] introduced the concept of interval-valued fuzzy sets and intuitionistic fuzzy sets, respectively. In traditional fuzzy logic, a number contained in the unit interval $[0,1]$ is used as a measure of expert confidence in other statements. However, it is often difficult for experts to accurately quantify their certainty. In other words, the probability of quantified figures being accurate is low. Therefore, it is necessary to increase the accuracy of measurement using the sub-interval of $[0,1]$, and there is an interval-value fuzzy set developed as a mathematical tool for this. Fuzzy sets or interval-valued fuzzy sets are a very useful tools for measuring against one factor, but it is not appropriate to measure two factors at the same time. Jun et al. [4] defined the notion of cubic sets, which is a kind of hybrid structure, by using a fuzzy set and an interval-valued fuzzy set. In addition, it is a good mathematical tool for evaluating both factors at the same time, and it is being applied in many places (see [5–9]). After the introduction of cubic set, various concepts related to it, i.e., cubic set, (generalized) cubic intuitionistic fuzzy set, cubic interval-valued intuitionistic fuzzy set, cubic picture fuzzy set, cubic hesitant fuzzy set, cubic bipolar fuzzy set, cubic Pythagorean fuzzy set, cubic soft set, etc., have emerged and are being applied in various ways. We can consider intuitionistic fuzzy set as a tool to measure both positive and negative factors for every outcome/assessment at the same time. We need the ability to handle three different tasks at the same time amid increasingly diverse social phenomena due to the development of science. In addition, mathematicians feel the need to develop mathematical tools to support this, and they have a desire to develop a wider hybrid structure. With a wider hybrid structure, Lee et al. [10] defined an octahedron set composed of an interval-valued fuzzy set, an intuitionistic fuzzy set and a fuzzy set that will provide more information about uncertainty. This structure allows point measurements, interval measurements, and positive and negative simultaneous measurements as event assessments at the same time. As mathematicians, the purpose of this paper is to carry out the study of applying octahedron sets to algebraic structures, in particular, groups and rings. From now on, we expect octahedron sets to be applied to several branches, including algebraic structures, topological structures, metric spaces, medical science, decision making systems, aggregation operators, expert systems, etc. The composition of this paper is as follows. In Section 2, we list some basic concepts needed in the next sections: for examples, an intuitionistic number, an intuitionistic fuzzy set, an interval number, an interval-valued fuzzy set, an octahedron number and an octahedron set. In Section 3, we define the i-product of two octahedron sets in a groupoid and introduce the concept of i-octahedron subgroupoids of a groupoid by using it. In particular, we obtain four characterizations of i-octahedron groupoids (See Theorems 2–4). Also, we define an i-OLI [resp., i-ORI and i-OI]] of a groupoid and study some of their properties. Moreover, we obtain some properties for the image and preimage of an i-octahedron subgroupoid [resp., i-OLI, i-ORI and i-OI] under groupoid homomorphism. In Section 4, we define an i-octahedron subgroup of a group and investigate some of its properties. In particular, we obtain two characterizations of i-octahedron subgroup and i-OLI [resp., i-ORI and i-OI] of a group (See Theorems 7 and 8). In Section 4, we introduce the concepts of i-octahedron subrings [resp., i-OLIs, i-ORIs and i-OIs] of a ring and obtain their characterizations (See Theorems 13 and 15). Furthermore, we find a sufficient condition for which a commutative ring with a unity e is a field (See Proposition 27).

2. Preliminaries

Let $I \oplus I = \{\bar{a} = (a^{\in}, a^{\notin}) \in I \times I : a^{\in} + a^{\notin} \leq 1\}$, where $I = [0,1]$. Then each member $\bar{a}$ of $I \oplus I$ is called an intuitionistic point or intuitionistic number. In particular, we denote $(0,1)$ and $(1,0)$ as $\bar{0}$ and $\bar{1}$, respectively. Refer to [11] for the definitions of the order ($\leq$) and the equality ($=$) of two intuitionistic numbers, and the infimum and the supremum of any intuitionistic numbers.

Definition 1 ([3])**.** *For a nonempty set X, a mapping $A : X \to I \oplus I$ is called an intuitionistic fuzzy set (briefly, IF set) in X, where for each $x \in X$, $A(x) = (A^{\in}(x), A^{\notin}(x))$, and $A^{\in}(x)$ and $A^{\notin}(x)$ represent the degree of membership and the degree of nonmembership of an element x to A, respectively. Let $(I \oplus I)^X$ denote the set of all IF sets in X and for each $A \in (I \oplus I)^X$, we write $A = (A^{\in}, A^{\notin})$. In particular, $\bar{\mathbf{0}}$ and $\bar{\mathbf{1}}$ denote the IF empty set and the IF whole set in X defined by, respectively: for each $x \in X$,*

$$\bar{\mathbf{0}}(x) = \bar{0} \text{ and } \bar{\mathbf{1}}(x) = \bar{1}.$$

For the definitions of the inclusion, the equality, the union and the intersection of two IF sets, the complement of an IF set, two operations [ ] and $\diamond$ on $(I \oplus I)^X$, refer to [3].

The set of all closed subintervals of I is denoted by $[I]$, and members of $[I]$ are called interval numbers and are denoted by $\widetilde{a}$, $\widetilde{b}$, $\widetilde{c}$, etc., where $\widetilde{a} = [a^-, a^+]$ and $0 \leq a^- \leq a^+ \leq 1$. In particular, if $a^- = a^+$, then we write as $\widetilde{a} = \mathbf{a}$ (See [12]).

For the definitions of the order and the equality of two interval numbers, and the infimum and the supremum of any interval numbers, refer to [13,14].

Definition 2 ([2,15])**.** *For a nonempty set X, a mapping $A : X \to [I]$ is called an interval-valued fuzzy set (briefly, an IVF set) in X. Let $[I]^X$ denote the set of all IVF sets in X. For each $A \in [I]^X$ and $x \in X$, $A(x) = [A^-(x), A^+(x)]$ is called the degree of membership of an element x to A, where $A^-, A^+ \in I^X$ are called a lower fuzzy set and an upper fuzzy set in X, respectively. For each $A \in [I]^X$, we write $A = [A^-, A^+]$. In particular, $\widetilde{0}$ and $\widetilde{1}$ denote the interval-valued fuzzy empty set and the interval-valued fuzzy empty whole set in X defined by, respectively: for each $x \in X$,*

$$\widetilde{0}(x) = \mathbf{0} \text{ and } \widetilde{1}(x) = \mathbf{1}.$$

For the definitions of the inclusion, the equality, the union, the intersection of two IV sets and the complement of an IV set, refer to [2,15].

Now members of $[I] \times (I \oplus I) \times I$ are written as $\widetilde{\bar{a}} = \langle \widetilde{a}, \bar{a}, a \rangle = \langle [a^-, a^-], (a^{\in}, a^{\notin}), a \rangle$, $\widetilde{\bar{b}} = \langle \widetilde{b}, \bar{b}, b \rangle = \langle [b^-, b^-], (b^{\in}, b^{\notin}), b \rangle$, etc. and are called octahedron numbers. Furthermore, we will define the following order relations in $[I] \times (I \otimes I) \times I$ (see [10]):

(Oi) (Equality) $\widetilde{\bar{a}} = \widetilde{\bar{b}} \Leftrightarrow \widetilde{a} = \widetilde{b},\ \bar{a} = \bar{b},\ a = b$,

(Oii) (Type 1-order) $\widetilde{\bar{a}} \leq_1 \widetilde{\bar{b}} \Leftrightarrow a^- \leq b^-,\ a^+ \leq b^+,\ a^{\in} \leq b^{\in},\ a^{\notin} \geq b^{\notin},\ a \leq b$,

(Oiii) (Type 2-order) $\widetilde{\bar{a}} \leq_2 \widetilde{\bar{b}} \Leftrightarrow a^- \leq b^-,\ a^+ \leq b^+,\ a^{\in} \leq b^{\in},\ a^{\notin} \geq b^{\notin},\ a \geq b$,

(Oiv) (Type 3-order) $\widetilde{\bar{a}} \leq_3 \widetilde{\bar{b}} \Leftrightarrow a^- \leq b^-,\ a^+ \geq b^+,\ a^{\in} \geq b^{\in},\ a^{\notin} \leq b^{\notin},\ a \leq b$,

(Ov) (Type 4-order) $\widetilde{\bar{a}} \leq_4 \widetilde{\bar{b}} \Leftrightarrow a^- \leq b^-,\ a^+ \leq b^+,\ a^{\in} \geq b^{\in},\ a^{\notin} \leq b^{\notin},\ a \geq b$.

Definition 3 ([10]). *Let X be a nonempty set and let $\mathbf{A} = [A^-, A^+] \in [I]^X$, $A = (A^{\in}, A^{\notin}) \in (I \oplus I)^X$, $\lambda \in I^X$. Then the triple $\mathcal{A} = \langle \mathbf{A}, A, \lambda \rangle$ is called an* octahedron *set in X. In fact, $\mathcal{A} : X \to [I] \times (I \oplus I) \times I$ is a mapping.*

We can consider following special octahedron sets in X:

$\left\langle \widetilde{0}, \bar{\mathbf{0}}, 0 \right\rangle = \ddot{0}$,

$\left\langle \widetilde{0}, \bar{\mathbf{0}}, 1 \right\rangle$, $\left\langle \widetilde{0}, \bar{\mathbf{1}}, 0 \right\rangle$, $\left\langle \widetilde{1}, \bar{\mathbf{0}}, 0 \right\rangle$,

$\left\langle \widetilde{0}, \bar{\mathbf{1}}, 1 \right\rangle$, $\left\langle \widetilde{1}, \bar{\mathbf{0}}, 1 \right\rangle$, $\left\langle \widetilde{1}, \bar{\mathbf{1}}, 0 \right\rangle$,

$\left\langle \widetilde{1}, \bar{\mathbf{1}}, 1 \right\rangle = \ddot{1}$.

In this case, $\ddot{0}$ [resp., $\ddot{1}$] is called an octahedron empty set [resp., octahedron whole set] in X. We denote the set of all octahedron sets as $\mathcal{O}(X)$.

It is obvious that for each $A \in 2^X$, $\chi_A = \langle [\chi_A, \chi_A], (\chi_A, \chi_{A^c}), \chi_A \rangle \in \mathcal{O}(X)$ and then $2^X \subset \mathcal{O}(X)$, where 2^X denotes the set of all subsets of X and χ_A denotes the characteristic function of A. Furthermore, we can easily see that for each $\mathbf{A} = \langle A, \lambda \rangle \in \mathcal{C}(X)$, $\mathbf{A} = \langle A, (A^-, A^+), \lambda \rangle$, $\mathbf{A} = \langle A, (\lambda, \lambda^c), \lambda \rangle \in \mathcal{O}(X)$ and then $\mathcal{C}(X) \subset \mathcal{O}(X)$. In this case, we denote $\langle A, (A^-, A^+), \lambda \rangle$ and $\langle A, (\lambda, \lambda^c), \lambda \rangle$ as $\mathcal{A}_A$ and $\mathcal{A}_\lambda$, respectively. In fact, we can consider octahedron sets as a generalization of cubic sets.

Definition 4 ([10]). *Let X be a nonempty set and let $\mathcal{A} = \langle \mathbf{A}, A, \lambda \rangle$, $\mathcal{B} = \langle \mathbf{B}, B, \mu \rangle \in \mathcal{O}(X)$. Then we can define following order relations between $\mathcal{A}$ and $\mathcal{B}$:*

(i) (Equality) $\mathcal{A} = \mathcal{B} \Leftrightarrow \mathbf{A} = \mathbf{B},\ A = B,\ \lambda = \mu$,

(ii) (Type 1-order) $\mathcal{A} \subset_1 \mathcal{B} \Leftrightarrow \mathbf{A} \subset \mathbf{B},\ A \subset B, \lambda \leq \mu$,

(iii) (Type 2-order) $\mathcal{A} \subset_2 \mathcal{B} \Leftrightarrow \mathbf{A} \subset \mathbf{B},\ A \subset B, \lambda \geq \mu$,

(iv) (Type 3-order) $\mathcal{A} \subset_3 \mathcal{B} \Leftrightarrow \mathbf{A} \subset \mathbf{B},\ A \supset B, \lambda \leq \mu$,

(v) (Type 4-order) $\mathcal{A} \subset_4 \mathcal{B} \Leftrightarrow \mathbf{A} \subset \mathbf{B},\ A \supset B, \lambda \geq \mu$.

Definition 5 ([10]). *Let X be a nonempty set and let $(\mathcal{A}_j)_{j \in J} = (\langle \mathbf{A}_j, A_j, \lambda_j \rangle)_{j \in J}$ be a family of octahedron sets in X. Then the Type i-union $\cup^i$ and Type i-intersection $\cap^i$ of $(\mathcal{A}_j)_{j \in J}$, $(i = 1,\ 2,\ ,3,4)$, are defined as follows, respectively:*

(i) (Type i-union) $\bigcup^1_{j \in J} \mathcal{A}_j = \left\langle \bigcup_{j \in J} \mathbf{A}_j, \bigcup_{j \in J} A_j, \bigcup_{j \in J} \lambda_j \right\rangle$,

$\bigcup^2_{j \in J} \mathcal{A}_j = \left\langle \bigcup_{j \in J} \mathbf{A}_j, \bigcup_{j \in J} A_j, \bigcap_{j \in J} \lambda_j \right\rangle$,

$\bigcup^3_{j \in J} \mathcal{A}_j = \left\langle \bigcup_{j \in J} \mathbf{A}_j, \bigcap_{j \in J} A_j, \bigcup_{j \in J} \lambda_j \right\rangle$,

$\bigcup^4_{j \in J} \mathcal{A}_j = \left\langle \bigcup_{j \in J} \mathbf{A}_j, \bigcap_{j \in J} A_j, \bigcap_{j \in J} \lambda_j \right\rangle$,

(ii) (Type i-intersection) $\bigcap^1_{j \in J} \mathcal{A}_j = \left\langle \bigcap_{j \in J} \mathbf{A}_j, \bigcap_{j \in J} A_j, \bigcap_{j \in J} \lambda_j \right\rangle$,

$\bigcap^2_{j \in J} \mathcal{A}_j = \left\langle \bigcap_{j \in J} \mathbf{A}_j, \bigcap_{j \in J} A_j, \bigcup_{j \in J} \lambda_j \right\rangle$,

$$\bigcap^3_{j\in J} \mathcal{A}_j = \left\langle \bigcap_{j\in J} \mathbf{A}_j, \bigcup_{j\in J} A_j, \bigcap_{j\in J} \lambda_j \right\rangle,$$
$$\bigcap^4_{j\in J} \mathcal{A}_j = \left\langle \bigcap_{j\in J} \mathbf{A}_j, \bigcup_{j\in J} A_j, \bigcup_{j\in J} \lambda_j \right\rangle.$$

Definition 6 ([10]). *Let X be a nonempty set and let $\mathcal{A} = \langle \mathbf{A}, A, \lambda \rangle$ be an octahedron set in X. Then the complement $\mathcal{A}^c$, operators $[\,]$ and $\diamond$ of $\mathcal{A}$ are defined as follows, respectively: for each $x \in X$,*
(i) $\mathcal{A}^c = \langle \mathbf{A}^c, A^c, \lambda^c \rangle$,
(ii) $[\,]\mathcal{A} = \langle \mathbf{A}, [\,]A, \lambda \rangle$,
(iii) $\diamond\mathcal{A} = \langle \mathbf{A}, \diamond A, \lambda \rangle$.

Definition 7 ([10]). *Let $\mathcal{A} = \langle \mathbf{A}, A, \lambda \rangle \in O(X)$, let $\tilde{\bar{a}} = \langle \tilde{a}, \bar{a}, a \rangle \in [I]$ be an octahedron number such that $a^+ > 0$, $\bar{a} \in I \oplus I$ with $\bar{a} \neq \bar{0}$, $a \in I$ with $a \neq 0$. Then A is called an octahedron point with the support $x \in X$ and the value $\tilde{\bar{a}}$, denoted by $A = x_{\tilde{\bar{a}}}$, if for each $y \in X$,*

$$x_{\tilde{\bar{a}}}(y) = \begin{cases} \tilde{\bar{a}} & \text{if } y = x \\ \left\langle \tilde{0}, \bar{0}, 0 \right\rangle & \text{otherwise.} \end{cases}$$

The set of all octahedron points in X is denoted by $\mathcal{O}_P(X)$.

Definition 8 ([16]). *Let $(X, \cdot)$ be a groupoid and let $\lambda,\ \mu \in I^X$. Then the product of λ and μ, denoted by $\lambda \circ_F \mu$, is a fuzzy set in X defined as follows: for each $x \in X$,*

$$(\lambda \circ_F \mu)(x) = \begin{cases} \bigvee_{yz=x,\ y,\ z\in X}[\lambda(y) \wedge \mu(z)] & \text{if } yz = x \\ 0 & \text{otherwise.} \end{cases}$$

Definition 9 ([17]). *Let $(X, \cdot)$ be a groupoid and let $A,\ B \in (I \oplus I)^X$. Then the product of A and B, denoted by $A \circ_{IF} B$, is an IF set in X defined as follows: for each $x \in X$,*

$$(A \circ_{IF} B)(x) = \begin{cases} (\bigvee_{yz=x,\ y,\ z\in X}[A^{\in}(y) \wedge B^{\in}(z)], \bigwedge_{yz=x,\ y,\ z\in X}[A^{\notin}(y) \wedge B^{\notin}(z)] & \text{if } yz = x \\ (0, 1) & \text{otherwise.} \end{cases}$$

Definition 10 ([18]). *Let $(X, \cdot)$ be a groupoid and let $\mathbf{A},\ \mathbf{B} \in [I]^X$. Then the product of $\mathbf{A}$ and $\mathbf{B}$, denoted by $\mathbf{A} \circ_{IV} \mathbf{B}$, is an IVF set in X defined as follows: for each $x \in X$,*

$$(\mathbf{A} \circ_{IV} \mathbf{B})(x) = \begin{cases} [\ \bigvee_{yz=x,\ y,\ z\in X}[A^-(y) \wedge B^-(z)], \bigvee_{yz=x,\ y,\ z\in X}[A^+(y) \wedge B^+(z)]\ ] & \text{if } yz = x \\ [0, 0] & \text{otherwise.} \end{cases}$$

3. Octahedron Subgroupoids

In this section, we list the product of fuzzy sets [resp., intuitionistic fuzzy sets and interval-valued fuzzy sets] and we define the product of octahedron sets by using each product. Next we introduce the concepts of octahedron subgroupoid and octahedron ideal in a groupoid X, and find some of their properties and give some examples.

Throughout this section and next section, for an octahedron set $\mathcal{A} = \langle \mathbf{A}, A, \lambda \rangle$ in a set X, $\mathcal{A} \neq \ddot{0}$ [resp., $\left\langle \tilde{0}, \bar{0}, 1 \right\rangle$, $\left\langle \tilde{0}, \bar{1}, 0 \right\rangle$ and $\left\langle \tilde{0}, \bar{1}, 1 \right\rangle$] means that
$\mathbf{A} \neq \tilde{0},\ A \neq \bar{0},\ \lambda \neq 0$ [resp., $\mathbf{A} \neq \tilde{0},\ A \neq \bar{0},\ \lambda \neq 1$; $\mathbf{A} \neq \tilde{0},\ A \neq \bar{1},\ \lambda \neq 0$ and $\mathbf{A} \neq \tilde{0},\ A \neq \bar{1},\ \lambda \neq 1$].

Based on the order relations (Oi), (Oii), (Oiii), (Oiv) and (Ov), we can define the inf and the sup of octahedron numbers as follows:

Definition 11. *Let $\tilde{\bar{a}}, \tilde{\bar{b}} \in [I] \times (I \oplus I) \times I$. Then*
(i) $\tilde{\bar{a}} \wedge^1 \tilde{\bar{b}} = \langle [a^- \wedge b^-, a^+ \wedge b^+], (a^{\in} \wedge b^{\in}, a^{\notin} \vee b^{\notin}), a \wedge b \rangle$,

$\widetilde{a} \wedge^2 \widetilde{b} = \langle [a^- \wedge b^-, a^+ \wedge b^+], (a^{\in} \wedge b^{\in}, a^{\notin} \vee b^{\notin}), a \vee b \rangle$,
$\widetilde{a} \wedge^3 \widetilde{b} = \langle [a^- \wedge b^-, a^+ \wedge b^+], (a^{\in} \vee b^{\in}, a^{\notin} \wedge b^{\notin}), a \wedge b \rangle$,
$\widetilde{a} \wedge^4 \widetilde{b} = \langle [a^- \wedge b^-, a^+ \wedge b^+], (a^{\in} \vee b^{\in}, a^{\notin} \wedge b^{\notin}), a \vee b \rangle$,
(ii) $\widetilde{a} \vee^1 \widetilde{b} = \langle [a^- \vee b^-, a^+ \vee b^+], (a^{\in} \vee b^{\in}, a^{\notin} \wedge b^{\notin}), a \vee b \rangle$,
$\widetilde{a} \vee^2 \widetilde{b} = \langle [a^- \vee b^-, a^+ \vee b^+], (a^{\in} \vee b^{\in}, a^{\notin} \wedge b^{\notin}), a \wedge b \rangle$,
$\widetilde{a} \vee^3 \widetilde{b} = \langle [a^- \vee b^-, a^+ \vee b^+], (a^{\in} \wedge b^{\in}, a^{\notin} \vee b^{\notin}), a \vee b \rangle$,
$\widetilde{a} \vee^4 \widetilde{b} = \langle [a^- \vee b^-, a^+ \vee b^+], (a^{\in} \wedge b^{\in}, a^{\notin} \vee b^{\notin}), a \wedge b \rangle$.

By using Definition 11, we can find the product of two octahedron sets as follows:

Definition 12. *Let $(X, \cdot)$ be a groupoid and let $\mathcal{A} = \langle \mathbf{A}, A, \lambda \rangle$, $\mathcal{B} = \langle \mathbf{B}, B, \mu \rangle \in \mathcal{O}(X)$. Then the i-product of $\mathcal{A}$ and $\mathcal{B}$, denoted by $\mathcal{A} \circ_i \mathcal{B}$ (i = 1, 2, ,3, 4), is an octahedron set in X defined as follows: for each $x \in X$,*

$$(\mathcal{A} \circ_1 \mathcal{B})(x) = \begin{cases} \bigvee^1_{yz=x,\ y,\ z\in X}[\mathcal{A}(y) \wedge^1 \mathcal{B}(z)] & \text{if } yz = x \text{ for some } y, z \in X \\ \ddot{0} & \text{otherwise,} \end{cases}$$

$$(\mathcal{A} \circ_2 \mathcal{B})(x) = \begin{cases} \bigvee^2_{yz=x,\ y,\ z\in X}[\mathcal{A}(y) \wedge^2 \mathcal{B}(z)] & \text{if } yz = x \text{ for some } y, z \in X \\ \langle [0,0], (0,1), 1 \rangle & \text{otherwise,} \end{cases}$$

$$(\mathcal{A} \circ_3 \mathcal{B})(x) = \begin{cases} \bigvee^3_{yz=x,\ y,\ z\in X}[\mathcal{A}(y) \wedge^3 \mathcal{B}(z)] & \text{if } yz = x \text{ for some } y, z \in X \\ \langle [0,0], (1,0), 0 \rangle & \text{otherwise,} \end{cases}$$

$$(\mathcal{A} \circ_4 \mathcal{B})(x) = \begin{cases} \bigvee^4_{yz=x,\ y,\ z\in X}[\mathcal{A}(y) \wedge^4 \mathcal{B}(z)] & \text{if } yz = x \text{ for some } y, z \in X \\ \langle [0,0], (1,0), 1 \rangle & \text{otherwise.} \end{cases}$$

Remark 1. *From Definitions 8–12, we can easily see that followings hold:*
(1) $\mathcal{A} \circ_1 \mathcal{B} = \langle \mathbf{A} \circ_{IV} \mathbf{B}, A \circ_{IF} B, \lambda \circ_F \mu \rangle$,
(2) $\mathcal{A} \circ_2 \mathcal{B} = \langle \mathbf{A} \circ_{IV} \mathbf{B}, A \circ_{IF} B, \lambda \circ_2 \mu \rangle$, *where*

$$(\lambda \circ_2 \mu)(x) = \begin{cases} \bigwedge_{yz=x,\ y,\ z\in X}[\lambda(y) \vee \mu(z)] & \text{if } yz = x \text{ for some } y, z \in X \\ 1 & \text{otherwise,} \end{cases}$$

(3) $\mathcal{A} \circ_3 \mathcal{B} = \langle \mathbf{A} \circ_{IV} \mathbf{B}, A \circ_3 B, \lambda \circ_F \mu \rangle$, *where*

$$(A \circ_3 B)(x) = \begin{cases} \left(\bigwedge_{yz=x,\ y,\ z\in X}[A^{\in}(y) \vee B^{\in}(z)], \bigvee_{yz=x,\ y,\ z\in X}[A^{\notin}(y) \wedge B^{\notin}(z)] \right) & \text{if } yz = x \\ & \text{for some } y, z \in X \\ (1,0) & \text{otherwise,} \end{cases}$$

(4) $\mathcal{A} \circ_4 \mathcal{B} = \langle \mathbf{A} \circ_{IV} \mathbf{B}, A \circ_3 B, \lambda \circ_2 \mu \rangle$.

Example 1. *Let $X = \{a, b, c\}$ be the groupoid with the following Cayley Table 1:*

Table 1. Caley.

$\cdot$	a	b	c
a	a	a	a
b	b	a	b
c	c	c	a

Consider two octahedron sets $\mathcal{A}$ and $\mathcal{B}$ in X, respectively given by:
$\mathcal{A}(a) = \langle [0.3, 0.6], (0.7, 0.2), 0.5 \rangle$, $\mathcal{A}(b) = \langle [0.2, 0.4], (0.6, 0.3), 0.7 \rangle$,

$\mathcal{A}(c) = \langle [0.4, 0.7], (0.5, 0.4), 0.3 \rangle$, $\mathcal{B}(a) = \langle [0.2, 0.6], (0.6, 0.3), 0.7 \rangle$,
$\mathcal{B}(b) = \langle [0.3, 0.5], (0.5, 0.2), 0.6 \rangle$, $\mathcal{B}(c) = \langle [0.4, 0.7], (0.7, 0.2), 0.8 \rangle$.
Then we can easily calculate $\mathcal{A} \circ_i \mathcal{B}$ with Tables 2 and 3:

Table 2. $(\mathcal{A} \circ_1 \mathcal{B})(t)$ and $(\mathcal{A} \circ_2 \mathcal{B})(t)$.

	$(\mathcal{A} \circ_1 \mathcal{B})(t)$	$(\mathcal{A} \circ_2 \mathcal{B})(t)$
a	$\langle [0.4, 0.7], (0.5, 0.2), 0.6 \rangle$	$\langle [0.4, 0.7], (0.5, 0.2), 0.6 \rangle$
b	$\langle [0.2, 0.4], (0.6, 0.3), 0.7 \rangle$	$\langle [0.2, 0.4], (0.6, 0.3), 0.7 \rangle$
c	$\langle [0.3, 0.6], (0.5, 0.4), 0.3 \rangle$	$\langle [0.3, 0.6], (0.5, 0.4), 0.7 \rangle$

Table 3. $(\mathcal{A} \circ_3 \mathcal{B})(t)$ and $(\mathcal{A} \circ_4 \mathcal{B})(t)$.

	$(\mathcal{A} \circ_3 \mathcal{B})(t)$	$(\mathcal{A} \circ_4 \mathcal{B})(t)$
a	$\langle [0.4, 0.7], (0.6, 0.2), 0.6 \rangle$	$\langle [0.4, 0.7], (0.6, 0.2), 0.6 \rangle$
b	$\langle [0.2, 0.4], (0.6, 0.2), 0.7 \rangle$	$\langle [0.2, 0.4], (0.6, 0.2), 0.7 \rangle$
c	$\langle [0.3, 0.6], (0.6, 0.3), 0.3 \rangle$	$\langle [0.3, 0.6], (0.6, 0.3), 0.7 \rangle$

Proposition 1. *Let $(X, \cdot)$ be a groupoid, let $\mathcal{A} = \langle \mathbf{A}, A, \lambda \rangle$, $\mathcal{B} = \langle \mathbf{B}, B, \mu \rangle \in \mathcal{O}(X)$ and let $x_{\tilde{\tilde{a}}} = x_{\langle \tilde{a}, \bar{a}, a \rangle}$, $y_{\tilde{\tilde{b}}} = y_{\langle \tilde{b}, \bar{b}, b \rangle} \in \mathcal{O}_P(X)$. Then we have*

(1) *$x_{\tilde{\tilde{a}}} \circ_i y_{\tilde{\tilde{b}}} = (xy)_{\tilde{\tilde{a}} \wedge^i \tilde{\tilde{b}}}$, for $i = 1, 2, 3, 4$, i.e.,*

$$x_{\tilde{\tilde{a}}} \circ_1 y_{\tilde{\tilde{b}}} = \langle (xy)_{\tilde{a} \wedge \tilde{b}}, (xy)_{\bar{a} \wedge \bar{b}}, (xy)_{a \wedge b} \rangle, \; x_{\tilde{\tilde{a}}} \circ_2 y_{\tilde{\tilde{b}}} = \langle (xy)_{\tilde{a} \wedge \tilde{b}}, (xy)_{\bar{a} \wedge \bar{b}}, (xy)_{a \vee b} \rangle,$$

$$x_{\tilde{\tilde{a}}} \circ_3 y_{\tilde{\tilde{b}}} = \langle (xy)_{\tilde{a} \wedge \tilde{b}}, (xy)_{\bar{a} \vee \bar{b}}, (xy)_{a \wedge b} \rangle, \; x_{\tilde{\tilde{a}}} \circ_4 y_{\tilde{\tilde{b}}} = \langle (xy)_{\tilde{a} \wedge \tilde{b}}, (xy)_{\bar{a} \vee \bar{b}}, (xy)_{a \vee b} \rangle,$$

(2) *$\mathcal{A} \circ_i \mathcal{B} = \bigcup^i_{x_{\tilde{\tilde{a}}} \in_i \mathcal{A},\, y_{\tilde{\tilde{b}}} \in_i \mathcal{B}} x_{\tilde{\tilde{a}}} \circ_i y_{\tilde{\tilde{b}}}$, for $i = 1, 2, 3, 4$.*

Proof. (1) The proofs are obvious from Definitions 7 and 12.

(2) Case 1: Let $i = 1$. Then the proof of the first part follows from Proposition 1.1 [16], Proposition 2.2 [17] and Proposition 3.2 [18].

Case 2: Let $i = 2$. From Remark 3.5 (2), it is sufficient to prove that $\lambda \circ_2 \mu = \bigcap_{x_a \in_2 \lambda,\, y_b \in_2 \mu} x_a \circ_2 y_b$. Let $C = \bigcap_{x_a \in_2 \lambda, y_b \in_2 \mu} x_a \circ_2 y_b$. For each $z \in X$, we may suppose that there are $u, v \in X$ such that $uv = z$, $x_a \neq 1$ and $y_b \neq 1$ without loss of generality. Then

$$\begin{aligned}
(\lambda \circ_2 \mu)(z) &= \textstyle\bigwedge_{z=uv}[\lambda(u) \vee \mu(v)] \\
&\leq \textstyle\bigwedge_{z=uv}(\bigwedge_{x_a \in_2 \lambda,\, y_b \in_2 \mu}[x_a(u) \vee y_b(v)]) \\
&= \textstyle\bigwedge_{x_a \in_2 \lambda,\, y_b \in_2 \mu}(\bigwedge_{z=uv}[x_a(u) \vee y_b(v)]) \\
&= (\textstyle\bigcap_{x_a \in_2 \lambda,\, y_b \in_2 \mu} x_a \circ_2 y_b)(z) \\
&= C.
\end{aligned}$$

Since $u_{\lambda(u)} \in_2 \lambda$ and $v_{\mu(v)} \in_2 \mu$,

$$\begin{aligned}
(\textstyle\bigcap_{x_a \in_2 \lambda,\, y_b \in_2 \mu} x_a \circ_2 y_b)(z) &= \textstyle\bigwedge_{x_a \in_2 \lambda,\, y_b \in_2 \mu} \bigwedge_{z=uv}[x_a(u) \vee y_b(v)] \\
&\leq \textstyle\bigwedge_{z=uv}[u_{\lambda(u)}(u) \vee v_{\mu(v)}(v)] \\
&= \textstyle\bigwedge_{z=uv}[\lambda(u) \vee \mu(v)] \\
&= (\lambda \circ_2 \mu)(z).
\end{aligned}$$

Thus, $(\lambda \circ_2 \mu)(z) = C(z)$. So $\mathcal{A} \circ_2 \mathcal{B} = \bigcup^2_{x_{\tilde{\tilde{a}}} \in_2 \mathcal{A},\, y_{\tilde{\tilde{b}}} \in_2 \mathcal{B}} x_{\tilde{\tilde{a}}} \circ_2 y_{\tilde{\tilde{b}}}$.

Case 3: Let $i = 3$. From Remark 1 (3), it is sufficient to prove that

$$A \circ_3 B = \Big(\bigcap_{x_a \in_3 A,\, y_b \in_3 B} x_a \circ_3 y_b, \bigcup_{x_a \in_3 A,\, y_b \in_3 B} x_a \circ_3 y_b \Big),$$

where $(A \circ_3 B)^{\in} = \bigcap_{x_a \in_3 A,\ y_b \in_3 B} x_a \circ_3 y_b$ and $(A \circ_3 B)^{\notin} = \bigcup_{x_a \in_3 A,\ y_b \in_3 B} x_a \circ_3 y_b$. Let $z \in X$. Then from the proof of Case 2 and Proposition 1.1 [16] (ii), we have

$$(A \circ_3 B)^{\in}(z) = (\bigcap_{x_a \in_3 A,\ y_b \in_3 B} x_a \circ_3 y_b)(z),\ (A \circ_3 B)^{\notin}(z) = (\bigcup_{x_a \in_3 A,\ y_b \in_3 B} x_a \circ_3 y_b)(z).$$

Thus, $\mathcal{A} \circ_3 \mathcal{B} = \bigcup^3_{x_{\tilde{a}} \in_3 \mathcal{A},\ y_{\tilde{b}} \in_3 \mathcal{B}} x_{\tilde{a}} \circ_3 y_{\tilde{b}}$.

For $i = 4$, from Cases 2 and 3, the proof is obvious. □

The followings are immediate results of Definition 12.

Proposition 2. *Let $(X, \cdot)$ be a groupoid and let $i = 1, 2, 3, 4$.*

(1) If "$\cdot$" is associative [resp., commutative] in X, then so is "$\circ_i$" in $\mathcal{O}(X)$.

(2) If "$\cdot$" has an identity $e \in X$, then we have

(2_a) $e_{\ddot{1}} \in \mathcal{O}_P(X)$ is an identity of "$\circ_1$" in $\mathcal{O}(X)$, i.e.,
$\mathcal{A} \circ e_{\ddot{1}} = e_{\ddot{1}} \circ \mathcal{A} = \mathcal{A}$, for each $\mathcal{A} \in \mathcal{O}(X)$,

(2_b) $e_{\langle \tilde{1}, \bar{\mathbf{1}}, 0 \rangle} \in \mathcal{O}_P(X)$ is an identity of "$\circ_2$" in $\mathcal{O}(X)$, i.e.,
$\mathcal{A} \circ e_{\langle \tilde{1}, \bar{\mathbf{1}}, 0 \rangle} = e_{\langle \tilde{1}, \bar{\mathbf{1}}, 0 \rangle} \circ \mathcal{A} = \mathcal{A}$, for each $\mathcal{A} \in \mathcal{O}(X)$,

(2_c) $e_{\langle \tilde{1}, \bar{\mathbf{0}}, 1 \rangle} \in \mathcal{O}_P(X)$ is an identity of "$\circ_3$" in $\mathcal{O}(X)$, i.e.,
$\mathcal{A} \circ e_{\langle \tilde{1}, \bar{\mathbf{0}}, 1 \rangle} = e_{\langle \tilde{1}, \bar{\mathbf{0}}, 1 \rangle} \circ \mathcal{A} = \mathcal{A}$, for each $\mathcal{A} \in \mathcal{O}(X)$,

(2_d) $e_{\langle \tilde{1}, \bar{\mathbf{0}}, 0 \rangle} \in \mathcal{O}_P(X)$ is an identity of "$\circ_4$" in $\mathcal{O}(X)$, i.e.,
$\mathcal{A} \circ e_{\langle \tilde{1}, \bar{\mathbf{0}}, 0 \rangle} = e_{\langle \tilde{1}, \bar{\mathbf{0}}, 0 \rangle} \circ \mathcal{A} = \mathcal{A}$, for each $\mathcal{A} \in \mathcal{O}(X)$.

Definition 13. *Let $(X, \cdot)$ be a groupoid and let $\mathcal{A} = \langle \mathbf{A}, A, \lambda \rangle \in \mathcal{O}(X)$. Then*

(i) $\ddot{0} \neq \mathcal{A}$ is called a 1-octahedron subgroupoid in X, if $\mathcal{A} \circ_1 \mathcal{A} \subset^1 \mathcal{A}$, i.e.,

$$\mathbf{A} \circ_{IV} \mathbf{A} \subset \mathbf{A},\ A \circ_{IF} A \subset A,\ \lambda \circ_F \lambda \subset \lambda,$$

(ii) $\left\langle \tilde{0}, \bar{\mathbf{0}}, 1 \right\rangle \neq \mathcal{A}$ is called a 2-octahedron subgroupoid in X, if $\mathcal{A} \circ_2 \mathcal{A} \subset^2 \mathcal{A}$, i.e.,

$$\mathbf{A} \circ_{IV} \mathbf{A} \subset \mathbf{A},\ A \circ_{IF} A \subset A,\ \lambda \circ_2 \lambda \supset \lambda,$$

(iii) $\left\langle \tilde{0}, \bar{\mathbf{1}}, 0 \right\rangle \neq \mathcal{A}$ is called a 3-octahedron subgroupoid in X, if $\mathcal{A} \circ_3 \mathcal{A} \subset^3 \mathcal{A}$, i.e.,

$$\mathbf{A} \circ_{IV} \mathbf{A} \subset \mathbf{A},\ A \circ_3 A \supset A,\ \lambda \circ_F \lambda \subset \lambda,$$

(iv) $\left\langle \tilde{0}, \bar{\mathbf{1}}, 1 \right\rangle \neq \mathcal{A}$ is called a 4-octahedron subgroupoid in X, if $\mathcal{A} \circ_4 \mathcal{A} \subset^4 \mathcal{A}$, i.e.,

$$\mathbf{A} \circ_{IV} \mathbf{A} \subset \mathbf{A},\ A \circ_3 A \supset A,\ \lambda \circ_2 \lambda \supset \lambda.$$

We will denote the set of all i-octahedron subgroupoids in X as $OGP_i(X)$ ($i = 1, 2, 3, 4$).

Let us denote the set of all fuzzy [resp., intuitionistic fuzzy, interval-valued fuzzy] subgroupoids in a groupoid X in the sense of Liu [16] [resp., Hur et al. [17], Kang and Hur [18]] as $FGP(X)$ [resp., $IFGP(X)$, $IVGP(X)$].

Remark 2. *Let $(X, \cdot)$ be a groupoid and let $\mathcal{A} = \langle \mathbf{A}, A, \lambda \rangle \in \mathcal{O}(X)$. Then*

(1) $\mathcal{A} \in OGP_1(X)$ if and only if $\mathbf{A} \in IVGP(X)$, $A \in IFGP(X)$, $\lambda \in FGP(X)$,

(2) $\mathcal{A} \in OGP_2(X)$ if and only if $\mathbf{A} \in IVGP(X)$, $A \in IFGP(X)$, $\lambda \circ_2 \lambda \supset \lambda$,

(3) $\mathcal{A} \in OGP_3(X)$ if and only if $\mathbf{A} \in IVGP(X)$, $A \circ_3 A \supset A$, $\lambda \in FGP(X)$,

(4) $\mathcal{A} \in OGP_3(X)$ if and only if $\mathbf{A} \in IVGP(X)$, $A \circ_3 A \supset A$, $\lambda \circ_2 \lambda \supset \lambda$.

Example 2. *(1) Let $(X,\cdot)$ be the subgroupoid and let $\mathcal{A}$ be the octahedron set in X given in Example 1. Then we can easily calculate that*

$$(\mathbf{A} \circ_{IV} \mathbf{A})(a) = [0.4, 0.7] \not\leq [0.3, 0.6] = \mathbf{A}(a),$$

$$(\lambda \circ_2 \lambda)(a) = 0.3 \not\geq 0.5 = \lambda(a),$$

$$(A \circ_3 A)(a) = (0.5, 0.4) \not\geq (0.7, 0.2) = A(a).$$

Thus, $\mathcal{A} \notin OGP_i(X)$, for $i = 1,\ 2,\ 3,\ 4$.

(2) Let $X = \{a, b, c\}$ be the groupoid with the following Cayley Table 4:

Table 4. Caley.

$\cdot$	a	b	c
a	a	a	a
b	b	a	b
c	c	c	c

Consider the octahedron set $\mathcal{A}$ in X given in Example 1. Then we can easily see that $\mathcal{A} \in OGP_i(X)$ for $i = 1,\ 2$ but $\mathcal{A} \notin OGP_i(X)$ for $i = 3,\ 4$.

(3) Let $(X, \cdot)$ be a groupoid and let $\mathbf{A} \in IVGP(X)$. Then clearly, $\mathcal{O}_{\mathbf{A}} \in OGP_1(X)$, where $\mathcal{O}_{\mathbf{A}}$ is the octahedron set in X induced by $\mathbf{A}$ (See Example 3.2 (3) in [10]).

(4) Let $(X, \cdot)$ be a groupoid and let $A \in IFGP(X)$. Then clearly, $\mathcal{O}_A \in OGP_1(X)$, where $\mathcal{O}_A$ is the octahedron set in X induced by A (See Example 3.2 (4) in [10]).

(5) Let $(X, \cdot)$ be a groupoid and let $\mathcal{A} \in OGP_i(X)$. Then clearly, $[\,]\mathcal{A},\ \diamond\mathcal{A} \in OGP_i(X)$ $(i = 1,\ 2,\ 3,\ 4)$.

The followings are immediate results of Definitions 11–13, Proposition 2 (2) and Remark 2 (1).

Theorem 1. *Let $(X, \cdot)$ be a groupoid and let $\ddot{0} \neq \mathcal{A} = \langle \mathbf{A}, A, \lambda \rangle \in \mathcal{O}(X)$. Then the followings are equivalent:*

(1) $\mathcal{A} \in OGP_1(X)$,

(2) *for every $x_{\tilde{\bar{a}}},\ y_{\tilde{\bar{b}}} \in \mathcal{A}$, $x_{\tilde{\bar{a}}} \circ_1 y_{\tilde{\bar{b}}} \in \mathcal{A}$, i.e., $(\mathcal{A}, \circ_1)$ is a groupoid,*

(3) *for every $x,\ y \in X$, $\mathcal{A}(xy) \geq \mathcal{A}(x) \wedge^1 \mathcal{A}(y)$, i.e.,*

(i) $A^-(xy) \geq A^-(x) \wedge A^-(y),\ A^+(xy) \geq A^+(x) \wedge A^+(y)$,

(ii) $A^{\in}(xy) \geq A^{\in}(x) \wedge A^{\in}(y),\ A^{\notin}(xy) \leq A^{\notin}(x) \vee A^{\notin}(y)$,

(iii) $\lambda(xy) \geq \lambda(x) \wedge \lambda(y)$.

From Definitions 8–10, Remark 1 (1) and the above proposition, it is obvious that $(\mathcal{A}, \circ_1)$ is a groupoid if and only if $(\mathbf{A}, \circ_{IV})$, $(A, \circ_{IF})$ and $(\lambda, \circ_F)$ are groupoids.

Proposition 3. *Let $(X, \cdot)$ be a groupoid and let $\mathcal{A} = \langle \mathbf{A}, A, \lambda \rangle \in OGP_1(X)$.*

(1) *If "$\cdot$" is associative in X, then so is "$\circ_1$" in $\mathcal{A}$, i.e., for every $x_{\tilde{\bar{a}}},\ y_{\tilde{\bar{b}}},\ z_{\tilde{\bar{c}}} \in \mathcal{A}$,*

$$(x_{\tilde{\bar{a}}} \circ_1 y_{\tilde{\bar{b}}}) \circ_1 z_{\tilde{\bar{c}}} = x_{\tilde{\bar{a}}} \circ_1 (y_{\tilde{\bar{b}}} \circ_1 z_{\tilde{\bar{c}}}),$$

(2) *If "$\cdot$" is commutative in X, then so is "$\circ_1$" in $\mathcal{A}$, i.e., for every $x_{\tilde{\bar{a}}},\ y_{\tilde{\bar{b}}} \in \mathcal{A}$,*

$$x_{\tilde{\bar{a}}} \circ_1 y_{\tilde{\bar{b}}} = y_{\tilde{\bar{b}}} \circ_1 x_{\tilde{\bar{a}}},$$

(3) *If "$\cdot$" has an identity $e \in X$, then for each $x_{\tilde{\bar{a}}} \in \mathcal{A}$,*

$$e_{\ddot{1}} \circ_1 x_{\tilde{\bar{a}}} = x_{\tilde{\bar{a}}} = x_{\tilde{\bar{a}}} \circ_1 e_{\ddot{1}}.$$

The followings are immediate consequences of Definitions 11–13, Proposition 2 (2) and Remark 2 (2).

Theorem 2. *Let $(X, \cdot)$ be a groupoid and let $\left\langle \widetilde{0}, \bar{\mathbf{0}}, 1 \right\rangle \neq \mathcal{A} = \langle \mathbf{A}, A, \lambda \rangle \in \mathcal{O}(X)$. Then the followings are equivalent:*

(1) $\mathcal{A} \in OGP_2(X)$,

(2) *for every $x_{\tilde{a}}$, $y_{\tilde{b}} \in \mathcal{A}$, $x_{\tilde{a}} \circ_2 y_{\tilde{b}} \in \mathcal{A}$, i.e., $(\mathcal{A}, \circ_2)$ is a groupoid,*

(3) *for every x, $y \in X$, $\mathcal{A}(xy) \geq \mathcal{A}(x) \wedge^2 \mathcal{A}(y)$, i.e.,*

(i) $A^-(xy) \geq A^-(x) \wedge A^-(y)$, $A^+(xy) \geq A^+(x) \wedge A^+(y)$,

(ii) $A^{\in}(xy) \geq A^{\in}(x) \wedge A^{\in}(y)$, $A^{\notin}(xy) \leq A^{\notin}(x) \vee A^{\notin}(y)$,

(iii) $\lambda(xy) \leq \lambda(x) \vee \lambda(y)$.

Proposition 4. *Let $(X, \cdot)$ be a groupoid and let $\mathcal{A} = \langle \mathbf{A}, A, \lambda \rangle \in OGP_2(X)$.*

(1) *If "$\cdot$" is associative in X, then so is "$\circ_2$" in $\mathcal{A}$, i.e., for every $x_{\tilde{a}}$, $y_{\tilde{b}}$, $z_{\tilde{c}} \in \mathcal{A}$,*

$$(x_{\tilde{a}} \circ_2 y_{\tilde{b}}) \circ_2 z_{\tilde{c}} = x_{\tilde{a}} \circ_2 (y_{\tilde{b}} \circ_2 z_{\tilde{c}}),$$

(2) *If "$\cdot$" is commutative in X, then so is "$\circ_2$" in $\mathcal{A}$, i.e., for every $x_{\tilde{a}}$, $y_{\tilde{b}} \in \mathcal{A}$,*

$$x_{\tilde{a}} \circ_2 y_{\tilde{b}} = y_{\tilde{b}} \circ_2 x_{\tilde{a}},$$

(3) *If "$\cdot$" has an identity $e \in X$, then for each $x_{\tilde{a}} \in \mathcal{A}$,*

$$e_{\langle \tilde{1}, \bar{\mathbf{1}}, 0 \rangle} \circ_2 x_{\tilde{a}} = x_{\tilde{a}} = x_{\tilde{a}} \circ_2 e_{\langle \tilde{1}, \bar{\mathbf{1}}, 0 \rangle}.$$

The followings are immediate consequences of Definitions 11–13, Proposition 2 (3) and Remark 2 (3).

Theorem 3. *Let $(X, \cdot)$ be a groupoid and let $\left\langle \widetilde{0}, \bar{\mathbf{1}}, 0 \right\rangle \neq \mathcal{A} = \langle \mathbf{A}, A, \lambda \rangle \in \mathcal{O}(X)$. Then the followings are equivalent:*

(1) $\mathcal{A} \in OGP_3(X)$,

(2) *for every $x_{\tilde{a}}$, $y_{\tilde{b}} \in \mathcal{A}$, $x_{\tilde{a}} \circ_3 y_{\tilde{b}} \in \mathcal{A}$, i.e., $(\mathcal{A}, \circ_3)$ is a groupoid,*

(3) *for every x, $y \in X$, $\mathcal{A}(xy) \geq \mathcal{A}(x) \wedge^3 \mathcal{A}(y)$, i.e.,*

(i) $A^-(xy) \geq A^-(x) \wedge A^-(y)$, $A^+(xy) \geq A^+(x) \wedge A^+(y)$,

(ii) $A^{\in}(xy) \leq A^{\in}(x) \vee A^{\in}(y)$, $A^{\notin}(xy) \geq A^{\notin}(x) \wedge A^{\notin}(y)$,

(iii) $\lambda(xy) \geq \lambda(x) \wedge \lambda(y)$.

Proposition 5. *Let $(X, \cdot)$ be a groupoid and let $\mathcal{A} = \langle \mathbf{A}, A, \lambda \rangle \in OGP_3(X)$.*

(1) *If "$\cdot$" is associative in X, then so is "$\circ_3$" in $\mathcal{A}$, i.e., for every $x_{\tilde{a}}$, $y_{\tilde{b}}$, $z_{\tilde{c}} \in \mathcal{A}$,*

$$(x_{\tilde{a}} \circ_3 y_{\tilde{b}}) \circ_3 z_{\tilde{c}} = x_{\tilde{a}} \circ_3 (y_{\tilde{b}} \circ_3 z_{\tilde{c}}),$$

(2) *If "$\cdot$" is commutative in X, then so is "$\circ_3$" in $\mathcal{A}$, i.e., for every $x_{\tilde{a}}$, $y_{\tilde{b}} \in \mathcal{A}$,*

$$x_{\tilde{a}} \circ_3 y_{\tilde{b}} = y_{\tilde{b}} \circ_3 x_{\tilde{a}},$$

(3) *If "$\cdot$" has an identity $e \in X$, then for each $x_{\tilde{a}} \in \mathcal{A}$,*

$$e_{\langle \tilde{1}, \bar{\mathbf{0}}, 1 \rangle} \circ_3 x_{\tilde{a}} = x_{\tilde{a}} = x_{\tilde{a}} \circ_3 e_{\langle \tilde{1}, \bar{\mathbf{0}}, 1 \rangle}.$$

The followings are immediate consequences of Definitions 11–13, Proposition 2 (4) and Remark 2 (4).

Theorem 4. *Let $(X, \cdot)$ be a groupoid and let $\left\langle \widetilde{0}, \bar{\mathbf{1}}, 1 \right\rangle \neq \mathcal{A} = \langle \mathbf{A}, A, \lambda \rangle \in \mathcal{O}(X)$. Then the followings are equivalent:*

(1) $\mathcal{A} \in OGP_4(X)$,
(2) *for every* $x_{\tilde{a}}$, $y_{\tilde{b}} \in \mathcal{A}$, $x_{\tilde{a}} \circ_3 y_{\tilde{b}} \in \mathcal{A}$, *i.e.*, $(\mathcal{A}, \circ_4)$ *is a groupoid*,
(3) *for every* x, $y \in X$, $\mathcal{A}(xy) \geq \mathcal{A}(x) \wedge^4 \mathcal{A}(y)$, *i.e.*,
(i) $A^-(xy) \geq A^-(x) \wedge A^-(y)$, $A^+(xy) \geq A^+(x) \wedge A^+(y)$,
(ii) $A^{\in}(xy) \leq A^{\in}(x) \vee A^{\in}(y)$, $A^{\notin}(xy) \geq A^{\notin}(x) \wedge A^{\notin}(y)$,
(iii) $\lambda(xy) \leq \lambda(x) \vee \lambda(y)$.

Proposition 6. *Let* $(X, \cdot)$ *be a groupoid and let* $\mathcal{A} = \langle \mathbf{A}, A, \lambda \rangle \in OGP_4(X)$.
(1) *If "$\cdot$" is associative in X, then so is "$\circ_4$" in* $\mathcal{A}$, *i.e., for every* $x_{\tilde{a}}$, $y_{\tilde{b}}$, $z_{\tilde{c}} \in \mathcal{A}$,

$$(x_{\tilde{a}} \circ_4 y_{\tilde{b}}) \circ_4 z_{\tilde{c}} = x_{\tilde{a}} \circ_4 (y_{\tilde{b}} \circ_4 z_{\tilde{c}}),$$

(2) *If "$\cdot$" is commutative in X, then so is "$\circ_4$" in* $\mathcal{A}$, *i.e., for every* $x_{\tilde{a}}$, $y_{\tilde{b}} \in \mathcal{A}$,

$$x_{\tilde{a}} \circ_4 y_{\tilde{b}} = y_{\tilde{b}} \circ_4 x_{\tilde{a}},$$

(3) *If "$\cdot$" has an identity* $e \in X$, *then for each* $x_{\tilde{a}} \in \mathcal{A}$,

$$e_{\langle \tilde{1}, \bar{0}, 0 \rangle} \circ_3 x_{\tilde{a}} = x_{\tilde{a}} = x_{\tilde{a}} \circ_4 e_{\langle \tilde{1}, \bar{0}, 0 \rangle}.$$

Remark 3. *Let* $(X, \cdot)$ *be a groupoid and let* $A \in 2^X$. *Then we have*

$$\chi_{\mathcal{A}} \in OGP_1(X) \iff A \text{ is a subgroupoid of } X.$$

Definition 14. *Let* $(X, \cdot)$ *be a groupoid,* $\mathcal{A} \in \mathcal{O}(X)$ *and let* $i = 1, 2, 3, 4$. *Then* $\mathcal{A}$ *is called a:*
(i) i-octahedron left ideal (briefly, i-OLI) of X, if for every x, $y \in X$,

$$\mathcal{A}(xy) \geq_i \mathcal{A}(y), \text{ i.e.,}$$

(ii) i-octahedron right ideal (briefly, i-ORI) of X, if for every x, $y \in X$,

$$\mathcal{A}(xy) \geq_i \mathcal{A}(x), \text{ i.e.,}$$

(iii) i-octahedron ideal (simply, i-OI) of X, if it is both an i-OLI and an i-ORI of X.

In this case, we will denote the set of all i-OIs [resp., i-OLIs and i-ORIs] of X as $OI_i(X)$ *[resp.,* $OLI_i(X)$ *and* $ORI_i(X)$*].*

Remark 4. *From the above Definition, we have the followings.*
(1) $\mathcal{A} \in OLI_1(X)$
$\iff A^-(xy) \geq A^-(y)$, $A^+(xy) \geq A^+(y)$, $A^{\in}(xy) \geq A^{\in}(y)$,
$A^{\notin}(xy) \leq A^{\notin}(y)$, $\lambda(xy) \geq \lambda(y)$,
$\mathcal{A} \in OLI_2(X)$
$\iff A^-(xy) \geq A^-(y)$, $A^+(xy) \geq A^+(y)$, $A^{\in}(xy) \geq A^{\in}(y)$,
$A^{\notin}(xy) \leq A^{\notin}(y)$, $\lambda(xy) \leq \lambda(y)$,
$\mathcal{A} \in OLI_3(X)$
$\iff A^-(xy) \geq A^-(y)$, $A^+(xy) \geq A^+(y)$, $A^{\in}(xy) \leq A^{\in}(y)$,
$A^{\notin}(xy) \geq A^{\notin}(y)$, $\lambda(xy) \geq \lambda(y)$,
$\mathcal{A} \in OLI_4(X)$
$\iff A^-(xy) \geq A^-(y)$, $A^+(xy) \geq A^+(y)$, $A^{\in}(xy) \leq A^{\in}(y)$,
$A^{\notin}(xy) \geq A^{\notin}(y)$, $\lambda(xy) \leq \lambda(y)$,
(2) $\mathcal{A} \in ORI_1(X)$
$\iff A^-(xy) \geq A^-(x)$, $A^+(xy) \geq A^+(x)$, $A^{\in}(xy) \geq A^{\in}(x)$,

$$A^{\notin}(xy) \leq A^{\notin}(x),\ \lambda(xy) \geq \lambda(x),$$
$$\mathcal{A} \in ORI_2(X)$$
$$\Longleftrightarrow A^{-}(xy) \geq A^{-}(x),\ A^{+}(xy) \geq A^{+}(x),\ A^{\in}(xy) \geq A^{\in}(x),$$
$$A^{\notin}(xy) \leq A^{\notin}(x),\ \lambda(xy) \leq \lambda(x),$$
$$\mathcal{A} \in ORI_3(X)$$
$$\Longleftrightarrow A^{-}(xy) \geq A^{-}(x),\ A^{+}(xy) \geq A^{+}(x),\ A^{\in}(xy) \leq A^{\in}(x),$$
$$A^{\notin}(xy) \geq A^{\notin}(x),\ \lambda(xy) \geq \lambda(x),$$
$$\mathcal{A} \in ORI_4(X)$$
$$\Longleftrightarrow A^{-}(xy) \geq A^{-}(x),\ A^{+}(xy) \geq A^{+}(x),\ A^{\in}(xy) \leq A^{\in}(x),$$
$$A^{\notin}(xy) \geq A^{\notin}(x),\ \lambda(xy) \leq \lambda(x),$$

(3) $\mathcal{A} \in OI_1(X)$
$$\Longleftrightarrow A^{-}(xy) \geq A^{-}(x) \vee A^{-}(y),\ A^{+}(xy) \geq A^{+}(x) \vee A^{+}(y),$$
$$A^{\in}(xy) \geq A^{\in}(x) \vee A^{\in}(y),\ A^{\notin}(xy) \leq A^{\notin}(x) \wedge A^{\notin}(x),\ \lambda(xy) \geq \lambda(x) \vee \lambda(y),$$
$$\mathcal{A} \in ORI_2(X)$$
$$\Longleftrightarrow A^{-}(xy) \geq A^{-}(x) \vee A^{-}(y),\ A^{+}(xy) \geq A^{+}(x) \vee A^{+}(y),$$
$$A^{\in}(xy) \geq A^{\in}(x) \vee A^{\in}(y),\ A^{\notin}(xy) \leq A^{\notin}(x) \wedge A^{\notin}(x),\ \lambda(xy) \leq \lambda(x) \wedge \lambda(y),$$
$$\mathcal{A} \in OI_3(X)$$
$$\Longleftrightarrow A^{-}(xy) \geq A^{-}(x) \vee A^{-}(y),\ A^{+}(xy) \geq A^{+}(x) \vee A^{+}(y),$$
$$A^{\in}(xy) \leq A^{\in}(x) \wedge A^{\in}(y),\ A^{\notin}(xy) \geq A^{\notin}(x) \vee A^{\notin}(x),\ \lambda(xy) \geq \lambda(x) \vee \lambda(y),$$
$$\mathcal{A} \in OI_4(X)$$
$$\Longleftrightarrow A^{-}(xy) \geq A^{-}(x) \vee A^{-}(y),\ A^{+}(xy) \geq A^{+}(x) \vee A^{+}(y),$$
$$A^{\in}(xy) \leq A^{\in}(x) \wedge A^{\in}(y),\ A^{\notin}(xy) \geq A^{\notin}(x) \vee A^{\notin}(x),\ \lambda(xy) \leq \lambda(x) \wedge \lambda(y).$$

Remark 5. *An i-octahedron left ideal [resp., right ideal and ideal] in a semigroup S, a group G and a ring G is defined as Definition 14.*

For a groupoid $(X, \cdot)$, let us denote the set of all fuzzy ideals [resp., left ideals and right ideals] (See [19]), the set of all IVIs [resp., IVLIs and IVRIs] (See [18]) and the set of all IFIs [resp., IFLIs, IFRIs] (See [17]) of X as $FI(X)$ [resp., $FLI(X)$ and $FRI(X)$], $IVI(X)$ [resp., $IVLI(X)$ and $IVRI(X)$] and $IFI(X)$ [resp., $IFLI(X)$ and $IFRI(X)$]. Then we can easily see that $\mathcal{A} = \langle \mathbf{A}, A, \lambda \rangle \in OI_1(X)$ [resp., $OLI_1(X)$ and $ORI_1(X)$] if and only if $\mathbf{A} \in IVI(X)$, $A \in IFI(X)$, $\lambda \in FI(X)$ [resp., $\mathbf{A} \in IVLI(X)$, $A \in IFLI(X)$, $\lambda \in FLI(X)$ and $\mathbf{A} \in IVRI(X)$, $A \in IFRI(X)$, $\lambda \in FRI(X)$]. Furthermore, it is obvious that $\mathcal{A} \in OGP_i(X)$, for each $\mathcal{A} \in OI_i(X)$ [resp., $OLI_i(X)$ and $ORI_i(X)$] ($i = 1, 2, 3, 4$) but the converse is not true in general (See Example 3 (1)).

Note that for every $\mathcal{A} \in OGP_i(X)$ ($i = 1, 2, 3, 4$), we have: for each $x \in X$,

$$\mathcal{A}(x^n) \geq_i \mathcal{A}(x), \text{ i.e.,}$$

where x^n is any composite of x's.

Example 3. *(1) Let $(X, \cdot)$ be the groupoid and $\mathcal{A} \in OGP_1(X)$ given in Example 2 (2). Then clearly, $\lambda(ab) = 0.5 \not\geq 0.7 = \lambda(b)$. Thus, $\lambda \notin FLI(X)$. So $\mathcal{A} \notin OLI_1(X)$.*

(2) Let $X = \{a, b, c\}$ be the groupoid with the following Cayley Table 5:

Table 5. Caley.

$\cdot$	a	b	c
a	a	a	a
b	a	a	c
c	a	b	c

Consider the octahedron set $\mathcal{A}$ in X given by:

$$\mathcal{A}(a) = \langle [0.4, 0.8], (0.7, 0.2), 0.8 \rangle,$$

$$\mathcal{A}(b) = \langle [0.3, 0.7], (0.6, 0.3), 0.7 \rangle,$$

$$\mathcal{A}(c) = \langle [0.2, 0.6], (0.5, 0.4), 0.6 \rangle.$$

Then we can easily calculate that $\mathcal{A} \in OLI_1(X)$. But $A^-(bc) = 0.2 \not\geq 0.3 = A^-(b)$. Thus, $\mathbf{A} \notin IVRI(X)$. So $\mathcal{A} \notin ORI_1(X)$.

(3) Let $X - \{a, b, c\}$ be the groupoid with the following Cayley Table 6:

Table 6. Caley.

·	a	b	c
a	a	a	a
b	b	b	a
c	c	a	c

Consider the octahedron set $\mathcal{A}$ in X given by:

$$\mathcal{A}(a) = \langle [0.4, 0.8], (0.7, 0.2), 0.9 \rangle,$$

$$\mathcal{A}(b) = \langle [0.3, 0.7], (0.6, 0.3), 0.7 \rangle,$$

$$\mathcal{A}(c) = \langle [0.2, 0.6], (0.5, 0.4), 0.8 \rangle.$$

Then we can easily calculate that $\mathcal{A} \in ORI_1(X)$. But $A^-(ba) = 0.3 \not\geq 0.4 = A^-(a)$. Thus, $\mathbf{A} \notin IVLI(X)$. So $\mathcal{A} \notin OLI_1(X)$.

From Proposition 3.2 in [19], we have the following result.

Theorem 5. *Let $(X, \cdot)$ be a groupoid and let $A \in 2^X$. Then $\chi_{\mathcal{A}} \in OLI_1(X)$ [resp., $ORI_1(X)$ and $OI_1(X)$] if and only if A is a left ideal [resp., a right ideal and an ideal] of X.*

Definition 15 ([10]). *Let X be a nonempty set, let $\tilde{\bar{a}} = \langle \tilde{a}, \bar{a}, a \rangle \in [I] \times (I \oplus I) \times I$ and let $\mathcal{A} = \langle \mathbf{A}, A, \lambda \rangle \in \mathcal{O}(X)$. Then two subsets $[\mathcal{A}]_{\tilde{\bar{a}}}$ and $[\mathcal{A}]^*_{\tilde{\bar{a}}}$ of X are defined as follows:*

$$[\mathcal{A}]_{\tilde{\bar{a}}} = \{x \in X : \mathbf{A}(x) \geq \tilde{a},\ A(x) \geq \bar{a},\ \lambda(x) \geq a\},$$

$$[\mathcal{A}]^*_{\tilde{\bar{a}}} = \{x \in X : \mathbf{A}(x) > \tilde{a},\ A(x) > \bar{a},\ \lambda(x) > a\}.$$

In this case, $[\mathcal{A}]_{\tilde{\bar{a}}}$ is called an $\tilde{\bar{a}}$-level set of $\mathcal{A}$ and $[\mathcal{A}]^*_{\tilde{\bar{a}}}$ is called a strong $\tilde{\bar{a}}$-level set of $\mathcal{A}$.

The following is an immediate consequence of Theorem 1, Definitions 13 and 14.

Proposition 7. *Let $(X, \cdot)$ be a groupoid and let $\mathcal{A} = \langle \mathbf{A}, A, \lambda \rangle \in \mathcal{O}(X)$. If $\mathcal{A} \in OGP_1(X)$ or $\mathcal{A} \in OLI_1(X)$ [resp., $ORI_1(X)$ and $OI_1(X)$], then $[\mathcal{A}]_{\tilde{\bar{a}}}$ is a subgroupoid or a left ideal [resp., a right ideal and an ideal] of X, for each $\tilde{\bar{a}} \in [I] \times (I \oplus I) \times I$.*

Proposition 8. *Let $(X, \cdot)$ be a groupoid and let $i = 1, 2, 3, 4$. If $(\mathcal{A}_j)_{j \in J} = (\langle \mathbf{A}_j, A_j, \lambda_j \rangle)_{j \in J} \subset OGP_i(X)$, then $\bigcap^i_{j \in J} \mathcal{A}_j \in OGP_i(X)$, where J denotes an index set.*

Proof. Case 1: Suppose $(\mathcal{A}_j)_{j\in J} = (\langle \mathbf{A}_j, A_j, \lambda_j\rangle)_{j\in J} \subset OGP_1(X)$. Then from Propositions 3.8 in [18], 3.9 in [17] and 3.1 in [19], we have

$$\bigcap_{j\in J} \mathbf{A}_j \in IVGP(X),\ \bigcap_{j\in J} A_j \in IFGP(X),\ \bigcap_{j\in J} \lambda_j \in FGP(X).$$

Thus, $\bigcap^1_{j\in J} \mathcal{A}_j \in OGP_1(X)$.

Case 2: Suppose $(\mathcal{A}_j)_{j\in J} = (\langle \mathbf{A}_j, A_j, \lambda_j\rangle)_{j\in J} \subset OGP_2(X)$ and let $x,\ y \in X$. Then by Definition 5 and Theorem 2,

$$\begin{aligned}(\textstyle\bigcup_{j\in J} \lambda_j)(xy) &= \textstyle\bigvee_{j\in J} \lambda_j(xy) \le \bigvee_{j\in J}(\lambda_j(x) \vee \lambda_j(y))\\ &= (\textstyle\bigvee_{j\in J} \lambda_j(x)) \vee (\bigvee_{j\in J} \lambda_j(y))\\ &= (\textstyle\bigcup_{j\in J} \lambda_j)(x) \vee (\bigcup_{j\in J} \lambda_j)(y).\end{aligned}$$

Thus, $(\bigcup_{j\in J} \lambda_j)$ satisfies the the condition (iii) of Theorem 2 (3). By the hypothesis and Case 1, $\bigcap_{j\in J} \mathbf{A}_j$ and $\bigcap_{j\in J} A_j$ satisfy the conditions (i) and (ii) of Theorem 2 (3). So $\bigcap^2_{j\in J} \mathcal{A}_j \in OGP_2(X)$.

Case 3: Suppose $(\mathcal{A}_j)_{j\in J} = (\langle \mathbf{A}_j, A_j, \lambda_j\rangle)_{j\in J} \subset OGP_3(X)$ and let $x,\ y \in X$. Then by Definition 5 and Theorem 3 (ii),

$$\begin{aligned}(\textstyle\bigcup_{j\in J} A^{\in}_j)(xy) &= \textstyle\bigvee_{j\in J} A^{\in}_j(xy) \le \bigvee_{j\in J}(A^{\in}_j(x) \vee A^{\in}_j(y))\\ &= (\textstyle\bigvee_{j\in J} A^{\in}_j(x)) \vee (\bigvee_{j\in J} A^{\in}_j(y))\\ &= (\textstyle\bigcup_{j\in J} A^{\in}_j)(x) \vee (\bigcup_{j\in J} A^{\in}_j)(y).\end{aligned}$$

Similarly, we have $(\bigcap_{j\in J} A^{\notin}_j)(xy) \ge (\bigcap_{j\in J} A^{\notin}_j)(x) \wedge (\bigcap_{j\in J} A^{\notin}_j)(y)$. Thus, $(\bigcup_{j\in J} A_j)$ satisfies the the condition Theorem 3 (ii). By the hypothesis and Case 1, $\bigcap_{j\in J} \mathbf{A}_j$ and $\bigcap_{j\in J} \lambda_j$ satisfy the conditions (i) and (iii) of Theorem 2. So $\bigcap^3_{j\in J} \mathcal{A}_j \in OGP_3(X)$.

Case 4: Suppose $(\mathcal{A}_j)_{j\in J} = (\langle \mathbf{A}_j, A_j, \lambda_j\rangle)_{j\in J} \subset OGP_3(X)$. Then by Case 2 and 3, we can easily see that $\bigcap^4_{j\in J} \mathcal{A}_j \in OGP_4(X)$. □

Remark 6. *for every* $\mathcal{A},\ \mathcal{B} \in OGP_i(X)$, $\mathcal{A} \cup^i \mathcal{B} \notin OGP_i(X)$ *in general* $(i = 1,\ 2,\ 3,\ 4)$.

Example 4. *Let* $(X, \cdot)$ *be the groupoid and* $\mathcal{A} \in OGP_1(X)$ *given in Example 2 (2). Consider the octahedron subgroupoid in X given by:*

$$\mathcal{B}(a) = \mathcal{B}(b) = \mathcal{B}(c) = \langle [0.1, 0.7], (0.5, 0.4), 0.6\rangle\,.$$

Then $(A \cup B)(ab) = (0.7, 0.4) \not\ge (0.6, 0.3) = (A \cup B)(a) \wedge (A \cup B)(b)$. *Thus,* $A \cup B \notin IFGP(X)$. *So* $\mathcal{A} \cup^1 \mathcal{B} \notin OGP_1(X)$.

Remark 7. *Let* $(X, \cdot)$ *be a groupoid and let* $(\mathcal{A}_j)_{j\in J} = (\langle \mathbf{A}_j, A_j, \lambda_j\rangle)_{j\in J} \subset OGP_i(X)$ $(i = 1,\ 2,\ 3,\ 4)$. *Then from Proposition 8, we can easily see that*

$$\bigcap^i\{\mathcal{A} \in OGP_i(X) : \bigcup^i_{j\in J} \mathcal{A}_j \subset_i \mathcal{A}\} \in OGP_i(X).$$

In this case, we will denote $\bigcap^i\{\mathcal{A} \in OGP_i(X) : \bigcup^i_{j\in J} \mathcal{A}_j \subset_i \mathcal{A}\}$ *as* $\bigvee^i_{j\in J} \mathcal{A}_j$.

In particular, it is obvious that $(OGP_1(X), \subset_1)$ *is a complete lattice with the least element* $\ddot{0}$ *and the greatest element* $\ddot{1}$, *where for each* $(\mathcal{A}_j)_{j\in J} \subset_1 OGP_1(X)$, *the inf and the sup of* $(\mathcal{A}_j)_{j\in J}$ *are* $inf_{j\in J}\mathcal{A}_j = \bigcap^1_{j\in J} \mathcal{A}_j$ *and* $sup_{j\in J}\mathcal{A}_j = \bigvee^1_{j\in J} \mathcal{A}_j$.

The following is an immediate result of Proposition 8.

Corollary 1. *Let* $(X, \cdot)$ *be the groupoid,* $\mathcal{A} \in \mathcal{O}(X)$ *and let*

$$(\mathcal{A}) = \bigcap^{1} \{\mathcal{B} \in OGP_1(X) : \mathcal{A} \subset_1 \mathcal{B}\}.$$

Then $(\mathcal{A}) \in OGP_1(X)$.

In this case, $(\mathcal{A})$ is called the octahedron subgroupoid in X generated by $\mathcal{A}$.

Proposition 9. *Let* $(X, \cdot)$ *be a groupoid, and let* (A) *be the subgroupoid generated by* A *and* $\chi_{(A)} = \left\langle [\chi_{(A)}, \chi_{(A)}], (\chi_{(A)}, \chi_{(A^c)}), \chi_{(A)} \right\rangle$ *for each* $A \in 2^X$. *Then*

$$(\chi_{\mathcal{A}}) = \chi_{(A)}.$$

Proof. From Remark 3 and Corollary 1, it is obvious that $\chi_{(A)} \in OGP_1(X)$. Let $\mathcal{B} \in OGP_1(X)$ such that $\mathcal{B} \supset_1 \chi_{(A)}$. Then clearly,

$$\mathcal{B}(x) = \langle [1,1], (1,0), 1 \rangle, \text{ for each } x \in A.$$

Since $\mathcal{B} \in OGP_1(X)$, $\mathcal{B}(xy) = \langle [1,1], (1,0), 1 \rangle$ for every $x,\ y \in A$. Thus, $\mathcal{B} \supset_1 \chi_{(A)}$. So

$$\chi_{(A)} \subset_1 \bigcap^{1} \{\mathcal{B} \in OGP_1(X) : \mathcal{B} \supset_1 \chi_{\mathcal{A}}\} = (\chi_{\mathcal{A}}).$$

We can easily prove that $(\chi_{\mathcal{A}}) \subset_1 \chi_{(A)}$. Hence $(\chi_{\mathcal{A}}) = \chi_{(A)}$. □

From the above Proposition, the subgoupoid lattice of X can be regarded as a sublattice of the octahedron subgroupoid lattice of X.

Proposition 10. *Let* $(X, \cdot)$ *be a groupoid and let* $i = 1,\ 2,\ 3,\ 4$. *Then the i-intersection or the i-union of any i-octahedron (left, right) ideals is an i-octahedron (left, right) ideal.*

Proof. Let $(\mathcal{A}_j)_{j \in J} \subset OLI_i(X)$ [resp., $ORI_i(X)$ and $OI_i(X)$], where $\mathcal{A}_j = \langle \mathbf{A}_j, A_j, \lambda_j \rangle$. We only prove that $\bigcup^{i}_{j \in J} \mathcal{A}_j \in OLI_i(X)$ and the remainder's proofs are omitted.

Case 1: $(\mathcal{A}_j)_{j \in J} \subset OLI_1(X)$ and let $x,\ y \in X$. Then by Definition 5 and Remark 4 (1), we have

$$(\bigcup_{j \in J} \mathbf{A}^-)(xy) = \bigvee_{j \in J} \mathbf{A}^-(xy) \geq \bigvee_{j \in J} \mathbf{A}^-(y) = (\bigcup_{j \in J} \mathbf{A}^-)(y).$$

Similarly, $(\bigcap_{j \in J} \mathbf{A}^+)(xy) \geq (\bigcap_{j \in J} \mathbf{A}^+)(y)$. From Proposition 3.3 in [19] and 3.10 in [17], we have

$$(\bigcup_{j \in J} \lambda)(xy) \geq (\bigcup_{j \in J} \lambda)(y),\ (\bigcup_{j \in J} A)(xy) \geq (\bigcup_{j \in J} A)(y).$$

Thus, $(\mathcal{A}_j)_{j \in J} \in OLI_1(X)$.

Case 2: $(\mathcal{A}_j)_{j \in J} \subset OLI_2(X)$ and let $x,\ y \in X$. Then by Definition 5 and Remark 4 (1), we have

$$(\bigcap_{j \in J} \lambda)(xy) = \bigwedge_{j \in J} \lambda(xy) \leq \bigwedge_{j \in J} \lambda(y) = (\bigcap_{j \in J} \lambda)(y).$$

Thus, by Case 1, $(\mathcal{A}_j)_{j \in J} \in OLI_2(X)$.

Case 3: $(\mathcal{A}_j)_{j \in J} \subset OLI_3(X)$ and let $x,\ y \in X$. Then by Definition 5 and Remark 4 (1), we have

$$(\bigcap_{j \in J} A^{\in})(xy) = \bigwedge_{j \in J} A^{\in}(xy) \leq \bigwedge_{j \in J} A^{\in}(y) = (\bigcap_{j \in J} A^{\in})(y).$$

Similarly, $(\bigcup_{j\in J} A^{\notin})(xy) \geq (\bigcup_{j\in J} A^{\notin})(y)$. Thus, by Case 1, $(\mathcal{A}_j)_{j\in J} \in OLI_3(X)$.

Case 4: $(\mathcal{A}_j)_{j\in J} \subset OLI_4(X)$. Then by Cases 2 and 3, $(\mathcal{A}_j)_{j\in J} \in OLI_4(X)$. This completes the proof. □

Definition 16 ([10]). *Let X, Y be two sets, let $f : X \to Y$ be a mapping and let $\mathcal{A} = \langle \mathbf{A}, A, \lambda \rangle \in \mathcal{O}(X)$, $\mathcal{B} = \langle \mathbf{B}, B, \mu \rangle \in \mathcal{O}(Y)$.*

(i) The preimage of $\mathcal{B}$ under f, denoted by $f^{-1}(\mathcal{B}) = \langle f^{-1}(\mathbf{B}), f^{-1}(B), f^{-1}(\mu) \rangle$, is the octahedron set in X defined as follows: for each $x \in X$,

$$f^{-1}(\mathcal{B})(x) = \left\langle [(B^{-} \circ f)(x), (B^{+} \circ f)(x))], ((B^{\in} \circ f)(x), (B^{\notin} \circ f)(x)), (\mu \circ f)(x) \right\rangle.$$

(ii) The image of $\mathcal{A}$ under f, denoted by $f(\mathcal{A}) = \langle f(\mathbf{A}), f(A), f(\lambda) \rangle$, is the octahedron set in Y defined as follows: for each $y \in Y$,

$$f(\mathbf{A})(y) = \begin{cases} [\bigvee_{x\in f^{-1}(y)} A^{-}(x), \bigvee_{x\in f^{-1}(y)} A^{+}(x)] & \text{if } f^{-1}(y) \neq \phi \\ \mathbf{0} & \text{otherwise,} \end{cases}$$

$$f(A)(y) = \begin{cases} (\bigvee_{x\in f^{-1}(y)} A^{\in}(x), \bigwedge_{x\in f^{-1}(y)} A^{\notin}(x)) & \text{if } f^{-1}(y) \neq \phi \\ \bar{0} & \text{otherwise,} \end{cases}$$

$$f(\lambda)(y) = \begin{cases} \bigvee_{x\in f^{-1}(y)} \lambda(x) & \text{if } f^{-1}(y) \neq \phi \\ 0 & \text{otherwise.} \end{cases}$$

It is obvious that $f(x_{\langle \tilde{a}, \bar{b}, \alpha \rangle}) = [f(x)]_{\langle \tilde{a}, \bar{b}, \alpha \rangle}$, for each $x_{\langle \tilde{a}, \bar{b}, \alpha \rangle} \in \mathcal{O}_P(X)$.

Proposition 11. *Let $f : X \to Y$ be a groupoid homomorphism, let $\mathcal{B} = \langle \mathbf{B}, B, \mu \rangle \in \mathcal{O}(Y)$ and let $i = 1, 2, 3, 4$.*

(1) If $\mathcal{B} \in OGP_i(Y)$, then $f^{-1}(\mathcal{B}) \in OGP_i(X)$.

(2) If $\mathcal{B} \in OLI_i(Y)$ [resp., $ORI_i(Y)$ and $OI_i(Y)$], then $f^{-1}(\mathcal{B}) \in OLI_i(X)$ [resp., $ORI_i(X)$ and $OI_i(X)$].

Proof. (1) Case 1: Suppose $\mathcal{B} = \langle \mathbf{B}, B, \mu \rangle \in OGP_1(Y)$. Then clearly, Propositions 3.9 in [18], 4.1 in [19] and 4.1 (1) in [17], $f^{-1}(\mu) \in FGP(X)$, $f^{-1}(\mathbf{B}) \in IVGP(X)$ and $f^{-1}(B) \in IFGP(X)$. Thus, $f^{-1}(\mathcal{B}) \in OGP_1(X)$.

Case 2: Suppose $\mathcal{B} \in OGP_2(Y)$ and let x, $y \in X$. Then by Theorem 2 (iii),

$$\begin{aligned} f^{-1}(\mu)(xy) &= (\mu \circ f)(xy) = \mu(f(xy)) \\ &= \mu(f(x)f(y)) \text{ [Since } f \text{ is a groupoid homomorphism]} \\ &\leq \mu(f(x)) \vee \mu(f(y)) \\ &= f^{-1}(\mu)(x) \vee f^{-1}(\mu)(y). \end{aligned}$$

Thus, $f^{-1}(\mu)$ satisfies the condition (iii) of Theorem 2 (3). So by Case 1, $f^{-1}(\mathcal{B}) \in OGP_2(X)$.

Case 3: Suppose $\mathcal{B} \in OGP_3(Y)$ and let x, $y \in X$. Then by Theorem 3 (ii),

$$\begin{aligned} f^{-1}(B^{\in})(xy) &= (B^{\in} \circ f)(xy) = B^{\in}(f(xy)) \\ &= B^{\in}(f(x)f(y)) \\ &\leq B^{\in}(f(x)) \vee B^{\in}(f(y)) \\ &= f^{-1}(B^{\in})(x) \vee f^{-1}(B^{\in})(y). \end{aligned}$$

Similarly, we have $f^{-1}(B^{\notin})(xy) \geq f^{-1}(B^{\notin})(x) \wedge f^{-1}(B^{\notin})(y)$. Thus, $f^{-1}(B)$ satisfies the condition (ii) of Theorem 3 (3). So by Case 1, $f^{-1}(\mathcal{B}) \in OGP_3(X)$.

Case 4: Suppose $\mathcal{B} \in OGP_4(Y)$. Then by Cases 2 and 3, $f^{-1}(\mathcal{B}) \in OGP_4(X)$.

(2) We only prove that $f^{-1}(\mathcal{B}) \in OLI_i(X)$ and the other proofs are omitted.

Case 1: Suppose $\mathcal{B} \in OLI_1(Y)$ and let x, $y \in X$. Then

$$f^{-1}(\mathbf{B})(xy) = [B^{-}(f(x)f(y)), B^{+}(f(x)f(y))]$$

$\geq [B^-(f(y)), B^+(f(y))]$ [Since $\mathbf{B} \in IVLI(Y)$]
$= f^{-1}(\mathbf{B})(y)$.

Thus, $f^{-1}(\mathbf{B}) \in IVLI(X)$. Moreover, from Propositions 4.1 in [19] and 4.1 in [17], $f^{-1}(\mu) \in FLL(X)$ and $f^{-1}(B) \in IFLI(X)$. So $f^{-1}(\mathcal{B}) \in OLI_1(X)$.

Case 2: Suppose $\mathcal{B} \in OLI_2(Y)$ and let $x,\ y \in X$. Then by Remark 4 (1),

$$f^{-1}(\mu)(xy) = \lambda(f(xy)) = \mu(f(x)f(y)) \leq \mu(f(y)) = f^{-1}(\mu)(y).$$

Thus, $f^{-1}(\mu)(xy) \leq f^{-1}(\mu)(y)$. So by Case 1, $f^{-1}(\mathcal{B}) \in OLI_2(X)$.

Case 3: Suppose $\mathcal{B} \in OLI_3(Y)$ and let $x,\ y \in X$. Then by Remark 4 (1),

$$f^{-1}(B^{\in})(xy) = B^{\in}(f(xy)) = B^{\in}(f(x)f(y)) \leq B^{\in}(f(y)) = f^{-1}(B^{\in})(y).$$

Similarly, we have $f^{-1}(B^{\notin})(xy) \geq f^{-1}(B^{\notin})(y)$. Thus, by Case 1, $f^{-1}(\mathcal{B}) \in OLI_3(X)$.

Case 4: Suppose $\mathcal{B} \in OLI_3(Y)$. Then by Cases 2 and 3, $f^{-1}(\mathcal{B}) \in OLI_4(X)$. This completes the proof. □

Definition 17. *Let X be a nonempty set and let $\mathcal{A} = \langle \mathbf{A}, A, \lambda \rangle \in \mathcal{O}(X)$. Then we say that $\mathcal{A}$ has the i-sup-property ($i = 1,\ 2,\ 3,\ 4$), if for each $T \in 2^X$, there is $t_0 \in T$ such that*

$$\mathcal{A}(t_0) = \bigvee_{t \in T}^{i} \mathcal{A}(t).$$

It is obvious that if $\mathcal{A}$ takes on only finitely many values, then it has the i-sup-property. In particular, $\mathcal{A} \in \mathcal{O}(X)$ has the 1-sup-property if and only if $\mathbf{A}$, A and λ have the sup-property.

Proposition 12. *Let $f : X \to Y$ be a groupoid homomorphism, let $\mathcal{A} = \langle \mathbf{A}, A, \lambda \rangle \in \mathcal{O}(X)$ has the i-sup-property and let $i = 1,\ 2,\ 3,\ 4$.*

(1) If $\mathcal{A} \in OGP_i(X)$, then $f(\mathcal{A}) \in OGP_i(Y)$.

(2) If $\mathcal{A} \in OLI_i(X)$ [resp., $ORI_i(X)$ and $OI_i(X)$], then $f(\mathcal{A}) \in OLI_i(Y)$ [resp., $ORI_i(Y)$ and $OI_i(Y)$].

Proof. (1) Case 1: Suppose $\mathcal{A} \in OGP_1(X)$. Then from Propositions 3.11 in [18], 4.2 in [19] and 4.4 (1) in [17],

$$f(\mathbf{A}) \in IVGP(X),\ f(\lambda) \in FGP(X),\ f(A) \in IFGP(X).$$

Thus, by Remark 2 (1), $f(\mathcal{A}) \in OGP_1(Y)$.

Case 2: Suppose $\mathcal{A} \in OGP_2(X)$. Since $f(\mathbf{A}) \in IVGP(X),\ f(A) \in IFGP(X)$ by Case 1, it is sufficient to show that $f(\lambda)$ satisfies the condition (iii) of Theorem 2. Let $y,\ y' \in Y$. Then we can consider four cases:

(i) $f^{-1}(y) \neq \emptyset,\ f^{-1}(y') \neq \emptyset$, (ii) $f^{-1}(y) \neq \emptyset,\ f^{-1}(y') = \emptyset$,
(iii) $f^{-1}(y) = \emptyset,\ f^{-1}(y') \neq \emptyset$, (iv) $f^{-1}(y) = \emptyset,\ f^{-1}(y') = \emptyset$.

We prove only case (i) and omit the other ones. Since $\mathcal{A}$ has the 2-sup-property, there are $x_0,\ x_0' \in X$ such that $\lambda(x_0) = \bigwedge_{t \in f^{-1}(y)} \lambda(t),\ \lambda(x_0') = \bigwedge_{t' \in f^{-1}(y')} \lambda(t')$. Then

$f(\lambda)(yy') = \bigwedge_{z \in f^{-1}(yy')} \lambda(z)$
$\leq \lambda(x_0x_0')$ [Since $f(x_0x_0') = f(x_0)f(x_0') = yy'$.]
$\leq \lambda(x_0) \vee \lambda(x_0')$ [Since by the hypothesis and Theorem 2 (iii)]
$= (\bigwedge_{t \in f^{-1}(y)} \lambda(t)) \vee (\bigwedge_{t' \in f^{-1}(y')} \lambda(t'))$
$= f(\lambda)(y) \vee f(\lambda)(y')$.

Thus, by Theorem 2, $f(\mathcal{A}) \in OGP_2(Y)$.

Case 3: Suppose $\mathcal{A} \in OGP_3(X)$. Since $f(\mathbf{A}) \in IVGP(X)$, $f(\lambda) \in FGP(X)$ by Case 1, it is sufficient to show that $f(A)$ satisfies the condition (ii) of Theorem 3. Let $y,\ y' \in Y$ and we show only case (i) of Case 2. Since $\mathcal{A}$ has the 3-sup-property, there are $x_0,\ x_0' \in X$ such that

$$A^{\in}(x_0) = \bigwedge_{t\in f^{-1}(y)} A^{\in}(t),\ A^{\in}(x_0') = \bigwedge_{t'\in f^{-1}(y')} A^{\in}(t')$$

and

$$A^{\notin}(x_0) = \bigvee_{t\in f^{-1}(y)} A^{\notin}(t),\ A^{\notin}(x_0') = \bigvee_{t'\in f^{-1}(y')} A^{\notin}(t').$$

Then

$$\begin{aligned}
[f(A)]^{\in}(yy') &= \textstyle\bigwedge_{z\in f^{-1}(yy')} A^{\in}(z)\\
&\leq A^{\in}(x_0x_0')\\
&\leq A^{\in}(x_0) \vee A^{\in}(x_0')\\
&\quad \text{[Since by the hypothesis and Theorem 3 (ii)]}\\
&= (\textstyle\bigwedge_{t\in f^{-1}(y)} A^{\in}(t)) \vee (\bigwedge_{t'\in f^{-1}(y')} A^{\in}(t'))\\
&= [f(A)]^{\in}(y) \vee [f(A)]^{\in}(y').
\end{aligned}$$

Similarly, we have $[f(A)]^{\notin}(yy') \geq [f(A)]^{\notin}(y) \wedge [f(A)]^{\notin}(y')$. Thus, by Theorem 3, $f(\mathcal{A}) \in OGP_3(Y)$.

Case 4: Suppose $\mathcal{A} \in OGP_4(X)$. Then by Cases 2, 3 and Theorem 4, we can easily prove that $f(\mathcal{A}) \in OGP_4(Y)$.

(2) We prove only that $f(\mathcal{A}) \in OLI_i(Y)$.

Case 1: Suppose $\mathcal{A} \in OLI_1(X)$. From Propositions 4.2 in [19] and 4.4 (2) in [17], $f(\lambda) \in FLI(Y)$ and $f(B) \in IFLI(Y)$. Then it is sufficient to show that $f(\mathbf{A}) \in IVLI(Y)$. Let $y,\ y' \in X$ and we prove only case (i) of Case 2 in (1). Since $\mathcal{A}$ has the 1-sup-property, there are $x_0 \in f^{-1}(y)$ and $x_0' \in f^{-1}(y')$ such that

$$\mathcal{A}(x_0) = \bigvee_{t\in f^{-1}(y)} \mathcal{A}(x) \text{ and } \mathcal{A}(x_0') = \bigvee_{t'\in f^{-1}(y')} \mathcal{A}(t').$$

Then

$$\begin{aligned}
f(\mathbf{A})^{-}(yy') &= \textstyle\bigvee_{z\in f^{-1}(yy')} A^{-}(z)\\
&\geq A^{-}(x_0x_0')\ [\text{Since } f(x_0x_0') = f(x_0)f(x_0') = yy']\\
&\geq A^{-}(x_0')\ [\text{Since } \mathbf{A} \in IVLI(X)]\\
&= \textstyle\bigvee_{t'\in f^{-1}(y')} A^{-}(t')\\
&= f(\mathbf{A})^{-}(y').
\end{aligned}$$

Similarly, we have $f(\mathbf{A})^{+}(yy') \geq f(\mathbf{A})^{+}(y')$. Thus, $f(\mathbf{A})(yy') \geq f(\mathbf{A})(y')$. So $f(\mathbf{A}) \in IVLI(Y)$. Hence by Remark 4 (1), $f(\mathcal{A}) \in OLI_1(Y)$.

Case 2: Suppose $\mathcal{A} \in OLI_2(X)$ and let $y,\ y' \in Y$. Since $f(\mathbf{A}) \in IVLI(Y)$, $f(A) \in IFLI(Y)$ by Case (1), it is sufficient to prove that $f(\lambda)(yy') \leq f(\lambda)(y')$. Since $\mathcal{A}$ has the 2-sup-property, there are $x_0,\ x_0' \in X$ such that

$$\lambda(x_0) = \bigwedge_{t\in f^{-1}(y)} \lambda(t),\ \lambda(x_0') = \bigwedge_{t'\in f^{-1}(y')} \lambda(t').$$

Then

$$\begin{aligned}
f(\lambda)(yy') &= \textstyle\bigwedge_{z\in f^{-1}(yy')} \lambda(z)\\
&\leq \lambda(x_0x_0')\\
&\leq \lambda(x_0')\ \text{[Since by the hypothesis and Remark 4 (1)]}\\
&= \textstyle\bigwedge_{t'\in f^{-1}(y')} \lambda(t')\\
&= f(\lambda)(y').
\end{aligned}$$

Thus, by Remark 4 (1), $f(\mathcal{A}) \in OLI_2(Y)$.

Case 3: Suppose $\mathcal{A} \in OLI_3(X)$ and let $y,\ y' \in Y$. Since $f(\mathbf{A}) \in IVLI(Y)$, $f(\lambda) \in FLI(Y)$ by Case (1), it is sufficient to prove that $[f(A)]^{\in}(yy') \leq [f(A)]^{\in}(y')$ and $[f(A)]^{\notin}(yy') \geq [f(A)]^{\notin}(y')$. Since $\mathcal{A}$ has the 3-sup-property, there are $x_0,\ x_0' \in X$ such that

$$A^{\in}(x_0) = \bigwedge_{t \in f^{-1}(y)} A^{\in}(t),\ A^{\in}(x_0') = \bigwedge_{t' \in f^{-1}(y')} A^{\in}(t')$$

and

$$A^{\notin}(x_0) = \bigvee_{t \in f^{-1}(y)} A^{\notin}(t),\ A^{\notin}(x_0') = \bigvee_{t' \in f^{-1}(y')} A^{\notin}(t').$$

Then

$[f(A)]^{\in}(yy') = \bigwedge_{z \in f^{-1}(yy')} A^{\in}(z)$
$\leq A^{\in}(x_0 x_0')$
$\leq A^{\in}(x_0')$
[Since by the hypothesis and Remark 4 (1)]
$= \bigwedge_{t' \in f^{-1}(y')} A^{\in}(t')$
$= [f(A)]^{\in}(y')$.

Similarly, we have $[f(A)]^{\notin}(yy') \geq [f(A)]^{\notin}(y')$. Thus, by Remark 4 (1), $f(\mathcal{A}) \in OLI_3(Y)$.

Case 4: Suppose $\mathcal{A} \in OLI_4(X)$. Then by Cases 2 and 3, we can easily show that $f(\mathcal{A}) \in OLI_4(Y)$. This completes the proof. □

Definition 18. *Let X, Y be sets, $f : X \to Y$ be a mapping and let $\mathcal{A} \in \mathcal{O}(X)$. Then $\mathcal{A}$ is said to be f-invariant, if for every $x,\ y \in X$, $f(x) = f(y)$ implies $\mathcal{A}(x) = \mathcal{A}(y)$.*

It is obvious that $\mathcal{A}$ is f-invariant if and only if $\mathbf{A}$, A and λ are f-invariant. Moreover, we can easily see that if $\mathcal{A}$ is f-invariant, then $f^{-1}(f(\mathcal{A})) = \mathcal{A}$.

The following is the immediate result of Definition 18.

Proposition 13. *Let X, Y be sets, let $f : X \to Y$ be a mapping and let*

$$\Omega = \{\mathcal{A} \in \mathcal{O}(X) : \mathcal{A} \text{ is } f - invariant\}.$$

Then there is a one-to-one correspondence between Ω and $\mathcal{O}(Imf)$, where Imf denotes the image of f.

The following is the immediate result of Propositions 12 (1) and 13.

Proposition 14. *Let $f : X \to Y$ be a groupoid homomorphism and let*

$$\Phi = \{\mathcal{A} \in OGP_i(X) : \mathcal{A} \text{ is } f - invariant \text{ and has the } i - sup\ property\},$$

where $i = 1,\ 2,\ 3,\ 4$. Then there is a one-to-one correspondence between Φ and $OGP_i(Imf)$.

4. Octahedron Subgroups

Unless stated otherwise in this section, G denotes a group and e is the identity of G.

Definition 19 ([19]). *Let $\lambda \in FGP(G)$. Then λ is called a fuzzy subgroup of G, if it satisfies the following condition:*

$$\lambda(x^{-1}) \geq \lambda(x), \text{ for each } x \in G.$$

We will denote the set of fuzzy subgroups of G as $FG(G)$.

Definition 20 ([20]). *Let $A \in IFGP(G)$. Then A is called an intuitionistic fuzzy subgroup (briefly, IFG) of G, if it satisfies the following condition: for each $x \in G$,*

$$A(x^{-1}) \geq A(x), \text{ i.e., } A^{\in}(x^{-1}) \geq A^{\in}(x),\ A^{\notin}(x^{-1}) \leq A^{\notin}(x).$$

We will denote the set of IFGs of G as $IFG(G)$.

Definition 21 ([18,21]). *Let $\mathbf{A} \in IVFGP(X)$. Then $\mathbf{A}$ is called an interval-valued fuzzy subgroup (briefly, IVG) of G, if it satisfies the following condition: for each $x \in G$,*

$$\mathbf{A}(x^{-1}) \geq A(x), \text{ i.e., } A^{-}(x^{-1}) \geq A^{-}(x),\ A^{+}(x^{-1}) \geq A^{+}(x).$$

We will denote the set of IVGs of G as $IVG(G)$.

Definition 22. *Let $\mathcal{A} \in OGP_i(G)$ $(i = 1, 2, 3, 4)$. Then $\mathcal{A}$ is called a i-octahedron subgroup of G, if it satisfies the following condition: for each $x \in G$,*

$$\mathcal{A}(x^{-1}) \geq_i \mathcal{A}(x).$$

We will denote the set of all i-octahedron subgroups of G as $OG_i(G)$.
In particular, if $\mathcal{A} \in OGP_1(G)$, then $\mathcal{A}$ will simply called an octahedron subgroup of G.

From Theorems 1–4 and Definition 22, we obtain easily the characterizations of i-octahedron subgroups of G.

Theorem 6. *Let $\mathcal{A} \in \mathcal{O}(G)$.*
(1) $\mathcal{A} \in OG_1(G) \iff \mathbf{A} \in IVG(G),\ A \in IFG(G),\ \lambda \in FG(G)$.
(2) $\mathcal{A} \in OG_2(G) \iff \mathbf{A} \in IVG(G),\ A \in IFG(G)$ *and λ satisfies the following conditions: for every $x, y \in G$,*

$$\lambda(xy) \leq \lambda(x) \vee \lambda(y),\ \lambda(x^{-1}) \leq \lambda(x).$$

(3) $\mathcal{A} \in OG_3(G) \iff \mathbf{A} \in IVG(G),\ \lambda \in FG(G)$ *and A satisfies the following conditions: for every $x, y \in G$,*
(i) $A^{\in}(xy) \leq A^{\in}(x) \vee A^{\in}(y),\ A^{\notin}(xy) \geq A^{\notin}(x) \wedge A^{\notin}(y)$,
(ii) $A^{\in}(x^{-1}) \leq A^{\in}(x),\ A^{\notin}(x^{-1}) \geq A^{\notin}(x)$.
(4) $\mathcal{A} \in OG_4(G) \iff \mathbf{A} \in IVG(G)$, *A and λ satisfies the following conditions: for every $x, y \in G$,*
(i) $\lambda(xy) \leq \lambda(x) \vee \lambda(y),\ \lambda(x^{-1}) \leq \lambda(x)$,
(ii) $A^{\in}(xy) \leq A^{\in}(x) \vee A^{\in}(y),\ A^{\notin}(xy) \geq A^{\notin}(x) \wedge A^{\notin}(y)$,
(iii) $A^{\in}(x^{-1}) \leq A^{\in}(x),\ A^{\notin}(x^{-1}) \geq A^{\notin}(x)$.

Example 5. *(1) Consider the additive group $(\mathbb{Z}, +)$. We define five mappings $\mathbf{A} = [A^{-}, A^{+}] : \mathbb{Z} \to [I]$, $A = (A^{\in}, A^{\notin})$, $B = (B^{\in}, B^{\notin}) : \mathbb{Z} \to I \oplus I$ and λ, $\mu : \mathbb{Z} \to I$, respectively as follows: for each $0 \neq n \in \mathbb{Z}$,*

$$\mathbf{A}(0) = [1,1],\ A(0) = (1,0),\ B(0) = (0,1),\ \lambda(0) = 1,\ \mu(0) = \frac{1}{6}, \tag{1}$$

$$\mathbf{A}(n) = \begin{cases} \left[\frac{1}{2}, \frac{2}{3}\right] & \text{if } n \text{ is odd} \\ \left[\frac{1}{3}, \frac{4}{5}\right] & \text{if } n \text{ is even,} \end{cases} \tag{2}$$

$$A(n) = \begin{cases} \left(\frac{1}{2}, \frac{1}{3}\right) & \text{if } n \text{ is odd} \\ \left(\frac{2}{3}, \frac{1}{5}\right) & \text{if } n \text{ is even,} \end{cases} \tag{3}$$

$$B(n) = \begin{cases} \left(\frac{2}{3}, \frac{1}{5}\right) & \text{if } n \text{ is odd} \\ \left(\frac{1}{2}, \frac{1}{3}\right) & \text{if } n \text{ is even,} \end{cases} \tag{4}$$

$$\lambda(n) = \begin{cases} \frac{1}{2} & \text{if } n \text{ is odd} \\ \frac{3}{5} & \text{if } n \text{ is even,} \end{cases} \tag{5}$$

$$\mu(n) = \begin{cases} \frac{3}{5} & \text{if } n \text{ is odd} \\ \frac{1}{2} & \text{if } n \text{ is even.} \end{cases} \tag{6}$$

Then we can easily check that $\lambda \in FG(\mathbb{Z})$. Moreover, $\mathbf{A} \in IVG(\mathbb{Z})$ and $A \in IFG(\mathbb{Z})$ from Example 4.1 in [18] and Example 2.1 in [20]. Thus, $\langle \mathbf{A}, A, \lambda \rangle$ is an octahedron subgroup of $\mathbb{Z}$.

On the other hand, we can easily check that μ and B satisfy the conditions of Theorem 6 (2) and (i) and (ii) of Theorem 6 (3), respectively. Hence by Theorem 6 (2) and (3), $\langle \mathbf{A}, A, \mu \rangle \in OG_2(X)$ and $\langle \mathbf{A}, B, \lambda \rangle \in OG_3(X)$. Thus, by Theorem 6 (2), $\mathcal{A} = \langle \mathbf{A}, A, \mu \rangle \in OG_2(X)$. Furthermore, from Theorem 6 (4), we can easily see that $\langle \mathbf{A}, B, \mu \rangle \in OG_4(X)$.

(2) If $\mathbf{A} \in IVGP(G)$, then $\mathcal{O}_{\mathbf{A}} \in OG_1(G)$. Also if $A \in IFGP(G)$, then $\mathcal{O}_A \in OG_1(G)$.

(3) If $\mathcal{A} \in OG_i(X)$, then clearly, $[\,]\mathcal{A}$, $\diamond \mathcal{A} \in OG_i(X)$ $(i = 1, 2, 3, 4)$.

Remark 8. *(1) If $\lambda \in FG(G)$, then we have*

$$\langle [\lambda, \lambda], (\lambda, \lambda^c), \lambda \rangle \in OG_1(G), \ \ \langle [\lambda, \lambda], (\lambda, \lambda^c), \lambda^c \rangle \in OG_2(G),$$

$$\left\langle [\lambda, \lambda], (\lambda^C, \lambda), \lambda \right\rangle \in OG_3(G), \ \ \left\langle [\lambda, \lambda], (\lambda^C, \lambda), \lambda^c \right\rangle \in OG_4(G).$$

(2) If $\mathbf{A} \in IVG(G)$, then we have

$$\langle \mathbf{A}, (A^-, (A^+)^c), A^- \rangle, \ \langle \mathbf{A}, (A^-, (A^+)^c), A^+ \rangle \in OG_1(G),$$

$$\langle \mathbf{A}, (A^+, (A^-)^c), A^+ \rangle, \ \langle \mathbf{A}, (A^+, (A^-)^c), A^- \rangle \in OG_1(G),$$

$$\langle \mathbf{A}, (A^-, (A^+)^c), (A^-)^c \rangle, \ \langle \mathbf{A}, (A^-, (A^+)^c), (A^+)^c \rangle \in OG_2(G),$$

$$\left\langle \mathbf{A}, (A^+, (A^-)^c), (A^+)^C \right\rangle, \ \langle \mathbf{A}, (A^+, (A^-)^c), (A^-)^c \rangle \in OG_2(G),$$

$$\langle \mathbf{A}, ((A^-)^c, A^+), A^- \rangle, \ \langle \mathbf{A}, ((A^-)^c, A^+), A^+ \rangle \in OG_3(G),$$

$$\langle \mathbf{A}, ((A^+)^c, A^-), A^+ \rangle, \ \langle \mathbf{A}, ((A^+)^c, A^-), A^- \rangle \in OG_3(G),$$

$$\langle \mathbf{A}, ((A^-)^c, A^+), (A^-)^c \rangle, \ \langle \mathbf{A}, ((A^-)^c, A^+), (A^+)^c \rangle \in OG_4(G),$$

$$\langle \mathbf{A}, ((A^+)^c, A^-), (A^+)^c \rangle, \ \langle \mathbf{A}, ((A^+)^c, A^-), (A^-)^c \rangle \in OG_4(G).$$

(3) If $A \in IFG(G)$, then we have

$$\left\langle [A^{\in}, (A^{\notin})^c], A, A^{\in} \right\rangle \in OG_1(G), \ \left\langle [A^{\in}, (A^{\notin})^c], A, A^{\notin} \right\rangle \in OG_2(G),$$

$$\left\langle [A^{\in}, (A^{\notin})^c], A^c, A^{\in} \right\rangle \in OG_3(G), \ \left\langle [A^{\in}, (A^{\notin})^c], A^c, A^{\notin} \right\rangle \in OG_4(G).$$

The following is an immediate result of Theorem 6 (1) and Remark 3.

Proposition 15. *For every $H \subset G$, H is a subgroup of G if and only if $\chi_{\mathcal{H}} \in OG_1(G)$.*

The following is an immediate result of Theorem 6 (1) and Proposition 8.

Proposition 16. *If $(\mathcal{A}_j)_{j\in J} = (\langle \mathbf{A}_j, A_j, \lambda_j \rangle)_{j\in J} \subset OG_1(G)$, then $\bigcap^1_{j\in J} \mathcal{A}_j \in OG_1(G)$, where J denotes an index set.*

The following is an immediate result of Proposition 16 and Corollary 1.

Corollary 2. *Let $\mathcal{A} \in \mathcal{O}(G)$ and let*

$$(\mathcal{A}) = \bigcap^1 \{\mathcal{B} \in OG_1(G) : \mathcal{A} \subset_1 \mathcal{B}\}.$$

Then $(\mathcal{A}) \in OG_1(G)$.
In this case, $(\mathcal{A})$ is called the octahedron subgroup of G generated by $\mathcal{A}$.

The following is an immediate result of Proposition 9 and Corollary 2.

Corollary 3. *For each $A \in 2^X$, let (A) be the subgroup generated by A and let $\chi_{(A)} = \left\langle [\chi_{(A)}, \chi_{(A)}], (\chi_{(A)}, \chi_{(A^c)}), \chi_{(A)} \right\rangle$. Then*

$$(\chi_{\mathcal{A}}) = \chi_{(A)}.$$

Proposition 17. *Let $\mathcal{A} \in OG_i(G)$ $(i = 1, 2, 3, 4)$. Then for each $x \in G$,*

$$\mathcal{A}(x^{-1}) = \mathcal{A}(x),\ \mathcal{A}(e) \geq_i \mathcal{A}(x).$$

Proof. Case 1: Suppose $\mathcal{A} \in OG_1(G)$. Then by Theorem 6 (1), we have

$$\mathbf{A} \in IVG(G),\ A \in IFG(G),\ \lambda \in FG(G).$$

Thus, by Propositions 3.1 in [21], 2.6 in [20] and 5.4 in [19], we have

$$\mathbf{A}(x^{-1}) = \mathbf{A}(x),\ A(x^{-1}) = A(x),\ \lambda(x^{-1}) = \lambda(x),$$

$$\mathbf{A}(e) \geq \mathbf{A}(x),\ A(e) \geq A(x),\ \lambda(e) \geq \lambda(x).$$

So $\mathcal{A}(x^{-1}) = \mathcal{A}(x)$ and $\mathcal{A}(e) \geq_1 \mathcal{A}(x)$.
Case 2: Suppose $\mathcal{A} \in OG_2(G)$. Then by Theorem 6 (2), we have

$$\mathbf{A} \in IVG(G),\ A \in IFG(G)$$

and

$$\lambda(xy) \leq \lambda(x) \vee \lambda(y),\ \lambda(x^{-1}) \leq \lambda(x), \text{ for every } x,\ y \in G.$$

Thus, $\lambda(x) = \lambda((x^{-1})^{-1}) \leq \lambda(x^{-1}) \leq \lambda(x)$, i.e., $\lambda(x^{-1}) = \lambda(x)$.
On the other hand, $\lambda(e) = \lambda(xx^{-1}) \leq \lambda(x) \vee \lambda(x^{-1}) = \lambda(x)$. By Case 1, we have

$$\mathbf{A}(x^{-1}) = \mathbf{A}(x),\ A(x^{-1}) = A(x) \text{ and } \mathbf{A}(e) \geq \mathbf{A}(x),\ A(e) \geq A(x).$$

So $\mathcal{A}(x^{-1}) = \mathcal{A}(x)$ and $\mathcal{A}(e) \geq_2 \mathcal{A}(x)$.

Case 3: Suppose $\mathcal{A} \in OG_2(G)$. Then by Case 1 and Theorem 6 (3), we have

$$\mathbf{A}(x^{-1}) = \mathbf{A}(x),\ \lambda(x^{-1}) = \lambda(x),\ \mathbf{A}(e) \geq \mathbf{A}(x),\ \lambda(e) \geq \lambda(x)$$

and A satisfies the conditions (i) and (ii). By (ii),

$$A^{\in}(x) = A^{\in}((x^{-1})^{-1}) \leq A^{\in}(x^{-1}) \leq A^{\in}(x), \text{ i.e., } A^{\in}(x) = A^{\in}(x^{-1}).$$

Similarly, we have $A^{\notin}(x) = A^{\notin}(x^{-1})$. Thus, $A(x) = A(x^{-1})$. By (i), we can easily prove that $A(e) \leq A(x)$.. So $\mathcal{A}(x^{-1}) = \mathcal{A}(x)$ and $\mathcal{A}(e) \geq_3 \mathcal{A}(x)$.

Case 4: Suppose $\mathcal{A} \in OG_2(G)$. Then by Cases 1 and 2, we have $\mathcal{A}(x^{-1}) = \mathcal{A}(x)$ and $\mathcal{A}(e) \geq_4 \mathcal{A}(x)$. This completes the proof. □

Theorem 7. *Let $\mathcal{A} \in \mathcal{O}(G)$ and let $i = 1, 2, 3, 4$. Then $\mathcal{A} \in OG_i(G)$ if and only if $\mathcal{A}(xy^{-1}) \geq_i \mathcal{A}(x) \wedge_i \mathcal{A}(x)(y)$,* for every $x,\ y \in G$.

Proof. We prove only the necessity of the condition.

Case 1: Suppose $\mathcal{A} \in OG_1(G)$ and let $x,\ y \in G$. Then by Theorem 6 (1), we have

$$\mathbf{A} \in IVG(G),\ A \in IFG(G),\ \lambda \in FG(G).$$

Thus, by Propositions 3.2 in [21] and 5.6 in [19], we have

$$\mathbf{A}(xy^{-1}) = \mathbf{A}(x) \wedge \mathbf{A}(y),\ \lambda(xy^{-1}) = \lambda(x) \wedge \lambda(y).$$

On the other hand, by Proposition 17,

$$A^{\in}(xy^{-1}) \geq A^{\in}(x) \wedge A^{\in}(y^{-1}) = A^{\in}(x) \wedge A^{\in}(y),$$

$$A^{\notin}(xy^{-1}) \leq A^{\notin}(x) \vee A^{\notin}(y^{-1}) = A^{\notin}(x) \vee A^{\notin}(y).$$

So $\mathcal{A}(xy^{-1}) \geq_1 \mathcal{A}(x) \wedge^1 \mathcal{A}(y)$.

Case 2: Suppose $\mathcal{A} \in OG_2(G)$ and let $x,\ y \in G$. Then by Theorem 6 (2) and Case 1, we have

$$\mathbf{A}(xy^{-1}) = \mathbf{A}(x) \wedge \mathbf{A}(y),\ A(xy^{-1}) = A(x) \wedge A(y)$$

and

$$\lambda(xy) \leq \lambda(x) \vee \lambda(y),\ \lambda(x^{-1}) \leq \lambda(x).$$

Thus, by Proposition 17,

$$\lambda(xy^{-1}) \leq \lambda(x) \vee \lambda(y^{-1}) = \lambda(x) \vee \lambda(y).$$

So $\mathcal{A}(xy^{-1}) \geq_2 \mathcal{A}(x) \wedge^2 \mathcal{A}(y)$.

Case 3: Suppose $\mathcal{A} \in OG_2(G)$ and let $x,\ y \in G$. Then by Case 1 and Theorem 6 (3), we have

$$\mathbf{A}(xy^{-1}) \geq \mathbf{A}(x) \wedge \mathbf{A}(x),\ \lambda(xy^{-1}) \geq \lambda(x) \wedge \lambda(y)$$

and A satisfies the conditions (i) and (ii). Thus, by (i) and Proposition 17,

$$A^{\in}(xy^{-1}) \leq A^{\in}(x) \vee A^{\in}(y^{-1}) = A^{\in}(x) \vee A^{\in}(y),$$

$$A^{\notin}(xy^{-1}) \geq A^{\notin}(x) \wedge A^{\in}(y^{-1}) = A^{\in}(x) \vee A^{\in}(y).$$

So $\mathcal{A}(xy^{-1}) \geq_3 \mathcal{A}(x) \wedge^3 \mathcal{A}(y)$.

Case 4: Suppose $\mathcal{A} \in OG_2(G)$. Then by Cases 1 and 2, we can easily prove that $\mathcal{A}(xy^{-1}) \geq_4 \mathcal{A}(x) \wedge^4 \mathcal{A}(y)$. This completes the proof. □

The following is an immediate consequence of Corollary in [19], Propositions 4.6 in [18] and 2.7 in [20].

Proposition 18. *If $\mathcal{A} \in OG_1(G)$. Then $G_\mathcal{A} = \{x \in G : \mathcal{A}(x) = \mathcal{A}(e)\}$ is a subgroup of G.*

Proposition 19. *Let $\mathcal{A} \in OG_1(G)$ $(i = 1, 2, 3, 4)$ and let $x, y \in G$. If $\mathcal{A}(xy^{-1}) = \mathcal{A}(e)$, then $\mathcal{A}(x) = \mathcal{A}(y)$.*

Proof. (1) Case 1: Suppose $\mathcal{A} \in OG_1(G)$ and let $x, y \in G$. Then by Theorem 6 (1), we have

$$\mathbf{A} \in IVG(G),\ A \in IFG(G),\ \lambda \in FG(G).$$

Thus, by Propositions 4.7 in [18], 2.8 in [20] and 5.5 in [19], we have

$$\mathbf{A}(x) = \mathbf{A}(y),\ A(x) = A(y)\ \lambda(x) = \lambda(y).$$

So $\mathcal{A}(x) = \mathcal{A}(y)$.

Case 2: Suppose $\mathcal{A} \in OG_2(G)$ and let $x, y \in G$. Then by Theorem 6 (2) and Case 1, we have

$$\mathbf{A}(x) = \mathbf{A}(y),\ A(x) = A(y)$$

and

$$\lambda(xy) \leq \lambda(x) \vee \lambda(y),\ \lambda(x^{-1}) \leq \lambda(x).$$

By Proposition 17,

$$\lambda(x) = \lambda((xy^{-1})y) \leq \lambda(xy^{-1}) \vee \lambda(y) = \lambda(e) \vee \lambda(y) = \lambda(y)$$
$$= \lambda((yx^{-1})x) \leq \lambda(e) \vee \lambda(x) = \lambda(x).$$

Thus, $\lambda(x) = \lambda(y)$. So $\mathcal{A}(x) = \mathcal{A}(y)$.

Case 3: Suppose $\mathcal{A} \in OG_2(G)$ and let $x, y \in G$. Then by Case 1 and Theorem 6 (3), we have $\mathbf{A}(x) = \mathbf{A}(y)$, $\lambda(x) = \lambda(y)$ and A satisfies the conditions (i) and (ii). By (i) and Proposition 17,

$$A^{\in}(x) = A^{\in}((xy^{-1})y) \leq A^{\in}(e) \vee A^{\in}(y) = A^{\in}(y)$$
$$= A^{\in}((yx^{-1})x) \leq A^{\in}(e) \vee A^{\in}(x) = A^{\in}(x).$$

Thus, $A^{\in}(x) = A^{\in}(y)$. Similarly, we have $A^{\notin}(x) = A^{\notin}(y)$.

$$A^{\notin}(xy^{-1}) \geq A^{\notin}(x) \wedge A^{\in}(y^{-1}) = A^{\in}(x) \vee A^{\in}(y).$$

So $\mathcal{A}(x) = \mathcal{A}(y)$.

Case 4: Suppose $\mathcal{A} \in OG_2(G)$. Then by Cases 1 and 2, we can easily prove that $\mathcal{A}(x) = \mathcal{A}(y)$. This completes the proof. □

The following is an immediate consequence of Corollary in [19], Corollaries 4.7-1 in [18] and 2.8-1 in [20].

Corollary 4. *Let $\mathcal{A} \in OG_1(G)$. If $G_\mathcal{A}$ is a normal subgroup of G, then $\mathcal{A}$ is constant on each coset of $G_\mathcal{A}$.*

The following is an immediate result of Corollary in [19], Corollaries 4.7-2 in [18] and 2.8-2 in [20].

Corollary 5. *Let $\mathcal{A} \in OG_1(G)$ and let $G_\mathcal{A}$ be a normal subgroup of G. If $G_\mathcal{A}$ has a finite index, then $\mathcal{A}$ has the sup property.*

The following is an immediate result of Propositions 5.7 in [19], 4.8 in [18] and 2.10 in [20].

Proposition 20. *A group G cannot be the* 1*-union of two proper* 1*-octahedron subgroups of G.*

Theorem 8. $\mathcal{A} \in OLI_i(G)$ *[resp.,* $\mathcal{A} \in ORI_i(G)$ *and* $\mathcal{A} \in OI_i(G)$*] if and only if* $\mathcal{A}$ *is a constant mapping* $(i = 1, 2, 3, 4)$*.*

Proof. We prove only that $\mathcal{A} \in OLI_i(G)$ if and only if $\mathcal{A}$ is a constant mapping. Suppose $\mathcal{A}$ is a constant mapping. Then we can easily show that $\mathcal{A} \in OLI_i(G)$. Thus, it is sufficient to prove only that the necessary condition holds.

Case 1: Suppose $\mathcal{A} \in OLI_1(G)$. Then by Propositions 4.14 in [18], 2.16 in [20] and 5.9 in [19], $\mathbf{A}$, A and λ are constant mappings. Thus, $\mathcal{A}$ is a constant mapping.

Case 2: Suppose $\mathcal{A} \in OLI_2(G)$. Since $\mathbf{A}$ and A are constant mappings by Case 1, it is enough to show that λ is a constant mapping. Let $x, y \in G$. Since $\mathcal{A} \in OLI_2(G)$, by Remark 4 (1), we have $\lambda(xy) \leq \lambda(x) \vee \lambda(y)$. Let $y = e$. Then by Proposition 17, we have

$$\lambda(x) \leq \lambda(x) \vee \lambda(e) = \lambda(e), \text{ for each } x \in G.$$

Now let $x = y^{-1}$. Then by Proposition 17, we have

$$\lambda(e) \leq \lambda(y^{-1}) \vee \lambda(y) = \lambda(y), \text{ for each } y \in G.$$

Thus, $\lambda(x) = \lambda(y) = \lambda(e)$. So λ is a constant mapping. Hence $\mathcal{A}$ is a constant mapping.

Case 3: Suppose $\mathcal{A} \in OLI_3(G)$. Then by Remark 4 (1) and Proposition 17, we can easily see that A is a constant mapping. Thus, $\mathcal{A}$ is a constant mapping.

Case 4: Suppose $\mathcal{A} \in OLI_4(G)$. Then by Cases (2) and (3), we can easily prove that $\mathcal{A}$ is a constant mapping. This completes the proof. □

Proposition 21. *Let* $f : G \to G'$ *be a group homomorphism, let* $\mathcal{A} \in OG_i(G)$ *and let* $\mathcal{B} \in OG_i(G')$ $(i = 1, 2, 3, 4)$*.*

(1) *If* $\mathcal{A}$ *has the i-sup-property,* $f(\mathcal{A}) \in OG_i(G')$*.*

(2) $f^{-1}(\mathcal{B}) \in OG_i(G)$*.*

Proof. (1) Since $f(\mathcal{A}) \in OGP_i(G')$ by Proposition 12 (1), It is sufficient to show that $f(\mathcal{A})(y^{-1}) \geq_i f(\mathcal{A})(y)$ for each $y \in f(G)$.

Case 1: Suppose $\mathcal{A} \in OG_1(G)$. Then by Propositions 4.11 in [18], 2.13 in [20] and 5.8 in [19], $f(\mathbf{A}) \in IVG(G')$, $f(A) \in IFG(G')$, $f(\lambda) \in FG(G')$. Thus, by Theorem 6 (1), $f(\mathcal{A}) \in OG_1(G')$.

Case 2: Suppose $\mathcal{A} \in OG_2(G)$ and let $y \in f(G)$. Since $f(\mathbf{A}) \in IVG(G')$ and $f(A) \in IFG(G')$ by Case 1, it is enough to prove that $f(\lambda)(y^{-1}) \leq f(\lambda)(y)$. Since $\mathcal{A}$ has the 2-sup-property, there is $x_0 \in f^{-1}(y)$ such that

$$\lambda(x_0) = \bigwedge_{t \in f^{-1}(y)} \lambda(t).$$

Then $f(\lambda)(y^{-1}) = \bigwedge_{t \in f^{-1}(y^{-1})} \lambda(t) \leq \lambda(x_0^{-1}) \leq \lambda(x_0) = f(\lambda)(y)$. Thus, by Theorem 6 (2), $f(\mathcal{A}) \in OG_2(G')$.

Case 3: Suppose $\mathcal{A} \in OG_3(G)$ and let $y \in f(G)$. Since $f(\mathbf{A}) \in IVG(G')$ and $f(\lambda) \in FG(G')$ by Case 1, it is sufficient to show that $[f(A)]^{\in}(y^{-1}) \leq [f(A)]^{\in}(y)$ and $[f(A)]^{\notin}(y^{-1}) \geq [f(A)]^{\in}(y)$. Since $\mathcal{A}$ has the 3-sup-property, there is $x_0 \in f^{-1}(y)$ such that

$$A^{\in}(x_0) = \bigwedge_{t \in f^{-1}(y)} A^{\in}(t), \; A^{\notin}(x_0) = \bigvee_{t \in f^{-1}(y)} A^{\notin}(t).$$

Then $[f(A)]^{\in}(y^{-1}) = \bigwedge_{t \in f^{-1}(y^{-1})} A^{\in}(t) \leq A^{\in}(x_0^{-1}) \leq A^{\in}(x_0) = [f(A)]^{\in}(y)$. Similarly, we have $[f(A)]^{\notin}(y^{-1}) \geq [f(A)]^{\notin}(y^{-1})$. Thus, by Theorem 6 (3), $f(\mathcal{A}) \in OG_3(G')$.

Case 4: Suppose $\mathcal{A} \in OG_4(G)$. Then from Cases (2), (3) and Theorem 6 (4), we can easily prove that $f(\mathcal{A}) \in OG_4(G^{'})$.

(2) Case 1: Suppose $\mathcal{B} \in OG_1(G^{'})$. Then Propositions 4.11 in [18], 2.13 in [20] and 5.8 in [19], $f^{-1}(\mathbf{B}) \in IVG(G)$, $f^{-1}(B) \in IFG(G)$, $f^{-1}(\mu) \in FG(G)$. Thus, $f^{-1}(\mathcal{B}) \in OG_1(G)$.

Case 2: Suppose $\mathcal{B} \in OG_2(G^{'})$ and let $x \in G$. Since $f^{-1}(\mathbf{B}) \in IVG(G)$, $f^{-1}(B) \in IFG(G)$ by Case 1, it is sufficient to prove that $f^{-1}(\mu)(x^{-1}) \leq f^{-1}(\mu)(x)$. Then

$$f^{-1}(\mu)(x^{-1}) = \mu(f(x^{-1})) = \mu(f(x)^{-1}) \leq \mu(f(x)) = f^{-1}(\mu)(x).$$

Thus, by Theorem 6 (2), $f(\mathcal{A}) \in OG_2(G^{'})$.

Case 3: Suppose $\mathcal{B} \in OG_3(G^{'})$ and let $x \in G$. Since $f^{-1}(\mathbf{B}) \in IVG(G)$, $f^{-1}(\mu) \in FG(G)$ by Case 1, it is enough to show that $[f^{-1}(A)]^{\in}(x^{-1}) \leq [f^{-1}(A)]^{\in}(x)$ and $[f^{-1}(A)]^{\notin}(x^{-1}) \geq [f^{-1}(A)]^{\notin}(x)$. Then

$$[f^{-1}(A)]^{\in}(x^{-1}) = A^{\in}(f(x^{-1})) = A^{\in}(f(x)^{-1}) \leq A^{\in}(f(x)) = [f^{-1}(A)]^{\in}(x).$$

Similarly, we have $[f^{-1}(A)]^{\notin}(x^{-1}) \geq [f^{-1}(A)]^{\notin}(x)$. Thus, by Theorem 6 (3), $f(\mathcal{A}) \in OG_3(G^{'})$.

Case 4: Suppose $\mathcal{B} \in OG_3(G^{'})$. Then from Cases (2), (3) and Theorem 6 (4), we can easily prove that $f(\mathcal{A}) \in OG_4(G^{'})$. This completes the proof. □

From Propositions 4.16 and 4.17 in [18], 2.18 and 2.19 in [20], and Theorems 2.1 and 2.2 in [22], we have the following result.

Theorem 9. *If $\mathcal{A} \in OG_1(G)$, then $[\mathcal{A}]_{\tilde{\tilde{a}}}$ is a subgroup of G, for each $\tilde{\tilde{a}} \in [I] \times (I \oplus I) \times I$ such that $\tilde{\tilde{a}} \leq_1 \mathcal{A}(e)$. Conversely, if $\mathcal{A} \in \mathcal{O}(G)$ such that $[\mathcal{A}]_{\tilde{\tilde{a}}}$ is a subgroup of G, for each $\tilde{\tilde{a}} \in [I] \times (I \oplus I) \times I$ such that $\tilde{\tilde{a}} \leq_1 \mathcal{A}(e)$, then $\mathcal{A} \in OG_1(G)$.*

Theorem 10. *Let G_p be the cyclic group of prime order p. Then $\mathcal{A} \in OG_i(G_p)$ if and only if $\mathcal{A}(x) = \mathcal{A}(1) \leq_i \mathcal{A}(0)$, for $i = 1,\ 2,\ 3,\ 4$.*

Proof. We prove only for $i = 1,\ 2$ and the proofs are omitted for $i = 3,\ 4$.

For $i = 1$, from Propositions 4.12 in [18], 2.14 in [20] and 5.10 in [19], we have for each $0 \neq x \in G_p$,

$$\mathbf{A} \in IVG(G_p) \text{ iff } \mathbf{A}^{-}(x) = \mathbf{A}^{-}(1) \leq \mathbf{A}^{-}(0) \text{ and } \mathbf{A}^{+}(x) = \mathbf{A}^{+}(1) \leq \mathbf{A}^{+}(0), \tag{7}$$

$$A \in IFG(G_p) \text{ iff } A^{\in}(x) = A^{\in}(1) \leq A^{\in}(0) \text{and } A^{\notin}(x) = A^{\notin}(1) \geq A^{\notin}(0), \tag{8}$$

$$\lambda \in FG(G_p) \text{ iff } \lambda(x) = \lambda(1) \leq \lambda(0). \tag{9}$$

Thus, by Theorem 6 (1), $\mathcal{A} \in OG_1(G_p)$ iff $\mathcal{A}(x) = \mathcal{A}(1) \leq_i \mathcal{A}(0)$.

For $i = 2$, suppose $\mathcal{A} \in OG_2(G_p)$ and let $y \in G_p$. Then by Theorem 6 (2), $\lambda(xy) \leq \lambda(x) \vee \lambda(y)$. Since G_p is the cyclic group of prime order p, $G_p = \{0, 1, 2, \ldots, p-1\}$. Since x is the sum of $i's$ and i is the sum of $x's$, $\lambda(x) \leq \lambda(1) \leq \lambda(x)$. Thus, $\lambda(x) = \lambda(1)$. Since 0 is the identity of G_p, $\lambda(0) \leq \lambda(x)$. Thus, $\lambda(x) = \lambda(1) \geq \lambda(0)$. So $\mathcal{A}(x) = \mathcal{A}(1) \leq_2 \mathcal{A}(0)$.

Conversely, suppose $\mathcal{A}(x) = \mathcal{A}(1) \leq_2 \mathcal{A}(0)$. Then by Theorem 7, $\mathcal{A} \in OG_2(G_p)$. This completes the proof. □

Definition 23. *Let $\mathcal{A} \in \mathcal{O}(G)$ and let $i = 1,\ 2,\ 3,\ 4$. Then $\mathcal{A}$ is called an i-octahedron normal subgroup (briefly, an i-ONG) of G, if it satisfies the following conditions:*

$$\mathcal{A} \in OG_i(G) \text{ and } \mathcal{A}(xy) = \mathcal{A}(yx), \text{ for every} x,\ y \in G.$$

We will denote the set of all i-ONGs of G as $ONG_i(G)$. It is obvious that if G is abelian, then $\mathcal{A} \in ONG_i(G)$, for each $\mathcal{A} \in OG_i(G)$. Furthermore,

$$\mathcal{A} \in OG_1(G) \iff \mathbf{A} \in IVNG(G),\ A \in IFNG(G),\ \lambda \in FNG(G),$$

where $IVNG(G)$ [resp., $IFNG(G)$ and $FNG(G)$] is denoted by the set of all interval-valued fuzzy normal subgroups [resp., intuitionistic fuzzy normal subgroups and fuzzy normal subgroups or fuzzy invariant subgroups] of G (See [18] [resp., See [16,20]]).

Example 6. *Let $GL(n,R)$ be the general linear group of degree n and let I_n be the unit matrix of $GL(n,R)$. Then clearly, $GL(n,R)$ is a non abelian group. We define the interval-valued fuzzy set $\mathbf{A}$, two intuitionistic fuzzy sets A, B and two fuzzy sets λ, μ in $GL(n,R)$ as follows: for each $I_n \neq M \in GL(n,R)$,*

$$\mathbf{A}(I_n) = [1,1],\ A(I_n) = (1,0),\ B(I_n) = (0,1),\ \lambda) = 1,\ \mu(I_n) = 0,$$

$$\mathbf{A}(M) = \begin{cases} \left[\frac{1}{5}, \frac{2}{3}\right] & \text{if } M \text{ is not a triangular matrix} \\ \left[\frac{1}{3}, \frac{1}{2}\right] & \text{if } M \text{ is a triangular matrix,} \end{cases}$$

$$A(M) = \begin{cases} \left[\frac{2}{3}, \frac{1}{5}\right] & \text{if } M \text{ is not a triangular matrix} \\ \left[\frac{1}{2}, \frac{1}{3}\right] & \text{if } M \text{ is a triangular matrix,} \end{cases}$$

$$A(M) = \begin{cases} \left[\frac{1}{5}, \frac{2}{3}\right] & \text{if } M \text{ is not a triangular matrix} \\ \left[\frac{1}{3}, \frac{1}{2}\right] & \text{if } M \text{ is a triangular matrix,} \end{cases}$$

$$\lambda(M) = \begin{cases} \frac{2}{3} & \text{if } M \text{ is not a triangular matrix} \\ \frac{1}{3} & \text{if } M \text{ is a triangular matrix,} \end{cases}$$

$$\mu(M) = \begin{cases} \frac{1}{3} & \text{if } M \text{ is not a triangular matrix} \\ \frac{2}{3} & \text{if } M \text{ is a triangular matrix.} \end{cases}$$

Then we can easily check that that the followings hold:

$$\langle \mathbf{A}, A, \lambda \rangle \in ONG_1(GL(n,R)),\ \langle \mathbf{A}, A, \mu \rangle \in ONG_2(GL(n,R)),$$

$$\langle \mathbf{A}, B, \lambda \rangle \in ONG_3(GL(n,R)),\ \langle \mathbf{A}, B, \mu \rangle \in ONG_4(GL(n,R)).$$

From Propositions 5.2 in [18], 3.2 in [20] and 2.1 (i) in [16], and Remark 1 (1), we have the following.

Proposition 22. *Let $\mathcal{A} \in \mathcal{O}(G)$ and let $\mathcal{B} \in ONG_1(G)$. Then $\mathcal{A} \circ_1 \mathcal{B} = \mathcal{B} \circ_1 \mathcal{A}$.*

Also from Propositions 5.3 in [18], 3.3 in [20] and 2.1 (ii) in [16], and Remark 1 (1), we have the following.

Proposition 23. *Let $\mathcal{A} \in ONG_1(G)$. if $\mathcal{B} \in OG_1(G)$, then $\mathcal{B} \circ_1 \mathcal{A} \in OG_1(G)$.*

Proposition 24. *If $\mathcal{A} \in ONG_1(G)$, then $G_{\mathcal{A}}$ is a normal subgroup of G.*

Proof. From Propositions 5.4 in [18], 3.5 in [20], 2.2 (ii) in [16] and Proposition 18, the proof is clear. □

It is obvious that A is a normal subgroup of G, then $\chi_A \in ONG_1(G)$ and $G_{\chi_A} = A$.

Definition 24. *Let $\mathcal{A} \in ONG_1(G)$. Then the quotient group $G/G_{\mathcal{A}}$ is called the octahedron quotient group (briefly, OQG) of G with respect to $\mathcal{A}$.*

Now let $\pi : G \to G/G_{\mathcal{A}}$ be the natural projection.

Proposition 25. *If $\mathcal{A} \in ONG_1(G)$ and $\mathcal{B} \in \mathcal{O}(G)$, then $\pi^{-1}(\pi(\mathcal{B})) = G/G_{\mathcal{A}} \circ_1 \mathcal{B}$.*

Proof. From Propositions 5.6 in [18], 3.7 in [20] and 2.3 (ii) in [16], the proof is obvious. □

5. Octahedron Ideals

Definition 25. *Let $(R, +, \cdot)$ be a ring and let $\mathcal{A} \in \mathcal{O}(R)$. Then*

(i) $\ddot{0} \neq \mathcal{A}$ is called a 1-octahedron subring of R, if it satisfies the following conditions: for every $x, y \in R$,

(a) $\mathcal{A}(x+y) \geq_1 \mathcal{A}(x) \wedge^1 \mathcal{A}(y)$,

(b) $\mathcal{A}(x^{-1}) \geq_1 \mathcal{A}(x)$,

(c) $\mathcal{A}(xy) \geq_1 \mathcal{A}(x) \wedge^1 \mathcal{A}(y)$,

(ii) $\left\langle \widetilde{0}, \bar{\mathbf{0}}, 1 \right\rangle \neq \mathcal{A}$ is called a 2-octahedron subring of R, if it satisfies the following conditions: for every $x, y \in R$,

(a) $\mathcal{A}(x+y) \geq_2 \mathcal{A}(x) \wedge^2 \mathcal{A}(y)$,

(b) $\mathcal{A}(x^{-1}) \geq_2 \mathcal{A}(x)$,

(c) $\mathcal{A}(xy) \geq_2 \mathcal{A}(x) \wedge^2 \mathcal{A}(y)$,

(iii) $\left\langle \widetilde{0}, \bar{\mathbf{1}}, 0 \right\rangle \neq \mathcal{A}$ is called a 3-octahedron subring of R, if it satisfies the following conditions: for every $x, y \in R$,

(a) $\mathcal{A}(x+y) \geq_3 \mathcal{A}(x) \wedge^3 \mathcal{A}(y)$,

(b) $\mathcal{A}(x^{-1}) \geq_3 \mathcal{A}(x)$,

(c) $\mathcal{A}(xy) \geq_3 \mathcal{A}(x) \wedge^3 \mathcal{A}(y)$,

(iv) $\left\langle \widetilde{0}, \bar{\mathbf{1}}, 1 \right\rangle \neq \mathcal{A}$ is called a 4-octahedron subring of R, if it satisfies the following conditions: for every $x, y \in R$,

(a) $\mathcal{A}(x+y) \geq_4 \mathcal{A}(x) \wedge^4 \mathcal{A}(y)$,

(b) $\mathcal{A}(x^{-1}) \geq_4 \mathcal{A}(x)$,

(c) $\mathcal{A}(xy) \geq_4 \mathcal{A}(x) \wedge^4 \mathcal{A}(y)$.

We will denote the set of all i-octahedron subrings of R as $OR_i(R)$ $(i = 1, 2, 3, 4)$.

It is clear that if A is a subring of R, then $\chi_A \in OR_1(R)$.

Example 7. *Consider the ring $(\mathbb{Z}_2, +, \cdot)$, where $\mathbb{Z}_2 = \{0, 1\}$. Let us define the interval-valued fuzzy set $\mathbf{A}$, two intuitionistic fuzzy sets A, B and two fuzzy sets λ, μ in $\mathbb{Z}_2$ as follows:*

$$\mathbf{A}(0) = [0.5, 0.8],\ \mathbf{A}(1) = [0.4, 0.6],$$

$$A(0) = (0.7, 0.2),\ A(1) = (0.5, 0.3),\ B(0) = (0.6, 0.3),\ B(1) = (0.8, 0.2),$$

$$\lambda(0) = 0.8,\ \lambda(1) = 0.5,\ \mu(0) = 0.6,\ \mu(1) = 0.7.$$

Then we can easily check that the followings hold:

$$\langle \mathbf{A}, A, \lambda \rangle \in OR_1(\mathbb{Z}_2),\ \langle \mathbf{A}, A, \mu \rangle \in OR_2(\mathbb{Z}_2),$$

$$\langle \mathbf{A}, B, \lambda \rangle \in OR_3(\mathbb{Z}_2),\ \langle \mathbf{A}, B, \mu \rangle \in OR_4(\mathbb{Z}_2).$$

From the definitions of orders of two octahedron numbers and Definition 11, and Theorems 1–4 and 6, we have the following.

Theorem 11. *Let $(R,+,\cdot)$ be a ring and let $\mathcal{A} \in \mathcal{O}(R)$. Then*
(1) $\ddot{0} \neq \mathcal{A} \in OR_1(R) \iff \mathcal{A} \in OG_1((R,+)), \mathcal{A} \in OGP_1((R,\cdot))$,
(2) $\left\langle \tilde{0}, \bar{\mathbf{0}}, 1 \right\rangle \neq \mathcal{A} \in OR_2(R) \iff \mathcal{A} \in OG_2((R,+)), \mathcal{A} \in OGP_2((R,\cdot))$,
(3) $\left\langle \tilde{0}, \bar{\mathbf{0}}, 1 \right\rangle \neq \mathcal{A} \in OR_3(R) \iff \mathcal{A} \in OG_3((R,+)), \mathcal{A} \in OGP_3((R,\cdot))$,
(4) $\left\langle \tilde{0}, \bar{\mathbf{0}}, 1 \right\rangle \neq \mathcal{A} \in OR_4(R) \iff \mathcal{A} \in OG_4((R,+)), \mathcal{A} \in OGP_4((R,\cdot))$.
The following is an immediate result of Theorems 7 and 11.

Corollary 6. *Let R be a ring and let $\mathcal{A} \in \mathcal{O}(R)$. Then $\mathcal{A} \in OR_i(R)$ if and only if it satisfies the following conditions: for every $x,\ y \in R$ and for every $i = 1,\ 2,\ 3,\ 4$,*
(i) $\mathcal{A}(x-y) \geq_i \mathcal{A}(x) \wedge^i \mathcal{A}(y)$,
(ii) $\mathcal{A}(xy) \geq_i \mathcal{A}(x) \wedge^i \mathcal{A}(y)$.
The following is an immediate result of Remark 3 and Proposition 15.

Theorem 12. *Let R be a ring. Then A be a subring of R if and only if $\chi_{\mathcal{A}} \in OR_1(R)$.*

Definition 26. *Let R be a ring and let $\mathcal{A} \in OR_i(R)$ $(I = 1,\ 2,\ 3,\ 4)$. Then*
(i) $\mathcal{A}$ is called an i-octahedron left ideal (briefly, i-OLI) if for every $x,\ y \in R$,

$$\mathcal{A}(xy) \geq_i \mathcal{A}(y),$$

(ii) $\mathcal{A}$ is called an i-octahedron right ideal (briefly, i-ORI), if for every $x,\ y \in R$,

$$\mathcal{A}(xy) \geq_i \mathcal{A}(x),$$

(iii) $\mathcal{A}$ is called an i-octahedron ideal (briefly, i-OI), if it is an i-octahedron left ideal and an i-octahedron right ideal of R.
we will denote the set of all i-OLIs [resp., i-ORIs and i-OIs] in R as $OLI_i(R)$ [resp., $ORI_i(R)$ and $OI_i(R)$].

Example 8. *Consider the ring $(\mathbb{Z}_4,+,\cdot)$, where $\mathbb{Z}_4 = \{0,1,2,3\}$. Let us define an interval-valued fuzzy set* **A**, **B**, *two intuitionistic fuzzy sets A, B and two fuzzy sets $\lambda,\ \mu,\ \eta,\ \delta$ in $\mathbb{Z}_2$ as follows:*

$$\mathbf{A}(0) = [0.6,0.8],\ \mathbf{A}(1) = \mathbf{A}(3) = [0.4,0.6],\ \mathbf{A}(2) = [0.5,0.7],$$

$$A(0) = (0.7,0.2),\ A(1) = A(3) = (0.5,0.4),\ A(2) = (0.6,0.3),$$

$$B(0) = (0.5,0.4),\ B(1) = B(3) = (0.7,0.2),\ B(2) = (0.6,0.3),$$

$$\lambda(0) = 0.9,\ \lambda(1) = \lambda(3) = 0.6,\ \lambda(2) = 0.7,$$

$$\mu(0) = 0.5,\ \mu(1) = \mu(3) = 0.8,\ \mu(2) = 0.6.$$

Then we can easily check that the followings hold:

$$\langle \mathbf{A}, A, \lambda \rangle \in OLI_1(\mathbb{Z}_4),\ \langle \mathbf{A}, A, \mu \rangle \in OLI_2(\mathbb{Z}_4),$$

$$\langle \mathbf{A}, B, \lambda \rangle \in OLI_3(\mathbb{Z}_4),\ \langle \mathbf{A}, B, \mu \rangle \in OLI_4(\mathbb{Z}_4).$$

Remark 9. *(1) Let R be a ring. If $\lambda \in FLI(R)$ [resp., $FRI(R)$ and $FI(R)$], then*
$\langle [\lambda,\lambda], (\lambda,\lambda^c), \lambda \rangle \in OLI_1(R)$ *[resp., $ORI_1(R)$ and $OI_1(R)$],*
$\langle [\lambda,\lambda], (\lambda,\lambda^c), \lambda^c \rangle \in OLI_2(R)$ *[resp., $ORI_2(R)$ and $OI_2(R)$],*

$\langle[\lambda,\lambda],(\lambda^c,\lambda),\lambda\rangle \in OLI_3(R)$ *[resp.,* $ORI_3(R)$ *and* $OI_3(R)$*],*
$\langle[\lambda,\lambda],(\lambda^c,\lambda),\lambda^c\rangle \in OLI_4(R)$ *[resp.,* $ORI_4(R)$ *and* $OI_4(R)$*].*
(2) Let R be a ring. If $A \in IFLI(R)$ *[resp.,* $IFRI(R)$ *and* $IFI(R)$*], then*
$\langle[A^{\in},(A^{\notin})^c],A,A^{\in}\rangle \in OLI_1(R)$ *[resp.,* $ORI_1(R)$ *and* $OI_1(R)$*],*
$\langle[A^{\in},(A^{\notin})^c],A,(A^{\notin})^c\rangle \in OLI_2(R)$ *[resp.,* $ORI_2(R)$ *and* $OI_2(R)$*],*
$\langle[A^{\in},(A^{\notin})^c],A^c,A^{\in}\rangle \in OLI_3(R)$ *[resp.,* $ORI_3(R)$ *and* $OI_3(R)$*],*
$\langle[A^{\in},(A^{\notin})^c],A^c,(A^{\notin})^c\rangle \in OLI_4(R)$ *[resp.,* $ORI_4(R)$ *and* $OI_4(R)$*].*

The following is an immediate result of Propositions 11, 12 and 21.

Proposition 26. *Let* $f: R \to R'$ *be a ring homomorphism and let* $i = 1, 2, 3, 4$.
(1) *If* $\mathcal{A} \in OR_i(R)$ *or* $\mathcal{A} \in OLI_i(R)$ *[resp.,* $ORI_i(R)$ *and* $OI_i(R)$*], then so is* $f(\mathcal{A})$.
(2) *If* $\mathcal{B} \in OR_i(R')$ *or* $\mathcal{B} \in OLI_i(R')$ *[resp.,* $ORI_i(R')$ *and* $OI_i(R')$*], then so is* $f^{-1}(\mathcal{B})$.

The following is an immediate result of Corollary 6 and Definition 26.

Theorem 13. *Let R be a ring,* $\mathcal{A} \in \mathcal{O}(R)$ *and let* $i = 1, 2, 3, 4$. *Then* $\mathcal{A} \in OI_i(R)$ *[resp.,* $\mathcal{A} \in OLI_i(R)$ *and* $\mathcal{A} \in ORI_i(R)$*] if and only if it satisfies the following conditions: for every* $x, y \in R$,
(i) $\mathcal{A}(x-y) \geq_i \mathcal{A}(x) \wedge^i \mathcal{A}(y)$,
(ii) $\mathcal{A}(xy) \geq_i \mathcal{A}(x) \vee^i \mathcal{A}(y)$ *[resp.,* $\mathcal{A}(xy) \geq_i \mathcal{A}(y)$ *and* $\mathcal{A}(xy) \geq_i \mathcal{A}(x)$*].*

The following is an immediate result of Theorems 12 and 13.

Theorem 14. *Let R be a ring. Then A an ideal [resp., a left ideal and a right ideal] of R if and only* $\chi_A \in OI_1(R)$ *[resp.,* $OLI_1(R)$ *and* $ORI_1(R)$*].*

Theorem 15. *Let R be a skew field (also division ring) and let* $\mathcal{A} \in \mathcal{O}(R)$*, where 0 and e denote the identity for "+" and "·". Then* $\mathcal{A} \in OI_i(R)$ *[resp.,* $OLI_i(R)$ *and* $ORI_i(R)$*] if and only if for each* $0 \neq x \in R$, $\mathcal{A}(x) = \mathcal{A}(e) \leq_i \mathcal{A}(0)$ *for* $i = 1, 2, 3, 4$.

Proof. We show only that $\mathcal{A} \in OLI_i(R)$ iff for each $0 \neq x \in R$, $\mathcal{A}(x) = \mathcal{A}(e) \leq_i \mathcal{A}(0)$ for $i = 1, 2$. The remainder's proofs are omitted.

Case 1: Let $i = 1$. Then from Propositions 6.6 in [18], 4.7 in [20] and 3.3 in [16],
$\mathcal{A} \in OLI_1(R)$ iff for each $0 \neq x \in R$, $\mathcal{A}(x) = \mathcal{A}(e) \leq_1 \mathcal{A}(0)$, i.e.,

$$\mathbf{A}(x) = \mathbf{A}(e) \leq \mathbf{A}(0),\ A(x) = A(e) \leq A(0),\ \lambda(x) = \lambda(e) \leq \lambda(0). \tag{10}$$

Case 2: Let $i = 2$. Suppose $\mathcal{A} \in OLI_2(R)$ and let $0 \neq x \in R$. Then by Definition 26, $\lambda(x) = \lambda(xe) \leq \lambda(e)$ and $\lambda(e) = \lambda(x^{-1}x) \leq \lambda(x)$. Thus, $\lambda(x) = \lambda(e)$. Since $\mathcal{A} \in OLI_2(R)$, $\mathcal{A} \in OR_2(R)$. By Corollary 6 (i) and Definition 2,

$$\lambda(0) = \lambda(e-e) \leq \lambda(e) \vee \lambda(e) = \lambda(e).$$

So $\lambda(x) = \lambda(e) \geq \lambda(0)$. Hence by the first and the second parts of (5.1),

$$\mathcal{A}(x) = \mathcal{A}(e) \leq_2 \mathcal{A}(0).$$

Conversely, suppose the necessary condition holds and let $x, y \in R$. Then we have four cases:
(i) $x \neq 0,\ y \neq 0,\ x \neq y$, (ii) $x \neq 0,\ y \neq 0,\ x = y$,
(iii) $x \neq 0,\ y = 0$, (iv) $x = 0,\ y \neq 0$.

Case (i). Suppose $x \neq 0,\ y \neq 0,\ x \neq y$. Then by the hypothesis, we have

$$\mathbf{A}(x) = \mathbf{A}(y) = \mathbf{A}(e) = \mathbf{A}(x-y) = \mathbf{A}(xy),$$

$$A(x) = A(y) = A(e) = A(x-y) = A(xy),$$

$$\lambda(x) = \lambda(y) = \lambda(e) = \lambda(x-y) = \lambda(xy).$$

Thus,

$$\mathbf{A}(x-y) \geq \mathbf{A}(x) \wedge \mathbf{A}(y),\ \mathbf{A}(xy) \geq \mathbf{A}(y),$$

$$A(x-y) \geq A(x) \wedge A(y),\ A(xy) \geq A(y),$$

$$\lambda(x-y) \leq \lambda(x) \vee \lambda(y),\ \lambda(xy) \leq \lambda(y).$$

So $\mathcal{A}(x-y) \geq_2 \mathcal{A}(x) \wedge^2 \mathcal{A}(y)$ and $\mathcal{A}(xy) \geq_2 \mathcal{A}(y)$.

Case (ii). Suppose $x \neq 0,\ y \neq 0,\ x = y$. Then by the hypothesis, we have

$$\mathbf{A}(x) = \mathbf{A}(y) = \mathbf{A}(e) = \mathbf{A}(xy) = \mathbf{A}(x-y) = \mathbf{A}(0),$$

$$A(x) = A(y) = A(e) = A(xy) = A(x-y) = A(0),$$

$$\lambda(x) = \lambda(y) = \lambda(e)) = \lambda(x-y) = \lambda(xy) = \lambda(0).$$

Thus, we have the same result in Case (i):

$$\mathcal{A}(x-y) \geq_2 \mathcal{A}(x) \wedge^2 \mathcal{A}(y) \text{ and } \mathcal{A}(xy) \geq_2 \mathcal{A}(y).$$

Case (iii). Suppose $x \neq 0,\ y = 0$. Then by the hypothesis, we have

$$\mathbf{A}(x-y) = \mathbf{A}(x) = \mathbf{A}(e) \geq \mathbf{A}(x) \wedge \mathbf{A}(y),\ \mathbf{A}(xy) = \mathbf{A}(0) \geq \mathbf{A}(y)$$

$$A(x-y) = A(x) = A(e) \geq A(x) \wedge A(y),\ A(xy) = A(0) \geq A(y),$$

$$\lambda(x-y) = \lambda(x) = \lambda(e)) \leq \lambda(x) \vee \lambda(y),\ \lambda(xy) = \lambda(0) \leq \lambda(y).$$

Thus, we have the same result in Case (i):

$$\mathcal{A}(x-y) \geq_2 \mathcal{A}(x) \wedge^2 \mathcal{A}(y) \text{ and } \mathcal{A}(xy) \geq_2 \mathcal{A}(y).$$

Case (iv). Suppose $x = 0,\ y \neq 0$. Then by the similar proof to Case (iii), we have

$$\mathcal{A}(x-y) \geq_2 \mathcal{A}(x) \wedge^2 \mathcal{A}(y) \text{ and } \mathcal{A}(xy) \geq_2 \mathcal{A}(y).$$

Hence in either cases, by Theorem 13, $\mathcal{A} \in OLI_2(R)$. □

Remark 10. *Theorem 15 shows that an i-OLI (ORI) is an i-OI in a skew field.*

The following gives a characteristic of a (usual) field by a 1-OI.

Proposition 27. *Let R be a commutative ring with a unity e. Suppose for each $\mathcal{A} \in OI_1(R)$,*

$$\mathcal{A}(x) = \mathcal{A}(e) \leq_1 \mathcal{A}(0) \text{ for each } 0 \neq x \in R.$$

Then R is a field.

Proof. Let A be an ideal of R such that $A \neq R$. Then clearly by Theorem 14, $\chi_A \in OI_1(R)$ such that $A \neq \ddot{1}$. Thus, there is $y \in R$ such that $y \notin A$. So $\chi_A(y) = \ddot{0}$. By the hypothesis, $\chi_A(x) = \chi_A(e) \leq_1 \chi_A(0)$. Hence $\chi_A(0) = \ddot{1}$, i.e., $A = \{0\}$. Therefore R is a field. □

6. Conclusions

By using the i-product of two octahedron sets, we introduce the concept of i-octahedron subgroupoids of a groupoid. In particular, we obtain four characterizations of i-octahedron groupoids. Also, we defined an i-OLI [resp., i-ORI and i-OI]] of a groupoid and investigated some of their properties. Moreover, we obtain some properties for the image and preimage of an i-octahedron subgroupoid [resp., i-OLI, i-ORI and i-OI] under groupoid homomorphism. Next, we define i-octahedron subgroups of a group and study some of their properties. In particular, we obtain two characterizations of i-octahedron subgroups and i-OLI [resp., i-ORI and i-OI] of a group. We introduce the concepts of i-octahedron subrings [resp., i-OLIs, i-ORIs and i-OIs] of a ring and obtain their characterizations. Furthermore, we found a sufficient condition for which a commutative ring with a unity e is a field.

In the future, we expect that one applies octahedron sets to BCI/BCK-algebras, topologies, decision-making, measures and entropy measures, etc.

Author Contributions: Created and conceptualized ideas, J.-G.L. and K.H.; writing—original draft preparation, J.-G.L. and K.H.; writing—review and editing, Y.B.J.; funding acquisition, J.-G.L. All authors have read and agreed to the published version of the manuscript.

Funding: This research was supported by Basic Science Research Program through the National Research Foundation of Korea (NRF) funded by the Ministry of Education (2018R1D1A1B07049321).

Acknowledgments: We are very grateful to the reviewers for their careful reading and their meaningful suggestions.

Conflicts of Interest: The authors declare no conflict of interest.

Abbreviations

The following abbreviations are used in this manuscript:

$\mathcal{O}(X)$	the set of all octahedron sets in X
$\mathcal{O}_P(X)$	the set of all octahedron points in X
$FGP(X)$	the set of all fuzzy subgroupoids in a groupoid X
$IFGP(X)$	the set of all intuitionistic fuzzy subgroupoids in a groupoid X
$IVGP(X)$	the set of all interval-valued fuzzy subgroupoids in a groupoid X
$OGP_i(X)$	the set of all i-octahedron subgroupoids in X
i-OLI	i-octahedron left ideal
i-ORI	i-octahedron right ideal
i-OI	i-octahedron ideal
$OLI_i(X)$	the set of all i-OLIs of X
$ORI_i(X)$	the set of all i-ORIs of X
$OI_i(X)$	the set of all i-OIs of X
$FI(X)$	the set of all fuzzy ideals of X
$FLI(X)$	the set of all fuzzy left ideals of X
$FRI(X)$	the set of all fuzzy right ideals of X
IVI	interval-valued fuzzy ideals
IVLI	interval-valued fuzzy left ideal
$IVRI$	interval-valued fuzzy right ideal
$IVI(X)$	the set of all IVIs of X
$IVLI(X)$	the set of all IVLIs of X
$IVRI(X)$	the set of all IVRIs of X
IFI	intuitionistic fuzzy ideal
$IFLI$	intuitionistic fuzzy left ideal

$IFRI$	intuitionistic fuzzy right ideal
$IFI(X)$	the set of all IFIs of X
$IFLI(X)$	the set of all IFLIs of X
$IFRI(X)$	the set of all IFRIs of X
$FGP(G)$	the set of all fuzzy subgroupoids in a group G
$IFGP(G)$	the set of all intuitionistic fuzzy subgroupoids in a group G
$OGP_i(X)$	the set of all i-octahedron groupoids in a groupoid X
$FG(G)$	the set of fuzzy subgroups of G
IFG	intuitionistic fuzzy subgroup
$IFG(G)$	the set of IFGs of G
IVG	interval-valued fuzzy subgroup
$IVG(G)$	the set of IVGs of G
$OG_i(G)$	the set of all i-octahedron subgroups of G
i-ONG	i-octahedron normal subgroup
$ONG_i(G)$	the set of all i-ONGs of G
$IVNG(G)$	the set of all interval-valued fuzzy normal subgroups
$IFNG(G)$	the set of all intuitionistic fuzzy normal subgroups
$FNG(G)$	the set of all fuzzy normal subgroups
$OR_i(R)$	the set of all i-octahedron subrings of R

References

1. Zadeh, L.A. Fuzzy sets. *Inf. Control* **1965**, *8*, 338–353. [CrossRef]
2. Zadeh, L.A. The concept of a linguistic variable and its application to approximate reasoning-I. *Inform. Sci.* **1975**, *8*, 199–249. [CrossRef]
3. Atanassov, K.T. Intuit, ionistic fuzzy sets. *Fuzzy Sets Syst.* **1986**, *20*, 87–96. [CrossRef]
4. Jun, Y.B.; Kim, C.S.; Yang, K.O. Cubic sets. *Ann. Fuzzy Math. Inform.* **2012**, *4*, 83–98.
5. Jun, Y.B. A novel extension of cubic sets and its applications in BCI/BCK-algebras. *Ann. Fuzzy Math. Inform.* **2017**, *14*, 475–486.
6. Jun, Y.B.; Khan, A. Cubic ideals in semigroups. *Honam Math. J.* **2013**, *35*, 607–623. [CrossRef]
7. Jun, Y.B.; Jung, S.T.; Kim, M.S. Cubic subgroups. *Ann. Fuzzy Math. Inform.* **2011**, *2*, 9–15.
8. Kaur, G.; Garg, H. Multi-attribute decision-making based on bonferroni mean operators under cubic intuitionistic fuzzy environment. *Entropy* **2018**, *20*, 65. [CrossRef]
9. Kaur, G.; Garg, H. Cubic intuitionistic fuzzy aggregation operators. *Int. J. Uncertain. Quantif.* **2018**, *8*, 405–428.
10. Lee, J.G.; Senel, G.; Lim, P.K.; Kim, J.; Hur, K. Octahedron sets. *Ann. Fuzzy Math. Inform.* **2020**, *19*, 211–238.
11. Cheong, M.; Hur, K. Intuitionistic interval-valued fuzzy sets. *J. Korean Inst. Intell. Syst.* **2010**, *20*, 864–874. [CrossRef]
12. Mondal, T.K.; Samanta, S.K. Topology of interval-valued fuzzy sets. *Indian J. Pure Appl. Math.* **1999**, *30*, 133–189.
13. Kim, J.; Lim, P.K.; Lee, J.G.; Hur, K. Cubic relations. *Ann. Fuzzy Math. Inform.* **2020**, *19*, 21–43. [CrossRef]
14. Kim, J.; Cheong, M.; Lee, J.G.; Hur, K.; Mostafa, S.M. Interval-valued intuitionistic cubic structures of medial ideals on BCI-algebras. *Ann. Fuzzy Math. Inform.* **2019**, *19*, 109–125.
15. Gorzalczany, M.B. A method of inference in approximate reasoning based on interval-valued fuzzy sets. *Fuzzy Sets Syst.* **1987**, *21*, 1–17. [CrossRef]
16. Liu, W.J. Fuzzy invariant subgroups and fuzzy ideals. *Fuzzy Sets Syst.* **1982**, *8*, 133–139. [CrossRef]
17. Hur, K.; Jang, S.Y.; Kang, H.W. Intutionistic fuzzy subgroupoids. *Int. J. Fuzzy Log. Intell. Syst.* **2002**, *2*, 92–147.
18. Kang, H.W.; Hur, K. Interval-valued fuzzy subgroups and rings. *Honam Math. J.* **2010**, *32*, 593–617. [CrossRef]
19. Rosenfeld, A. Fuzzy groups. *J. Math. Anal. Appl.* **1071**, *35*, 512–517. [CrossRef]
20. Hur, K.; Kang, H.W.; Song, H.K. Intutionistic fuzzy subgroups and subrings. *Honam Math. J.* **2003**, *25*, 19–41.

21. Biswas, R. Rosenfeld's fuzzy subgroups with interval-valued membership functions. *Fuzzy Sets Syst.* **1995**, *63*, 87–90. [CrossRef]
22. Das Sivaramakrishna, P. Fuzzy groups and level subgroups. *J. Math. Anal. Appl.* **1981**, *84*, 264–269.

mathematics

MDPI

Article

Strong Tolerance and Strong Universality of Interval Eigenvectors in a Max-Łukasiewicz Algebra

Martin Gavalec [1,*], Zuzana Němcová [1] and Ján Plavka [2]

1 Faculty of Informatics and Management, University of Hradec Králové, 50003 Hradec Králové, Czech Republic; zuzana.nemcova@uhk.cz

2 Faculty of Electrical Engineering and Informatics, Technical University of Košice, 04200 Košice, Slovakia; jan.plavka@tuke.sk

* Correspondence: martin.gavalec@uhk.cz; Tel.: +420-493-332-248

Received: 9 August 2020; Accepted: 2 September 2020; Published: 4 September 2020

Abstract: The Łukasiewicz conjunction (sometimes also considered to be a logic of absolute comparison), which is used in multivalued logic and in fuzzy set theory, is one of the most important t-norms. In combination with the binary operation 'maximum', the Łukasiewicz t-norm forms the basis for the so-called max-Łuk algebra, with applications to the investigation of systems working in discrete steps (discrete events systems; DES, in short). Similar algebras describing the work of DES's are based on other pairs of operations, such as max-min algebra, max-plus algebra, or max-T algebra (with a given t-norm, T). The investigation of the steady states in a DES leads to the study of the eigenvectors of the transition matrix in the corresponding max-algebra. In real systems, the input values are usually taken to be in some interval. Various types of interval eigenvectors of interval matrices in max-min and max-plus algebras have been described. This paper is oriented to the investigation of strong, strongly tolerable, and strongly universal interval eigenvectors in a max-Łuk algebra. The main method used in this paper is based on max-Ł linear combinations of matrices and vectors. Necessary and sufficient conditions for the recognition of strong, strongly tolerable, and strongly universal eigenvectors have been found. The theoretical results are illustrated by numerical examples.

Keywords: max-Łukasiewicz algebra; interval matrix; interval eigenvector; strong interval eigenvector

MSC: 90C15

---

1. Introduction

A max-Łukasiewicz algebra (max-Łuk algebra, in short), is one of the so-called max-T fuzzy algebras, which are defined for various triangular norms.

A max-T fuzzy algebra contains values in the unit interval $\mathcal{I} = \langle 0, 1 \rangle$ and uses the binary operation of maximum and one of the triangular norms, T, instead of the conventional operations of addition and multiplication. Thus, by a max-T fuzzy algebra we understand a triplet $(\mathcal{I}, \oplus, \otimes_T)$, where $\mathcal{I}$ is the interval $\langle 0, 1 \rangle$ and $\oplus = \max$, $\otimes_T = T$ are binary operations on $\mathcal{I}$. The symbol $\mathcal{I}(m, n)$, respectively, $\mathcal{I}(n)$, denotes the set of all matrices (respectively, vectors) of the given dimensions over $\mathcal{I}$. The operations $\oplus$, $\otimes_T$ are extended to matrices and vectors in the standard way. The linear ordering on $\mathcal{I}$ induces partial orderings on $\mathcal{I}(m, n)$ and $\mathcal{I}(n)$.

The triangular norms (t-norms, in short) were introduced in [1], in the context of probabilistic metric spaces. The t-norms are interpreted as the conjunction in multi-valued fuzzy logics, or as the intersection of fuzzy sets. These functions are used in many fields, such as statistics and game theory, information and data fusion, decision making support, risk management, and probability theory. The t-norms (and the corresponding t-conorms) play an important role in fuzzy set theory. Many t-norms can be found in [2]).

The Łukasiewicz norm is often considered to be a logic of absolute (or metric) comparison. The Łukasiewicz conjunction is defined by

$$x \otimes_L y = \max\{x + y - 1, 0\}. \tag{1}$$

The simplest norm is the Gödel norm, and the conjunction is defined as the minimum of the entries: the truth degrees of the constituents. Gödel logic is considered to be a logic of relative comparison.

$$x \otimes_G y = \min(x, y) \tag{2}$$

In the particular case when $T = \min$ is the Gödel t-norm, we get an important max-min algebra which has useful applications to optimization and scheduling problems. Max-min algebras belong to the so-called tropical mathematics, with a wide scope of applications and interesting contributions to mathematical theory. Several monographs [3–6] and collections of papers [7–13] have dealt with tropical mathematics and related problems.

Tropical algebras can be naturally used for the study of systems working in discrete time (DES). The state of the system at time t is described by a vector $x(t)$. The transitions of the system from one state to another are described by the transition matrix A. The next state $x(t+1)$ is obtained by multiplying the transition matrix and the state vector; in matrix notation we write $A \otimes x(t) = x(t+1)$. When a DES reaches a steady state, after some time of operation, then the state vectors of the steady states are eigenvectors of A. In any tropical algebra, the eigenproblem for a given matrix $A \in \mathcal{I}(n,n)$ consists of finding an eigenvalue $\lambda \in \mathcal{I}$ and an eigenvector $x \in \mathcal{I}(n)$ fulfilling $A \otimes x = \lambda \otimes x$.

The eigenproblem in tropical algebra has been described in many papers, see [14]. Interesting results describing the structure of the eigenspace and several algorithms for computing the largest eigenvector of a given matrix have been published, for example, in [15,16]. The eigenvectors in a max-T algebra, for various triangular norms T, are useful in fuzzy set theory. Such eigenvectors have been studied in [17–19]. The eigenvalues and eigenvectors are interesting characteristics of the DES in fuzzy algebras. The eigenspace structures for the drastic and t-norm have been studied in [18,19]. Finally, [17] describes the case of Łukasiewicz fuzzy algebra.

2. Strong Types of Interval Eigenvectors in Max-Łuk Algebras

The investigation in this paper will be started by a simple numerical example.

Example 1. *(Numerical illustration: Steady state vector).*

Assume

$$A = \begin{pmatrix} 0.5 & 0.4 & 0.1 & 0.1 & 0.5 \\ 0.8 & 0.8 & 0.2 & 0.1 & 0.7 \\ 0.6 & 0.6 & 0.6 & 0.2 & 0 \\ 0.4 & 0.5 & 0.5 & 0.2 & 0.4 \\ 0.3 & 0.1 & 0.4 & 0 & 0.8 \end{pmatrix}, \quad x = \begin{pmatrix} 0.5 \\ 0.7 \\ 0.5 \\ 0.4 \\ 0.8 \end{pmatrix}, \quad \lambda = 0.8\,.$$

Then

$$A \otimes_L x = \begin{pmatrix} 0.3 \\ 0.5 \\ 0.3 \\ 0.2 \\ 0.6 \end{pmatrix} = 0.8 \otimes_L x\,.$$

That is, x is a max-Łuk eigenvector of A with the eigenvalue $\lambda = 0.8$.

In practical applications, the matrix entries usually are not exact numbers, but are contained in some intervals. Interval arithmetic is an efficient way to represent matrix operations on a computer. Similarly, matrices and vectors with interval coefficients are studied in a max-Łuk algebra (or a max-min algebra, or some other tropical algebra), see [9,17,20–23]. The classification of various types of the interval eigenvectors in a max-min algebra has been investigated in [24,25].

Let n be a given natural number. We define $N = \{1, 2, \ldots, n\}$. Similarly to [21,25–27], we define the interval matrix with bounds $\underline{A}, \overline{A} \in \mathcal{I}(n,n)$ and the interval vector with bounds $\underline{x}, \overline{x} \in \mathcal{I}(n)$ as

$$[\underline{A}, \overline{A}] = \{\, A \in \mathcal{I}(n,n);\ \underline{A} \le A \le \overline{A} \,\}, \quad [\underline{x}, \overline{x}] = \{\, x \in \mathcal{I}(n);\ \underline{x} \le x \le \overline{x} \,\}.$$

Let us assume that an interval matrix $\mathbf{A} = [\underline{A}, \overline{A}]$ and an interval vector $\mathbf{X} = [\underline{x}, \overline{x}]$ have been fixed. The interval max-Łuk eigenproblem for $\mathbf{A}$ and $\mathbf{X}$ aims at recognizing whether $A \otimes_L x = \lambda \otimes_L x$ holds true for $A \in \mathbf{A}$, $x \in \mathbf{X}$, $\lambda \in \mathcal{I}$, with suitable quantifiers (e.g., for all $A \in \mathbf{A}$, for some $A \in \mathbf{A}$, for all $x \in \mathbf{X}$, for some $x \in \mathbf{X}$) and their various combinations. Various types of interval max-Łuk eigenvectors are defined, using various choices of quantifiers and their order (see [25] for the further classification types).

Definition 1. *Assume that an interval matrix* $\mathbf{A}$ *and an interval vector* $\mathbf{X}$ *are given. Then,* $\mathbf{X}$ *is called:*

- *A* strong max-Łuk eigenvector *of* $\mathbf{A}$
  if $(\exists \lambda \in \mathcal{I})(\forall A \in \mathbf{A})(\forall x \in \mathbf{X})[\, A \otimes_L x = \lambda \otimes_L x]$;
- *A* strongly tolerable max-Łuk eigenvector *of* $\mathbf{A}$
  if $(\exists \lambda \in \mathcal{I})(\exists A \in \mathbf{A})(\forall x \in \mathbf{X})[\, A \otimes_L x = \lambda \otimes_L x]$;
- *A* strongly universal max-Łuk eigenvector *of* $\mathbf{A}$
  if $(\exists \lambda \in \mathcal{I})(\exists x \in \mathbf{X})(\forall A \in \mathbf{A})[\, A \otimes_L x = \lambda \otimes_L x]$.

Remark 1. *In general, an interval vector* $\mathbf{X}$ *is called a* tolerable *max-Łuk eigenvector of* $\mathbf{A}$ *if there is an eigenvalue* $\lambda \in \mathcal{I}$ *such that every* $x \in \mathbf{X}$ *preserves the state vector up to a multiple by* λ*, for some* $A \in \mathbf{A}$ *(in other words: A* tolerates *x with eigenvalue* λ*).*

In the case when there is one common *tolerating matrix* $A \in \mathbf{A}$ *for all of the vectors* $x \in \mathbf{X}$*, the interval vector* $\mathbf{X}$ *is called* strongly tolerable. *Otherwise, the tolerating matrix A depends on x. If also* the eigenvalue λ *depends on x, then the interval eigenvector* $\mathbf{X}$ *is usually called* weakly tolerable.

Remark 2. *Similarly, an interval vector* $\mathbf{X}$ *is called a* universal *max-Łuk eigenvector of* $\mathbf{A}$ *if there is an eigenvalue* $\lambda \in \mathcal{I}$ *such that for every* $A \in \mathbf{A}$*, some* $x \in \mathbf{X}$ *preserves the state vector up to a multiple by* λ*, (in other words: x is* universal *for A with eigenvalue* λ*).*

In the case when there is one common *universal vector* $x \in \mathbf{X}$ *for all matrices* $A \in \mathbf{A}$*, the interval vector* $\mathbf{X}$*, as well as the common universal x, are called* strongly universal. *Otherwise, the universal vector x depends on A. If also* the eigenvalue λ *depends on A, then the interval eigenvector* $\mathbf{X}$ *is usually called* weakly universal.

In this paper, we study in more detail the strong max-Łuk interval eigenvectors, the strongly tolerable and strongly universal max-Łuk interval eigenvectors (the remaining types of max-Łuk interval eigenvector are not considered here). Necessary and sufficient conditions are described for recognizing whether a given interval vector is a strong (strongly tolerable, strongly universal) eigenvector of a given interval matrix in a max-Łuk algebra. The results are illustrated by numerical examples.

3. Strong Interval Eigenvectors in a Max-Łuk Algebra

In this section, we assume that an interval matrix $\mathbf{A} = [\underline{A}, \overline{A}]$ and an interval vector $\mathbf{X} = [\underline{x}, \overline{x}]$ are given. For each pair $i, j \in N$, define $\tilde{A}^{(ij)} \in \mathcal{I}(n,n)$ and $\tilde{x}^{(i)} \in \mathcal{I}(n)$ by putting for every $k, l \in N$,

$$\tilde{a}_{kl}^{(ij)} = \begin{cases} \overline{a}_{ij}, & \text{for } k = i, l = j \\ \underline{a}_{kl}, & \text{otherwise} \end{cases}, \qquad \tilde{x}_k^{(i)} = \begin{cases} \overline{x}_i, & \text{for } k = i \\ \underline{x}_k, & \text{otherwise} \end{cases}.$$

It is shown in the following lemma that every $A \in \mathbf{A}$ can be written as a max-Łuk linear combination of *generators* $\tilde{A}^{(ij)}$ with $i, j \in N$. Similarly, every $x \in \mathbf{X}$ is equal to a max-Łuk linear combination of generators $\tilde{x}^{(i)}$ with $i \in N$.

Lemma 1. *Let $x \in \mathcal{I}(n)$ and $A \in \mathcal{I}(n,n)$. Then,*

(i) $x \in \mathbf{X}$ *if and only if* $x = \bigoplus_{i \in N} \beta_i \otimes_L \tilde{x}^{(i)}$ *for some* $\beta_i \in \mathcal{I}$ *with* $\underline{x}_i - \overline{x}_i + 1 \leq \beta_i \leq 1$,

(ii) $A \in \mathbf{A}$ *if and only if* $A = \bigoplus_{i,j \in N} \alpha_{ij} \otimes_L \tilde{A}^{(ij)}$ *for some* $\alpha_{ij} \in \mathcal{I}$ *with* $\underline{a}_{ij} - \overline{a}_{ij} + 1 \leq \alpha_{ij} \leq 1$.

Proof. For the proof of statement (i), assume that $x \in \mathbf{X}$: that is, $\underline{x}_i \leq x_i \leq \overline{x}_i$ for every $i \in N$. Put $\beta_i = x_i - \overline{x}_i + 1$ for each $i \in N$. It is easy to see that the β_i's satisfy the inequalities in assertion (i). Moreover, for every $j \in N$

$$\begin{aligned} \left(\bigoplus_{i \in N} \beta_i \otimes_L \tilde{x}^{(i)}\right)_j &= \bigoplus_{i \in N} \left((x_i - \overline{x}_i + 1) \otimes_L \tilde{x}^{(i)}\right)_j \\ &= \left((x_j - \overline{x}_j + 1) \otimes_L \tilde{x}_j^{(j)}\right) \oplus \bigoplus_{i \in N \setminus \{j\}} \left((x_i - \overline{x}_i + 1) \otimes_L \tilde{x}_j^{(i)}\right) \\ &= \left((x_j - \overline{x}_j + 1) \otimes_L \overline{x}_j\right) \oplus \bigoplus_{i \in N \setminus \{j\}} \left((x_i - \overline{x}_i + 1) \otimes_L \underline{x}_j\right). \end{aligned} \tag{3}$$

In particular, $(x_j - \overline{x}_j + 1) \otimes_L \overline{x}_j = (x_j - \overline{x}_j + 1) + \overline{x}_j - 1 = x_j$, since $x_j \geq 0$. On the other hand, for $i \neq j$ we have $(x_i - \overline{x}_i + 1) \otimes_L \underline{x}_j = (x_i - \overline{x}_i + 1) + \underline{x}_j - 1 \leq x_j$, because $x_i - \overline{x}_i \leq 0$.

For the converse implication, assume that $\underline{x}_i - \overline{x}_i + 1 \leq \beta_i \leq 1$ for every $i \in N$ and $x = \bigoplus_{i \in N} \beta_i \otimes_L \tilde{x}^{(i)}$. For every $j \in N$

$$\begin{aligned} x_j &= \left(\bigoplus_{i \in N} \beta_i \otimes_L \tilde{x}^{(i)}\right)_j \leq \left(\bigoplus_{i \in N} 1 \otimes_L \tilde{x}^{(i)}\right)_j \\ &= \bigoplus_{i \in N} \tilde{x}_j^{(i)} = \overline{x}_j \oplus \bigoplus_{i \in N \setminus \{j\}} \underline{x}_j = \overline{x}_j, \end{aligned} \tag{4}$$

$$\begin{aligned} x_j &= \left(\bigoplus_{i \in N} \beta_i \otimes_L \tilde{x}^{(i)}\right)_j \geq \bigoplus_{i \in N} \left((\underline{x}_i - \overline{x}_i + 1) \otimes_L \tilde{x}^{(i)}\right)_j \\ &= \left((\underline{x}_j - \overline{x}_j + 1) + \overline{x}_j - 1\right) \oplus \bigoplus_{i \in N \setminus \{j\}} \left((\underline{x}_i - \overline{x}_i + 1) + \underline{x}_j - 1\right) \\ &\geq \underline{x}_j \oplus \bigoplus_{i \in N \setminus \{j\}} (\underline{x}_i - \overline{x}_i + \underline{x}_j) = \underline{x}_j. \end{aligned} \tag{5}$$

We have shown that $\underline{x} \leq x \leq \overline{x}$. That is, $x \in [\underline{x}, \overline{x}]$. The proof of (ii) is analogous. □

Theorem 1. *The interval vector $\mathbf{X} = [\underline{x}, \overline{x}]$ is a strong max-Łuk eigenvector of the interval matrix $\mathbf{A} = [\underline{A}, \overline{A}]$ if and only if there exists $\lambda \in \mathcal{I}$, $\lambda > 0$, such that for every $i \in N$.*

$$\underline{A} \otimes_L \tilde{x}^{(i)} = \lambda \otimes_L \tilde{x}^{(i)}, \tag{6}$$

$$\overline{A} \otimes_L \tilde{x}^{(i)} = \lambda \otimes_L \tilde{x}^{(i)}. \tag{7}$$

Proof. Assume that $\lambda \in \mathcal{I}$ fulfills conditions (6) and (7), and that $x \in \mathbf{X}$ is given. Then x is a max-Łuk linear combination $x = \bigoplus_{i\in N} \beta_i \otimes_L \tilde{x}^{(i)}$ for some coefficients $\beta_i \in \mathcal{I}, i \in N$ with $\underline{x}_i - \overline{x}_i + 1 \le \beta_i \le 1$, according to Lemma 1(i). In view of (6) we get

$$\begin{aligned} \underline{A} \otimes_L x = \underline{A} \otimes_L \left(\bigoplus_{i\in N} \beta_i \otimes_L \tilde{x}^{(i)} \right) &= \bigoplus_{i\in N} \left(\underline{A} \otimes_L \beta_i \otimes_L \tilde{x}^{(i)} \right) \\ = \bigoplus_{i\in N} \beta_i \otimes_L \left(\underline{A} \otimes_L \tilde{x}^{(i)} \right) &= \bigoplus_{i\in N} \beta_i \otimes_L \left(\lambda \otimes_L \tilde{x}^{(i)} \right) \\ = \lambda \otimes_L \bigoplus_{i\in N} \left(\beta_i \otimes_L \tilde{x}^{(i)} \right) &= \lambda \otimes_L x. \end{aligned} \tag{8}$$

Using (7) we analogously get

$$\underline{A} \otimes_L x = \overline{A} \otimes_L \left(\bigoplus_{i\in N} \beta_i \otimes_L \tilde{x}^{(i)} \right) = \lambda \otimes_L \bigoplus_{i\in N} \left(\beta_i \otimes_L \tilde{x}^{(i)} \right) = \lambda \otimes_L x. \tag{9}$$

From (8) and (9) it easily follows that

$$\begin{aligned} \lambda \otimes_L x = \underline{A} \otimes_L x \le A \otimes_L x \le \overline{A} \otimes_L x = \lambda \otimes_L x, \\ A \otimes_L x = \lambda \otimes_L x, \end{aligned} \tag{10}$$

for every $A \in \mathbf{A}$. That is, $\mathbf{X}$ is a strong max-Łuk eigenvector of $\mathbf{A}$. The converse implication is trivial. □

Example 2. *(Numerical illustration: Strong max-Luk eigenvector)*
Assume lower and upper bounds for $A \in [\underline{A}, \overline{A}]$ and for $x \in [\underline{x}, \overline{x}]$

$$\underline{A} = \begin{pmatrix} 0.6 & 0.2 & 0.1 & 0.1 & 0.5 \\ 0.2 & 0.2 & 0.1 & 0 & 0.4 \\ 0.1 & 0.3 & 0.2 & 0.4 & 0.2 \\ 0.3 & 0.5 & 0.4 & 0.6 & 0.2 \\ 0 & 0.1 & 0.3 & 0 & 0.3 \end{pmatrix}, \quad \overline{A} = \begin{pmatrix} 0.6 & 0.6 & 0.3 & 0.1 & 0.6 \\ 0.2 & 0.6 & 0.3 & 0.1 & 0.6 \\ 0.5 & 0.9 & 0.6 & 0.4 & 0.9 \\ 0.7 & 1 & 0.8 & 0.6 & 0.9 \\ 0.2 & 0.6 & 0.3 & 0.1 & 0.6 \end{pmatrix}$$

$$\underline{x} = \begin{pmatrix} 0 \\ 0 \\ 0.7 \\ 0.9 \\ 0 \end{pmatrix}, \quad \overline{x} = \begin{pmatrix} 0.8 \\ 0.4 \\ 0.7 \\ 0.9 \\ 0.4 \end{pmatrix}.$$

Then

$$\tilde{x}^{(1)} = \begin{pmatrix} 0.8 \\ 0 \\ 0.7 \\ 0.9 \\ 0 \end{pmatrix}, \quad \tilde{x}^{(2)} = \begin{pmatrix} 0 \\ 0.4 \\ 0.7 \\ 0.9 \\ 0 \end{pmatrix}, \quad \tilde{x}^{(3)} = \begin{pmatrix} 0 \\ 0 \\ 0.7 \\ 0.9 \\ 0 \end{pmatrix}, \quad \tilde{x}^{(4)} = \begin{pmatrix} 0 \\ 0 \\ 0.7 \\ 0.9 \\ 0 \end{pmatrix}, \quad \tilde{x}^{(5)} = \begin{pmatrix} 0 \\ 0 \\ 0.7 \\ 0.9 \\ 0.4 \end{pmatrix}.$$

The following equations hold for $\lambda = 0.6$

$$\underline{A} \otimes_L \tilde{x}^{(1)} = \begin{pmatrix} 0.4 \\ 0 \\ 0.3 \\ 0.5 \\ 0 \end{pmatrix} = \overline{A} \otimes_L \tilde{x}^{(1)} = 0.6 \otimes_L \tilde{x}^{(1)},$$

$$\underline{A} \otimes_L \tilde{x}^{(2)} = \begin{pmatrix} 0 \\ 0 \\ 0.3 \\ 0.5 \\ 0 \end{pmatrix} = \overline{A} \otimes_L \tilde{x}^{(2)} = 0.6 \otimes_L \tilde{x}^{(2)},$$

$$\underline{A} \otimes_L \tilde{x}^{(3)} = \begin{pmatrix} 0 \\ 0 \\ 0.3 \\ 0.5 \\ 0 \end{pmatrix} = \overline{A} \otimes_L \tilde{x}^{(3)} = 0.6 \otimes_L \tilde{x}^{(3)},$$

$$\underline{A} \otimes_L \tilde{x}^{(4)} = \begin{pmatrix} 0 \\ 0 \\ 0.3 \\ 0.5 \\ 0 \end{pmatrix} = \overline{A} \otimes_L \tilde{x}^{(4)} = 0.6 \otimes_L \tilde{x}^{(4)},$$

$$\underline{A} \otimes_L \tilde{x}^{(5)} = \begin{pmatrix} 0 \\ 0 \\ 0.3 \\ 0.5 \\ 0 \end{pmatrix} = \overline{A} \otimes_L \tilde{x}^{(5)} = 0.6 \otimes_L \tilde{x}^{(5)}.$$

Hence, $\mathbf{X} = [\underline{x}, \overline{x}]$ *is a strong max-Łuk eigenvector of A with the eigenvalue* $\lambda = 0.6$.

Theorem 1 leads to the following recognition problem: given **A** and **X**, recognize whether there is, or is no value $\lambda \in \mathcal{I} \setminus \{0\}$ such that (6) and (7) hold for every $k \in N$. If the answer is positive, then find all (or at least one) such values.

If $i, k \in N$, then we write, for brevity,

$$z_{ik} = \left(\underline{A} \otimes_L \tilde{x}^{(i)}\right)_k, \quad z'_{ik} = \left(\overline{A} \otimes_L \tilde{x}^{(i)}\right)_k. \tag{11}$$

Furthermore, we write

$$Z^0 = \{(i,k) \in N \times N;\ z_{ik} = 0\} \tag{12}$$

$$Z^> = \{(i,k) \in N \times N;\ z_{ik} > 0\} \tag{13}$$

$$I = \left(0, \min_{(i,k) \in N \times N} \left(1 - \tilde{x}_k^{(i)}\right)\right\rangle \tag{14}$$

Theorem 2. *The interval vector* $\mathbf{X} = [\underline{x}, \overline{x}]$ *is a strong max-Łuk eigenvector of the interval matrix* $\mathbf{A} = [\underline{A}, \overline{A}]$ *if and only if*

(i) $(\forall (i,k) \in N \times N)\quad z_{ik} = z'_{ik}$,

(ii) $(\forall (i,k) \in Z^0)\quad \tilde{x}_k^{(i)} < 1$,

(iii) $(\forall (i,k),(j,l) \in Z^>)\quad \tilde{x}_k^{(i)} - z_{ik} = \tilde{x}_l^{(j)} - z_{jl} < 1$,

(iv) $(\forall (i,k) \in Z^>, (j,l) \in Z^0)\quad \tilde{x}_l^{(j)} \leq \tilde{x}_k^{(i)} - z_{ik}$.

Proof. Assume that $\mathbf{X}$ is a strong interval eigenvector of $\mathbf{A}$. That is, there exists $\lambda \in \mathcal{I}, \lambda > 0$ fulfilling conditions (6) and (7). The statement (i) then follows immediately. For $(i,k) \in Z^0$ we have $z_{ik} = 0$, which gives $\lambda \otimes_L \tilde{x}_k^{(i)} = 0$, in view of (6). Then, by definition of $\otimes_L$, we have $\lambda + \tilde{x}_k^{(i)} - 1 \leq 0$, which implies $\lambda \leq 1 - \tilde{x}_k^{(i)}$. Now, statement (ii) easily follows, in view of the assumption that $0 < \lambda$.

For $(i,k),(j,l) \in Z^>$, we have $z_{ik} > 0$, which gives $\lambda \otimes_L \tilde{x}_k^{(i)} > 0$, in view of (6). Consequently, $z_{ik} = \lambda + \tilde{x}_k^{(i)} - 1$. That is, $\lambda = -\tilde{x}_k^{(i)} + z_{ik} + 1$. Similarly, $\lambda = -\tilde{x}_l^{(j)} + z_{jl} + 1$. In view of the assumption that $\lambda > 0$, we get (iii) by a simple computation. Finally, assume $(i,k) \in Z^0, (j,l) \in Z^>$. By the same arguments as above, we get $\lambda = -\tilde{x}_k^{(i)} + z_{ik} + 1 \leq 1 - \tilde{x}_l^{(j)}$. Then (iv) follows directly.

For the converse implication, assume that statements (i)–(iv) hold. We shall show that then a $\lambda > 0$ can be found such that (6) and (7) are satisfied. We distinguish two cases.

Case 1. $Z^> = \emptyset$. Then $Z^0 = N \times N$, and by (ii) we have $0 < 1 - \tilde{x}_k^{(i)}$ for every $(i,k) \in N \times N$. That is, the interval I is non-empty. Choose an arbitrary $\lambda \in I$. Then, for every $(i,k) \in N \times N$, we have $\lambda \leq 1 - \tilde{x}_k^{(i)}$ which gives $\lambda + \tilde{x}_k^{(i)} - 1 \leq 0$. That is, $\lambda \otimes_L \tilde{x}_k^{(i)} = 0 = z_{ik}$. As (i,k) is arbitrary, (6) has been demonstrated. Then, (7) follows by statement (i) .

Case 2. $Z^> \neq \emptyset$. Let $(i,k) \in Z^>$ be fixed. By (iii) we have $\tilde{x}_k^{(i)} - z_{ik} < 1$, which gives $0 < z_{ik} + 1 - \tilde{x}_k^{(i)}$. Choosing $\lambda = z_{ik} + 1 - \tilde{x}_k^{(i)}$, we get $\lambda > 0$ and $z_{ik} = \lambda + \tilde{x}_k^{(i)} - 1$. Then, the assumption that $(i,k) \in Z^>$ implies $z_{ik} > 0$ and $\lambda + \tilde{x}_k^{(i)} - 1 > 0$. That is, $\lambda + \tilde{x}_k^{(i)} - 1 = \lambda \otimes_L \tilde{x}_k^{(i)}$, which implies $z_{ik} = \lambda \otimes_L \tilde{x}_k^{(i)}$.

Consider an arbitrary $(j,l) \in Z^>$. We have $\lambda = z_{ik} + 1 - \tilde{x}_k^{(i)} = z_{jl} + 1 - \tilde{x}_l^{(j)}$, in view of (iii). That is, $z_{jl} = \lambda \otimes_L \tilde{x}_l^{(j)}$, similarly as above. On the other hand, for every $(j,l) \in Z^0$ we have $\tilde{x}_l^{(j)} \leq \tilde{x}_k^{(i)} - z_{ik}$, in view of (iv). Consequently, we get $z_{ik} + 1 - \tilde{x}_k^{(i)} \leq 1 - \tilde{x}_l^{(j)}$. That is, $\lambda \leq 1 - \tilde{x}_l^{(j)}$, which gives $\lambda + \tilde{x}_l^{(j)} - 1 \leq 0$. This implies $\lambda \otimes_L \tilde{x}_l^{(j)} = 0$, i.e., $z_{jl} = \lambda \otimes_L \tilde{x}_l^{(j)}$. As $(j,l) \in Z^> \cup Z^0 = N \times N$ was arbitrary, we have shown that (6) is satisfied. By (i), (7) holds as well. □

Remark 3. The proof of Theorem 2 contains a description of the set $S(\mathbf{A},\mathbf{X}) = \{\lambda > 0;\ (\forall\, A \in \mathbf{A})(\forall\, x \in \mathbf{X})\ A \otimes_L x = \lambda \otimes_L x\}$. Namely

(i) if some of statements (i)–(iv) in Theorem 2 are not satisfied, then $S(\mathbf{A},\mathbf{X}) = \emptyset$,

(ii) if $(i,k) \in Z^>$, then $S(\mathbf{A},\mathbf{X}) = \{\lambda\} = \{z_{ik} + 1 - \tilde{x}_k^{(i)}\}$,

(iii) if $Z^> = \emptyset$, then $S(\mathbf{A},\mathbf{X}) = \left(0, \min_{i,k \in N} (1 - \tilde{x}_k^{(i)})\right\rangle$.

Theorem 3. *The recognition problem of whether a given interval vector* $\mathbf{X}$ *is a strong max-Łuk eigenvector of the interval matrix* $\mathbf{A}$ *is solvable in* $O(n^3)$ *time.*

Proof. According to Theorem 2, the problem can be solved by verifying conditions (i)–(iv). Each of them can be verified in $O(n^3)$ time. Therefore, the computational complexity is $O(n^3)$. □

4. Strongly Tolerable Interval Eigenvectors in a Max-Łuk Algebra

Theorem 4. *The interval vector* $\mathbf{X} = [\underline{\mathbf{x}}, \overline{\mathbf{x}}]$ *is a strongly tolerable max-Łuk eigenvector of the interval matrix* $\mathbf{A} = [\underline{A}, \overline{A}]$ *if and only if there exist an* $A \in \mathbf{A}$ *and* $\lambda \in \mathcal{I}$ *such that*

$$A \otimes_L \tilde{x}^{(k)} = \lambda \otimes_L \tilde{x}^{(k)} \quad \text{for every } k \in N. \tag{15}$$

Proof. Let us assume that $A \in \mathbf{A}$ and $\lambda \in \mathcal{I}$ fulfill condition (15). If $x \in \mathcal{I}(n)$ is an arbitrary vector in $\mathbf{X}$, then x is a max-Łuk linear combination $x = \bigoplus_{k\in N} \beta_k \otimes_L \tilde{x}^{(k)}$ for some coefficients $\beta_k \in \mathcal{I}, k \in N$ with $\underline{x}_k - \overline{x}_k + 1 \le \beta_i \le 1$. According to Lemma 1 (i),

$$\begin{aligned} A \otimes_L x &= A \otimes_L \left(\bigoplus_{k\in N} \beta_k \otimes_L \tilde{x}^{(k)}\right) = \bigoplus_{k\in N} \left(A \otimes_L \beta_k \otimes_L \tilde{x}^{(k)}\right) \\ &= \bigoplus_{k\in N} \beta_k \otimes_L \left(A \otimes_L \tilde{x}^{(k)}\right) = \bigoplus_{k\in N} \beta_k \otimes_L \left(\lambda \otimes_L \tilde{x}^{(k)}\right) \\ &= \lambda \otimes_L \bigoplus_{k\in N} \left(\beta_k \otimes_L \tilde{x}^{(k)}\right) = \lambda \otimes_L x. \end{aligned} \quad (16)$$

By (3), $\mathbf{X}$ is a strongly tolerable eigenvector of $\mathbf{A}$. The converse implication follows immediately. □

Remark 4. *The property (15) can be briefly expressed in words: A is a λ-*certificate *for the strong tolerance max-Łuk problem* $(\mathbf{A}, \mathbf{X})$.

Example 3. *(Numerical illustration: Strongly tolerable max-Luk eigenvector)*
Assume the lower and upper bounds for $A \in [\underline{A}, \overline{A}]$ and for $x \in [\underline{x}, \overline{x}]$ are

$$\underline{A} = \begin{pmatrix} 0.2 & 0.2 & 0.2 & 0.3 & 0 \\ 0.1 & 0.3 & 0.2 & 0.3 & 0 \\ 0 & 0.1 & 0.2 & 0.3 & 0.1 \\ 0.1 & 0.2 & 0.1 & 0.2 & 0.2 \\ 0.1 & 0.1 & 0.2 & 0.1 & 0.4 \end{pmatrix}, \quad \overline{A} = \begin{pmatrix} 0.8 & 0.8 & 0.9 & 0.7 & 0.5 \\ 0.5 & 0.8 & 0.9 & 0.8 & 0.5 \\ 0.5 & 0.7 & 0.5 & 0.5 & 0.8 \\ 0.5 & 0.5 & 0.5 & 0.8 & 0.8 \\ 0.5 & 0.2 & 0.9 & 0.9 & 0.8 \end{pmatrix}$$

$$\underline{x} = \begin{pmatrix} 0.6 \\ 0 \\ 0.6 \\ 0 \\ 0.7 \end{pmatrix}, \quad \overline{x} = \begin{pmatrix} 0.8 \\ 0.7 \\ 0.6 \\ 0.4 \\ 0.7 \end{pmatrix}.$$

Then

$$\tilde{x}^{(1)} = \begin{pmatrix} 0.8 \\ 0 \\ 0.6 \\ 0 \\ 0.7 \end{pmatrix}, \quad \tilde{x}^{(2)} = \begin{pmatrix} 0.6 \\ 0.7 \\ 0.6 \\ 0 \\ 0.7 \end{pmatrix}, \quad \tilde{x}^{(3)} = \begin{pmatrix} 0.6 \\ 0 \\ 0.6 \\ 0 \\ 0.7 \end{pmatrix}, \quad \tilde{x}^{(4)} = \begin{pmatrix} 0.6 \\ 0 \\ 0.6 \\ 0.4 \\ 0.7 \end{pmatrix}, \quad \tilde{x}^{(5)} = \begin{pmatrix} 0.6 \\ 0 \\ 0.6 \\ 0 \\ 0.7 \end{pmatrix}.$$

For $\lambda = 0.7$ and for given $A_1 \in \mathbf{A}$

$$A_1 = \begin{pmatrix} 0.7 & 0.6 & 0.7 & 0.6 & 0.1 \\ 0.2 & 0.7 & 0.4 & 0.6 & 0.1 \\ 0.2 & 0.2 & 0.3 & 0.4 & 0.6 \\ 0.2 & 0.3 & 0.4 & 0.7 & 0.3 \\ 0.2 & 0.1 & 0.8 & 0.8 & 0.4 \end{pmatrix}$$

the following equations hold

$$A_1 \otimes_L \tilde{x}^{(1)} = \begin{pmatrix} 0.7 & 0.6 & 0.7 & 0.6 & 0.1 \\ 0.2 & 0.7 & 0.4 & 0.6 & 0.1 \\ 0.2 & 0.2 & 0.3 & 0.4 & 0.6 \\ 0.2 & 0.3 & 0.4 & 0.7 & 0.3 \\ 0.2 & 0.1 & 0.8 & 0.8 & 0.4 \end{pmatrix} \otimes \begin{pmatrix} 0.8 \\ 0 \\ 0.6 \\ 0 \\ 0.7 \end{pmatrix} = \begin{pmatrix} 0.5 \\ 0 \\ 0.3 \\ 0 \\ 0.4 \end{pmatrix} = 0.7 \otimes_L \tilde{x}^{(1)},$$

$$A_1 \otimes_L \tilde{x}^{(2)} = \begin{pmatrix} 0.7 & 0.6 & 0.7 & 0.6 & 0.1 \\ 0.2 & 0.7 & 0.4 & 0.6 & 0.1 \\ 0.2 & 0.2 & 0.3 & 0.4 & 0.6 \\ 0.2 & 0.3 & 0.4 & 0.7 & 0.3 \\ 0.2 & 0.1 & 0.8 & 0.8 & 0.4 \end{pmatrix} \otimes \begin{pmatrix} 0.6 \\ 0.7 \\ 0.6 \\ 0 \\ 0.7 \end{pmatrix} = \begin{pmatrix} 0.3 \\ 0.4 \\ 0.3 \\ 0 \\ 0.4 \end{pmatrix} = 0.7 \otimes_L \tilde{x}^{(2)},$$

$$A_1 \otimes_L \tilde{x}^{(3)} = \begin{pmatrix} 0.7 & 0.6 & 0.7 & 0.6 & 0.1 \\ 0.2 & 0.7 & 0.4 & 0.6 & 0.1 \\ 0.2 & 0.2 & 0.3 & 0.4 & 0.6 \\ 0.2 & 0.3 & 0.4 & 0.7 & 0.3 \\ 0.2 & 0.1 & 0.8 & 0.8 & 0.4 \end{pmatrix} \otimes \begin{pmatrix} 0.6 \\ 0 \\ 0.6 \\ 0 \\ 0.7 \end{pmatrix} = \begin{pmatrix} 0.3 \\ 0 \\ 0.3 \\ 0 \\ 0.4 \end{pmatrix} = 0.7 \otimes_L \tilde{x}^{(3)},$$

$$A_1 \otimes_L \tilde{x}^{(4)} = \begin{pmatrix} 0.7 & 0.6 & 0.7 & 0.6 & 0.1 \\ 0.2 & 0.7 & 0.4 & 0.6 & 0.1 \\ 0.2 & 0.2 & 0.3 & 0.4 & 0.6 \\ 0.2 & 0.3 & 0.4 & 0.7 & 0.3 \\ 0.2 & 0.1 & 0.8 & 0.8 & 0.4 \end{pmatrix} \otimes \begin{pmatrix} 0.6 \\ 0 \\ 0.6 \\ 0.4 \\ 0.7 \end{pmatrix} = \begin{pmatrix} 0.3 \\ 0 \\ 0.3 \\ 0.1 \\ 0.4 \end{pmatrix} = 0.7 \otimes_L \tilde{x}^{(4)},$$

$$A_1 \otimes_L \tilde{x}^{(5)} = \begin{pmatrix} 0.7 & 0.6 & 0.7 & 0.6 & 0.1 \\ 0.2 & 0.7 & 0.4 & 0.6 & 0.1 \\ 0.2 & 0.2 & 0.3 & 0.4 & 0.6 \\ 0.2 & 0.3 & 0.4 & 0.7 & 0.3 \\ 0.2 & 0.1 & 0.8 & 0.8 & 0.4 \end{pmatrix} \otimes \begin{pmatrix} 0.6 \\ 0 \\ 0.6 \\ 0 \\ 0.7 \end{pmatrix} = \begin{pmatrix} 0.3 \\ 0 \\ 0.3 \\ 0 \\ 0.4 \end{pmatrix} = 0.7 \otimes_L \tilde{x}^{(5)},$$

Hence, $\mathbf{X} = [\underline{x}, \overline{x}]$ *is a strongly tolerable max-Łuk eigenvector of* $\mathbf{A}$ *with the eigenvalue* $\lambda = 0.7$*, and* A_1 *is the* λ*-certificate for the strong tolerance max-Łuk problem* $(\mathbf{A}, \mathbf{X})$.

Remark 5. *In general, not every matrix* $A \in \mathbf{A}$ *is a* λ*-certificate for* $(\mathbf{A}, \mathbf{X})$*, for some* λ*. Take, e.g.,*

$$A_2 = \begin{pmatrix} 0.8 & 0.8 & 0.9 & 0.7 & 0.5 \\ 0.5 & 0.8 & 0.9 & 0.8 & 0.5 \\ 0.5 & 0.7 & 0.5 & 0.5 & 0.8 \\ 0.5 & 0.5 & 0.5 & 0.8 & 0.8 \\ 0.5 & 0.2 & 0.9 & 0.9 & 0.8 \end{pmatrix}$$

and $\tilde{x}^{(1)}$, $\tilde{x}^{(2)}$, $\tilde{x}^{(3)}$, $\tilde{x}^{(4)}$, $\tilde{x}^{(5)}$ *from Example 3. Then*

$$A_2 \otimes_L \tilde{x}^{(1)} = A_2 \otimes \begin{pmatrix} 0.8 \\ 0 \\ 0.6 \\ 0 \\ 0.7 \end{pmatrix} = \begin{pmatrix} 0 \\ 0 \\ 0 \\ 0 \\ 0.1 \end{pmatrix} \neq \lambda \otimes_L \tilde{x}^{(1)}.$$

It is easy to see that the equality in the last position cannot hold for any $\lambda \in \mathcal{I}$*. That is,* A_2 *is not a* λ*-certificate in Example 3 with any* $\lambda \in \mathcal{I}$.

In Example 3, the certificate A_1 was given. Now the question arises of how to find a certificate (or to show that no certificate exists) for a given instance $(\mathbf{A}, \mathbf{X})$. In other words, how do we recognize whether or not $\mathbf{X}$ is a strongly tolerable interval eigenvector of $\mathbf{A}$?

A method for solving the strong tolerance interval eigenproblem in a max-Łuk algebra for instances with a natural additional condition is described in the rest of this section. We start with a simple lemma.

Lemma 2. *Assume $u, v, w \in \mathcal{I}$.*

(i) *If $u + v + w > 2$, then $(u \otimes_L v) \otimes_L w = u + v + w - 2 > 0$,*

(ii) *If $u + v + w \le 2$, then $(u \otimes_L v) \otimes_L w = 0$.*

Proof. Let $u + v + w > 2$. Then $u + v > 2 - w \ge 2 - 1 = 1$, since $0 \le w \le 1$. Hence, $u \otimes_L v = u + v - 1 > 0$, and $(u \otimes_L v) \otimes_L w = (u + v - 1) + w - 1 = u + v + w - 2 > 0$.

On the other hand, if $u + v + w \le 2$, then $(u + v - 1) + (w - 1) \le 0$. We consider two subcases.

Subcase 1. Suppose $u + v - 1 > 0$. Then $u \otimes_L v = u + v - 1$ and $u \otimes_L v + w - 1 \le 0$. Hence, $(u \otimes_L v) \otimes_L w = 0$.

Subcase 2. Suppose $u + v - 1 \le 0$. Then $u \otimes_L v = 0$ and $u \otimes_L v + w - 1 = 0 + w - 1 \le 0$, since $w \le 1$. That is, $(u \otimes_L v) \otimes_L w = 0$. □

Remark 6. *It is easy to see directly from the definition that the Łukasiewicz conjunction $\otimes_L$ is commutative. As a consequence of Lemma 2, $\otimes_L$ is associative, as well.*

Namely due to Lemma 2(i), we have, for any $u, v, w \in \mathcal{I}$ with $u + v + w > 2$, that $(u \otimes_L v) \otimes_L w = u + v + w - 2$, and, by the commutative law, $u \otimes_L (v \otimes_L w) = (v \otimes_L w) \otimes_L u = v + w + u - 2$. That is, $(u \otimes_L v) \otimes_L w = u \otimes_L (v \otimes_L w)$. Similar reasoning is used when $u + v + w \le 2$.

To recognize the existence of a certificate $A \in \mathbf{A}$ satisfying the conditions (15) from Theorem 4, the unknown A will be written as a max-Łuk linear combination of generators $\tilde{A}^{(ij)}$ as in Lemma 1 (ii).

The coefficients in the linear combination will be found as the solution to a system of max-Łuk linear equations with parameter λ, and the variables $\alpha_{(ij)}$ in the bounds $\underline{a}_{ij} = \underline{\alpha}_{(ij)} \le \alpha_{(ij)} \le \overline{\alpha}_{(ij)} = \overline{a}_{ij}$. The form of the system will require that every solution of the system, for some parameter value λ, gives coefficients for such a max-Łuk linear combination of generators which is a λ-certificate matrix for the given instance. Then the recognition of strong tolerability is equivalent to the recognition of whether there is a value λ for which the system is solvable. On the other hand, if the system is unsolvable for every $\lambda \in \mathcal{I}$, then no certificate exists for the given instance.

Formally, we consider the bounded max-Łuk linear system

$$\tilde{C} \otimes_L \alpha = \lambda \otimes_L \tilde{b} \tag{17}$$

$$\underline{\alpha} \le \alpha \le \overline{\alpha} \tag{18}$$

with parameter $\lambda \in \mathcal{I}$, where the columns $\tilde{C}^{(ij)}$ of $\tilde{C} \in \mathcal{I}(n^2, n^2)$ are constructed blockwise from $\tilde{A}^{(ij)} \otimes \tilde{x}^{(k)}$, $k \in N$. The right-hand side vector $\tilde{b} \in \mathcal{I}(n^2)$ is constructed blockwise from the generators $\tilde{x}^{(k)}$ for $k \in N$, and the bounds $\underline{\alpha}$, $\overline{\alpha} \in \mathcal{I}(n^2)$ for the variable vector $\alpha \in \mathcal{I}(n^2)$ are constructed from the columns of $\underline{A}$, $\overline{A}$, according to Lemma 1 (ii). That is, we have

$$\tilde{C}^{(ij)} = \begin{pmatrix} \tilde{A}^{(ij)} \otimes_L \tilde{x}^{(1)} \\ \tilde{A}^{(ij)} \otimes_L \tilde{x}^{(2)} \\ \vdots \\ \tilde{A}^{(ij)} \otimes_L \tilde{x}^{(n)} \end{pmatrix}, \quad \tilde{b} = \begin{pmatrix} \tilde{x}^{(1)} \\ \tilde{x}^{(2)} \\ \vdots \\ \tilde{x}^{(n)} \end{pmatrix}, \tag{19}$$

$$\underline{\alpha}_{(ij)} = \underline{a}_{ij} - \overline{a}_{ij} + 1 \le \alpha_{(ij)} \le 1 = \overline{\alpha}_{(ij)}. \tag{20}$$

Theorem 5. *The interval vector* $\mathbf{X} = [\underline{x}, \overline{x}]$ *is a strongly tolerable eigenvector of the interval matrix* $\mathbf{A} = [\underline{A}, \overline{A}]$ *if and only if there is a* $\lambda \in \mathcal{I}$ *such that the linear system* (17) *and* (18), *has a solution* $\alpha \in \mathcal{I}(n^2)$. *In the positive case, the max-Łuk linear combination*

$$A = \bigoplus_{(ij) \in N \times N} \alpha_{(ij)} \otimes_L \tilde{A}^{(ij)} \tag{21}$$

is a λ*-certificate for the given instance.*

Proof. Assume that there exists a $\lambda \in \mathcal{I}$ such that α satisfies (17), (18) with (19), (20). Then, $A \in \mathcal{I}(n,n)$ as defined in (21) belongs to $[\underline{A}, \overline{A}]$, in view of Lemma 1(ii). Moreover, we have the following block equations, for every $k \in N$

$$\bigoplus_{i,j \in N} \left(\tilde{A}^{(ij)} \otimes_L \tilde{x}^{(k)}\right) \otimes_L \alpha_{(ij)} = \lambda \otimes_L \tilde{x}^{(k)}, \tag{22}$$

$$\left(\bigoplus_{i,j \in N} \alpha_{(ij)} \otimes_L \tilde{A}^{(ij)}\right) \otimes_L \tilde{x}^{(k)} = \lambda \otimes_L \tilde{x}^{(k)}, \tag{23}$$

$$A \otimes_L \tilde{x}^{(k)} = \lambda \otimes_L \tilde{x}^{(k)}. \tag{24}$$

We will prove that the block Equations (22) and (23) are equivalent. In particular, we show that the left-hand sides of (22) and (23) in every row h and in every block row k are equal.

Assume $k, h \in N$ are fixed. Then

$$\left(\bigoplus_{i,j \in N} \left(\tilde{A}^{(ij)} \otimes_L \tilde{x}^{(k)}\right) \otimes_L \alpha_{(ij)}\right)_h = \bigoplus_{i,j \in N} \left(\tilde{A}^{(ij)} \otimes_L \tilde{x}^{(k)}\right)_h \otimes_L \alpha_{(ij)} \tag{25}$$

$$= \bigoplus_{i,j \in N} \left(\bigoplus_{g \in N} \tilde{A}^{(ij)}_{hg} \otimes_L \tilde{x}^{(k)}_g\right) \otimes_L \alpha_{(ij)} = \bigoplus_{i,j \in N} \left(\bigoplus_{g \in N} \left(\tilde{A}^{(ij)}_{hg} \otimes_L \tilde{x}^{(k)}_g\right) \otimes_L \alpha_{(ij)}\right) \tag{26}$$

$$= \bigoplus_{i,j \in N} \left(\bigoplus_{g \in N} \alpha_{(ij)} \otimes_L \left(\tilde{A}^{(ij)}_{hg} \otimes_L \tilde{x}^{(k)}_g\right)\right) = \bigoplus_{i,j \in N} \left(\bigoplus_{g \in N} \left(\alpha_{(ij)} \otimes_L \tilde{A}^{(ij)}_{hg}\right) \otimes_L \tilde{x}^{(k)}_g\right) \tag{27}$$

$$= \bigoplus_{i,j \in N} \left(\alpha_{(ij)} \otimes_L \tilde{A}^{(ij)}\right)_h \otimes_L \tilde{x}^{(k)} = \left(\bigoplus_{i,j \in N} \left(\alpha_{(ij)} \otimes_L \tilde{A}^{(ij)}\right) \otimes_L \tilde{x}^{(k)}\right)_h \tag{28}$$

Please note that the associative law has been used in (27). That is,

$$\alpha_{(ij)} \otimes_L \left(\tilde{A}^{(ij)}_{hg} \otimes_L \tilde{x}^{(k)}_g\right) = \left(\alpha_{(ij)} \otimes_L \tilde{A}^{(ij)}_{hg}\right) \otimes_L \tilde{x}^{(k)}_g,$$

according to Remark 6. The remaining equalities (25), (26) and (28) are consequences of standard arithmetic rules in max-Łuk algebras.

Now, in view of the fact that (22) means that $\alpha \in \mathcal{I}(n^2)$ is a solution of (19), while (23) says that (21) satisfies (15), we obtain, due to Theorem 4 that $\mathbf{X}$ is a strongly tolerable eigenvector of $\mathbf{A}$. The converse implication follows from the converse implication in Theorem 4. □

Theorem 5 reduces the recognition problem of whether $\mathbf{X}$ is a strongly tolerable eigenvector of $\mathbf{A}$ to the solvability problem of the bounded parametric system (17), (18) with dimension $n^2 \times n^2$ for some $\lambda \in \mathcal{I}$. The latter problem is a particular case of the bounded parametric solvability problem with general dimension $m \times n$. The recognition algorithm can be briefly described by the following steps (for details and notation, see [28]):

(i) permute the equations in the system so that the right-hand side will be decreasing, that is

$$0 \le 1 - b_1 \le 1 - b_2 \le \cdots \le 1 - b_m \le 1, \tag{29}$$

(ii) recognize the solvability for some λ with $1 - b_m < \lambda \le 1$, according to [28]/Theorem 3 (case a), by verifying $C \otimes_L y^\star(\lambda^m_{\max}) = \lambda^m_{\max} \otimes_L b$,

(iii) recognize the solvability for some λ with $0 \le \cdots \le 1 - b_h < \lambda \le 1 - b_{h+1} \le \ldots 1$, according to [28]/Theorem 4 (case b), by verifying $C \otimes_L y^\star(\lambda^h_{\max}) = \lambda^h_{\max} \otimes_L b$. This step may be repeated, if necessary, with different indices $h \le m$,

(iv) recognize the solvability for some λ with $0 \le \lambda \le 1 - b_1$, according to [28]/Theorem 5 (case c), by verifying $\underline{y}_j \le \bigwedge_{i \in M}(1 - c_{ij})$, for every $j \in N$.

(v) the system is solvable if the answer is positive at least once in steps 2, 3 or 4. Otherwise, the system is unsolvable for any value of λ.

Theorem 6. *The recognition problem of whether a given interval vector* **X** *is a strongly tolerable eigenvector of a given interval matrix* **A** *in a max-Łuk algebra is solvable in* $O(n^6)$ *time.*

Proof. According to [28], the parametric solvability problem with dimension $m \times n$ has the computational complexity $O(m\, n^2)$. Therefore, the computational complexity of the strong tolerance problem with dimension $n^2 \times n^2$ is $O\big((n^2)^3\big) = O(n^6)$. □

Example 4. *(Numerical illustration: Computing a certificate)*

Assume that the lower and upper bounds for $A \in [\underline{A}, \overline{A}]$ *and* $x \in [\underline{x}, \overline{x}]$ *in the interval eigenproblem are*

$$\underline{A} = \begin{pmatrix} 0.9 & 0.7 & 0.6 \\ 0.7 & 0.9 & 0.6 \\ 0.8 & 0.8 & 0.9 \end{pmatrix}, \qquad \overline{A} = \begin{pmatrix} 1 & 0.8 & 0.8 \\ 0.7 & 0.9 & 0.8 \\ 1 & 0.9 & 1 \end{pmatrix}$$

$$\underline{x} = \begin{pmatrix} 0.7 \\ 0.8 \\ 0.9 \end{pmatrix}, \qquad \overline{x} = \begin{pmatrix} 0.9 \\ 0.8 \\ 0.9 \end{pmatrix}.$$

If we wish to recognize whether **X** *is a strongly tolerable max-Łuk eigenvector of* **A**, *then, according to Theorem 4, we must recognize the existence of a* λ*-certificate for* $(\mathbf{A}, \mathbf{X})$. *In view of Theorem 5, we must recognize the solvability of the max-Łuk linear system* $\tilde{C} \otimes_L \alpha = \lambda \otimes_L \tilde{b}$ *with bounds* $\underline{\alpha} \le \alpha \le \overline{\alpha}$, *for some* $\lambda \in \mathcal{I}$. *The vector (matrix) generators are*

$$\tilde{x}^{(1)} = \begin{pmatrix} 0.9 \\ 0.8 \\ 0.9 \end{pmatrix}, \qquad \tilde{x}^{(2)} = \begin{pmatrix} 0.7 \\ 0.8 \\ 0.9 \end{pmatrix}, \qquad \tilde{x}^{(3)} = \begin{pmatrix} 0.7 \\ 0.8 \\ 0.9 \end{pmatrix}.$$

$$\tilde{A}^{(11)} = \begin{pmatrix} 1 & 0.7 & 0.6 \\ 0.7 & 0.9 & 0.6 \\ 0.8 & 0.8 & 0.9 \end{pmatrix}, \qquad \tilde{A}^{(12)} = \begin{pmatrix} 0.9 & 0.8 & 0.6 \\ 0.7 & 0.9 & 0.6 \\ 0.8 & 0.8 & 0.9 \end{pmatrix},$$

$$\tilde{A}^{(13)} = \begin{pmatrix} 0.9 & 0.7 & 0.8 \\ 0.7 & 0.9 & 0.6 \\ 0.8 & 0.8 & 0.9 \end{pmatrix}, \qquad \tilde{A}^{(21)} = \begin{pmatrix} 0.9 & 0.7 & 0.6 \\ 0.7 & 0.9 & 0.6 \\ 0.8 & 0.8 & 0.9 \end{pmatrix},$$

$$\tilde{A}^{(22)} = \begin{pmatrix} 0.9 & 0.7 & 0.6 \\ 0.7 & 0.9 & 0.6 \\ 0.8 & 0.8 & 0.9 \end{pmatrix}, \qquad \tilde{A}^{(23)} = \begin{pmatrix} 0.9 & 0.7 & 0.6 \\ 0.7 & 0.9 & 0.8 \\ 0.8 & 0.8 & 0.9 \end{pmatrix},$$

$$\tilde{A}^{(31)} = \begin{pmatrix} 0.9 & 0.7 & 0.6 \\ 0.7 & 0.9 & 0.6 \\ 1 & 0.8 & 0.9 \end{pmatrix}, \qquad \tilde{A}^{(32)} = \begin{pmatrix} 0.9 & 0.7 & 0.6 \\ 0.7 & 0.9 & 0.6 \\ 0.8 & 0.9 & 0.9 \end{pmatrix},$$

$$\tilde{A}^{(33)} = \begin{pmatrix} 0.9 & 0.7 & 0.6 \\ 0.7 & 0.9 & 0.6 \\ 0.8 & 0.8 & 1 \end{pmatrix}.$$

The columns of the matrix $\tilde{C} \in \mathcal{I}(9,9)$ and the right-hand side vector $\tilde{b} \in \mathcal{I}(9)$ are computed blockwise according to (19), as follows.

$$\tilde{C}^{(11)} = \begin{pmatrix} \tilde{A}^{(11)} \otimes_L \tilde{x}^{(1)} \\ \tilde{A}^{(11)} \otimes_L \tilde{x}^{(2)} \\ \tilde{A}^{(11)} \otimes_L \tilde{x}^{(3)} \end{pmatrix} = \begin{pmatrix} 0.9 \\ 0.7 \\ 0.8 \\ 0.7 \\ 0.7 \\ 0.8 \\ 0.7 \\ 0.7 \\ 0.8 \end{pmatrix}, \quad \tilde{C}^{(12)} = \begin{pmatrix} \tilde{A}^{(12)} \otimes_L \tilde{x}^{(1)} \\ \tilde{A}^{(12)} \otimes_L \tilde{x}^{(2)} \\ \tilde{A}^{(12)} \otimes_L \tilde{x}^{(3)} \end{pmatrix} = \begin{pmatrix} 0.8 \\ 0.7 \\ 0.8 \\ 0.6 \\ 0.7 \\ 0.8 \\ 0.6 \\ 0.7 \\ 0.8 \end{pmatrix},$$

$$\tilde{C}^{(13)} = \begin{pmatrix} \tilde{A}^{(13)} \otimes_L \tilde{x}^{(1)} \\ \tilde{A}^{(13)} \otimes_L \tilde{x}^{(2)} \\ \tilde{A}^{(13)} \otimes_L \tilde{x}^{(3)} \end{pmatrix} = \begin{pmatrix} 0.8 \\ 0.7 \\ 0.8 \\ 0.7 \\ 0.7 \\ 0.8 \\ 0.7 \\ 0.7 \\ 0.8 \end{pmatrix}, \quad \tilde{C}^{(21)} = \begin{pmatrix} \tilde{A}^{(21)} \otimes_L \tilde{x}^{(1)} \\ \tilde{A}^{(21)} \otimes_L \tilde{x}^{(2)} \\ \tilde{A}^{(21)} \otimes_L \tilde{x}^{(3)} \end{pmatrix} = \begin{pmatrix} 0.8 \\ 0.7 \\ 0.8 \\ 0.6 \\ 0.7 \\ 0.8 \\ 0.6 \\ 0.7 \\ 0.8 \end{pmatrix},$$

$$\tilde{C}^{(22)} = \begin{pmatrix} \tilde{A}^{(22)} \otimes_L \tilde{x}^{(1)} \\ \tilde{A}^{(22)} \otimes_L \tilde{x}^{(2)} \\ \tilde{A}^{(22)} \otimes_L \tilde{x}^{(3)} \end{pmatrix} = \begin{pmatrix} 0.8 \\ 0.7 \\ 0.8 \\ 0.6 \\ 0.7 \\ 0.8 \\ 0.6 \\ 0.7 \\ 0.8 \end{pmatrix}, \quad \tilde{C}^{(23)} = \begin{pmatrix} \tilde{A}^{(23)} \otimes_L \tilde{x}^{(1)} \\ \tilde{A}^{(23)} \otimes_L \tilde{x}^{(2)} \\ \tilde{A}^{(23)} \otimes_L \tilde{x}^{(3)} \end{pmatrix} = \begin{pmatrix} 0.8 \\ 0.7 \\ 0.8 \\ 0.6 \\ 0.7 \\ 0.8 \\ 0.6 \\ 0.7 \\ 0.8 \end{pmatrix},$$

$$\tilde{C}^{(31)} = \begin{pmatrix} \tilde{A}^{(31)} \otimes_L \tilde{x}^{(1)} \\ \tilde{A}^{(31)} \otimes_L \tilde{x}^{(2)} \\ \tilde{A}^{(31)} \otimes_L \tilde{x}^{(3)} \end{pmatrix} = \begin{pmatrix} 0.8 \\ 0.7 \\ 0.9 \\ 0.6 \\ 0.7 \\ 0.8 \\ 0.6 \\ 0.7 \\ 0.8 \end{pmatrix}, \quad \tilde{C}^{(32)} = \begin{pmatrix} \tilde{A}^{(32)} \otimes_L \tilde{x}^{(1)} \\ \tilde{A}^{(32)} \otimes_L \tilde{x}^{(2)} \\ \tilde{A}^{(32)} \otimes_L \tilde{x}^{(3)} \end{pmatrix} = \begin{pmatrix} 0.8 \\ 0.7 \\ 0.8 \\ 0.6 \\ 0.7 \\ 0.8 \\ 0.6 \\ 0.7 \\ 0.8 \end{pmatrix},$$

$$\tilde{C}^{(33)} = \begin{pmatrix} \tilde{A}^{(33)} \otimes_L \tilde{x}^{(1)} \\ \tilde{A}^{(33)} \otimes_L \tilde{x}^{(2)} \\ \tilde{A}^{(33)} \otimes_L \tilde{x}^{(3)} \end{pmatrix} = \begin{pmatrix} 0.8 \\ 0.7 \\ 0.9 \\ 0.6 \\ 0.7 \\ 0.9 \\ 0.6 \\ 0.7 \\ 0.9 \end{pmatrix}, \quad \tilde{b} = \begin{pmatrix} \tilde{x}^{(1)} \\ \tilde{x}^{(2)} \\ \tilde{x}^{(3)} \end{pmatrix} = \begin{pmatrix} 0.9 \\ 0.8 \\ 0.9 \\ 0.7 \\ 0.8 \\ 0.9 \\ 0.7 \\ 0.8 \\ 0.9 \end{pmatrix}.$$

Hence, we wish to recognize the solvability of the system

$$\tilde{C} \otimes_L \alpha = \begin{pmatrix} 0.9 & 0.8 & 0.8 & 0.8 & 0.8 & 0.8 & 0.8 & 0.8 & 0.8 \\ 0.7 & 0.7 & 0.7 & 0.7 & 0.7 & 0.7 & 0.7 & 0.7 & 0.7 \\ 0.8 & 0.8 & 0.8 & 0.8 & 0.8 & 0.8 & 0.9 & 0.8 & 0.9 \\ 0.7 & 0.6 & 0.7 & 0.6 & 0.6 & 0.6 & 0.6 & 0.6 & 0.6 \\ 0.7 & 0.7 & 0.7 & 0.7 & 0.7 & 0.7 & 0.7 & 0.7 & 0.7 \\ 0.8 & 0.8 & 0.8 & 0.8 & 0.8 & 0.8 & 0.8 & 0.8 & 0.9 \\ 0.7 & 0.6 & 0.7 & 0.6 & 0.6 & 0.6 & 0.6 & 0.6 & 0.6 \\ 0.7 & 0.7 & 0.7 & 0.7 & 0.7 & 0.7 & 0.7 & 0.7 & 0.7 \\ 0.8 & 0.8 & 0.8 & 0.8 & 0.8 & 0.8 & 0.8 & 0.8 & 0.9 \end{pmatrix} \otimes_L \begin{pmatrix} \alpha_1 \\ \alpha_2 \\ \alpha_3 \\ \alpha_4 \\ \alpha_5 \\ \alpha_6 \\ \alpha_7 \\ \alpha_8 \\ \alpha_9 \end{pmatrix} = \lambda \otimes_L \begin{pmatrix} 0.9 \\ 0.8 \\ 0.9 \\ 0.7 \\ 0.8 \\ 0.9 \\ 0.7 \\ 0.8 \\ 0.9 \end{pmatrix} = \lambda \otimes_L \tilde{b}.$$

The problem is a particular case of the bounded parametric solvability problem, with dimension $n^2 \times n^2$, and can be solved by the algorithm suggested in [28] (see also a brief description in this paper, before Theorem 6).

Depending on the permuted entries of $\tilde{b}$, we distinguish the following four cases: (a) $\lambda \in \langle 0.3, 1\rangle$, (b) $\lambda \in \langle 0.1, 0.2\rangle$, $\lambda \in \langle 0.2, 0.3\rangle$ and (c) $\lambda \in \langle 0, 0.1\rangle$. We can verify that for $\lambda = 0.9$ the system $\tilde{C} \otimes_L \alpha = \lambda \otimes_L \tilde{b}$ has a solution $\alpha = (0.9, 0.7, 0.9, 0.8, 0.9, 0.9, 0.8, 0.8, 1)^T$ fulfilling the inequalities $\underline{a}_{ij} - \overline{a}_{ij} + 1 \leq \alpha_{(ij)} \leq 1$, for every $(i,j) \in N \times N$.

Using the coefficients $\alpha_{(ij)}$ we get, by Theorem 5, that $\mathbf{X}$ is a strongly tolerable eigenvector of $\mathbf{A}$, with certificate

$$A = \bigoplus_{(ij) \in N \times N} \alpha_{(ij)} \otimes_L \tilde{A}^{(ij)} = \begin{pmatrix} 0.9 & 0.7 & 0.7 \\ 0.7 & 0.9 & 0.7 \\ 0.8 & 0.8 & 1 \end{pmatrix}.$$

Example 5. *(Numerical illustration: Computing a certificate - no certificate exists)*

We assume the same lower and upper bounds for $A \in [\underline{A}, \overline{A}]$ as in Example 4 and take different bounds for $x \in [\underline{x}, \overline{x}]$.

$$\underline{A} = \begin{pmatrix} 0.9 & 0.7 & 0.6 \\ 0.7 & 0.9 & 0.6 \\ 0.8 & 0.8 & 0.9 \end{pmatrix}, \quad \overline{A} = \begin{pmatrix} 1 & 0.8 & 0.8 \\ 0.7 & 0.9 & 0.8 \\ 1 & 0.9 & 1 \end{pmatrix},$$

$$\underline{x} = \begin{pmatrix} 0.7 \\ 0.7 \\ 0.7 \end{pmatrix}, \quad \overline{x} = \begin{pmatrix} 1 \\ 1 \\ 1 \end{pmatrix}.$$

Then the generators of A stay the same and

$$\tilde{x}^{(1)} = \begin{pmatrix} 1 \\ 0.7 \\ 0.7 \end{pmatrix}, \quad \tilde{x}^{(2)} = \begin{pmatrix} 0.7 \\ 1 \\ 0.7 \end{pmatrix}, \quad \tilde{x}^{(3)} = \begin{pmatrix} 0.7 \\ 0.7 \\ 1 \end{pmatrix}.$$

The matrix $\tilde{C}$ and the right-hand side $\tilde{b}$ then are

$$\tilde{C} = \begin{pmatrix} 1 & 0.9 & 0.9 & 0.9 & 0.9 & 0.9 & 0.9 & 0.9 & 0.9 \\ 0.7 & 0.7 & 0.7 & 0.7 & 0.7 & 0.7 & 0.7 & 0.7 & 0.7 \\ 0.8 & 0.8 & 0.8 & 0.8 & 0.8 & 0.8 & 1 & 0.8 & 0.8 \\ 0.7 & 0.8 & 0.7 & 0.7 & 0.7 & 0.7 & 0.7 & 0.7 & 0.7 \\ 0.9 & 0.9 & 0.9 & 0.9 & 0.9 & 0.9 & 0.9 & 0.9 & 0.9 \\ 0.8 & 0.8 & 0.8 & 0.8 & 0.8 & 0.8 & 0.8 & 0.9 & 0.8 \\ 0.7 & 0.6 & 0.8 & 0.6 & 0.6 & 0.6 & 0.6 & 0.6 & 0.6 \\ 0.6 & 0.6 & 0.6 & 0.6 & 0.6 & 0.8 & 0.6 & 0.6 & 0.6 \\ 0.9 & 0.9 & 0.9 & 0.9 & 0.9 & 0.9 & 0.9 & 0.9 & 1 \end{pmatrix}, \quad \tilde{b} = \begin{pmatrix} 1 \\ 0.7 \\ 0.7 \\ 0.7 \\ 1 \\ 0.7 \\ 0.7 \\ 0.7 \\ 1 \end{pmatrix}.$$

Similarly as in the previous example we distinguish, depending on the permuted entries of $\tilde{b}$, the following three cases: (a) $\lambda \in \langle 0.3, 1 \rangle$, (b) $\lambda \in \langle 0, 0.3 \rangle$ and (c) $\lambda = 0$. It can be verified that the system $\tilde{C} \otimes_L \alpha = \lambda \otimes_L \tilde{b}$ has no solution in any of these cases.

*Consequently, the considered system is not solvable for any value of λ. That is, no certificate for strong tolerability exists and the given **X** is not a strongly tolerable eigenvector of **A**.*

5. Strongly Universal Interval Eigenvectors in a Max-Łuk Algebra

In this section, we present two necessary and sufficient conditions for characterizing a strongly universal eigenvector. The first condition is based on the generators of **A**, while the second one uses the lower bound and the upper bound of **A**.

Theorem 7. *Let **A** and **X** be given such that $\underline{a}_{ij} = \overline{a}_{ij}$ for some $i, j \in N$. The interval vector **X** is a strongly universal max-Łuk eigenvector of the interval matrix $\mathbf{A} = [\underline{A}, \overline{A}]$ if and only if there exists an $x \in \mathbf{X}$ and a $\lambda \in \mathcal{I}$ such that*

$$\tilde{A}^{(ij)} \otimes_L x = \lambda \otimes_L x \quad \textit{for every } i, j \in N. \tag{30}$$

Proof. Let us assume that there are $x \in \mathbf{X}$ and $\lambda \in \mathcal{I}$ fulfilling condition (30). If $A \in \mathcal{I}(n^2)$ is an arbitrary matrix in **A**, then A is a max-Łuk linear combination $A = \bigoplus_{ij \in N} \alpha_{ij} \otimes_L \tilde{A}^{(ij)}$ for some coefficients $\alpha_{ij} \in \mathcal{I}, i, j \in N$ with $\underline{a}_{ij} - \overline{a}_{ij} + 1 \leq \alpha_{ij} \leq 1$. According to Lemma 1(ii),

$$\begin{aligned} A \otimes_L x &= \left(\bigoplus_{ij \in N} \alpha_{ij} \otimes_L \tilde{A}^{(ij)} \right) \otimes_L x = \bigoplus_{ij \in N} \alpha_{ij} \otimes_L \left(\tilde{A}^{(ij)} \otimes_L x \right) \\ &= \bigoplus_{ij \in N} \alpha_{ij} \otimes_L (\lambda \otimes_L x) = \left(\bigoplus_{ij \in N} \alpha_{ij} \otimes_L \lambda \right) \otimes_L x = \lambda \otimes_L x \end{aligned}$$

because $\underline{a}_{ij} = \overline{a}_{ij}$ for some $i, j \in N$ implies $\bigoplus_{ij \in N} \alpha_{ij} = 1$ and $\bigoplus_{ij \in N} \alpha_{ij} \otimes_L \lambda = \lambda$. By Definition 1, **X** is a strongly universal eigenvector of **A**. The converse implication follows immediately. □

Theorem 8. *Suppose given an interval matrix* $\mathbf{A} = [\underline{A}, \overline{A}]$ *and interval vector* $\mathbf{X}$ *with bounds* $\underline{x}$, $\overline{x}$. *Then* $\mathbf{X}$ *is a strongly universal eigenvector of* $\mathbf{A}$ *if and only if there are* $\lambda \in \mathcal{I}$ *and* $x \in \mathbf{X}$ *such that* $\underline{A} \otimes_L x = \lambda \otimes_L x$ *and* $\overline{A} \otimes_L x = \lambda \otimes_L x$.

Proof. Let us suppose that there are λ and $x \in \mathbf{X}$ such that $\underline{A} \otimes_L x = \overline{A} \otimes x = \lambda \otimes x$. From the monotonicity of the operations $\oplus$ and $\otimes_L$ we get $\lambda \otimes_L x = \underline{A} \otimes_L x \leq A \otimes_L x \leq \overline{A} \otimes_L x = \lambda \otimes_L x$ for every $A \in \mathbf{A}$. The converse implication is trivial. □

The condition described in Theorem 8 can be verified by solving a two-sided max-Łuk system defined as follows. Define the block matrices $C \in \mathcal{I}(2n, n)$, $D \in \mathcal{I}(2n, n)$ by

$$C = \begin{pmatrix} \underline{A} \otimes_L \tilde{x}^{(1)} \ldots \underline{A} \otimes_L \tilde{x}^{(n)} \\ \overline{A} \otimes_L \tilde{x}^{(1)} \ldots \overline{A} \otimes_L \tilde{x}^{(n)} \end{pmatrix}, \qquad D = \begin{pmatrix} \tilde{x}^{(1)} & \ldots & \tilde{x}^{(n)} \\ \tilde{x}^{(1)} & \ldots & \tilde{x}^{(n)} \end{pmatrix}. \tag{31}$$

Theorem 9. *Assume that an interval matrix* $\mathbf{A} = [\underline{A}, \overline{A}]$ *and an interval vector* $\mathbf{X} = [\underline{x}, \overline{x}]$ *are given. Then* $\mathbf{X}$ *is a strongly universal eigenvector of* $\mathbf{A}$ *if and only if the bounded two-sided max-Łuk linear system with variable* $\beta \in \mathcal{I}(n)$

$$C \otimes_L \beta = \lambda \otimes_L D \otimes_L \beta \tag{32}$$

$$\underline{x} - \overline{x} + 1 \leq \beta \leq 1 \tag{33}$$

is solvable for some value of the parameter $\lambda \in \mathcal{I}$. *If* $\beta \in \mathcal{I}(n)$ *is a solution to the system, then* $x = \bigoplus_{i=1}^{n} \beta_i \otimes \tilde{x}^i$ *satisfies the condition in Theorem 8.*

Proof. Let us suppose that there is a λ such that the two-sided system $C \otimes \beta = \lambda \otimes D \otimes \beta$ has a solution β. Put $x = \bigoplus_{i=1}^{n} \beta_i \otimes \tilde{x}^i$. Then the following formulas are equivalent

$$C \otimes \beta = \lambda \otimes D \otimes \beta \tag{34}$$

$$\bigoplus_{i=1}^{n} \underline{A} \otimes \tilde{x}^{(i)} \otimes \beta_i = \lambda \otimes \bigoplus_{i=1}^{n} \tilde{x}^{(i)} \otimes \beta_i \text{ and } \bigoplus_{i=1}^{n} \overline{A} \otimes \tilde{x}^{(i)} \otimes \beta_i = \lambda \otimes \bigoplus_{i=1}^{n} \tilde{x}^{(i)} \otimes \beta_i \tag{35}$$

$$\bigoplus_{i=1}^{n} \underline{A} \otimes \tilde{x}^{(i)} \otimes \beta_i = \lambda \otimes \bigoplus_{i=1}^{n} \tilde{x}^{(i)} \otimes \beta_i \text{ and } \bigoplus_{i=1}^{n} \overline{A} \otimes \tilde{x}^{(i)} \otimes \beta_i = \lambda \otimes \bigoplus_{i=1}^{n} \tilde{x}^{(i)} \otimes \beta_i \tag{36}$$

$$\underline{A} \otimes \bigoplus_{i=1}^{n} \beta_i \otimes \tilde{x}^{(i)} = \lambda \otimes \bigoplus_{i=1}^{n} \beta_i \otimes \tilde{x}^{(i)} \text{ and } \overline{A} \otimes \bigoplus_{i=1}^{n} \beta_i \otimes \tilde{x}^{(i)} = \lambda \otimes \bigoplus_{i=1}^{n} \beta_i \otimes \tilde{x}^{(i)} \tag{37}$$

$$\underline{A} \otimes x = \lambda \otimes x \text{ and } \overline{A} \otimes x = \lambda \otimes x. \tag{38}$$

The assertion follows by Theorem 8. □

By Theorem 9, the verification of whether $\mathbf{X}$ is a strongly universal eigenvector of $\mathbf{A}$ is reduced to the verification of the solvability of the system $C \otimes_L x = \lambda \otimes_L D \otimes_L x$. A similar situation in max-min algebra is solved by a polynomial algorithm for the solvability of such a system, with a complexity equal to $O(n^3)$ [29].

In max-plus algebra, the solvability of the considered system has been generally shown to be polynomially equivalent to solving a mean-payoff game [30]. That is, there exist efficient pseudopolynomial algorithms for this problem. On the other hand, the existence of a polynomial algorithm is a long-standing open question. Similarly, for a max-Łuk algebra, the existence of a polynomial algorithm for the solvability recognition problem remains open.

Example 6. *(Numerical illustration: Strongly universal interval eigenvector)*

Assume that $A \in [\underline{A}, \overline{A}]$ *and* $x \in [\underline{x}, \overline{x}]$

$$\underline{A} = \begin{pmatrix} 0.6 & 0.3 & 0.4 \\ 0.1 & 0 & 0.2 \\ 0.2 & 0.2 & 0.6 \end{pmatrix}, \qquad \overline{A} = \begin{pmatrix} 0.6 & 0.7 & 0.5 \\ 0.4 & 0.6 & 0.5 \\ 0.5 & 0.7 & 0.6 \end{pmatrix}$$

$$\underline{x} = \begin{pmatrix} 0.1 \\ 0.2 \\ 0.1 \end{pmatrix}, \qquad \overline{x} = \begin{pmatrix} 0.7 \\ 0.6 \\ 0.7 \end{pmatrix}.$$

It is easy to verify that for a given $\lambda = 0.4$, $x = (0.1, 0.2, 0.1)^T$ *and for every matrix generator* $\tilde{A}^{ij}$, $i, j \in N$,

$$\tilde{A}^{ij} \otimes x = \tilde{A}^{ij} \otimes \begin{pmatrix} 0.1 \\ 0.2 \\ 0.1 \end{pmatrix} = \begin{pmatrix} 0 \\ 0 \\ 0 \end{pmatrix} = 0.4 \otimes \begin{pmatrix} 0.1 \\ 0.2 \\ 0.1 \end{pmatrix} = \lambda \otimes_L x.$$

Hence, in view of Theorem 7, the interval vector $\mathbf{X}$ *is a strongly universal max-Łuk eigenvector of the interval matrix* $\mathbf{A}$.

Theorem 9 offers a more systematic approach: find a solution to the two-sided system (32) and (33) with unknown coefficients β, *for some* $\lambda' \in \mathcal{I}$:

$$\begin{pmatrix} 0.3 & 0 & 0.1 \\ 0 & 0 & 0 \\ 0 & 0 & 0.3 \\ 0.3 & 0.3 & 0.2 \\ 0.1 & 0.2 & 0.2 \\ 0.2 & 0.3 & 0.3 \end{pmatrix} \otimes_L \begin{pmatrix} \beta_1 \\ \beta_2 \\ \beta_3 \end{pmatrix} = \lambda' \otimes_L \begin{pmatrix} 0.7 & 0.1 & 0.1 \\ 0.2 & 0.6 & 0.2 \\ 0.1 & 0.1 & 0.7 \\ 0.7 & 0.1 & 0.1 \\ 0.2 & 0.6 & 0.2 \\ 0.1 & 0.1 & 0.7 \end{pmatrix} \otimes_L \begin{pmatrix} \beta_1 \\ \beta_2 \\ \beta_3 \end{pmatrix},$$

$$\begin{pmatrix} 0.4 \\ 0.6 \\ 0.4 \end{pmatrix} \le \begin{pmatrix} \beta_1 \\ \beta_2 \\ \beta_3 \end{pmatrix} \le \begin{pmatrix} 1 \\ 1 \\ 1 \end{pmatrix}.$$

In this particular instance, it is easy to verify that $\beta' = (0.9, 0.8, 0.8)^T$ *is a solution to the system with* $\lambda' = 0.6$. *Then the corresponding linear combination of generators,* $x' = (0.9 \otimes_L \tilde{x}^{(1)}) \oplus (0.8 \otimes_L \tilde{x}^{(2)}) \oplus (0.8 \otimes_L \tilde{x}^{(3)}) = (0.6, 0.4, 0.5)^T$, *satisfies the condition from Theorem 8. It is worth noticing that we have found two different universal eigenvectors* x *and* x' *for two different values* λ *and* λ' *in this example. Hence, we have shown that neither the "strongly universal" eigenvalue nor the strongly universal eigenvector are uniquely determined.*

6. Conclusions

Strong versions of the notion of an interval eigenvector of an interval matrix in a max-Łuk algebra have been investigated in this paper. The steady states of a given discrete events system (DES) correspond to eigenvectors of the transition matrix of the system under consideration. When the entries of the state vectors and transition matrix are supposed to be contained in some intervals, then several types of interval eigenvector can be defined, according to the choice of the quantifiers used in the definition. Three of the main important types of interval eigenvectors of a given interval matrix in a max-Łuk algebra have been studied: the strong eigenvector, the strongly tolerable eigenvector, and the strongly universal eigenvector.

Using vector generators and matrix generators belonging to given intervals, the structure of the eigenspace for each of the above mentioned types has been described, and necessary and sufficient conditions for the existence of an interval eigenvector have been formulated. Moreover, recognition algorithms have been suggested for the recognition of these conditions for the first two types: strong and strongly tolerable eigenvector. The existence of an efficient recognition algorithm for the strongly universal type has not been shown. This question remains as a challenge for future research.

These results can be useful in practical applications aimed at the construction of real DES working with Łukasiewicz fuzzy logic. The results have been illustrated by numerical examples.

Author Contributions: Investigation, M.G., Z.N. and J.P.; Writing–review and editing, M.G., Z.N. and J.P. All authors contributed equally to this manuscript. All authors have read and agreed to the published version of the manuscript.

Funding: This research was funded by the Czech Science Foundation (GAČR) #18-01246S and by the Faculty of Informatics and Management UHK, specific research project 2107 Computer Networks for Cloud, Distributed Computing, and Internet of Things III.

Conflicts of Interest: The authors declare no conflict of interest. The funders had no role in the design of the study; in the collection, analyses, or interpretation of data; in the writing of the manuscript, or in the decision to publish the results.

References

1. Schweizer, B.; Sklar, A. Statistical metric spaces. *Pac. J. Math.* **1960**, *10*, 313–334. [CrossRef]
2. Gottwald, S. *A Treatise on Many-Valued Logics*; Studies in Logic and Computation, Research Studies Press: Baldock, Great Britain, 2001.
3. Gavalec, M. *Periodicity in Extremal Algebras*; Gaudeamus: Hradec Králové, Czech Republic, 2004.
4. Golan, J.S. *Semirings and Their Applications*; Springer: Berlin, Germany, 2013.
5. Gondran, M.; Minoux, M. *Graphs, Dioids and Semirings: New Models and Algorithms*; Springer: Berlin, Germany, 2008; Volume 41.
6. Kolokoltsov, V.; Maslov, V.P. *Idempotent Analysis and Its Applications*; Springer: Berlin, Germany, 1997; Volume 401.
7. Mysková, H. Interval eigenvectors of circulant matrices in fuzzy algebra. *Acta Electrotech. Inform.* **2012**, *12*, 57. [CrossRef]
8. Mysková, H. Weak stability of interval orbits of circulant matrices in fuzzy algebra. *Acta Electrotech. Inform.* **2012**, *12*, 51. [CrossRef]
9. Plavka, J. On the weak robustness of fuzzy matrices. *Kybernetika* **2013**, *49*, 128–140.
10. Tan, Y.J. Eigenvalues and eigenvectors for matrices over distributive lattices. *Linear Algebra Appl.* **1998**, *283*, 257–272. [CrossRef]
11. Zimmermann, K. *Extremální Algebra*; Útvar vědeckých Informací Ekonomického Ústavu ČSAV: Praha, Czech Republic, 1976.
12. Saleem, N.; Abbas, M.; De la Sen, M. Optimal Approximate Solution of Coincidence Point Equations in Fuzzy Metric Spaces. *Mathematics* **2019**, *7*, 327. [CrossRef]
13. Alolaiyan, H.; Saleem, N.; Abbas, M. A natural selection of a graphic contraction transformation in fuzzy metric spaces. *J. Nonlinear Sci. Appl.* **2018**, *11*, 218–227. [CrossRef]
14. Butkovic, P. *Max-Linear Systems: Theory and Algorithms*; Springer: Berlin, Germany, 2010; p. 272. [CrossRef]
15. Cuninghame-Green, R.A. *Minimax Algebra*; Springer: Berlin, Germany, 2012; Volume 166.
16. Gavalec, M. Monotone eigenspace structure in max-min algebra. *Linear Algebra Appl.* **2002**, *345*, 149–167. [CrossRef]
17. Gavalec, M.; Němcová, Z.; Sergeev, S. Tropical linear algebra with the Łukasiewicz T-norm. *Fuzzy Sets Syst.* **2015**, *276*, 131–148. [CrossRef]
18. Gavalec, M.; Rashid, I.; Cimler, R. Eigenspace structure of a max-drast fuzzy matrix. *Fuzzy Sets Syst.* **2014**, *249*, 100–113. [CrossRef]
19. Rashid, I.; Gavalec, M.; Cimler, R. Eigenspace structure of a max-prod fuzzy matrix. *Fuzzy Sets Syst.* **2016**, *303*, 136–148. [CrossRef]
20. Collins, P.; Niqui, M.; Revol, N. A validated real function calculus. *Math. Comput. Sci.* **2011**, *5*, 437–467. [CrossRef]
21. Fiedler, M.; Nedoma, J.; Ramík, J.; Rohn, J.; Zimmermann, K. *Linear Optimization Problems with Inexact Data*; Springer: Berlin, Germany, 2006.

22. Gavalec, M.; Ramík, J.; Zimmermann, K. Interval eigenproblem in max-min algebra. In *Decision Making and Optimization*; Springer: Berlin, Germany, 2015; pp. 163–181.
23. Litvinov, G.L.; Sobolevskiī, A.N. Idempotent interval analysis and optimization problems. *Reliab. Comput.* **2001**, *7*, 353–377. [CrossRef]
24. Gavalec, M.; Plavka, J. Monotone interval eigenproblem in max–min algebra. *Kybernetika* **2010**, *46*, 387–396.
25. Gavalec, M.; Plavka, J.; Tomášková, H. Interval eigenproblem in max-min algebra. *Liner Algebra Appl.* **2014**, *440*, 24–33. [CrossRef]
26. Gavalec, M.; Plavka, J.; Ponce, D. Tolerance types of interval eigenvectors in max-plus algebra. *Inf. Sci.* **2017**, *367*, 14–27. [CrossRef]
27. Gavalec, M.; Zimmermann, K. Classification of solutions to systems of two-sided equations with interval coefficients. *Int. J. Pure Appl. Math.* **2008**, *45*, 533.
28. Gavalec, M.; Němcová, Z. Solvability of a Bounded Parametric System in Max-Łukasiewicz Algebra. *Mathematics* **2020**, *8*, 1026. [CrossRef]
29. Gavalec, M.; Zimmermann, K. Solving systems of two sided (max,min) linear equations. *Kybernetika* **2010**, *46*, 405–414.
30. Allamigeon, X.; Fahrenberg, U.; Gaubert, S.; Katz, R.D.; Legay, A. Tropical Fourier–Motzkin elimination, with an application to real-time verification. *Int. J. Algebra Comput.* **2014**, *24*, 569–607. [CrossRef]

mathematics

MDPI

Article

Which Alternative for Solving Dual Fuzzy Nonlinear Equations Is More Precise?

Joanna Kołodziejczyk, Andrzej Piegat and Wojciech Sałabun *

Research Team on Intelligent Decision Support Systems, Department of Artificial Intelligence Methods and Applied Mathematics, Faculty of Computer Science and Information Technology, West Pomeranian University of Technology in Szczecin ul. Żołnierska 49, 71-210 Szczecin, Poland; joanna.kolodziejczyk@zut.edu.pl (J.K.); apiegat@wi.zut.edu.pl (A.P.)

* Correspondence: wojciech.salabun@zut.edu.pl; Tel.: +48-91-449-5580

Received: 14 August 2020; Accepted: 3 September 2020; Published: 4 September 2020

Abstract: To answer the question stated in the title, we present and compare two approaches: first, a standard approach for solving dual fuzzy nonlinear systems (DFN-systems) based on Newton's method, which uses 2D FN representation and second, the new approach, based on multidimensional fuzzy arithmetic (MF-arithmetic). We use a numerical example to explain how the proposed MF-arithmetic solves the DFN-system. To analyze results from the standard and the new approaches, we introduce an imprecision measure. We discuss the reasons why imprecision varies between both methods. The imprecision of the standard approach results (roots) is significant, which means that many possible values are excluded.

Keywords: fuzzy nonlinear systems; fuzzy arithmetic; fuzzy calculus; multidimensional fuzzy arithmetic; RDM fuzzy arithmetic; fuzzy parametric form

1. Introduction

Solving linear and nonlinear fuzzy equations is a difficult task. This is evidenced by the fact that scientists are constantly developing new and better methods, because previous methods are not satisfactory and scientists have recognized the need to improve them. In the case of fuzzy nonlinear equations, one of the first methods was published in 1990 [1].

Newton's numerical methods for solving DFN-systems are widespread in the literature. Abbasbandy and Asady proposed the numerical parametric approach to find a fuzzy nonlinear equation using Newton's method. The efficiency of the algorithm was shown based on some numerical examples [2]. The continuation of their work is found in [3]. Ramli et al. show that the disadvantages of Newton's method arise from the need to calculate and invert the Jacobian matrix at each iteration. They proposed an eight-step algorithm to solve fuzzy nonlinear equations [4]. In [5], Kajani et al. have applied Newton's method for solving a DFN-system, which cannot be replaced by a fuzzy nonlinear system according to fuzzy arithmetic. Newton's method was also used to solve dual fuzzy polynomial equations, where the modified Adomian decomposition method was applied in the numerical algorithm [6]. Waziri and Majid proposed a new approach for solving DFN-equations by combining Newton's method for initial iteration and Broyden's method for the rest of the iterations [7]. The paper [8] presents a method of solving DFN-systems based on Chord Newton's method as an improvement of Newton's iterative method published in [5]. An unquestionable advantage of the method [8] is that it requires the Jacobian matrix to be calculated only once for all iterations whereas in Newton's method from [5] the matrix has to be calculated in each subsequent iteration, which is connected with a high computational effort. Wang et al. introduced the general family of n-point Newton-type iterative methods for solving nonlinear equations using direct Hermite interpolation [9]. All these approaches have one thing in common, as will be shown, which is they give imprecise results.

In the paper, we compare two approaches to solve DFN-systems first, a standard based on Newton's method and a 2D FN representation. The second is a novel alternative approach based on multidimensional fuzzy arithmetic. We introduce imprecision to measure how different a solution is from the universal algebraic solution to analyze both methods' solutions. The universal algebraic solution is a set of all solutions that satisfied the equations. We show that the standard approach finds a narrower set of solutions than the universal algebraic solution, which means that they omit many possible values. Our proposal's primary motivation is the advantage of determining all possible results when looking for solutions, not only the most narrow subset. The full universal algebraic solution is especially desired in decision support systems, where the imprecision of traditional approaches leads to limitations of possible states and, therefore, may constitute decision variants.

The paper is organized as follows: first, we give some essential basic definitions in a standard approach and proposed MF-arithmetic in Section 2. The ideas for both approaches are compared in Section 3. The imprecision measure is defined in Section 4. In Section 5, we solve the benchmark DFN-system using both proposed approaches. Finally, in Section 6, we discuss the most important outcomes.

2. Preliminaries

Definition 1. *[5] The dual fuzzy nonlinear system [8] is understood to be the system (1):*

$$Q(x) = R(x) + c \tag{1}$$

where all parameters are fuzzy numbers.

The standard approach to solving DFN-systems applies Newton's algorithm and a parametric form of a fuzzy number. Our approach uses MF-arithmetic based on a horizontal fuzzy number definition.

2.1. Basic Definitions in the Standard Approach

The definitions presented in the subsection are fundamental standard approaches using Newton's methods for solving DFN-systems and comes from articles: [5,8].

Definition 2. *A fuzzy number is a fuzzy set like $u : \mathbb{R} \to I = [0,1]$ which satisfies:*

1. *u is upper semicontinuous,*
2. *$u(x) = 0$ outside some interval $[c, d]$,*
3. *There are real numbers a, b such that $c \leq a \leq b \leq d$ and*
   (a) *$u(x)$ is monotonic increasing on $[c, a]$*
   (b) *$u(x)$ is monotonic decreasing on $[b, d]$*
   (c) *$u(x) = 1, a \leq x \leq b$*

Definition 3. *A fuzzy number in parametric form is a pair $\underline{u}, \overline{u}$ of functions $\underline{u}(r), \overline{u}(r)$, $0 \leq r \leq 1$ which satisfies the following:*

1. *$\underline{u}(r)$ is a bounded monotonic increasing left continuous function,*
2. *$\overline{u}(r)$ is a bounded monotonic decreasing left continuous function,*
3. *$\underline{u}(r) \leq \overline{u}(r), 0 \leq x \leq 1$.*

where the variable r represents the membership level.

A popular triangular fuzzy number with the parameters (a, b, c) (shown in Figure 1 (left graph)) in its parametric form is:

$$\underline{u}(r) = a + (b-a)r, \quad \bar{u}(r) = c - (c-b)r \tag{2}$$

In [8] the described Chord–Newton's approach was implemented according to the following steps:

Step 1 Transform the dual fuzzy nonlinear equations into parametric form.
Step 2 Determine the initial point x_0 by solving the parametric equations for $r = 0$ and $r = 1$.
Step 3 Compute the Jacobian matrix $J\left(\underline{x}_0, \overline{x}_0, r\right)$
Step 4 Compute (3)

$$x_n(r) = x_{n-1}(r) + \left(sa_{\theta_n}^{\gamma_n}\right) J^{-1}\left(\underline{x}_0, \overline{x}_0, r\right) \quad n = 1, 2, \ldots \tag{3}$$

Step 5 Repeat steps from 3 to 4 and continue with the next n keeping Jacobian until tolerance $\epsilon \leq 10^{-5}$ is satisfied.

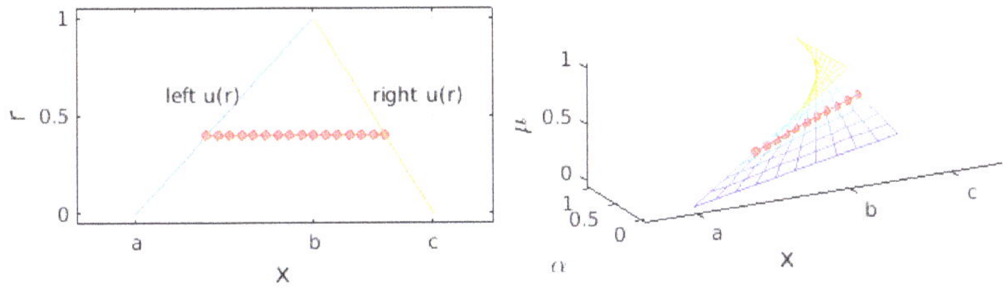

Figure 1. Triangle membership function of a fuzzy number X with the cut at membership level $r = \mu = 0.4$: left graph—parametric fuzzy number (FN), right graph—horizontal FN.

2.2. Multidimensional Fuzzy Arithmetic

The MF-arithmetic theoretical foundations are linked with the horizontal fuzzy numbers (HFN) [10]. The HFN with linear or/and nonlinear left and right borders can be defined based on the FN parametric definition.

Definition 4. *A horizontal fuzzy number, of an arbitrary fuzzy number* $(\underline{u}(\mu), \bar{u}(\mu))$ *is defined as follows:*

$$U = \left\{u^{gr} \in \mathbb{R} : u^{gr} = u\left(\mu, \alpha_u\right) = \underline{u}(\mu) + \alpha_u(\bar{u}(\mu) - \underline{u}(\mu))\right\}, \tag{4}$$

where $\mu, \alpha_u \in [0,1]$ *and* μ *represents the membership level.*

The variable $\alpha_u \in [0,1]$ is the Relative Distance Measure (RDM) [11,12] and ensures obtaining any FN value between the left border $\underline{u}(\mu)$ and right border $\overline{u}(\mu)$. For $\alpha_u = 0$ value of the left endpoint $\underline{u}(\mu)$ and for $\alpha_u = 1$ the right endpoint $\overline{u}(\mu)$ of FN.

The fuzzy number U represents only an approximate knowledge about the precise but unknown true value of the variable u^{gr} (epistemic approach [13]).

If we have an approximate knowledge of two uncertain values of the variables x and y in the form of horizontal membership function: $x^{gr}(\mu, \alpha_x)$ and $y^{gr}(\mu, \alpha_y)$ then the basic arithmetic operations $(* \in \{+, -, \times, /\})$ can be expressed as Equation (5).

$$x^{gr}\left(\mu, \alpha_x\right) * y^{gr}\left(\mu, \alpha_y\right) = z^{gr}\left(\mu, \alpha_x, \alpha_y\right), \quad \mu, \alpha_x, \alpha_y \in [0,1]. \tag{5}$$

The division operation guarantees a result in the form of a single granule if the following condition is satisfied: $0 \notin y^{gr}\left(\mu, \alpha_y\right)$. However, if $0 \in y^{gr}\left(\mu, \alpha_y\right)$ then the quotient will be multi-granular [14]. It should be noted that the result $z^{gr}\left(\mu, \alpha_x, \alpha_y\right)$ of any arithmetic operation is not defined in 2D space $\mu \times z$ but in 4D space $\mu \times \alpha_x \times \alpha_y \times z$.

Definition 5. [10] *The direct solution of an operation on n horizontal fuzzy numbers* $u_1(\mu, \alpha_{u_1}), \ldots, u_n(\mu, \alpha_{u_n})$*, where* $\mu, \alpha_{u_1}, \ldots, \alpha_{u_n} \in [0,1]$ *is a set of numbers expressed in the form of multidimensional formula* $Z = z(\mu, \alpha_{u_1}, \ldots, \alpha_{u_n})$ *with up to n HFN variables* $\alpha_{u_1}, \ldots, \alpha_{u_n} \in [0,1]$.

Definition 6. [10] *For the direct solution* $Z = z(\mu, \alpha_{u_1}, \ldots, \alpha_{u_n})$ *of the basic arithmetic operations on n horizontal fuzzy numbers span is a fuzzy number:*

$$s(Z^r) = \Big(\min_{\alpha_{u_1},\ldots,\alpha_{un}} z(\mu, \alpha_{u_1}, \ldots, \alpha_{u_n}) \quad \max_{\alpha_{u_1},\ldots,\alpha_{un}} z(\mu, \alpha_{u_1}, \ldots, \alpha_{u_n}) \Big) \tag{6}$$

Multidimensional fuzzy arithmetic has such important mathematical properties as (assuming X, Y, Z are HFN):

- Additive inverse element $X - X = 0$,
- Multiplicative inverse element $X \cdot 1/X = 1$,
- Distributive law $X(Y + Z) = XY + XZ$,
- Cancellation law for multiplication $XZ = YZ \Rightarrow X = Y$, and others.

These properties allow solving any problem by transforming fuzzy equations from one mathematical form to another. In other words, without these properties, transformations are impossible. 2D-fuzzy arithmetic methods do not have such properties, which causes calculation difficulties.

The horizontal triangle membership functions corresponding to the vertical triangle membership functions with the parameters (a, b, c), shown in Figure 1 (right graph), are represented by Equation (7) [14].

$$x^r = a + (b - a)\,\mu + (c - a)\,(1 - \mu)\,\alpha_x, \tag{7}$$

where $\mu, \alpha_x \in [0,1]$ and $X : x^r \in X$.

3. Multidimensional Fuzzy Arithmetic Approach

What we called a standard approach are two-dimensional (2D) versions of the fuzzy arithmetic, and calculus based on it is presented in the scheme in Figure 2.

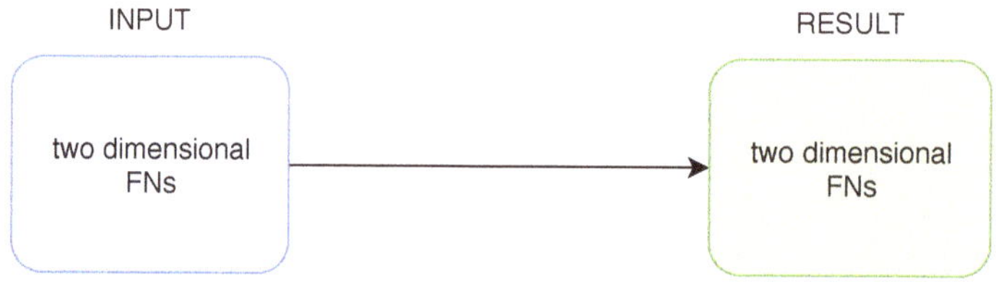

Figure 2. A standard approach (2D) scheme in fuzzy arithmetic.

This approach assumes that the result of calculations on 2D-FNs is also a 2D-FN. Reference [12] considers this assumption incorrect. Such the approach delivers: sometimes precise, sometimes imprecise, sometimes paradoxical results or sometimes they are not able to solve the problem at all [15] or sometimes they can only solve it partially, as in the case of the benchmark (Equation (9)).

On the other hand, multidimensional fuzzy arithmetic works according to a different scheme than standard fuzzy 2D arithmetics. The diagram is given in Figure 3.

Figure 3. A proposed multidimensional fuzzy (MF)-arithmetic approach scheme.

In 2D fuzzy calculation methods, what is assumed to be a "result" in MF-arithmetic is only a 2D indicator of a multidimensional result, which at most can be called "secondary result". There are three basic indicators for the multidimensional fuzzy result:

1. Span,
2. Cardinality distribution,
3. Center of gravity.

What is interpreted as the "result" in 2D methods is actually only the span of the multidimensional fuzzy result [16,17]. Until now, MF-arithmetic has been described in about 40 papers [15].

4. Imprecision Measure

Before numerical example presentation and further discussion, let us define a DFN-systems solution quality measure named "imprecision". This quality is related to the concept of a universal algebraic solution. Let:

1. UA-solution—universal algebraic solution always satisfies the given nonlinear equation regardless of its mathematical form (right and left sides of the equation are equal),
2. A—UA-solution delivered by the method in the form of an MF-arithmetic,
3. B—a solution delivered by the method in the form of a fuzzy number (FN),
4. $Supp(B)$—the support of B FN,
5. $Supp(s(A))$—the support of the A's span $s(A)$.

Definition 7. *The imprecision of a solution B is a measure of how different a solution B is from the A which is UA-solution:*

$$Imprecision(B) = \frac{Supp(B) - Supp(s(A))}{Supp(s(A))} \cdot 100\%, \tag{8}$$

The imprecision (B) calculates the percentage of values excluded (negative Imprecision for the case $Supp(B) \subset Supp(s(A))$) or values included (positive Imprecision for the case $Supp(s(A)) \subset Supp(B)$) from or to the UA-solution.

5. Solving the Benchmark Equation

The best proof of methods qualitative correctness and precision is the correct solution to real problems. To show the performance of Newton's approach, authors in [7] present solutions to some DFN-systems benchmark problems. To compare the standard and proposed MF-arithmetic approaches we take the benchmark from [5,7] called Problem 1 .

Problem 1 ([8]). *Consider the dual fuzzy nonlinear equation:*

$$(6,2,2)x^2 + (2,1,1)x = (2,1,1)x^2 + (2,1,1) \tag{9}$$

In this problem $(6, 2, 2)$ denotes triangle FN with a core $\sigma = 6$ and with a left and right deviation equal to 2 and $(2, 1, 1)$ denotes a FN with a core equal to $\sigma = 2$ and both deviations equal to 1. The above FNs are shown in Figure 4.

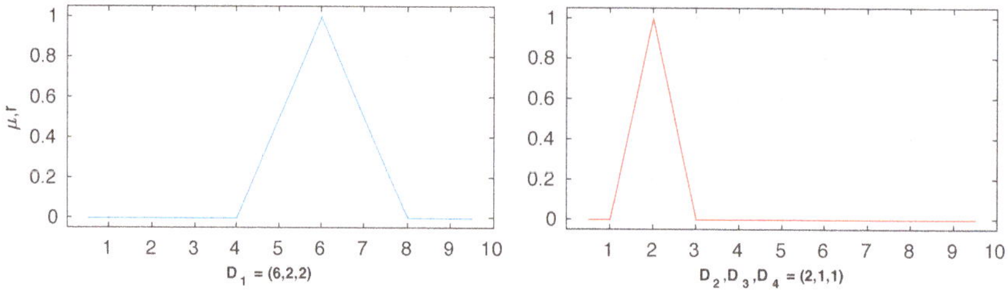

Figure 4. Triangle fuzzy numbers D_1, D_2, D_3, D_4 used as uncertain coefficients in the discussed benchmark Problem 1.

5.1. Standard Approach

Using the standard approach using Newton's method and parametric FN representation, the authors of [5,8] achieved only one positive solution (positive root) of Equation (9) in form of the fuzzy number X_2 shown in Figure 5.

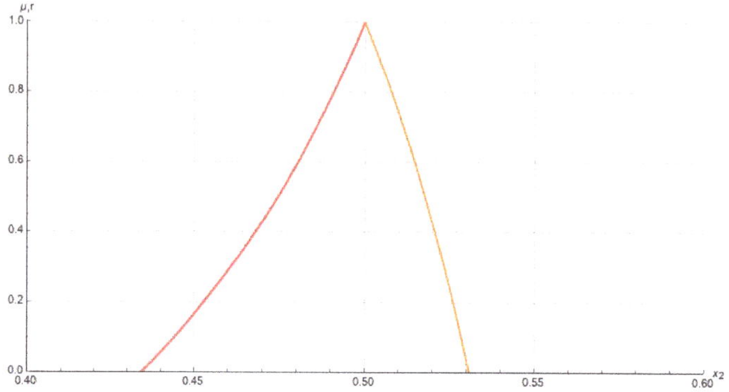

Figure 5. Solution X_2 of the benchmark Equation (9) achieved with the use of a standard approach based on Newton's methods presented in [5,8].

As will be shown, the positive solution X_2 delivered by the standard approach is, to a considerable degree, imprecise due to the FN definition and followed by its assumptions. Additionally, according to the authors [5,8] the negative solution X_1 of the benchmark Equation (9) does not exist at all and the explanation for this fact is given in [5].

5.2. MF-Arithmetic Approach

The benchmark equation (Equation (9)) represents a stationary, real system (Equation (10)),

$$d_1 x^2 + d_2 x = d_3 x^2 + d_4, \tag{10}$$

where d_1, d_2, d_3, d_4 are true, crisp coefficients describing properties in the real system. In the analysed benchmark these true values are not known exactly. However, we have approximate knowledge about them: $d_1 \in D_1 = (6,2,2)$, $d_2 \in D_2 = (2,1,1)$, $d_3 \in D_3 = (2,1,1)$, $d_4 \in D_4 = (2,1,1)$. The fact that

$$D_2 = D_3 = D_4 = (2,1,1) \tag{11}$$

does not imply that the true coefficient values are equal, i.e., $d_2 = d_3 = d_4$. The probability of such a case occurring is equal to zero, because it is the probability of drawing a single point from a cuboid (geometric probability).

For true coefficient values, the equation (Equation (10)) can be converted to (Equation (12)).

$$(d_1 - d_3)x^2 + d_2 x - d_4 = 0. \tag{12}$$

Well known Vieta's formulas (Equation (13)) are applied to solve the equation (Equation (12)).

$$\begin{aligned} & x_1 = \frac{-d_2 - \sqrt{\Delta}}{2(d_1 - d_3)}, \quad x_2 = \frac{-d_2 + \sqrt{\Delta}}{2(d_1 - d_3)}, \\ & \Delta = {d_2}^2 + 4(d_1 - d_3)d_4. \end{aligned} \tag{13}$$

In terms of multidimensional fuzzy arithmetic the mathematical models of true and possible coefficient values are described as follows (Equation (14)):

$$\begin{aligned} {d_1}^{gr} &= (4 + 2\mu) + 4(1-\mu)\alpha_{d_1}, \ \mu, \alpha_{d_1} \in [0,1], \\ {d_2}^{gr} &= (1 + \mu) + 2(1-\mu)\alpha_{d_2}, \ \mu, \alpha_{d_2} \in [0,1], \\ {d_3}^{gr} &= (1 + \mu) + 2(1-\mu)\alpha_{d_3}, \ \mu, \alpha_{d_3} \in [0,1], \\ {d_4}^{gr} &= (1 + \mu) + 2(1-\mu)\alpha_{d_4}, \ \mu, \alpha_{d_4} \in [0,1]. \end{aligned} \tag{14}$$

To achieve benchmark (Equation (9)) solutions, the expression obtained in (Equation (14)) is substituted in equations (Equation (13)):

$$\begin{aligned} x_1^{gr} &= \frac{-[(1+\mu)+2(1-\mu)\alpha_{d_2}] - \sqrt{\Delta}}{2[(3+\mu)+(1-\mu)(4\alpha_{d_1} - 2\alpha_{d_3})]}, \\ x_2^{gr} &= \frac{-[(1+\mu)+2(1-\mu)\alpha_{d_2}] + \sqrt{\Delta}}{2[(3+\mu)+(1-\mu)(4\alpha_{d_1} - 2\alpha_{d_3})]}, \\ \Delta &= [(1+\mu) + 2(1-\mu)\alpha_{d_2}]^2 + 4[(3+\mu) + \\ &+ (1-\mu)(4\alpha_{d_1} - 2\alpha_{d_3})][(1+\mu) + 2(1-\mu)\alpha_{d_4}], \\ & \mu, \alpha_{d_1}, \alpha_{d_2}, \alpha_{d_3}, \alpha_{d_4} \in [0,1]. \end{aligned} \tag{15}$$

The x_1^{gr} and x_2^{gr} possible values satisfy the following conditions: $x_1^{gr} < 0$ and $x_2^{gr} > 0$ and their denominators do not include zero. Hence, the solutions are single information granules. Each information granule is multidimensional (6D) because $x_1^{gr} = f_1(\mu, \alpha_{d_1}, \alpha_{d_2}, \alpha_{d_3}, \alpha_{d_4})$ and $x_2^{gr} = f_1(\mu, \alpha_{d_1}, \alpha_{d_2}, \alpha_{d_3}, \alpha_{d_4})$.

They are universal, algebraic solutionsof the benchmark, that can be easily verified by substitution in the Equation (9). The solutions ensure left and right-hand side equality regardless of the mathematical representation of the benchmark equation, which means that they are universal [11,12,15]. The multidimensional, algebraic solutions x_1^{gr} and x_2^{gr} are hypersurface fragments in 6D spaces and cannot be visualised. For simplified (2D) information about the solutions, low-dimensional indicators can be used such as: span, cardinality distribution and center of gravity of multidimensional fuzzy solution [11,12,15,17]. Spans of multidimensional solutions $s(x_{1,2}^{gr})$ can be obtained with Equation (16).

$$s(x_{1,2}^{gr}) = \left[\min_{\alpha_{d_1},\ldots,\alpha_{d_4}} x_{1,2}^{gr}(\mu, \alpha_{d_1}, \ldots, \alpha_{d_4}), \right. \\ \left. \max_{\alpha_{d_1},\ldots,\alpha_{d_4}} x_{1,2}^{gr}(\mu, \alpha_{d_1}, \ldots, \alpha_{d_4}) \right]. \quad (16)$$

To determine the span of the negative root x_1^{gr} (Equation (17)) equation, Equation (15) is substituted to Equation (16).

$$s(x_1^{gr}) = \left[\frac{-3+\mu-\sqrt{21+26\mu-11\mu^2}}{2+6\mu}, \right. \\ \left. \frac{1+\mu+\sqrt{29+18\mu-11\mu^2}}{-14+6\mu} \right]. \quad (17)$$

According to the authors of [8] the negative root does not exist. The lack of the negative solution is the consequence of the parametric FN form and the low-dimension fuzzy calculus. Newton's methods [5,8] found the second root that does not satisfy the third condition in Definition 3 and therefore was rejected. We observe that the low-dimensional fuzzy calculus is imprecise. It is strange that the negative root was rejected, because negative crisp benchmark solutions can easily be found. The span function of the negative root $s(x_1^{gr})$, which is the benchmark (Equation (9)) solution, is shown in Figure 6.

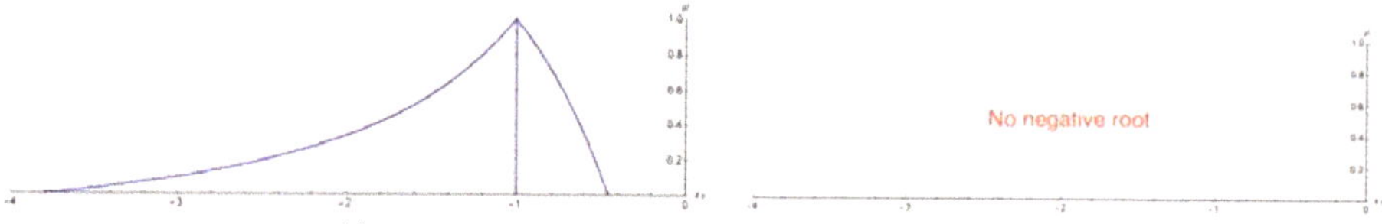

(a) (b)

Figure 6. The span function $s(x_1^{gr})$ of the negative multidimensional solution (root) of the benchmark (9) determined using MF-arithmetic (**a**) and solution found with a standard approach based on Newton's method [5,8] (**b**). (**a**) MF-arithmetic approach; (**b**) standard approach based on Newton's method.

The minimum value of the negative root,

$$\min_{\alpha_{d_1},\ldots,\alpha_{d_4}} x_1^{gr}(\mu, \alpha_{d_1}, \alpha_{d_2}, \alpha_{d_3}, \alpha_{d_4}), \quad (18)$$

was obtained for following values of RDM-variables:

$$\alpha_{d_1} = 0, \alpha_{d_2} = \alpha_{d_3} = \alpha_{d_4} = 1, \quad (19)$$

and the membership level $\mu = 0$ and is equal to $\min x_1^{gr} = -3.79129$.

The maximum value of the negative root,

$$\max_{\alpha_{d_1},\ldots,\alpha_{d_4}} x_1^{gr}(\mu, \alpha_{d_1}, \alpha_{d_2}, \alpha_{d_3}, \alpha_{d_4}), \quad (20)$$

was derived for the set of RDM-variables:

$$\alpha_{d_1} = 1, \alpha_{d_2} = \alpha_{d_3} = \alpha_{d_4} = 0, \quad (21)$$

and the $\mu = 0$ level and is equal to $\max x_1^{gr} = -0.456083$.

These results can easily be verified with a method described in [17]. Solving equations (Equation (14)) by substituting RDM-variables $\alpha_{d_1} = 0, \alpha_{d_2} = \alpha_{d_3} = \alpha_{d_4} = 1$ for $\min x_1^{gr}$ and $\mu = 0$ results in a set of corresponding coefficient values:

$$d_1 = 4, d_2 = 3, d_3 = 3, d_4 = 3. \tag{22}$$

All these values are possible because they are contained in supports of fuzzy coefficients D_1, D_2, D_3, D_4 in the benchmark (Equation (9)). For these coefficients the equation (Equation (9)) takes form:

$$4x^2 + 3x = 3x^2 + 3, \tag{23}$$

and have two roots: $x_{neg} = -3.79129$ and $x_{pos} = 0.791288$. The negative root is the same as that found using multidimensional fuzzy arithmetic.

Now we check the correctness of the $\max x_1^{gr}$ solution. Coefficient values corresponding to RDM-variables $\alpha_{d_1} = 1, \alpha_{d_2} = \alpha_{d_3} = \alpha_{d_4} = 0$ and $\mu = 0$ are

$$d_1 = 8, d_2 = 1, d_3 = 1, d_4 = 1. \tag{24}$$

They generate a possible form of the Equation (9):

$$8x^2 + x = x^2 + 1, \tag{25}$$

which has a negative root $x_{neg} = -0.456083$ and also a positive $x_{pos} = 0.313226$.

Other combinations of RDM-variables

$$\alpha_{d_1}, \alpha_{d_2}, \alpha_{d_3}, \alpha_{d_4} \in [0, 1] \tag{26}$$

generate intermediate values of the negative root $x_1 \in [-3.79129, -0.456083]$ as shown in Figure 6, which can be easily checked.

The above is an empirical proof that the negative solution (root) of the benchmark (Equation (9)) exists.

The positive solution (root) x_2 of the benchmark (Equation (9)) can be determined on the basis of Equations (15) and (16) and its span function $s(x_{gr}^2)$ is as follows:

$$s(x_2^{gr}) = \left[\frac{-3 + \mu + \sqrt{37 + 10\mu - 11\mu^2}}{14 - 6\mu}, \frac{-1 - \mu + \sqrt{13 + 34\mu - 11\mu^2}}{2 + 6\mu} \right]. \tag{27}$$

The minimum, left border of this function is evaluated for the RDM-variable set $\alpha_{d_1} = \alpha_{d_2} = 1, \alpha_{d_3} = \alpha_{d_4} = 0$ and the maximum, right border corresponds to the set: $\alpha_{d_1} = \alpha_{d_2} = 0, \alpha_{d_3} = \alpha_{d_4} = 1$. Span functions determined by multidimensional fuzzy arithmetic and the standard approach are shown in Figure 7.

The result in Figure 7b delivered by Newton's method seems to be of better quality (less uncertain), as its support is narrower than the support of the span function calculated using multidimensional fuzzy arithmetic (Figure 7a). If we expect the correct result to be precise and true, such a conclusion would be considered incorrect. The correct result does not necessarily mean less uncertain (narrower).

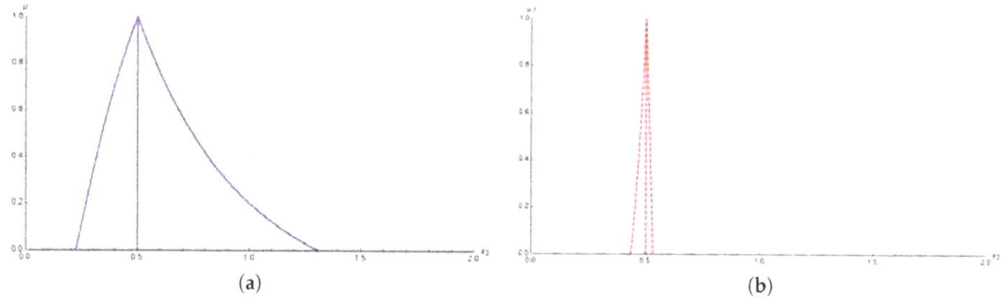

Figure 7. The span function $s(x_2^{gr})$ of the positive root determined by MF-arithmetic (a) and with standard approach using Newton's method [5,8] (b). (a) MF-arithmetic approach; (b) standard approach.

Let us check whether the standard approach result is possible and true. According to [8] possible solutions x_2 (only the positive root) for the benchmark (9) on the membership level $\mu = 0$ are in the range $[0.4343, 0.5307]$. As a consequence every x_2 lying outside of the interval is not possible. The validity of this conclusion can be easily verified by the following reasoning. The values of RDM-variables $\alpha_{d_1} = \alpha_{d_2} = 1, \alpha_{d_3} = \alpha_{d_4} = 0$ and $\mu = 0$ correspond to coefficients:

$$d_1 = 8, d_2 = 2, d_3 = 1, d_4 = 1, \tag{28}$$

which generate the possible benchmark (9) form:

$$8x^2 + 3x = x^2 + 1. \tag{29}$$

Such quadratic equation has two roots and the positive root is $x_2 = 0.220197$. It means that such root value is possible. However, according to the standard approach using Newton's method, this value is impossible.

Moreover, RDM-variable values $\alpha_{d_1} = \alpha_{d_2} = 0, \alpha_{d_3} = \alpha_{d_4} = 1$ and $\mu = 0$ resulted in

$$d_1 = 4, d_2 = 1, d_3 = 3, d_4 = 3 \tag{30}$$

coefficients that transforms the benchmark Equation (9) into the form

$$4x^2 + x = 3x^2 + 3. \tag{31}$$

This quadratic equation has a positive root $x_2 = 1.30278$, which according to the standard approach (Figure 7b), is impossible. Other intermediate, possible values of the root x_2 can be achieved for other combinations of RDM-variables $\alpha_{d_1}, \alpha_{d_2}, \alpha_{d_3}, \alpha_{d_4} \in [0, 1]$.

5.3. Final Comparison

The total comparison of all solutions (roots) obtained with multidimensional fuzzy arithmetic and with the standard approach using Newton's method and parametric FN representation is shown in Figure 8.

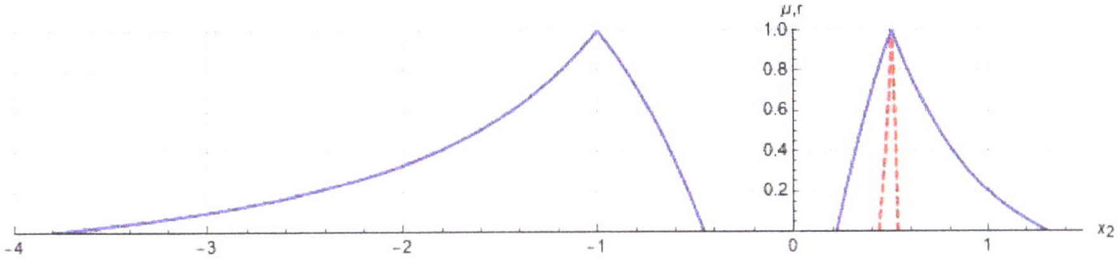

Figure 8. All solutions (roots) calculated with multidimensional fuzzy arithmetic (continuous line) and with the standard approach using Newton's method (dashed line).

We claim that the results of the standard approach are, to some degree, imprecise. Let us calculate the imprecision based on Definition 7. The standard Newton method excludes over 91% possible positive root solutions compared to possible solutions delivered by multidimensional fuzzy arithmetic:

$$Imprecision(standardApproach) = \frac{(0.0964-1.08258)}{1.08258} \cdot 100\% \\ = -91.095\%. \tag{32}$$

The number of solutions excluded from the negative root is 100%.

The experimental verification shows that roots: $x_1 \in [-3.39179, -0.4560838]$ and $x_2 \in [0.220197, 1.30278]$ computed by MF-arithmetic are true and possible. On this basis, we can conclude that the standard Newton method solution $x_2 \in [0.4343, 0.5307]$ is significantly imprecise.

6. Conclusions

Using an example and solving the DFN-equation benchmark, we compare the standard approach based on Newton's method, which uses 2D FN representation to the MF-arithmetic approach. Calculating the imprecision measure, we show that the standard approach results are significantly imprecise, i.e., differs from the universal algebraic solution (UA-solution). According to the standard approach, the benchmark equation's negative root does not exist at all, and the positive root is significantly (over 91%) imprecise (underestimated). The reason for such imprecision is twofold. First is the fuzzy numbers representation and, consequently, the low-dimensional fuzzy calculus, i.e., striving for a direct calculation of results without prior determination of the multidimensional, universal, algebraic fuzzy result. Second, the assumptions in the fuzzy number representation reject the negative root.

The precise solution of DFN-systems can be achieved with the MF-arithmetic, which works according to the scheme:

1. Get 2D fuzzy inputs,
2. Calculate a multidimensional direct fuzzy result,
3. Calculate the 2D secondary result.

Based on the presented example, we stated that without the multidimensional fuzzy calculation, in general, true and precise solutions for DFN-systems are impossible to obtain. The standard approach has limited applications for solving dual fuzzy nonlinear systems due to a high imprecision, i.e., by rejecting a range of possible values.

In future work, we explore MF-arithmetic applicability for solving other types of equations and systems of equations. We plan to apply our approach to real decision-making problems.

Author Contributions: Conceptualization, A.P. and J.K.; methodology, A.P., J.K.; software, J.K.; validation, A.P., J.K. and W.S.; formal analysis, A.P. and J.K.; investigation, J.K.; resources, J.L. and W.S.; data curation, J.K.; writing—original draft preparation, J.K., A.P. and W.S.; writing—review and editing, J.K. and W.S.; visualization, J.K.; supervision, A.P. and W.S.; project administration, J.K.; funding acquisition, W.S. All authors have read and agreed to the published version of the manuscript.

Funding: The work was supported by statutory funds of the Research Team on Intelligent Decision Support Systems, Department of Artificial Intelligence and Applied Mathematics, Faculty of Computer Science and Information Technology, West Pomeranian University of Technology in Szczecin.

Acknowledgments: The authors would like to thank the editor and the anonymous reviewers, whose insightful comments and constructive suggestions helped us to significantly improve the quality of this paper.

Conflicts of Interest: The authors declare no conflict of interest.

Abbreviations

The following abbreviations are used in this manuscript:

DFN-system	Dual Fuzzy Nonlinear System
FN	Fuzzy Number
RDM	Relative Distance Measure
MF-arithmetic	Multidimensional Fuzzy Arithmetic

References

1. Buckley, J.J.; Qu, Y. Solving Linear and Quadratic Fuzzy Equations. *Fuzzy Sets Syst.* **1990**, *38*, 43–59. [CrossRef]
2. Abbasbandy, S.; Asady, B. Newton's Method for Solving Fuzzy Nonlinear Equations. *Appl. Math. Comput.* **2004**, *159*, 349–356. [CrossRef]
3. Abbasbandy, S.; Ezzati, R. Newton's method for solving a system of fuzzy nonlinear equations. *Appl. Math. Comput.* **2006**, *175*, 1189–1199. [CrossRef]
4. Ramli, A.; Abdullah, M.L.; Mamat, M. Broyden's method for solving fuzzy nonlinear equations. *Adv. Fuzzy Syst.* **2010**, *2010*, 763270. [CrossRef]
5. Kajani, M.T.; Asady, B.; Vencheh, A.H. An Iterative Method for Solving Dual Fuzzy Nonlinear Equations. *Appl. Math. Comput.* **2005**, *167*, 316–323. [CrossRef]
6. Mosleh, M. Solution of dual fuzzy polynomial equations by modified Adomian decomposition method. *Fuzzy Inf. Eng.* **2013**, *5*, 45–56. [CrossRef]
7. Waziri, M.Y.; Majid, Z.A. A New Approach for Solving Dual Fuzzy Nonlinear Equations Using Broyden's and Newton's Methods. *Adv. Fuzzy Syst.* **2012**, *2012*, 682087. [CrossRef]
8. Waziri, M.Y.; Moyi, A.U. An Alternative Approach for Solving Dual Fuzzy Nonlinear Equations. *Int. J. Fuzzy Syst.* **2016**, *18*, 103–107. [CrossRef]
9. Wang, X.; Qin, Y.; Qian, W.; Zhang, S.; Fan, X.A. Family of Newton Type Iterative Methods for Solving Nonlinear Equations. *Algorithms* **2015**, *8*, 786–798. [CrossRef]
10. Landowski, M. Method with horizontal fuzzy numbers for solving real fuzzy linear systems. *Soft Comput.* **2019**, *23*, 3921–3933. [CrossRef]
11. Piegat, A.; Landowski, M. Fuzzy Arithmetic Type 1 with Horizontal Membership Functions. In *Uncertainty Modeling: Dedicated to Professor Boris Kovalerchuk on his Anniversary*; Springer International Publishing: Cham, Switzerland, 2017; pp. 233–250.
12. Piegat, A.; Landowski, M. Is Fuzzy Number the Right Result of Arithmetic Operations on Fuzzy Numbers? In *Advances in Fuzzy Logic and Technology 2017*; Kacprzyk, J., Szmidt, E., Zadrożny, S., Atanassov, K.T., Krawczak, M., Eds.; Springer International Publishing: Cham, Switzerland, 2018; pp. 181–194.
13. Lodwick, W.A.; Dubois, D. Interval Linear Systems As a Necessary Step in Fuzzy Linear Systems. *Fuzzy Sets Syst.* **2015**, *281*, 227–251. [CrossRef]
14. Piegat, A.; Pluciński, M. Fuzzy number division and the multi-granularity phenomenon. *Bull. Pol. Acad. Sci. Tech. Sci.* **2017**, *65*, 497–511. [CrossRef]
15. Piegat, A.; Landowski, M. Why Multidimensional Fuzzy Arithmetic? In *Proceedings of the 13th International Conference on Theory and Application of Fuzzy Systems and Soft Computing—ICAFS-2018, Istanbul, Turkey, 28–30 October 2018*; Aliev, R.A., Kacprzyk, J., Pedrycz, W., Jamshidi, M., Sadikoglu, F.M., Eds.; Springer International Publishing: Cham, Switzerland, 2019; pp. 16–23.

16. Piegat, A.; Landowski, M. Two Interpretations of Multidimensional RDM Interval Arithmetic-Multiplication and Division. *Int. J. Fuzzy Syst.* **2013**, *15*, 488–496.
17. Piegat, A.; Landowski, M. Correctness-checking of uncertain-equation solutions on example of the interval-modal method. In *Modern Approach in Fuzzy Sets, Intuitionistic Fuzzy Sets, Generalized Nets and Related Topics*; IBS PAN: Warswa, Poland, 2014; pp. 159–170.

mathematics

Article
Construction of Fuzzy Measures over Product Spaces

Fernando Reche, María Morales * and Antonio Salmerón

Department of Mathematics and Center for the Development and Transfer of Mathematical Research to Industry (CDTIME), University of Almería, 04120 Almería, Spain; fernando.reche@ual.es (F.R.); antonio.salmeron@ual.es (A.S.)
* Correspondence: maria.morales@ual.es

Received: 26 August 2020; Accepted: 15 September 2020; Published: 17 September 2020

Abstract: In this paper, we study the problem of constructing a fuzzy measure over a product space when fuzzy measures over the marginal spaces are available. We propose a definition of independence of fuzzy measures and introduce different ways of constructing product measures, analyzing their properties. We derive bounds for the measure on the product space and show that it is possible to construct a single product measure when the marginal measures are capacities of order 2. We also study the combination of real functions over the marginal spaces in order to produce a joint function over the product space, compatible with the concept of marginalization, paving the way for the definition of statistical indices based on fuzzy measures.

Keywords: fuzzy measures; monotone measures; product spaces

1. Introduction

Fuzzy measures, also known as capacities [1], non-additive measures, or monotone measures [2] emerged as an extension of classic probabilistic measure theory by relaxing the additivity property. Fuzzy measures started to receive significant interest from the scientific community due to Choquet's work on capacities [1], but it was Sugeno [3] who first used the term fuzzy measure in relation to non-additive measures on finite domains.

Examples of fuzzy measures can also be found in contexts related to probability theory, dating back to the works by Dempster [4] and Shafer [5], who studied the substitution of the additivity property by superadditivity and subadditivity, resulting in the so-called belief and plausibility measures, respectively, and showed that they can be regarded as probability intervals.

In this paper, we are interested in the problem of constructing a fuzzy measure over a product space when fuzzy measures over the marginal spaces are available. From a practical point of view, this can be regarded as the extension of particular information, given by the fuzzy measures on the marginal spaces, to a more general setting, determined by the product space. We also study the combination of real functions defined on the marginal spaces, in order to obtain a function over the product space coherent with the initial functions.

The problem of composing fuzzy measures has received remarkable attention in the last years, but most of the works consider measures defined over the same space, and typically over the same σ-algebra [6]. With the motivation of reducing the number of parameters involved in the definition of a fuzzy measure, a variety of particular types of fuzzy measures have been studied, like m-separable fuzzy measures [7], that take advantage of the structure of the space where the measure is defined in order to obtain a compact representation. k-maxitive fuzzy measures [8] have been recently proposed as a way of encoding the interactions between the subsets of the reference set. Previously, the internal structure of fuzzy measures showing partial additivity was studied in [9]. The combination of the elements in the decomposition is carried out using different aggregation measures, including copulas [10].

The problem of combining fuzzy measures from marginal spaces in order to obtain a fuzzy measure over a product space has been approached from different perspectives, fundamentally based on the concept of conditioning [11–18]. Recently, the problem has been studied within the context of game theory as a way of representing coalitions between agents [19]. The measure over the product space is obtained by combining a fuzzy measure over a given σ-algebra $\mathcal{A}$ with a Lebesgue measure over the Borel σ-algebra on the interval $[0,1]$.

In this paper, we consider a more general setting, in which the measures to be combined are general fuzzy measures over potentially different spaces. The rest of the paper is organized as follows. Section 2 is devoted to give the necessary basic definitions and preliminaries. The original contributions in this paper are presented in Sections 3 and 4, covering, respectively, the combination of fuzzy measures and real functions. The paper ends with conclusions in Section 5.

2. Preliminaries

Definition 1. *Let X be a set and $\mathcal{A}$ be a non-empty class of subsets of X so that $X \subset \mathcal{A}$ and $\emptyset \subset \mathcal{A}$. We say that $\mu : \mathcal{A} \longrightarrow [0,1]$ is a fuzzy measure if the following conditions hold.*

1. $\mu(\emptyset) = 0$.
2. $\mu(X) = 1$.
3. $\forall A, B \in \mathcal{A}$ *such that* $A \subseteq B$ *it holds that* $\mu(A) \leq \mu(B)$.
4. *If* $\{A_n\}_{n\in\mathbb{N}} \in \mathcal{A}$ *such that* $A_1 \subseteq A_2 \subseteq \ldots$ *and* $\bigcup_{n=1}^{\infty} A_n \in \mathcal{A}$, *then*

$$\lim_n \mu(A_n) = \mu\left(\bigcup_{n=1}^{\infty} A_n\right). \tag{1}$$

5. *If* $\{A_n\}_{n\in\mathbb{N}} \in \mathcal{A}$ *such that* $A_1 \supseteq A_2 \supseteq \ldots$ *and* $\bigcap_{n=1}^{\infty} A_n \in \mathcal{A}$, *then*

$$\lim_n \mu(A_n) = \mu\left(\bigcap_{n=1}^{\infty} A_n\right). \tag{2}$$

The triplet $(X, \mathcal{A}, \mu)$ is a *measurable space*, and X is called the *reference set*.

In this paper, we will only consider finite spaces that are sufficient to cover a wide range of applications domains [20]. Note that, in this case, the continuity conditions in Equations (1) and (2) always hold. Furthermore, we will also assume that $\mathcal{A}$ is the power set of X, i.e., the set of all subsets of X. In order to simplify the notation, from now on we will write μ_i for $\mu(\{x_i\})$, and μ_A for $\mu(A)$.

It can be shown [21] that a fuzzy measure over a reference set of cardinality n is equivalent to $n!$ probability functions, each one of them associated with one possible permutation of the elements in the reference set. We will denote by X^σ the ordering of the elements of X according to permutation σ, so that $X^\sigma = \{x_{(1)}, \ldots, x_{(n)}\}$.

Definition 2. *[21] Let $(X, \mathcal{A}, \mu)$ be a measurable space. The probability functions associated with μ and X^σ are defined as the set $P_\sigma = \{p_\sigma(x_{(1)}), \ldots, p_\sigma(x_{(n)})\}$ such that*

$$p_\sigma(x_{(i)}) = \begin{cases} \mu(A_{(i)}) - \mu(A_{(i+1)}) & \text{if } i < n, \\ \mu(x_{(n)}) & \text{if } i = n, \end{cases} \tag{3}$$

where $A_{(i)} = \{x_{(i)}, \ldots, x_{(n)}\}$.

It is straightforward to show that $0 \leq p_\sigma(x_i) \leq 1$ and $\sum_{i=1}^{n} p_\sigma(x_i) = 1$ for any σ.

Definition 3. [21] *Let $(X, \mathcal{A}, \mu)$ be a measurable space and let P_σ be the probability function associated with μ and X^σ. The probability measure generated by μ and X^σ is*

$$P_\sigma(A) = \sum_{x \in A} p_\sigma(x). \tag{4}$$

When it is clear from the context, we will use P_σ for both the probability function and probability measure.

Note that there are as many probability measures generated by μ as there are possible permutations of the element of the reference set. However, not all those measures are necessarily different.

An important property of the generated probability measure is that it bounds the underlying fuzzy measure in the extreme cases, as stated in the following proposition [22].

Proposition 1. *Let $(X, \mathcal{A}, \mu)$ be a measurable space and let $\{P_\sigma\}_{\sigma \in S_n}$ be the set of all the probability measures generated by μ. Then,*

$$\min_\sigma P_\sigma(A) \le \mu(A) \le \max_\sigma P_\sigma(A). \tag{5}$$

Two types of fuzzy measures that are specially relevant for this paper are belief functions [4,5] and capacities of order 2 [1,23].

Definition 4. *Let $(X, \mathcal{A}, \mu)$ be a measurable space. Given a function $m : \mathcal{P}(X) \to [0,1]$ such that $m(\emptyset) = 0$ and $\sum_{A \in \mathcal{P}(X)} m(A) = 1$, we say that μ is a belief function if*

$$\mu(A) = \sum_{B \in \mathcal{P}(X) | B \subseteq A} m(B), \quad \forall A \in \mathcal{P}(X), \tag{6}$$

where $\mathcal{P}(X)$ denotes the power set of X.

From now on, if μ is a belief function we will denote $\mu(A)$ as $\mathrm{Bel}\,(A)$.

Definition 5. *Let $(X, \mathcal{A}, \mu)$ be a measurable space. We say that μ is a monotone capacity of order 2 if*

$$\mu(A \cup B) + \mu(A \cap B) \ge \mu(A) + \mu(B), \quad \forall A, B \in \mathcal{A}. \tag{7}$$

Definition 6. *Let $(X, \mathcal{A}, \mu)$ be a measurable space. We say that μ is an alternating capacity of order 2 if*

$$\mu(A \cup B) + \mu(A \cap B) \le \mu(A) + \mu(B), \quad \forall A, B \in \mathcal{A}. \tag{8}$$

The following result [22] links capacities of order 2 with the probability measures generated by the fuzzy measure and will be key in the proof of one of the results in this paper.

Theorem 1. *A fuzzy measure μ is a monotone (alternating) capacity of order 2 if and only if*

$$\mu(A) = \min_\sigma P_\sigma(A) \quad \left(\mu(A) = \max_\sigma P_\sigma(A)\right) \tag{9}$$

for all $A \in \mathcal{A}$, being $\{P_\sigma\}_{\sigma \in S_n}$ the set of probability measures generated by μ (see Definition 3).

We will consider two possible scenarios related to a given product space $X_1 \times X_2$, where $\times$ denotes the Cartesian product:

- A fuzzy measure defined over the product space is available.

- Two fuzzy measures, respectively, defined over X_1 and X_2 are available, but no measure over $X_1 \times X_2$ is known.

In the first scenario, we are interested in particularizing the information contained in the fuzzy measure over the unidimensional spaces. Therefore, we need to define a marginalization operation over the measure on the product space.

In the second scenario, we focus on building a fuzzy measure over the product space, by combining the two measures over the marginal spaces. Thus, we need to define an appropriate way of combining fuzzy measures.

Likewise marginal spaces, in a measurable product space we will assume the product class $\mathcal{A}_{X_1 \times X_2}$ to be the power set of $X_1 \times X_2$, i.e., $\mathcal{A}_{X_1 \times X_2} = \mathcal{P}(X_1 \times X_2)$, which is not the same as $\mathcal{P}(X_1) \times \mathcal{P}(X_2)$.

Among the possible elements of a product class, we are particularly interested in those that can be obtained from sets in the marginal space. They are called *rectangles* and are formally defined as follows.

Definition 7. *Let $(X_1, \mathcal{A}_{X_1})$ and $(X_2, \mathcal{A}_{X_2})$ be two spaces where $\mathcal{A}_{X_1}$ and $\mathcal{A}_{X_2}$ are classes defined on X_1 and X_2, respectively. We define the class of rectangles of $\mathcal{A}_{X_1 \times X_2}$ as*

$$\mathcal{R} = \{H \in \mathcal{A}_{X_1 \times X_2} |\ H = A \times B, \ \textit{where } A \in \mathcal{A}_{X_1}, B \in \mathcal{A}_{X_2}\}. \tag{10}$$

Taking into account that we are assuming $\mathcal{A}_{X_1 \times X_2} = \mathcal{P}(X_1 \times X_2)$, it is easy to show that $\mathcal{R}$ is closed for intersections, but not for unions.

We will make use of the concept of triangular norm and conorm. Both are operators that raised within the context of probabilistic metric spaces [24]. They have also been widely used by the theory of fuzzy sets [25–29] as an extension of classic operations over sets.

Definition 8 ([24]). *An operator $T : [0,1]^2 \longrightarrow [0,1]$ is a triangular norm or t-norm for short, if it satisfies the following conditions.*

1. *$T(0, a) = 0$, $T(a, 1) = a$ for all $a \in [0,1]$. (Boundary conditions)*
2. *$T(a, b) = T(b, a)$. (Commutativity)*
3. *If $a \leq c$ and $b \leq d$, then $T(a, b) \leq T(c, d)$. (Monotonicity)*
4. *$T(T(a, b), c) = T(a, T(b, c))$. (Asocciativity)*

Example 1. *Some examples of t-norms are*

1. $T_0(x_1, x_2) = \begin{cases} \min\{x_1, x_2\} & \textit{if } \max\{x_1, x_2\} = 1, \\ 0 & \textit{otherwise.} \end{cases}$
2. $T_1(x_1, x_2) = \max\{x_1 + x_2 - 1, 0\}$.
3. $T_2(x_1, x_2) = x_1 x_2$.
4. $T_3(x_1, x_2) = \min\{x_1, x_2\}$.

Note that any t-norm T is always bounded by T_0 and T_3 in the following way.

$$T_0(x_1, x_2) \leq T(x_1, x_2) \leq T_3(x_1, x_2). \tag{11}$$

Definition 9 ([24]). *An operator $T : [0,1]^2 \longrightarrow [0,1]$ is a triangular conorm or t-conorm for short, if it satisfies the following properties.*

1. *$S(1, a) = 1$, $S(a, 0) = a$ for all $a \in [0,1]$. (Boundary conditions)*
2. *$S(a, b) = S(b, a)$. (Commutativity)*
3. *If $a \leq c$ and $b \leq d$, then $S(a, b) \leq S(c, d)$. (Monotonicity)*
4. *$S(S(a, b), c) = S(a, S(b, c))$. (Asocciativity)*

Given any t-norm T, a t-conorm S can always be constructed as

$$S(a,b) = 1 - T(1-a, 1-b). \tag{12}$$

Example 2. *Applying Equation (12) to the t-norms in Example 1, we obtain the following t-conorms.*

1. $S_0(x_1,x_2) = \begin{cases} \max\{x_1,x_2\} & \text{if } \min\{x_1,x_2\} = 0, \\ 1 & \text{otherwise.} \end{cases}$
2. $S_1(x_1,x_2) = \min\{x_1+x_2, 1\}$.
3. $S_2(x_1,x_2) = x_1 + x_2 - x_1x_2$.
4. $S_3(x_1,x_2) = \max\{x_1,x_2\}$.

A similar boundary condition as expressed in Equation (11) for t-norms, holds for t-conorms:

$$S_3(x_1,x_2) \leq S(x_1,x_2) \leq S_0(x_1,x_2). \tag{13}$$

A thorough study of the use of t-norms and t-conorms in the context of fuzzy measures and fuzzy sets can be found in [28].

Functions can be integrated with respect to a fuzzy measure using Choquet integral, which is a generalization of Lebesgue integral to non-additive monotone measures [1]. In the particular case of additive measures, Choquet and Lebesgue integrals coincide. It is formally defined as follows.

Definition 10. *Let $(X, \mathcal{A}, \mu)$ be a measurable space, and let h be a measurable real function of X. The Choquet integral of h with respect to μ is*

$$\oint_A h \circ \mu = \int_{-\infty}^{0} (\mu(H_\alpha \cap A) - 1)\, d\alpha + \int_0^{\infty} \mu(H_\alpha \cap A)\, d\alpha \tag{14}$$

where $A \in \mathcal{A}$ and H_α are the α-cuts *of h, defined as*

$$H_\alpha = \{x \in X / h(x) \geq \alpha\}. \tag{15}$$

If the reference set is finite, the integral can be expressed as

$$\oint h \circ \mu = h(x_{(1)})\mu(A_{(1)}) + \sum_{i=2}^{n} \mu(A_{(i)})[h(x_{(i)}) - h(x_{(i-1)})], \tag{16}$$

where X^σ is an ordering such that $h(x_{(1)}) \leq h(x_{(2)}) \leq \ldots \leq h(x_{(n)})$ and the sets $A_{(i)}$ are of the form $\{x_{(i)}, x_{(i+1)}, \ldots, x_{(n)}\}$.

3. Combining Fuzzy Measures

The main difficulty when combining fuzzy measures from marginal spaces in order to obtain a fuzzy measure over a product space is that, unlike probability measures, we cannot follow a procedure based on extending the measures, as additivity is required [11–18].

For instance, for sets of the form $A \times B$, where $A \in \mathcal{A}_{X_1}$ and $B \in \mathcal{A}_{X_2}$, we could define $\mu(A \times B) = \mu_{X_1}(A) \otimes \mu_{X_2}(B)$ for some appropriate operator $\otimes$. In the case of probability measures, this would suffice as, due to additivity, the measure can easily be extended to arbitrary sets of $X_1 \times X_2$ using integrals [30]. More precisely, in the case of additive measures, the product measure for sets of the form $A \times B$, is given by $\mu_{X_1}(A)\mu_{X_2}(B)$, while for the rest of sets $Q \subseteq X_1 \times X_2$, the product measure is computed as

$$\mu(Q) = \int \mu_{X_2}(Q_{x_1})\mu_{X_1}(dx_1) = \int \mu_{X_1}(Q_{x_2})\mu_{X_2}(dx_2) \tag{17}$$

where $Q_{x_1} = \{x_2 \in X_2 | \ (x_1, x_2) \in Q\}$ and $Q_{x_2} = \{x_1 \in X_1 | \ (x_1, x_2) \in Q\}$.

The same construction is not always possible In the case of non-additive measures, because the integrals in Equation (17) can be different [14]. It happens, for instance, if we use Choquet integral [1].

3.1. $\odot$-Independent Measures

Consider two measurable spaces $(X_1, \mathcal{A}_{X_1}, \mu_1)$ and $(X_2, \mathcal{A}_{X_2}, \mu_2)$. Our goal is to construct a fuzzy measure over the product space in a sensible way. We start off by defining a fuzzy measure over the product space, compatible with the marginal measures:

Definition 11. *A product fuzzy measure of μ_1 and μ_2 is a function $\mu_{12} : \mathcal{A}_{X_1 \times X_2} \longrightarrow [0,1]$ satisfying*

1. $\mu_{12}(\emptyset) = 0$, $\mu_{12}(X_1 \times X_2) = 1$.
2. *For all $A, B \in \mathcal{A}_{X_1 \times X_2}$ such that $A \subseteq B$ it holds that $\mu_{12}(A) \leq \mu_{12}(B)$.*
3. *For all $A \in \mathcal{A}_{X_1}$, it holds that $\mu_{12}(A \times X_2) = \mu_1(A)$.*
4. *For all $B \in \mathcal{A}_{X_2}$, it holds that $\mu_{12}(X_1 \times B) = \mu_2(B)$.*

The next step is to guarantee that the composition of measures using the product in Definition 11 is compatible with the concept of independence. More precisely, assuming independence between two fuzzy measures, their product fuzzy measure should be possible to be obtained using exclusively the two original fuzzy measures. In this work, we will assume that two fuzzy measures are independent if they can be composed resulting in a product fuzzy measure within the class $\mathcal{R}$. This is formally defined through the concept of $\odot$-independence.

Definition 12. *Let $(X_1, \mathcal{A}_{X_1}, \mu_1)$ and $(X_2, \mathcal{A}_{X_2}, \mu_2)$ be measurable spaces. We say that μ_1 and μ_2 are $\odot$-independent fuzzy measures if there exists a product fuzzy measure $\mu_{12}^{\odot}$ satisfying that for any $H \in \mathcal{R}$ it holds that*

$$\mu_{12}^{\odot}(H) = \mu_1(A) \odot \mu_2(B), \tag{18}$$

where $H = A \times B$ and $\odot$ is a t-norm.

From now on we will refer to this measure as the *$\odot$-independent product* of μ_1 and μ_2.

The next proposition shows that the $\odot$-independent product results in a well defined fuzzy measure on $\mathcal{R}$.

Proposition 2. *Let $(X_1, \mathcal{A}_{X_1}, \mu_1)$ and $(X_2, \mathcal{A}_{X_2}, \mu_2)$ be measurable spaces. The $\odot$-independent product of μ_1 and μ_2, $\mu_{12}^{\odot}$, is a fuzzy measure on $\mathcal{R}$.*

Proof. Let $H \in \mathcal{R}$ such that $H = A \times B$ with $A \in \mathcal{A}_{X_1}$ and $B \in \mathcal{A}_{X_2}$. We have to show that the conditions in Definition 11 are satisfied by $\mu_{12}^{\odot}$.

1. It is clear that $\emptyset \in \mathcal{R}$ as it can be expressed as $\emptyset = A \times B$ if at least one of them (A or B) is the empty set. In such case, $\mu_1(A)$, $\mu_2(B)$, or both are equal to zero (since they are fuzzy measures). Hence, $\mu_{12}^{\odot}(\emptyset) = 0$ because $\odot$ is a t-norm.
2. $X_1 \times X_2$ trivially belongs to $\mathcal{R}$, and thus $\mu_{12}^{\odot}(X_1 \times X_2) = \mu_1(X_1) \odot \mu_2(X_2) = 1 \odot 1 = 1$.
3. Let $H_1 \subseteq H_2 \in \mathcal{R}$. Assume that $H_1 = A_1 \times B_1$ and $H_2 = A_2 \times B_2$. Then, $A_1 \subseteq A_2$ and $B_1 \subseteq B_2$. As the t-norm $\odot$ is a monotone operator, it holds that

$$\mu_{12}^{\odot}(H_1) = \mu_1(A_1) \odot \mu_2(B_1) \leq \mu_1(A_2) \odot \mu_2(B_2) = \mu_{12}^{\odot}(H_2),$$

which means that $\mu_{12}^{\odot}$ is monotone.

4. Given $A \in \mathcal{A}_{X_1}$, it holds that $A \times X_2 \in \mathcal{R}$ and therefore

$$\mu_{12}^{\odot}(A \times X_2) = \mu_1(A) \odot \mu_2(X_2) = \mu_1(A) \odot 1 = \mu_1(A).$$

5. Analogously, we can see that $\mu_{12}^{\odot}(X_1 \times B) = \mu_2(B)$.

□

Example 3. *Let $(X_1, \mathcal{A}_{X_1}, \mu_1)$ and $(X_2, \mathcal{A}_{X_2}, \mu_2)$ be probabilistic spaces. Therefore, μ_1 and μ_2 are additive measures and $\mathcal{A}_{X_1}$ and $\mathcal{A}_{X_2}$ are algebras. By letting $\odot$ be equal to the product between real numbers, denoted by $\times$, we can define the $\times$-independent product of μ_1 and μ_2 as*

$$\mu_{12}^{\times}(H) = \mu_1(A) \times \mu_2(B) \tag{19}$$

with $H = A \times B \in \mathcal{R}$. It is known that, in this case, there exists a unique additive product measure defined over the smaller algebra generated by the corresponding rectangles [30].

Similar examples can be found for different types of fuzzy measures, as for instance, possibility measures [31,32], illustrating that, given a $\odot$-independent fuzzy measure, it is possible to find a unique product measure compatible with the initial ones. However, uniqueness does not hold in general, so we will instead pursue the idea of defining bounds where the product fuzzy measure lies.

Definition 13. *Let $(X_1, \mathcal{A}_{X_1}, \mu_1)$ and $(X_2, \mathcal{A}_{X_2}, \mu_2)$ be measurable spaces. We define the $\odot$-exterior product measure for any $H \in \mathcal{A}_{X_1 \times X_2}$ as*

$$\overline{\mu}_{12}^{\odot}(H) = \min_{A \times B \supseteq H} \mu_1(A) \odot \mu_2(B) \tag{20}$$

where $\odot$ is a t-norm.

Definition 14. *Let $(X_1, \mathcal{A}_{X_1}, \mu_1)$ and $(X_2, \mathcal{A}_{X_2}, \mu_2)$ be measurable spaces. We define the $\odot$-interior product measure for any $H \in \mathcal{A}_{X_1 \times X_2}$ as*

$$\underline{\mu}_{12}^{\odot}(H) = \max_{A \times B \subseteq H} \mu_1(A) \odot \mu_2(B) \tag{21}$$

where $\odot$ is a t-norm.

These definitions are more general than the ones introduced in [13], where the product t-norm is used instead. Figure 1 shows a representation of the $\odot$-interior and exterior product measures corresponding, respectively, to the contained rectangle of larger measure and the containing rectangle of lower measure.

The next results shows that both $\odot$-interior and exterior product measures are indeed product fuzzy measures.

Proposition 3. *Let $(X_1, \mathcal{A}_{X_1}, \mu_1)$ and $(X_2, \mathcal{A}_{X_2}, \mu_2)$ be measurable spaces. Then, the $\odot$-interior (resp. exterior) product measure is a product fuzzy measure of μ_1 and μ_2.*

Proof. We will show that $\underline{\mu}_{12}^{\odot}(H)$ satifies Definition 11. The proof for the $\odot$-exterior product measure is analogous. Note that

$$\underline{\mu}_{12}^{\odot}(\emptyset) = \max_{A \times B \subseteq \emptyset} \mu_1(A) \odot \mu_2(B) = 0,$$

as $A \times B \subseteq \emptyset$, and therefore at least one of them will be equal to $\emptyset$. We have also used that, as μ_1 and μ_2 are fuzzy measures and $\odot$ is a t-norm, the value of the $\odot$-interior product measure for the empty set is zero.

The value of the $\odot$-interior product measure for $X_1 \times X_2$ is

$$\begin{aligned} \underline{\mu}_{12}^{\odot}(X_1 \times X_2) &= \max_{A \times B \subseteq X_1 \times X_2} \mu_1(A) \odot \mu_2(B) \\ &= \mu_1(X_1) \odot \mu_2(X_2) \\ &= 1 \odot 1 = 1. \end{aligned}$$

Now we will show that monotonicity also holds. Let $H_1 \subseteq H_2 \subseteq X_1 \times X_2$, then

$$\begin{aligned} \underline{\mu}_{12}^{\odot}(H_1) &= \max_{A_1 \times B_1 \subseteq H_1} \mu_1(A_1) \odot \mu_2(B_1) \\ &= \max_{A_1 \times B_1 \subseteq H_1 \subseteq H_2} \mu_1(A_1) \odot \mu_2(B_1) \\ &\leq \max_{A_2 \times B_2 \subseteq H_2} \mu_1(A_2) \odot \mu_2(B_2) \\ &= \underline{\mu}_{12}^{\odot}(H_2). \end{aligned}$$

That is, as $H_1 \subseteq H_2$ and $\odot$ is a t-norm (and thus a monotone operator), in the worst case, the interior rectangle of larger measure in H_2 has, at least, the same measure as the one in H_1.

Let us check now the compatibility with marginalization. As $A \times X_2$ and $X_1 \times B$ belong to $\mathcal{R}$, and taking into account that $\odot$ is a t-norm, it holds that

$$\begin{aligned} \underline{\mu}_{12}^{\odot}(A \times X_2) &= \mu_1(A) \odot \mu_2(X_2) = \mu_1(A), \\ \underline{\mu}_{12}^{\odot}(X_1 \times B) &= \mu_1(X_1) \odot \mu_2(B) = \mu_2(B). \end{aligned}$$

□

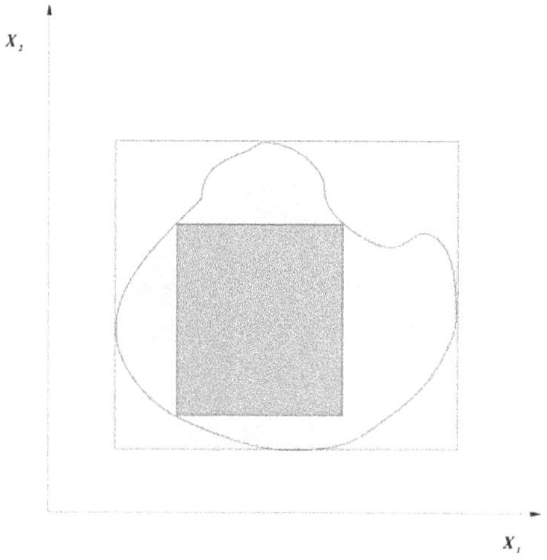

Figure 1. A representation of internal and external product measures.

Note that, for the particular case of the class $\mathcal{R}$, both measures turn out to be the same, i.e., for all $H \in \mathcal{R}$, it holds that

$$\underline{\mu}_{12}^{\odot}(H) = \mu_{12}^{\odot}(H) = \overline{\mu}_{12}^{\odot}(H). \quad (22)$$

Finally, the next result states how, in the general case, $\underline{\mu}_{12}^{\odot}$ and $\overline{\mu}_{12}^{\odot}$ conform lower and upper bounds, respectively, for any $\odot$-independent product fuzzy measure.

Proposition 4. *Let $(X_1, \mathcal{A}_{X_1}, \mu_1)$ and $(X_2, \mathcal{A}_{X_2}, \mu_2)$ be measurable spaces. Given any $\odot$-independent product of μ_1 and μ_2, it holds that for all $C \in \mathcal{A}_{X_1 \times X_2}$,*

$$\underline{\mu}_{12}^{\odot}(C) \leq \mu_{12}^{\odot}(C) \leq \overline{\mu}_{12}^{\odot}(C). \tag{23}$$

Proof. We will only develop the proof for the lower bound. The upper bound case is analogous.

As $\mu_{12}^{\odot}$ is a monotone measure, for any $A \times B \in \mathcal{R}$, with $A \times B \subseteq C \in \mathcal{A}_{X_1 \times X_2}$ it holds that $\mu_{12}^{\odot}(A \times B) \leq \mu(C)$, and therefore

$$\mu_{12}^{\odot}(C) \geq \max_{A \times B \subseteq C} \mu_{12}^{\odot}(A \times B) = \max_{A \times B \subseteq C} \mu_1(A) \odot \mu_2(B) = \underline{\mu}_{12}^{\odot}(C),$$

taking into account that $\mu_{12}^{\odot}$ satisfies that $\mu_{12}^{\odot}(A \times B) = \mu_1(A) \odot \mu_2(B)$. □

Up to this point, we have only been able to provide a way to obtain a combined fuzzy measure when working within the class $\mathcal{R}$. Outside it only bounds have been obtained. In the next section we solve this problem for some particular kinds of fuzzy measures.

3.2. Maximin Product

We will elaborate on an idea introduced in [17] based on an alternative representation of general fuzzy measures. Given any fuzzy measure, it is always possible to construct classes of measures bounding it. Such is the case of

$$\mathcal{M}_P(\mu) = \{P \in \mathfrak{P} |\ P(A) \geq \mu(A)\ \forall A \in \mathcal{A}\} \tag{24}$$

and

$$\mathcal{M}_{Bel}(\mu) = \{Bel(A) \in \mathfrak{B} |\ Bel(A) \leq \mu(A)\ \forall A \in \mathcal{A}\}, \tag{25}$$

where $\mathfrak{P}$ denotes the set of all probability measures and $\mathfrak{B}$ is the set of all belief measures.

It can be shown [17,18] that any fuzzy measure μ can be represented using elements of the classes defined above as

$$\mu(A) = \max_{\beta \in \mathcal{M}_{Bel}(\mu)} \min_{P \in \mathcal{M}_P(\beta)} P(A), \tag{26}$$

and the Choquet integral of any function h with respect to μ can also be computed as

$$\oint h \circ \mu = \max_{\beta \in \mathcal{M}_{Bel}(\mu)} \min_{P \in \mathcal{M}_P(\beta)} \int h\, dP. \tag{27}$$

In other words, given any fuzzy measure, there is always a probability measure whose value matches it for a given subset of the reference set.

Taking this representation as a basis, we have the conditions to propose a product fuzzy measure over $X_1 \times X_2$ as follows.

Definition 15. *Let $(X_1, \mathcal{A}_{X_1}, \mu_1)$ and $(X_2, \mathcal{A}_{X_2}, \mu_2)$ be measurable spaces. We define the maximin product measure as*

$$\mu_1 \otimes \mu_2(A) = \max_{\substack{\beta_i \in \mathcal{M}_{Bel}(\mu_i) \\ i=1,2}} \min_{\substack{P_i \in \mathcal{M}_P(\beta_i) \\ i=1,2}} P_1 \otimes P_2(A), \tag{28}$$

where $P_1 \otimes P_2$ is the standard product for probability measures.

The main drawback of this definition, from a practical point of view, is that it involves solving an optimization problem over very general classes of fuzzy measures. However, we will show that an approximation can be easily obtained taking advantage of the representation of a fuzzy measure by means of a set of probability measures. Remarkably, for some particular types of fuzzy measures, we will also show that, rather than just bounds, we can obtain a precise product fuzzy measure.

Definition 16. *Let $(X_1, \mathcal{A}_{X_1}, \mu_1)$ and $(X_2, \mathcal{A}_{X_2}, \mu_2)$ be measurable spaces and $P^{\mu_1}_{\sigma_1}$ and $P^{\mu_2}_{\sigma_2}$ be the probability functions associated with $X_1^{\sigma_1}$ and $X_2^{\sigma_2}$, respectively. We define the lower product p-measure as*

$$\underline{\mathfrak{m}}_{12}(C) = \min_{\sigma_1,\sigma_2} \left[P^{\mu_1}_{\sigma_1} \otimes P^{\mu_2}_{\sigma_2}(C) \right], \tag{29}$$

where $\otimes$ is the standard probabilistic product.

Definition 17. *Given the conditions in Definition 16, we define the upper product p-measure as*

$$\overline{\mathfrak{m}}_{12}(C) = \max_{\sigma_1,\sigma_2} \left[P^{\mu_1}_{\sigma_1} \otimes P^{\mu_2}_{\sigma_2}(C) \right], \tag{30}$$

where $\otimes$ is the standard probabilistic product.

The construction of these measures comprises the following steps, assuming measurable spaces $(X_1, \mathcal{A}_{X_1}, \mu_1)$ and $(X_2, \mathcal{A}_{X_2}, \mu_2)$ with cardinalities n_1 and n_2, respectively.

- Compute the $n_1!$ probability measures associated with μ_1 and the $n_2!$ corresponding to μ_2.
- Compute the $n_1! \times n_2!$ product probability measures using the standard product.
- The product p-measures are computed, for any given set, using the smaller and larger product probability measure.

In fact, there is no need to know the value of the probabilities for all the possible subsets for all the permutations, as the probability measures are additive, and therefore it suffices to know the values for the elementary events (singletones).

Example 4. *Consider two measurable spaces, $(X_1, \mathcal{A}_{X_1}, \mu_1)$ and $(X_2, \mathcal{A}_{X_2}, \mu_2)$, both with cardinality 3. Table 1 shows an example of a fuzzy measure with all its possible values specified, as well as the associated probability measures corresponding to each permutation of the subsets of the reference set.*

As both reference sets have cardinality 3, each one of them has three proper sets (i.e., excluding the total and empty sets). Thus, $X_1 \times X_2$ contains 9 elements and $2^9 - 2 = 510$ proper subsets. For the sake of readability, we will use the notation $(x_{1i}, x_{2j}) = z_{ij}$, meaning that

$$X_1 \times X_2 = \{z_{11}, z_{12}, z_{13}, z_{21}, z_{22}, z_{23}, z_{31}, z_{32}, z_{33}\}.$$

Let us see how to compute, from the data in Table 1, the product p-measures for some subsets:

- *Consider the unitary subset z_{11}.*

$$\underline{\mathfrak{m}}_{12}(\{z_{11}\}) = \min_{\sigma_1,\sigma_2} P^{\mu_1}_{\sigma_1}(\{x_{11}\}) P^{\mu_2}_{\sigma_2}(\{x_{21}\}) = 0.1 \cdot 0.1 = 0.01.$$

The calculation for unitary subsets is easy, as it requires just to search the permutation with lower value in both components, which in this case are $P^{\mu_1}_{(3,1,2)}$ and $P^{\mu_2}_{(2,3,1)}$.

- *Now we will consider a set from class $\mathcal{R}$, namely, $\{z_{22}, z_{23}\}$,*

$$\begin{aligned}\underline{\mathfrak{m}}_{12}(\{z_{22}, z_{23}\}) &= \min_{\sigma_1,\sigma_2}[P^{\mu_1}_{\sigma_1}(\{x_{12}\})P^{\mu_2}_{\sigma_2}(\{x_{22}\}) + P^{\mu_1}_{\sigma_1}(\{x_{12}\})P^{\mu_2}_{\sigma_2}(\{x_{23}\})] \\ &= \min_{\sigma_1,\sigma_2}[P^{\mu_1}_{\sigma_1}(\{x_{12}\})[P^{\mu_2}_{\sigma_2}(\{x_{22}\}) + P^{\mu_2}_{\sigma_2}(\{x_{23}\})]] \\ &= \min_{\sigma_1,\sigma_2}[P^{\mu_1}_{\sigma_1}(\{x_{12}\})P^{\mu_2}_{\sigma_2}(\{x_{22}, x_{23}\})] \\ &= 0.2 \cdot 0.6 = 0.12.\end{aligned}$$

The calculations here are analogous to the unitary set case, as it is enough to find the permutations returning the minimum value for the projections. In this case, they are $P^{\mu_1}_{(3,2,1)}$ for $\{x_{12}\}$ and $P^{\mu_2}_{(1,3,2)}$ for $\{x_{22}, x_{23}\}$.

- *Consider now a subset outside the class $\mathcal{R}$, for instance $\{z_{11}, z_{22}\}$. We obtain*

$$\begin{aligned}\underline{\mathfrak{m}}_{12}(\{z_{11}, z_{22}\}) &= \min_{\sigma_1,\sigma_2}[P^{\mu_1}_{\sigma_1}(\{x_{11}\})P^{\mu_2}_{\sigma_2}(\{x_{21}\}) + P^{\mu_1}_{\sigma_1}(\{x_{12}\})P^{\mu_2}_{\sigma_2}(\{x_{22}\})] \\ &= 0.3 \cdot 0.1 + 0.2 \cdot 0.6 = 0.15.\end{aligned}$$

This case is more complicated, as it requires exploring all the possible combination of products and finding the permutation returning the lowest value. In this case, they are $P^{\mu_1}_{(3,2,1)}$ and $P^{\mu_2}_{(2,3,1)}$.

Table 1. Probability measures generated by two sample fuzzy measures.

$\mathcal{P}(X_1)$	μ_1	$P^{\mu_1}_{(1,2,3)}$	$P^{\mu_1}_{(1,3,2)}$	$P^{\mu_1}_{(2,1,3)}$	$P^{\mu_1}_{(2,3,1)}$	$P^{\mu_1}_{(3,1,2)}$	$P^{\mu_1}_{(3,2,1)}$
x_{11}	0.3	0.2	0.2	0.15	0.3	0.1	0.3
x_{12}	0.4	0.35	0.4	0.4	0.4	0.4	0.2
x_{13}	0.45	0.45	0.4	0.45	0.3	0.5	0.5
x_{11}, x_{12}	0.5	0.55	0.6	0.55	0.7	0.5	0.5
x_{11}, x_{13}	0.6	0.65	0.6	0.6	0.6	0.6	0.8
x_{12}, x_{13}	0.8	0.8	0.8	0.85	0.7	0.9	0.7
$\mathcal{P}(X_2)$	μ_2	$P^{\mu_2}_{(1,2,3)}$	$P^{\mu_2}_{(1,3,2)}$	$P^{\mu_2}_{(2,1,3)}$	$P^{\mu_2}_{(2,3,1)}$	$P^{\mu_2}_{(3,1,2)}$	$P^{\mu_2}_{(3,2,1)}$
x_{21}	0.1	0.4	0.4	0.2	0.1	0.2	0.1
x_{22}	0.5	0.4	0.5	0.6	0.6	0.5	0.6
x_{23}	0.2	0.2	0.1	0.2	0.3	0.3	0.3
x_{21}, x_{22}	0.7	0.8	0.9	0.8	0.7	0.7	0.7
x_{21}, x_{23}	0.4	0.6	0.5	0.4	0.4	0.5	0.4
x_{22}, x_{23}	0.6	0.6	0.6	0.8	0.9	0.8	0.9

The calculations are similar for the upper product p-measure, resulting in

$$\begin{aligned}\overline{\mathfrak{m}}_{12}(\{z_{11}\}) &= 0.12, \\ \overline{\mathfrak{m}}_{12}(\{z_{22}, z_{23}\}) &= 0.36, \\ \overline{\mathfrak{m}}_{12}(\{z_{11}, z_{22}\}) &= 0.32.\end{aligned}$$

Note that the product p-measures provide an interval of measure over the product space, but they are obtained in a rather different way than the exterior and interior product measures defined in Equations (20) and (21). We will analyze now some remarkable properties of product p-measures, first of all checking that they are actually fuzzy measures.

Proposition 5. *$\underline{\mathfrak{m}}_{12}$ and $\overline{\mathfrak{m}}_{12}$ are fuzzy measures.*

Proof. Consider $\underline{\mathfrak{m}}_{12}(C) = \min_{\sigma_1,\sigma_2} P^{\mu_1}_{\sigma_1} \otimes P^{\mu_2}_{\sigma_2}(C)$. As $\otimes$ is the standard probabilistic product and we are assuming the reference set to be finite, it holds that there exist two permutations σ_1 and σ_2 such that

$$\underline{\mathfrak{m}}_{12}(C) = \sum_{(x_{1i},x_{2j})\in C} p_{\sigma_1}(x_{1i})p_{\sigma_2}(x_{2j}).$$

Given $C \subset H$, it follows from Definition 16 that there exist two permutations τ_1 and τ_2 such that

$$\begin{aligned}\underline{\mathfrak{m}}_{12}(H) &= \sum_{(x_{1i},x_{2j})\in H} p_{\tau_1}(x_{1i})p_{\tau_2}(x_{2j}) \\ &= \sum_{(x_{1i},x_{2j})\in C} p_{\tau_1}(x_{1i})p_{\tau_2}(x_{2j}) + \sum_{(x_{1i},x_{2j})\in H-C} p_{\tau_1}(x_{1i})p_{\tau_2}(x_{2j}).\end{aligned}$$

As σ_1 and σ_2 are the permutations that minimize the product probability of C, it holds that

$$\begin{aligned}\underline{\mathfrak{m}}_{12}(H) &\geq \sum_{(x_{1i},x_{2j})\in C} p_{\sigma_1}(x_{1i})p_{\sigma_2}(x_{2j}) + \sum_{(x_{1i},x_{2j})\in H-C} p_{\tau_1}(x_{1i})p_{\tau_2}(x_{2j}) \\ &= \underline{\mathfrak{m}}_{12}(C) + \sum_{(x_{1i},x_{2j})\in H-C} p_{\tau_1}(x_{1i})p_{\tau_2}(x_{2j})\end{aligned}$$

and thus the measure is monotone. The proof for the upper product p-measure is analogous. □

However, $\underline{\mathfrak{m}}_{12}$ and $\overline{\mathfrak{m}}_{12}$ are not, in general, product fuzzy measures of μ_1 and μ_2 since they can often fail to be consistent with the marginalization, i.e.

$$\underline{\mathfrak{m}}_{12}(A \times X_2) = \min_{\sigma_1,\sigma_2} P^{\mu_1}_{\sigma_1} \otimes P^{\mu_2}_{\sigma_2}(A \times X_2) = \min_{\sigma_1} P^{\mu_1}_{\sigma_1}(A),$$

that is not guaranteed to be equal to $\mu_1(A)$. The same happens to $\overline{\mathfrak{m}}_{12}$. However, these measures conform a bound of a product measure, as stated in the next proposition.

Proposition 6. *Let $(X_1, \mathcal{A}_{X_1}, \mu_1)$ and $(X_2, \mathcal{A}_{X_2}, \mu_2)$ be measurable spaces and let $\mu^{\times}_{12}$ be the $\times$-independent product of μ_1 and μ_2. Then, for any $H = A \times B \in \mathcal{R}$ it holds that*

$$\underline{\mathfrak{m}}_{12}(H) \leq \mu^{\times}_{12}(H) \leq \overline{\mathfrak{m}}_{12}(H). \tag{31}$$

Moreover, for any $C \in \mathcal{A}_{X_1 \times X_2}$,

$$\begin{aligned}\underline{\mu}^{\times}_{12}(C) &\leq \overline{\mathfrak{m}}_{12}(C), \\ \underline{\mathfrak{m}}_{12}(C) &\leq \overline{\mu}^{\times}_{12}(C).\end{aligned} \tag{32}$$

Proof. According to Proposition 1,

$$\min_{\sigma_1} P^{\mu_1}_{\sigma_1}(A) \leq \mu_1(A) \leq \max_{\sigma_1} P^{\mu_1}_{\sigma_1}(A)$$

and

$$\min_{\sigma_2} P^{\mu_2}_{\sigma_2}(B) \leq \mu_2(B) \leq \max_{\sigma_2} P^{\mu_2}_{\sigma_2}(B).$$

Multiplying both inequalities, we obtain

$$\min_{\sigma_1} P_{\sigma_1}^{\mu_1}(A) \min_{\sigma_2} P_{\sigma_2}^{\mu_2}(B) \le \mu_1(A)\mu_2(B) \le \max_{\sigma_1} P_{\sigma_1}^{\mu_1}(A) \max_{\sigma_2} P_{\sigma_2}^{\mu_2}(B)$$
$$\Rightarrow \min_{\sigma_1,\sigma_2} P_{\sigma_1}^{\mu_1}(A) P_{\sigma_2}^{\mu_2}(B) \le \mu_1(A)\mu_2(B) \le \max_{\sigma_1,\sigma_2} P_{\sigma_1}^{\mu_1}(A) P_{\sigma_2}^{\mu_2}(B)$$
$$\Rightarrow \underline{\mathfrak{m}}_{12}(H) \le \mu_{12}^{\times}(H) \le \overline{\mathfrak{m}}_{12}(H),$$

as $H = A \times B$ and $\mu_{12}^{\times}(H) = \mu_1(A)\mu_2(B)$. This proves Equation (31).

Now consider two sets $C \in \mathcal{A}_{X_1 \times X_2}$ and $G \in \mathcal{R}$ with $G \subseteq C$. As $\overline{\mathfrak{m}}_{12}$ is a fuzzy measure, and thus monotone, it holds that $\overline{\mathfrak{m}}_{12}(G) \le \overline{\mathfrak{m}}_{12}(C)$, which together with Equation (31) yields

$$\mu_{12}^{\times}(G) \le \overline{\mathfrak{m}}_{12}(G) \le \overline{\mathfrak{m}}_{12}(C). \tag{33}$$

As Equation (33) holds for any $G \in \mathcal{R}$ with $G \subseteq C$, in particular we can write

$$\max_{\substack{G\in\mathcal{R} \\ G\subseteq C}} \mu_{12}^{\times}(G) \le \overline{\mathfrak{m}}_{12}(C). \tag{34}$$

Note that, according to Definition 14, the left hand side of inequality (34) is the interior product measure constructed with the product t-norm, i.e., $\underline{\mu}_{12}^{\times}(C)$, which means that $\underline{\mu}_{12}^{\times}(C) \le \overline{\mathfrak{m}}_{12}(C)$. The remaining inequality is proven in a similar way. □

The next proposition relates the concept of product p-measure with standard probabilistic product.

Proposition 7. *Let $(X_1, \mathcal{A}_{X_1}, P_1)$ and $(X_2, \mathcal{A}_{X_2}, P_2)$ be probabilistic spaces. Then, $\underline{\mathfrak{m}}_{12}$ and $\overline{\mathfrak{m}}_{12}$ are equal to the standard probabilistic product.*

Proof. If P_1 and P_2 are probability measures (and thus fuzzy measures after all), it follows from Definitions 2 and 3 that all the associated probability measures are the same, and therefore $\underline{\mathfrak{m}}_{12} = \overline{\mathfrak{m}}_{12}$ and they are equal to the standard probabilistic product. □

The next theorem states that, for a particular class of fuzzy measures, constructing a single product measure from two marginal measures is indeed possible.

Theorem 2. *Let $(X_1, \mathcal{A}_{X_1}, \mu_1)$ and $(X_2, \mathcal{A}_{X_2}, \mu_2)$ be measurable spaces such that μ_1 and μ_2 are monotone (alternating) capacities of order 2, then $\underline{\mathfrak{m}}_{12}$ ($\overline{\mathfrak{m}}_{12}$) is a product fuzzy measure of μ_1 and μ_2.*

Proof. According to Proposition 5, both $\underline{\mathfrak{m}}_{12}$ and $\overline{\mathfrak{m}}_{12}$ are fuzzy measures. We only have to prove that they are consistent with the marginalization, which is straightforward taking into account that, if μ_1 and μ_2 are capacities of order 2, it follows from Theorem 1 that

$$\mu_i(A) = \min_{\sigma} P_{\sigma}^{\mu_i}(A), \quad i = 1,2 \qquad \text{(for monotone capacities)},$$
$$\mu_i(A) = \max_{\sigma} P_{\sigma}^{\mu_i}(A), \quad i = 1,2 \qquad \text{(for alternating capacities)}.$$

Thus, $\underline{\mathfrak{m}}_{12}(A \times X_2) = \min_{\sigma} P_{\sigma}^{\mu_1}(A) = \mu_1(A)$ and $\underline{\mathfrak{m}}_{12}(X_1 \times B) = \min_{\sigma} P_{\sigma}^{\mu_2}(B) = \mu_2(B)$. The result for alternating capacities is analogously obtained. □

Corollary 1. *If μ_1 and μ_2 are monotone (alternating) capacities of order 2 and $A \times B \in \mathcal{R}$, then*

$$\begin{aligned} \underline{\mathfrak{m}}_{12}(A \times B) &= \mu_1(A)\mu_2(B) \qquad \textit{(for monotone capacities)}, \\ \overline{\mathfrak{m}}_{12}(A \times B) &= \mu_1(A)\mu_2(B) \qquad \textit{(for alternating capacities)}. \end{aligned} \tag{35}$$

These results indicate that the composition of capacities of order 2 can be constructed using the product p-measures, so that if both marginals are monotone capacities of order 2, the product measure is $\underline{\mathfrak{m}}_{12}$, while it is $\overline{\mathfrak{m}}_{12}$ if the marginals are alternating capacities of order 2.

4. Functions over Product Spaces

After studying the problem of composing fuzzy measures, the next step is to consider the construction of functions over product spaces, from functions defined on the marginal spaces.

4.1. Composition of Functions

Our starting point consists of two non-negative functions h_1 and h_2 defined over the reference sets X_1 and X_2, respectively. We propose the use of t-norms for carrying out the composition of the functions, and therefore we will assume, without loss of generality, that they take values on $[0,1]$.

Definition 18. *Let h_1 and h_2 be functions defined on X_1 and X_2, respectively, and taking values on interval $[0,1]$. We say that $h^{\star}_{12} : X_1 \times X_2 \longrightarrow [0,1]$ is the $\star$-composition of h_1 and h_2 if $\forall (x_1, x_2) \in X_1 \times X_2$,*

$$h^{\star}_{12}(x_1, x_2) = h_1(x_1) \star h_2(x_2), \tag{36}$$

where $\star$ is a t-norm.

Example 5. *Consider the functions h_1 and h_2 defined on X_1 and X_2 given in Table 2. Using the t-norm* min, *we find that the* min*-composition of h_1 and h_2, denoted as $h^{\min}_{12}$, is the function specified in Table 3.*

Table 2. An example of two functions defined over the marginal spaces.

X_1	h_1	X_2	h_2
x_{11}	0.3	x_{21}	0.4
x_{12}	0.6	x_{22}	0.9
x_{13}	0.5	x_{23}	0.2

Table 3. Min-composition of the functions in Table 2.

	x_{21}	x_{22}	x_{23}
x_{11}	0.3	0.3	0.2
x_{12}	0.4	0.6	0.2
x_{13}	0.4	0.5	0.2

Note that $h^{\star}_{12}$ is not symmetric. A possible interpretation of the composed function is the worst value of a pair (x_{1i}, x_{2j}) regarding both marginal spaces simultaneously. For instance, a value of $h^{\star}_{12}(x_{12}, x_{21})$ equal to 0.4 would indicate that both $h_1(x_{12}) \geq 0.4$ and $h_2(x_{21}) \geq 0.4$.

Proposition 8. *Let h_1 and h_2 be functions defined on X_1 and X_2, respectively, and taking values on $[0,1]$. Then, the α-cuts generated by $h^{\min}_{12}$, belong to the class of rectangles, $\mathcal{R}$.*

Proof. The α-cuts of any function h in $X_1 \times X_2$ are (see Equation (15))

$$H_\alpha = \{(x_1, x_2) \in X_1 \times X_2 | h(x_1, x_2) \geq \alpha\}.$$

Considering the min t-norm, and thus $h = h_{12}^{\min}$, we find that

$$\begin{aligned} H_\alpha &= \{(x_1, x_2) \in X_1 \times X_2 | h_{12}^{\min}(x_1, x_2) \geq \alpha\} \\ &= \{(x_1, x_2) \in X_1 \times X_2 | \min\{h_1(x_1), h(x_2)\} \geq \alpha\} \\ &= \{(x_1, x_2) \in X_1 \times X_2 | h_1(x_1) \geq \alpha \text{ y } h_2(x_2) \geq \alpha\} \\ &= \{x_1 \in X_1 | h_1(x_1) \geq \alpha\} \times \{x_2 \in X_2 | h_2(x_2) \geq \alpha\} \in \mathcal{R}. \end{aligned}$$

□

Proposition 8 guarantees that, when composing functions using the min t-norm, the resulting function generates α-cuts belonging to class $\mathcal{R}$, which facilitates the calculation of a product measure as the interior and exterior product measures are the same in this case.

4.2. Marginalization

The marginalization operation acts in the opposite direction to composition, i.e., from a function defined on a product space, by applying marginalization we should obtain functions defined over the marginal spaces. In what follows, we pursue the definition of a well founded marginalization process.

Definition 19. *Let h be a function defined on $X_1 \times X_2$ and taking values on $[0,1]$. We define the $\oplus$-marginals of h as*

$$h_{X_1}^{\oplus}(x_{1i}) = \bigoplus_{x_{2j} \in X_2} h(x_{1i}, x_{2j}) = h(x_{1i}, x_{21}) \oplus h(x_{1i}, x_{22}) \oplus \ldots \oplus h(x_{1i}, x_{2m}), \quad (37)$$

$$h_{X_2}^{\oplus}(x_{2j}) = \bigoplus_{x_{1i} \in X_1} h(x_{1i}, x_{2j}) = h(x_{11}, x_{2j}) \oplus h(x_{12}, x_{2j}) \oplus \ldots \oplus h(x_{1n}, x_{2j}), \quad (38)$$

where $\oplus$ is a t-conorm, n is the cardinality of X_1 and m is the cardinality of X_2.

For instance, if we consider the max t-conorm, the marginalization process would result in

$$h_{X_1}^{\max}(x_{1i}) = \max\{h(x_{1i}, x_{21}), h(x_{1i}, x_{22}), \ldots, h(x_{1i}, x_{2m})\}.$$

Example 6. *Consider a function h defined on $X_1 \times X_2$ as specified in Table 4. The corresponding marginals when the max t-conorm is used are shown on the last column and row. Table 5 illustrates the opposite process, where the marginals have been combined usind the* min *t-conorm.*

Table 4. Specification of a function $h(x_{1i}, x_{2j})$ and its marginals using the max t-conorm.

	x_{21}	x_{22}	x_{23}	$h_{X_1}^{\max}$
x_{11}	0.5	0.1	0.2	0.5
x_{12}	0.7	0.6	0.4	0.7
x_{13}	0.3	0.5	0.2	0.5
$h_{X_2}^{\max}$	0.7	0.6	0.4	

Table 5. Composition of the marginals in Table 4 using the min t-conorm.

	x_{21}	x_{22}	x_{23}	$h_{X_1}^{\max}$
x_{11}	0.5	0.5	0.4	0.5
x_{12}	0.7	0.6	0.4	0.7
x_{13}	0.5	0.5	0.4	0.5
$h_{X_2}^{\max}$	0.7	0.6	0.4	

We can notice how, in Example 6, the function obtained by composing the marginals bounds from above the original function over the product space. The next proposition shows that this property holds in rather general settings.

Proposition 9. *Let h be a function defined on $X_1 \times X_2$ and taking values on $[0,1]$, and let $h^{\oplus}_{X_1}$ and $h^{\oplus}_{X_2}$ be the $\oplus$-marginals of h. Then, for any arbitrary t-conorm $\oplus$ it holds that*

$$h(x_{1i}, x_{2j}) \leq \min\{h^{\oplus}_{X_1}(x_{1i}), h^{\oplus}_{X_2}(x_{2j})\}, \quad \forall (x_{1i}, x_{2j}) \in X_1 \times X_2. \tag{39}$$

Proof. According to Definition 19,

$$h^{\oplus}_{X_1}(x_{1i}) = h(x_{1i}, x_{21}) \oplus h(x_{1i}, x_{22}) \oplus \ldots \oplus h(x_{1i}, x_{2m}). \tag{40}$$

As the max t-conorm bounds any other conorm from below,

$$h^{\oplus}_{X_1}(x_{1i}) \geq \max\{h(x_{1i}, x_{21}), h(x_{1i}, x_{22}), \ldots, h(x_{1i}, x_{2m})\} = h^{\max}_{X_1}(x_{1i}). \tag{41}$$

Likewise, it holds that

$$h^{\oplus}_{X_2}(x_{2j}) \geq h^{\max}_{X_2}(x_{2j}). \tag{42}$$

On the other hand, it is clear that

$$h(x_{1i}, x_{2j}) \leq \min\{h^{\max}_{X_1}(x_{1i}), h^{\max}_{X_2}(x_{2j})\}, \tag{43}$$

as $h^{\max}_{X_1}(x_{1i})$ is the maximum value of $h(x_{1i}, x_{2j})$ for a fixed value x_{1i} of X_1, and $h^{\max}_{X_2}(x_{2j})$ is the maximum value of $h(x_{1i}, x_{2j})$ for a fixed value x_{2j} of X_2. Therefore, combining Equation (43) with Equations (41) and (42), we obtain Equation (39). □

Corollary 2. *Assuming the conditions in Proposition 9, given an arbitrary t-conorm $\oplus$ it holds that, for any $(x_{1i}, x_{2j}) \in X_1 \times X_2$*

$$h(x_{1i}, x_{2j}) \leq h^{\min}_{12}(x_{1i}, x_{2j}). \tag{44}$$

5. Conclusions

In this paper, we have studied the problem of constructing fuzzy measures over product domains, when fuzzy measures over the marginal spaces are available. We have proposed a definition of independence of fuzzy measures and different ways of constructing product measures that are consistent with the defined concept of independence. Even though, in general, we have only been able to give bounds for the measure on the product space when we work outside the class of rectangles $\mathcal{R}$, we show in Theorem 2 that it is possible to construct a single product measure if the marginal measures are capacities of order 2.

Our proposal for combining real functions over the marginal spaces in order to produce a joint function over the product space satisfies that the resulting function yields α-cuts within the class of rectangles $\mathcal{R}$, if the min t-norm is used. The importance of this property is that within the class $\mathcal{R}$, we are able to compute a unique product fuzzy measure, as the interior and exterior product measures, which conform the bound of the product fuzzy measure, are the same in this case.

The results in the paper show that we are able to handle marginal spaces endowed with a fuzzy measure and a real function, and work on the product space with product measures and functions containing the information in the marginal case. Likewise, the marginal functions can be measured using, for instance, Choquet integral; the joint function can also be measured in the same way, by integrating with respect to the product measure. This provides the basic tools for defining statistical indices, as for instance indices of association, based on fuzzy measures.

Author Contributions: Investigation, F.R., M.M. and A.S.; Writing–original draft, F.R., M.M. and A.S.; Writing–review–editing, F.R., M.M. and A.S. All authors have read and agreed to the published version of the manuscript.

Funding: This research was funded by the Spanish Ministry of Science and Innovation through grant TIN2016-77902-C3-3-P and by ERDF-FEDER funds.

Conflicts of Interest: The authors declare no conflicts of interest.

References

1. Choquet, G. Theory of capacities. *Ann. L'Institut Fourier* **1954**, *5*, 131–295. [CrossRef]
2. Li, J. On Null-Continuity of Monotone Measures. *Mathematics* **2020**, *8*, 205. [CrossRef]
3. Sugeno, M. Theory of Fuzzy Integrals and Its Applications. Ph.D. Thesis, Tokyo Institute of Technology, Tokyo, Japan, 1974.
4. Dempster, A.P. Upper and lower probabilities induced by a multivalued mapping. *Ann. Math. Stat.* **1967**, *38*, 325–339. [CrossRef]
5. Shafer, G. *A Mathematical Theory of Evidence*; Princeton University Press: Princeton, NJ, USA, 1976.
6. Nedović, L.; Pap, E. Aggregation of sequence of fuzzy measures. *Iran. J. Fuzzy Syst.* **2020**, *17*, 39–55.
7. Narukawa, Y.; Torra, V. On distorted probabilities and m-separable fuzzy measures. *Int. J. Approx. Reason.* **2011**, *52*, 1325–1336. [CrossRef]
8. Murillo, J.; Guillaume, S.; Bulacio, P. k-maxitive fuzzy measures: A scalable approach to model interactions. *Fuzzy Sets Syst.* **2017**, *324*, 33–48. [CrossRef]
9. Combarro, E.; Miranda, P. On the structure of the k-additive fuzzy measures. *Fuzzy Sets Syst.* **2010**, *161*, 2314–2327. [CrossRef]
10. Navarro, J.; Spizzichino, F. Aggregation and signature based comparisons of multi-state systems via decompositions of fuzzy measures. *Fuzzy Sets Syst.* **2020**, *396*, 115–137. [CrossRef]
11. Denneberg, D. Conditioning (updating) non-additive measure. *Ann. Oper. Res.* **1994**, *52*, 21–42. [CrossRef]
12. Denneberg, D. *Non-Additive Measure and Integral*; Theory and Decision Library B; Springer: Dordrecht, Netherlands, 1994.
13. Hendon, E.; Jacobsen, H.J.; Sloth, B.; Tranæs, T. The product of capacities and belief functions. *Math. Soc. Sci.* **1996**, *32*, 95–108. [CrossRef]
14. Ghirardato, P. On idependence for Non-Additive measures, with Fubini theorem. *J. Econ. Theory* **1997**, *73*, 261–291. [CrossRef]
15. Koshevoy, G.A. Distributive lattice and product of capacities. *J. Math. Anal. Appl.* **1998**, *219*, 427–441. [CrossRef]
16. Denneberg, D. Totally monotone core and products of monotone measures. *Int. J. Approx. Reason.* **2000**, *24*, 273–281. [CrossRef]
17. Denneberg, D. Conditional expectation for monotone measure, the discrete case. *J. Math. Econ.* **2002**, *37*, 105–121. [CrossRef]
18. Brüning, M.; Denneberg, D. Max-min σ-additive representation of monotone measure. *Stat. Pap.* **2002**, *43*, 23–35. [CrossRef]
19. Candeloro, D.; Mesiar, R.; Sambucini, A. A special class of fuzzy measures: Choquet integral and applications. *Fuzzy Sets Syst.* **2019**, *355*, 83–99. [CrossRef]
20. Beliakov, G.; James, S.; Wu, J. Discrete Fuzzy Measures. In *Studies in Fuzziness and Soft Computing*; Springer International Publishing: Cham, Switzerland, 2020; Volume 382.
21. De Campos, L.M.; Bolaños, M.J. Representation of fuzzy measures through probabilities. *Fuzzy Sets Syst.* **1989**, *31*, 23–36. [CrossRef]
22. De Campos, L.M. Caracterización y Estudio de Medidas Difusas a Partir de Probabilidades. Ph.D. Thesis, Universidad de Granada, Granada, Spain, 1987.
23. Jia, G.; Zhang, N. New proofs for several properties of capacities. *arXiv* **2013**, arXiv:1307.0913v1.
24. Schweizer, B.; Sklar, A. *Probability Metric Spaces*; Elsevier: North-Holland, NY, USA, 1983.
25. Zadeh, L.A. Fuzzy sets. *Inf. Control* **1965**, *8*, 338–353. [CrossRef]
26. Dubois, D.; Prade, H. Fuzzy Sets and Systems: Theory and Applications. In *Mathematics in Science and Engineering*; Academic Press: Cambridge, MA, USA, 1980; Volume 114.

27. Grabisch, M.; Nguyen, H.T.; Walker, E.A. *Fundamentals of Uncertainty Calculi with Applications to Fuzzy Inference*; Theory and Decision Library B; Springer: Dordrecht, The Netherlands, 1995.
28. Klement, E.P. Some mathematical aspects of fuzzy sets: Triangular norms, fuzzy logic, and generalized measures. *Fuzzy Sets Syst.* **1997**, *90*, 133–140. [CrossRef]
29. Grabisch, M.; Murofushi, T.; Sugeno, M., Eds. Fuzzy Measure and Integrals. In *Studies in Fuzziness and Soft Computing*; Physica Verlag: Heidelberg, Germany, 2000; Volume 40.
30. Bauer, H. *Probability Theory and Elements of Measure Theory*, 2nd ed.; Academic Press: London, UK, 1981.
31. De Cooman, G. Possibility theory I: The measure and integral theoretic groundwork. *Int. J. Gen. Syst.* **1997**, *25*, 291–323. [CrossRef]
32. De Cooman, G.; Zhang, G.; Kerre, E. Possibility measures and possibility integrals defined on a complete lattice. *Fuzzy Sets Syst.* **2001**, *120*, 459–467. [CrossRef]

mathematics

MDPI

Article

Schauder-Type Fixed Point Theorem in Generalized Fuzzy Normed Linear Spaces

S. Chatterjee [1], T. Bag [1] and Jeong-Gon Lee [2,*]

1 Department of Mathematics, Siksha-Bhavana, Visva-Bharati, Santiniketan, Birbhum, West-Bengal 731235, India; shayani.mathvb10@gmail.com (S.C.); tarapadavb@gmail.com (T.B.)
2 Division of Applied Mathematics, Wonkwang University, 460, Iksan-daero, Iksan-Si, Jeonbuk 54538, Korea
* Correspondence: jukolee@wku.ac.kr

Received: 12 August 2020; Accepted: 19 September 2020; Published: 23 September 2020

Abstract: In the present article, the Schauder-type fixed point theorem for the class of fuzzy continuous, as well as fuzzy compact operators is established in a fuzzy normed linear space (fnls) whose underlying t-norm is left-continuous at $(1,1)$. In the fuzzy setting, the concept of the measure of non-compactness is introduced, and some basic properties of the measure of non-compactness are investigated. Darbo's generalization of the Schauder-type fixed point theorem is developed for the class of ψ-set contractions. This theorem is proven by using the idea of the measure of non-compactness.

Keywords: Schauder fixed point theorem; fuzzy normed linear space; t-norm; measure of non-compactness

MSC: 03B52; 03E72; 46B20; 46B99; 46A19; 03E70; 15A03; 54H25

1. Introduction

In 1930, Schauder established an important theorem in the field of fixed point theory. The theorem stated that "If B is a compact, convex subset of a Banach space X and $f : B \to B$ is a continuous function then f has a fixed point." However, to develop more results in functional analysis, Schauder relaxed the compactness by closedness. The theorem has an enormous influence on the theory of differential equations. At first, the Schauder-type fixed point theorem was applied to Peano's existence theorem for the first order differential equations. After that, many interesting applications of this theorem were given to differential equations. For example, in 2007, Chu and Torres [1] proved the existence of positive solutions to the second order singular differential equations with the help of this fixed point theorem. In 2009, A. F. Dizaji et al. [2] determined the sufficient condition for the existence of periodic solution of the initial value problems, which correspond to the Duffing's oscillator with time varying coefficients as an application of the Schauder-type fixed point theorem. Recently, in 2019, Shengjun Li et al. [3] established the existence of the periodic orbits of rapidly symmetric systems with a repulsive singularity. The line of proof of this existence problem is based on the use of Schauder's fixed point theorem. Moreover, the global existence of the solution for a class of functional equations is also studied using the Schauder fixed point theorem, which arises in various types of neural networks such as the Hopfield neural network, the Cohen–Grossberg neural network, cellular networks, etc. For the references, please see [4–6].

Due to its huge application in real-life problems, much scientific attention has been drawn towards the generalization of this theorem. In 1935, A. N. Tychonoff [7] extended Schauder's theorem to locally convex spaces. In 1950, M. Hukuhara [8] unified both the theorem of Schauder and Tychonoff. In 1955 [9], G. Darbo extended the Schauder theorem to a more general class of mappings, the so-called α-set contractions, which contain compact, as well as continuous mappings. Darbo proved this theorem using the concept of Kuratowski's measure of non-compactness. In 1961 [10], Ky Fan generalized both Schauder's and Tychonoff's theorem for the class of continuous set-valued mappings. In recent years, a significant contribution has been made towards the generalization of Schauder's fixed point theorem. For example, in 2012, R. L. Pouso [11] introduced a new version of Schauder's theorem for the class of discontinuous operators. In 2013, R. P. Agarawal et al. [12] established this theorem in semilinear Banach spaces. In 2016, Wei-Shih Du [13] generalized this theorem in an another direction, i.e., the compactness assumption is replaced by the finite open cover, and the continuity condition is totally removed.

On the other hand, several authors, viz. Xio and Zhu, Bag and Samanta, and Zhang and Guo, have played important roles in the process of the formulation of the Schauder-type fixed point theorem in the fuzzy setting. For the references, please see [14–16]. However, all of them considered the underlying t-norm as the continuous t-norm. Therefore, naturally, a question may arise: Is it possible to prove the Schauder-type fixed point theorem in a fuzzy normed linear space (fnls) w.r.t. the general t-norm?

In this paper, we try to give an affirmative answer to this question.

In this paper, we develop the Schauder-type fixed point theorem for a fuzzy continuous, as well as a fuzzy compact operator in an fnls whose underlying t-norm is left-continuous only at $(1,1)$. We also establish Darbo's generalization of the Schauder-type fixed point theorem in the fuzzy setting for the class of ψ-set contraction mappings using the properties of the measure of non-compactness.

This article is divided into three parts. Section 2 deals with preliminary results, which are used in the subsequent sections. In Section 3, the Schauder-type fixed point theorem for the class of fuzzy continuous, as well as fuzzy compact mappings is established in generalized fnls. In Section 4, the definition of the measure of non-compactness is given, and some basic properties are studied to prove Darbo's generalization of the Schauder-type fixed point theorem.

2. Preliminaries

Definition 1 ([17]). *Let X be a linear space over the field $\mathcal{F}$ ($\mathbb{C}$ or $\mathbb{R}$). A fuzzy subset N of $X \times \mathbb{R}$ ($\mathbb{R}$ is the set of all real numbers) is called a fuzzy norm on X if:*

(N1) $\forall t \in \mathbb{R}$ *with* $t \leq 0,\ N(x,t) = 0$;

(N2) $(\forall t \in \mathbb{R},\ t > 0,\ N(x,t) = 1)\ iff\ x = \theta$;

(N3) $\forall t \in \mathbb{R},\ t > 0,\ N(cx,t) = N(x, \frac{t}{|c|})\ if\ c \neq 0$;

(N4) $\forall s,t \in \mathbb{R};\ x,u \in X;\ N(x+u, s+t) \geq N(x,s) * N(u,t)$;

(N5) $N(x,.)$ *is a non-decreasing function of* $\mathbb{R}$ *and* $\lim\limits_{t\to\infty} N(x,t) = 1$.

The triplet $(X, N, *)$ is referred to as an fnls.

Throughout the paper, we assume the following conditions:

1. For each $x \neq \theta$, $N(x,t)$ is a left-continuous function w.r.t. t.
2. The t-norm $*$ is left-continuous at one with respect to the first or second component.

Theorem 1 ([17]). *Let $(X, N, *)$ be a finite-dimensional fnls in which the underlying t-norm $*$ is continuous at $(1,1)$. Then, a subset A is compact iff A is closed and bounded.*

Lemma 1 ([18]). *Let $(X, N, *)$ be an fnls. Then:*

$$\lim_{n\to\infty} N(x_n - x, t) = 1\ \forall t > 0 \Leftrightarrow \lim_{n\to\infty} \wedge\{t > 0 : N(x_n - x, t) > 1 - \alpha\} = 0\ \forall \alpha \in (0,1)$$

Proposition 1 ([18]). *Let $(X, N, *)$ be an fnls. Then, the function $M : X \times X \times [0,\infty) \to [0,1]$ defined by $M_N(x,y,t) = N(x-y,t)$ is a fuzzy metric space defined by H. Wu [19]. Thus, the family $\mathcal{B}$ (the collection of* *all* (α, t) *neighborhoods* $B_N(x, \alpha, t)$, $x \in X$, $0 < \alpha < 1$, $t > 0$) *induces a Hausdorff topology* τ *such that* $\mathcal{B}$ *is a base for* τ *and* τ *also satisfies the first countability axiom, where* $B_N(x, \alpha, t) = \{y \in X : N(x - y, t) > 1 - \alpha\}$.

Definition 2 ([20]). *A fuzzy metric space* $(X, M, *)$ *is called compact if* (X, τ_M) *is compact.*

Theorem 2 ([20]). *A fuzzy metric space* $(X, M, *)$ *is fuzzy totally bounded iff every sequence has a Cauchy subsequence.*

Note 1. *The above result is also true if* $(X, M, *)$ *is the H. Wu-type fuzzy metric space.*

Definition 3 ([21]). *Let* $(X, N, *)$ *be an fnls. Let* $\{x_n\}$ *be a sequence in* X. *Then,* $\{x_n\}$ *is said to be convergent if* $\exists\, x \in X$ *such that:*

$$\lim_{n \to \infty} N(x_n - x, t) = 1\ \forall t > 0.$$

In that case, x *is called the limit of the sequence* $\{x_n\}$ *and is denoted by* $\lim x_n$.

Definition 4 ([21]). *A subset* A *of an fnls is said to be fuzzy bounded if for each* α, $0 < \alpha < 1\ \exists\ t(\alpha) > 0$ *such that* $N(x, t) > 1 - \alpha\ \forall x \in A$.

Definition 5 ([21]). *Let* $(X, N, *)$ *be an fnls. A subset* F *of* X *is said to be closed if for any sequence* $\{x_n\}$ *in* F*, it converges to* x*, i.e.,*

$$\lim_{n \to \infty} N(x_n - x, t) = 1\ \forall t > 0$$

implies that $x \in F$.

Definition 6 ([21]). *Let* $(X, N, *)$ *be an fnls. A subset* B *of* X *is said to be the closure of* F *if for any* $x \in B$, $\exists$, *a sequence* $\{x_n\}$ *in* F *such that:*

$$\lim_{n \to \infty} N(x_n - x, t) = 1\ \forall t > 0.$$

We denote the set B *by* $\overline{F}$.

Definition 7 ([21]). *Let* $(X, N, *)$ *be an fnls. A subset* A *of* X *is said to be compact if any sequence* $\{x_n\}$ *in* A *has a subsequence converging to an element of* A.

Definition 8 ([21]). *A sequence* $\{x_n\}$ *is said to be Cauchy if* $\lim_{n \to \infty} N(x_n - x_{n+p}, t) = 1$, $\forall t > 0$, $p = 1, 2, 3...$

This definition of a Cauchy sequence is equivalent to $\lim_{n,m \to \infty} N(x_n - x_m, t) = 1$, $\forall t > 0$. Throughout the paper, we use this as the definition of the Cauchy sequence.

Lemma 2 ([22]). *Let* $(X, N, *)$ *be an fnls. If* $A \subseteq X$ *is fuzzy bounded, then* $\overline{A}$ *is also.*

Definition 9 ([22]). *Let* $(X, N_1, *_1)$ *and* $(Y, N_2, *_2)$ *be two fnlss. A linear operator* $T : (X, N_1, *_1) \to (Y, N_2, *_2)$ *is called a fuzzy compact linear operator if for every fuzzy bounded subset* M *of* X*, the subset* $T(M)$ *of* Y *is relatively compact, i.e.,* $\overline{T(M)}$ *is a compact set w.r.t.* τ_{N_2}.

Theorem 3 ([22]). *Let* $T : (X, N_1, *_1) \to (Y, N_2, *_2)$ *be a linear operator and* $*_2$ *be continuous at* $(1, 1)$. *Then,* T *is a fuzzy compact linear operator iff it maps every bounded sequence* $\{x_n\}$ *in* $(X, N_1, *_1)$ *onto a sequence* $\{T(x_n)\}$ *in* $(Y, N_2, *_2)$*, which has a convergent subsequence.*

Lemma 3 ([23]). *A fuzzy metric space* $(X, M, *)$ *is sequentially compact iff it is compact.*

Note 2. *By Lemma 3, in an fnls, Definition 2 and Definition 7 are equivalent.*

Theorem 4 ([24]). *In an fnls* $(X, N, *)$*, a subset* A *of* X *is fuzzy bounded iff* A *is bounded in topology* τ_N.

Theorem 5 ([24]). *In an fnls* $(X, N, *)$*, the following statements are equivalent:*

(i) A *is fuzzy totally bounded.*

(ii) $\forall \alpha \in (0,1),\ \forall t > 0\ \exists \{x_1, x_2, \cdots x_n\} \subseteq A : A \subseteq \bigcup_{i=1}^{n}(x_i + B(\theta, \alpha, t))$

Theorem 6 ([24]). *Let* $(X, N, *)$ *be an fnls and* $K \subseteq X$ *be a compact set in* (X, τ_N)*. Then,* K *is fuzzy totally bounded.*

Definition 10 ([25]). *An fnls* (X, N) *is a fuzzy Banach space if its induced fuzzy metric is complete.*

Definition 11 ([26]). *A subset* A *of an fnls* $(X, N, *)$ *is called fuzzy totally bounded if:*

$$\forall \alpha \in (0,1),\ \exists \{x_1, x_2, \cdots x_n\} \subseteq X : A \subseteq \bigcup_{i=1}^{n}(x_i + B(\theta, \alpha, \alpha)).$$

Theorem 7 ([26]). *Let* $T : (X, N_1) \to (Y, N_2)$ *be a mapping where* (X, N_1) *and* (Y, N_2) *are fnlss. Then, the following statements are equivalent:*

(i) T *is fuzzy continuous on* X.

(ii) T *is continuous on* X.

(iii) T *maps a fuzzy bounded set to a fuzzy bounded set.*

Theorem 8. *In a fuzzy Banach space* $(X, N, *)$*, if a subset* A *of* X *is fuzzy totally bounded, then it is compact in* (X, τ_N).

Proof. Consider a sequence $\{x_n\}$ in A. By Theorem 2, $\{x_n\}$ has a Cauchy subsequence. Since $(X, N, *)$ is fuzzy Banach space, then the Cauchy subsequence of $\{x_n\}$ is convergent in $(X, N, *)$. Therefore, by Definition 7, A is compact in $(X, N, *)$. □

Definition 12 ([27]). *(Fuzzy continuous) A mapping* T *from* (X, N_1) *to* (Y, N_2) *is said to be fuzzy continuous at* $x_0 \in X$ *if for given* $\epsilon > 0,\ \alpha \in (0,1)\ \exists\, \delta(\alpha, \epsilon) > 0,\ \beta(\alpha, \epsilon) \in (0,1)$ *such that* $\forall x \in X$:

$$N_1(x - x_0, \delta) > \beta \implies N_2(Tx - Tx_0, \epsilon) > \alpha$$

If T *is fuzzy continuous at each,*
$x \in X$*, then* T *is fuzzy continuous on* X.

Definition 13 ([27]). *(Sequentially fuzzy continuous) A mapping* T *from* (X, N_1) *to* (Y, N_2) *is said to be sequentially fuzzy continuous at* $x_0 \in X$ *if for any sequence* $\{x_n\}$, $x_n \in X$ *with* $x_n \to x_0$ *implies* $Tx_n \to Tx_0$*, i.e.,* $\lim_{n\to\infty} N_1(x_n - x_0, t) = 1\ \forall t > 0, \implies \lim_{n\to\infty} N_2(Tx_n - Tx_0, t) = 1\ \forall t > 0$

Theorem 9 ([27]). *Let* $T : (X, N_1) \to (Y, N_2)$ *be a mapping where* (X, N_1) *and* (Y, N_2) *are fnlss. Then,* T *is fuzzy continuous iff it is sequentially fuzzy continuous.*

Note 3. *From Definition 9, it is clear that if* T *is a fuzzy compact linear operator, then* T *maps bounded sets of* X *to bounded sets of* Y *by Theorem 4. Thus,* T *is a continuous mapping from* (X, τ_{N_1}) *to* (Y, τ_{N_2}).

3. Schauder-Type Fixed Point Theorem

In this section, we first define the uniformly fuzzy convergence and pointwise fuzzy convergence for a sequence of functions and investigate the relation between them. After that, we propound three types of Schauder-type fixed point theorems for the fuzzy compact class, as well as the fuzzy continuous linear operator in a generalized fnls and try to prove them.

Definition 14. *Let $f_n : (X, N_1, *_1) \to (Y, N_2, *_2)$ be a family of functions.*

(i) *$\{f_n\}$ is said to be uniformly fuzzy convergent to a function f on a subset A of X if for each $\alpha \in (0,1)$,*

$$\lim_{n\to\infty} \bigvee_{x\in A} \wedge\{t > 0 : N_2(f_n(x) - f(x), t) > 1 - \alpha\} = 0$$

i.e., for each $\alpha \in (0,1)$ and for each $\epsilon > 0 \; \exists \; N_0(\alpha, \epsilon) \in \mathbb{N}$ such that:

$$\bigvee_{x\in A} \wedge \{t > 0 : N_2(f_n(x) - f(x), t) > 1 - \alpha\} < \epsilon \; \forall n \geq N_0$$

(ii) *$\{f_n\}$ is said to be pointwise fuzzy convergent to a function f on a subset A of X if for each $\alpha \in (0,1)$, for each $x \in Y \; \exists \; N_0(\alpha, \epsilon, x) \in \mathbb{N}$ such that:*

$$\wedge\{t > 0 : N_2(f_n(x) - f(x), t) > 1 - \alpha\} < \epsilon \; \forall n \geq N_0$$

From the definition, it is obvious that (i) implies (ii), but (ii) does not imply (i).We verify this by the following example.

Example 1. *Let us consider a real nls (normed linear space) $(\mathbb{R}, \| \, \|)$, where $\mathbb{R}$ is the set of all real numbers and $\|x\| = |x|, \; \forall x \in \mathbb{R}$. Define two functions as follows:*

$$N_1(x,t) = \begin{cases} \frac{t}{t+\|x\|}, \; t > 0 \\ 0, \quad t \leq 0 \end{cases} \qquad N_2(x,t) = \begin{cases} 1, \; t \geq \|x\| \\ 0, \quad t < \|x\| \end{cases}$$

Define $f_n : (\mathbb{R}, N_1, \wedge) \to (\mathbb{R}, N_2, \wedge)$ by $f_n(x) = x^n$. Now, if we consider $f_n : [0,1] \to [0,1]$, then f_n is pointwise fuzzy convergent, but not uniformly fuzzy convergent.

Lemma 4. *Let f be self-mapping defined on a fuzzy Banach space $(X, N, *)$ and f also be a fuzzy compact linear operator on a subset M of X. Then, there exists a sequence of continuous mappings $\{f_n\}$ such that:*

(i) $\{f_n\}$ is uniformly fuzzy convergent to f.
(ii) $\{f_n(M)\}$ generates a finite-dimensional subspace of X.

Proof. Since f is a fuzzy compact linear operator, thus the set $\overline{\{f(x); x \in M\}}$ is a fuzzy compact set, i.e., $\overline{\{f(x); x \in M\}}$ is a compact set w.r.t. τ_N. Now, by Theorem 6, $\overline{\{f(x); x \in M\}}$ is fuzzy totally bounded. Let $\alpha_0 \in (0,1)$ and $\{t_n\}$ be a strictly decreasing sequence that tends to 0. Then, for each t_n, we can find a finite No. of elements $y_1^n, y_2^n, \cdots y_m^n \in f(M)$ such that:

$$\begin{aligned} & f(M) \subseteq \bigcup_{i=1}^{m} y_i^n + B_N(\theta, \alpha_0, t_n) \\ \implies & N(f(x) - y_i^n, t_n) > 1 - \alpha_0 \; \forall x \in M \\ \implies & \wedge\{t > 0 : N(f(x) - y_i^n, t) > 1 - \alpha_0\} \leq t_n \end{aligned} \tag{1}$$

We now define f_n on $f(M)$ for each $y \in f(M)$, by:

$$f_n(y) = \sum_{i=1}^{m} g_i^n(y) y_i^n / \sum_{i=1}^{m} g_i^n(y)$$
$$where\ g_i^n(y) = max\{0, t_n - \wedge\{t > 0 : N(y - y_i^n, t) > 1 - \alpha_0\}\}$$

Since the family $\wedge\{t > 0 : N(x,t) > 1 - \alpha\}$ is a continuous function on X for each $\alpha \in (0,1)$ and f is continuous on M by Note 3, so $g_i^n(x)$ is continuous on M. Thus, each f_n is a continuous function on $f(M)$. Now,

$$\begin{aligned}
&\wedge\{t > 0 : N(f_n(y) - y, t) > 1 - \alpha_0\} \\
&= \wedge\{t > 0 : N(\sum_{i=1}^{m} g_i^n(y) y_i^n / \sum_{i=1}^{m} g_i^n(y) - y, t) > 1 - \alpha_0\} \\
&= \frac{1}{\sum_{i=1}^{m} g_i^n(y)} \wedge \{t > 0 : N(\sum_{i=1}^{m} g_i^n(y)\{y_i^n - y\}, t) > 1 - \alpha_0\} \\
&= \frac{\sum_{i=1}^{m} g_i^n(y)}{\sum_{i=1}^{m} g_i^n(y)} \wedge \{\frac{t}{\sum_{i=1}^{m} g_i^n(y)} > 0 : N(y_i^n - y, \frac{t}{\sum_{i=1}^{m} g_i^n(y)}) > 1 - \alpha_0\} \\
&= \wedge\{t' > 0 : N(y_i^n - y, t') > 1 - \alpha_0\} \leq t_n
\end{aligned} \tag{2}$$

Now, define $\tilde{f}_n : M \to M$ by $\tilde{f}_n(x) = f_n(f(x))$. Thus, by Inequality (1),

$$\begin{aligned}
&\wedge\{t > 0 : N(f_n(f(x)) - f(x), t) > 1 - \alpha_0\} \leq t_n\ \forall x \in M \\
&\implies \bigvee_{x \in M} \wedge \{t > 0 : N(f_n(f(x)) - f(x), t) > 1 - \alpha_0\} \leq t_n
\end{aligned}$$

Thus, $\lim_{n\to\infty} \bigvee_{x \in M} \wedge\{t > 0 : N(f_n(f(x)) - f(x), t) > 1 - \alpha_0\} = 0$.

Since $\alpha_0 \in (0,1)$ is arbitrary, then the above relation is true for each $\alpha_0 \in (0,1)$. Thus, $\tilde{f}_n$ uniformly fuzzy converges to f. Condition (ii) is automatically valid by the construction of f_n. □

Remark 1. *In Lemma 4, each $\{f_n\}$ contains a fixed point, say x_n. This can be shown in the following way: Now, the sequence $\{f_n\}$, which is uniformly fuzzy convergent to f, is of the form:*

$$f_n(x) = \sum_{i=1}^{m} g_i^n(f(x)) y_i^n / \sum_{i=1}^{m} g_i^n(f(x))$$
$$where\ g_i^n(f(x)) = max\{0, t_n - \wedge\{t > 0 : N(f(x) - y_i^n, t) > 1 - \alpha_0\}\}$$

Now, if we choose $C_n = \overline{Co}\{y_i^n\}_{i=1}^m$ (convex closure of $\{y_i^n\}_{i=1}^m$), $Y_n = Span\{y_i^n\}_{i=1}^m$, then C_n is a closed, bounded, convex subset of the finite-dimensional subspace Y_n of X and $f_n(C_n) \subseteq C_n$ (by the definition of f_n). Each f_n is continuous. Now, by the Brouwer fixed point theorem, $\exists$ a point $x_n \in C_n$ such that $f_n(x_n) = x_n$.

Remark 2. *If $\{f_n\}$ is uniformly fuzzy convergent to f on X, then for each $x \in X$,*

$$\lim_{n\to\infty} N_2(f_n(x) - f(x), t) = 1\ \forall t > 0.$$

Proof. Since $\{f_n\}$ is uniformly fuzzy convergent to f, then $\{f_n\}$ pointwise fuzzy converges to f. Thus:

$$\lim_{n\to\infty} \wedge \{t > 0 : N_2(f_n(x) - f(x), t) > 1 - \alpha\} = 0\ \forall x \in X,\ \forall \alpha \in (0,1).$$

Now, from Lemma 1, the required result follows immediately. □

Lemma 5. *Let $\{T_n\}$ be a sequence of fuzzy compact linear operators defined on $E \subseteq X$, where $(X, N, *)$ is an fnls. Again, $\{T_n\}$ is uniformly fuzzy convergent on E. Then, the set $\tilde{E} = \bigcup_{i=1}^{\infty} \overline{T_i E}$ is a fuzzy compact set, i.e., compact w.r.t. the topology τ_N.*

Proof. We show that $\tilde{E}$ is fuzzy totally bounded. Then, by Theorem 8, the assertion of the lemma is automatically valid. Let $\epsilon > 0$ be an arbitrary No. and $\alpha_0 \in (0,1)$ be given. Then, by the left-continuity of $*$ at $(1,1)$ $\exists$ $\beta_0 \in (0,1)$ such that:

$$(1-\beta_0) * (1-\beta_0) * (1-\beta_0) > 1-\alpha_0$$

Since $\{T_n\}$ uniformly fuzzy converges to T, then $\exists N_0(\epsilon/4, \beta_0) \in \mathbb{N}$ such that:

$$\begin{aligned}
&\bigvee_{x\in E} \wedge\{t > 0 : N(T_n(x) - T(x), t) > 1-\beta_0\} < \epsilon/4 \;\forall n \geq N_0 \\
&\Longrightarrow \wedge\{t > 0 : N(T_n(x) - T(x), t) > 1-\beta_0\} < \epsilon/4 \;\forall n \geq N_0,\ \forall x \in E \\
&\Longrightarrow N(T_n(x) - T(x), \epsilon/4) > 1-\beta_0 \;\forall n \geq N_0,\ \forall x \in E
\end{aligned} \tag{3}$$

Again, the sets $\overline{T_0 E}, \overline{T_1 E}, \cdots, \overline{T_{N_0} E}$ are fuzzy compact sets, i.e., compact w.r.t τ_N by the definition of the fuzzy compact linear operator. Therefore, $\bigcup_{i=0}^{N_0} \overline{T_i E}$ is compact w.r.t. τ_N. By Theorem 6, $\bigcup_{i=0}^{N_0} \overline{T_i E}$ is fuzzy totally bounded. Now, by the definition of the fuzzy total boundedness, we can find $y_1, y_2, \cdots, y_n$ such that:

$$T_j(x) \in \bigcup_{i=0}^{n} B_N(y_i, \epsilon/2, \beta_0) \;\forall\; T_j(x) \in \bigcup_{i=0}^{N_0} \overline{T_i E}$$

Now, for any $T_m(x) \in \bigcup_{i=0}^{\infty} \overline{T_i E}$, if $m \leq N_0$, we have:

$$T_m(x) \in \bigcup_{i=0}^{n} B_N(y_i, \epsilon/2, \beta_0) \subseteq \bigcup_{i=0}^{n} B_N(y_i, \epsilon, \alpha_0) \tag{4}$$

If $m > N_0$, then:

$$\begin{aligned}
N(T_m(x) - y_i, \epsilon) &\geq N(T_m(x) - T(x), \epsilon/4) * N(T(x) - T_{N_0}(x), \epsilon/4) * N(T_{N_0}(x) - y_i, \epsilon/2) \\
&\geq (1-\beta_0) * (1-\beta_0) * (1-\beta_0) > (1-\alpha_0)
\end{aligned}$$

$$\therefore T_m(x) \in \bigcup_{i=0}^{n} B_N(y_i, \epsilon, \alpha_0) \tag{5}$$

Thus, $\bigcup_{i=0}^{\infty} \overline{T_i E}$ is fuzzy totally bounded. This completes the proof. □

Lemma 6. *Let T be a continuous self-mapping on $(X, N, *)$ and $\dim T(X) < \infty$. Then, T is a fuzzy compact linear operator.*

Proof. Let $\{y_n\}$ be a fuzzy bounded sequence. Then, by Theorem 7, $\{Ty_n\}$ is a fuzzy bounded sequence. Again, the range set of $\{Ty_n\}$ say $R(Ty_n : n \in \mathbb{N})$ is fuzzy bounded. Now, by Lemma 2, $\overline{R(Ty_n : n \in \mathbb{N})}$ is fuzzy bounded. Since $T(X)$ is finite-dimensional, thus $\overline{R(Ty_n : n \in \mathbb{N})}$ is fuzzy compact. Therefore, $\{Ty_n\}$ has a fuzzy convergent subsequence. Thus, T is a fuzzy compact linear operator by Theorem 3. □

Lemma 7. *Let* $(X, N, *)$ *be an fnls. For each* $\alpha \in (0,1)$ $\exists\, \beta \in (0, \alpha/3)$ *such that:*

$$B(\theta, \beta, \beta) + B(\theta, \beta, \beta) + B(\theta, \beta, \beta) \subseteq B(\theta, \alpha, \alpha)$$

Proof. Suppose $\alpha_0 \in (0,1)$. Then, $\alpha_0/3$ also belongs to $(0,1)$. By the left-continuity of '$*$' at $(1,1)$, $\exists \beta_0 \in (0, \alpha/3]$ such that:

$$(1-\beta_0) * (1-\beta_0) * (1-\beta_0) > 1 - \alpha_0/3$$

Let $y \in B(\theta, \beta_0, \beta_0) + B(\theta, \beta_0, \beta_0) + B(\theta, \beta_0, \beta_0)$. Thus, $y = y_1 + y_2 + y_3$, where $y_1, y_2, y_3 \in B(\theta, \beta_0, \beta_0)$. Now:

$$\begin{aligned}
&N(y, \alpha_0) \\
&\geq N(y_1, \alpha_0/3) * N(y_2, \alpha_0/3) * N(y_3, \alpha_0/3) \\
&\geq N(y_1, \beta) * N(y_2, \beta) * N(y_3, \beta) \\
&\geq (1-\beta) * (1-\beta) * (1-\beta) \\
&> (1-\alpha_0) \\
&\therefore y \in B(\theta, \alpha_0, \alpha_0)
\end{aligned}$$

This completes the proof. □

Theorem 10. *(Schauder-type fixed point theorem) Let* $(X, N, *)$ *be an fnls, C be a bounded, closed, convex subset in X w.r.t.* τ_N*, and* $f : C \to C$ *be a fuzzy compact linear operator. Then, there exists a point* $x_0 \in C$ *such that* $f(x_0) = x_0$.

Proof. Since f is a fuzzy compact linear operator, then by Lemma 4 and Remark 1, there exists a sequence of continuous mappings $\{f_n\}$, which is uniformly fuzzy convergent to f, and each $\{f_n\}$ contains a fixed point, say x_n, i.e., $f_n(x_n) = x_n$, $\forall n \in \mathbb{N}$.

Since each $x_n \in \tilde{C}$, then by Lemma 5, $\{x_n\}$ has a fuzzy convergent subsequence, say $\{x_{n_k}\}$, i.e., $x_{n_k} \to x_0$. Now, for any $t > 0$, $\alpha_0 \in (0,1)$ with
$(1-\beta_0) * (1-\beta_0) * (1-\beta_0) > 1 - \alpha_0$, we have:

$$\begin{aligned}
N(f(x_0) - x_0, t) \geq & N(f(x_0) - f_{n_k}(x_0), t/4) * N(f_{n_k}(x_0) - f_{n_k}(x_{n_k}), t/4) * \\
& N(f_{n_k}(x_{n_k}) - x_{n_k}, t/4) * N(x_{n_k} - x_0, t/4)
\end{aligned} \tag{6}$$

Since $\{f_{n_k}\}$ uniformly fuzzy converges to f, then by Remark 2, for $x_0 \in X$,

$$\lim_{k \to \infty} N(f(x_0) - f_{n_k}(x_0), t/4) = 1$$

Again, each $\{f_{n_k}\}$ is continuous, so $x_{n_k} \to x_0 \implies f_{n_k}(x_{n_k}) \to f_{n_k}(x_0)$,

$$i.e., \lim_{k \to \infty} N(f_{n_k}(x_0) - f_{n_k}(x_0), t/4) = 1$$

Taking $lim\ k \to \infty$ in both sides of Inequality 6, we get,

$$\begin{aligned}
&N(f(x_0) - x_0, t) \geq 1 \\
&\Rightarrow N(f(x_0) - x_0, t) = 1\ \forall t > 0 \\
&\Rightarrow f(x_0) = x_0\ (by\ N2)
\end{aligned}$$

This completes the proof. □

Theorem 11. *Let $(X, N, *)$ be an fnls. Let C be a convex, compact subset of X and f be a continuous operator from C into C. Then, there exists $x_0 \in C$ such that $f(x_0) = x_0$.*

Proof. Since C is compact w.r.t. τ_N, thus by Theorem 6, C is fuzzy totally bounded. Now, consider a strictly decreasing sequence $\{\alpha_n\}$ with $\alpha_n \to 0$, then $\exists \{x_n^1, x_n^2, \cdots x_n^m\} \subseteq C$ such that,

$$C \subseteq \bigcup_{i=1}^{m} \{x_n^i\} + B(\theta, \alpha_n, \alpha_n)$$

Now, define a family of functions such that:

$$f_n(x) = \frac{\sum_{i=1}^{m} \beta_n^i(x) x_n^i}{\sum_{i=1}^{m} \beta_n^i(x)}$$

where $\sum_{i=1}^{m} \beta_n^i(x) = max\{0, \alpha_n - \wedge\{t > 0 : N(f(x) - x_n^i, t) > 1 - \alpha_n\}\}$

Let $x \in C$ and $\alpha_0 \in (0,1)$.

Since $\alpha_n \to 0$, so $\exists N_0 \in \mathbb{N}$ such that $\alpha_n < \alpha_0 \; \forall n \geq N_0$. Now:

$$\begin{aligned}
&\wedge \{t > 0 : N(f_n(x) - f(x), t) > 1 - \alpha_0\} \\
&= \wedge\{t > 0 : N(\frac{\sum_{i=1}^{m} \beta_n^i(x) x_n^i}{\sum_{i=1}^{m} \beta_n^i(x)} - f(x), t) > 1 - \alpha_0\} \\
&= \wedge\{t > 0 : N(x_n^i - f(x), t) > 1 - \alpha_0\} \\
&\leq \wedge\{t > 0 : N(x_n^i - f(x), t) > 1 - \alpha_n\} \leq \alpha_n \; \forall n \geq N_0 \\
&\Rightarrow \bigvee_{x \in C} \wedge \{t > 0 : N(f_n(x) - f(x), t) > 1 - \alpha_0\} \leq \alpha_n \; \forall n \geq N_0 \\
&\Rightarrow \bigvee_{x \in C} \wedge \{t > 0 : N(f_n(x) - f(x), t) > 1 - \alpha_0\} \to 0 \; as \; n \to \infty
\end{aligned}$$

Since $\alpha_0 \in (0,1)$ is arbitrary, thus $\{f_n\}$ uniformly fuzzy converges to f. Again, $\{f_n\}$ is a family of continuous functions from (X, τ_N) to itself. For each $n \in \mathbb{N}$, f_n maps from C to the closed convex hull C_n of $\{x_n^i, i = 1, 2, 3, \cdots\}$. Since C is convex, then $C_n \subseteq C$. We constrict the restricted mapping $f_n : C_n \to C_n$, and it turns out that it maps the compact, convex subset of a finite-dimensional set C_n of $Y_n =$ the span of $\{x_n^i, i = 1, 2, \cdots, m(n)\}$ into itself. Thus, by the Browder fixed point theorem, $\exists x_n \in C_n \subseteq C$ such that $f_n(x_n) = x_n$, $\forall n \in \mathbb{N}$. Since C is compact w.r.t. τ_N, $\{x_n\}$ has a convergent subsequence, say $\{x_{n_k}\}$ w.r.t. fuzzy norm N and $\{x_{n_k}\} \to x_0$. Now, consider $t_0(> 0) \in \mathbb{R}$.

We have $N(f(x_0) - x_0, t_0)$

$$\geq N(f(x_0) - f(x_{n_k}), t_0/3) * N(f(x_{n_k}) - x_{n_k}, t_0/3) * N(x_{n_k} - x_0, t_0/3)$$

Taking $k \to \infty$ on both sides, we get $N(f(x_0) - x_0, t_0) = 1$. Again, $t_0 > 0$ is arbitrary.

$$\begin{aligned}
&\text{So, } N(f(x_0) - x_0, t) = 1, \; \forall t > 0 \\
&\Rightarrow f(x_0) = x_0.
\end{aligned}$$

□

Theorem 12. *Let $(X, N, *)$ be a fuzzy Banach space, C be a closed and convex subset of X, and $f : C \to C$ be a continuous mapping such that the image of C is contained ina compact set. Then, $\exists x_0 \in C$ such that $f(x_0) = x_0$.*

Proof. Let $B = f(C)$. Consider $K = \overline{Co(f(C))}$ (where $Co(f(C))$ is the convex combination of the element of $f(C)$). It is clear that K is a convex subset of X. We show that K is compact w.r.t. τ_N. We have $B(\subseteq C)$, a compact subset of X w.r.t. τ_N. Therefore, B is fuzzy totally bounded.
Let $\alpha_0 \in (0,1)$. Then, $\exists \beta_0 \in (0, \alpha_0/3)$ such that $(1-\beta_0) * (1-\beta_0) * (1-\beta_0) > (1-\alpha_0)$. Again, since $\beta_0 \in (0,1)$, $\exists \{x_1, x_2, \cdots, x_n\} \subseteq B$ such that:

$$B \subseteq \bigcup_{i=1}^{n} \{x_i\} + B(\theta, \beta_0, \beta_0) \tag{7}$$

Let $x \in Co(B)$. Thus, x is of the form $\sum_{j-1}^{m} \alpha_j y_j$, where $\sum_{j-1}^{m} \alpha_j = 1$. Again, each $y_j \in \bigcup_{i=1}^{n} \{x_i\} + B(\theta, \beta_0, \beta_0)$, $j = \{1, 2, \cdots, m(\leq)n\}$, Therefore, for each y_j, $\exists\, x_i$ for some $i \in \{1, 2, \cdots, n\}$ such that:

$$N(x_i - y_j, \beta_0) > 1 - \beta_0$$

Here:

$$\begin{aligned} &N(\sum_{j=1}^{m} \alpha_j x_j - \sum_{j=1}^{m} \alpha_j y_j, \beta_0) \\ &= N(x_j - y_j, \beta_0) > 1 - \beta_0 \\ &\therefore x \in \sum_{j=1}^{m} \alpha_j x_j + B(\theta, \beta_0, \beta_0) \end{aligned}$$

$\therefore x \in C_j + B(\theta, \beta_0, \beta_0)$, where $C_j = \sum_{j=1}^{m} \alpha_j x_j$. Since each fnls is a topological vector space:

$$\text{so, } \overline{Co(B)} = \bigcap_{\alpha \in (0,1)} Co(B) + B(\theta, \alpha, \alpha)$$

Thus:

$$\begin{aligned} \overline{Co(B)} &\subseteq Co(B) + B(\theta, \beta_0, \beta_0) \\ &\subseteq C_j + B(\theta, \beta_0, \beta_0) + B(\theta, \beta_0, \beta_0) \\ &\subseteq \overline{C_j} + B(\theta, \beta_0, \beta_0) + B(\theta, \beta_0, \beta_0) \end{aligned}$$

Here, $\overline{C_j}$ is a closed bounded subset of $Y_j = Span\{x_j\}_{j=1}^{m}$. Therefore, $\overline{C_j}$ is compact w.r.t. τ_N. Thus, $\exists \{p_k\}_{k=1}^{r} \in \overline{C_j} \subseteq \overline{Co(B)}$ such that:

$$\overline{C_j} \subseteq \bigcup_{k=1}^{r} \{p_k\} + B(\theta, \beta_0, \beta_0)$$

$$\begin{aligned} \therefore \overline{Co(B)} &\subseteq \bigcup_{k=1}^{r} \{p_k\} + B(\theta, \beta_0, \beta_0) + B(\theta, \beta_0, \beta_0) + B(\theta, \beta_0, \beta_0) \\ &\subseteq \bigcup_{k=1}^{r} \{p_k\} + B(\theta, \alpha, \alpha) \end{aligned}$$

Thus, we get that $K = \overline{Co(B)}$ is totally bounded and complete, i.e., compact w.r.t. τ_N. Again, $f(K) \subseteq f(C) \subseteq K$. By theorem 11, $\exists x_0 \in K$ such that $f(x_0) = x_0$. □

4. Darbo's Generalization of the Schauder-Type Fixed Point Theorem Using the Concept of the Measure of Non-Compactness

In this section, we first consider two types of fuzzy bounded subsets of a KM-type fuzzy metric space (i.e., M is a left-continuous function w.r.t. t, and $*$ is left-continuous at $(1,1)$). We renamed them as strongly and weakly and studied the relation between them. After that, the measure of the non-compactness of a strongly fuzzy bounded subset of the fuzzy metric space is defined. Using this concept, a family of ψ-set contraction mapping is specified, and Darbo's generalization of the Schauder-type fixed point theorem is established for these types of contraction mappings.

Definition 15. *(Strongly fuzzy boundedness) Let $(X, M, *)$ be a fuzzy metric space. A subset Q of X is said to be strongly fuzzy bounded if $\exists\, t > 0$ such that for each $\alpha \in (0,1)$:*

$$x \in B_M(y, \alpha, t)\ \forall\, x, y \in Q$$

i.e., fuzzy diameter of Q $(f - \delta(Q))$ less than ∞ where
$f - \delta(Q) = \bigvee_{\alpha \in (0,1)} \bigvee_{x,y \in Q} \wedge\{t > 0 : M(x,y,t) > 1 - \alpha\}$ *(defined by Bag and Samanta in the paper [28].)*

An example is presented to understand the strongly fuzzy boundedness more clearly.

Example 2. *Let $X = \mathbb{R}^2$ (the set of all ordered pairs of the elements of the set of all real numbers) and $\|x\|' = |x_1| + |x_2|, \|x\|'' = (|x_1|^2 + |x_2|^2)^{1/2}$, where $x = (x_1, x_2)$ are two norms on X. Clearly, $\|x\|' \geq \|x\|'', \forall x \in X$. Define a function $M : X \times X \times \mathbb{R} \to [0,1]$ by:*

$$M(x,y,t) = \begin{cases} 1, \ t \geq \|x-y\|' \\ 1/2, \ \|x-y\|'' \leq t < \|x-y\|' \\ 0, \ t < \|x-y\|'' \end{cases}$$

Then, M is a fuzzy metric on X w.r.t. the min t-norm. Clearly,

$$\wedge\{t > 0 : M(x,y,t) > 1 - \alpha\} = \begin{cases} \|x-y\|', \ 0 < \alpha < 1/2 \\ \|x-y\|'', \ 1/2 \leq \alpha < 1 \end{cases}$$

Consider $A = B(0,1)$. Now:

$$\begin{aligned} f - \delta(A) &= \bigvee_{\alpha \in (0,1)} \bigvee_{x,y \in A} \wedge\{t > 0 : M(x,y,t) > 1 - \alpha\} \\ &= diam A \ w.r.t. \| \ \|' \\ &= 1 < \infty \end{aligned}$$

$\therefore$ A is strongly fuzzy bounded.

The fuzzy boundedness defined in Definition 4 is renamed as the weakly fuzzy bounded subset of a fuzzy metric space $(X, M, *)$. From the two definitions, it is clear that strongly fuzzy bounded implies the weakly fuzzy boundedness, but the converse may not be. This can be justified by the following example.

Example 3. *Consider the fuzzy metric:*

$$M(x,y,t) = \begin{cases} \frac{t}{t+|x-y|}, t > |x-y| \\ 0, t \leq |x-y| \end{cases}$$

Now, $\wedge\{t > 0 : M(x,y,t) > 1-\alpha\} = \frac{\alpha}{1-\alpha}|x-y|$. *Let* $A = [0,1]$. *Since* $|x-y| \leq 1 < 2$, *so* A *is weakly fuzzy bounded. However,* $\bigvee_{\alpha\in(0,1)} \frac{\alpha}{1-\alpha} = \infty$. *Thus,* A *is not strongly fuzzy bounded.*

The weakly fuzzy bounded subset of a fuzzy metric space can also be defined as if $\beta(A) = 1$, then A is weakly fuzzy bounded where $\beta(A) = \underset{t>0}{Sup}\ \underset{x,y\in A}{inf}\ M(x,y,t)$. The equivalence between these two definitions was already proved in the paper [29].

Definition 16. *(Kuratowski's measure of non-compactness) Let* $(X, M, *)$ *be a fuzzy metric space and* Q *be a strongly fuzzy bounded subset of* X. *Then, Kuratowski's measure of the non-compactness of* Q *denoted by* $\psi(Q)$ *is defined as:*

$$\psi(Q) = inf\{\epsilon > 0;\ Q \subseteq \bigcup_{i=1}^{n} S_i,\ S_i \subseteq X,\ f-\delta(S_i) < \epsilon\ \forall i \in \{1,2,\cdots,n\}\}$$

From the definition, it is clear that $\psi(Q) < f-\delta(Q)$, for each strongly fuzzy bounded subset Q of X.

Definition 17. *(α-level Kuratowski measure of non-compactness) Let* $(X, M, *)$ *be a fuzzy metric space and* Q *be a weakly fuzzy bounded subset (or strongly fuzzy bounded subset) of* X. *Then, for each* $\alpha \in (0,1)$, *the* α*-level Kuratowski measure of the non-compactness of* Q *denoted by* $\psi_\alpha(Q)$ *is defined as:*

$$\psi_\alpha(Q) = inf\{\epsilon > 0;\ Q \subseteq \bigcup_{i=1}^{n} S_i,\ S_i \subseteq X,\ \alpha-\delta(S_i) < \epsilon\ \forall i \in \{1,2,\cdots,n\}\}$$

where $\alpha-\delta(S_i) = \bigvee_{x,y\in S_i} \wedge\{t > 0 : M(x,y,t) > 1-\alpha\}$, *defined by Bag and Samanta [28].*

From the definition of $\psi(Q)$ and $\psi_\alpha(Q)$, it is clear that if Q is a strongly fuzzy bounded subset, then $\psi_\alpha(Q) \leq \psi(Q)\ \forall\alpha \in (0,1)$, i.e., $\bigvee_{\alpha\in(0,1)} \psi_\alpha(Q) \leq \psi(Q)$.

Lemma 8. *Let* Q, Q_1, Q_2 *be strong fuzzy bounded subsets of a complete fuzzy metric space* $(X, M, *)$. *Then:*

(i) $\psi(Q) = 0 \iff \bar{Q}$ *is compact w.r.t.* τ_M.
(ii) $\psi(Q) = \psi(\bar{Q})$
(iii) $Q_1 \subseteq Q_2 \implies \psi(Q_1) \leq \psi(Q_2)$
(iv) $\psi(Q_1 \cup Q_2) = max\{\psi(Q_1), \psi(Q_2)\}$

Again, if $(X, N, *)$ *is an fnls, then the followings properties also hold.*

(v) $\psi(Q_1 + Q_2) \leq \psi(Q_1) + \psi(Q_2)$
(vi) $\psi(Q + x_0) = \psi(Q)$
(vii) $\psi(rQ_1) = |r|\psi(Q_1)$
(viii) $\psi(ConvQ) = \psi(Q)$

Proof. (i) First, we suppose that $\psi(Q) = 0$. Then, for each $\epsilon > 0\ \exists\ \{S_i\}_{i=1}^{n}$ with $f-\delta(S_i) < \epsilon$ such that $Q \subseteq \bigcup_{i=1}^{n} S_i$. Now, if Q is totally bounded, then $\bar{Q}$ is also, and we get the required result. Let $\alpha_0 \in (0,1)$ and $\epsilon_0 > 0$. Consider a fixed $x_i \in S_i$ for each $i = \{1,2,\cdots,n\}$. Then, it is clear that $S_i \subseteq B_M(x_i, \alpha_0, \epsilon_0)$.

$$\therefore Q \subseteq \bigcup_{i=1}^{n} S_i \subseteq \bigcup_{i=1}^{n} B_M(x_i, \alpha_0, \epsilon_0)$$

Thus, Q is totally bounded. Conversely, suppose that $\bar{Q}$ is compact w.r.t. τ_M. Then, Q is totally bounded. Let $\epsilon > 0$ be given. Then, for any $\alpha \in (0,1)$, and for, $\epsilon > 0$, $\exists\ \{x_1, x_2, \cdots, x_n\}$ such that

$Q \subseteq \bigcup_{i=1}^{n} B_M(x_i, \alpha, \epsilon/2)$. Consider:

$$S_i = \bigvee_{\alpha \in (0,1)} \{y \in X, \wedge\{t > 0 : M(x_i, y, t) > 1 - \alpha < \epsilon/2\}, \forall i = \{1, 2, \cdots, n\}.$$

Then, $Q \subseteq \bigcup_{i=1}^{n} S_i$, where $f - \delta(S_i) < \epsilon$ for each $i = \{1, 2, \cdots, n\}$. Since $\epsilon > 0$ is arbitrary, thus $\psi(Q) = 0$.

(ii) We first prove that $f - \delta(Q) = f - \delta(\bar{Q})$. Then, the required result follows immediately. Obviously, $f - \delta(Q) \leq f - \delta(\bar{Q})$. For the reverse part, let $x, y \in \bar{Q}$. Then, $\exists \{x_n\}$ and $\{y_n\}$ in Q such that:

$$\lim_{n\to\infty} M(x_n, x, t) = 1 \, \forall t > 0$$
$$\lim_{n\to\infty} M(y_n, y, t) = 1 \, \forall t > 0$$

Here:

$$\begin{aligned}
&\wedge \{t > 0 : M(x, y, t) > 1 - \alpha\} \\
&\leq \wedge\{t > 0 : M(x, x_n, t) > 1 - \beta\} + \wedge\{t > 0 : M(x_n, y_n, t) > 1 - \beta\} + \wedge\{t > 0 : M(y_n, y, t) > 1 - \beta\} \\
&(\beta \leq \alpha) \\
&\implies \wedge\{t > 0 : M(x, y, t) > 1 - \alpha\} \leq \liminf_{n\to\infty} \wedge \{t > 0 : M(x_n, y_n, t) > 1 - \beta\} \leq f - \delta(Q) \\
&\therefore f - \delta(\bar{Q}) \leq f - \delta(Q)
\end{aligned}$$

Thus, we arrive at the required conclusion.

(iii) For $Q_1 \subseteq Q_2$, the set $\{\epsilon > 0;\ Q_2 \subseteq \bigcup_{i=1}^{n} S_i,\ S_i \subseteq X,\ f - \delta(S_i) < \epsilon$ $\forall i \in \{1, 2, \cdots, n\}\} \subseteq \{\epsilon > 0;\ Q_1 \subseteq \bigcup_{i=1}^{n} S_i,\ S_i \subseteq X,\ f - \delta(S_i) < \epsilon$ $\forall i \in \{1, 2, \cdots, n\}\}. \therefore \psi(Q_1) \leq \psi(Q_2)$.

(iv) From (iii), $\psi(Q_1 \cup Q_2) \leq max\{\psi(Q_1), \psi(Q_2)\}$ follows. The reverse part is similar to a crisp set. For the references, please see [30].

For (v), (vi), (vii), and (viii), we first prove that in an fnls, the following properties hold.

(1) $f - \delta(Q_1 + Q_2) \leq f - \delta(Q_1) + f - \delta(Q_2)$.
(2) $f - \delta(Q + x_0) = f - \delta(Q)$
(3) $f - \delta(rQ) = |r| f - \delta(Q)$
(4) $f - \delta(ConvQ) = f - \delta(Q)$

Then rest of the proof of (v), (vi), (vii), and (viii) is similar to the classical version of this theorem.

(1) Now:

$$\begin{aligned}
f - \delta(Q_1 + Q_2) &= \bigvee_{\alpha \in (0,1)} \bigvee_{x, y \in Q_1 + Q_2} \wedge\{t > 0 : N(x - y, t) > 1 - \alpha\} \\
&= \bigvee_{\alpha \in (0,1)} \bigvee_{x_1, x_2 \in Q_1, y_1, y_2 \in Q_2} \wedge\{t > 0 : N(x_1 + y_1 - x_2 - y_2, t) > 1 - \alpha\} \\
&\leq \bigvee_{\alpha \in (0,1)} \bigvee_{x_1, x_2 \in Q_1} \wedge\{t > 0 : N(x_1 - x_2, t) > 1 - \alpha\} \\
&+ \bigvee_{\alpha \in (0,1)} \bigvee_{y_1, y_2 \in Q_2} \wedge\{t > 0 : N(y_1 - y_2, t) > 1 - \alpha\} \\
&= f - \delta(Q_1) + f - \delta(Q_2)
\end{aligned}$$

(2) Again:

$$\begin{aligned} f-\delta(Q) &= \bigvee_{\alpha\in(0,1)} \bigvee_{x,y\in Q} \wedge\{t>0 : N(x-y,t)>1-\alpha\} \\ &= \bigvee_{\alpha\in(0,1)} \bigvee_{x,y\in Q} \wedge\{t>0 : N(x+x_0-y-x_0,t)>1-\alpha\} \\ &= f-\delta(Q+x_0) \end{aligned}$$

(3)

$$\begin{aligned} f-\delta(rQ) &= \bigvee_{\alpha\in(0,1)} \bigvee_{x,y\in rQ} \wedge\{t>0 : N(x-y,t)>1-\alpha\} \\ &= \bigvee_{\alpha\in(0,1)} \bigvee_{x_1,y_1\in Q} \wedge\{t>0 : N(rx_1-ry_1,t)>1-\alpha\} \\ &= \bigvee_{\alpha\in(0,1)} \bigvee_{x_1,y_1\in Q} \wedge\{|r|t/|r|>0 : N(x_1-y_1,t/|r|)>1-\alpha\} \\ &= |r| \bigvee_{\alpha\in(0,1)} \bigvee_{x_1,y_1\in Q} \wedge\{t>0 : N(x_1-y_1,t)>1-\alpha\} \\ &= |r| f-\delta(Q) \end{aligned}$$

(4) $f-\delta(Q)\leq f-\delta(ConvQ)$ is obvious as $Q\subseteq ConvQ$. We only show that for a fixed $\alpha_0\in(0,1)$ and for a fixed $x_0(\neq\theta)\in X$,
$\bigvee_{y\in Conv(Q)} \wedge\{t>0 : N(x_0-y,t)>1-\alpha_0\} = \bigvee_{y\in Q} \wedge\{t>0 : N(x_0-y,t)>1-\alpha_0\}$.

Since α_0 and x_0 are arbitrary, thus
$f-\delta(ConvQ)\leq f-\delta(Q)$. Consider $y\in Conv(Q)$. Thus, $y=\sum_{i=1}^{n}\lambda_i x_i,\ x_i\in Q,\ \sum_{i=1}^{n}\lambda_i=1$.

$$\begin{aligned} \therefore\ & \bigvee_{y\in Conv(Q)} \wedge\{t>0 : N(x_0-y,t)>1-\alpha_0\} \\ &= \bigvee_{x_i\in Q} \wedge\{t>0 : N(\sum_{i=1}^{n}\lambda_i x_0-\sum_{i=1}^{n}\lambda_i x_i,t)>1-\alpha\} \\ &= \bigvee_{x_i\in Q} \wedge\{t>0 : N(x_0-x_i,t/\sum_{i=1}^{n}\lambda_i)>1-\alpha\} \\ &= \bigvee_{x_i\in Q} \wedge\{t>0 : N(x_0-x_i,t)>1-\alpha\} \\ &\leq \bigvee_{y\in Q} \wedge\{t>0 : N(x_0-y,t)>1-\alpha\} \end{aligned}$$

$\therefore$ we arrive at the required conclusion. □

Definition 18. *(Axiomatic approach) Let $(X,M,*)$ be a complete fuzzy metric space and $\mathcal{B}$ the family of strongly fuzzy bounded subsets of X. A map $\psi:\mathcal{B}\to[0,\infty)$ is called a measure of non-compactness if it satisfies the following properties:*

(1) $\psi(B)=0 \iff B$ is fuzzy totally bounded, $\forall B\in\mathcal{B}$.
(2) $\psi(B)=\psi(\bar{B}), \forall B\in\mathcal{B}$.
(3) $\psi(B_1\cup B_2)=max\{\psi(B_1),\psi(B_2)\}, \forall B_1,\ B_2\in\mathcal{B}$

Using this axiomatic approach, we give some examples of the measure of the non-compactness in a fuzzy metric space.

Example 4. *Let $X=\mathbb{R}^2$ (the set of all ordered pairs of the elements of the set of all real numbers) and $\|x\|'=|x_1|+|x_2|, \|x\|''=(|x_1|^2+|x_2|^2)^{1/2}$, where $x=(x_1,x_2)$ are two norms on X. Clearly,*

$\|x\|' \geq \|x\|'', \forall x \in X$. *Define a function* $M : X \times X \times \mathbb{R} \to [0,1]$ *by:*

$$M(x,y,t) = \begin{cases} 1, \ t \geq \|x-y\|' \\ 1/2, \ \|x-y\|'' \leq t < \|x-y\|' \\ 0, \ t < \|x-y\|'' \end{cases}$$

Then, M is a fuzzy metric on X w.r.t. the min t-norm. Clearly,

$$\wedge\{t > 0 : M(x,y,t) > 1-\alpha\} = \begin{cases} \|x-y\|', \ 0 < \alpha < 1/2 \\ \|x-y\|'', \ 1/2 \leq \alpha < 1 \end{cases}$$

Define functions ψ_1 *and* ψ_2 *from the set of all strongly fuzzy bounded subsets of X to* $[0,\infty)$ *by:*

$$\psi_1(B) = \begin{cases} 0, \ if B \text{ is totally fuzzy bounded} \\ 1, otherwise \end{cases}$$

and $\psi_2(B) = f - \delta(B)$.

Both ψ_1 *and* ψ_2 *satisfy all the conditions of Definition 18. Therefore, both are the measure of the non-compactness of fuzzy metric space* (X, M, min).

Theorem 13. *Let* $(X, M, *)$ *be a complete fuzzy metric space. If* $\{F_n\}$ *is a decreasing sequence of non-empty closed, strongly fuzzy bounded subsets of X such that* $\lim_{n\to\infty} \psi(F_n) = 0$, *then the intersection* $F_\infty = \bigcap_{n=1}^{\infty} F_n$ *is a non-empty compact subset of X w.r.t.* τ_M.

Proof. Here, $\psi(F_\infty) \leq \lim_{n\to\infty} \psi(F_n) = 0$. Thus, by Lemma 8, F_∞ is compact w.r.t. τ_M, as F_∞ is closed. Now, we will show that F_∞ is non-empty. Since $\lim_{n\to\infty} \psi(F_n) = 0$, so $\lim_{n\to\infty} \psi_\alpha(F_n) = 0$, $\forall \alpha \in (0,1)$. Let $\{x_n\} \subseteq X$ and $x_n \in F_n$, i.e., $\{x_n\} \in F_1$; $\{x_n\}_{n=2}^{\infty} \in F_2$, and so on. Consider $\alpha_0 \in (0,1)$. Thus, $\lim_{n\to\infty} \psi_{\alpha_0}(F_n) = 0$.

By Definition 17, for every $n \in \mathbb{N}$, $F_n \subseteq \bigcup_{i=1}^{k_n} F_i^n$ such that $\alpha_0 - \delta(F_i^n) < \psi_{\alpha_0}(F_n) + 1/n$. Since $\{x_n\} \subseteq F_1$ $\exists \{x_n^1\} \subseteq F_i^1 \cap F_2 \subseteq F_1$ for some $i = \{1,2,\cdots,k_n\}$, so $\alpha_0 - \delta(x_n^1) < \psi_{\alpha_0}(F_1) + 1$. Consider a subsequence $\{x_n^2\}$ of $\{x_n^1\}$ with $\{x_n^1\} \subseteq F_i^2 \cap F_2$, for some $i = \{1,2,\cdots,k_n\}$. Thus, $\alpha_0 - \delta(x_n^2) < \psi_{\alpha_0}(F_2) + 1/2$.

Similarly, we get a subsequence of $\{x_n^j\}$ of $\{x_n^{j-1}\}$ with $\alpha_0 - \delta(x_n^j) < \psi_{\alpha_0}(F_j) + 1/j$, i.e., $\lim_{j\to\infty} \alpha_0 - \delta(x_n^j) = 0$. This is true for any $\alpha_0 \in (0,1)$.

Thus, $\lim_{j\to\infty} \alpha - \delta(x_n^j) = 0$, $\forall \alpha \in (0,1)$.

For any $\alpha \in (0,1)$ and $\epsilon > 0$ $\forall j_1, j_2 \in \mathbb{N}$, $\exists N_0(\alpha, \epsilon) \in \mathbb{N}$ such that:

$$\wedge \{t > 0 : M(x_n^{j_1}, x_n^{j_2}, t) > 1-\alpha\} < \epsilon \ \forall j_1, j_2 > N_0$$
$$\implies M(x_n^{j_1}, x_n^{j_2}, \epsilon) > 1-\alpha \ \forall j_1, j_2 > N_0$$
$$\implies \lim_{j_1, j_2 \to\infty} M(x_n^{j_1}, x_n^{j_2}, t) = 1 \ \forall t > 0.$$

$\therefore\ \{x_n^j\}_j$ is a Cauchy sequence w.r.t. $(X, M, *)$, i.e., converges to $x \in F_n\ \forall n \in \mathbb{N}$.
i.e. $x \in \bigcap_{n=1}^{\infty} F_n = F_\infty$. Thus, F_∞ is non-empty. □

Definition 19. *Let $(X, M, *)$ be a complete fuzzy metric space and $f : X \to X$ be a fuzzy continuous mapping. Then, f is called a ψ-set contraction if there exists $k \in [0,1)$ such that for all strongly fuzzy bounded subsets C of X, the following relation holds, $\psi(f(C)) \le k\psi(C)$, where ψ is the measure of the noncompactness of C.*

This definition is inspired by the α-set contraction in the classical set theory. For the references, please see the book [9].

Theorem 14. *(Darbo's generalization of the Schauder-type fixed point theorem) Let $(X, N, *)$ be a fuzzy Banach space and C be a closed, strongly fuzzy bounded, and convex subset of X. If $f : C \to C$ is a ψ-set contraction, then f has a fixed point in C.*

Proof. For each $n \in \mathbb{N}$, consider $C_n = \overline{Convf(C_{n-1})}$. Clearly, $C_{n+1} \subseteq C_n\ \forall n$. Now, $C_\infty = \bigcap_{n=1}^{\infty} C_n$, which is a closed and convex set, and $\psi(C) \le \lim_{n\to\infty} \psi(C_n)$. Again, $\lim_{n\to\infty} \psi(C_n) = 0$ ($\because \psi(f(C_n)) \le k^n \psi(C)$).

Furthermore, $f(C_\infty) = f(\bigcap_{n=1}^{\infty} C_n) \subseteq \bigcap_{n=1}^{\infty} f(C_n) \subseteq \bigcap_{n=1}^{\infty} C_{n+1} = C_\infty$. By Theorem 13, C_∞ is compact and non-empty. Thus, $f : C_\infty \to C_\infty$ is a continuous mapping from a compact, convex set to itself. Thus, by Theorem 11, $\exists\ x_0 \in C_\infty$ such that $f(x_0) = x_0$. This completes the proof. □

Example 5. *Let $X = C[0,1]$ (the set of all continuous functions over [0,1]) and $\|x\|' = \underset{0\le t\le 1}{Sup} |x(t)|$, $\|x\|'' = \int_0^1 |x(t)|dt$ be two norms on X. Clearly, $\|x\|' \ge \|x\|'', \forall x \in X$.*

Define a function $N : X \times \mathbb{R} \to [0,1]$ by:

$$N(x,t) = \begin{cases} 1, \ t \ge \|x\|' \\ 1/2, \ \|x\|'' \le t < \|x\|' \\ 0, \ t < \|x\|'' \end{cases}$$

Then, (X, N, min) is fuzzy Banach space. Clearly,

$$\wedge\{t > 0 : N(x,t) > 1-\alpha\} = \begin{cases} \|x\|', \ 0 < \alpha < 1/2 \\ \|x\|'', \ 1/2 \le \alpha < 1 \end{cases}$$

Define a function $f : \overline{B(\theta, 1/2, 1)} \to \overline{B(\theta, 1/2, 1)}$ with $\|f(x) - f(y)\|' \le k\|x-y\|''$, $\forall x, y \in \overline{B(\theta, 1/2, 1)}$ and $\psi(C) = f - \delta(C) = \bigvee_{x,y\in C} \|x-y\|'$, where C is a strongly fuzzy bounded subset of $\overline{B(\theta, 1/2, 1)}$. Clearly, $\psi(f(C)) = \bigvee_{x,y\in C} \|f(x)-f(y)\|' \le k \bigvee_{x,y\in C} \|x-y\|'' \le k \bigvee_{x,y\in C} \|x-y\|' = k\psi(C)$.
$\therefore f$ is a ψ-set contraction mapping. By Theorem 14, f has a fixed point in C.

Remark 3. *In Example 5, $\overline{B(\theta, 1/2, 1)} = \{y \in X, \|x-y\|'' \le 1\}$. It is a closed, convex, bounded subset in $(X, \| \|'')$, where $\|x\|'' = \int_0^1 |x(t)|dt$. However, $(X, \| \|'')$ is not a Banach space. Therefore, the classical version of Darbo's generalization of the Schauder-type fixed point theorem will not be able to give the existence result of a fixed point of f, which is defined in Example 5. In this scene, our theorem is more general than its classical form.*

5. Conclusions

Schauder's fixed point theorem and its generalizations play a pivotal role in this context of nonlinear functional analysis. The aim of this paper is to study different types of Schauder's fixed point theorems in the context of fuzzy settings. For this reason, two types of fuzzy convergence are defined for a sequence of linear operators whose domain and range space are the fnlss. Moreover, the notion of two types of fuzzy bounded subsets of a fuzzy metric space is formulated, and the relation between them is studied. Further, the concept of Kuratowski's measure of non-compactness in a fuzzy metric and an fnls are introduced for both fuzzy bounded subsets. This concept is used as a tool to prove Darbo's generalization of the Schauder-type fixed point theorem. This is the first instance of studying the measure of the non-compactness in fuzzy settings. There is a huge scope of further research in this area, and many fixed point theorems can be developed by using these types of measures of non-compactness. Schauder's fixed point theorem has various applications in the theory of differential equations such as Peano's existence theorem for the first-order differential equations, the existence of the positive solution to the second-order singular differential equations, the existence of periodic orbits of rapidly symmetric systems, and so on. The theorems developed in this manuscript will promote future studies on the fuzzified area of the above-mentioned differential equations, as well as in the fuzzy neural networks.

Author Contributions: Conceptualization, S.C. and J.-G.L.; funding acquisition, J.-G.L.; writing, original draft, S.C.; Writing, review and editing, T.B. and J.-G.L. All authors have read and agreed to the published version of the manuscript.

Funding: This research is supported by the Basic Science Research Program through the National Research Foundation of Korea (NRF) funded by the Ministry of Education (2018R1D1A1B07049321).

Acknowledgments: The authors are grateful to the reviewers for their valuable suggestions and comments in rewriting the article in the present form. The authors are also thankful to the Editor-in-Chief of the journal for his valuable comments.

Conflicts of Interest: The authors declare no conflict of interest.

Reference

1. Chu, J.; Torres, P.J. Applications of Schauder's fixed point theorem to singular differential equations. *Bull. Lond. Math. Soc.* **2007**, *39*, 653–660. [CrossRef]
2. Dizaji, A.F.; Sepiani, H.A.; Ebrahimi, F.; Allahverdizadeh, A.; Sepiani, H.A. Schauder fixed point theorem based existence of periodic solution for the response of Duffing's oscillator. *J. Mech. Sci. Technol.* **2009**, *23*, 2299–2307. [CrossRef]
3. Li, S.; Tang, X.; Luo, H.; Applications of Schauder's fixed point theorem to singular radially symmetric systems. *J. Fixed Point Theory Appl.* **2019**, *21*, 46. [CrossRef]
4. Zhu, H.; Feng, C. Existence and global uniform asymptotic stability of pseudo almost periodic solutions for Cohen-Grossberg neural networks with discrete and distributed delays. *Math. Probl. Eng.* **2014**. [CrossRef]
5. Liang, X.; Wang, L.; Wang, R. Random Attractor of Reaction-Diffusion Hopfield Neural Networks Driven by Wiener Processes. *Math. Probl. Eng.* **2018**. [CrossRef]
6. Li, Y.; Qin, J.; Li, B. Periodic solutions for quaternion-valued fuzzy cellular neural networks with time-varying delays, *Adv. Differ. Equations* **2019**, *2019*, 63.
7. Tychonoff, A.N. Ein Fixpunktsatz. *Math. Ann.* **1935**, *111*, 767–776. [CrossRef]
8. Hukuhara, M. Sur i'existence des points invariants d'une transformation dans l'espace fonctionnel. *Jpn. J. Math.* **1950**, *20*, 1–4. [CrossRef]
9. Istratescu, V.I. *Fixed Point Theory: An Introduction*; Springer: Berlin/Heidelberg, Germany, 1981.
10. Fan, Ky. A Generalization of Tychonoff's Fixed Point Theorem. *Math. Ann.* **1961**, *142*, 305–310. [CrossRef]
11. Pouso, R.L. Schauder's fixed-point theorem: New applications and a new version for discontinuous operators. *Bound. Value Probl.* **2012**, *2012*, 92.
12. Agarwal, R.P.; Arshad, S.; O'Regan, D.; Lupulescu, V. A Schauder fixed point theorem in semilinear spaces and applications. *Fixed Point Theory Appl.* **2013**, *2013*, 306. [CrossRef]

13. Du, W.-S. A Short Note on a Simple Proof of Schauder's Fixed Point Theorem and its Generalization without Continuity and Compactness Assumptions. *Int. J. Math. Anal.* **2016**, *10*, 933–938. [CrossRef]
14. Xiao, J.-Z.; Zhu, X.-H. Fixed point theorems in generating spaces of quasi-norm family and applications. *Fixed Point Theory Appl.* **2006**, *1*, 61623. [CrossRef]
15. Zhang, S.S; Guo, J.L. Probability integrals, Gateaux differentials and the Schauder principle in probabilistic normed linear spaces. *J. Sichuan Univ.* **1989**, *26*, 127–135.
16. Bag, T.; Samanta, S.K. Some fixed point theorems in fuzzy normed linear spaces. *Inf. Sci.* **2007**, *177*, 3271–3289. [CrossRef]
17. Bag, T.; Samanta, S.K. Finite dimensional fuzzy normed linear spaces. *Ann. Fuzzy Math. Inform.* **2013**, *6*, 271–283.
18. Chatterjee, S.; Bag, T.; Samanta, S.K. Uniform boundedness principle in generalized fuzzy normed linear spaces. *Afrika Matematika* **2020**. [CrossRef]
19. Wu, H.-C. Hausdorff Topology Induced by the Fuzzy Metric and the Fixed Point Theorems in Fuzzy Metric Spaces, *J. Korean Math. Soc.* **2015**, *52*, 1287–1303. [CrossRef]
20. Gregori, V.; Romaguera, S. Some properties of fuzzy metric spaces. *Fuzzy Sets Syst.* **2000**, *115*, 485–489. [CrossRef]
21. Bag, T.; Samanta, S.K. Finite dimensional fuzzy normed linear spaces. *J. Fuzzy Math.* **2003**, *11*, 687–705.
22. Chatterjee, S.; Bag, T.; Samanta, S.K. Fuzzy compact linear operator. *Adv. Fuzzy Math.* **2017**, *12*, 215–228.
23. Adhya, S.; Ray, A. D. On Lebesgue Property for Fuzzy Metric Spaces. *arXiv* **2018**, arXiv:1812.03093.
24. Bînzar, T.; Pater, F.; Nuaduaban, S. A study of boundedness in fuzzy normed linear spaces. *Symmetry* **2019**, *11*, 923. [CrossRef]
25. Sadeqi, I. Some fixed point theorems in fuzzy reflexive Banach spaces. *Chaos Solitons Fractals* **2009**, *41*, 2606–2612. [CrossRef]
26. Sadeqi, I.; Kia, F.S. Fuzzy normed linear space and its topological structure. *Chaos Solitons Fractals* **2009**, *40*, 2576–2589. [CrossRef]
27. Bag, T.; Samanta, S.K. Fuzzy bounded linear operators. *Fuzzy Sets Syst.* **2005**, *151*, 513–547. [CrossRef]
28. Bag, T.; Samanta, S.K. Fixed point theorems on fuzzy normed linear spaces. *Inf. Sci.* **2006**, *176*, 2910–2931. [CrossRef]
29. Xiao, J.-Z.; Zhu, X.-H.; Zhou, H. On the topological structure of KM fuzzy metric spaces and normed spaces. *IEEE Trans. Fuzzy Syst.* **2019**, *28*, 1575–1584. [CrossRef]
30. Malkowsky, E.; Rakocević, V. An introduction into the theory of sequence spaces and measures of noncompactness. *Zb. Rad.* **2000**, *17*, 143–234.

mathematics

Article

On the Generalized Cross-Law of Importation in Fuzzy Logic

Yifan Zhao and Kai Li *

School of Cyber Security and Computer, Hebei University, Baoding 071002, China; wmk6905@sina.com
* Correspondence: likai@hbu.cn

Received: 28 August 2020; Accepted: 27 September 2020; Published: 1 October 2020

Abstract: Recently, Baczyński et al. introduced two pexider-type generalisations of the law of importation in fuzzy logic, i.e., $I(C(x, \alpha), y) = I(x, J(\alpha, y)$ (GLI) and $I(C(x, \alpha), y) = J(x, I(\alpha, y)$ (CLI), where C is a fuzzy conjunction and I, J are fuzzy implications. However, (CLI) has not been adequately investigated so far. In this paper, we firstly show that (CLI) can be derived from the α-migrativity of an R-implication obtained from an α-migrative t-norm. Secondly, the relationships between the satisfaction of the law of importation (LI) by the pairs (C, I) or (C, J) and the satisfaction of (CLI) by the triple (C, I, J) are studied. Moreover, some necessary conditions of (CLI) are given. Finally, we study (CLI) under three different perspectives.

Keywords: fuzzy logic connectives; fuzzy implication; law of importation; α-migrativity; t-norm

1. Introduction

1.1. On the Laws of Importation in Fuzzy Logic

Fuzzy logic connectives attract a good deal of attention for research because of their interesting properties and wide range of applications, not only in approximate reasoning and fuzzy control, but also in many other research area they have proved to be valuable like composition of fuzzy relations [1,2], fuzzy relational equations [3,4], fuzzy mathematical morphology [5], fuzzy neural networks [6], fuzzy rough sets [7–9] and data mining [10]. This fact has led more and more people to a systematic research of many fuzzy logic connectives in theory, analyze additional properties of fuzzy implications and solve functional equations involving this kind of operators (see the recent survey [11–13]).

In the framework of fuzzy logic, the law of importation on fuzzy implications plays an important role in fuzzy relational inference mechanisms (FRIM), since one can generate an equivalent multi-layered scheme that markedly improves the computational efficiency of the whole system (see [14]). Furthermore, some applications of the law of importation dealing with Zadeh's compositional rule of inference (CRI) have been studied in [15], and most of them believed it is necessary to theoretically study this law before applying it.

In classical two-valued logic, one of the most important tautologies is the following law of importation:

$$(p \wedge q) \rightarrow r \equiv (p \rightarrow (q \rightarrow r)). \tag{1}$$

The general form of the above equivalence is the well-known law of important (LI, for short):

$$I(T(x, \alpha), y) = I(x, I(\alpha, y)), \quad \alpha, x, y \in [0, 1], \tag{2}$$

where T is a t-norm and I is a fuzzy implication. Moreover, Mas et al. [16] extended the above equation to the following form:

$$I(U_c(x, \alpha), y) = I(x, I(\alpha, y)), \quad \alpha, x, y \in [0, 1], \tag{3}$$

where U_c is a conjunctive uninorm and I is a fuzzy implication derived from uninorms. There are some results already known about this property. Specifically, in A-implications defined by Türkşena et al. [17], the Equation (2) with T as the product t-norm $T_P(x, y) = xy$ was taken as one of the axioms. Later, Mas et al. [18] studied the law of important for (S, N)-, R-, QL- and D-implications derived from smooth discrete t-norms and t-conorms. In [19], Massanet and Torrens introduced a weaker version of Equation (2), called the weak law of important (WLI, for short):

$$I(F(x, \alpha), y) = I(x, I(\alpha, y)), \quad \alpha, x, y \in [0, 1], \tag{4}$$

where F is a commutative, conjunctive, and non-decreasing function. Moreover, they have made new characterizations of (S, N)-implications, R-implications, and their counterparts for uninorms based on Equation (4). Therefore, it seems interesting and important to study various laws of importation in fuzzy logic.

1.2. Motivation of This Work

Recently, Baczyński et al. [20] have generalized Equation (2) to the following functional equations through the α-migrativity of fuzzy implications, called generalized laws of importation, as can be seen below:

$$I(C(x, \alpha), y) = I(x, J(\alpha, y)), \quad \alpha, x, y \in [0, 1], \tag{5}$$

$$I(C(x, \alpha), y) = J(x, I(\alpha, y)), \quad \alpha, x, y \in [0, 1], \tag{6}$$

where I and J are fuzzy implications and C is a fuzzy conjunction. Note that when $C = T$ and $J = I$, then both Equations (5) and (6) reduce to Equation (2). In other words, Equation (2) is a special case of Equations (5) and (6).

However, the generalized cross-law of importation (6) has not been investigated yet. A meaningful way of establishing the connection between (6) and the law of α-migrativity is desired. Moreover, there are two questions that we want to study:

Are there different implication functions I and J such that (6) holds for some fuzzy conjunction C? If the answer is yes, then the next question arises: Are there triples (C, I, J) that satisfy both (5) and (6)? Both of those questions are answered in this paper.

In this work, we want to study the recently proposed property of generalized cross-law of importation in fuzzy logic. Furthermore, we go on to investigate the relationship with α-migrativity and some conditions to satisfy the studied functional equations. On the one hand, we hope that our work give a chance for better understanding of a connection between the law of importation and the law of α-migrativity. On the other hand, we believe that the connection will be useful in results and applications wherein (LI) plays a key role (see [15,21,22] for details).

1.3. Novelties of This Work

The novelties of this work are threefold:

(i) Showing the generalized cross-law of importation can be derived from the α-migrativity of an R-implication obtained from an α-migrative t-norm.
(ii) Discussing the relationship between Equations (2), (5) and (6) under three different perspectives.
(iii) Extending Equations (5) and (6) to a more generalized version which depend on four functions.

The structure of the paper is organized as follows: In Section 2, we recall some basic definitions and provide several examples that are useful in further considerations. In Section 3, we give some necessary conditions of (6), and study the generalized cross-law of importation from three different perspectives. In Section 4, we discuss with different results given in the previous section and present some issues worth further investigate. Finally, Section 5 covers some conclusions.

2. Preliminaries

In order to help the reader ger familiar with the theory, we recall here some basic definitions and facts which are necessary for the development of this article. More details about t-norms, fuzzy negations, fuzzy implications, and fuzzy conjunctions can be found in [23–25].

Definition 1 ([23]). *A binary function* $T : [0, 1]^2 \to [0, 1]$ *is called a t-norm, if it satisfies, for all* $x, y, z \in [0, 1]$, *the following conditions:*

$$T(x, y) = T(y, x), \tag{7}$$

$$T(T(x, y), z) = T(x, T(y, z)), \tag{8}$$

$$T(x, y) \leq T(x, z) \text{ for } y \leq z, \tag{9}$$

$$T(x, 1) = x. \tag{10}$$

A t-norm T is called α-migrative (see [24]) if it satisfies the condition (11) for a fixed $\alpha \in (0, 1)$ and for all $x, y \in [0, 1]$

$$T(\alpha x, y) = T(x, \alpha y). \tag{11}$$

Note that if T is α-migrative for all $\alpha \in (0, 1)$, then T is said to be migrative. However, not all t-norms are migrative, for instance, the Gödel t-norm $T_G(x, y) = \min(x, y)$.

Definition 2 ([25]). *A decreasing function* $N : [0, 1] \to [0, 1]$ *is called a fuzzy negation, if* $N(0) = 1$ *and* $N(1) = 0$. *Furthermore, a fuzzy negation N is called*

(i) strict, if it is continuous and strictly decreasing,
(ii) strong, if it satisfies $N(N(x)) = x$ for all $x \in [0, 1]$.

Definition 3 ([25]). *A binary function* $I : [0, 1]^2 \to [0, 1]$ *is called a fuzzy implication if it satisfies, the following conditions:*

$$I(x, z) \geq I(y, z) \text{ when } x \leq y, \text{ for all } z \in [0, 1], \tag{12}$$

$$I(x, y) \leq I(x, z) \text{ when } y \leq z, \text{ for all } z \in [0, 1], \tag{13}$$

$$I(0, 0) = I(1, 1) = 1 \text{ and } I(1, 0) = 0. \tag{14}$$

Remark 1. *Note that from Definition 3, we can deduce that* $I(x, 1) = 1$ *and* $I(0, y) = 1$ *for all* $x, y \in [0, 1]$, *whereas the symmetric values* $I(1, x)$ *and* $I(y, 0)$ *are not derived from the above definition. Moreover, the family of all fuzzy implications will be denoted by* $\mathcal{FI}$.

A fuzzy implication I is called α-migrative (see [20]), if it satisfies the condition (15) for a fixed $\alpha \in (0, 1)$ and for all $x, y \in [0, 1]$

$$I(\alpha x, y) = I(x, 1 - \alpha + \alpha y). \tag{15}$$

Note that if I is α-migrative for all $\alpha \in (0, 1)$, then I is said to be migrative. Every fuzzy implication is α-migrative when $\alpha = 0$ or $\alpha = 1$.

There are many other properties usually required for fuzzy implications (see [25]). We present here several properties that are used in this paper.

Definition 4 ([25]). *A fuzzy implication I is said to satisfy*

(i) The boundary property if:

$$I(x, 0) = N(x), \quad x \in [0, 1], \tag{16}$$

where N is a (continuous, strict, strong) fuzzy negation.

(ii) The exchange principle if

$$I(x, I(y, z)) = I(y, I(x, z)), \ x, y, z \in [0, 1]. \tag{17}$$

Example 1 ([25]). *Examples of fuzzy implications that are used in this paper are:*

1. The least I_{Lt} and the greatest I_{Gt} fuzzy implications:

$$I_{\mathrm{Lt}}(x, y) = \begin{cases} 1, & if\, x = 0\, or\, y = 1, \\ 0, & if\, x > 0\, and\, y < 1, \end{cases} \quad I_{\mathrm{Gt}}(x, y) = \begin{cases} 1, & if\, x < 1\, or\, y > 0, \\ 0, & if\, x = 1\, and\, y = 0. \end{cases}$$

2. R-implications derived from a left-continuous t-norm T:

$$I_T(x, y) = \sup\{z \in [0, 1] \big| T(x, z) \le y\}, \quad x, y \in [0, 1].$$

3. (S, N)-implications derived from a t-conorm S and a fuzzy negation N:

$$I_{S,N}(x, y) = S(N(x), y), \quad x, y \in [0, 1].$$

Theorem 1 ([19]). *Let $I \in \mathcal{FI}$. If I satisfies (16) with a continuous fuzzy negation N, then*

$$I \,\text{satisfies}\, (2) \Leftrightarrow I \,\text{satisfies}\, (4).$$

Theorem 2 ([26]). *For a function $I : [0, 1]^2 \to [0, 1]$ the following statements are equivalent.*

(i) I is an (S, N)-implication generated from some t-conorm S and some continuous (strict, strong) fuzzy negation N.

(ii) I satisfies (12), (17) and $I(x, 0) = N(x)$ is a continuous (strict, strong) fuzzy negation.

Moreover, the representation $I(x, y) = S(N(x), y)$ is unique in this case.

Definition 5 ([25]). *Let $I \in \mathcal{FI}$. The function N_I defined by $N_I(x) = I(x, 0)$ for all $x \in [0, 1]$, is called the natural negation of I.*

Remark 2. *According to the above definition, from any fuzzy negation one can defined a fuzzy implication, but according to Equation (16) it only happens for some fuzzy implications.*

Definition 6 ([20]). *A binary function $C : [0, 1]^2 \to [0, 1]$ is called a fuzzy conjunction if it satisfies the following conditions:*

$$C \,\text{is increasing in each variable}, \tag{18}$$

$$C(x, 0) = C(0, x) = 0 \text{ for all } x \in [0, 1], \tag{19}$$

$$C(1, 1) = 1. \tag{20}$$

Remark 3. *A fuzzy conjunction C which satisfies (7), (8) and (10) is a t-norm. Every t-norm is a fuzzy conjunction, but the converse is not true. The left- and right-neutral elements of C will be denoted by e_l and e_r, respectively. The family of all fuzzy conjunctions will be denoted by $\mathcal{C}$.*

Example 2 ([20]). *Here are some fuzzy conjunctions which will be used in this paper:*

$$C_x^0(x,y)=\begin{cases}0, & if\ y=0,\\ x, & otherwise,\end{cases}\quad C_y^0(x,y)=\begin{cases}0, & if\ x=0,\\ y, & otherwise,\end{cases}$$

$$C_m^y(x,y)=\begin{cases}y, & if \min\{x,y\}>0.5,\\ \min\{x,y\}, & otherwise,\end{cases}\quad C_p(x,y)=\begin{cases}1, & if\ x=1\ and\ y=1,\\ \frac{xy}{2}, & otherwise.\end{cases}$$

3. The Main Results

3.1. Solutions of Equation (6)—Some Necessary Conditions

Remark 4. *It can be shown that the generalized cross-law of importation (6) can be derived from the α-migrativity of an R-implication obtained from an α-migrative t-norm. Now, we shall consider the following cases:*

(1) *If $\alpha \in [0,1]$, then we have*

$$I_T(\alpha x,y)=\sup\{z\in[0,1]\big|T(\alpha x,z)\le y\}=\sup\{z\in[0,1]\big|T(x,\alpha z)\le y\},\quad x,y\in[0,1]. \tag{21}$$

Note that as z varies over $[0,1]$, αz varies over $[0,\alpha]$ and substituting $\beta=\alpha z$ in to (21), we obtain

$$\begin{aligned} I_T(\alpha x,y) &= \sup\{\tfrac{\beta}{\alpha}\in[0,1]\big|T(x,\beta)\le y\}\\ &=\tfrac{1}{\alpha}\sup\{\beta\in[0,\alpha]\big|T(x,\beta)\le y\}\\ &=\min\{\tfrac{1}{\alpha}\sup\{t\in[0,1]\big|T(x,t)\le y\},1\}\\ &=\min\{\tfrac{I_T(x,y)}{\alpha},1\}. \end{aligned} \tag{22}$$

Note that (22) can be expressed as $I_T(\alpha x,y)=I_{GG}(\alpha,I_T(x,y))$, where I_{GG} is the Goguen implication.

(2) *If $\alpha=0$, then we have* LHS of (22) $=I_T(0,y)=1=I_{GG}(0,I_T(x,y))=$ RHS of (22).

Finally, substituting $\alpha=x$ (and $x=\alpha$) in to (22), then we obtain $I_T(x\alpha,y)=I_{GG}(x,I_T(\alpha,y))$ which is a special case of (6) with the triplet $(C=T_P, I=I_T, J=I_{GG})$ being fixed.

Remark 5. *Let $C\in\mathcal{C}$ andI, $J\in\mathcal{FI}$. Of course, there are many other triples (C,I,J) that satisfy (6). For example, consider the least and the greatest fuzzy implications (see Example 1-1).*

(i) If C satisfies that $=C(x,y)=0\Leftrightarrow x=0$ or $y=0$, then the triplet $(C,I_{Lt},J=I_{Lt})$ satisfies (6). To see this, note that we have the following equivalences:

$$\begin{cases}\text{LHS of (6)}=0\Leftrightarrow I_{Lt}(C(x,\alpha),y)=0\Leftrightarrow C(x,\alpha)>0\text{ and }y<1\Leftrightarrow \alpha,x>0\text{ and }y<1,\\ \text{RHS of (6)}=0\Leftrightarrow I_{Lt}(x,I_{Lt}(\alpha,y))=0\Leftrightarrow x>0\text{ and }I_{Lt}(\alpha,y)<1\Leftrightarrow \alpha,x>0\text{ and }y<1.\end{cases}$$

$$\begin{cases}\text{LHS of (6)}=1\Leftrightarrow I_{Lt}(C(x,\alpha),y)=1\Leftrightarrow C(x,\alpha)=0\text{ or }y=1\Leftrightarrow \alpha=0\text{ or }x=0\text{ or }y=1,\\ \text{RHS of (6)}=1\Leftrightarrow I_{Lt}(x,I_{Lt}(\alpha,y))=1\Leftrightarrow x=0\text{ or }I_{Lt}(\alpha,y)=1\Leftrightarrow \alpha=0\text{ or }x=0\text{ or }y=1.\end{cases}$$

(ii) Similarly, if C satisfies that $C(x,y)=1\Leftrightarrow x=y=1$, then the triplet $(C,I_{Gt},J=I_{Gt})$ satisfies (6).

$$\begin{cases}\text{LHS of (6)}=0\Leftrightarrow I_{Gt}(C(x,\alpha),y)=0\Leftrightarrow C(x,\alpha)=1\text{ and }y=0\Leftrightarrow \alpha=x=1\text{ and }y=0,\\ \text{RHS of (6)}=0\Leftrightarrow I_{Gt}(x,I_{Gt}(\alpha,y))=0\Leftrightarrow x=1\text{ and }I_{Gt}(\alpha,y)=0\Leftrightarrow \alpha=x=1\text{ and }y=0.\end{cases}$$

$$\begin{cases} \text{LHS of (6)} = 1 \Leftrightarrow I_{\text{Gt}}(C(x, \alpha), y) = 1 \Leftrightarrow C(x, \alpha) < 1 \text{ or } y > 0 \Leftrightarrow \alpha < 1 \text{ or } x < 1 \text{ or } y > 0, \\ \text{RHS of (6)} = 1 \Leftrightarrow I_{\text{Gt}}(x, I_{\text{Gt}}(\alpha, y)) = 1 \Leftrightarrow x < 1 \text{ or } I_{\text{Gt}}(\alpha, y) > 0 \Leftrightarrow \alpha < 1 \text{ or } x < 1 \text{ or } y > 0. \end{cases}$$

Example 3. *Consider* $C_y^0(x, y) = \begin{cases} 0, & \text{if } x = 0, \\ y, & \text{otherwise} \end{cases}$ *and the fuzzy implication* $J_1(x, y) = \begin{cases} 1, & \text{if } x = 0, \\ y, & \text{otherwise.} \end{cases}$ *It is easy to verify that the triple* $(C_y^0, I_{\text{Lt}}, J_1)$ *satisfies (6). On the other hand, the pair* (C_y^0, I_{Lt}) *satisfies (2), as can be seen below:*

$$\text{LHS of (2)} = I_{\text{Lt}}(C_y^0(x, \alpha), y) = I_{\text{Lt}}\left(\begin{cases} 0, & x = 0, \\ \alpha, & x \neq 0, \end{cases} y\right) - \begin{cases} 0, & \alpha, x > 0 \text{ and } y < 1, \\ 1, & \text{otherwise,} \end{cases}$$

$$\begin{aligned} \text{RHS of (2)} &= I_{\text{Lt}}(x, I_{\text{Lt}}(\alpha, y)) = I_{\text{Lt}}\left(x, \begin{cases} 0, & \alpha > 0 \text{ and } y < 1, \\ 1, & \text{otherwise,} \end{cases}\right) \\ &= \begin{cases} 0, & \alpha, x > 0 \text{ and } y < 1, \\ 1, & \text{otherwise,} \end{cases} = \text{LHS of (2)}. \end{aligned}$$

Interestingly, the pair (C_y^0, J_1) also satisfies Equation (2), as shown below:

$$\text{LHS of (2)} = J_1(C_y^0(x, \alpha), y) = J_1\left(\begin{cases} 0, & x = 0, \\ \alpha, & x > 0, \end{cases} y\right) = \begin{cases} 1, & \alpha = 0 \text{ or } x = 0, \\ y, & \alpha, x \in (0, 1], \end{cases}$$

$$\begin{aligned} \text{RHS of (2)} &= J_1(x, J_1(\alpha, y)) = J_1\left(x, \begin{cases} 1, & \alpha = 0, \\ y, & \alpha > 0, \end{cases}\right) \\ &= \begin{cases} 1, & \alpha = 0 \text{ or } x = 0, \\ y, & \alpha, x \in (0, 1], \end{cases} = \text{LHS of (2)}. \end{aligned}$$

If we consider $C_y^1(x, y) = \begin{cases} y, & \text{if } x = 1, \\ 0, & \text{otherwise} \end{cases}$ and its residual $J_2(x, y) = \begin{cases} y, & \text{if } x = 1, \\ 1, & \text{otherwise,} \end{cases}$ then the triple $(C_y^1, I_{\text{Gt}}, J_2)$ satisfies (6), as shown below:

$$\text{LHS of (6)} = I_{\text{Gt}}(C_y^1(x, \alpha), y) = I_{\text{Gt}}\left(\begin{cases} \alpha, & x = 1, \\ 0, & x \neq 1, \end{cases} y\right) = \begin{cases} 0, & \alpha = x = 1 \text{ and } y = 0, \\ 1, & \text{otherwise,} \end{cases}$$

$$\begin{aligned} \text{RHS of (6)} &= J_2(x, I_{\text{Gt}}(\alpha, y)) = J_2\left(x, \begin{cases} 0, & \alpha = 1 \text{ and } y = 0, \\ 1, & \text{otherwise,} \end{cases}\right) \\ &= \begin{cases} 0, & \alpha = x = 1, y = 0, \\ 1, & \text{otherwise,} \end{cases} = \text{LHS of (6)}. \end{aligned}$$

On the other hand, the pairs (C_y^1, I_{Gt}) and (C_y^1, J_2) also satisfy (2). Thus, a natural question arises: If the triplet (C, I, J) satisfies (6), do the corresponding pairs (C, I) and (C, J) necessarily satisfy (2)? Unfortunately, the answer is negative. However, there are some sufficient conditions such that the pair (C, I) satisfies (2). For more details we refer the readers to [19].

Remark 6. *Let* $C \in \mathcal{C}$ *and* $I, J \in \mathcal{FI}$. *It can be shown that the satisfaction of (2) by either/both the pairs* (C, I) *and* (C, J) *is neither sufficient nor necessary for the triplet* (C, I, J) *to fulfill (6). Consider the following fuzzy implications and the result is presented in Table 1.*

$$I_1(x, y) = \begin{cases} 1, & \text{if } y = 1, \\ 1 - x, & \text{otherwise,} \end{cases} \quad I_2(x, y) = \begin{cases} 1, & \text{if } y = 1, \\ 1 - x^2, & \text{otherwise,} \end{cases}$$

$$I_3(x, y) = \begin{cases} 0, & if\, x > 0.5\, and\, y < 1, \\ 1, & otherwise, \end{cases} \quad I_4(x, y) = \begin{cases} 1, & if\, x \leq 0.5, \\ y, & otherwise, \end{cases}$$

$$I_5(x, y) = \begin{cases} y, & if\, x = 1, \\ 0, & if\, x > 0\, and\, y = 0, \\ 1, & otherwise, \end{cases} \quad I_D(x, y) = \begin{cases} 1, & if\, x = 0\, or\, y = 1, \\ y, & otherwise. \end{cases}$$

Table 1. The relationships between (2) and (6).

C	I	J	**(C, I) Satisfies (2)**	**(C, J) Satisfies (2)**	**(C, I, J) Satisfies (6)**
C_x^0	I_{Lt}	I_2	√	√	×
C_m^y	I_4	I_3	√	×	×
C_y^0	I_{Gt}	I_D	×	√	√
C_p	I_1	I_D	×	√	×
C_m^y	I_2	I_3	×	×	×
C_y^0	I_{Gt}	I_5	×	×	√
C_x^0	I_{Lt}	I_5	√	×	√
C_y^1	I_{Gt}	J_2	√	√	√

Finally, we want to finish this subsection with the following result, some necessary conditions on the triple (C, I, J) to satisfy (6).

Proposition 1. *Let the triple (C, I, J) satisfy (6), $e_l, e_r \in (0, 1]$ be the left- and right-neutral elements of C and $f_l \in (0, 1]$ be the left-neutral elements of I. Let $\Re_I = Ran(I)$.*

(i) $J(e_l, r) = r$ on $\Re_I$. Further, if $\Re_I = [0, 1]$, then J has left-neutral element e_l.
(ii) If f_l is the right-neutral element of C, i.e., $f_l = e_r$, then $I = J$.
(iii) If $I(e_r, 0) = 0$, then $N_I = N_J = N$ and $N_I(x) = 0$ whenever $x \in [e_r, 1]$.

Proof. First, to prove item (i), substituting $x = e_l$ in to (6), we obtain

$$I(C(e_l, \alpha), y) = J(e_l, I(\alpha, y)), \quad \alpha, e_l, y \in [0, 1]. \tag{23}$$

Since $C(e_l, \alpha) = \alpha$ for all $\alpha \in [0, 1]$ and hence $I(\alpha, y) = J(e_l, I(\alpha, y))$. Now, let $I(\alpha, y) = r \in \Re_I = Ran(I)$, it is sufficient to see that $J(e_l, r) = r$ on $\Re_I$. Furthermore, if $\Re_I = [0, 1]$, then $J(e_l, r) = r$ for all $r \in [0, 1]$, which implies e_l is also the left-neutral element of J.

For item (ii), substituting $\alpha = e_r$ we obtain from (6),

$$I(x, y) = I(C(x, e_r), y) = J(x, I(e_r, y)), \quad x, e_r, y \in [0, 1]. \tag{24}$$

Obviously, if $f_l = e_r$, then $J(x, I(e_r, y)) = J(x, I(f_l, y)) = J(x, y)$ and thus $I = J$.

To prove item (iii), substituting $\alpha = e_r$ and $y = 0$ in to (6), after a simple rearrangement we obtain

$$I(x, 0) = J(x, I(e_r, 0)), \quad x, e_r \in [0, 1]. \tag{25}$$

If $I(e_r, 0) = 0$, then $I(x, 0) = J(x, 0)$ for all $x \in [0, 1]$ and by Definition 5, we obtain $N_I = N_J = N$. Finally, if $x \in [e_r, 1]$, then by (12) we deduce that $N_I(x) = 0$. □

3.2. Perspective One: The Pair (C, I) Satisfies (2)

Let us start with the first perspective when the pair (C, I) satisfies (2). Specifically, we have the following result.

Theorem 3. *Let $C \in \mathcal{C}$ and $I, J \in \mathcal{FI}$ and consider the following items:*

(i) The pair (C, I) satisfies (2).
(ii) The triple (C, I, J) satisfies (6).
(iii) $I = J$.

Then one has the following items:

(1) If $I(f_l, y) = y$ for all $y \in [0, 1]$, then (i) and (ii) $\Rightarrow$ (iii).
(2) Without any further assumption, (i) and (iii) $\Rightarrow$ (ii); (ii) and (iii) $\Rightarrow$ (i).

Proof.

(1) It is clear that (i) and (ii) imply RHS of (2) = RHS of (6), i.e., $I(x, I(\alpha, y)) = J(x, I(\alpha, y))$ for all $\alpha, x, y \in [0, 1]$. Let $f_l \in (0, 1]$. If $I(f_l, y) = y$ for all $y \in [0, 1]$. Now, substituting $\alpha = f_l$ in the above equation, then we obtain $I = J$.
(2) It is trivially true that (i) and (iii) $\Rightarrow$ (ii) and (ii) and (iii) $\Rightarrow$ (i). □

Remark 7. *Note that in Theorem 3.(1), even if I does not have left-neutral element, we can still have the implication (i) and* (ii) $\Rightarrow$ (iii). *To see this consider the pair (C_x^0, I_6) (see Example 2), where*

$$I_6(x, y) = \begin{cases} 1, & if\, y = 1, \\ 1 - x^6, & otherwise. \end{cases} \tag{26}$$

It is clear that I_6 does not have any left-neutral element. On the other hand, the pair (C_x^0, I_6) satisfies Equation (2) as can be seen below:

$$\text{LHS of (2)} = I_6(C_x^0(x, \alpha), y) = I_6\left(\begin{cases} 0, & \alpha = 0, \\ x, & \alpha \neq 0, \end{cases} y\right) = \begin{cases} 1, & \alpha = 0\, or\, y = 1, \\ 1 - x^6, & \alpha > 0\, and\, y < 1, \end{cases}$$

$$\begin{aligned} \text{RHS of (2)} &= I_6(x, I_6(\alpha, y)) = I_6\left(x, \begin{cases} 1, & y = 1, \\ 1 - \alpha^6, & y \neq 1, \end{cases}\right) \\ &= \begin{cases} 1, & \alpha = 0\, or\, y = 1, \\ 1 - x^6, & \alpha > 0\, and\, y < 1, \end{cases} = \text{LHS of (2)}. \end{aligned}$$

Now, assume that the triple (C_x^0, I_6, J) satisfies (6), then we have $J = I_6$, as shown below:
Let $x, y \in [0, 1]$ and $J(x, y) = \mu$. We divide our argument in two cases:

Case 1. If $y = 1$, then $J(x, 1) = \mu = 1$.
Case 2. If $y \in [0, 1)$ and $\alpha = 0$, then LHS of (6) $= I_6(C_x^0(x, 0), y) = 1 = J(x, 1) = J(x, I_6(0, y)) =$ RHS of (6); If $y \in [0, 1)$ and $\alpha \in (0, 1]$, then

$$\text{LHS of (6)} = I_6(C_x^0(x, \alpha), y) = I_6(x, y) = 1 - x^6,$$

$$\text{RHS of (6)} = J(x, I_6(\alpha, y)) = J(x, 1 - \alpha^6).$$

Combining the above two equations, we have $J(x, 1 - \alpha^6) = 1 - x^6$. As α varies over $(0, 1]$, $1 - \alpha^6$ varies over $[0, 1)$. Thus, substituting $1 - \alpha^6$ by y we obtain $J(x, y) = \mu = 1 - x^6$, when $y < 1$. Therefore, we conclude that $J = I_6$.

3.3. Perspective Two: The Pair (C, J) Satisfies (2)

In this subsection, we focus on the second perspective when the pair (C, J) satisfies (2).

Theorem 4. *Let $C \in \mathcal{C}$ and $I, J \in \mathcal{FI}$ and consider the following items:*

(i) The pair (C, J) satisfies (2).
(ii) The triple (C, I, J) satisfies (6).
(iii) J satisfies (17).
(iv) $J = I$.

Moreover, consider the following two properties with respect to C and J.

(a) f_l is the right-neutral element of C, i.e., $f_l = e_r$.
(b) N_J is a continuous fuzzy negation.

Then the following items hold:

(1) If (a) is true, then $(\text{ii}) \Rightarrow (\text{iv})$.
(2) If (b) is true, then $(\text{i}) \Leftrightarrow (\text{iii})$.
(3) Without any further assumption, (i) and $(\text{iv}) \Rightarrow (\text{ii})$; (ii) and $(\text{iv}) \Rightarrow (\text{i})$.

Proof.

(1) If $f_l = e_r$, then $(\text{ii}) \Rightarrow (\text{iv})$ follows from Proposition 1 (ii).
(2) Necessity. Assume that the pair (C, J) satisfies (2). We already know that every t-norm is a fuzzy conjunction. By the commutativity of a t-norm, if an implication J satisfies (2) with respect to any t-norm T, then J satisfies (17).

Sufficiency. Assume that J satisfies (17). Note that J is a fuzzy implication it satisfies (12) and then Theorem 2 implies that J is an (S, N)-implication derived from a t-conorm S and a continuous fuzzy negation N. Moreover, $J_{S,N}$ satisfies (4) with the function $F(x, y) = N_1(S(N(x), N(y)))$ where N_1 such that $N \circ N_1 = id_{[0,1]}$. Finally, Theorem 1 ensures that the pair (C, J) satisfies (2).

(3) The implications (i) and $(\text{iv}) \Rightarrow (\text{ii})$ and (ii) and $(\text{iv}) \Rightarrow (\text{i})$ are trivially true. □

Remark 8. *Now, let us consider the necessity of the distinct conditions used in Theorem 4.*

(i) Note that in Theorem 4, we have considered two properties on C and J. In fact, the assumption (a) is not necessary. To see this, consider the pair (C_m^y, I_4) (see Remark 6), it is easy to see that C_m^y does not have any right-neutral element but I_4 has left-neutral element whenever $f_l \in [0.5, 1]$ and thus (a) is not valid. Now, assume that the triple (C_m^y, I_4, J) satisfies (6), then we have $J = I_4$, as shown below:

Let $x, y \in [0, 1]$ and $J(x, y) = v$. If $x \in [0, 0.5]$, then

$$\text{LHS of (6)} = I_4(C_m^y(x, \alpha), y) = I_4(\min\{x, \alpha\}, y) = 1,$$

$$\text{RHS of (6)} = J(x, I_4(\alpha, y)) = J(x, \begin{cases} 1, & \alpha \le 0.5, \\ y, & \alpha > 0.5, \end{cases}) = J(x, y),$$

and thus $J(x, y) = v = 1$, when $x \le 0.5$. If $x \in (0.5, 1]$, then

$$\text{LHS of (6)} = I_4(C_m^y(x, \alpha), y) = I_4(\begin{cases} \alpha, & \alpha \le 0.5, \\ \alpha, & \alpha > 0.5, \end{cases} y) = \begin{cases} 1, & \alpha \le 0.5, \\ y, & \alpha > 0.5, \end{cases}$$

$$\text{RHS of } (6) = J(x, I_4(\alpha, y)) = J(x, \begin{cases} 1, & \alpha \le 0.5, \\ y, & \alpha > 0.5, \end{cases}) = \begin{cases} 1, & \alpha \le 0.5, \\ J(x, y), & \alpha > 0.5, \end{cases}$$

which implies that $J(x, y) = v = y$ if $x > 0.5$ and thus $J = I_4$.

(ii) Similarly, one can show that the assumption (b) is not necessary, viz., even if N_J is not continuous, there exists a fuzzy implication J satisfies (17) with the pair (C, J) satisfies (2). To see this, consider the following pair $(C_\lambda, J = I_{\mathrm{WB}})$, where

$$C_\lambda(x, y) = \begin{cases} 1, & if\, x = 1\, and\, y = 1, \\ (xy)^\lambda, \lambda \ge 1, & otherwise, \end{cases} \quad I_{\mathrm{WB}}(x, y) = \begin{cases} 1, & if\, x < 1, \\ y, & otherwise. \end{cases}$$

As is well known that I_{WB} satisfies the exchange principle (17) but the natural negation of I_{WB} is not continuous. However, the pair $(C_\lambda, I_{\mathrm{WB}})$ satisfies (2) as can be seen below:

$$\text{LHS of } (2) = I_{\mathrm{WB}}(C_\lambda(x, \alpha), y) = I_{\mathrm{WB}}\left(\begin{cases} 1, & \alpha = x = 1, \\ (\alpha x)^\lambda, & \alpha < 1\, or\, x < 1, \end{cases} y\right) = \begin{cases} 1, & \alpha < 1\, or\, x < 1, \\ y, & \alpha = x = 1, \end{cases}$$

$$\text{RHS of } (2) = I_{\mathrm{WB}}(x, I_{\mathrm{WB}}(\alpha, y)) = I_{\mathrm{WB}}\left(x, \begin{cases} 1, & \alpha < 1, \\ y, & \alpha = 1, \end{cases}\right) = \begin{cases} 1, & \alpha < 1\, or\, x < 1, \\ y, & \alpha = x = 1, \end{cases} = \text{LHS of } (2).$$

(iii) Finally, note that the assumption (iv) is not necessary, i.e., even if $J \neq I$, there can exist a triple (C, I, J) satisfies (6) with the corresponding pair (C, J) satisfies (2). See for instance Example 3.

3.4. *Perspective Three: The Triple (C, I, J) Satisfies (5) and (6)*

In this subsection, we want to discuss the second question as we have mentioned in the introduction, i.e., are there exist triples $(C, I, J \neq I)$ that satisfy both (5) and (6)?

Remark 9.

(i) *We already know that the triples $(C_y^0, I_{\mathrm{Lt}}, J_1)$ and $(C_y^1, I_{\mathrm{Gt}}, J_2)$ satisfy (6) (see Example 3). However, the triples also satisfy (5) as can be seen below:*

$$\text{LHS of } (5) = I_{\mathrm{Lt}}(C_y^0(x, \alpha), y) = I_{\mathrm{Lt}}\left(\begin{cases} 0, & x = 0, \\ \alpha, & x \neq 0, \end{cases} y\right) = \begin{cases} 0, & \alpha, x > 0\, and\, y < 1, \\ 1, & otherwise, \end{cases}$$

$$\begin{aligned} \text{RHS of } (5) &= I_{\mathrm{Lt}}(x, J_1(\alpha, y)) = I_{\mathrm{Lt}}\left(x, \begin{cases} 1, & \alpha = 0, \\ y, & \alpha > 0, \end{cases}\right) = \begin{cases} 0, & \alpha, x > 0\, and\, y < 1, \\ 1, & otherwise, \end{cases} \\ &= \text{LHS of } (5). \end{aligned}$$

$$\text{LHS of } (5) = I_{\mathrm{Gt}}(C_y^1(x, \alpha), y) = I_{\mathrm{Gt}}\left(\begin{cases} \alpha, & x = 1, \\ 0, & x \neq 1, \end{cases} y\right) = \begin{cases} 0, & \alpha = x = 1\, and\, y = 0, \\ 1, & otherwise, \end{cases}$$

$$\begin{aligned} \text{RHS of } (5) &= I_{\mathrm{Gt}}(x, J_2(\alpha, y)) = I_{\mathrm{Gt}}\left(x, \begin{cases} y, & \alpha = 1, \\ 1, & \alpha < 1, \end{cases}\right) = \begin{cases} 0, & \alpha = x = 1\, and\, y = 0, \\ 1, & otherwise, \end{cases} \\ &= \text{LHS of } (5). \end{aligned}$$

(ii) In a similar way as in Remark 6, one can show that the satisfaction of Equation (5) by the triple (C, I, J) is neither sufficient nor necessary to satisfy Equation (6). The result is presented in Table 2.

(iii) Note that if the triples (C, I, J) that both satisfy (5) and (6), then

$$I(x, J(\alpha, y)) = J(x, I(\alpha, y)), \quad \alpha, x, y \in [0, 1], \tag{27}$$

but the converse is not true. For instance, consider the following triple $(C_{\prod}, I_5, J_1^0)$(see Remark 6), where

$$C_{\prod}(x, y) = \begin{cases} 0, & if\, x = 0\, or\, y = 0, \\ xy, & otherwise, \end{cases} \quad J_0^1(x, y) = \begin{cases} 1, & if\, x = 0\, or\, y > 0, \\ 0, & otherwise. \end{cases}$$

Table 2. The relationships between (5) and (6).

C	I	J	**(C, I, J) Satisfies (5)**	**(C, I, J) Satisfies (6)**
C_p	I_1	I_D	×	×
C_m^y	I_4	I_3	√	×
C_x^0	I_{Lt}	I_5	×	√
C_y^1	I_{Gt}	J_2	√	√

As is shown below, the corresponding pair (I_5, J_1^0) satisfies (27).

$$\text{LHS of (27)} = I_5(x, J_1^0(\alpha, y)) = I_5\left(x, \begin{cases} 1, & \alpha = 0\, or\, y > 0, \\ 0, & otherwise, \end{cases}\right) = \begin{cases} 1, & \alpha = 0\, or\, x = 0\, or\, y > 0, \\ 0, & otherwise, \end{cases}$$

$$\begin{aligned} \text{RHS of (27)} &= J_1^0(x, I_5(\alpha, y)) = J_1^0\left(x, \begin{cases} y, & \alpha = 1, \\ 0, & \alpha > 0\, and\, y = 0, \\ 1, & otherwise, \end{cases}\right) = \begin{cases} J_1^0(x, y), & \alpha = 1, \\ J_1^0(x, 0), & \alpha > 0\, and\, y = 0, \\ 1, & otherwise, \end{cases} \\ &= \begin{cases} 1, & \alpha = 0\, or\, x = 0\, or\, y > 0, \\ 0, & otherwise, \end{cases} = \text{LHS of (27)}. \end{aligned}$$

On the other hand, it is easy to verify that the triple $(C_{\prod}, I_5, J_1^0)$ does not satisfy (5) and (6), since

$$I_5(C_{\prod}(1, 1), 0.5) = 0.5 \neq 1 = I_5(1, J_1^0(1, 0.5)) = J_1^0(1, I_5(1, 0.5)).$$

(iv) Finally, observe that Equations (5) and (6) can be extend a more generalized version which depend on four functions:

$$I(C(x, \alpha), y) = J(x, K(\alpha, y)), \quad \alpha, x, y \in [0, 1], \tag{28}$$

where $I, J, K \in \mathcal{FI}$ and $C \in \mathcal{C}$. Of course, there exists a quadruple (C, I, J, K) such that the above equation holds. Let us give an example of this.

Example 4. *Consider the following quadruple of functions* $(C_0^1, I_{Lt}, J = I_{Gt}, K_n^1)$, where

$$C_0^1(x, y) = \begin{cases} 1, & if\, (x, y) = (1, 1), \\ 0, & otherwise, \end{cases} \quad K_n^1(x, y) = \begin{cases} 1, & if\, y = 1, \\ 1 - x^n,\ n \geq 1, & otherwise. \end{cases}$$

Then the quadruple $(C_0^1, I_{Lt}, J = I_{Gt}, K_n^1)$ satisfies (28), as shown below:

$$\text{LHS of (28)} = I_{Lt}(C_0^1(x, \alpha), y) = I_{Lt}\left(\begin{cases} 1, & \alpha = x = 1, \\ 0, & \alpha < 1\, or\, x < 1, \end{cases} y\right) = \begin{cases} 1, & \alpha < 1\, or\, x < 1\, or\, y = 1, \\ 0, & \alpha = x = 1,\ y < 1, \end{cases}$$

$$\begin{aligned} \text{RHS of (28)} &= I_{Gt}(x, K_n^1(\alpha, y)) = I_{Gt}\left(x, \begin{cases} 1, & y = 1, \\ 1 - \alpha^n, & otherwise, \end{cases}\right) = \begin{cases} 1, & y = 1, \\ I_{Gt}(x, 1 - \alpha^n), & otherwise, \end{cases} \\ &= \begin{cases} 1, & \alpha < 1\, or\, x < 1\, or\, y = 1, \\ 0, & \alpha = x = 1,\ y < 1, \end{cases} = \text{LHS of (28)}. \end{aligned}$$

Theorem 5. *Let $C \in \mathcal{C}$ and $I,\ J,\ K \in \mathcal{FI}$ and consider the following items:*

(i) The quadruple (C, I, J, K) satisfies (28).
(ii) The triple (C, I, K) satisfies (5).
(iii) The triple (C, I, J) satisfies (6).
(iv) $I = J$.
(v) $I = K$.

Then the following items hold:

(1) Without any further assumption, (i) and (iv) $\Rightarrow$ (ii).
(2) Without any further assumption, (i) and (v) $\Rightarrow$ (iii).

Proof. Assume that the quadruple (C, I, J, K) satisfies (28). The first item is clear because if $I = J$, then $I(C(x, \alpha), y) = I(x, K(\alpha, y))$ which implies that the triple (C, I, K) satisfies (5). Similarly, if $I = K$, then we obtain $I(C(x, \alpha), y) = J(x, I(\alpha, y))$, which implies that the triple (C, I, K) satisfies (6), this completes the proof. □

Remark 10. *Next, let us discuss the necessity of the distinct conditions used in Theorem 5.*

(i) Note that the assumption (iv) is not necessary, i.e., even if $I \neq J$, there can exist the quadruple (C, I, J, K) satisfies (28) with the corresponding triple (C, I, K) satisfying (5). So, to see this we need to search among those triples (I, J, K) satisfy that $I(x, K(\alpha, y)) = J(x, K(\alpha, y))$ for all $\alpha, x, y \in [0, 1]$. However, the above equation holds rather rarely when $I \neq J$. For example, consider the following quadruple $(C_0^1, I_{\mathrm{Gt}}, J = I_{\mathrm{WB}}, K = I_{\mathrm{Gt}})$. Clearly, it satisfies (28) as can be seen below:

$$\text{LHS of (28)} = I_{\mathrm{Gt}}(C_0^1(x, \alpha), y) = I_{\mathrm{Gt}}\left(\begin{cases} 1, & \alpha = x = 1, \\ 0, & \alpha < 1\,or\,x < 1, \end{cases} y\right) = \begin{cases} 1, & \alpha < 1\,or\,x < 1\,or\,y > 0, \\ 0, & \alpha = x = 1,\ y = 0, \end{cases}$$

$$\begin{aligned} \text{RHS of (28)} &= I_{\mathrm{WB}}(x, I_{\mathrm{Gt}}(\alpha, y)) = I_{\mathrm{WB}}\left(x, \begin{cases} 1, & \alpha < 1\,or\,y > 0, \\ 0, & otherwise, \end{cases}\right) \\ &= \begin{cases} 1, & \alpha < 1\,or\,x < 1\,or\,y > 0, \\ 0, & \alpha = x = 1,\ y = 0 \end{cases} = \text{LHS of (28)}. \end{aligned}$$

On the other hand, the corresponding triple $(C_0^1, I_{\mathrm{Gt}}, K = I_{\mathrm{Gt}})$ satisfies (5), as shown below:

$$\text{RHS of (5)} = I_{\mathrm{Gt}}(x, I_{\mathrm{Gt}}(\alpha, y)) = \begin{cases} I_{\mathrm{Gt}}(x, 1), & \alpha < 1\,or\,y > 0, \\ I_{\mathrm{Gt}}(x, 0), & \alpha = 1\,and\,y = 0, \end{cases} = \text{LHS of (28)} = \text{LHS of (5)}.$$

(ii) Similarly, the assumption (v) is not necessary, i.e., even if $I \neq K$, we can still have that the quadruple (C, I, J, K) satisfies (28) with the corresponding triple (C, I, J) satisfying (6). To see this, let us consider the quadruple $(C_{x^2}^0, I_1, J = I_2, K = I_2)$ (see Remark 6), where

$$C_{x^2}^0(x, y) = \begin{cases} 0, & if\ y = 0, \\ x^2, & otherwise. \end{cases} \tag{29}$$

Observe that the quadruple $(C_{x^2}^0, I_1, J = I_2, K = I_2)$ satisfies (28), as shown below:

$$\text{LHS of (28)} = I_1(C_{x^2}^0(x, \alpha), y) = I_1\left(\begin{cases} 0, & \alpha = 0, \\ x^2, & \alpha > 0, \end{cases} y\right) = \begin{cases} 1, & \alpha = 0\,or\,y = 1, \\ 1 - x^2, & otherwise, \end{cases}$$

$$\begin{aligned}\text{RHS of } (28) &= I_2(x, I_2(\alpha, y)) = I_2\left(x, \begin{cases} 1, & y = 1, \\ 1-\alpha^2, & y < 1, \end{cases}\right) = \begin{cases} 1, & \alpha = 0 \text{ or } y = 1, \\ 1-x^2, & \text{otherwise}, \end{cases} \\ &= \text{LHS of } (16).\end{aligned}$$

Moreover, the corresponding triple $(C^0_{x^2}, I_1, K = I_2)$ satisfies (6) as follows:

$$\text{RHS of } (6) = I_2(x, I_1(\alpha, y)) = I_2\left(x, \begin{cases} 1, & y = 1, \\ 1-\alpha, & y < 1, \end{cases}\right) = \text{LHS of } (28) = \text{LHS of } (6).$$

4. Discussion

In the previous section, we have studied the relationship between (2), (5) and (6) under three different perspectives. It is shown that the satisfaction of (2) by either/both the pairs (C, I) and (C, J) is neither sufficient nor necessary for the triplet (C, I, J) to fulfill (6). In a similar way, it is shown that the satisfaction of (5) by the triple (C, I, J) is neither sufficient nor necessary to satisfy (6). Thus, a natural question arises: If the triple (C, I, J) satisfies (5) and (6), do the corresponding pairs (C, I) and (C, J) necessarily satisfy (2)? However, we have established a connection between the cross-law of importation (6) and the law of α-migrativity and discussed some conditions to satisfy the studied functional equations. However, there are still some issues worth further investigation, such as

- We have found some cases when the triples (C, I, J) satisfy both (5) and (6) (see Example 3), but to characterize all the cases is still an open problem.
- The sufficient and necessary conditions under which (6) holds for α-migrative fuzzy implications.
- Fixed a concrete fuzzy conjunction C, for which triples (I, J, K) such that Equation (28) holds? For instance, which triples (I, J, K) satisfy the following functional equation

$$I(x\alpha, y) = J(x, K(\alpha, y)), \quad \alpha, x, y \in [0, 1],$$

  that comes from (28) with $C = C_\prod$?

  We intend to study the above issues in a future work.

5. Conclusions

The generalized cross-law of importation (6) has not been investigated so far. In this work, we have shown that Equation (6) can be derived from the α-migrativity of an R-implication with respect to an α-migrative t-norm (Remark 4). Another important fact is that the satisfaction of (2) by either/both the pairs (C, I) and (C, J) is neither sufficient nor necessary for the triplet (C, I, J) to satisfy (6) (Remark 6). In addition, some necessary conditions for solutions to Equation (6) are given (Proposition 1). Following this, we have discussed the relationship between Equations (2), (5) and (6) under three different perspectives. In particular, note that both Equations (5) and (6) can be further generalized as mentioned in Remark 9 (iv).

We believe that our work provides an opportunity for better understanding of a connection between the laws of importation and the laws of α-migrativity.

Author Contributions: Supervision, K.L.; writing—original draft, Y.Z.; writing—review and editing, K.L. All authors have read and agreed to the published version of the manuscript.

Funding: This research was partially supported by the Natural Science Foundation of Hebei Province (Grant NO. F2018201060).

Acknowledgments: The authors are extremely grateful to the Editor and anonymous reviewers for their very valuable comments and suggestions.

Conflicts of Interest: The authors declare no conflict of interest.

References

1. Bošnjak, I.; Madarász, R. On the composition of fuzzy power relations. *Fuzzy Sets Syst.* **2015**, *271*, 81–87. [CrossRef]
2. Elkano, M.; Sanz, J.A.; Galar, M.; Pękala, B.; Bentkowska, U.; Bustince, H. Composition of interval-valued fuzzy relations using aggregation functions. *Inf. Sci.* **2016**, *369*, 690–703. [CrossRef]
3. Jayaram, B.; Mesiar, R. I-fuzzy equivalence relations and I-fuzzy partitions. *Inf. Sci.* **2009**, *179*, 1278–1297. [CrossRef]
4. Khorram, E.; Ezzati, R.; Valizadeh, Z. Linear fractional multi-objective optimization problems subject to fuzzy relational equations with a continuous Archimedean triangular norm. *Inf. Sci.* **2014**, *267*, 225–239. [CrossRef]
5. Deng, T.D.; Heijmans, H. Grey-scale morphology based on fuzzy logic. *J. Math. Imaging Vis.* **2002**, *16*, 155–171. [CrossRef]
6. Valle, M.E.; Sussner, P. A general framework for fuzzy morphological associative memories. *Fuzzy Sets Syst.* **2008**, *159*, 747–768. [CrossRef]
7. Wang, C.Y. Topological characterizations of generalized fuzzy rough sets. *Fuzzy Sets Syst.* **2017**, *312*, 109–125. [CrossRef]
8. Wang, C.Y. Single axioms for lower fuzzy rough approximation operators determined by fuzzy implications. *Fuzzy Sets Syst.* **2018**, *336*, 116–147. [CrossRef]
9. Fang, B.W.; Hu, B.Q. Granular fuzzy rough sets based on fuzzy implicators and coimplicators. *Fuzzy Sets Syst.* **2019**, *359*, 112–139. [CrossRef]
10. Yan, P.; Chen, G.Q. Discovering a cover set of ARsi with hierarchy from quantitative databases. *Inf. Sci.* **2005**, *173*, 319–336. [CrossRef]
11. Mas, M.; Monserrat, M.; Torrens, J.; Trillas, E. A survey on fuzzy implication functions. *IEEE Trans. Fuzzy Syst.* **2007**, *15*, 1107–1121. [CrossRef]
12. Baczyński, M.; Jayaram, B. (S, N)-and R-implications: A state-of-the-art survey. *Fuzzy Sets Syst.* **2008**, *159*, 1836–1859. [CrossRef]
13. Baczyński, M.; Jayaram, B.; Massanet, S.; Torrens, J. Fuzzy implications: Past, present, and future. In *Springer Handbook of Computational Intelligence*; Springer: Berlin/Heidelberg, Germany, 2015; pp. 183–202. [CrossRef]
14. Mandal, S.; Jayaram, B. SISO fuzzy relational inference systems based on fuzzy implications are universal approximators. *Fuzzy Sets Syst.* **2015**, *277*, 1–21. [CrossRef]
15. Jayaram, B. On the Law of Importation $(x \wedge y) \longrightarrow z \equiv (x \longrightarrow (y \longrightarrow z))$ in Fuzzy Logic. *IEEE Trans. Fuzzy Syst.* **2008**, *16*, 130–144. [CrossRef]
16. Mas, M.; Monserrat, M.; Torrens, J. On the law of importation for some kinds of fuzzy implications derived from uninorms. In Proceedings of the IPMU'08, Torremolinos, Spain, 22–27 June 2008; pp. 1303–1310.
17. Türksen, I.; Kreinovich, V.; Yager, R. A new class of fuzzy implications. Axioms of fuzzy implication revisited. *Fuzzy Sets Syst.* **1998**, *100*, 267–272. [CrossRef]
18. Mas, M.; Monserrat, M.; Torrens, J. The law of importation for discrete implications. *Inf. Sci.* **2009**, *179*, 4208–4218. [CrossRef]
19. Massanet, S.; Torrens, J. The law of importation versus the exchange principle on fuzzy implications. *Fuzzy Sets Syst.* **2011**, *168*, 47–69. [CrossRef]
20. Baczyński, M.; Jayaram, B.; Mesiar, R. Fuzzy implications: Alpha migrativity and generalised laws of importation. *Inf. Sci.* **2020**, *531*, 87–96. [CrossRef]
21. Mas, M.; Monserrat, M.; Torrens, J. A characterization of (U, N), RU, QL and D-implications derived from uninorms satisfying the law of importation. *Fuzzy Sets Syst.* **2010**, *161*, 1369–1387. [CrossRef]
22. Mandal, S.; Jayaram, B. Bandler-Kohout subproduct with Yager's classes of fuzzy implications. *IEEE Trans. Fuzzy Syst.* **2014**, *22*, 469–482. [CrossRef]
23. Klement, E.P.; Mesiar, R.; Pap, E. *Triangular Norms*; Kluwer: Dordrecht, The Netherlands, 2000. [CrossRef]
24. Durante, F.; Sarkoci, P. A note on the convex combinations of triangular norms. *Fuzzy Sets Syst.* **2008**, *159*, 77–80. [CrossRef]
25. Baczyński, M.; Jayaram, B. *Fuzzy Implications*; Springer: Berlin/Heidelberg, Germany, 2008. [CrossRef]
26. Baczyński, M.; Jayaram, B. On the characterization of (S, N)-implications. *Fuzzy Sets Syst.* **2007**, *158*, 1713–1727. [CrossRef]

mathematics

MDPI

Article

Distance and Similarity Measures for Octahedron Sets and Their Application to MCGDM Problems

Güzide Şenel [1], Jeong-Gon Lee [2,*] and Kul Hur [2]

[1] Department of Mathematics, University of Amasya, 05100 Amasya, Turkey; g.senel@amasya.edu.tr
[2] Division of Applied Mathematics, Wonkwang University, 460, Iksan-daero, Iksan-Si, Jeonbuk 54538, Korea; kulhur@wku.ac.kr
* Correspondence: jukolee@wku.ac.kr

Received: 12 August 2020; Accepted: 27 September 2020; Published: 1 October 2020

check for updates

Abstract: In this paper, in order to apply the concept of octahedron sets to multi-criteria group decision-making problems, we define several similarity and distance measures for octahedron sets. We present a multi-criteria group decision-making method with linguistic variables in octahedron set environment. We give a numerical example for multi-criteria group decision-making problems.

Keywords: octahedron set; distance measure; similarity measure

MSC: 46S40; 03E72; 68T37

1. Introduction

In real world, we frequently encounter with decision-making problems with uncertainty and vagueness that can be difficult to solve with the classical methods. A number of techniques have been developed to solve uncertinities; similarity measures are one of tools solving decision-making problems. Chen and Hsiao [1] studied some similarity measures for fuzzy sets introduced by Zadeh [2]. Pramanik and Mondal [3] defined the concept of weighted fuzzy similarity measure (called a tangent similarity measure) and applied it to medical diagnosis. Hwang and Yang [4] made a new similarity measure for intuitionistic fuzzy sets proposed by Atanassov [5]. Pramanik and Mondal [6] proposed intuitionistic fuzzy similarity measure based on tangent function and applied it to multi–attribute decision. Ren and Wang [7] introduced the notion of similarity measures for interval-valued intuitionstic fuzzy sets proposed by Atanassov and Gargov [8]. Baroumi and Smarandache [9] dealt with several similarity measures between neutrosophic sets and applied them to decision-making problems introduced by Smarandache [10]. Ye [11] defined a similarity measure for interval neutrosophic sets and applied it to decision-making method. Sahin and Liu [12] introduced various distance and similarity measures between single-valued neutrosophic hesitant fuzzy sets (see Reference [13,14]) and discussed MADM problems based on the single-valued neutrosophic hesitant fuzzy information. Kaur and Garg [15,16] applied it to decision-making and studied cubic intuitionistic fuzzy aggregation operators (see Reference [17,18]). Pramanik et al. [19] proposed a similarity measure for cubic neutrosophic sets and applied it a multi-criteria group decision-making (MCGDM) method.

Recently, Kim et al. [20] defined an octahedron set composed of an interval-valued fuzzy set, an intuitionistic set and a fuzzy set that will provide more information about uncertainty and vagueness. The purpose of this paper is to review recent research into the octahedron set applications with MCGDM method. This paper proposes a new methodology for the theory of octahedron set with defining several distance and similarity measures between octahedron sets. Moreover, we prove that each is distance and similarity measure. This new application offers some important insights into octahedron sets with MCGDM method based on a similarity measure for octahedron sets. The findings reported here shed new light on an application area of octahedron sets. This approach will prove useful in expanding our understanding of how the application of octahedron sets can be done. The findings of this study could have a number of important implications for future practice with analyzing our numerical example for MCGDM problems that is given in this paper. In order to apply the concept of octahedron sets to MCGDM problems, this paper is presented as follows: In Section 2, we list some basic notions that are needed in next section. In Section 3, we propose several distance and similarity measures between octahedron sets and prove that each is distance and similarity measure. In Section 4, we present a MCGDM method based on a similarity measure for octahedron sets environments. In addition, we give a numerical example for MCGDM problems to demonstrate the usefulness and applicability of our proposed method. There is a relatively small body of literature that is concerned with the application of soft octahedron sets. In recent years, there has been an increasing amount of literature on the application of set theories. One of the aim of this research was to improve the works on the application of octahedron sets by proposing suitable examples in this paper. All these findings will provide a base to researchers who want to work in the field of the application of octahedron sets and will help to strengthen the foundations of the other MCGDM problems in octahedron set environment, such as economic policy, foreign policy between countries, trade policy, financial policy, etc., by using big data.

2. Preliminaries

Let $I \oplus I = \{\bar{a} = (a^{\in}, a^{\notin}) \in I \times I : a^{\in} + a^{\notin} \leq 1\}$, where $I = [0,1]$. Then, each member $\bar{a}$ of $I \oplus I$ is called an intuitionistic point or intuitionistic number. In particular, we denote $(0,1)$ and $(1,0)$ as $\bar{0}$ and $\bar{1}$, respectively. Refer to Reference [21] for the definitions of the order $(\leq)$ and the equality $(=)$ of two intuitionistic numbers, and the infimum and the supremum of any intuitionistic numbers.

Definition 1 (Reference [5]). *For a nonempty set X, a mapping $A : X \to I \oplus I$ is called an intuitionistic fuzzy set (briefly, IF set) in X, where, for each $x \in X$, $A(x) = (A^{\in}(x), A^{\notin}(x))$, and $A^{\in}(x)$ and $A^{\notin}(x)$ represent the degree of membership and the degree of nonmembership of an element x to A, respectively. Let $(I \oplus I)^X$ denote the set of all IF sets in X and for each $A \in (I \oplus I)^X$, we write $A = (A^{\in}, A^{\notin})$. In particular, $\bar{\mathbf{0}}$ and $\bar{\mathbf{1}}$ denote the IF empty set and the IF whole set in X defined by, respectively:*

For each $x \in X$,

$$\bar{\mathbf{0}}(x) = \bar{0} \text{ and } \bar{\mathbf{1}}(x) = \bar{1}.$$

Refer to Reference [5] for the definitions of the inclusion, the equality, the intersection, and the union of intuitionistic fuzzy set and operations ^{c}, $[\,]$, $\diamond$ on $(I \oplus I)^X$.

The set of all closed subintervals of I is denoted by $[I]$, and members of $[I]$ are called interval numbers and are denoted by $\widetilde{a}$, $\widetilde{b}$, $\widetilde{c}$,, etc., where $\widetilde{a} = [a^-, a^+]$ and $0 \leq a^- \leq a^+ \leq 1$. In particular, if $a^- = a^+$, then we write as $\widetilde{a} = \mathbf{a}$. Refer to Reference [20] for the definitions of the order and the equality of two interval numbers, as well as the infimum and the supremum of any interval numbers.

Definition 2 (Reference [22,23]). *For a nonempty set X, a mapping $A : X \to [I]$ is called an interval-valued fuzzy set (briefly, an IVF set) in X. Let $[I]^X$ denote the set of all IVF sets in X. For each $A \in [I]^X$ and $x \in X$, $A(x) = [A^-(x), A^+(x)]$ is called the degree of membership of an element x to A, where $A^-, A^+ \in I^X$ are called a lower fuzzy set and an upper fuzzy set in X, respectively. For each $A \in [I]^X$, we write $A = [A^-, A^+]$. In particular, $\widetilde{0}$ and $\widetilde{1}$ denote the interval-valued fuzzy empty set and the interval-valued fuzzy empty whole set in X defined by, respectively: for each $x \in X$,*

$$\widetilde{0}(x) = \mathbf{0} \text{ and } \widetilde{1}(x) = \mathbf{1}.$$

Refer to Reference [22,23] for the definitions of the inclusion, the equality, the intersection, and the union of intuitionistic fuzzy set and operation ^{c} on $[I]^X$.

Now, members of $[I] \times (I \oplus I) \times I$ are written by $\tilde{\bar{a}} =< \tilde{a}, \bar{a}, a >=< [a^-, a^-], (a^{\in}, a^{\notin}), a >$, $\tilde{\bar{b}} =< \tilde{b}, \bar{b}, b >=< [b^-, b^-], (b^{\in}, b^{\notin}), b >$, etc., and are called octahedron numbers. Furthermore, we define the following order relations between $\tilde{\bar{a}}$ and $\tilde{\bar{b}}$ (see Reference [20]):

(i) (Equality) $\tilde{\bar{a}} = \tilde{\bar{b}} \Leftrightarrow \tilde{a} = \tilde{b},\ \bar{a} = \bar{b},\ a = b$,
(ii) (Type 1-order) $\tilde{\bar{a}} \leq_1 \tilde{\bar{b}} \Leftrightarrow a^- \leq b^-,\ a^+ \leq b^+,\ a^{\in} \leq b^{\in},\ a^{\notin} \geq b^{\notin},\ a \leq b$,
(iii) (Type 2-order) $\tilde{\bar{a}} \leq_2 \tilde{\bar{b}} \Leftrightarrow a^- \leq b^-,\ a^+ \leq b^+,\ a^{\in} \leq b^{\in},\ a^{\notin} \geq b^{\notin},\ a \geq b$,
(iv) (Type 3-order) $\tilde{\bar{a}} \leq_3 \tilde{\bar{b}} \Leftrightarrow a^- \leq b^-,\ a^+ \geq b^+,\ a^{\in} \geq b^{\in},\ a^{\notin} \leq b^{\notin},\ a \leq b$,
(v) (Type 4-order) $\tilde{\bar{a}} \leq_4 \tilde{\bar{b}} \Leftrightarrow a^- \leq b^-,\ a^+ \leq b^+,\ a^{\in} \geq b^{\in},\ a^{\notin} \leq b^{\notin},\ a \geq b$.

From the above orders, we can define the inf and the sup of octahedron numbers as follows.

Definition 3 (Reference [20]). *Let $\tilde{\bar{a}},\ \tilde{\bar{b}} \in [I] \times (I \oplus I) \times I$. Then,*

(i)
$$\tilde{\bar{a}} \wedge^1 \tilde{\bar{b}} = \left\langle [a^- \wedge b^-, a^+ \wedge b^+], (a^{\in} \wedge b^{\in}, a^{\notin} \vee b^{\notin}), a \wedge b \right\rangle,$$
$$\tilde{\bar{a}} \wedge^2 \tilde{\bar{b}} = \left\langle [a^- \wedge b^-, a^+ \wedge b^+], (a^{\in} \wedge b^{\in}, a^{\notin} \vee b^{\notin}), a \vee b \right\rangle,$$
$$\tilde{\bar{a}} \wedge^3 \tilde{\bar{b}} = \left\langle [a^- \wedge b^-, a^+ \wedge b^+], (a^{\in} \vee b^{\in}, a^{\notin} \wedge b^{\notin}), a \wedge b \right\rangle,$$
$$\tilde{\bar{a}} \wedge^4 \tilde{\bar{b}} = \left\langle [a^- \wedge b^-, a^+ \wedge b^+], (a^{\in} \vee b^{\in}, a^{\notin} \wedge b^{\notin}), a \vee b \right\rangle,$$
(ii)
$$\tilde{\bar{a}} \vee^1 \tilde{\bar{b}} = \left\langle [a^- \vee b^-, a^+ \vee b^+], (a^{\in} \vee b^{\in}, a^{\notin} \wedge b^{\notin}), a \vee b \right\rangle,$$
$$\tilde{\bar{a}} \vee^2 \tilde{\bar{b}} = \left\langle [a^- \vee b^-, a^+ \vee b^+], (a^{\in} \vee b^{\in}, a^{\notin} \wedge b^{\notin}), a \wedge b \right\rangle,$$
$$\tilde{\bar{a}} \vee^3 \tilde{\bar{b}} = \left\langle [a^- \vee b^-, a^+ \vee b^+], (a^{\in} \wedge b^{\in}, a^{\notin} \vee b^{\notin}), a \vee b \right\rangle,$$
$$\tilde{\bar{a}} \vee^4 \tilde{\bar{b}} = \left\langle [a^- \vee b^-, a^+ \vee b^+], (a^{\in} \wedge b^{\in}, a^{\notin} \vee b^{\notin}), a \wedge b \right\rangle.$$

Definition 4 (Reference [20]). *Let X be a nonempty set, and let $\mathbf{A} = [A^-, A^+] \in [I]^X$, $A = (A^{\in}, A^{\notin}) \in (I \oplus I)^X$, $\lambda \in I^X$. Then, the triple $\mathcal{A} = \langle \mathbf{A}, A, \lambda \rangle$ is called an octahedron set in X. In fact, $\mathcal{A} : X \to [I] \times (I \oplus I) \times I$ is a mapping. In particular, the octahedron empty (resp. whole) set in X, denoted by $\ddot{0}$ (resp. $\ddot{1}$), is an octahedron set in X defined by:*

$$\ddot{0} = \left\langle \tilde{0}, \bar{\mathbf{0}}, 0 \right\rangle,\ \ddot{1} = \left\langle \tilde{1}, \bar{\mathbf{1}}, 1 \right\rangle.$$

It is obvious that, for each $A \in 2^X$, $\chi_A = \langle [\chi_A, \chi_A], (\chi_A, \chi_{A^c}), \chi_A \rangle \in \mathcal{O}(X)$ and then $2^X \subset \mathcal{O}(X)$, where 2^X denotes the set of all subsets of X and χ_A denotes the characteristic function of A. Furthermore, we can easily see that, for each $\mathbf{A} = \langle A, \lambda \rangle \in \mathcal{C}(X)$, $\mathbf{A} = \langle A, (A^-, A^+), \lambda \rangle$, $\mathbf{A} = \langle A, (\lambda, \lambda^c), \lambda \rangle \in \mathcal{O}(X)$ and then $\mathcal{C}(X) \subset \mathcal{O}(X)$. In this case, we denote $\langle A, (A^-, A^+), \lambda \rangle$ and $\langle A, (\lambda, \lambda^c), \lambda \rangle$ as $\mathcal{A}_A$ and $\mathcal{A}_\lambda$, respectively. In fact, we can consider octahedron sets as a generalization of cubic sets.

Definition 5 (Reference [20]). *Let X be a nonempty set, and let $\mathcal{A} = \langle \mathbf{A}, A, \lambda \rangle$, $\mathcal{B} = \langle \mathbf{B}, B, \mu \rangle \in \mathcal{O}(X)$. Then, we can define following order relations between $\mathcal{A}$ and $\mathcal{B}$:*

(i) (Equality) $\mathcal{A} = \mathcal{B} \Leftrightarrow \mathbf{A} = \mathbf{B},\ A = B,\ \lambda = \mu$,
(ii) (Type 1-order) $\mathcal{A} \subset_1 \mathcal{B} \Leftrightarrow \mathbf{A} \subset \mathbf{B},\ A \subset B,\ \lambda \leq \mu$,
(iii) (Type 2-order) $\mathcal{A} \subset_2 \mathcal{B} \Leftrightarrow \mathbf{A} \subset \mathbf{B},\ A \subset B,\ \lambda \geq \mu$,
(iv) (Type 3-order) $\mathcal{A} \subset_3 \mathcal{B} \Leftrightarrow \mathbf{A} \subset \mathbf{B},\ A \supset B,\ \lambda \leq \mu$,
(v) (Type 4-order) $\mathcal{A} \subset_4 \mathcal{B} \Leftrightarrow \mathbf{A} \subset \mathbf{B},\ A \supset B,\ \lambda \geq \mu$.

Definition 6 (Reference [20]). *Let X be a nonempty set, and let* $(\mathcal{A}_j)_{j\in J} = (\langle \mathbf{A}_j, A_j, \lambda_j \rangle)_{j\in J}$ *be a family of octahedron sets in X. Then, the Type i-union* $\cup^i$ *and Type i-intersection* $\cap^i$ *of* $(\mathcal{A}_j)_{j\in J}$, $(i = 1,\ 2,\ ,3,4)$, *are defined as follows, respectively:*

(i) (Type i-union) $\bigcup^1_{j\in J} \mathcal{A}_j = \left\langle \bigcup_{j\in J} \mathbf{A}_j, \bigcup_{j\in J} A_j, \bigcup_{j\in J} \lambda_j \right\rangle,$

$$\bigcup^2_{j\in J} \mathcal{A}_j = \left\langle \bigcup_{j\in J} \mathbf{A}_j, \bigcup_{j\in J} A_j, \bigcap_{j\in J} \lambda_j \right\rangle,$$

$$\bigcup^3_{j\in J} \mathcal{A}_j = \left\langle \bigcup_{j\in J} \mathbf{A}_j, \bigcap_{j\in J} A_j, \bigcup_{j\in J} \lambda_j \right\rangle,$$

$$\bigcup^4_{j\in J} \mathcal{A}_j = \left\langle \bigcup_{j\in J} \mathbf{A}_j, \bigcap_{j\in J} A_j, \bigcap_{j\in J} \lambda_j \right\rangle,$$

(ii) (Type i-intersection) $\bigcap^1_{j\in J} \mathcal{A}_j = \left\langle \bigcap_{j\in J} \mathbf{A}_j, \bigcap_{j\in J} A_j, \bigcap_{j\in J} \lambda_j \right\rangle,$

$$\bigcap^2_{j\in J} \mathcal{A}_j = \left\langle \bigcap_{j\in J} \mathbf{A}_j, \bigcap_{j\in J} A_j, \bigcup_{j\in J} \lambda_j \right\rangle,$$

$$\bigcap^3_{j\in J} \mathcal{A}_j = \left\langle \bigcap_{j\in J} \mathbf{A}_j, \bigcup_{j\in J} A_j, \bigcap_{j\in J} \lambda_j \right\rangle,$$

$$\bigcap^4_{j\in J} \mathcal{A}_j = \left\langle \bigcap_{j\in J} \mathbf{A}_j, \bigcup_{j\in J} A_j, \bigcup_{j\in J} \lambda_j \right\rangle.$$

Let $\mathbb{R}$ be the real space. Then, for any intervals $A = [a_1, a_2]$ and $B = [b_1, b_2]$ of $\mathbb{R}$, the Hausdorff distance $d_H(A, B)$ between A and B is defined by:

$$d_H(A, B) = \max(\mid a_1 - b_1 \mid, \mid a_2 - b_2 \mid). \tag{1}$$

Throughout this paper, let $X = \{x_1, x_2, x_3, \cdots, x_n\}$ be a universal set.

3. Distance and Similarity Measures between Octahedron Sets

Definition 7. *A mapping* $d : \mathcal{O}(X) \times \mathcal{O}(X) \to I$ *is called a distance measure on* $\mathcal{O}(X)$, *if it satisfies the following conditions: For any* $\mathcal{A},\ \mathcal{B},\ \mathcal{C} \in \mathcal{O}(X)$,

(DM_1) $0 \le d(\mathcal{A}, \mathcal{B}) \le 1$,
(DM_2) $d(\mathcal{A}, \mathcal{B}) = 0$ *if and only if* $\mathcal{A} = \mathcal{B}$,
(DM_3) $d(\mathcal{A}, \mathcal{B}) = d(\mathcal{B}, \mathcal{A})$,
(DM_4) *if* $\mathcal{A} \subset_1 \mathcal{B} \subset_1 \mathcal{C}$, *then* $d(\mathcal{A}, \mathcal{C}) \ge d(\mathcal{A}, \mathcal{B}) \vee d(\mathcal{B}, \mathcal{C})$.

In this case, $d(\mathcal{A}, \mathcal{B})$ *is called the distance measure between* $\mathcal{A}$ *and* $\mathcal{B}$.

Definition 8. *A mapping* $s : \mathcal{O}(X) \times \mathcal{O}(X) \to I$ *is called a similarity measure on* $\mathcal{O}(X)$, *if it satisfies the following conditions: For any* $\mathcal{A},\ \mathcal{B},\ \mathcal{C} \in \mathcal{O}(X)$,

(DM_1) $0 \le s(\mathcal{A}, \mathcal{B}) \le 1$,
(DM_2) $s(\mathcal{A}, \mathcal{B}) = 1$ *if and only if* $\mathcal{A} = \mathcal{B}$,
(DM_3) $s(\mathcal{A}, \mathcal{B}) = s(\mathcal{B}, \mathcal{A})$,
(DM_4) *if* $\mathcal{A} \subset_1 \mathcal{B} \subset_1 \mathcal{C}$, *then* $s(\mathcal{A}, \mathcal{C}) \le s(\mathcal{A}, \mathcal{B}) \wedge s(\mathcal{B}, \mathcal{C})$.

In this case, $d(\mathcal{A}, \mathcal{B})$ *is called the similarity measure between* $\mathcal{A}$ *and* $\mathcal{B}$.

In fact, distance measure and similarity measure from Definitions 7 and 8, we can easily see that $s(\mathcal{A}, \mathcal{B}) = 1 - d(\mathcal{A}, \mathcal{B})$.

Now, we give some types of distance measures between two octahedron sets in the following:

Example 1. *(1) (Generalized normalized distance measure) Let* $\delta > 0$. *We define* $d_{GN} : \mathcal{O}(X) \times \mathcal{O}(X) \to I$ *is the mapping defined as: For any* $\mathcal{A}, \mathcal{B} \in \mathcal{O}(X)$,

$$\begin{aligned} d_{GN}(\mathcal{A},\mathcal{B}) &= [\tfrac{1}{5n}\Sigma_{i=1}^{n}(\mid A^{-}(x_i) - B^{-}(x_i) \mid^{\delta} + \mid A^{+}(x_i) - B^{+}(x_i) \mid^{\delta} \\ &+ \mid A^{\in}(x_i) - B^{\in}(x_i) \mid^{\delta} + \mid A^{\notin}(x_i) - B^{\notin}(x_i) \mid^{\delta} \\ &+ \mid \lambda(x_i) - \mu(x_i) \mid^{\delta})]^{\frac{1}{\delta}}. \end{aligned}$$

Then, d_{GN} *is a distance measure on* $\mathcal{O}(X)$ *(see Propositions 1 and 2).*

In particular, if $\delta = 1$, *then* d_{GN} *reduces an octahedron normalized Hamming distance and denoted by* d_{NH}:

$$\begin{aligned} d_{NH}(\mathcal{A},\mathcal{B}) &= \tfrac{1}{5n}\Sigma_{i=1}^{n}(\mid A^{-}(x_i) - B^{-}(x_i) \mid + \mid A^{+}(x_i) - B^{+}(x_i) \mid \\ &+ \mid A^{\in}(x_i) - B^{\in}(x_i) \mid + \mid A^{\notin}(x_i) - B^{\notin}(x_i) \mid \\ &+ \mid \lambda(x_i) - \mu(x_i) \mid). \end{aligned}$$

If $\delta = 2$, *then* d_{GN} *reduces an octahedron normalized Euclidean distance and denoted by* d_{NE}:

$$\begin{aligned} d_{NE}(\mathcal{A},\mathcal{B}) &= [\tfrac{1}{5n}\Sigma_{i=1}^{n}(\mid A^{-}(x_i) - B^{-}(x_i) \mid^{2} + \mid A^{+}(x_i) - B^{+}(x_i) \mid^{2} \\ &+ \mid A^{\in}(x_i) - B^{\in}(x_i) \mid^{2} + \mid A^{\notin}(x_i) - B^{\notin}(x_i) \mid^{2} \\ &+ \mid \lambda(x_i) - \mu(x_i) \mid^{2})]^{\frac{1}{2}}. \end{aligned}$$

In fact, d_{GN} *can be viewed as a most generalized case of distance measures.*

(2) (Generalized octahedron normalized Hausdorff distance) We define $d_{GNH} : \mathcal{O}(X) \times \mathcal{O}(X) \to I$ *is the mapping defined as: for any* $\mathcal{A}, \mathcal{B} \in \mathcal{O}(X)$,

$$\begin{aligned} d_{GNH}(\mathcal{A},\mathcal{B}) &= [\tfrac{1}{n}\Sigma_{i=1}^{n}\max(\mid A^{-}(x_i) - B^{-}(x_i) \mid^{\delta}, \mid A^{+}(x_i) - B^{+}(x_i) \mid^{\delta}, \\ &\mid A^{\in}(x_i) - B^{\in}(x_i) \mid^{\delta}, \mid A^{\notin}(x_i) - B^{\notin}(x_i) \mid^{\delta}, \\ &\mid \lambda(x_i) - \mu(x_i) \mid^{\delta})]^{\frac{1}{\delta}}. \end{aligned}$$

Then, d_{GNH} *is a distance measure on* $\mathcal{O}(X)$ *(see Propositions 3 and 4).*

In particular, if $\delta = 1$, *then* d_{GNH} *reduces an octahedron normalized Hamming-Hausdorff distance and denoted by* d_{NHH}:

$$\begin{aligned} d_{NHH}(\mathcal{A},\mathcal{B}) &= \tfrac{1}{n}\Sigma_{i=1}^{n}\max(\mid A^{-}(x_i) - B^{-}(x_i) \mid, \mid A^{+}(x_i) - B^{+}(x_i) \mid, \\ &\mid A^{\in}(x_i) - B^{\in}(x_i) \mid, \mid A^{\notin}(x_i) - B^{\notin}(x_i) \mid, \\ &\mid \lambda(x_i) - \mu(x_i) \mid). \end{aligned}$$

If $\delta = 2$, *then* d_{GNH} *reduces an octahedron normalized Euclidean-Hausdorff distance and denoted by* d_{NEH}:

$$\begin{aligned} d_{NEH}(\mathcal{A},\mathcal{B}) &= [\tfrac{1}{n}\Sigma_{i=1}^{n}\max(\mid A^{-}(x_i) - B^{-}(x_i) \mid^{2}, \mid A^{+}(x_i) - B^{+}(x_i) \mid^{2}, \\ &\mid A^{\in}(x_i) - B^{\in}(x_i) \mid^{2}, \mid A^{\notin}(x_i) - B^{\notin}(x_i) \mid^{2}, \\ &\mid \lambda(x_i) - \mu(x_i) \mid^{2})]^{\frac{1}{2}}. \end{aligned}$$

In many practical situations, the weight of each element $x_i \in X$ should be taken into account. For example, in MADM problems, since the considered attribute has different importance in general, we need to be assigned with different weights. Since an octahedron set has three types of degree (an interval-valued fuzzy membership degree, an intuitionistic fuzzy membership degree and a fuzzy membership degree) and each degree may have different importance according to a decision-maker, different weights can be assigned to each element in each degree. We assume that the weights

$\omega = (\omega_1, \omega_2, \cdots, \omega_n)^T$ with $\omega_i \in I$, $\Sigma_{i=1}^n \omega_i = 1$; $\eta = (\eta_1, \eta_2, \cdots, \eta_n)^T$ with $\eta_i \in I$, $\Sigma_{i=1}^n \eta_i = 1$; $\xi = (\xi_1, \xi_2, \cdots, \xi_n)^T$ with $\xi_i \in I$, $\Sigma_{i=1}^n \xi_i = 1$ denote the weights assigned to interval-valued fuzzy membership degree, intuitionistic fuzzy membership degree and fuzzy membership degree of an octahedron set.

Now, we give some types of weighted distance measures between two octahedron sets in the following:

Example 2. *(1) (Generalized octahedron weighted distance measure) Let $\delta > 0$. We define $d_{GW} : \mathcal{O}(X) \times \mathcal{O}(X) \to I$ is the mapping defined as: for any $\mathcal{A}$, $\mathcal{B} \in \mathcal{O}(X)$,*

$$\begin{aligned} d_{GW}(\mathcal{A}, \mathcal{B}) &= [\tfrac{1}{5n}\Sigma_{i=1}^n(\omega_i \mid A^-(x_i) - B^-(x_i) \mid^\delta + \omega_i \mid A^+(x_i) - B^+(x_i) \mid^\delta \\ &+ \eta_i \mid A^\in(x_i) - B^\in(x_i) \mid^\delta + \eta_i \mid A^\notin(x_i) - B^\notin(x_i) \mid^\delta \\ &+ \xi_i \mid \lambda(x_i) - \mu(x_i) \mid^\delta)]^{\frac{1}{\delta}}. \end{aligned}$$

Then, d_{GW} is a weighted distance measure on $\mathcal{O}(X)$ (The proof is omitted).

In particular, if $\delta = 1$, then d_{GW} reduces an octahedron weighted Hamming distance and denoted by d_{WH}:

$$\begin{aligned} d_{WH}(\mathcal{A}, \mathcal{B}) &= \tfrac{1}{5n}\Sigma_{i=1}^n(\omega_i \mid A^-(x_i) - B^-(x_i) \mid + \omega_i \mid A^+(x_i) - B^+(x_i) \mid \\ &+ \eta_i \mid A^\in(x_i) - B^\in(x_i) \mid + \eta_i \mid A^\notin(x_i) - B^\notin(x_i) \mid \\ &+ \xi_i \mid \lambda(x_i) - \mu(x_i) \mid). \end{aligned}$$

If $\delta = 2$, then d_{GW} reduces an octahedron weighted Euclidean distance and denoted by d_{WE}:

$$\begin{aligned} d_{WE}(\mathcal{A}, \mathcal{B}) &= [\tfrac{1}{5n}\Sigma_{i=1}^n(\omega_i \mid A^-(x_i) - B^-(x_i) \mid^2 + \omega_i \mid A^+(x_i) - B^+(x_i) \mid^2 \\ &+ \eta_i \mid A^\in(x_i) - B^\in(x_i) \mid^2 + \eta_i \mid A^\notin(x_i) - B^\notin(x_i) \mid^2 \\ &+ \xi_i \mid \lambda(x_i) - \mu(x_i) \mid^2)]^{\frac{1}{2}}. \end{aligned}$$

(2) (Generalized octahedron weighted Hausdorff distance) We define $d_{GWH} : \mathcal{O}(X) \times \mathcal{O}(X) \to I$ is the mapping defined as: For any $\mathcal{A}$, $\mathcal{B} \in \mathcal{O}(X)$,

$$\begin{aligned} d_{GWH}(\mathcal{A}, \mathcal{B}) &= [\tfrac{1}{n}\Sigma_{i=1}^n \max(\omega_i \mid A^-(x_i) - B^-(x_i) \mid^\delta, \omega_i \mid A^+(x_i) - B^+(x_i) \mid^\delta, \\ &\eta_i \mid A^\in(x_i) - B^\in(x_i) \mid^\delta, \eta_i \mid A^\notin(x_i) - B^\notin(x_i) \mid^\delta, \\ &\xi_i \mid \lambda(x_i) - \mu(x_i) \mid^\delta)]^{\frac{1}{\delta}}. \end{aligned}$$

Then, d_{GWH} is a weighted distance measure on $\mathcal{O}(X)$ (The proof is omitted).

In particular, if $\delta = 1$, then d_{GWH} reduces an octahedron weighted Hamming-Hausdorff distance and denoted by d_{WHH}:

$$\begin{aligned} d_{WHH}(\mathcal{A}, \mathcal{B}) &= \tfrac{1}{n}\Sigma_{i=1}^n \max(\omega_i \mid A^-(x_i) - B^-(x_i) \mid, \omega_i \mid A^+(x_i) - B^+(x_i) \mid, \\ &\eta_i \mid A^\in(x_i) - B^\in(x_i) \mid, \eta_i \mid A^\notin(x_i) - B^\notin(x_i) \mid, \\ &\xi_i \mid \lambda(x_i) - \mu(x_i) \mid). \end{aligned}$$

If $\delta = 2$, then d_{GWH} reduces an octahedron weighted Euclidean-Hausdorff distance and denoted by d_{WEH}:

$$\begin{aligned} d_{WEH}(\mathcal{A}, \mathcal{B}) &= [\tfrac{1}{n}\Sigma_{i=1}^n \max(\omega_i \mid A^-(x_i) - B^-(x_i) \mid^2, \omega_i \mid A^+(x_i) - B^+(x_i) \mid^2, \\ &\eta_i \mid A^\in(x_i) - B^\in(x_i) \mid^2, \eta_i \mid A^\notin(x_i) - B^\notin(x_i) \mid^2, \\ &\xi_i \mid \lambda(x_i) - \mu(x_i) \mid^2)]^{\frac{1}{2}}. \end{aligned}$$

Proposition 1. *The mapping d_{NH} defined in Example 1 (1) is a distance measure on $\mathcal{O}(X)$.*

Proof. Since the proofs of (DM_2) and (DM_3) are easy from the definition of d_{NH}, we will show only (DM_1) and (DM_4).

(DM_1) Let $\mathcal{A}$, $\mathcal{B} \in \mathcal{O}(X)$. Then, by the definition of d_{NH},

$$\begin{aligned}&\mid A^-(x_i) - B^-(x_i) \mid \geq 0, \ \mid A^+(x_i) - B^+(x_i) \mid \geq 0,\\ &\mid A^{\in}(x_i) - B^{\in}(x_i) \mid \geq 0, \ \mid A^{\notin}(x_i) - B^{\notin}(x_i) \mid \geq 0, \mid \lambda(x_i) - \mu(x_i) \mid \geq 0.\end{aligned}$$

Thus, we have

$$\begin{aligned} d_{NH}(\mathcal{A},\mathcal{B}) &= \tfrac{1}{5n}\Sigma_{i=1}^{n}(\mid A^-(x_i) - B^-(x_i) \mid + \mid A^+(x_i) - B^+(x_i) \mid \\ &+ \mid A^{\in}(x_i) - B^{\in}(x_i) \mid + \mid A^{\notin}(x_i) - B^{\notin}(x_i) \mid \\ &+ \mid \lambda(x_i) - \mu(x_i) \mid) \\ &\geq 0. \end{aligned}$$

In addition, by the definition of d_{NH},

$$\begin{aligned}&\mid A^-(x_i) - B^-(x_i) \mid \leq 1, \ \mid A^+(x_i) - B^+(x_i) \mid \leq 1,\\ &\mid A^{\in}(x_i) - B^{\in}(x_i) \mid \leq 1, \ \mid A^{\notin}(x_i) - B^{\notin}(x_i) \mid \leq 1, \mid \lambda(x_i) - \mu(x_i) \mid \leq 1.\end{aligned}$$

So, we get

$$\begin{aligned} d_{NH}(\mathcal{A},\mathcal{B}) &= \tfrac{1}{5n}\Sigma_{i=1}^{n}(\mid A^-(x_i) - B^-(x_i) \mid + \mid A^+(x_i) - B^+(x_i) \mid \\ &+ \mid A^{\in}(x_i) - B^{\in}(x_i) \mid + \mid A^{\notin}(x_i) - B^{\notin}(x_i) \mid \\ &+ \mid \lambda(x_i) - \mu(x_i) \mid) \\ &\leq 1. \end{aligned}$$

Hence, $0 \leq d_{NH}(\mathcal{A},\mathcal{B}) \leq 1$.

(DM_4) Suppose $\mathcal{A} = \langle \mathbf{A}, A, \lambda \rangle$, $\mathcal{B} = \langle \mathbf{B}, B, \mu \rangle$, $\mathcal{C} = \langle \mathbf{C}, C, \nu \rangle \in \mathcal{O}(X)$ such that $\mathcal{A} \subset_1 \mathcal{B} \subset_1 \mathcal{C}$, and let $x_i \in X$. Then, we have

$$\mathcal{A}(x_i) \leq_1 \mathcal{B}(x_i) \leq_1 \mathcal{C}(x_i), \text{ i.e.,}$$

$$\begin{aligned}&A^-(x_i) \leq B^-(x_i) \leq C^-(x_i), \ A^+(x_i) \leq B^+(x_i) \leq C^+(x_i),\\ &A^{\in}(x_i) \leq B^{\in}(x_i) \leq C^{\in}(x_i), \ A^{\notin}(x_i) \geq B^{\notin}(x_i) \geq C^{\notin}(x_i),\\ &\lambda(x_i) \leq \mu(x_i) \leq \nu(x_i).\end{aligned}$$

Thus, we get

$$\begin{aligned} d_{NH}(\mathcal{A},\mathcal{C}) &= \tfrac{1}{5n}\Sigma_{i=1}^{n}(\mid A^-(x_i) - C^-(x_i) \mid + \mid A^+(x_i) - C^+(x_i) \mid \\ &+ \mid A^{\in}(x_i) - C^{\in}(x_i) \mid + \mid A^{\notin}(x_i) - C^{\notin}(x_i) \mid \\ &+ \mid \lambda(x_i) - \nu(x_i) \mid) \\ &\geq \tfrac{1}{n}\Sigma_{i=1}^{n}(\mid A^-(x_i) - B^-(x_i) \mid + \mid A^+(x_i) - B^+(x_i) \mid \\ &+ \mid A^{\in}(x_i) - B^{\in}(x_i) \mid + \mid A^{\notin}(x_i) - B^{\notin}(x_i) \mid \\ &+ \mid \lambda(x_i) - \mu(x_i) \mid) \\ &= d_{NH}(\mathcal{A},\mathcal{B}). \end{aligned}$$

Similarly, we have $d_{NH}(\mathcal{A},\mathcal{C}) \geq d_H(\mathcal{B},\mathcal{C})$. So, $d_{NH}(\mathcal{A},\mathcal{C}) \geq d_{NH}(\mathcal{A},\mathcal{C}) \vee d_{NH}(\mathcal{B},\mathcal{C})$. This completes the proof. □

Proposition 2. *Two mappings d_{GN} and d_{NE} defined in Example 1 (1) are distance measure on $\mathcal{O}(X)$.*

Proof. The proofs are similar to the proof of Proposition 1. □

Proposition 3. *A mapping d_G and d_{GNH} defined in Example 1 (2) are distance measure on $\mathcal{O}(X)$.*

Proof. Since the proofs of (DM_2) and (DM_3) are easy from the definition of d_{GNH}, we will show only (DM_1) and (DM_4).

(DM_1) Let $\mathcal{A}$, $\mathcal{B} \in \mathcal{O}(X)$. Then, by the definition of d_{GNH}, we can easily obtain the following:

$$\max(| A^-(x_i) - B^-(x_i) |^\delta, \ | A^+(x_i) - B^+(x_i) |^\delta, \ | A^\in(x_i) - B^\in(x_i) |^\delta, \\ | A^\notin(x_i) - B^\notin(x_i) |^\delta, \ | \lambda(x_i) - \mu(x_i) |^\delta) \geq 0.$$

Thus, $d_{GNH}(\mathcal{A}, \mathcal{B}) \geq 0$. Similarly, we can easily prove that $d_{GNH}(\mathcal{A}, \mathcal{B}) \leq 1$. So, $0 \leq d_{GNH}(\mathcal{A}, \mathcal{B}) \leq 1$.

(DM_4) Suppose $\mathcal{A}$, $\mathcal{B}$, $\mathcal{C} \in \mathcal{O}(X)$ such that $\mathcal{A} \subset_1 \mathcal{B} \subset_1 \mathcal{C}$, and let $x_i \in X$. Then, we have

$$A^-(x_i) \leq B^-(x_i) \leq C^-(x_i), \ A^+(x_i) \leq B^+(x_i) \leq C^+(x_i), \\ A^\in(x_i) \leq B^\in(x_i) \leq C^\in(x_i), \ A^\notin(x_i) \geq B^\notin(x_i) \geq C^\notin(x_i), \\ \lambda(x_i) \leq \mu(x_i) \leq \nu(x_i).$$

Thus, we get

$$\max(| A^-(x_i) - C^-(x_i) |^\delta, \ | A^+(x_i) - C^+(x_i) |^\delta, \ | A^\in(x_i) - C^\in(x_i) |^\delta, \\ | A^\notin(x_i) - C^\notin(x_i) |^\delta, \ | \lambda(x_i) - \nu(x_i) |^\delta) \\ \geq \max(| A^-(x_i) - B^-(x_i) |^\delta, \ | A^+(x_i) - B^+(x_i) |^\delta, \ | A^\in(x_i) - B^\in(x_i) |^\delta, \\ | A^\notin(x_i) - B^\notin(x_i) |^\delta, \ | \lambda(x_i) - \mu(x_i) |^\delta).$$

So, $d_{GNH}(\mathcal{A}, \mathcal{C}) \geq d_H(\mathcal{A}, \mathcal{B})$. Similarly, we get $d_{GNH}(\mathcal{A}, \mathcal{C}) \geq d_{GNH}(\mathcal{B}, \mathcal{C})$. Hence, $d_{GNH}(\mathcal{A}, \mathcal{C}) \geq d_{GNH}(\mathcal{A}, \mathcal{B}) \vee d_{GNH}(\mathcal{B}, \mathcal{C})$. This completes the proof. □

Proposition 4. *Two mappings d_{NHH} and d_{NEH} defined in Example 1 (2) are distance measure on $\mathcal{O}(X)$.*

Proof. The proofs are similar to the proof of Proposition 3. □

Remark 1. *In Definition 7, although the condition (DM_4) is changed into the following:*
(DM_4') if $\mathcal{A} \subset_i \mathcal{B} \subset_i \mathcal{C}$ ($i = 1, 2, 3, 4$), then $d(\mathcal{A}, \mathcal{C}) \geq d(\mathcal{A}, \mathcal{B}) \vee d(\mathcal{B}, \mathcal{C})$,
we can easily see that all the distance measures given in Examples 1 and 2 satisfy the condition (DM_4').

Now, from the relationships between distance measures and similarity measures, we can give some examples of similarity measures on $\mathcal{O}(X)$.

Example 3. *Let $\mathcal{A}$, $\mathcal{B} \in \mathcal{O}(X)$.*

(1) (Generalized octahedron similarity measure corresponding to d_{GN})

$$s_{GN}(\mathcal{A}, \mathcal{B}) = 1 - d_{GN}(\mathcal{A}, \mathcal{B}). \tag{2}$$

In fact,

$$s_{GN}(\mathcal{A}, \mathcal{B}) = 1 - [\tfrac{1}{5n}\Sigma_{i=1}^n(| A^-(x_i) - B^-(x_i) |^\delta + | A^+(x_i) - B^+(x_i) |^\delta \\ + | A^\in(x_i) - B^\in(x_i) |^\delta + | A^\notin(x_i) - B^\notin(x_i) |^\delta \\ + | \lambda(x_i) - \mu(x_i) |^\delta)]^{\frac{1}{\delta}}.$$

(Octahedron similarity measure corresponding to d_{NH})

$$s_{NH}(\mathcal{A}, \mathcal{B}) = 1 - d_{NH}(\mathcal{A}, \mathcal{B}). \tag{3}$$

In fact,

$$\begin{aligned} s_{NH}(\mathcal{A},\mathcal{B}) &= \tfrac{1}{n}\Sigma_{i=1}^{n}[1-\tfrac{1}{5}(\mid A^{-}(x_i)-B^{-}(x_i)\mid + \mid A^{+}(x_i)-B^{+}(x_i)\mid \\ &+ \mid A^{\in}(x_i)-B^{\in}(x_i)\mid + \mid A^{\notin}(x_i)-B^{\notin}(x_i)\mid \\ &+ \mid \lambda(x_i)-\mu(x_i)\mid)]. \end{aligned}$$

(Octahedron similarity measure corresponding to d_{NE})

$$s_{NE}(\mathcal{A},\mathcal{B}) = 1 - d_{NE}(\mathcal{A},\mathcal{B}). \quad (4)$$

In fact,

$$\begin{aligned} s_{NE}(\mathcal{A},\mathcal{B}) &= 1-[\tfrac{1}{5n}\Sigma_{i=1}^{n}(\mid A^{-}(x_i)-B^{-}(x_i)\mid^2 + \mid A^{+}(x_i)-B^{+}(x_i)\mid^2 \\ &+ \mid A^{\in}(x_i)-B^{\in}(x_i)\mid^2 + \mid A^{\notin}(x_i)-B^{\notin}(x_i)\mid^2 \\ &+ \mid \lambda(x_i)-\mu(x_i)\mid^2)]^{\frac{1}{2}}. \end{aligned}$$

(2) *(Generalized octahedron similarity measure corresponding to d_{GNH})*

$$s_{GNH}(\mathcal{A},\mathcal{B}) = 1 - d_{GNH}(\mathcal{A},\mathcal{B}). \quad (5)$$

In fact,

$$\begin{aligned} s_{GNH}(\mathcal{A},\mathcal{B}) &= 1-[\tfrac{1}{n}\Sigma_{i=1}^{n}\max(\mid A^{-}(x_i)-B^{-}(x_i)\mid^{\delta}, \mid A^{+}(x_i)-B^{+}(x_i)\mid^{\delta}, \\ &\mid A^{\in}(x_i)-B^{\in}(x_i)\mid^{\delta}, \mid A^{\notin}(x_i)-B^{\notin}(x_i)\mid^{\delta} \\ &\mid \lambda(x_i)-\mu(x_i)\mid^{\delta})]^{\frac{1}{\delta}}. \end{aligned}$$

(Octahedron similarity measure corresponding to d_{NHH})

$$s_{NHH}(\mathcal{A},\mathcal{B}) = 1 - d_{NHH}(\mathcal{A},\mathcal{B}). \quad (6)$$

In fact,

$$\begin{aligned} s_{NHH}(\mathcal{A},\mathcal{B}) &= \tfrac{1}{n}\Sigma_{i=1}^{n}[1-\max(\mid A^{-}(x_i)-B^{-}(x_i)\mid, \mid A^{+}(x_i)-B^{+}(x_i)\mid, \\ &\mid A^{\in}(x_i)-B^{\in}(x_i)\mid, \mid A^{\notin}(x_i)-B^{\notin}(x_i)\mid, \\ &\mid \lambda(x_i)-\mu(x_i)\mid)]. \end{aligned}$$

(Octahedron similarity measure corresponding to d_{NEH})

$$s_{NEH}(\mathcal{A},\mathcal{B}) = 1 - d_{NEH}(\mathcal{A},\mathcal{B}). \quad (7)$$

In fact,

$$\begin{aligned} s_{NEH}(\mathcal{A},\mathcal{B}) &= 1-[\tfrac{1}{n}\Sigma_{i=1}^{n}\max(\mid A^{-}(x_i)-B^{-}(x_i)\mid^2, \mid A^{+}(x_i)-B^{+}(x_i)\mid^2, \\ &\mid A^{\in}(x_i)-B^{\in}(x_i)\mid^2, \mid A^{\notin}(x_i)-B^{\notin}(x_i)\mid^2, \\ &\mid \lambda(x_i)-\mu(x_i)\mid^2)]^{\frac{1}{2}}. \end{aligned}$$

(3) *(Generalized octahedron similarity measure corresponding to d_{GW})*

$$s_{GW}(\mathcal{A},\mathcal{B}) = 1 - d_{GW}(\mathcal{A},\mathcal{B}). \quad (8)$$

In fact,

$$\begin{aligned} s_{GW}(\mathcal{A},\mathcal{B}) &= 1-[\tfrac{1}{5n}\Sigma_{i=1}^{n}(\omega_i \mid A^-(x_i)-B^-(x_i)\mid^\delta + \omega_i \mid A^+(x_i)-B^+(x_i)\mid^\delta \\ &+\eta_i \mid A^\in(x_i)-B^\in(x_i)\mid^\delta + \eta_i \mid A^\notin(x_i)-B^\notin(x_i)\mid^\delta \\ &+\xi_i \mid \lambda(x_i)-\mu(x_i)\mid^\delta)]^{\frac{1}{\delta}}. \end{aligned}$$

(Octahedron similarity measure corresponding to d_{WH})

$$s_{WH}(\mathcal{A},\mathcal{B}) = 1-d_{WH}(\mathcal{A},\mathcal{B}). \tag{9}$$

In fact,

$$\begin{aligned} s_{WH}(\mathcal{A},\mathcal{B}) &= \tfrac{1}{n}\Sigma_{i=1}^{n}[1-\tfrac{1}{5}(\omega_i \mid A^-(x_i)-B^-(x_i)\mid + \omega_i \mid A^+(x_i)-B^+(x_i)\mid \\ &+\eta_i \mid A^\in(x_i)-B^\in(x_i)\mid + \eta_i \mid A^\notin(x_i)-B^\notin(x_i)\mid \\ &+\xi_i \mid \lambda(x_i)-\mu(x_i)\mid)]. \end{aligned}$$

(Octahedron similarity measure corresponding to d_{WE})

$$s_{WE}(\mathcal{A},\mathcal{B}) = 1-d_{WE}(\mathcal{A},\mathcal{B}). \tag{10}$$

In fact,

$$\begin{aligned} s_{WE}(\mathcal{A},\mathcal{B}) &= 1-[\tfrac{1}{5n}\Sigma_{i=1}^{n}(\omega_i \mid A^-(x_i)-B^-(x_i)\mid^2 + \omega_i \mid A^+(x_i)-B^+(x_i)\mid^2 \\ &+\eta_i \mid A^\in(x_i)-B^\in(x_i)\mid^2 + \eta_i \mid A^\notin(x_i)-B^\notin(x_i)\mid^2 \\ &+\xi_i \mid \lambda(x_i)-\mu(x_i)\mid^2)]^{\frac{1}{2}}. \end{aligned}$$

(4) *(Generalized octahedron similarity measure corresponding to d_{GWH})*

$$s_{GWH}(\mathcal{A},\mathcal{B}) = 1-d_{GWH}(\mathcal{A},\mathcal{B}). \tag{11}$$

In fact,

$$\begin{aligned} s_{GWH}(\mathcal{A},\mathcal{B}) &= 1-[\tfrac{1}{n}\Sigma_{i=1}^{n}\max(\omega_i \mid A^-(x_i)-B^-(x_i)\mid^\delta, \omega_i \mid A^+(x_i)-B^+(x_i)\mid^\delta, \\ &\eta_i \mid A^\in(x_i)-B^\in(x_i)\mid^\delta, \eta_i \mid A^\notin(x_i)-B^\notin(x_i)\mid^\delta, \\ &\xi_i \mid \lambda(x_i)-\mu(x_i)\mid^\delta)]^{\frac{1}{\delta}}. \end{aligned}$$

(Octahedron similarity measure corresponding to d_{WHH})

$$s_{WHH}(\mathcal{A},\mathcal{B}) = 1-d_{WHH}(\mathcal{A},\mathcal{B}). \tag{12}$$

In fact,

$$\begin{aligned} s_{WHH}(\mathcal{A},\mathcal{B}) &= \tfrac{1}{n}\Sigma_{i=1}^{n}[1-\max(\omega_i \mid A^-(x_i)-B^-(x_i)\mid, \omega_i \mid A^+(x_i)-B^+(x_i)\mid, \\ &\eta_i \mid A^\in(x_i)-B^\in(x_i)\mid, \eta_i \mid A^\notin(x_i)-B^\notin(x_i)\mid, \\ &\xi_i \mid \lambda(x_i)-\mu(x_i)\mid)]. \end{aligned}$$

(Octahedron similarity measure corresponding to d_{WEH})

$$s_{WEH}(\mathcal{A},\mathcal{B}) = 1-d_{WEH}(\mathcal{A},\mathcal{B}). \tag{13}$$

In fact,

$$\begin{aligned} s_{WEH}(\mathcal{A},\mathcal{B}) &= 1-[\tfrac{1}{n}\Sigma_{i=1}^{n}\max(\omega_i \mid A^-(x_i)-B^-(x_i)\mid^2, \omega_i \mid A^+(x_i)-B^+(x_i)\mid^2,\\ &\eta_i \mid A^{\in}(x_i)-B^{\in}(x_i)\mid^2, \eta_i \mid A^{\notin}(x_i)-B^{\notin}(x_i)\mid^2,\\ &\xi_i \mid \lambda(x_i)-\mu(x_i)\mid^2)]^{\frac{1}{2}}. \end{aligned}$$

From Propositions 1–4, and the duality between distance measures and similarity measures, we can prove that (2)–(13) are similarity measures. But, we will show directly that (3) satisfies the conditions that are defined in Definition 8.

Proposition 5. *s_{NH} is a similarity measure for two octahedron sets $\mathcal{A}$ and $\mathcal{B}$.*

Proof. Let $\mathcal{A}$, $\mathcal{B} \in \mathcal{O}(X)$, and, for each $i = 1, 2, \cdots, n$, let

$$D_i = (\mid A^-(x_i)-B^-(x_i)\mid + \mid A^+(x_i)-B^+(x_i)\mid + \mid A^{\in}(x_i)-B^{\in}(x_i)\mid .$$

(i) $0 \le s_{NH}(\mathcal{A},\mathcal{B}) \le 1$.

Case 1. Suppose $D_i = 0$ or $D_i = 5$. Then, clearly, we have

$$s_{NH}(\mathcal{A},\mathcal{B}) = 1 \text{ or } s_{NH}(\mathcal{A},\mathcal{B}) = 0. \quad (14)$$

Case 2. Suppose $0 < D_i < 5$. Then, clearly, $0 < \frac{D_i}{5} < 1$. Thus, $0 < 1 - \frac{D_i}{5} < 1$. So,

$$0 = \frac{1}{n}\Sigma_{i=1}^{n}0 < \frac{1}{n}\Sigma_{i=1}^{n}(1-\frac{D_i}{5}) < \frac{1}{n}\Sigma_{i=1}^{n}1 = 1.$$

Hence,

$$0 < s_{NH}(\mathcal{A},\mathcal{B}) < 1. \quad (15)$$

Therefore, from (14) and (15), we get $0 \le s_{NH}(\mathcal{A},\mathcal{B}) \le 1$.

(ii) $s_{NH}(\mathcal{A},\mathcal{B}) = 1$ iff $\mathcal{A} = \mathcal{B}$.

$$\begin{aligned} &s_{NH}(\mathcal{A},\mathcal{B}) = 1\\ &\Leftrightarrow \tfrac{1}{n}\Sigma_{i=1}^{n}(1-\tfrac{D_i}{5}) = 1\\ &\Leftrightarrow D_i = 0\\ &\Leftrightarrow \mid A^-(x_i)-B^-(x_i)\mid = 0, \mid A^+(x_i)-B^+(x_i)\mid = 0,\\ &\Leftrightarrow \mid A^{\in}(x_i)-B^{\in}(x_i)\mid = 0, \mid A^{\notin}(x_i)-B^{\notin}(x_i)\mid = 0, \mid \lambda(x_i)-\mu(x_i)\mid = 0\\ &\Leftrightarrow \mathcal{A} = \mathcal{B}. \end{aligned}$$

(iii) $s_{NH}(\mathcal{A},\mathcal{B}) = s_{NH}(\mathcal{B},\mathcal{A})$. The proof is obvious from the property of "| |".

(iv) Let $\mathcal{A}$, $\mathcal{B}$, $\mathcal{B} \in \mathcal{O}(X)$ such that $\mathcal{A} \subset_1 \mathcal{B} \subset_1 \mathcal{C}$. Then,

$$s_{NH}(\mathcal{A},\mathcal{C}) \le s_{NH}(\mathcal{A},\mathcal{B}) \text{ and } s_{NH}(\mathcal{A},\mathcal{C}) \le s_{NH}(\mathcal{B},\mathcal{C}).$$

For each $i = 1, 2, \cdots, n$, let $x_i \in X$. Since $\mathcal{A} \subset_1 \mathcal{B} \subset_1 \mathcal{C}$, we have

$$\begin{aligned} &A^-(x_i) \le B^-(x_i) \le C^-(x_i),\ A^+(x_i) \le B^+(x_i) \le C^+(x_i),\\ &A^{\in}(x_i) \le B^{\in}(x_i) \le C^{\in}(x_i),\ A^{\notin}(x_i) \ge B^{\notin}(x_i) \ge C^{\notin}(x_i),\\ &\lambda(x_i) \le \mu(x_i) \le \nu((x_i)). \end{aligned}$$

Then, we get $D_i(\mathcal{A},\mathcal{C}) \geq D_i(\mathcal{A},\mathcal{B})$,
where

$$\begin{aligned} D_i(\mathcal{A},\mathcal{B}) &= (\mid A^-(x_i) - B^-(x_i) \mid + \mid A^+(x_i) - B^+(x_i) \mid \\ &+ \mid A^{\in}(x_i) - B^{\in}(x_i) \mid + \mid A^{\notin}(x_i) - B^{\notin}(x_i) \mid + \mid \lambda(x_i) - \mu(x_i) \mid) \end{aligned}$$

and

$$\begin{aligned} D_i(\mathcal{A},\mathcal{C}) &= (\mid A^-(x_i) - C^-(x_i) \mid + \mid A^+(x_i) - C^+(x_i) \mid \\ &+ \mid A^{\in}(x_i) - C^{\in}(x_i) \mid + \mid A^{\notin}(x_i) - C^{\notin}(x_i) \mid + \mid \lambda(x_i) - \nu(x_i) \mid). \end{aligned}$$

Then, clearly we can easily see that $D_i(\mathcal{A},\mathcal{C}) \geq D_i(\mathcal{A},\mathcal{B})$. Thus, we have

$$1 - \frac{1}{5}D_i(\mathcal{A},\mathcal{C}) \leq 1 - \frac{1}{5}D_i(\mathcal{A},\mathcal{B}).$$

So, we get $\frac{1}{n}\Sigma_{i=1}^{n}[1 - \frac{1}{5}D_i(\mathcal{A},\mathcal{C})] \leq \frac{1}{n}\Sigma_{i=1}^{n}[1 - \frac{1}{5}D_i(\mathcal{A},\mathcal{B})]$. Hence, we get

$$s_{NH}(\mathcal{A},\mathcal{C}) \leq s_{NH}(\mathcal{A},\mathcal{B}).$$

Similarly, we can prove that $s_{NH}(\mathcal{A},\mathcal{C}) \leq s_{NH}(\mathcal{B},\mathcal{C})$. Therefore, s_{NH} a similarity measure on $\mathcal{O}(X)$.

□

4. MCGDM Method Based on Similarity Measure in Octahedron Set Environment

In this section, we give a new method based on similarity measure in octahedron set environment. Assume that $\alpha = \{\alpha_1, \alpha_2, \cdots, \alpha_n\}$ is a set of n alternatives with criteria $\beta = \{\beta_1, \beta_2, \cdots, \beta_m\}$, and let $\gamma = \{\gamma_1, \gamma_2, \cdots, \gamma_r\}$ be the r decision-makers. Let $\delta = \{\delta_1, \delta_2, \cdots, \delta_r\}$ be the weight vector of decision-makers such that $\delta_k > 0$ and $\Sigma_{k=1}^{r}\delta_k = 1$. We propose MCGDM method presented using the following steps.

Step 1. Formation of ideal octahedron set decision matrix. Ideal octahedron set decision matrix is an important matrix for similarity measure of MCGDM given in the following form:

$$\begin{bmatrix} & \beta_1 & \beta_2 & \cdots & \beta_m \\ \alpha_1 & \mathcal{A}_{11} & \mathcal{A}_{12} & \cdots & \mathcal{A}_{1m} \\ \alpha_2 & \mathcal{A}_{21} & \mathcal{A}_{22} & \cdots & \mathcal{A}_{2m} \\ \cdot & \cdot & \cdot & \cdots & \cdot \\ \alpha_n & \mathcal{A}_{n1} & \mathcal{A}_{n2} & \cdots & \mathcal{A}_{nm} \end{bmatrix}, \tag{16}$$

where $\mathcal{A}_{ij} = \langle \mathbf{A}_{ij}, A_{ij}, \lambda_{ij} \rangle$, $i = 1, 2, \cdots, n, j = 1, 2, \cdots, m$.

Step 2. Construction of octahedron set decision matrix. Since r decision-makers are involved in the decision-making process, the k-th ($k = 1, 2, \cdots, r$) decision-maker gives the evaluation information of the alternative α_i ($i = 1, 2, \cdots, n$) with respect to criteria β_j ($j = 1, 2, \cdots, m$) in terms of octahedron set. The k-th decision matrix, denoted by M^k, is constructed by the following matrix:

$$M^k = \left\langle \mathcal{A}_{ij}^k \right\rangle = \begin{bmatrix} & \beta_1 & \beta_2 & \cdots & \beta_m \\ \alpha_1 & \mathcal{A}_{11}^k & \mathcal{A}_{12}^k & \cdots & \mathcal{A}_{1m}^k \\ \alpha_2 & \mathcal{A}_{21}^k & \mathcal{A}_{22}^k & \cdots & \mathcal{A}_{2m}^k \\ \cdot & \cdot & \cdot & \cdots & \cdot \\ \alpha_n^k & \mathcal{A}_{n1}^k & \mathcal{A}_{n2}^k & \cdots & \mathcal{A}_{nm}^k \end{bmatrix}, \tag{17}$$

where $k = 1, 2, \cdots, r$, $i = 1, 2, \cdots, n$ and $j = 1, 2, \cdots, m$.

Step 3. Determination of attribute weight. All attributes are not equally important in a decision-making situation. Every decision-maker provides their own opinion regarding to the attribute weight in terms of linguistic variables that can be converted into octahedron set. Let $w_k(\beta_j)$ denote the attribute weight for the attribute β_j given by the k-th decision-maker in terms of octahedron set. We convert into $w_k(\beta_j)$ into fuzzy number as follows:

$$w_k^F(\beta_j) = \begin{cases} [1 - (\frac{V_{kj}}{5})^{\frac{1}{2}}] & \text{if } \beta_j \in \beta \\ 0 & \text{otherwise,} \end{cases} \tag{18}$$

where

$$\begin{aligned} V_{kj} &= [(1 - A^-(\beta_j))^2 + (1 - A^+(\beta_j))^2 + (1 - A^{\in}(\beta_j))^2 + (A^{\notin}((\beta_j))^2 \\ &+ (1 - \lambda(\beta_j))^2]^{\frac{1}{2}} \end{aligned},$$

and each of the above values denote the value of the octahedron set corresponding to (k, β_j).

Then, aggregate weight for the criteria β_j can be determined as follows:

$$W_j = \frac{[1 - \Pi_{k=1}^r (1 - w_k^F(\beta_j))]}{\Sigma_{k=1}^r [1 - \Pi_{k=1}^r (1 - w_k^F(\beta_j))]}, \tag{19}$$

where $\Sigma_{k=1}^r W_j = 1$.

Step 4. Calculation of weighted similarity measure. We calculate weighted similarity measure between the ideal matrix M and the k-th decision matrix M^k as follows:

$$s_{NH}^W(M, M^k) = \left\langle \lambda_i^k \right\rangle = (\lambda_1^k, \lambda_2^k, \cdots, \lambda_n^k)^T = \left[\frac{1}{m}\Sigma_{j=1}^m (1 - \frac{D_{ij}^k}{5}) W_j\right]_{i=1}^n, \tag{20}$$

where $D_{ij}^k = | A_{ij}^-(x_r) - A_{ij}^{k,-}(x_r) | + | A_{ij}^+(x_r) - A_{ij}^{k,+}(x_r) | + | A_{ij}^{\in}(x_r) - A_{ij}^{k,\in}(x_r) |$
$+ | A_{ij}^{\notin}(x_r) - A_{ij}^{k,\notin}(x_r) | + | \lambda(x_r) - \lambda(x_r) |$ for each $x_r \in X$ and $k = 1, 2, \cdots, r$.

Step 5. Ranking of alternatives. In order to rank alternatives, we give the following formula:

$$\rho_i = \Sigma_{k=1}^r \delta_k \lambda_i^k, \tag{21}$$

where $i = 1, 2, \cdots, n$.

We can arrange alternatives according to the descending order values of ρ_i. The highest value of ρ_i reflects the best alternative.

Example 4 (Numerical example). *In order to solve a MCGDM problem, we adapt "Illustrative example" given by Ye [11] to demonstrate the applicablility and effectiveness of the proposed method. Assume that an investment company wants to invest a sum of money in the best option. The investment company is composed of a decision-making committee comprised of three members, say k_1, k_2, k_3 to make a panel of four alternatives to invest money. The alternatives Car company (α_1), Food company (α_2), Computer company (α_3), and Arm company (α_4). Decision-makers take decision based on the criteria, namely risk analysis (β_1), growth analysis (β_2), environment impact (β_3), and criteria weights, which are given by the decision-makers in terms of linguistic variables that can be converted into octahedron set (see Table 1).*

Table 1. Linguistic term for rating of attribute/criterion.

Linguistic Terms	Octahedron Set
Very important (VI)	$\langle [0.7, 0.9], (0.7, 0.2), 0.9 \rangle$
Important (I)	$\langle [0.6, 0.8], (0.6, 0.3), 0.6 \rangle$
Medium (M)	$\langle [0.4, 0.5], (0.5, 0.4), 0.5 \rangle$
Unimportant (UI)	$\langle [0.2, 0.4], (0.3, 0.6), 0.4 \rangle$
Very unimportant (VUI)	$\langle [0.1, 0.2], (0.2, 0.7), 0.2 \rangle$

Step 1. Formation of ideal octahedron set decision matrix. *Ideal octahedron set decision matrix M is given as follows:*

$$M = \begin{bmatrix} & \beta_1 & \beta_2 & \beta_3 \\ \alpha_1 & \langle[1,1],(1,0),1\rangle & \langle[1,1],(1,0),1\rangle & \langle[1,1],(1,0),1\rangle \\ \alpha_2 & \langle[1,1],(1,0),1\rangle & \langle[1,1],(1,0),1\rangle & \langle[1,1],(1,0),1\rangle \\ \alpha_3 & \langle[1,1],(1,0),1\rangle & \langle[1,1],(1,0),1\rangle & \langle[1,1],(1,0),1\rangle \\ \alpha_4 & \langle[1,1],(1,0),1\rangle & \langle[1,1],(1,0),1\rangle & \langle[1,1],(1,0),1\rangle \end{bmatrix} \quad (22)$$

Step 2. Construction of octahedron set decision matrix. *The k_i-th decision matrix M^{k_i} $(i = 1, 2, 3)$ in octahedron set form is constructed for four alternatives with respect to the three criteria by the following matrix:*

$$M^{k_1} = \begin{bmatrix} & \beta_1 & \beta_2 & \beta_3 \\ \alpha_1 & \langle[0.7,0.9],(0.7,0.2),0.9\rangle & \langle[0.7,0.9],(0.7,0.2),0.9\rangle & \langle[0.4,0.5],(0.5,0.4),0.5\rangle \\ \alpha_2 & \langle[0.6,0.8],(0.6,0.3),0.8\rangle & \langle[0.4,0.5],(0.5,0.4),0.5\rangle & \langle[0.7,0.9],(0.7,0.2),0.9\rangle \\ \alpha_3 & \langle[0.4,0.5],(0.5,0.4),0.5\rangle & \langle[0.6,0.8],(0.6,0.3),0.8\rangle & \langle[0.4,0.5],(0.5,0.4),0.5\rangle \\ \alpha_4 & \langle[0.3,0.4],(0.4,0.5),0.4\rangle & \langle[0.4,0.5],(0.5,0.4),0.5\rangle & \langle[0.7,0.9],(0.7,0.2),0.9\rangle \end{bmatrix},$$

$$M^{k_2} = \begin{bmatrix} & \beta_1 & \beta_2 & \beta_3 \\ \alpha_1 & \langle[0.3,0.4],(0.4,0.5),0.4\rangle & \langle[0.4,0.5],(0.5,0.4),0.5\rangle & \langle[0.7,0.9],(0.7,0.2),0.9\rangle \\ \alpha_2 & \langle[0.4,0.5],(0.5,0.4),0.5\rangle & \langle[0.4,0.5],(0.5,0.4),0.5\rangle & \langle[0.7,0.9],(0.7,0.2),0.9\rangle \\ \alpha_3 & \langle[0.7,0.9],(0.7,0.2),0.9\rangle & \langle[0.7,0.9],(0.7,0.2),0.9\rangle & \langle[0.4,0.5],(0.5,0.4),0.5\rangle \\ \alpha_4 & \langle[0.6,0.8],(0.6,0.3),0.8\rangle & \langle[0.4,0.5],(0.5,0.4),0.5\rangle & \langle[0.7,0.9],(0.7,0.2),0.9\rangle \end{bmatrix},$$

$$M^{k_3} = \begin{bmatrix} & \beta_1 & \beta_2 & \beta_3 \\ \alpha_1 & \langle[0.4,0.5],(0.5,0.4),0.5\rangle & \langle[0.4,0.5],(.5,.4),0.5\rangle & \langle[0.7,0.9],(0.7,0.2),0.9\rangle \\ \alpha_2 & \langle[0.4,0.5],(0.5,0.4),0.5\rangle & \langle[0.7,0.9],(.7,.2),0.9\rangle & \langle[0.4,0.5],(0.5,0.4),0.5\rangle \\ \alpha_3 & \langle[0.7,0.9],(0.7,0.2),0.9\rangle & \langle[0.6,0.8],(0.6,0.3),0.8\rangle & \langle[0.6,0.8],(0.6,0.3),0.8\rangle \\ \alpha_4 & \langle[0.7,0.9],(0.7,0.2),0.9\rangle & \langle[0.4,0.5],(0.5,0.4),0.5\rangle & \langle[0.3,0.4],(0.4,0.5),0.4\rangle \end{bmatrix}.$$

Step 3. Determination of attribute weight. *Linguistic terms given in Table 1 are used to evaluate each attribute. The importance of each attribute for every decision-maker is rated with linguistic terms (see Table 2). Moreover, each linguistic term is converted into octahedron set (see Table 3).*

Table 2. Attribute rating linguistic variables.

	β_1	β_2	β_3
k_1	VI	M	I
k_2	VI	VI	M
k_3	M	VI	M

Table 3. Attribute rating in octahedron set.

	β_1	β_2	β_3
k_1	$\langle[0.7,0.9],(0.7,0.2),0.9\rangle$	$\langle[0.4,0.5],(0.5,0.4),0.5\rangle$	$\langle[0.6,0.8],(0.6,0.3),0.8\rangle$
k_2	$\langle[0.7,0.9],(0.7,0.2),0.9\rangle$	$\langle[0.7,0.9],(0.7,0.2),0.9\rangle$	$\langle[0.4,0.5],(0.5,0.4),0.5\rangle$
k_3	$\langle[0.4,0.5],(0.5,0.4),0.5\rangle$	$\langle[0.7,0.9],(0.7,0.2),0.9\rangle$	$\langle[0.4,0.5],(0.5,0.4),0.5\rangle$

By using Equations (18) and (19), we get the following attribute weights:

$$W_1 = W_2 = W_3 = 0.33. \quad (23)$$

Step 4. Calculation of weighted similarity measures. *By using Formula (20), we obtain weighted similarity measures between the ideal matrix M and the k_s-th decision matrix M^{k_s} ($s = 1, 2, 3$) as follows:*

$$s^W_{NH}(M, M^{k_1}) = \begin{bmatrix} 0.205 \\ 0.207 \\ 0.187 \\ 0.178 \end{bmatrix}, s^W_{NH}(M, M^{k_2}) = \begin{bmatrix} 0.178 \\ 0.185 \\ 0.218 \\ 0.220 \end{bmatrix}, s^W_{NH}(M, M^{k_3}) = \begin{bmatrix} 0.211 \\ 0.211 \\ 0.229 \\ 0.187 \end{bmatrix}, \quad (24)$$

Step 5. Ranking of alternatives. *In order to rank the alternatives according to the descending value of ρ_i, by using Equations (22)–(24), we obtain ρ_i ($i = 1, 2, 3, 4$):*

$$\rho_1 = 0.196, \ \rho_2 = 0.199, \ \rho_3 = 0.232, \ \rho_4 = 0.193.$$

Then, $\rho_3 > \rho_2 > \rho_1 > \rho_4$. Thus, the ranking order is as follows:

$$\alpha_3 > \alpha_2 > \alpha_1 > \alpha_4.$$

So, we can see that Computer company (α_3) is the best alternative for money investment.

5. Conclusions

With this paper, we wished to renew an interest in the systematic study of the relationships between multi-criteria group decision-making (MCGDM) method with respect to octahedron set theory. For this purpose, various distance and similarity measures for octahedron sets were defined, and some of their properties were proved. The usefulness and interest of this correspondence of new defined distance and similarity measures will of course be enhanced if there is a way of returning from the transforms, that is to say, if there is a new method that characterize our proposed similarity measure. In Section 4, all the studies came to fruition, and we took up a result, MCGDM method based on a similarity measure for octahedron sets environments, which plays a pivotal role for demonstrating the usefulness of giving numerical examples. The detailed application of MCGDM method was carried out by introducing a a numerical example in the closing of Section 4. It considered some of the new results and consequences, which could be useful from the point of view of octahedron set theory, which were not studied at all. All these findings will provide a base to researchers who want to work in the field of the application of octahedron sets and will help to strengthen the foundations of the other MCGDM problems in octahedron set environment, such as economic policy, foreign policy between countries, trade policy, financial policy, etc., by using big data.

Author Contributions: Conceptualization, J.-G.L. and K.H.; funding acquisition, J.-G.L.; writing, original draft, J.-G.L. and K.H.; Writing, review and editing, G.Ş. All authors have read and agreed to the published version of the manuscript.

Funding: This research was supported by Basic Science Research Program through the National Research Foundation of Korea (NRF) funded by the Ministry of Education (2018R1D1A1B07049321).

Acknowledgments: The authors are grateful to the reviewers for their valuable suggestions and comments in rewriting the article in the present form. The authors are also thankful to the Editor-in-Chief of the journal for his valuable comments.

Conflicts of Interest: The authors declare no conflict of interest.

References

1. Chen, S.M.; Hsiao, P.H. A comparison of similarity measures of fuzzy values. *Fuzzy Sets Syst.* **1995**, *72*, 79–89. [CrossRef]
2. Zadeh, L.A. Fuzzy sets. *Inf. Control.* **1965**, *8*, 338–353. [CrossRef]
3. Pramanik, S.; Mondal, K. Weighted fuzzy similarity measure based on tangent function and its application to medical diagnosis. *Int. J. Innov. Res. Sci. Eng. Technol.* **2015**, *4*, 158–164.

4. Hwang, C.M.; Yang, S.M. A new construction for similarity measures between intuitionistic fuzzy sets based on lower, upper and middle fuzzy fuzzy sets. *Int. J. Fuzzy Syst.* **2013**, *15*, 371–378.
5. Atanassov, K.T. Intuitionistic fuzzy sets. *Fuzzy Sets Syst.* **1986**, *20*, 87–96. [CrossRef]
6. Pramanik, S.; Mondal, K. Intuitionistic fuzzy similarity measure based on tangent function and its application to multi–attribute decision. *Glob. J. Adv. Res.* **2015**, *2*, 464–471.
7. Ren, H.; Wang, G. An interval-valued intuitionistic fuzzy MADM method based on a new similarity measure. *Information* **2015**, *6*, 880–894. [CrossRef]
8. Atanassov, K.T.; Gargov, G. Interval-valued intuitionistic fuzzy sets. *Fuzzy Sets Syst.* **1989**, *31*, 343–349. [CrossRef]
9. Broumi, L.; Smarandache, F. Several similarity measures of neutrosophic sets and their decision-making. *arXiv* **2013**, arXiv:1301.0456vI.
10. Smarandache, F. *Neutrosophy, Neutrosophic Probability, and Logic*; American Research Press: Rehoboth, DE, USA, 1998.
11. Ye, J. Similarity measures between interval neutrosophic sets and their multi-criteria decision-making method. *J. Intell. Fuzzy Syst.* **2014**, *26*, 167–172.
12. Sahin, R.; Liu, P. Distance and similarity measures for multiple attribute decision-making with single-valued neutrosophic hesitant fuzzy information. *New Trends Neutrosophic Theory Appl.* **2016**, 35–54.
13. Torra, V. Hesitant fuzzy sets. *Int. J. Intell. Syst.* **2010**, *25*, 529–539. [CrossRef]
14. Wang, H.; Smarandache, F.; Zhang, Y.Q.; Sunderraman, R. Single valued neutrosophic sets. *Multispace Multistruct.* **2010**, *4*, 410–413.
15. Kaur, G.; Garg, H. Multi-attribute decision-making based on bonferroni mean operators under cubic intuitionistic fuzzy environment. *Entropy* **2018**, *20*, 65. [CrossRef]
16. Kaur, G.; Garg, H. Cubic intuitionistic fuzzy aggregation operators. *Int. J. Uncertain. Quantif.* **2018**, *8*, 405–428. [CrossRef]
17. Jun, Y.B. A novel extension of cubic sets and its applications in BCI/BCK-algebras. *Ann. Fuzzy Math. Inform.* **2017**, *14*, 475–486.
18. Jun, Y.B.; Kim, C.S.; Yang, K.O. Cubic sets. *Ann. Fuzzy Math. Inform.* **2012**, *4*, 83–98.
19. Pramanik, S.; Dalapati, S.; Alam, S.; Roy, T.K.; Smarandach, F. Neutrosophic cubic MCGDM method based on similarity measure. *Neutrosophic Sets Syst.* **2017**, *16*, 44–56.
20. Kim, J.; Senel, G.; Lim, P.K.; Lee, J.G.; Hur, K. Octahedron sets. *Ann. Fuzzy Math. Inform.* **2020**, *19*, 211–238.
21. Cheong, M.; Hur, K. Intuitionistic interval-valued fuzzy sets. *J. Korean Inst. Intell. Syst.* **2010**, *20*, 864–874. [CrossRef]
22. Gorzalczany, M.B. A method of inference in approximate reasoning based on interval-valued fuzzy sets. *Fuzzy Sets Syst.* **1987**, *21*, 1–17. [CrossRef]
23. Zadeh, L.A. The concept of a linguistic variable and its application to approximate reasoning-I. *Inform. Sci.* **1975**, *8*, 199–249. [CrossRef]

mathematics

Article

A New Continuous-Discrete Fuzzy Model and Its Application in Finance

Hoang Viet Long [1,2], Haifa Bin Jebreen [3,*] and Y. Chalco-Cano [4]

1 Division of Computational Mathematics and Engineering, Institute for Computational Science, Ton Duc Thang University, Ho Chi Minh City 70000, Vietnam; hoangvietlong@tdtu.edu.vn
2 Faculty of Mathematics and Statistics, Ton Duc Thang University, Ho Chi Minh City 758307, Vietnam
3 Department of Mathematics, College of Science, King Saud University, P.O. Box 2455, Riyadh 11451, Saudi Arabia
4 Departamento de Matemática, Universidad de Tarapacá, Casilla 7D, Arica 09010, Chile; ychalco@uta.cl
* Correspondence: hjebreen@ksu.edu.sa

Received: 19 September 2020; Accepted: 5 October 2020; Published: 16 October 2020

Abstract: In this paper, we propose a fuzzy differential-difference equation for modeling of mixed continuous-discrete phenomena. In the special case, we present the general solution of linear fuzzy differential-difference equations. The dynamical process in the intervals is presented by the corresponding fuzzy differential equation and with impulsive jumps in some points. We illustrate the applicability of the model to study the time value of money.

Keywords: fuzzy differential equations; fuzzy difference equations; mixed continuous-discrete model; strongly generalized Hukuhara differentiability; time value of money

1. Introduction

Differential and difference equations play a relevant role in modeling problems that arise in physics, engineering, biology, economics, finance, and many other areas. However, in some cases, these equations are restricted in their ability to describe phenomena due to the imprecise or incomplete information about the parameters, variables and initial conditions available. This can result from errors in measurement, observation, or experimental data; application of different operating conditions; or maintenance induced errors [1]. To overcome uncertainties or lack of precision, one can use a fuzzy environment in parameters, variables and initial conditions in place of exact (fixed) ones, by turning general differential and difference equations into fuzzy differential and difference equations, respectively [1–7]. These uncertainties may be modeled by fuzzy set theory when an abundance of data is not available. Accordingly, there is often a need to model, solve and interpret the problems one encounters in the world of uncertainty. The governing differential and difference equations will then become uncertain. Therefore, recently many researchers have studied fuzzy differential equations and fuzzy difference equations in different approaches [8–12].

Fuzzy set theory refers to the uncertainty when we have a lack of knowledge or incomplete information about the variables and parameters. In the financial markets there are elements of uncertainties and lack of precision associates to fluctuation and votality of financial markets. We cannot make forecasts easily, we have incomplete information or some type of uncertainty, about the values of financial factors such as taxes, inflation, interest rate, price change rate among other factors [4,13,14]. In this direction, some problems of the financial field can be approached via fuzzy difference equations [4]. In particular, Papadopoulos at al. in Reference [4] demonstrated the applicability of fuzzy difference equations to the problems of time value of money. The results obtained in the article [4] were motivated by models introduced by Kwapisz in Reference [15], where several difference equations to study the basic problems of finance such as capital deposits and capital investments were presented.

Although small discrete systems are easy to work with, the continuous models are easier to deal with than large discrete systems. Whether or not nature is fundamentally discrete, the most useful models are often continuous because the discreteness can only occur in very small scales. Discrete models are probably useful if nature has genuinely discrete structure. But on larger scales a discrete model would contain some parameters that we cannot measure and might not even be interested in. This is related to the observation that continuous models often work well for large discrete systems. Discreteness is useful to include in the model if it occurs in the situation we are interested in. Therefore, a mixed continuous-discrete model and, in a special case, a differential-difference equation can possess the inherent properties and advantages of both discrete and continuous models which are useful for modeling of real-world phenomena. In this direction, Kwapisz in Reference [15] introduced and studied a general mixed continuous-discrete model describing dynamical processes in some problems that arise in finance. In this model, the dynamical process in each interval is presented by the corresponding differential equation and it displays impulsive jumps, obtained by the corresponding difference equation. Therefore, proposing a mixed continuous-discrete model is a natural way to study a phenomenon which is continuous on some sub-intervals and it has discontinuities in some points.

Motived by the results obatined in Reference [15], on continuous-discrete models, and the recent advances on fuzzy differential equations [2,8], in this article we introduce fuzzy differential-difference equations. We present some results on existence and uniqueness of solutions for this class of models. Finally, we give an example on the time value of money to demonstrate the effectiveness of theoretical results.

2. Preliminaries

We start by recalling some preliminaries about the fuzzy sets defined on $\mathbb{R}$. A fuzzy set on $\mathbb{R}$ is a mapping $u\colon \mathbb{R} \to [0,1]$, where the value $u(x)$ denotes the degree of membership of the element x to the fuzzy set u. For $0 < \alpha \leq 1$, the α-level of u is defined by the set $[u]^\alpha = \{x \in \mathbb{R} \mid u(x) \geq \alpha\}$. For $\alpha = 0$, the support of u is defined as the set $[u]^0 = \text{supp}(u) = \overline{\{x \in \mathbb{R} \mid u(x) > 0\}}$. We denote

$$\mathbb{R}_F = \{u : \mathbb{R} \to [0,1] | u \text{ satisfies } (i) - (iv) \text{ below}\},$$

where

(i) u is normal, that is, there exists $x_0 \in \mathbb{R}$ such that $u(x_0) = 1$.
(ii) u is fuzzy convex, that is, $u(\lambda x + (1-\lambda)y) \geq \min\{u(x), u(y)\}$, for any $x, y \in \mathbb{R}$ and $0 \leq \lambda \leq 1$.
(iii) u is upper semicontinuous.
(iv) $[u]^0$ is compact.

If $u \in \mathbb{R}_F$, we say that u is a fuzzy number.

According to Zadeh's Extension Principle [2], operations of addition and scalar multiplication on $\mathbb{R}_F$ are defined as:

$$(u+v)(x) = \sup_{y+z=x} \min\{u(y), v(z)\}, \qquad \text{and} \qquad (\lambda u)(x) = \begin{cases} u(\frac{x}{\lambda}) & \lambda \neq 0, \\ \chi_{\{0\}}(x) & \lambda = 0, \end{cases}$$

where $\chi_{\{0\}}$ is the characteristic function of $\{0\}$. Moreover, the following relations hold:

$$[u+v]^\alpha = [u]^\alpha + [v]^\alpha, \qquad \text{and} \qquad [\lambda u]^\alpha = \lambda [u]^\alpha, \quad \forall u, v \in \mathbb{R}_F, \quad \forall \alpha \in [0,1].$$

Definition 1. *Let $u, v, w \in \mathbb{R}_F$. An element w is called the Hukuhara difference (H-difference, for short) of u and v, if it verifies the equation $u = v + w$. If the H-difference exists, it will be denoted by $u \ominus_H v$. Clearly, $u \ominus_H u = \{0\}$, and if $u \ominus_H v$ exists, it is unique.*

The space $\mathbb{R}_F$ is a complete metric space with the distance $D(u,v)$ given by

$$D(u,v) = \sup_{\alpha\in[0,1]} d([u]^\alpha,[v]^\alpha), \quad \forall u,v \in \mathbb{R}_F,$$

where $d(\cdot,\cdot)$ is the well known Pompeiu-Hausdorff distance on the space $\mathcal{K}_c^n$ of all nonempty, compact and convex subsets of the n-dimensional Euclidean space $\mathbb{R}^n$.

We need the following theorem in this paper.

Theorem 1 ([2]). *(i) For any $u,v,w \in \mathbb{R}_F$, we have*

$$[(u+v)w]^\alpha \subseteq [uw]^\alpha + [vw]^\alpha, \ \forall \alpha \in [0,1],$$

and, in general, distributivity does not hold.

(ii) For any $u,v,w \in \mathbb{R}_F$ such that none of the supports of u,v,w contain 0, we have

$$u(vw) = (uv)w.$$

In the sequel, we fix $I = (0,T)$, for $T \in \mathbb{R}$. There are several approaches to study fuzzy differential equations [10,12,13,16–19]. In the following, we use the generalized Hukuhara differentiability concept of fuzzy functions [3,8].

Definition 2. *Let $F : I \to \mathbb{R}_F$ and $t_0 \in I$ be fixed. Then, we say that F is differentiable at t_0 if there exists an element $F'(t_0) \in \mathbb{R}_F$ such that either*

(i) For all $h > 0$ sufficiently small, the H-differences $F(t_0+h) \ominus F(t_0), F(t_0) \ominus F(t_0-h)$ exist and the limits (in the metric D)

$$\lim_{h\to 0^+} \frac{F(t_0+h) \ominus F(t_0)}{h} = \lim_{h\to 0^+} \frac{F(t_0) \ominus F(t_0-h)}{h} = F'(t_0),$$

or

(ii) For all $h > 0$ sufficiently small, the H-differences $F(t_0) \ominus F(t_0+h), F(t_0-h) \ominus F(t_0)$ exist and the limits (in the metric D)

$$\lim_{h\to 0^+} \frac{F(t_0) \ominus F(t_0+h)}{-h} = \lim_{h\to 0^+} \frac{F(t_0-h) \ominus F(t_0)}{-h} = F'(t_0).$$

We say that F is (i)-differentiable on I if F is differentiable in the sense (i) of Definition 2. Similarly, we say that F is (ii)-differentiable on I if F is differentiable in the sense (ii) of Definition 2. In this paper, we make use of the following theorem [10].

Theorem 2 ([10]). *Let $F : I \to \mathbb{R}_F$ be a fuzzy function such that $[F(t)]^\alpha = [f_\alpha(t), g_\alpha(t)]$ for each $\alpha \in [0,1]$. Then, we have*

(i) If F is (i)-differentiable, then f_α and g_α are differentiable functions and we have

$$[F'(t)]^\alpha = [f_\alpha'(t), g_\alpha'(t)].$$

(ii) If F is (ii)-differentiable, then f_α and g_α are differentiable functions and we have

$$[F'(t)]^\alpha = [g_\alpha'(t), f_\alpha'(t)].$$

Let us consider the initial value problem to fuzzy differential equation

$$\begin{cases} y'(t) &= f(t, y(t)), \qquad t \in I, \\ y(0) &= y_0, \end{cases} \tag{1}$$

where $f : I \times \mathbb{R}_F \to \mathbb{R}_F$ is a continuous fuzzy mapping and y_0 is a fuzzy number. It is well known from Reference [10] that the sufficient conditions for the existence and uniqueness of the (i)-differentiable solution to the initial value problem (1) are

(a) The fuzzy mapping f is continuous on $I \times \mathbb{R}_F$;
(b) The fuzzy mapping f satisfies Lipschitz condition

$$D(f(t,u), f(t,v) \leq LD(u,v), \;\; L > 0, \; \forall u, v \in \mathbb{R}_F, t \in I.$$

In Reference [2], the sufficient conditions for the unique existence of the (ii)-differentiable solution to the initial value problem (1) are presented.

Now, we consider the first-order fuzzy linear differential equation

$$\begin{cases} y'(t) &= a(t)y(t) + b(t), \qquad t \in I, \\ y(0) &= y_0, \end{cases} \tag{2}$$

where a, $b : I \to \mathbb{R}$ are fuzzy mappings and $y_0 \in \mathbb{R}_F$ is the fuzzy initial condition. The initial value problem (2) was studied in Reference [8] by Bede, Rudas and Bencsik. They have presented the general solution of the problem in some special cases. Later, in Reference [20], the authors have presented the solution of the problem (2) with general conditions.

Theorem 3 ([8]). *Consider the initial value problem (2). Then, we have*

(a) If $a > 0$, then the (i)-differentiable solution to the problem (2) is given by

$$y(t) = e^{\int_0^t a(u)du} \left(y_0 + \int_0^t b(s) e^{-\int_0^s a(u)du} ds \right).$$

(b) If $a < 0$, then the (ii)-differentiable solution to the problem (2) is given by

$$y(t) = e^{\int_0^t a(u)du} \left(y_0 \ominus (-1) \int_0^t b(s) e^{-\int_0^s a(u)du} ds \right),$$

provided the H-difference exists.

Fuzzy difference equation is a difference equation whose parameters or initial data are fuzzy numbers and its solutions are given in the form of fuzzy number sequences. Due to the applicability of fuzzy difference equations in the analysis of phenomena where imprecision is inherent, this class of difference equations is an interesting topic from theoretical point of view. Deeba et al. [9] have studied the first-order fuzzy difference equation $x_{n+1} = wx_n + q$, $n = 0, 1, \ldots$ to investigate the population genetics, where $\{x_n\}$ is a sequence of positive fuzzy numbers and $w, q, x_0 \in \mathbb{R}_F^+$. In Reference [21], Papaschinopoulos et. al. studied the existence and some related properties of the positive solutions of the fuzzy difference equation $x_{n+1} = A + B/x_n, n = 0, 1, \ldots,$ where $\{x_n\}$ is a sequence of positive fuzzy numbers and $A, B \in \mathbb{R}_F^+$.

In the following, we consider the first-order fuzzy difference equation

$$z_{n+1} = \mu_n z_n + \nu_n, \;\; n = 0, 1, 2, \ldots \tag{3}$$

where $\{\mu_n\}$ and $\{\nu_n\}$ are sequences of positive fuzzy numbers and $z_0 \in \mathbb{R}_F^+$. The difference Equation (3) is a generalization of the following fuzzy difference equations, studied in Reference [4]

$$F_{n+1} = F_n + IF_0, \text{ and } F_{n+1} = F_n(I' + 1) + b_n, \ n = 0, 1, 2, \ldots$$

where $I, I', b_n \in \mathbb{R}_F^+$. Furthermore, the Equation (3) is also a generalization of the fuzzy difference equation $x_{n+1} = wx_n + q$, which was studied in Reference [5,9]. In the following, we study the existence of positive solution to the difference Equation (3).

Since μ_n, ν_n, z_0 are positive fuzzy numbers for each $n = 0, 1, \ldots$, then the $\alpha-$cuts of z_{n+1} is given by

$$[\underline{z}_{n+1}, \overline{z}_{n+1}]^\alpha = [\underline{\mu}_n \underline{z}_n + \underline{\nu}_n, \overline{\mu}_n \overline{z}_n + \overline{\nu}_n], \qquad n = 0, 1, 2, \ldots$$

Then, we have two classical difference equations

$$\underline{z}_{n+1} = \underline{\mu}_n \underline{z}_n + \underline{\nu}_n, \ n = 0, 1, 2, \ldots$$

and

$$\overline{z}_{n+1} = \overline{\mu}_n \overline{z}_n + \overline{\nu}_n, \ n = 0, 1, 2, \ldots$$

Therefore, by using the results of classic difference equations [22], we have

$$\underline{z}_n = \underline{z}_0 \prod_{i=0}^{n-1} \underline{\mu}_i + \sum_{i=0}^{n-1} \underline{\nu}_i \prod_{j=i+1}^{n-1} \underline{\mu}_j, \qquad n = 0, 1, \ldots$$

and

$$\overline{z}_n = \overline{z}_0 \prod_{i=0}^{n-1} \overline{\mu}_i + \sum_{i=0}^{n-1} \overline{\nu}_i \prod_{j=i+1}^{n-1} \overline{\mu}_j, \ n = 0, 1, \ldots$$

Consequently, we obtain

$$[\underline{z}_n, \overline{z}_n]^\alpha = \left[\underline{z}_0 \prod_{i=0}^{n-1} \underline{\mu}_i + \sum_{i=0}^{n-1} \underline{\nu}_i \prod_{j=i+1}^{n-1} \underline{\mu}_j, \overline{z}_0 \prod_{i=0}^{n-1} \overline{\mu}_i + \sum_{i=0}^{n-1} \overline{\nu}_i \prod_{j=i+1}^{n-1} \overline{\mu}_j \right].$$

Additionally, since μ_n, ν_n, z_0 are positive fuzzy numbers, we have the following result

Theorem 4. *For each $n \in \mathbb{N}$, let $z_0, \nu_n, \mu_n \in \mathbb{R}_F^+$. Then, the positive solution of the first-order fuzzy difference Equation (3) is given by*

$$z_n = z_0 \prod_{i=0}^{n-1} \mu_i + \sum_{i=0}^{n-1} \nu_i \prod_{j=i+1}^{n-1} \mu_j, \qquad n = 0, 1, \ldots$$

There are various methods to compare and arrange fuzzy numbers. In the theoretical point of view, the set of fuzzy numbers can only be partially ordered and hence, it cannot be compared. However, in practical applications such as decision making, scheduling, market analysis or optimization with fuzzy uncertainties, the comparison of fuzzy numbers becomes crucial [23]. In this study, we use the following definition for ordering fuzzy numbers.

Definition 3. *For each $u, v \in \mathbb{R}_F$, we say that the fuzzy number u is greater than the fuzzy number v, denoted by $u \gg v$, if and only if $\underline{u}^\alpha > \underline{v}^\alpha$ and $\overline{u}^\alpha > \overline{v}^\alpha$ for all $\alpha \in [0, 1]$.*

It is well-known that $u \in \mathbb{R}_F^+$ if and only if $u \gg \tilde{0}$, that is, $\overline{u}^\alpha \geq \underline{u}^\alpha > 0$ for all $\alpha \in [0,1]$, where $\tilde{0} = \chi_{\{0\}}$. Similarly, $u \in \mathbb{R}_F^-$ if and only if $\tilde{0} \gg u$, that is, $\underline{u}^\alpha \leq \overline{u}^\alpha < 0$ for all $\alpha \in [0,1]$.

Proposition 1. *Let $u, v, w \in \mathbb{R}_F^+$ and $v \ominus w$ exist such that $v \gg w$. Then, we have*

$$u(v \ominus w) = uv \ominus uw.$$

Proof. Since $v \ominus w$ exists such that $v \gg w$, it implies that

$$\begin{cases} \underline{v} > \underline{w}, \\ \overline{v} > \overline{w} \\ \underline{v} - \underline{w} < \overline{v} - \overline{w}. \end{cases}$$

Thus, it implies that $0 < \underline{v} - \underline{w} < \overline{v} - \overline{w}$ and hence, $v \ominus w \gg \tilde{0}$.

Finally, for each $\alpha \in [0,1]$, we have

$$\begin{aligned} [u(v \ominus w)]^\alpha &= [\underline{u}(\underline{v} - \underline{w}), \overline{u}(\overline{v} - \overline{w})] \\ &= [\underline{uv} - \underline{uw}, \overline{uv} - \overline{uw}] \\ &= [\underline{uv}, \overline{uv}] \ominus [\underline{uw}, \overline{uw}] \\ &= [uv \ominus uw]^\alpha. \end{aligned}$$

□

We have the following lemma.

Lemma 1. *Let $u, v, w \in \mathbb{R}_F$ and the H-differences $u \ominus v$, $(u \ominus v) \ominus w$ exist. Then, the H-difference $u \ominus (v + w)$ exist and we have $(u \ominus v) \ominus w = u \ominus (v + w)$.*

Proof. Let $u \ominus v = \tau_1$ and $(u \ominus v) \ominus w = \tau_2$. Then, $\tau_2 + w = u \ominus v$. So, we have $\tau_2 + w + v = u$. Therefore, $\tau_2 = u \ominus (v + w)$. □

Theorem 5. *Assume that the numbers $\mu_n, \nu_n, z_0 \in \mathbb{R}_F^+$ be such that the H-differences $\mu_n z_n \ominus \nu_n$ exist and $\mu_n z_n \gg \nu_n$ for all $n \geq 0$. Then, the fuzzy solution of the fuzzy difference equation*

$$z_{n+1} = \mu_n z_n \ominus \nu_n, \qquad n = 0, 1, \ldots \tag{4}$$

is given by

$$z_n = z_0 \prod_{i=0}^{n-1} \mu_i \ominus \sum_{i=0}^{n-1} \nu_i \prod_{j=i+1}^{n-1} \mu_j, \qquad n = 0, 1, \ldots$$

Proof. By the assumption that $z_0 \in \mathbb{R}_F^+$, $\mu_n z_n \gg \nu_n$ for each $n \in \mathbb{N}$ and the H-differences $\mu_n z_n \ominus \nu_n$ exist, it implies that $\mu_n z_n \ominus \nu_n \gg \tilde{0}$ and hence, we have $z_{n+1} \in \mathbb{R}_F^+$ for each $n \in \mathbb{N}$. On the other hand, since μ_n and ν_n are positive fuzzy numbers for $n = 0, 1, \ldots$, the α-cuts of z_{n+1} are given by

$$[\underline{z}_{n+1}, \overline{z}_{n+1}]^\alpha = \left[\underline{\mu}_n \underline{z}_n - \underline{\nu}_n, \overline{\mu}_n \overline{z}_n - \overline{\nu}_n\right], \qquad \text{for } n = 0, 1, \ldots \text{ and } \alpha \in [0,1].$$

Then, we have two classical difference equations

$$\underline{z}_{n+1} = \underline{\mu}_n \underline{z}_n - \underline{\nu}_n, \qquad n = 0, 1, \ldots$$

and

$$\overline{z}_{n+1} = \overline{\mu}_n \overline{z}_n - \overline{\nu}_n, \qquad n = 0, 1, \ldots$$

Therefore, by using the results of classic difference equations [22], we have

$$\underline{z}_n = \underline{z}_0 \prod_{i=0}^{n-1} \underline{\mu}_i - \sum_{i=0}^{n-1} \underline{\nu}_i \prod_{j=i+1}^{n-1} \underline{\mu}_j, \quad n = 0, 1, \ldots$$

and

$$\overline{z}_n = \overline{z}_0 \prod_{i=0}^{n-1} \overline{\mu}_i - \sum_{i=0}^{n-1} \overline{\nu}_i \prod_{j=i+1}^{n-1} \overline{\mu}_j, \quad n = 0, 1, \ldots$$

Therefore, we obtain

$$z_n = z_0 \prod_{i=0}^{n-1} \mu_i \ominus \sum_{i=0}^{n-1} \nu_i \prod_{j=i+1}^{n-1} \mu_j, \quad n = 0, 1, \ldots \tag{5}$$

It is easy to check that the H-difference in (5) exists. Indeed, the corresponding H-differences $z_1 = \mu_0 z_0 \ominus \nu_0$ and $z_2 = \mu_1 z_1 \ominus \nu_1 = \mu_1(\mu_0 z_0 \ominus \nu_0) \ominus \nu_1$ exist. Therefore, by using Lemma 1, the H-difference $\mu_1 \mu_0 z_0 \ominus (\mu_1 \nu_0 + \nu_1)$ exists and we have

$$z_2 = \mu_1 \mu_0 z_0 \ominus (\mu_1 \nu_0 + \nu_1).$$

By mathematical induction principle, we can see that the H-difference in (5) exists and hence, the proof is complete. □

3. General Mixed Continuous-Discrete Fuzzy Model

The fuzzy difference equations introduced in Reference [4] are the special cases of the following linear fuzzy difference equation

$$F_{n+1} = a_n F_n + b_n F_{\gamma_n} + f_n, \; n = 0, 1, \ldots \tag{6}$$

where $\{a_n\}, \{b_n\}, \{f_n\}$ are given sequences of fuzzy numbers and $\gamma_n = k\delta_n$ with $\delta_n = [\frac{n}{k}]$ for some integer $k > 0$.

In the following, we consider a positive increasing sequence $\{t_n\}$ satisfying $t_n \to +\infty$, a sequence of fuzzy functions $\{f_n\} \subset C(\overline{J_n} \times \mathbb{R}_F, \mathbb{R}_F)$, $J_n = (t_n, t_{n+1}]$, the fuzzy functions $d_n : \mathbb{R}_F \times \mathbb{R}_F \times \mathbb{R}_F \to \mathbb{R}_F$ for each $n = 0, 1, \ldots$ and the initial value $F_0 \in \mathbb{R}_F$. We introduce a dynamical process as a mixed continuous-discrete fuzzy model by the set of fuzzy differential equations and fuzzy difference equations as follows:

$$\begin{cases} y_n'(t; F_n) = f_n(t, y_n(t; F_n)), & t \in J_n, \\ y_n(t_n^+; F_n) = F_n, & n = 0, 1, \ldots \end{cases} \tag{7}$$

and

$$F_{n+1} = d_n(F_n, F_{\gamma_n}, y_n(t_{n+1}; F_n)). \tag{8}$$

Here, we assume that

$$y_n(t_n^+; F_n) = \lim_{h \to 0^+} y_n(t_n + h; F_n).$$

In order to get the uniqueness of the process, we assume that the sufficient conditions for the existence of (i)-differentiable and (ii)-differentiable solutions are fulfilled (see References [2,10]). Then the unique process in each type of differentiability is defined as

$$y(t; F_0) = y_n(t; F_n), \;\; t \in (t_n, t_{n+1}], \; n = 0, 1, \ldots$$

Therefore, the dynamical process in interval J_n is presented by the corresponding fuzzy differential equation and it displays impulsive jumps in the points of the sequence $\{t_n\}$. It is easy to see that when the functions d_n do not depend on the third variable or $f_n(t,y) \equiv 0,\ n = 0,1,\ldots,$ we have a purely discrete process described by the fuzzy difference equations. For instance, if we assume

$$d_n(x,y,z) = a_n x + b_n y + f_n,$$

then the fuzzy differential-difference Equations (7) and (8) are transformed into (6). On the other hand, if we assume $f_n(t,x) \equiv f(t,x)$ and $d_n(x,y,z) = z$, then we obtain a continuous dynamical process formulated by the fuzzy initial value problem

$$\begin{cases} y'(t) = f(t,y(t)), & t \geq t_0, \\ y(t_0) = F_0. \end{cases}$$

4. Linear Fuzzy Differential-Difference Equations

In this section, we consider the equation of linear form

$$f(t,y) = a(t)y + b(t), \tag{9}$$

where $a : [0,+\infty) \to \mathbb{R}$ and $b : [0,+\infty) \to \mathbb{R}_F$ are continuous functions and

$$d_n(x,y,z) = d_n z + e_n,\ n = 0,1,\ldots \tag{10}$$

is a fuzzy difference equation w.r.t. the sequences $\{d_n\}$ and $\{e_n\}$ of positive fuzzy numbers. Hence, the problem (7)–(8) is transformed into the following fuzzy model

$$\begin{cases} y_n'(t;F_n) = a(t)y_n(t;F_n) + b(t), & t \in (t_n, t_{n+1}], \\ y_n(t_n^+;F_n) = F_n, & n = 0,1,\ldots \end{cases} \tag{11}$$

$$F_{n+1} = d_n y_n(t_{n+1};F_n) + e_n, \qquad n = 0,1,\ldots \tag{12}$$

where the initial value $F_0 \in \mathbb{R}_F^+$. In the following, we will present an explicit formula for the solution $y_n(t;F_n)$ on each interval $J_n = (t_n, t_{n+1}]$. For this aim, we consider three different cases of the real function $a(t)$.

Theorem 6. *Consider the linear mixed continuous-discrete fuzzy model (11)–(12) where $d_n, e_n, F_0 \in \mathbb{R}_F^+$ and $a : [0,+\infty) \to \mathbb{R}^+$, $b : [0,+\infty) \to \mathbb{R}_F^+$ are continuous functions. Then, the (i)-differentiable solution to the model (11)–(12) is given by*

$$\begin{aligned} y_n(t;F_n) = {} & F_0 e^{\int_{t_0}^t a(u)du} \prod_{i=0}^{n-1} d_i + \sum_{i=0}^{n-1} \left[\left(\prod_{j=i}^{n-1} d_j \right) \int_{t_i}^{t_{i+1}} b(s) e^{\int_s^t a(u)du} ds \right] \\ & + \sum_{i=0}^{n-1} \left[e_i e^{\int_{t_{i+1}}^t a(u)du} \prod_{j=i+1}^{n-1} d_j \right] + \int_{t_n}^t b(s) e^{\int_s^t a(u)du} ds, \end{aligned} \tag{13}$$

for each $t \in (t_n, t_{n+1}]$ and each $n = 0,1,\ldots$

Proof. If the function $a(t)$ is positive, then according to Theorem 3, the (i)-differentiable solution of the fuzzy differential-difference Equations (11) and (12) is given by

$$y_n(t;F_n) = e^{\int_{t_n}^t a(u)du} \left(F_n + \int_{t_n}^t b(s) e^{-\int_{t_n}^s a(u)du} ds \right),$$

or equivalently,

$$y_n(t;F_n) = F_n e^{\int_{t_n}^t a(u)du} + \int_{t_n}^t b(s) e^{\int_s^t a(u)du} ds, \tag{14}$$

for all $t_n < t \le t_{n+1}$ and $n = 0, 1, \ldots$ Therefore, by using the difference Equation (12), we directly have

$$F_{n+1} = d_n \left[F_n e^{\int_{t_n}^{t_{n+1}} a(u)du} + \int_{t_n}^{t_{n+1}} b(s) e^{\int_s^{t_{n+1}} a(u)du} ds \right] + e_n.$$

For each $n \ge 1$, since the terms $d_n, F_n e^{\int_{t_n}^{t_{n+1}} a(u)du}$ and $\int_{t_n}^{t_{n+1}} b(s) e^{\int_s^{t_{n+1}} a(u)du} ds$ are in $\mathbb{R}_F^+$, then Theorem 1 implies that

$$F_{n+1} = F_n d_n e^{\int_{t_n}^{t_{n+1}} a(u)du} + d_n \int_{t_n}^{t_{n+1}} b(s) e^{\int_s^{t_{n+1}} a(u)du} ds + e_n,$$

or equivalently,

$$F_{n+1} = A_n F_n + B_n, \qquad n = 0, 1, \ldots, \tag{15}$$

where

$$A_n = d_n e^{\int_{t_n}^{t_{n+1}} a(u)du},$$
$$B_n = d_n \int_{t_n}^{t_{n+1}} b(s) e^{\int_s^{t_{n+1}} a(u)du} ds + e_n.$$

Therefore, according to Theorem 4, the solution of the fuzzy difference Equation (15) is

$$F_n = F_0 \prod_{i=0}^{n-1} A_i + \sum_{i=0}^{n-1} B_i \prod_{j=i+1}^{n-1} A_j, \qquad n = 0, 1, \ldots$$

and the proof is complete. □

Remark 1. *It is well-known that the Hukuhara differentiable functions have increasing length of support, that is, when the time goes by, the diameter of the fuzzy functions increases, see Reference [10]. Therefore, the solution of the model (11)–(12) has increasing length of support with some impulsive jumps in the points of sequence $\{t_n\}$.*

For $a < 0$, we have the following result.

Theorem 7. *Consider the linear mixed continuous-discrete fuzzy model (11)–(12) where $d_n, e_n, F_0 \in \mathbb{R}_F^+$ and $a : [0, +\infty) \to \mathbb{R}^-$, $b : [0, +\infty) \to \mathbb{R}_F^-$ are continuous functions. Assume that for each $n \ge 0$, the H-differences*

$$F_n e^{\int_{t_n}^{t} a(u)du} \ominus (-1) \int_{t_n}^{t} b(s) e^{\int_s^{t_{n+1}} a(u)du} ds$$

exist and

$$F_n e^{\int_{t_n}^{t} a(u)du} \gg (-1) \int_{t_n}^{t} b(s) e^{\int_s^{t_{n+1}} a(u)du} ds.$$

Then, the (ii)-differentiable solution to the model (11)–(12) is given by

$$y_n(t; F_n) = F_0 e^{\int_{t_0}^{t} a(u)du} \prod_{i=0}^{n-1} d_i \ominus \sum_{i=0}^{n-1} \left[\left(\prod_{j=i}^{n-1} d_j \right) \int_{t_i}^{t_{i+1}} b(s) e^{\int_s^{t} a(u)du} ds \right]$$
$$+ \sum_{i=0}^{n-1} \left[e_i e^{\int_{t_{i+1}}^{t} a(u)du} \prod_{j=i+1}^{n-1} d_j \right] + \int_{t_n}^{t} b(s) e^{\int_s^{t} a(u)du} ds, \tag{16}$$

for each $t_n < t \le t_{n+1}$ and each $n = 0, 1, \ldots$

Proof. According to Theorem 3, the (ii)-differentiable solution of (11) is given by

$$y_n(t;F_n) = e^{\int_{t_n}^{t} a(u)du}\left(F_n \ominus (-1)\int_{t_n}^{t} b(s)e^{-\int_{t_n}^{s} a(u)du}ds\right),$$

or equivalently,

$$y_n(t;F_n) = F_n e^{\int_{t_n}^{t} a(u)du} \ominus (-1)\int_{t_n}^{t} b(s)e^{\int_{s}^{t_{n+1}} a(u)du}ds, \tag{17}$$

for each $t_n < t \leq t_{n+1}$, and each $n = 0,1,\ldots$. Therefore, by using the difference Equation (12), we have

$$F_{n+1} = d_n\left[F_n e^{\int_{t_n}^{t_{n+1}} a(u)du} \ominus (-1)\int_{t_n}^{t_{n+1}} b(s)e^{\int_{s}^{t_{n+1}} a(u)du}ds\right] + e_n.$$

Then, since $d_n, F_n \in \mathbb{R}_F^+$ and $b : [0,\infty) \to \mathbb{R}_F^-$, Proposition 1 follows that

$$F_{n+1} = F_n d_n e^{\int_{t_n}^{t_{n+1}} a(u)du} \ominus (-1)d_n\int_{t_n}^{t_{n+1}} b(s)e^{\int_{s}^{t_{n+1}} a(u)du}ds + e_n,$$

or equivalently

$$F_{n+1} = A_n F_n \ominus B_n, \quad n = 0,1,\ldots \tag{18}$$

where

$$A_n = d_n e^{\int_{t_n}^{t_{n+1}} a(u)du}, \quad n = 0,1,\ldots$$
$$B_n = (-1)d_n\int_{t_n}^{t_{n+1}} b(s)e^{\int_{s}^{t_{n+1}} a(u)du}ds + e_n.$$

Thus, by Theorem 5, we obtain the solution as

$$F_n = F_0\prod_{i=0}^{n-1} A_i \ominus \sum_{i=0}^{n-1} B_i \prod_{j=i+1}^{n-1} A_j, \quad n = 0,1,\ldots$$

which completes the proof. □

Remark 2. *It is well-known that the (ii)-differentiable functions have non-increasing length of support, that is, when the time goes by, the diameter of the fuzzy functions decrease, see Reference [2]. Therefore, the solution of the model (11)–(12) under the differentiability in type (ii) has non-increasing length of support with some impulsive jumps in the points of sequence $\{t_n\}$.*

In the case $a(t) = 0$, we have the mixed continuous-discrete fuzzy model

$$\begin{cases} y_n'(t;F_n) = b(t), & t \in (t_n, t_{n+1}], \\ y_n(t_n^+;F_n) = F_n, & n = 0,1,\ldots \end{cases} \tag{19}$$

$$F_{n+1} = d_n y_n(t_{n+1};F_n) + e_n, \qquad n = 0,1,\ldots \tag{20}$$

where $F_0 \in \mathbb{R}_F^+$, $b : [0,+\infty) \to \mathbb{R}_F$ is a continuous function and $\{d_n\}$, $\{e_n\}$ are sequences of positive fuzzy numbers. We have the following results for $a(t) = 0$.

Theorem 8. *Consider the mixed continuous-discrete fuzzy model (19)–(20), where the parameters $d_n, e_n, F_0 \in \mathbb{R}_F^+$ and $b : [0,+\infty) \to \mathbb{R}_F$ is a continuous function. Then,*

(i) If the function $b : [0,+\infty) \to \mathbb{R}_F^+$ is continuous, then the (i)-differentiable solution of the fuzzy model (19)–(20) is given by

$$y_n(t;F_n) = F_0\prod_{i=0}^{n-1} d_i + \sum_{i=0}^{n-1}\left[\prod_{j=i}^{n-1} d_j \int_{t_i}^{t_{i+1}} b(s)ds\right] + \sum_{i=0}^{n-1}\left[e_i \prod_{j=i+1}^{n-1} d_j\right] + \int_{t_n}^{t} b(s)ds,$$

for each $t_n < t \leq t_{n+1}$ *and each* $n = 0, 1, \ldots$

(ii) *If* $b : [0, +\infty) \to \mathbb{R}_F^-$ *is a continuous function such that the H-difference* $F_n \ominus (-1)\int_{t_n}^{t} b(s)ds$ *exists and the following term holds*

$$F_n \gg (-1)\int_{t_n}^{t} b(s)ds,$$

then, the (ii)-differentiable solution of the model (19)–(20) is given by

$$y_n(t;F_n) = F_0\prod_{i=0}^{n-1} d_i \ominus \sum_{i=0}^{n-1}\left[\prod_{j=i}^{n-1} d_j \int_{t_i}^{t_{i+1}} b(s)ds\right] + \sum_{i=0}^{n-1}\left[e_i \prod_{j=i+1}^{n-1} d_j\right] + \int_{t_n}^{t} b(s)ds,$$

for each $t_n < t \leq t_{n+1}$ *and each* $n = 0, 1, \ldots$

Proof. By similar arguments as in the case $a(t) > 0$, we obtain the (i)-differentiable solution if $b : [0, +\infty) \to \mathbb{R}_F^+$, while in the case $a(t) < 0$, we receive the (ii)-differentiable solution of the model with $b : [0, +\infty) \to \mathbb{R}_F^-$. □

Example 1. *Consider the following fuzzy differential-difference equation*

$$\begin{cases} y_n'(t, F_n) = y_n(t, F_n) + [\alpha + 1, 3 - \alpha]t, & t \in (t_n, t_{n+1}], \\ y_n(t_n^+, F_n) = F_n, & n = 0, 1, \ldots \end{cases} \tag{21}$$

$$F_{n+1} = 1.1 y_n(t_{n+1}, F_n) + 0.1(n+1)[\alpha + 1, 3 - \alpha], \quad n = 0, 1, \ldots \tag{22}$$

where $F_0 = [2 + \alpha, 4 - \alpha]$, $t_n = 0.2n$, $n = 0, 1, \ldots$ *Then, the solution of the fuzzy differential-difference Equations (21) and (22) is determined by the formula (13) and its graphical representation is given in Figure 1 for* $\alpha = 0, 1$. *As we see in Figure 1, the length of the support of the solution is increasing. Starting from the triangular fuzzy initial value* $(2, 3, 4)$, *the diameter of the solution increases as time goes by and in the point* $t_1 = 0.2$, *according to Equation (22), we have a jump. Again, starting from the point* t_1 *and using FDE (21), we obtain the solution on* $(0.2, 0.4]$. *We can follow this procedure for* $(0.4, 0.6]$.

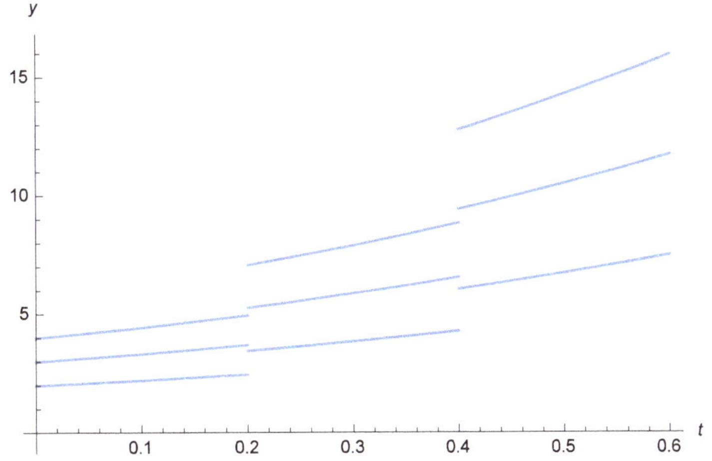

Figure 1. The solution of the fuzzy differential-difference Equations (21) and (22) for $\alpha = 0.1$.

5. Application: Time Value of Money

It is a fact that a fixed amount of money to get after some years is worth less than the same amount today. The main reason is that money due in the future or locked in a fixed term account cannot be spent right away. Meanwhile, prices may rise and the amount will not have the same purchasing power as it would have at present. In addition, there is always a risk that the money will never be received. Therefore, whenever a future payment is uncertain, its value today will be reduced to compensate for the risk. We mention that in this paper, we shall consider situations free from such risk. Bank deposit and bond are generic examples of risk-free assets [24]. A bank deposit is a specific sum of money taken and held on account by a bank, as a service to its customers. Some banks pay the customer through the interest of the funds deposited while others may charge a fee for this service. Therefore a bank deposit is a type of asset. There are many ways that a bank can pay interest on the funds deposited, see for example Reference [24].

Many mathematical methodologies have been developed to study the uncertainty in the estimation of the time value of money. An important effort has been made by Buckley [13] where he has developed fuzzy analogues of the elementary compound interest problem in financial mathematics. Later, in Reference [4], the authors have presented an alternative methodology using fuzzy difference equations. Their method has some advantages such as simplicity and capacity in studying the uncertain factors which cause the change of value of money in different time periods. In this paper, we consider this topic in three following cases:

Case I. Simple Interest: Chrysasif et al. [4] considered a simple capitalization problem. Let us assume that an amount of money is deposited in a bank account to obtain the interest. Then, the future value of this investment consists of the initial value of deposit P, namely the principal, plus all the interest earned during the period of investment. The authors considered the case when the interest is received only by the principal. This motivates the following fuzzy difference equation of simple interest [4]

$$V_{n+1} = V_n + IP, \;\; n = 0, 1, \ldots$$

where I is the rate of interest and $V_0 = P$.

Case II. Periodic Compounding: Let us assume that an amount of money P is deposited in a bank account to receive interest at a constant rate I. Here, in contrast to the case of simple interest, we assume that the interest earned will be added to the initial principal periodically. Consequently, the interest will be received not only by the principal, but also by all the interest earned so far. This motivates the following fuzzy difference equation [4]

$$V_{n+1} = V_n(1 + I), \;\; V_0 = P, \;\; n = 0, 1, \ldots$$

The authors of Reference [4] have studied the compound interest problem considering a new factor e_n, which is added into the equation, denoting the deposits realized during the life of the account

$$V_{n+1} = V_n(1 + I) + e_n, \;\; V_0 = P, \;\; n = 0, 1, \ldots$$

It is natural to use fuzzy number for the extra deposits because we do not know certainly the number of deposits that the customer will make during the period of investment.

Case III. Continuous Compounding: In this case, the rate of growth of the deposit is proportional to the current wealth. In the periodic compounding, if we consider limit case as $n \to \infty$, we get $V(t) = e^{tI}P$, which is the solution of the following Cauchy problem [24]

$$V'(t) = IV(t), \; V(0) = P.$$

This is known as the continuous compounding, where the corresponding growth factor is e^{tI}.

Remark 3 ([24]). *For the fixed principal P and interest rate I, the continuous compounding produces the higher future value than periodic compounding with any frequency n.*

Example 2. *In the following, by using the results of Section 4, we introduce a new mixed continuous-discrete fuzzy model to study the future value of money. Consider the following fuzzy differential-difference equation*

$$\begin{cases} V_n'(t;F_n) = IV_n(t;F_n), & t \in (t_n, t_{n+1}], \\ V_n(t_n^+;F_n) = F_n, & n = 0,1,\ldots \end{cases} \tag{23}$$

$$F_{n+1} = V_n(t_{n+1};F_n) + e_n, \quad n = 0,1,\ldots \tag{24}$$

where the initial value $F_0 = P$. *Let us consider triangular fuzzy numbers* $F_0 = \left(\phi, \frac{(\phi+\rho)}{2}, \rho\right)$ *and* $e_n = a_n\left(s, \frac{s+t}{2}, t\right)$, *where their membership functions are given by*

$$F_0(x) = \begin{cases} \frac{-2x+2\phi}{\phi-\rho}, & x \in [\phi, \frac{(\phi+\rho)}{2}), \\ \frac{-2x+2\rho}{\rho-\phi}, & x \in [\frac{(\phi+\rho)}{2}, \rho), \\ 0, & otherwise, \end{cases}$$

and

$$e_n(x) = \begin{cases} \frac{-2x+2a_n s}{a_n(s-t)}, & x \in [a_n s, \frac{a_n(s+t)}{2}), \\ \frac{-2x+2a_n t}{a_n(t-s)}, & x \in [\frac{a_n(s+t)}{2}, a_n t), \\ 0, & otherwise. \end{cases}$$

Hence, their level sets are given by

$$[F_0]^\alpha = \left[\frac{\alpha(\rho-\phi)+2\rho}{2}, \frac{\alpha(\phi-\rho)+2\phi}{2}\right],$$
$$[e_n]^\alpha = \frac{a_n}{2}\left[\alpha(t-s)+2s, \alpha(s-t)+2t\right],$$

for all $\alpha \in [0,1]$. *Then, according to the Formula (13), we obtain the solution of the fuzzy differential-difference Equations (23) and (24) is*

$$V_n(t;F_n) = F_0 e^{I(t-t_0)} + \sum_{i=0}^{n-1} e_i e^{I(t-t_{i+1})}. \tag{25}$$

In particular, we consider $I = 3.5$, $t_n = 0.2n, n = 0,1,\ldots$ *and* F_0, e_n *are fuzzy numbers whose level sets are given by*

$$[F_0]^\alpha = 50000 + 5000[-1+\alpha, 1-\alpha],$$
$$[e_n]^\alpha = 200(n+1)[9+\alpha, 11-\alpha].$$

Finally, the solution of fuzzy differential-difference Equations (23) and (24) is determined by the Formula (25) and its graphical representation with $\alpha = 0,1$ *is shown in Figure 2.*

Here, the initial value of the deposit is the triangular fuzzy number $(45,000, 50,000, 55,000)$. *Using the FDE (23), we obtain the solution on* $(0, 0.2]$. *At* $t = 0.2$, *we have an impulsive jump such that we can obtain the value at* t_1^+ *by (24). To obtain the solution on* $(0.2, 0.4]$, *we use the FDE (23) with initial value* F_1. *By following this procedure, we obtain the solution on* $[0, 0.6]$ *in Figure 2.*

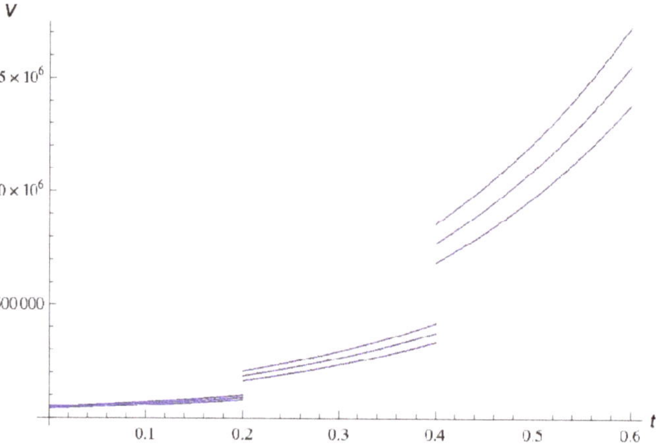

Figure 2. The solution of fuzzy differential-difference Equations (23) and (24) on the interval $[0, 0.6]$ with $\alpha = 0, 1$.

6. Conclusions

In the present paper, a fuzzy differential-difference equation is proposed to model the mixed continuous-discrete phenomena. We presented the dynamical process in the intervals by a fuzzy differential equation and impulsive jumps in some points by the corresponding fuzzy difference equation. This study generalizes the results of Reference [15] to the fuzzy set theory to consider the uncertain factors in differential equations and difference equations. By this approach, we modeled the uncertainty in initial values and parameters of the differential-difference equations. The general solution of linear fuzzy differential-difference equations is presented, too. Finally, the applicability of the model is illustrated by studying the time value of money in finance.

For further research, we propose to extend these results to study the existence of both (i)-differentiable and (ii)-differentiable solutions of the mixed continuous-discrete fuzzy model corresponding to each case of a. The current work opens up many potential results in studying control problems or numerical algorithms for the fuzzy differential-difference equations, that are inspired by pioneer works [25–29].

Author Contributions: All authors contributed equally. All authors have read and agreed to the published version of the manuscript.

Funding: This project was supported by Researchers Supporting Project number (RSP-2020/210), King Saud University, Riyadh, Saudi Arabia.

Conflicts of Interest: The authors declare no conflict of interest. The writers state that they have no known personal relationships or competing financial interests that could have appeared to affect the work reported in this work.

References

1. Chakraverty, S.; Tapaswini, S.; Behera, D. *Fuzzy Differential Equations and Applications for Engineers and Scientists*; Taylor& Francis: Oxfordshire, UK, 2016.
2. Bede, B. *Mathematics of Fuzzy Sets and Fuzzy Logic*; Springer: London, UK, 2013.
3. Chalco-Cano, Y.; Román-Flores, H. On new solutions of fuzzy differential equations. *Chaos Solitons Fractals* **2008**, *38*, 112–119. [CrossRef]
4. Chrysafis, K.A.; Papadopoulos, B.K.; Papaschinopoulos, G. Papaschinopoulos, On the fuzzy difference equations of finance. *Fuzzy Sets Syst.* **2008**, *159*, 3259–3270. [CrossRef]
5. Khastan, A. New solutions for first order linear fuzzy difference equations. *J. Comput. Appl. Math.* **2017**, *312*, 156–166. [CrossRef]
6. Papaschinopoulos, G.; Papadopoulos, B.K. On the fuzzy difference equation $x_{n+1} = A + \frac{x_n}{x_{n-m}}$. *Fuzzy Sets Syst.* **2002**, *129*, 73–81. [CrossRef]

7. Villamizar-Roa, E.J.; Angulo-Castillo, V.; Chalco-Cano, Y. Existence of solutions to fuzzy differential equations with generalized Hukuhara derivative via contractive-like mapping principles. *Fuzzy Sets Systems* **2015**, *265*, 24–38. [CrossRef]
8. Bede, B.; Rudas, I.J.; Bencsik, A.L. First order linear fuzzy differential equations under generalized differentiability. *Inform. Sci.* **2007**, *177*, 1648–1662. [CrossRef]
9. Deeba, E.Y.; de Korvin, A. Analysis by fuzzy difference equations of a model of CO_2 level in the blood. *Appl. Math. Lett.* **1999**, *12*, 33–40. [CrossRef]
10. Kaleva, O. Fuzzy differential equations. *Fuzzy Sets Syst.* **1987**, *24*, 301–317. [CrossRef]
11. Papaschinopoulos, G.; Stefanidou, G. Boundedness and asymptotic behaviour of the solutions of a fuzzy difference equation. *Fuzzy Sets Syst.* **2003**, *140*, 523–539. [CrossRef]
12. Rodríguez-López, R. On the existence of solutions to periodic boundary value problems for fuzzy linear differential equations. *Fuzzy Sets Syst.* **2013**, *219*, 1–26. [CrossRef]
13. Buckley, J.J. The fuzzy mathematics of finance. *Fuzzy Sets Syst.* **1987**, *21*, 257–273. [CrossRef]
14. Córdova, J.D.; Molina, E.C.; López, P.N. Fuzzy logic and financial risk. A proposed classification of financial risk to the cooperative sector. *Contaduría Adm.* **2017**, *62*, 1687–1703. [CrossRef]
15. Kwapisz, M. On difference equations arising in mathematics of finance. *Nonlinear Anal. Theory Methods Appl.* **1997**, *30*, 1207–1218. [CrossRef]
16. Diamond, P.; Kloeden, P. *Metric Spaces of Fuzzy Sets*; World Scientific: Singapore, 1994.
17. Gasilov, N.; Amrahov, S.E.; Fatullayev, A.G. Solution of linear differential equations with fuzzy boundary values. *Fuzzy Sets Syst.* **2014**, *257*, 169–183. [CrossRef]
18. Nieto, J.J.; Rodríguez-López, R.; Franco, D. Linear first order fuzzy differential equations. *Int. J. Uncertain. Fuzziness Knowl.-Based Syst.* **2006**, *14*, 687–709. [CrossRef]
19. Nieto, J.J.; Rodríguez-López, R.; Georgiou, D.N. Fuzzy differential systems under generalized metric spaces approach. *Dyn. Syst. Appl.* **2008**, *17*, 1–24.
20. Khastan, A.; Rodríguez-López, R. On the solutions to first order linear fuzzy differential equations. *Fuzzy Sets Syst.* **2016**, *295*, 114–135. [CrossRef]
21. Papaschinopoulos, G.; Papadopoulos, B.K. On the fuzzy difference equation $x_{n+1} = A + \frac{B}{x_n}$. *Soft Comput.* **2002**, *6*, 456–461. [CrossRef]
22. Lakshmikantham, V.; Trigiante, D. *Theory of Difference Equations: Numerical Methods and Applications*; Academic Press: New York, NY, USA, 1988.
23. Kacprzyk, J.; Fedrizzi, M. *Fuzzy Regression Analysis*; Physica-Verlag: Heidelberg, Germany, 1992.
24. Capinski, M.; Zastawniak, T. *Mathematics for Finance: An Introduction to Financial Engineering*; Springer: London, UK, 2003.
25. Dong, N.P.; Long, H.V.; Khastan, A. Optimal control of a fractional order model for granular SEIR epidemic model. *Commun. Nonlinear Sci. Numer. Simulat.* **2020**, *88*, 105312. [CrossRef]
26. Mazandarani, M.; Kamyad, A.V. Modified fractional Euler method for solving fuzzy fractional initial value problem. *Commun. Nonlinear Sci. Numer. Simul.* **2013**, *18*, 12–21. [CrossRef]
27. Mazandarani, M.; Zhao, Y. Fuzzy Bang-Bang control problem under granular differentiability. *J. Franklin Inst.* **2018**, *355*, 4931–4951. [CrossRef]
28. Son, N.T.K.; Dong, N.P.; Son, L.H.; Abdel-Basset, M.; Manogaran, G.; Long, H.V. On the stabilizability for a class of linear time-invariant systems under uncertainty. *Circ. Syst. Signal Process.* **2020**, *39*, 919–960. [CrossRef]
29. Son, N.T.K.; Dong, N.P.; Long, H.V.; Son, L.H.; Khastan, A. Linear quadratic regulator problem governed by granular neutrosophic fractional differential equations. *ISA Trans.* **2019**, *97*, 296–316. [CrossRef] [PubMed]

mathematics

MDPI

Article

Eigen Fuzzy Sets and their Application to Evaluate the Effectiveness of Actions in Decision Problems

Ferdinando Di Martino [1,2,*] **and Salvatore Sessa** [1,2]

[1] Dipartimento di Architettura, Università degli Studi di Napoli Federico II, Via Toledo 402, 80134 Napoli, Italy; sessa@unina.it
[2] Centro Interdipartimentale di Ricerca "Alberto Calza Bini",Università degli Studi di Napoli Federico II, Via Toledo 402, 80134 Napoli, Italy
* Correspondence: fdimarti@unina.it; Tel.: +39-0812538908; Fax: +39-081238905

Received: 21 October 2020; Accepted: 6 November 2020; Published: 9 November 2020

Abstract: We propose a new method based on the greatest (resp., smallest) eigen fuzzy set (GEFS, resp., SEFS) of a fuzzy relation R with respect to the max–min (resp., min–max) composition in order to implement the actions of a decisor. Using information derived from judgments of the evaluators on how much a characteristic is improved with respect to others, we construct the fuzzy relations, RMAX (resp., RMIN), where any entry $RMAX_{ij}$j (resp., $RMIN_{ij}$) expresses how much the efficacy produced on the *i*th characteristic is equal to or greater (resp., lesser) than that one produced by the *j*th characteristic. The GEFS of RMAX (resp., SEFS of RMIN) are calculated in order to improve the performances of each characteristic. In the wake of previous applications based on GEFS and SEFS, we propose a method to evaluate the tourism enhancement policies in the historical center of an important Italian city. This method is new and different from those known in the literature so far. It is applied to evaluate benefits brought about by locals in order to enhance tourism in a historical center Comparison tests show that the results obtained are consistent with those expressed by the tourists interviewed

Keywords: GEFS; SEFS; fuzzy relations: fuzzy sets; max–min composition; min–max composition

1. Introduction

The greatest eigen fuzzy set (for short, GEFS) of a fuzzy relation with respect to the max–min composition and the smallest eigen fuzzy set (SEFS) of it with respect to the min–max decomposition have been studied in [1–4].

GEFS and SEFS have been applied to problems of image information retrieval [5,6], image analysis [7], and image reconstruction [8,9]. In [9–11], a hybrid method is proposed in which GEFS and SEFS are applied to construct a fitness function of a genetic algorithm used for image reconstruction.

In [12–15], GEFS and SEFS are applied to evaluate the effectiveness of pharmaceutical treatments prescribed to patients in the presence of specific symptoms, considering the influence that the symptoms have on each other. In [16], GEFS and SEFS are applied in decision-making problems.

In [14], the authors proposed a method based on GEFS and SEFS to evaluate the effectiveness of Bayer's aspirin in making throat inflammation symptoms disappear in patients after treatment.

Two NxN fuzzy relations R_{MAX} and R_{MIN} are constructed where the entry R_{MAXij} measures how the action of the drug on the *i*th symptom is considered equal to or stronger than the *j*th one, and R_{MINij} measures how much the action of the drug on the *i*th symptom is considered equal to or weaker than the *j*th one. Let A_{MAX} and B_{MIN} be the GEFS of R_{MAX} and the SEFS of R_{MIN}, respectively. The authors of [13,14] conclude that Bayer's aspirin removes the *i*th symptom in a range between B_{MINi} and A_{MAXi}.

We propose to extend this model to analyze a generic problem in which we need to evaluate the effectiveness of an action applied on entities for the purpose of improving their performance.

By this term, we mean an evaluation of the real benefit brought by an action performed on an entity in terms of protection and/or performance improvement. In this context, we generalize the terms entity and action, meaning by entity any object or set of objects described by a set of characteristics and with action any generic action performed on the entity whose possible benefits are to be assessed.

For example, if it is intended to evaluate the effectiveness of a drug on patients who present a specific disease, the entity is constituted by the patients and the action consists of the use of the drug by the patients; the measure of the effectiveness of the action is carried out by analyzing the variation of the patient's characteristics, consisting of the symptoms of the disease.

Another example refers to the evaluation of the performance improvement of the services provided by an infrastructure. In this case, the entity is constituted by the infrastructure and the action is constituted by the works carried out to make the infrastructure more efficient. The characteristics of the entity consist of those attributes that determine its performance.

The aim of this research was to extend the method proposed in [14] in order to apply it in various contexts and to extract an evaluation of both the effectiveness of the action in improving the characteristic of the entity and the uncertainty of this evaluation.

We have tested our method on a problem of tourist enhancement of a historic center. In specific interviews with tourists who had visited the place during the previous year, they were asked to which extent they felt that one of the tourist attractions of the historic center had improved compared to the other ones. In this way, R_{MAX} and R_{MIN} relations were built, and consequently, A_{MAX} and B_{MIN} were determined.

The mean value of the *i*th component of A_{MAX} and B_{MIN} provides an estimate of the effectiveness of the action: the difference between the *i*th components of A_{MAX} and B_{MIN} provide the uncertainty of this estimate.

In Section 2, we discuss the concepts of the eigen fuzzy set of a fuzzy relation with respect to the max–min and min–max operators, and we show how GEFS and SEFS can be found as well. In Section 3, we introduce the proposed method based on GEFS and SEFS applied to evaluate the effectiveness of an action on entities. In Section 4, we show the results of our tests. In Section 5, final considerations are given for future studies.

2. Preliminaries

The theory of eigen fuzzy sets of fuzzy relations is prevalent in the literature. For example, in [17], eigen fuzzy sets are determined via evolutionary algorithms and neural nets for solving fuzzy relation equations. In [18], subsystems of a fuzzy transition system are characterized in terms of eigen fuzzy sets, in [19], some properties of nilpotent fuzzy matrices are determined in terms of eigen fuzzy sets, and in [20], this concept is used for application to linear differential equations. Fuzzy matrices are useful in various fuzzy systems, with products usually determined by the max–min rule, which is well known in fuzzy set theory. The min–max rule is the duality of the max–min rule, and it is also used in fuzzy systems theory. On the other hand, sufficient conditions for convergence under max–min (and consequently under min–max, for duality) products are well known (e.g., see [21]). In [22], a generalization of the greatest eigen fuzzy sets is proposed.

For making this paper self-contained, we recall well-known results from [1,2]. Let $X = \{x_1, x_2, \ldots, x_n\}$ be a universe of discourse given by a finite set and let R be a fuzzy relation defined on X × X, $R: X \times X \rightarrow [0, 1]$. Furthermore, let A be a fuzzy set of X, that is A: X® [0, 1], such that

$$R \circ A = A \quad (1)$$

where the symbol "∘" denotes the well-known max–min composition operator. A is called *eigen fuzzy set of R with respect to the max–min composition*. In terms of membership functions, Equation (1) is read as

$$A(y) = \max\{\min\{A(x),R(x,y)\}: x \in X\}. \quad (2)$$

We define the fuzzy set A_0 of X where $A_0(y) = a = \min\{\max R(x,y): x \in X\}$ for every $y \in X$. A_0 is an eigen fuzzy set as it satisfies Equation (2). In fact, we obtain for every $y \in X$:

$$A_0(y) = \max\{\min\{A_0(x),R(x,y)\}: x \in X\} = \max\{\min\{a,R(x,y)\}: x \in X\} = \min\{\{a, \max R(x,y)\}: x \in X\} = a.$$

Now, let A_i i = 1,2, ... be fuzzy sets of X defined recursively by

$$A_1(y) = \max\{R(x,y): x \in X\} \quad \forall y \in X, \quad A_2 = R \circ A_1, \ldots, A_{n+1} = R \circ A_n, \ldots \tag{3}$$

The following theorem holds:

Theorem 1. *([2]): $A_{i+1} \subseteq A_i$ for every i = 1,2, ... ,n, ...*

Proof. We have $A_2 \subseteq A_1$, since $A_2(y) = \max\{\min\{A_1(x),R(x,y)\}: x \in X\} \le \max\{R(x,y): x \in X\} = A_1(y)$ for every $y \in X$. Then, we suppose that $A_n \subseteq A_{n-1}$ and prove that $A_{n+1} \subseteq A_n$ by induction. Indeed, we have every $y \in X$ that $A_{n+1}(y) = \max\{\min\{A_n(x),R(x,y)\}: x \in X\} \le \max\{\min\{A_{n-1}(x),R(x,y)\}: x \in X\}$ as $A_n \subseteq A_{n-1}$.

It is easy to see that $A_0 \subseteq A_1$, since $A_0(y) = \max\{\min\{a, R(x,y)\}: x \in X\} \le \max\{R(x,y)\}: x \in X\} = A_1(y)$ for every $y \in X$. Moreover, being $A_0 \subseteq A_1$ and thus $A_0 = R \circ A_0 \subseteq R \circ A_1 = A_2$. By Theorem 1, then we have that

$$A_0 \subseteq \ldots \subseteq A_n \subseteq A_{n-1} \subseteq \ldots \subseteq A_1. \tag{4}$$

We search the greatest eigen fuzzy set of R with respect to the max–min composition. In accordance to the known literature (e.g., [1,12,18,19]), there exists the smallest integer $p \in \{1, \ldots, \text{card}X\}$ such that $A_{p+1} = R \circ A_p = A_p$; furthermore, $A_p(x) \ge A(x)$ for every $x \in X$ and $A \in F(X)$ satisfying Equation (1), that is, A_p is the greatest eigen fuzzy set of R. The following illustrative example make the above concepts clear.□

Example. *We consider the following fuzzy relation:*

$$R = \begin{pmatrix} 0.3 & 0.7 & 0.6 & 0.5 & 0.8 \\ 0.9 & 0.3 & 0.1 & 0.2 & 0.4 \\ 0.6 & 0.5 & 0.4 & 1.0 & 0.7 \\ 0.2 & 0.8 & 0.2 & 0.1 & 0.6 \\ 1.0 & 0.4 & 0.4 & 0.6 & 0.3 \end{pmatrix}.$$

As $\min\{\{\max_{x \in X} R(x,z)\}: z \in X\} = \min\{1.0, 0.8, 0.6, 1.0, 0.8\} = 0.6$, then
$A_0 = = \min\{\max R(x,z): z \in X\} = \{0.6, 0.6, 0.6, 0.6, 0.6\}$.
Furthermore, by recursion, we obtain that
$A_1 = \{1.0, 0.8, 0.6, 1.0, 0.8\}$,
$A_2 = R \circ A_1 = \{0.8, 0.8, 0.6, 0.6, 0.8\}$,
$A_3 = R \circ A_2 = A_2$.
Then, $p = 2$ and A_2 is the GEFS of R.
Let B be a fuzzy set of X. The dual operator of the max–min composition (1) is given by

$$R \bullet B = B \tag{5}$$

where "•" denotes the min–max composition and B is said an *eigen fuzzy set of R with respect to the min–max composition*. In terms of membership functions, Equation (5) is read for every $y \in X$ as

$$B(y) = \min\{\max\{B(x),R(x,y)\} : x \in X\}. \tag{6}$$

We define the fuzzy set B_0 of X where $B_0(y) = b = \max\{\min R(x,z): z \in X\}$ for every $y \in X$. B_0 is an eigen fuzzy set satisfying Equation (6). In fact, we obtain for every $y \in X$:

$$B_0(y) = \min_{x \in X} \{\max\{B_0(x), R(x,y)\}\} = \min_{x \in X} (\max\{b, R(x,y)\}) = \max\{b, \min_{x \in X} R(x,y)\} = b = B_0(y).$$

Let B_i $i = 1,2, \ldots$ be fuzzy sets of X defined recursively by

$$B_1(y) = \min_{x \in X} R(x,y) \quad \forall y \in X, \quad B_2 = R \bullet B_1, \ldots, B_{n+1} = R \bullet B_n, \ldots \tag{7}$$

For the principle of duality, the following theorem holds:

Theorem 2. *([2]): $B_{i+1} \supseteq B_i$ $\forall$ $i = 1,2, \ldots, n, \ldots$*

Since $B_0 \supseteq B_1$ and B_0 is an eigen fuzzy set of R with respect to the min–max composition, we deduce that $B_0 = R \circ B_0 \supseteq R \circ B_1 = B_2$.

Then, we obtain by Theorem 2:

$$B_0 \supseteq \ldots \supseteq B_n \supseteq B_{n-1} \supseteq \ldots \supseteq B_1. \tag{8}$$

We search the smallest eigen fuzzy set of R with respect to the min–max composition. By the principle of duality, it is easily seen that there exists the smallest integer $q \in \{1, \ldots, \text{card } X\}$ such that $B_{q+1} = R \bullet B_q = B_q$; furthermore, $B_q(x) \le B(x)$ for any $x \in X$ and fuzzy set B of X satisfying Equation (5), that is, B_q is the smallest eigen fuzzy set (SEFS) of R.

Returning to the above example, by using the sequence defined from (7), we have that $q = 1$, since

$B_1 = \min_{x \in X} R(x,z) = (0.2, 0.3, 0.1, 0.1, 0.3)$

$B_2 = R \bullet B_1 = (0.2, 0.3, 0.2, 0.1, 0.3)$

$B_3 = R \bullet B_2 = B_2 = (0.2, 0.3, 0.1, 0.1, 0.3)$.

Then, $q = 2$ and B_2 is the SEFS of R.

The sequences defined from (3) and (6) are used in our tests, where N = cardX = 256.

3. The Proposed Method

We propose to apply the GEFS and SEFS to study the effect produced by an action performed on entities to safeguard them or improve their performances (for example, a restoration or maintenance intervention on a damaged or degraded building, or a medical treatment prescribed to a patient to eradicate a disease). We intend to generalize the method proposed in [14] to evaluate the effectiveness of a drug in making a symptom of a disease disappear in patients.

Let $X = \{x_1, x_2, \ldots, x_n\}$ be an universe of discourse given by a set of positive or negative characteristics of the entity that highlights its good condition (or its degradation). We consider a fuzzy relation R_{MAX} whose entry R_{MAXij} is a value in [0, 1], representing how much the action performed on the entity has enhanced the positive characteristic (or has attenuated the negative characteristic) x_i more or in the same way as x_j. The values of R_{MAXij} are obtained by investing N evaluators who, due to the action performed, consider that the characteristic x_i is enhanced (or attenuated) in a manner equal to or greater than x_j. If n_{ij} is the number of evaluators considering $x_i$enhanced (or attenuated) in a manner equal to or greater than x_j, we compute

$$R_{MAXij} = \frac{n_{ij}}{N}. \tag{9}$$

The element R_{MAXii} represents the percentage of evaluators considering x_i enhanced (or attenuated) and not modified after the intervention performed. The GEFS of R_{MAX} is given by the fuzzy set A_{MAX} satisfying the equation:

$$R_{MAX} \circ A_{MAX} = A_{MAX}. \tag{10}$$

We can interpret A_{MAXi} as the maximum effect produced by the intervention enhancing (or mitigating) the *i*th characteristics. Furthermore, we consider a fuzzy relation R_{MIN} whose entry R_{MINij} is a value in [0, 1] representing how much the intervention produced on the entities has attenuated the negative characteristic S_i less than or at least in the same way as x_j. If m_{ij} is the number of evaluators considering x_i enhanced (or attenuated) to a lesser or identical extent with respect to x_j, we compute as

$$R_{MINij} = \frac{m_{ij}}{N}. \tag{11}$$

The SEFS of R_{MIN} is given by the fuzzy set B_{MIN} satisfying the equation:

$$R_{MIN} \circ B_{MIN} = B_{MIN}. \tag{12}$$

We can interpret B_{MINi} as the minimum effect produced by the action performed enhancing (or mitigating) the *i*th characteristics. Then, the interval $[B_{MINi}, A_{MAXi}]$ can represent the range of effectiveness of the action in enhancing (or mitigating) the *i*th characteristics. We consider the mean value given by

$$E_i = \frac{(A_{MAXi} + B_{MINi})}{2} \tag{13}$$

as the mean effectiveness of the action in enhancing (mitigating) the *i*th characteristics. The value

$$U_i = \frac{(A_{MAXi} - B_{MINi})}{2} \tag{14}$$

is considered as the mean uncertainty in the evaluation of the effectiveness of the action in enhancing (or mitigating) the *i*th characteristics. The effectiveness of the action in enhancing (or mitigating) the *i*th characteristics will be evaluated as $E_i \pm U_i$.

The block diagram of the proposed algorithm is shown in Figure 1.

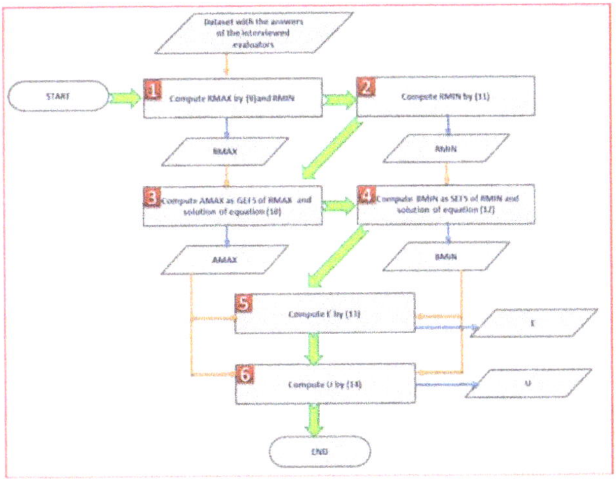

Figure 1. Block diagram schematizing the proposed algorithm.

The processes are represented with rectangles, and the input and output data to the processes are represented with parallelograms. Processes are numbered according to the sequential order in which they are executed in the algorithm. The green arrows link each process to the next process. The yellow arrows connect input data to the process, and the blue arrows show the output data obtained after the process execution is complete.

The pseudocode of the proposed Algorithm 1 is shown below.

Algorithm 1 Eigen Fuzzy Sets Action Effectiveness Evaluation

Input: Dataset with the answers of the interviewed evaluators
Output: Mean effectiveness E and mean uncertainty U
For i: = 1 to n
For j: = 1 to n
Compute R_{MAXij} by (9)
Compute R_{MINij} by (11)
Next j
Next i
Compute A_{MAX} obtaining the GEFS of R_{MAX} by solving (10)
Compute B_{MIN} obtaining the SEFS of R_{MIN} by solving (12)
For i: = 1 to n
Compute E_i by (13)
Compute U_i by (14)
Next i
Return E, U

---

4. An Illustrative Example

We applied our method to analyze any benefits brought about by local policies over a year in order to enhance tourism in a historical center. The calculation processes are developed in Microsoft Excel, and the results of this test are included in our shared repository https://drive.google.com/file/d/17t3QWbu06xX3B63A10o6imMsCtSj2Iz6/view?usp=sharing.

In this example, the universe of discourse is given by six positive and negative characteristics of a historic urban center that represents a pole of cultural tourist attraction. The six characteristics (the first three negative and the last three positive) are the following:

x_1: "poor variety and quality of museum services as information and booking points"
x_2: "poor reachability of museums by public transport"
x_3: "high state of decay of some monuments and churches"
x_4: "discrete presence of equipped hotel facilities in the historic city center"
x_5: "good presence of restaurants and entertainment venues in the historical city center"
x_6: "discrete presence in the historic center of furnished housing units for tourist use".

We carried out our tests, made recently, considering the historical center of the municipality of Naples, in Italy. The evaluators were about 150 tourists who visited the historical center of the city of Naples in 2019 and who were asked to evaluate how much these characteristics had improved between them after a year because of policies to enhance the cultural tourism heritage pursued by the local administration. We construct the fuzzy relations R_{MAX} and R_{MIN} by (9) and (11), respectively.

The component R_{MAXij} represents how much the ith characteristic is improved better or to the same extent as the jth one. For example, the entry $R_{MAX1,2}$ contains as information to what extent the interviewed tourists assess that the action implemented has reduced the poor variety and quality of museum services as information and booking points in an equal or more effective way than the poor reachability of museums by public transport. The R_{MAX14} entry contains as information to what extent the interviewed tourists assess that the action implemented has reduced the poor variety and quality of museum services as information and booking points in an equal or greater manner than the improvement of the discrete presence of equipped hotel facilities in the historical city center.

We apply the proposed algorithm to evaluate the effectiveness of the implementation of the policies applied to enhance the cultural tourism heritage and to improve the six characteristics.

The two matrices R_{MAX}, R_{MIN}, GEFS, and SEFS are shown in Tables 1–3, respectively.

Table 1. R_{MAX}.

R_{MAX}	x_1	x_2	x_3	x_4	x_5	x_6
x_1	0.5	0.4	0.6	0.6	0.5	0.5
x_2	0.7	0.8	0.7	0.8	0.6	0.5
x_3	0.6	0.4	0.6	0.6	0.5	0.4
x_4	0.5	0.2	0.4	0.3	0.4	0.2
x_5	0.5	0.4	0.6	0.7	0.5	0.5
x_6	0.6	0.5	0.6	0.8	0.6	0.7

Table 2. R_{MIN}.

R_{MIN}	x_1	x_2	x_3	x_4	x_5	x_6
x_1	0.5	0.7	0.6	0.5	0.5	0.6
x_2	0.4	0.8	0.4	0.2	0.4	0.5
x_3	0.6	0.7	0.6	0.4	0.6	0.6
x_4	0.6	0.8	0.6	0.3	0.7	0.8
x_5	0.5	0.6	0.5	0.4	0.5	0.6
x_6	0.5	0.5	0.4	0.2	0.5	0.7

Table 3. Greatest eigen fuzzy set (GEFS) of R_{MAX} and SEFS of R_{MIN}.

	x_1	x_2	x_3	x_4	x_5	x_6
A_{MAX}	0.7	0.8	0.6	0.6	0.6	0.7
B_{MIN}	0.5	0.5	0.5	0.3	0.5	0.5
E	0.60	0.65	0.55	0.45	0.55	0.60
U	0.10	0.15	0.05	0.15	0.05	0.10

The characteristic x_2 "poor reachability of museums by public transport" is that one on which the policies of the cultural tourism heritage are most affected ($E_2 = 0.65$). On the contrary, the characteristic x_4 "discrete presence of equipped hotel facilities in the historic city center" is that one on which the implementation of these policies has had the least impact, even if worsening the characteristic x_4 ($E_4 = 0.45$). The characteristics x_3 and x_5 are those for which the uncertainty in evaluating the effectiveness produced by the implementation of these policies is the smallest one ($U_3 = U_5 = 0.05$).

To perform a comparison analysis of our method, we compare the results in Table 3 with the percentage of preferences assigned to each characteristic by the tourists. Each tourist interviewed was also asked to indicate which of the six characteristics had improved more significantly compared with last year.

Table 4 shows for each characteristic the number of preferences for which the characteristic improved more significantly and its percentage value calculated with respect to the total number of interviews.

The characteristic that improved most significantly for the largest number of respondents (approximately 23%) was characteristic x_2. About 20% of the respondents believed that the characteristics that improved most significantly were characteristics x_1 and x_6; 14% of them thought that the characteristic that improved most significantly was characteristic x_5 and 13% of them thought that the characteristic that improved most significantly was characteristic x_3. Finally, only about 10% of respondents believed that the characteristic that improved most significantly was characteristic x_4.

The results in Table 4 are consistent with those obtained by applying the proposed method, as the higher the value of the mean effectiveness of the action in enhancing a characteristic of the entity, the higher the percentage of preferences assigned by the tourists interviewed to that characteristic.

Table 4. Preferences expressed by tourists about the characteristics that improved most significantly.

	x_1	x_2	x_3	x_4	x_5	x_6	Tot
Number of preferences	30	35	20	15	21	31	152
Percentage of preferences	19.74%	23.03%	13.16%	9.87%	13.82%	20.39%	100.00%

5. Conclusions

We propose a method based on GEFS and SEFS to evaluate the effect produced by an action performed on entities to safeguard them or improve their performances. These performances are measured considering their characteristics: a set of evaluators express judgments on how much a characteristic is improved in comparison to others. This information allows creating two fuzzy relations, R_{MAX} and R_{MIN}, where R_{MAXij} expresses the percentage evaluating how the efficacy produced on the ith characteristic is equal to or greater than that produced by the jth characteristic. Conversely, R_{MINij} expresses the percentage evaluating how the efficacy produced on the ith characteristic is equal to or less than that produced by the jth characteristic. We compute the GEFS of R_{MAX} and the SEFS of R_{MIN} to assess how effective the action produced was regarding increasing the performances of each characteristic.

In the future, we intend to experiment with our method the extraction of information relating to the judgments given by evaluators expressed essentially in social groups in order to increase the number of judgments and therefore to obtain less uncertain assessments of the effectiveness of the action taken.

Author Contributions: Conceptualization, F.D.M. and S.S.; methodology, F.D.M. and S.S.; software, F.D.M. and S.S.; validation, F.D.M. and S.S.; formal analysis, F.D.M. and S.S.; investigation, F.D.M. and S.S.; resources, F.D.M. and S.S.; data curation, F.D.M. and S.S.; writing—original draft preparation, F.D.M. and S.S.; writing—review and editing, F.D.M. and S.S.; visualization, F.D.M. and S.S.; supervision, F.D.M. and S.S. All authors have read and agreed to the published version of the manuscript.

Funding: This research received no external funding.

Conflicts of Interest: The authors declare no conflict of interest.

References

1. Sanchez, E. Resolution of eigen fuzzy sets equation. *Fuzzy Sets Syst.* **1978**, *1*, 69–74. [CrossRef]
2. Sanchez, E. Eigen fuzzy sets and fuzzy relations. *J. Math. Anal. Appl.* **1981**, *81*, 399–421. [CrossRef]
3. Wagenknecht, M.; Hartmann, K. On the construction of fuzzy eigen solutions in given regions. *Fuzzy Sets Syst.* **1986**, *20*, 55–65. [CrossRef]
4. Bourke, M.M.; Grant Fisher, D. Convergence, eigen fuzzy sets and stability analysis of relation matrices. *Fuzzy Sets Syst.* **1996**, *8*, 227–234. [CrossRef]
5. Di Martino, F.; Nobuhara, H.; Sessa, S. Eigen fuzzy sets and image information retrieval. In Proceedings of the 2004 IEEE International Conference on Fuzzy Systems, Budapest, Hungary, 25–29 July 2004; pp. 1285–1390. [CrossRef]
6. Di Martino, F.; Nobuhara, H.; Sessa, S. Eigen fuzzy sets and image information retrieval. In *Handbook of Granular Computing*; Pedrycz, W., Skowron, A., Kreinovich, V., Eds.; Wiley: New York, NY, USA, 2008; pp. 863–872, ISBN 9780470035542.
7. Nobuhara, H.; Hirota, K. A solution for eigen fuzzy sets of adjoint max-min decomposition and its application to image analysis. In Proceedings of the IEEE International Symposium on Intelligent Signal Processing, Budapest, Hungary, 4–6 September 2003; pp. 27–30.
8. Nobuhara, H.; Bede, B.; Hirota, K. On various eigen fuzzy sets and their application to image reconstruction. *Inf. Sci.* **2006**, *176*, 2988–3010. [CrossRef]
9. Di Martino, F.; Sessa, S. A Genetic Algorithm Based on Eigen Fuzzy Sets for Image Reconstruction. In *Applications of Fuzzy Sets Theory*; Lecture Notes in Computer Science; Masulli, F., Mitra, S., Pasi, G., Eds.; Springer: Berlin/Heidelberg, Germany, 2007; Volume 4578, pp. 342–348. [CrossRef]

10. Nobuhara, H.; Iyoda, E.M.; Bede, B.; Hirota, K. A solution for generalized eigen fuzzy sets equations by genetic algorithms and its application to image analysis. In Proceedings of the 2nd International IEEE Conference on Intelligent Systems, Varna, Bulgaria, 22–24 June 2004; Volume 1, pp. 208–212. [CrossRef]
11. Di Martino, F.; Sessa, S. Image matching by using fuzzy transforms. *Adv. Fuzzy Syst.* **2013**, *2013*, 760704. [CrossRef]
12. Gestenkorn, T.; Rakus-Andersson, E. An application of fuzzy set theory to differentiating the effectiveness of drugs in treatment of inflammation of genital organs. *Fuzzy Sets Syst.* **1994**, *68*, 327–333. [CrossRef]
13. Rakus-Andersson, E. An application of fuzzy numbers in eigen fuzzy set problem to differentiating the effectiveness of drugs. In Proceedings of the International Conference on Fuzzy Information Processing: Theories and Applications, Bejing, China, 1–4 March 2003; Tsinghua University Press: Bejing, China; Springer: Berlin, Germany, 2003; pp. 85–90.
14. Rakus-Andersson, E. The greatest and the least eigen fuzzy sets in evaluation of the drug effectiveness levels. In *Artificial Intelligence and Soft Computing—ICAISC 2006*; Lecture Notes in Computer Science; Rutkowski, L., Tadeusiewicz, R., Zadeh, L.A., Żurada, J.M., Eds.; Springer: Berlin/Heidelberg, Germany, 2006; Volume 402. [CrossRef]
15. Rakus-Andersson, E. Fuzzy and rough techniques in medical diagnosis and medication. In *Studies in Fuzziness and Soft Computing*; Springer: Berlin/Heidelberg, Germany, 2007; Volume 212, ISBN 103540497072.
16. Guleria, A.; Bajaj, R.K. Eigen spherical fuzzy sets and its application in decision making problem, Scientia Iranica. *Int. J. Sci. Technol.* **2019**, *2019*, 29. [CrossRef]
17. Buckley, J.; Feuring, T.; Hayashi, Y. Solving fuzzy equations using evolutionary algorithms and neural nets. *Soft Comput.* **2002**, *6*, 116–123. [CrossRef]
18. Ignjatović, J.; Ćirić, M.; Simović, V. Fuzzy relation equations and subsystems of fuzzy transition systems. *Knowl. Based Syst.* **2013**, *38*, 48–61. [CrossRef]
19. Lur, Y.-Y.; Pang, C.-T.; Guu, S.-M. On nilpotent fuzzy matrices. *Fuzzy Sets Syst.* **2004**, *145*, 287–299. [CrossRef]
20. Mizukoshi, M.T.; Lodwick, W.A. The interval eigenvalue problem using constraint interval analysis with an application to linear differential equations. *Fuzzy Sets Syst.* **2020**, in press. [CrossRef]
21. Thomason, M.G. Convergence of powers of a fuzzy matrix. *J. Math. Anal. Appl.* **1977**, *57*, 476–480. [CrossRef]
22. Fernández, M.J.; Suárez, F.; Gil, P. T-eigen fuzzy sets. *Inf. Sci.* **1993**, *75*, 63–80. [CrossRef]

mathematics

MDPI

Article

Statistical Parameters Based on Fuzzy Measures

Fernando Reche, María Morales * and Antonio Salmerón

Department of Mathematics and Center for the Development and Transfer of Mathematical Research to Industry (CDTIME), University of Almería, 04120 Almería, Spain; fernando.reche@ual.es (F.R.); antonio.salmeron@ual.es (A.S.)

* Correspondence: maria.morales@ual.es

Received: 21 October 2020; Accepted: 9 November 2020; Published: 12 November 2020

Abstract: In this paper, we study the problem of defining statistical parameters when the uncertainty is expressed using a fuzzy measure. We extend the concept of monotone expectation in order to define a monotone variance and monotone moments. We also study parameters that allow the joint analysis of two functions defined over the same reference set. Finally, we propose some parameters over product spaces, considering the case in which a function over the product space is available and also the case in which such function is obtained by combining those in the marginal spaces.

Keywords: monotone statistical parameters; fuzzy measures; monotone measures; product spaces; fuzzy statistics

1. Introduction

Fuzzy measures [1], also known as capacities [2], non-additive measures or monotone measures [3], have shown to be a valuable tool for representing uncertainty, since they are able to cope with more general scenarios than probability measures do. Even though fuzzy measures have been successfully applied in a wide range of applications [4], no theory analogous to mathematical statistics has emerged around them in the general case, due to the difficulty of defining statistical parameters with a clear interpretation when additivity is replaced by monotonicity.

A remarkable exception is the case of the so-called imprecise probabilities [5,6], characterized by upper and lower expectations that provide rich semantics and interpretability. Dempster–Shafer belief functions [7,8], for instance, can be formulated as special cases of imprecise probabilities.

The field of fuzzy probability and statistics [9–14] has received significant attention during the last two decades. The contributions in this field can be classified into two basic groups according to the underlying approach they follow [15]. One of the groups include the methods that deal with the analysis of classical (non-fuzzy) data using methods based on fuzzy set theory, while the other group focuses on analyzing fuzzy data using statistical methods. In this context, fuzzy data refers to data in which the values correspond to fuzzy numbers [16], characterized by a membership function that returns a value between 0 and 1 indicating to which extent a given real number matches a given fuzzy number.

Examples within the first group include fuzzy clustering [17], fuzzy linear regression [18], testing fuzzy hypothesis from non-fuzzy data [19], fuzzy statistical quality control [20], time series forecasting based on fuzzy logic [21] and making statistical decisions with fuzzy utilities [22].

The second group includes methods for maximum likelihood estimation from fuzzy data [23], classification when data are labeled with Dempster–Shafer belief functions [24], distance-based statistical analysis [25], statistical hypothesis testing from fuzzy data [26], principal component analysis [27], discriminant analysis [28] and clustering [29].

In this paper, we are interested in the definition of statistical parameters when the uncertainty is represented by a general fuzzy measure. More precisely, our starting point is a measurable space and a measurable real-valued function defined on the reference set of the space. We also assume that the measurable space is endowed with a fuzzy measure, and we will study the definition of statistical parameters over the measurable function, in a similar way as statistical parameters over a random variable can be defined from a probability measure. In this way, we attempt to handle more general scenarios than the ones covered by probability measures. To achieve this, we rely on the concept of monotone expectation [30]. We consider the case of marginal spaces as well as product spaces, and take advantage of recent advances in the construction of fuzzy measures over product spaces [31]. Our study is restricted to discrete reference sets.

The rest of the paper is organized as follows. Section 2 establishes the basic notation and definitions, and highlights the fundamental properties of product measures that are used throughout the paper. Section 3 contains the original contributions in this paper, in what concerns the definition of parameters in a marginal measurable space, while Section 4 describes our proposals for product spaces. The paper ends with conclusions in Section 5.

2. Preliminaries and Notation

Definition 1. *[1] Let X be a set and $\mathcal{A}$ be a non-empty class of subsets of X so that $X \subset \mathcal{A}$ and $\emptyset \subset \mathcal{A}$. A function $\mu : \mathcal{A} \longrightarrow [0,1]$ is a* fuzzy measure *if:*

1. $\mu(\emptyset) = 0$.
2. $\mu(X) = 1$.
3. $\forall A, B \in \mathcal{A}$ *such that* $A \subseteq B$ *it holds that* $\mu(A) \leq \mu(B)$.
4. *If* $\{A_n\}_{n\in\mathbb{N}} \in \mathcal{A}$ *such that* $A_1 \subseteq A_2 \subseteq \ldots$ *and* $\bigcup_{n=1}^{\infty} A_n \in \mathcal{A}$, *then* $\lim_n \mu(A_n) = \mu\left(\bigcup_{n=1}^{\infty} A_n\right)$.
5. *If* $\{A_n\}_{n\in\mathbb{N}} \in \mathcal{A}$ *such that* $A_1 \supseteq A_2 \supseteq \ldots$ *and* $\bigcap_{n=1}^{\infty} A_n \in \mathcal{A}$, *then* $\lim_n \mu(A_n) = \mu\left(\bigcap_{n=1}^{\infty} A_n\right)$.

The triplet $(X, \mathcal{A}, \mu)$ is a *measurable space*, and X is called the *reference set*. We will only work with finite reference sets [4] in this paper. By default, we will assume that $\mathcal{A}$ is the power set of X.

Example 1 (Modified from [32]). *Imagine there is a vehicle covering the connection between the harbor and the railway station in a city. This vehicle has four compartments: one for a car, one for a van, one for a motor-bike and another one for a bike. Assume that the gas tank of this vehicle has exactly the capacity necessary to carry the vehicle, with the four compartments busy, from the harbor to the railway station. Then we can regard this capacity to be equal to 1 unit. In this example, $X = \{c, v, m, b\}$, where c stands for* car compartment busy, *v for* van compartment busy, *m for* motor-bike compartment busy *and b for* bike compartment busy. *Assume also that the vehicle does not start the trip unless at least one of the compartments is busy. All the possible transportation situations are then the elements in $\mathcal{A} = \mathcal{P}(X)$ ($\mathcal{P}(X)$ stands for the power set of (X). In these conditions, for every $A \subseteq X$, $\mu(A)$ can be interpreted as the proportion of gas consumed if A happens. A possible specification of a fuzzy measure for this problem is as follows.*

$$\mu(\{b\}) = 0.1,\ \mu(\{v\}) = 0.4,\ \mu(\{c\}) = 0.3,\ \mu(\{m\}) = 0.2,$$

$$\mu(\{c, v\}) = 0.6,\ \mu(\{c, b\}) = 0.35,\ \mu(\{c, m\}) = 0.45,$$

$$\mu(\{b, v\}) = 0.42,\ \mu(\{b, m\}) = 0.21,\ \mu(\{v, m\}) = 0.68,$$

$$\mu(\{c, v, b\}) = 0.7,\ \mu(\{c, v, m\}) = 0.75,\ \mu(\{c, b, m\}) = 0.5,\ \mu(\{v, b, m\}) = 0.69.$$

Note how the fuzzy measure in Example 1 is non-additive. Therefore, the same information cannot be represented by a single probability distribution.

Every fuzzy measure over a reference set of cardinality n can be characterized by $n!$ probability functions (not necessarily different) [33], each one of them corresponding to one possible permutation of the reference set. Given a permutation σ of the set of indices $\{1,\ldots,n\}$, we will denote by X^σ the ordering of the elements of X according to permutation σ, i.e., $X^\sigma = \{x_{\sigma(1)},\ldots,x_{\sigma(n)}\}$. When it is clear from the context, we will drop σ from the subscripts and write $X^\sigma = \{x_{(1)},\ldots,x_{(n)}\}$.

Definition 2. *[33] Let $(X,\mathcal{A},\mu)$ be a measurable space. The* probability function associated with *μ and X^σ is defined as the set $P_\sigma = \{p_\sigma(x_{(1)}),\ldots,p_\sigma(x_{(n)})\}$ such that*

$$p_\sigma(x_{(i)}) = \begin{cases} \mu(A_{(i)}) - \mu(A_{(i+1)}) & \text{if } i < n, \\ \mu(x_{(n)}) & \text{if } i = n, \end{cases} \tag{1}$$

where $A_{(i)} = \{x_{(i)},\ldots,x_{(n)}\}$.

Definition 3. *[33] Let $(X,\mathcal{A},\mu)$ be a measurable space and let P_σ be the probability function associated with μ and X^σ. The* probability measure generated by *μ and X^σ is*

$$P_\sigma(A) = \sum_{x\in A} p_\sigma(x), \quad \forall A \in \mathcal{A}. \tag{2}$$

We will use P_σ for both the probability function and the probability measure when it is clear from the context.

We will consider measures over marginal spaces $(X,\mathcal{A})$ as well as product spaces $(X_1 \times X_2, \mathcal{A}_{X_1\times X_2})$ resulting from composing the marginal spaces $(X_1,\mathcal{A}_{X_1})$ and $(X_2,\mathcal{A}_{X_2})$, with $\mathcal{A}_{X_1\times X_2} = \mathcal{P}(X_1 \times X_2)$, which is not the same as $\mathcal{P}(X_1) \times \mathcal{P}(X_2)$.

Of particular interest are the elements of a product class that can be obtained from sets in the marginal space. They are called *rectangles* and are formally defined as follows:

Definition 4. *Let $(X_1,\mathcal{A}_{X_1})$ and $(X_2,\mathcal{A}_{X_2})$ be two spaces where $\mathcal{A}_{X_1}$ and $\mathcal{A}_{X_2}$ are classes defined on X_1 and X_2, respectively. The class of* rectangles *of $\mathcal{A}_{X_1\times X_2}$ is*

$$\mathcal{R} = \{H \in \mathcal{A}_{X_1\times X_2} \mid H = A \times B, \text{ where } A \in \mathcal{A}_{X_1}, B \in \mathcal{A}_{X_2}\}. \tag{3}$$

Our proposals in this paper will be based on the product measures described in [31], which make use of the concept of triangular norm and conorm.

Definition 5. *[34] An operator $T : [0,1]^2 \longrightarrow [0,1]$ is a* triangular norm *or* t-norm *for short, if it satisfies the following conditions:*

1. *$T(0,a) = 0$, $T(a,1) = a$ for all $a \in [0,1]$. (Boundary conditions)*
2. *$T(a,b) = T(b,a)$. (Commutativity)*
3. *If $a \le c$ and $b \le d$, then $T(a,b) \le T(c,d)$. (Monotonicity)*
4. *$T(T(a,b),c) = T(a,T(b,c))$. (Associativity)*

Definition 6. *[34] An operator $T : [0,1]^2 \longrightarrow [0,1]$ is a* triangular conorm *or* t-conorm *for short, if it satisfies the following properties:*

1. *$S(1,a) = 1$, $S(a,0) = a$ for all $a \in [0,1]$. (Boundary conditions)*
2. *$S(a,b) = S(b,a)$. (Commutativity)*
3. *If $a \le c$ and $b \le d$, then $S(a,b) \le S(c,d)$. (Monotonicity)*
4. *$S(S(a,b),c) = S(a,S(b,c))$. (Assocciativity)*

The usual way of integrating real functions with respect to a fuzzy measure is by means of the so-called Choquet integral, which is a generalization of Lebesgue integral to monotone measures.

Definition 7. [2] *Let* $(X, \mathcal{A}, \mu)$ *be a measurable space, and let* h *be a measurable real function of* X. *The* Choquet integral *of* h *with respect to* μ *is*

$$\oint_A h \circ \mu = \int_{-\infty}^{0} (\mu(H_\alpha \cap A) - 1)\, d\alpha + \int_0^{\infty} \mu(H_\alpha \cap A)\, d\alpha \tag{4}$$

where $A \in \mathcal{A}$ *and* H_α *are the* α-cuts *of* h, *defined as*

$$H_\alpha = \{x \in X / h(x) \geq \alpha\}. \tag{5}$$

If the reference set is finite, the integral can be expressed as

$$\oint h \circ \mu = h(x_{(1)})\mu(A_{(1)}) + \sum_{i=2}^{n} \mu(A_{(i)})[h(x_{(i)}) - h(x_{(i-1)})], \tag{6}$$

where X^σ is an ordering such that $h(x_{(1)}) \leq h(x_{(2)}) \leq \ldots \leq h(x_{(n)})$ and the sets $A_{(i)}$ are of the form $\{x_{(i)}, x_{(i+1)}, \ldots, x_{(n)}\}$. Furthermore, if h is non-negative, it can be computed as

$$\oint h \circ \mu = \sum_{i=1}^{n} h(x_{(i)}) p_\sigma(x_{(i)}), \quad p_\sigma \in P_h, \tag{7}$$

where P_h is the probability function associated with the ordering X^σ induced by h (see Definition 2).

Given two measurable spaces $(X_1, \mathcal{A}_{X_1}, \mu_1)$ and $(X_2, \mathcal{A}_{X_2}, \mu_2)$, the concept of product fuzzy measure is defined as follows.

Definition 8. [31] *A* product fuzzy measure *of* μ_1 *and* μ_2 *is a function* $\mu_{12} : \mathcal{A}_{X_1 \times X_2} \longrightarrow [0, 1]$ *satisfying:*

1. $\mu_{12}(\emptyset) = 0$, $\mu_{12}(X_1 \times X_2) = 1$.
2. *For all* $A, B \in \mathcal{A}_{X_1 \times X_2}$ *such that* $A \subseteq B$ *it holds that* $\mu_{12}(A) \leq \mu_{12}(B)$.
3. *For all* $A \in \mathcal{A}_{X_1}$, *it holds that* $\mu_{12}(A \times X_2) = \mu_1(A)$.
4. *For all* $B \in \mathcal{A}_{X_2}$, *it holds that* $\mu_{12}(X_1 \times B) = \mu_2(B)$.

The next definitions particularize the concept of a fuzzy measure product, so that it is guaranteed to be compatible with the intuitive idea of independence, in the sense that if two fuzzy measures are independent, their fuzzy measure product should be possible to be obtained using exclusively the two original fuzzy measures.

Definition 9. [31] *Let* $(X_1, \mathcal{A}_{X_1}, \mu_1)$ *and* $(X_2, \mathcal{A}_{X_2}, \mu_2)$ *be measurable spaces.* μ_1 *and* μ_2 *are* $\odot$-independent fuzzy measures *if there exists a product fuzzy measure* $\mu_{12}^{\odot}$ *such that for any* $H \in \mathcal{R}$,

$$\mu_{12}^{\odot}(H) = \mu_1(A) \odot \mu_2(B), \tag{8}$$

where $H = A \times B$ *and* $\odot$ *is a t-norm.* $\mu_{12}^{\odot}$ *is called the* $\odot$*-independent product of* μ_1 *and* μ_2.

Definition 10. [31] *Let* $(X_1, \mathcal{A}_{X_1}, \mu_1)$ *and* $(X_2, \mathcal{A}_{X_2}, \mu_2)$ *be measurable spaces. The* $\odot$-exterior product measure *for any* $H \in \mathcal{A}_{X_1 \times X_2}$ *is defined as*

$$\overline{\mu}_{12}^{\odot}(H) = \min_{A \times B \supseteq H} \mu_1(A) \odot \mu_2(B), \tag{9}$$

where $\odot$ *is a t-norm.*

Definition 11. [31] *Let* $(X_1, \mathcal{A}_{X_1}, \mu_1)$ *and* $(X_2, \mathcal{A}_{X_2}, \mu_2)$ *be measurable spaces. The* $\odot$-interior product measure *for any* $H \in \mathcal{A}_{X_1 \times X_2}$ *is defined as*

$$\underline{\mu}^{\odot}_{12}(H) = \max_{A\times B\subseteq H} \mu_1(A) \odot \mu_2(B), \tag{10}$$

where $\odot$ *is a t-norm.*

Both measures conform to lower and upper bounds for any $\odot$-independent product fuzzy measure.

Proposition 1. *[31] Let* $(X_1, \mathcal{A}_{X_1}, \mu_1)$ *and* $(X_2, \mathcal{A}_{X_2}, \mu_2)$ *be measurable spaces. Given any* $\odot$*-independent product of* μ_1 *and* μ_2*, it holds that for all* $C \in \mathcal{A}_{X_1\times X_2}$*,*

$$\underline{\mu}^{\odot}_{12}(C) \leq \mu^{\odot}_{12}(C) \leq \overline{\mu}^{\odot}_{12}(C). \tag{11}$$

Note that, for the particular case of the class $\mathcal{R}$, both measures coincide [31], i.e., for all $H \in \mathcal{R}$,

$$\underline{\mu}^{\odot}_{12}(H) = \mu^{\odot}_{12}(H) = \overline{\mu}^{\odot}_{12}(H). \tag{12}$$

Product fuzzy measures can also be defined in terms of the associated probability measures [31].

Definition 12. *[31] Let* $(X_1, \mathcal{A}_{X_1}, \mu_1)$ *and* $(X_2, \mathcal{A}_{X_2}, \mu_2)$ *be measurable spaces and* $P^{\mu_1}_{\sigma_1}$ *and* $P^{\mu_2}_{\sigma_2}$ *be the probability functions associated with* $X_1^{\sigma_1}$ *and* $X_2^{\sigma_2}$*, respectively. The* lower product p-measure *is defined as*

$$\underline{\mathfrak{m}}_{12}(C) = \min_{\sigma_1,\sigma_2}\left[P^{\mu_1}_{\sigma_1} \otimes P^{\mu_2}_{\sigma_2}(C)\right], \tag{13}$$

for all $C \in \mathcal{A}_{X_1\times X_2}$*, where* $\otimes$ *is the standard probabilistic product, i.e.,* $P^{\mu_1}_{\sigma_1} \otimes P^{\mu_2}_{\sigma_2}(C) = P^{\mu_1}_{\sigma_1}(C)P^{\mu_2}_{\sigma_2}(C)$.

Definition 13. *[31] Given the conditions in Definition 12, the* upper product p-measure *is defined as*

$$\overline{\mathfrak{m}}_{12}(C) = \max_{\sigma_1,\sigma_2}\left[P^{\mu_1}_{\sigma_1} \otimes P^{\mu_2}_{\sigma_2}(C)\right], \tag{14}$$

where $\otimes$ *is the standard probabilistic product.*

3. Parameters over One Measurable Space

In this section we propose statistical parameters aimed to characterize the behavior of functions defined on a measurable space endowed with a fuzzy measure. We will separately address the case of analyzing a single function and the case of simultaneously analyzing two functions.

3.1. The Case of Only One Function

Our proposals rely on the extension of the concept of mathematical expectation associated with probability measures, to the more general case of fuzzy measures. Consider a measurable space $(X, \mathcal{A}, \mu)$ where μ is a fuzzy measure, and the class $\mathfrak{P}$ of all the additive measures over X. One way to extend the concept of mathematical expectation [5,35] is based on defining the set

$$\mathcal{M}_P(\mu) = \{P \in \mathfrak{P} | P(A) \geq \mu(A),\ \forall A \in \mathcal{A}\} \tag{15}$$

of all the probability measures that dominate the fuzzy measure μ.

Since all the elements in $\mathcal{M}_P(\mu)$ are additive measures, the expectation of a function h with respect to a fuzzy measure μ can be defined as

$$E_\mu(h) = \min_{P\in\mathcal{M}_P(\mu)} E_P(h), \tag{16}$$

where $E_P(h)$ is the mathematical expectation of h with respect to the probability measure P.

The problem of this definition is that it is not always well defined, since there can exist a fuzzy measure μ for which $\mathcal{M}_P(\mu) = \emptyset$. It happens, for instance, when the sum of the fuzzy measure μ over the unitary subsets of X is greater than 1, as it is not possible to find a probability measure bounding μ from above.

A class of fuzzy measures that are compatible with the definition of expectation in Equation (16) are those that conform a lower envelope of a set of probability measures [6], i.e., $\mu(A) = \min\{P(A) | P \in \mathcal{M} \subseteq \mathfrak{P}\}$, because in that case $\mathcal{M}_P(\mu) \neq \emptyset$.

A more general definition of expectation, based on Choquet integral [2], was given in [30] with the aim of extending the probabilistic concept of expectation to non-additive settings.

Definition 14. *[30] Let $(X, \mathcal{A}, \mu)$ be a measurable space and let h be a non-negative, real valued measurable function of X. The monotone expectation of h with respect to the fuzzy measure μ is defined as*

$$E_\mu(h) = \oint h \circ \mu. \tag{17}$$

Since a fuzzy measure can always be characterized by a set of probability measures, it is clear from Definition 14 and Equation (7) that the monotone expectation is equal to the mathematical expectation obtained with the probability function associated with the fuzzy measure μ and the ordering induced by the function h (see Definition 2), i.e.,

$$E_\mu(h) = E_{P_{\mu,h}}(h), \tag{18}$$

where $P_{\mu,h}$ denotes the probability function associated with μ and the ordering induced by h. In the particular case of considering a finite reference set, the monotone expectation can be expressed as

$$E_\mu(h) = \sum_{i=1}^{n} h(x_{(i)}) p_\sigma(x_{(i)}), \quad p_\sigma \in P_{\mu,h}. \tag{19}$$

The relation between the monotone expectation and the mathematical expectation is also illustrated in Proposition 2.

Proposition 2. *[30] Let $(X, \mathcal{A}, \mu)$ be a measurable space and let $\{P_\sigma, \sigma \in S_n\}$ be the set of all the probability functions associated with the fuzzy measure μ. Then, for any non-negative real valued, measurable function h of X it holds that*

$$\min_\sigma E_{P_\sigma}(h) \leq E_\mu(h) \leq \max_\sigma E_{P_\sigma}(h). \tag{20}$$

3.1.1. Monotone Variance

In the same way as the monotone expectation extends in a natural way the concept of mathematical expectation to non-additive measures, we will pursue the extension of other statistical parameters in a similar way.

We will start off considering the extension of the concept of variance to a non-monotone context. A direct approach is to define an extension of the variance using Choquet integral, as in the case of the monotone expectation, which yields

$$\mathrm{Var}_\mu(h) = E_\mu[(h - E_\mu(h))^2]. \tag{21}$$

However, the definition of variance in Equation (21) is problematic, since the distribution associated with μ and the ordering induced by h is not, in general, the same as the one induced by $(h - E_\mu(h))^2$. The reason is that functions h and $(h - E_\mu(h))^2$ are not comonotone, and therefore they may induce different orderings of the reference set. Hence, the monotone variance defined in this way could not be considered as a measure of dispersion with respect to the monotone expectation, as the underlying probability distribution can be different (see Definition 2).

Taking this into account, we propose a definition of monotone variance that preserves the underlying probability measure associated with μ and the ordering induced by h.

Definition 15. *Let $(X, \mathcal{A}, \mu)$ be a measurable space and let h be a non-negative real valued measurable function of X. We define the monotone variance of h with respect to the fuzzy measure μ as*

$$Var_\mu(h) = Var_{P_{\mu,h}}(h), \tag{22}$$

where $P_{\mu,h}$ is the probability function associated with μ and the ordering induced by h.

It is clear from the definition that $\text{Var}_\mu(h) \geq 0$ and that it is equal to the traditional variance when μ is a probability measure.

Example 2. *Consider the fuzzy measure over the reference set $X = \{x_1, x_2, x_3\}$ and its associated probability distributions in Table 1, and the function h defined as $h(x_1) = 0.4, h(x_2) = 0.1$ and $h(x_3) = 0.7$. The ordering of X induced by h is thus (x_2, x_1, x_3), i.e., the ordering induced by permutation $\sigma = (2, 1, 3)$, which corresponds to the probability distribution $P_{(2,1,3)}$. Therefore, according to Equation (22), the monotone variance of h is just the variance of h computed using probability distribution $P_{(2,1,3)}$, resulting in*

$$Var_\mu(h) = 0.0621.$$

Table 1. A fuzzy measure and the associated probability distributions corresponding to all the possible permutations of the indices $(1, 2, 3)$.

$\mathcal{A}$	μ	$P_{(1,2,3)}$	$P_{(1,3,2)}$	$P_{(2,1,3)}$	$P_{(2,3,1)}$	$P_{(3,1,2)}$	$P_{(3,2,1)}$
x_1	0.2	0.6	0.6	0.3	0.2	0.4	0.2
x_2	0.1	0.1	0.1	0.4	0.4	0.1	0.3
x_3	0.3	0.3	0.3	0.3	0.4	0.5	0.5
x_1, x_2	0.5	0.7	0.7	0.7	0.6	0.5	0.5
x_1, x_3	0.6	0.9	0.9	0.6	0.6	0.9	0.7
x_2, x_3	0.4	0.4	0.4	0.7	0.8	0.6	0.8

Our definition of monotone variance preserves some properties of the traditional variance, likewise the monotone expectation preserves some properties of the mathematical expectation. In particular, the result in Theorem 1 is of practical value as it simplifies the calculation, and it is also of interest because it links the concepts of monotone variance and monotone expectation.

Theorem 1. *Let $(X, \mathcal{A}, \mu)$ be a measurable space and let h be a non-negative real valued measurable function of X, then it holds that*

$$Var_\mu(h) = E_\mu(h^2) - E^2_\mu(h). \tag{23}$$

Proof. According to Equation (22),

$$\text{Var}_\mu(h) = \text{Var}_{P_{\mu,h}}(h),$$

i.e., the variance of h computed according to probability distribution $P_{\mu,h}$, which can be calculated as

$$\text{Var}_{P_{\mu,h}}(h) = E_{P_{\mu,h}}(h^2) - \left[E_{P_{\mu,h}}(h)\right]^2,$$

and thus

$$\mathrm{Var}_\mu(h) = E_{P_{\mu,h}}(h^2) - \left[E_{P_{\mu,h}}(h)\right]^2. \quad (24)$$

The functions h and h^2 are comonotone, and therefore they induce the same ordering of the reference set and hence yield the same associated probability distribution (see Definition 2). Thus, it holds that $P_{\mu,h} = P_{\mu,h^2}$ and therefore,

$$E_{P_{\mu,h}}(h^2) = E_{P_{\mu,h^2}}(h^2).$$

In addition, according to Equation (18), $E_\mu(h^2) = E_{P_{\mu,h^2}}(h^2) = E_{P_{\mu,h}}(h^2)$ and $E_\mu(h) = E_{P_{\mu,h}}(h)$. Now, replacing $E_{P_{\mu,h}}(h^2)$ by $E_\mu(h^2)$ and $E_{P_{\mu,h}}(h)$ by $E_\mu(h)$ in Equation (24) we obtain Equation (23). □

Example 3. *As a continuation of Example 2, we will compute $Var_\mu(h)$ using Equation (23).*

$$\begin{aligned} E_\mu(h^2) &= E_{P_{(2,1,3)}}(h^2) \\ &= 0.3 \cdot 0.4^2 + 0.4 \cdot 0.1^2 + 0.3 \cdot 0.7^2 = 0.199. \end{aligned}$$

$$\begin{aligned} E_\mu(h) &= E_{P_{(2,1,3)}}(h) \\ &= 0.3 \cdot 0.4 + 0.4 \cdot 0.1 + 0.3 \cdot 0.7 = 0.37. \end{aligned}$$

Hence,

$$Var_\mu(h) = 0.199 - 0.37^2 = 0.0621.$$

The next result shows that the monotone variance behaves in a similar way as traditional variance in relation to affine transformations.

Proposition 3. *Assume the conditions in Theorem 1 and let t be a function defined as $t = ah + b$ with $a \in \mathbb{R}_0^+$ and $b \in \mathbb{R}$. It holds that*

$$Var_\mu(t) = a^2 Var_\mu(h). \quad (25)$$

Proof. First, we have to show that t and h are comonotone, i.e., that for all $x, y \in X$, $(h(x) - h(y))$ and $(t(x) - t(y))$ have the same sign:

$$(h(x) - h(y))(t(x) - t(y)) = (h(x) - h(y))(ah(x) + b - ah(y) - b) = a(h(x) - h(y))^2 \geq 0,$$

since $a \in \mathbb{R}_0^+$. Therefore, the probability distribution associated with the measure μ is the same for both functions, i.e., $P_{\mu,t} = P_{\mu,h}$ and thus

$$\mathrm{Var}_\mu(t) = \mathrm{Var}_{P_{\mu,t}}(t) = \mathrm{Var}_{P_{\mu,h}}(t) = a^2 \mathrm{Var}_{P_{\mu,h}}(h) = a^2 \mathrm{Var}_\mu(h).$$

□

The next results analyze when the monotone variance is equal to 0.

Theorem 2. *Let $(X, \mathcal{A}, \mu)$ be a measurable space and let h be a non-negative real valued measurable function of X. Let $P_{\mu,h}$ be the probability function associated with μ and h. Then, the following three conditions are equivalent:*

1. $Var_\mu(h) = 0$.
2. $\exists! i \ (1 \leq i \leq n)$ *such that* $p_\sigma(x_i) = 1$ *and* $p_\sigma(x_j) = 0, \forall j \neq i$, *with* $p_\sigma \in P_{\mu,h}$.
3. $\exists i \ (1 \leq i \leq n)$ *such that*

$$\mu(H_{\alpha_j}) = 1, \quad \forall j \leq i \quad \text{and} \quad \mu(H_{\alpha_j}) = 0, \quad \forall j > i$$

where $H_{\alpha_i} = \{x \in X | h(x) \geq h(x_i)\}, i = 1, \ldots, n.$

Proof. Let us assume without loss of generality that

$$h(x_1) \leq h(x_2) \leq \cdots \leq h(x_n). \tag{26}$$

$(1) \Longrightarrow (2)$

Since $p_\sigma(x_i) \geq 0$, $i = 1, \ldots, n$, and $\sum_{i=1}^{n} p_\sigma(x_i) = 1$, there must be at least one $i \in \{1, \ldots, n\}$ such that $p_\sigma(x_i) \neq 0$.

Suppose that $\mathrm{Var}_\mu(h) = 0$ and there exist two different $j, k \in \{1, \ldots, n\}, j < k$, such that $p_\sigma(x_j) \neq 0$ and $p_\sigma(x_k) \neq 0$. Then it holds that

$$\mathrm{Var}_\mu(h) = \mathrm{Var}_{p_\sigma}(h) = p_\sigma(x_j)(h(x_j) - E_\mu(h))^2 + p_\sigma(x_k)(h(x_k) - E_\mu(h))^2 = 0,$$

which means that $h(x_j) = E_\mu(h)$ and $h(x_k) = E_\mu(h)$. However, according to the assumption in Equation (26), it holds that $h(x_j) \leq h(x_{j+1}) \leq \cdots \leq h(x_{k-1}) \leq h(x_k)$. Hence,

$$E_\mu(h) = h(x_j) \leq h(x_{j+1}) \leq \cdots \leq h(x_{k-1}) \leq h(x_k) = E_\mu(h),$$

which means that

$$\begin{aligned} h(x_j) = h(x_{j+1}) = \cdots = h(x_k) = E_\mu(h) \quad &\Rightarrow \quad H_{\alpha_j} = H_{\alpha_{j+1}} = \cdots = H_{\alpha_k} \\ &\Rightarrow \quad p_\sigma(x_j) = p_\sigma(x_{j+1}) = \cdots = p_\sigma(x_{k-1}) = 0, \end{aligned}$$

which is a contradiction with the assumption that $p_\sigma(x_j) \neq 0$. Thus, there is only one $p_\sigma(x_i) \neq 0$ and furthermore, $p_\sigma(x_i) = 1$.

$(2) \Longrightarrow (3)$

Assume $\exists! i$ such that $p_\sigma(x_i) \neq 0$. Then,

$$p_\sigma(x_1) = p_\sigma(x_2) = \cdots = p_\sigma(x_{i-1}) = p_\sigma(x_{i+1}) = \cdots = p_\sigma(x_n) = 0$$

and therefore

$$\mu(H_{\alpha_1}) = \mu(H_{\alpha_2}) = \cdots = \mu(H_{\alpha_i})$$

and

$$\mu(H_{\alpha_{i+1}}) = \mu(H_{\alpha_{i+2}}) = \cdots = \mu(H_{\alpha_n}).$$

On the other hand, since $\mu(H_{\alpha_i}) - \mu(H_{\alpha_{i+1}}) = 1$, it follows that $\mu(H_{\alpha_j}) = 1$ if $j \leq i$ and $\mu(H_{\alpha_j}) = 0$ if $j > i$. $(3) \Longrightarrow (1)$

It is straightforward from the definition of monotone variance. □

Corollary 1. *If h is constant, then* $Var_\mu(h) = 0$, *for any fuzzy measure* μ.

Example 4. *Assume a function h defined on* $X = \{x_1, x_2, x_3\}$ *as* $h(x_1) = 0.4$, $h(x_2) = 0.1$ *and* $h(x_3) = 0.7$, *and an associated probability distribution* p_σ *such that* $p_\sigma(x_1) = 1$, $p_\sigma(x_2) = 0$ *and* $p_\sigma(x_3) = 0$. *We will see how the monotone variance is equal to 0. However, first we need to calculate the monotone expectation.*

$$E_\mu(h) = 1 \cdot 0.4 + 0 \cdot 0.1 + 0 \cdot 0.7 = 0.4.$$

Thus,

$$Var_\mu(h) = 1 \cdot (0.4 - 0.4)^2 + 0 \cdot (0.1 - 0.4)^2 + 0 \cdot (0.7 - 0.4)^2 = 0.$$

Now we will calculate the value of the measure μ over the sets $H_{\alpha_i} = \{x \in X | h(x) \geq h(x_i)\}, i = 1,2,3$, i.e., $H_{\alpha_1} = \{x_1, x_3\}$, $H_{\alpha_2} = \{x_1, x_2, x_3\}$ and $H_{\alpha_3} = \{x_3\}$.

We can obtain the values of μ from p_σ using Definition 2. The result is

$$\begin{aligned} \mu(\{x_3\}) &= p_\sigma(x_3) = 0, \\ \mu(\{x_1, x_3\}) &= p_\sigma(x_1) + \mu(\{x_3\}) = 1 + 0 = 1, \\ \mu(\{x_1, x_2, x_3\}) &= p_\sigma(x_2) + \mu(\{x_1, x_3\}) = 0 + 1 = 1. \end{aligned}$$

Therefore, $\mu(H_{\alpha_1}) = 1$, $\mu(H_{\alpha_2}) = 1$ and $\mu(H_{\alpha_3}) = 0$.

3.1.2. Monotone Moments

Following the same idea underlying the definition of monotone variance, we can extend the concepts of central and non-central moments from a probabilistic setting to a monotone one.

Definition 16. *Let $(X, \mathcal{A}, \mu)$ be a measurable space and let h be a non-negative real valued measurable function of X. We define the k-th non-central monotone moment of h with respect to μ as*

$$g_\mu^k(h) = E_\mu(h^k). \tag{27}$$

Note that Equation (27) is well defined, since h and h^k are comonotone, and therefore the corresponding probability function is the same for both of them, regardless of the value of k.

The definition of central monotone moments is, however, more problematic. If we follow the same idea as in Definition 16, and define the central monotone moment as $E_\mu(h - E_\mu(h))^k$, we find the problem that functions h and $(h - E_\mu(h))^k$ are not comonotone, and that would mean that different underlying probability distributions would be used to compute $E_\mu(h)$ and $E_\mu(h - E_\mu(h))^k$. We will therefore generalize the definition of monotone variance to values of $k \neq 2$, utilizing the probability function associated with μ and h.

Definition 17. *Let $(X, \mathcal{A}, \mu)$ be a measurable space and let h be a non-negative real valued measurable function of X. We define the k-th central monotone moment of h with respect to μ as*

$$\gamma_\mu^k(h) = E_{P_{\mu,h}}\left[(h - E_\mu(h))^k\right], \tag{28}$$

where $P_{\mu,h}$ is the probability function associated with μ and h.

The following result establishes the relation between central and non-central monotone moments.

Proposition 4. *Let $(X, \mathcal{A}, \mu)$ be a measurable space and let h be a non-negative real valued measurable function of X. It holds that*

$$\gamma_\mu^k(h) = \sum_{j=0}^{k} (-1)^j \binom{k}{j} \left[g_\mu(h)\right]^j g_\mu^{k-j}(h). \tag{29}$$

Proof. Assume $X = \{x_1, \ldots, x_n\}$.

$$\begin{aligned}\gamma^k_\mu(h) &= E_{P_{\mu,h}}\left[(h - E_\mu(h))^k\right] = \sum_{i=1}^{n}(h(x_i) - E_\mu(h))^k P_{\mu,h}(x_i) \\ &= \sum_{i=1}^{n}\sum_{j=0}^{k}(-1)^j \binom{k}{j} E^j_\mu(h) h^{k-j}(x_i) P_{\mu,h}(x_i) \\ &= \sum_{j=0}^{k}(-1)^j \binom{k}{j} E^j_\mu(h) \sum_{i=1}^{n} h^{k-j}(x_i) P_{\mu,h}(x_i) \\ &= \sum_{j=0}^{k}(-1)^j \binom{k}{j} E^j_\mu(h) E_\mu(h^{k-j}) = \sum_{j=0}^{k}(-1)^j \binom{k}{j} \left[g_\mu(h)\right]^j g_\mu^{k-j}(h).\end{aligned}$$

□

3.2. The Case of Two Functions

In this section we approach the simultaneous analysis of two functions h_1 and h_2 over the same reference set, X. Our goal is to model the information that both functions have in common, or the way in which they interact with one another.

Generalizing the concept of covariance, for instance, by using $E_\mu[(h_1 - E_\mu(h_1))(h_2 - E_\mu(h_2))]$, raises the problem that the underlying probability distribution used to compute the monotone expectation is not the one induced by h_1 nor by h_2 for the same fuzzy measure μ, and therefore it is not clear that this monotone covariance in fact measures the relationship between both functions at all. We will therefore explore a different approach, in which we will model the degree of similarity between h_1 and h_2, by measuring the common region determined by both functions.

Definition 18. *Let $(X, \mathcal{A}, \mu)$ be a measurable space and let h_1 and h_2 be non-negative real valued measurable functions of X. We define the common expectation of h_1 and h_2 with respect to μ as*

$$\psi_\mu(h_1, h_2) = E_\mu[\min\{h_1, h_2\}]. \tag{30}$$

The concept of common expectation is illustrated in Figure 1. More precisely, the value of the common expectation of h_1 and h_2 is the measure, according to μ, of the function under which the shaded area is.

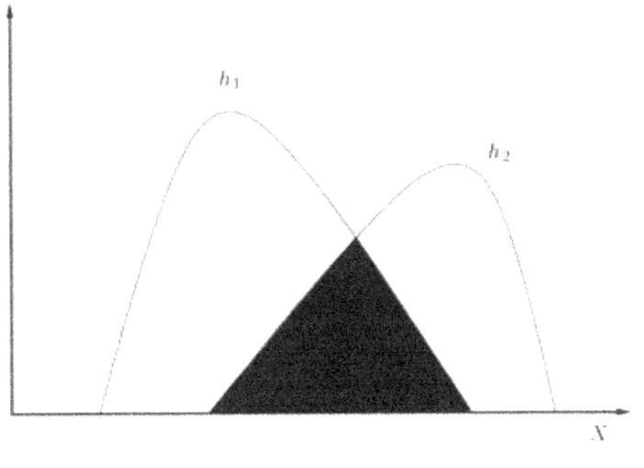

Figure 1. An illustration of the concept of common expectation of h_1 and h_2.

Example 5. *We want to obtain the global grade for two students out of the individual grades they obtained in four different courses $\{x_1, x_2, x_3, x_4\}$. In the final grade we want to reflect if a student shows a good performance in the two scientific courses, $\{x_1, x_2\}$, the humanistic ones, $\{x_3, x_4\}$, or in the combination*

$\{x_2, x_3\}$, corresponding to a social sciences profile. These criteria are encoded in the fuzzy measure in Table 2, while the grades obtained by both students (between 0 and 1) in each of the courses are shown in Table 3.

Table 2. A fuzzy measure matching the criteria in Example 5.

Reference Subsets	Measure
$\{x_1\}, \{x_2\}, \{x_3\}, \{x_4\}$	0.2
$\{x_1, x_2\}$	0.6
$\{x_1, x_3\}, \{x_1, x_4\}$	0.3
$\{x_2, x_3\}$	0.5
$\{x_2, x_4\}$	0.4
$\{x_3, x_4\}$	0.7
$\{x_1, x_2, x_3\}$	0.9
$\{x_1, x_2, x_4\}$	0.6
$\{x_1, x_3, x_4\}$	0.7
$\{x_2, x_3, x_4\}$	0.8

Table 3. Grades obtained by the students in Example 5 in the individual courses.

Student	x_1	x_2	x_3	x_4
h_1	0.9	0.8	0.3	0.2
h_2	0.2	0.3	0.8	0.9

The calculation of the respective monotone expectations and variances result in

$$E_\mu(h_1) = 0.61, \qquad E_\mu(h_2) = 0.65,$$
$$Var_\mu(h_1) = 0.0769, \qquad Var_\mu(h_2) = 0.0765,$$

which are quite similar, while the common expectation is $\psi_\mu(h_1, h_2) = 0.25$.

The next proposition states the basic properties of the common expectation.

Proposition 5. *Let $(X, \mathcal{A}, \mu)$ be a measurable space and let h_1 and h_2 be non-negative real valued measurable functions of X. Then, ψ_μ satisfies the following properties:*

1. $\psi_\mu(h_1, h_2) = \psi_\mu(h_2, h_1)$.
2. $\psi_\mu(h_1, h_2) \leq \min\{E_\mu(h_1), E_\mu(h_2)\}$.
3. *If $\forall x \in X$, $h_1(x) \leq h_2(x)$, then for any non-negative real valued measurable function h of X, $\psi_\mu(h_1, h) \leq \psi_\mu(h_2, h)$.*
4. *If $\forall x \in X$, $h_1(x) \leq h_2(x)$, then $\psi_\mu(h_1, h_2) = E_\mu(h_1)$.*
5. $\psi_\mu(h_1, h_2) = 0 \iff \{x \in X | h_1(x) > 0\} \cap \{x \in X | h_2(x) > 0\} = \emptyset$.

Proof.

1. It is straightforward from Equation (30).
2. It follows from the facts that E_μ is a monotone functional and that $\min\{h_1, h_2\}$ is bounded from above by both h_1 and h_2.
3. It is a direct consequence of the monotonicity of operator min and functional E_μ.
4. If $h_1 \leq h_2$, then $\min\{h_1, h_2\} = h_1$, and therefore both expectations are the same.
5. If $\{x \in X | h_1(x) > 0\} \cap \{x \in X | h_2(x) > 0\} = \emptyset$ then the minimum of both functions is the identically null function, which is known to be the only one that has null monotone expectation [30].

□

The common expectation is not normalized, and therefore its value alone is not enough to determine if it can be regarded as high or low. For instance, in Example 5 we obtained $\psi_\mu(h_1, h_2) = 0.25$, but that value does not tell us if it is high or low. However, it is clear that the common expectation can be bounded from above, since it is known that for any positive real numbers a and b, it holds that $\min\{a, b\} \leq \sqrt{a \cdot b} \leq \max\{a, b\}$ and the equality is reached only when $a = b$. Hence, we can normalize the common expectation using these bounds, which yields three possible definitions of coefficients of concordance between h_1 and h_2.

Definition 19. *Let $(X, \mathcal{A}, \mu)$ be a measurable space and let h_1 and h_2 be non-negative real valued measurable functions of X. We define the coefficients of concordance ρ_1, ρ_2 and ρ_3 between h_1 and h_2 with respect to μ as*

$$\rho_1^\mu(h_1, h_2) = \frac{\psi_\mu(h_1, h_2)}{\sqrt{E_\mu(h_1)E_\mu(h_2)}}, \tag{31}$$

$$\rho_2^\mu(h_1, h_2) = \frac{\psi_\mu(h_1, h_2)}{E_\mu(\max\{h_1, h_2\})}, \tag{32}$$

$$\rho_3^\mu(h_1, h_2) = \frac{\psi_\mu(h_1, h_2)}{\min\{E_\mu(h_1), E_\mu(h_2)\}}. \tag{33}$$

The next proposition shows the basic properties of the three concordance coefficients (when it is clear from the context, we will drop the measure and the functions, thus denoting $\rho_i^\mu(h_1, h_2)$ by ρ_i).

Proposition 6. *Assume the conditions in Definition 19. The coefficients of concordance satisfy the following conditions:*

1. $0 \leq \rho_i \leq 1, i = 1, 2, 3.$
2. $h_1 = h_2 \Rightarrow \rho_1 = \rho_2 = \rho_3 = 1.$
3. $\rho_2 \leq \rho_1 \leq \rho_3.$
4. $\rho_1 = \rho_2 = \rho_3 = 0$ *iff h_1 and h_2 have empty intersection, i.e., $\{x \in X / h_1(x) > 0\} \cap \{x \in X / h_2(x) > 0\} = \emptyset$.*
5. *If $h_1 \leq h_2$, then*

$$\rho_1 = \sqrt{\frac{E_\mu(h_1)}{E_\mu(h_2)}},$$
$$\rho_2 = \frac{E_\mu(h_1)}{E_\mu(h_2)},$$
$$\rho_3 = 1.$$

6. *If $h_1 = kh_2$, with $k > 1$, then*

$$\rho_1 = \frac{1}{\sqrt{k}},$$
$$\rho_2 = \frac{1}{k},$$
$$\rho_3 = 1.$$

Proof.

1. It is clear that

$$\min\{E_\mu(h_1), E_\mu(h_2)\} \leq \sqrt{E_\mu(h_1)E_\mu(h_2)} \leq \max\{E_\mu(h_1), E_\mu(h_2)\}.$$

Furthermore, since E_μ is a monotone functional, $\max\{E_\mu(h_1), E_\mu(h_2)\} \leq E_\mu(\max\{h_1, h_2\})$. According to property 2 in Proposition 5, $\psi_\mu(h_1, h_2) \leq \min\{E_\mu(h_1), E_\mu(h_2)\}$, and thus

$$\psi_\mu(h_1, h_2) \leq \min\{E_\mu(h_1), E_\mu(h_2)\} \leq \sqrt{E_\mu(h_1)E_\mu(h_2)} \leq \max\{E_\mu(h_1), E_\mu(h_2)\}.$$

Therefore

$$\begin{aligned} \rho_1^\mu(h_1, h_2) &= \frac{\psi_\mu(h_1, h_2)}{\sqrt{E_\mu(h_1)E_\mu(h_2)}} \leq 1, \\ \rho_2^\mu(h_1, h_2) &= \frac{\psi_\mu(h_1, h_2)}{E_\mu(\max\{h_1, h_2\})} \leq 1, \\ \rho_3^\mu(h_1, h_2) &= \frac{\psi_\mu(h_1, h_2)}{\min\{E_\mu(h_1), E_\mu(h_2)\}} \leq 1. \end{aligned}$$

On the other hand, since h_1 and h_2 are non-negative, so it is $\psi_\mu(h_1, h_2)$, which means that $\rho_1^\mu(h_1, h_2) \geq 0$, $\rho_2^\mu(h_1, h_2) \geq 0$ and $\rho_3^\mu(h_1, h_2) \geq 0$.

2. If $h_1 = h_2 = h$, then $\min\{h_1, h_2\} = \max\{h_1, h_2\} = h$ and $\psi_\mu(h_1, h_2) = E_\mu(h)$; hence, the three coefficients are equal to 1.
3. From the proof of property 1, we know that

$$\min\{E_\mu(h_1), E_\mu(h_2)\} \leq \sqrt{E_\mu(h_1)E_\mu(h_2)} \leq E_\mu(\max\{h_1, h_2\}) \Rightarrow$$

$$\frac{\psi_\mu(h_1, h_2)}{E_\mu(\max\{h_1, h_2\})} \leq \frac{\psi_\mu(h_1, h_2)}{\sqrt{E_\mu(h_1)E_\mu(h_2)}} \leq \frac{\psi_\mu(h_1, h_2)}{\min\{E_\mu(h_1), E_\mu(h_2)\}} \Rightarrow$$

$$\rho_2 \leq \rho_1 \leq \rho_3.$$

4. $\rho_1 = \rho_2 = \rho_3 = 0$ iff $\psi_\mu(h_1, h_2)$ which, according to property 5 in Proposition 5, can only happen if $\{x \in X / h_1(x) > 0\} \cap \{x \in X / h_2(x) > 0\} = \emptyset$.
5. If $h_1 \leq h_2$, then $\min\{h_1, h_2\} = h_1$, $\max\{h_1, h_2\} = h_2$ and $E_\mu(h_1) \leq E_\mu(h_2)$. Therefore,

$$\rho_1 = \frac{E_\mu(h_1)}{\sqrt{E_\mu(h_1)E_\mu(h_2)}} = \sqrt{\frac{E_\mu(h_1)}{E_\mu(h_2)}}, \quad \rho_2 = \frac{E_\mu(h_1)}{E_\mu(h_2)} \quad \text{and } \rho_3 = \frac{E_\mu(h_1)}{\min\{E_\mu(h_1), E_\mu(h_2)\}} = 1.$$

6. If $h_1 = kh_2$ with $k > 0$, then $\min\{h_1, h_2\} = h_2$, $\max\{h_1, h_2\} = kh_2$ and $E_\mu(h_1) = kE_\mu(h_2)$. Therefore,

$$\rho_1 = \frac{E_\mu(h_2)}{\sqrt{kE_\mu(h_2)E_\mu(h_2)}} = \frac{1}{\sqrt{k}}, \quad \rho_2 = \frac{E_\mu(h_2)}{E_\mu(kh_2)} = \frac{1}{k} \quad \text{and } \rho_3 = \frac{E_\mu(h_2)}{\min\{kE_\mu(h_2), E_\mu(h_2)\}} = 1.$$

□

Example 6. *As a continuation of Example 5, we can use the data in Tables 2 and 3 to compute the coefficients of concordance, obtaining*

$$\rho_1(h_1, h_2) = 0.397, \quad \rho_2(h_1, h_2) = 0.301, \quad \rho_3(h_1, h_2) = 0.410.$$

Note how the three coefficients have low values, which is consistent with the data in the example, as in spite of the similar values for the monotone expectation and variance corresponding to both students, they have a clearly different profile, scientific in the case of h_1 and humanistic in the case of h_2.

4. Parameters Defined over Product Spaces

In this section we explore scenarios where we have two measurable spaces each of them equipped with a different fuzzy measure. We will consider the definition of statistical parameters on the product space.

Likewise, in Section 3, we will separately study the case of one or two real functions. In both cases, it is necessary to obtain a fuzzy measure over the product space. We will rely on the proposals in [31] to obtain the product measures.

4.1. The Case of One Function

The methods proposed in [31] for constructing fuzzy measures over product spaces, rather than single measures, usually yield a set of them, bounded by an upper and lower measure. Similarly, our proposals here will consist of intervals of parameters rather than a single one.

We will start defining the concept of joint expectation making use of the interior and exterior product measures (see Definitions 10 and 11).

Definition 20. *Let $(X_1, \mathcal{A}_{X_1}, \mu_1)$ and $(X_2, \mathcal{A}_{X_2}, \mu_2)$ be measurable spaces, $h : X_1 \times X_2 \to [0,1]$ and $\underline{\mu}_{12}^{\odot}, \overline{\mu}_{12}^{\odot}$ the $\odot$-interior and $\odot$-exterior product measures. We define the joint lower and upper $\odot$-expectations as*

$$\underline{E}_{12}^{\odot}(h) = \oint h \circ \underline{\mu}_{12}^{\odot} \quad \textit{(lower)}, \tag{34}$$

$$\overline{E}_{12}^{\odot}(h) = \oint h \circ \overline{\mu}_{12}^{\odot} \quad \textit{(upper)}. \tag{35}$$

Proposition 7. *Let $(X_1, \mathcal{A}_{X_1}, \mu_1)$ and $(X_2, \mathcal{A}_{X_2}, \mu_2)$ be measurable spaces, $h : X_1 \times X_2 \to [0,1]$ and $\underline{\mu}_{12}^{\odot}, \overline{\mu}_{12}^{\odot}$ the $\odot$-interior and $\odot$-exterior product measures. It holds that*

$$\underline{E}_{12}^{\odot}(h) \leq \overline{E}_{12}^{\odot}(h). \tag{36}$$

Furthermore, if $\mu_{12}^{\odot}$ is any $\odot$-independent product measure of μ_1 and μ_2 (see Definition 9), it also holds that

$$\underline{E}_{12}^{\odot}(h) \leq E_{12}^{\odot}(h) \leq \overline{E}_{12}^{\odot}(h), \tag{37}$$

where $E_{12}^{\odot}(h) = \oint h \circ \mu_{12}^{\odot}$.

Proof. Note that $\underline{E}_{12}^{\odot}, \overline{E}_{12}^{\odot}$ and $E_{12}^{\odot}$ are monotone expectations, namely, $E_{\underline{\mu}_{12}^{\odot}}, E_{\overline{\mu}_{12}^{\odot}}$ and $E_{\mu_{12}^{\odot}}$ respectively. Therefore, Equations (36) and (37) are a direct consequence of the monotonicity of the monotone expectation and Proposition 1. □

The concept of joint $\odot$-expectations is analogous to the concept of monotone expectation in a marginal space, with the difference that, in the case of the product space, the underlying fuzzy measure is not known, but instead we have an interval of measures bounded by the interior and exterior $\odot$-product measures.

We can define joint expectations using other product measures, as the p-measures given in Definitions 12 and 13.

Definition 21. *Let $(X_1, \mathcal{A}_{X_1}, \mu_1)$ and $(X_2, \mathcal{A}_{X_2}, \mu_2)$ be measurable spaces, $h : X_1 \times X_2 \to [0,1]$ and $\underline{\mathfrak{m}}_{12}$, $\overline{\mathfrak{m}}_{12}$ the lower and upper product p-measures respectively. We define the lower and upper joint probabilistic expectations as*

$$E_{\underline{\mathfrak{m}}_{12}}(h) = \oint h \circ \underline{\mathfrak{m}}_{12} \quad (lower), \tag{38}$$

$$E_{\overline{\mathfrak{m}}_{12}}(h) = \oint h \circ \overline{\mathfrak{m}}_{12} \quad (upper). \tag{39}$$

Since we have a function defined over the product space and fuzzy measures defined over the marginal spaces, it is natural to define marginal expectations. We will utilize the concept of $\oplus$-marginal of a function [31].

Definition 22. *[31] Let h be a function defined on $X_1 \times X_2$ and taking values on $[0,1]$. We define the* $\oplus$-marginals *of h as*

$$h^{\oplus}_{X_1}(x_{1i}) = \bigoplus_{x_{2j}\in X_2} h(x_{1i}, x_{2j}) = h(x_{1i}, x_{21}) \oplus h(x_{1i}, x_{22}) \oplus \ldots \oplus h(x_{1i}, x_{2m}), \tag{40}$$

$$h^{\oplus}_{X_2}(x_{2j}) = \bigoplus_{x_{1i}\in X_1} h(x_{1i}, x_{2j}) = h(x_{11}, x_{2j}) \oplus h(x_{12}, x_{2j}) \oplus \ldots \oplus h(x_{1n}, x_{2j}), \tag{41}$$

where $\oplus$ is a t-conorm (see Definition 6), n is the cardinality of X_1 and m is the cardinality of X_2.

Definition 23. *Let $(X_1, \mathcal{A}_{X_1}, \mu_1)$ and $(X_2, \mathcal{A}_{X_2}, \mu_2)$ be measurable spaces and let h be a function defined on $X_1 \times X_2$ and taking values on $[0,1]$. We define the marginal $\oplus$-expectations as*

$$E^{\oplus}_{X_i}(h) = \oint h^{\oplus}_{X_i} \circ \mu_i, \quad i = 1, 2, \tag{42}$$

where $h^{\oplus}_{X_i}$ are the $\oplus$-marginals of h.

4.2. The Case of Two Functions

We will now assume that we have two different functions, one for each marginal space, and define parameters that combine the information provided by the marginal spaces.

Definition 24. *Let $(X_1, \mathcal{A}_{X_1}, \mu_1)$ and $(X_2, \mathcal{A}_{X_2}, \mu_2)$ be measurable spaces, and let h_1, h_2 be functions defined on X_1 and X_2 respectively, taking values on $[0,1]$. We define the upper and lower global expectation of h_1 and h_2 as*

$$\underline{\phi}^{\star}_{\odot}(h_1, h_2) = \oint h^{\star}_{12} \circ \underline{\mu}^{\odot}_{12} \quad (lower), \tag{43}$$

$$\overline{\phi}^{\star}_{\odot}(h_1, h_2) = \oint h^{\star}_{12} \circ \overline{\mu}^{\odot}_{12} \quad (upper). \tag{44}$$

where $\star$ and $\odot$ are arbitrary t-norms (see Definition 5), $h^{\star}_{12}(x_1, x_2) = h_1(x_1) \star h_2(x_2), \forall (x_1, x_2) \in X_1 \times X_2$ and $\underline{\mu}^{\odot}_{12}$ and $\overline{\mu}^{\odot}_{12}$ are the interior and exterior product measures of μ_1 and μ_2 respectively.

The next proposition shows that both expectations coincide when $\star$ is the min t-norm.

Proposition 8. *Assume the conditions in Definition 24. If $\star$ is the* min *t-norm, it holds that*

$$\underline{\phi}^{\star}_{\odot}(h_1, h_2) = \overline{\phi}^{\star}_{\odot}(h_1, h_2). \tag{45}$$

Proof. According to ([31], Proposition 8), the α-cuts generated by $h^{\star}$ belong to $\mathcal{R}$ when $\star$ is the min t-norm. Furthermore, Equation (12) establishes that $\underline{\mu}^{\odot}_{12} = \overline{\mu}^{\odot}_{12}$ for the elements of $\mathcal{R}$, which proves the result. □

As a consequence of Proposition 8, when using the min t-norm we will just write $\phi^{\star}_{\odot}$ for both $\underline{\phi}^{\star}_{\odot}$ and $\overline{\phi}^{\star}_{\odot}$.

The global expectation is in fact an extension of the monotone expectation in the sense expressed by the next theorem.

Theorem 3. *Let $(X, \mathcal{A}, \mu)$ be a measurable space and let h be a function defined on X and taking values on $[0,1]$. Consider the product space $X \times X$ and let both $\star$ and $\odot$ be the* min *t-norm. Then,*

$$\phi^{\star}_{\odot}(h,h) = \phi^{\min}_{\min}(h,h) = E_{\mu}(h). \tag{46}$$

Proof. Assume $X = \{x_1, x_2, \ldots, x_n\}$. Then

$$\forall (x_1, x_2) \in X \times X \quad \Rightarrow \quad h^{\min}(x,y) = \min\{h(x_1), h(x_2)\}.$$

Without loss of generality, we can assume that

$$h(x_1) < h(x_2) < \ldots < h(x_n),$$

in which case the α-cuts generated by $h^{\min}$ are of the form

$$H_{\alpha_i} = \{(x_k, x_l) \in X \times X |\, k, l \geq i\},$$

which are elements of the class $\mathcal{R}$ with their two projections being identical.

Since we are using the min t-norm to construct the product measure, it turns out that the measure in each α-cut of the product space is equal to the measure assigned by μ to the α-cuts in the marginal space, and thus

$$\begin{aligned}\phi^{\min}_{\min}(h,h) &= \oint h^{\min} \circ \mu^{\min}_{12} = \sum_{i=1}^{n \times n} \mu^{\min}_{12}(H_{\alpha_i})(\alpha_i - \alpha_{i-1}) \\ &= \sum_{i=1}^{n} \left[\mu(H^{\downarrow X}_{\alpha_i})\right](\alpha_i - \alpha_{i-1}) = E_{\mu}(h).\end{aligned}$$

□

Example 7. *Consider a reference set $X = \{x_1, x_2, x_3\}$ and the function defined as $h(x_1) = 0.1, h(x_2) = 0,4, h(x_3) = 0.7$. The function $h^{\min}$ is displayed in Table 4.*

Table 4. Values of the function $h^{\min}$.

	x_1	x_2	x_3
x_1	0.1	0.1	0.1
x_2	0.1	0.4	0.4
x_3	0.1	0.4	0.7

It can be seen how the diagonal contains the original values of h, and its α-cuts are

$$\begin{aligned}
H_{0.1} &= X \times X, & \mu^{\min}_{12}(H_{0.1}) &= \min\{\mu(X), \mu(X)\}, \\
H_{0.4} &= \{x_2, x_3\} \times \{x_2, x_3\}, & \mu^{\min}_{12}(H_{0.4}) &= \min\{\mu(\{x_2, x_3\}), \mu(\{x_2, x_3\})\}, \\
H_{0.7} &= \{x_3\} \times \{x_3\}, & \mu^{\min}_{12}(H_{0.1}) &= \min\{\mu(\{x_3\}), \mu(\{x_3\})\},
\end{aligned}$$

and therefore

$$\begin{aligned}\phi_{\min}^{\min}(h,h) &= \mu_{12}^{\min}(H_{0.1})(0.1-0)+\mu_{12}^{\min}(H_{0.4})(0.4-0.1)+\mu_{12}^{\min}(H_{0.7})(0.7-0.4)\\ &= \mu(X)(0.1-0)+\mu(\{x_2,x_3\})(0.4-0.1)+\mu(\{x_3\})(0.7-0.4)\\ &= E_\mu(h).\end{aligned}$$

Likewise for the common expectation, the global expectation is not normalized, but it can be easily normalized in the same way as we did for the common expectation case, as stated in the next definition.

Definition 25. *Let $(X_1,\mathcal{A}_{X_1},\mu_1)$ and $(X_2,\mathcal{A}_{X_2},\mu_2)$ be measurable spaces and let h_1 and h_2 be functions defined on X_1 and X_2 respectively and taking values on $[0,1]$. We define the global coefficients of concordance of h_1 and h_2 as*

$$\Phi_1(h_1,h_2) = \frac{\phi_{\min}^{\min}(h_1,h_2)}{\min\{E_{\mu_1}(h_1),E_{\mu_2}(h_2)\}}, \tag{47}$$

$$\Phi_2(h_1,h_2) = \frac{\phi_{\min}^{\min}(h_1,h_2)}{\sqrt{E_{\mu_1}(h_1)E_{\mu_2}(h_2)}}. \tag{48}$$

Example 8. *(Continuation of Example 5)*

Using the data in Table 3 we can obtain the function $h_{12}^{\min}$, the values of which are given in Table 5.

Table 5. Values of the function $h_{12}^{\min}$.

Course	x_1	x_2	x_3	x_4
x_1	0.2	0.3	0.8	0.9
x_2	0.2	0.3	0.8	0.8
x_3	0.2	0.3	0.3	0.3
x_4	0.2	0.2	0.2	0.2

Using the fuzzy measure in Table 2, we find that $\phi_{\min}^{\min}=0.6$ and the global concordance coefficients are $\Phi_1(h_1,h_2)=0.953$ and $\Phi_2(h_1,h_2)=0.984$.

The value of the global expectation (0.6) is very close to the values of the monotone expectations for each student in Example 5 (0.61 and 0.65 respectively). It can be interpreted as the fact that the grades of both students are acceptable individually and also globally, which is reflected in high values of the global coefficients of concordance. Note how the global expectation is not detecting the fact that both students have different profiles (scientific and humanistic), while the common expectation detected this fact yielding a much lower value (0.25) resulting in lower values of the coefficients of concordance as well.

5. Conclusions

With the introduction of the concept of monotone variance, we have complemented the already known concept of monotone expectation. It can be regarded as a measure of dispersion with respect to a central position measure. We have also introduced the concepts of central and non-central monotone moments, that can serve as a vehicle to define further statistical parameters based on fuzzy measures as, for instance, shape measures. The potential application scope is certainly wide, as it covers non-additive scenarios like the ones described in the examples in this paper, and just to mention some of them, such scenarios can be found in Engineering and Social Sciences applications.

The common expectation and concordance coefficients can be interpreted as measures of match between the functions, and in that sense can provide information about to which extent one function explains the other one. A possible application of these concepts is the development of prediction models when the measures are not additive.

Thanks to the developments in [31] we have been able to extend the concept of monotone expectation to product spaces, where, in addition, we have shown how to marginalize the information provided by a function over a product space using the marginal $\oplus$-expectations.

All the developments in this paper are restricted to finite reference sets. Even though it covers a wide variety of practical applications, it is worth exploring the formulation of the results obtained here to uncountable reference sets, which seems to be a promising research line. The first step in this direction would be the extension of the results in [31] to continuous domains.

Author Contributions: Investigation, F.R., M.M. and A.S.; writing—original draft, F.R., M.M. and A.S.; writing—review and editing, F.R., M.M. and A.S. All authors have read and agreed to the published version of the manuscript.

Funding: This research was funded by the Spanish Ministry of Science and Innovation through grants TIN2016-77902-C3-3-P, PID2019-106758GB-C32 and by ERDF-FEDER funds.

Conflicts of Interest: The authors declare no conflict of interest.

References

1. Sugeno, M. Theory of Fuzzy Integrals and Its Applications. Ph.D. Thesis, Tokyo Institute of Technology, Tokyo, Japan, 1974.
2. Choquet, G. Theory of capacities. *Annales de l'Institut Fourier* **1954**, *5*, 131–295. [CrossRef]
3. Li, J. On null-continuity of monotone measures. *Mathematics* **2020**, *8*, 205. [CrossRef]
4. Beliakov, G.; James, S.; Wu, J. *Discrete Fuzzy Measures*; Studies in Fuzziness and Soft Computing; Springer: Berlin, Germany, 2020; Volume 382.
5. Walley, P. *Statistical Reasoning with Imprecise Probabilities*; Chapman and Hall: London, UK, 1991.
6. Walley, P. *BI Statistical Methods. Volume I: Foundations*; Prescience Press: New York, NY, USA, 2015.
7. Dempster, A.P. Upper and lower probabilities induced by a multivalued mapping. *Ann. Math. Stat.* **1967**, *38*, 325–339. [CrossRef]
8. Shafer, G. *A Mathematical Theory of Evidence*; Princeton University Press: Princeton, NJ, USA, 1976.
9. Buckley, J. *Fuzzy Probability and Statistics*; Studies in Fuzziness and Soft Computing; Springer: Berlin, Germany, 2006; Volume 196.
10. Coppi, R.; Gil, M.A.; Kiers, H.A.L. The fuzzy approach to statistical analysis. *Comput. Stat. Data Anal.* **2006**, *51*, 1–14. [CrossRef]
11. D'Urso, P.; Gil, M.A. Fuzzy data analysis and classification. *Adv. Data Anal. Classif.* **2017**, *11*, 645–657. [CrossRef]
12. Intan, R.; Mukaidono, M. Fuzzy conditional probability relations and their applications in fuzzy information systems. *Knowl. Inf. Syst.* **2004**, *6*, 345–365. [CrossRef]
13. Nguyen, H.; Wu, B. *Fundamentals of Statistics with Fuzzy Data*; Studies in Fuzziness and Soft Computing; Springer: Berlin, Germany, 2006; Volume 198.
14. Vierti, R. *Statistical Methods for Fuzzy Data*; Wiley: Chichester, UK, 2011.
15. Kruse, R.; Held, P.; Moewes, C. On fuzzy data analysis. *On Fuzziness—A Homage to Lotfi A. Zadeh, Volume 1*; Studies in Fuzziness and Soft Computing; Springer: Heidelberg, Germany, 2013; Volume 298, pp. 343–347.
16. Dijkman, J.G.; van Haeringen, H.; Lange, S.J. Fuzzy numbers. *J. Math. Anal. Appl.* **1983**, *92*, 301–341. [CrossRef]
17. D'Urso, P. Informational paradigm, management of uncertainty and theoretical formalisms in the clustering framework: A review. *Inf. Sci.* **2017**, *400–401*, 30–62. [CrossRef]
18. Tanaka, H.; Uejima, S.; Asai, K. Linear regression analysis with fuzzy model. *IEEE Trans. Syst. Man Cybern.* **1982**, *12*, 903–907.
19. Parchami, A.; Taheri, S.M.; Mashinchi, M. Fuzzy *p*-value in testing fuzzy hypotheses with crisp data. *Stat. Pap.* **2010**, *51*, 209. [CrossRef]
20. Grzegorzewski, P.; Hryniewicz, O. Soft methods in statistical quality control. *Control Cybern.* **2000**, *29*, 119–140.
21. Zhang, R.; Ashuri, B.; Deng, Y. A novel method for forecasting time series based on fuzzy logic and visibility graph. *Adv. Data Anal. Classif.* **2017**, *11*, 759–783. [CrossRef]

22. Gil, M.A.; López-Díaz, M. Fundamentals and Bayesian analyses of decision problems with fuzzy-valued utilities. *Int. J. Approx. Reason.* **1996**, *15*, 95–115. [CrossRef]
23. Denoeux, T. Maximum likelihood estimation from fuzzy data using the EM algorithm. *Fuzzy Sets Syst.* **2011**, *183*, 72–91. [CrossRef]
24. Quost, B.; Denoeux, T.; Li, S. Parametric classification with soft labels using the evidential EM algorithm: Linear discriminant analysis versus logistic regression. *Adv. Data Anal. Classif.* **2017**, *11*, 659–690. [CrossRef]
25. Blanco-Fernández, A.; Casals, M.R.; Colubi, A.; Corral, N.; García-Bárzana, M.; Gil, M.A.; González-Rodríguez, G.; López, M.T.; Lubiano, M.A.; Montenegro, M.; et al. A distance-based statistical analysis of fuzzy number-valued data. *Int. J. Approx. Reason.* **2014**, *55*, 1487–1501. [CrossRef]
26. Wu, H.C. Statistical hypotheses testing for fuzzy data. *Inf. Sci.* **2005**, *279*, 446–459. [CrossRef]
27. Calcagnì, A.; Lombardi, L.; Pascali, E. A dimension reduction technique for two-mode non-convex fuzzy data. *Soft Comput.* **2016**, *20*, 749–762. [CrossRef]
28. Colubi, A.; González-Rodríguez, G.; Gil, M.A.; Trutschnig, W. Nonparametric criteria for supervised classification of fuzzy data. *Int. J. Approx. Reason.* **2011**, *52*, 1272–1282. [CrossRef]
29. Coppi, R.; D'Urso, P.; Giordani, P. Fuzzy and possibilistic clustering for fuzzy data. *Comput. Stat. Data Anal.* **2012**, *56*, 915–927. [CrossRef]
30. Bolaños, M.J.; De Campos, L.M.; González, A. Convergence properties on monotone expectation and its applications to the extension of fuzzy measures. *Fuzzy Sets Syst.* **1989**, *33*, 201–212. [CrossRef]
31. Reche, F.; Morales, M.; Salmerón, A. Construction of fuzzy measures over product spaces. *Mathematics* **2020**, *8*, 1605. [CrossRef]
32. Reche, F.; Salmerón, A. Operational approach to general fuzzy measures. *Int. J. Uncertain. Fuzziness -Knowl.-Based Syst.* **2000**, *8*, 369–382. [CrossRef]
33. De Campos, L.M.; Bolaños, M.J. Representation of fuzzy measures through probabilities. *Fuzzy Sets Syst.* **1989**, *31*, 23–36. [CrossRef]
34. Schweizer, B.; Sklar, A. *Probability Metric Spaces*; Elsevier North Holland: New York, NY, USA, 1983.
35. Huber, P.H. *Robust Statistics*; Wiley Series in Probability and Mathematical Statistics; John Wiley and Sons: Hoboken, NJ, USA, 1981.

mathematics

MDPI

Article

Application of Hexagonal Fuzzy MCDM Methodology for Site Selection of Electric Vehicle Charging Station

Arijit Ghosh [1], Neha Ghorui [2], Sankar Prasad Mondal [3,*], Suchitra Kumari [4], Biraj Kanti Mondal [5], Aditya Das [6] and Mahananda Sen Gupta [4]

1 Department of Mathematics, St. Xavier's College (Autonomous), Kolkata 700016, India; arijitghosh@sxccal.edu
2 Department of Mathematics, Prasanta Chandra Mahalanobis Mahavidyalaya, Kolkata 700108, India; neha.mundhra@thebges.edu.in
3 Department of Applied Science, Maulana Abul Kalam Azad University of Technology, West Bengal, Haringhata 741249, India
4 Department of Commerce, St. Xavier's College (Autonomous), Kolkata 700016, India; suchitra@sxccal.edu (S.K.); mahanandasg@gmail.com (M.S.G.)
5 Department of Geography, Netaji Subhas Open University, Kolkata 700016, India; birajmondal.kolkata@gmail.com
6 Department of Commerce, University of Calcutta, Kolkata 700073, India; dasaditya635@gmail.com
* Correspondence: sankarprasad.mondal@makautwb.ac.in; Tel.: +91-96-3557-8078

Abstract: In this paper, the application of hexagonal fuzzy multiple-criteria decision-making (MCDM) methodology for the site selection of electric vehicle charging stations is considered. In this regard, four factors and thirteen sub-factors have been taken into consideration for E-vehicle charging site selection. In this research, the geographic information system (GIS) has been incorporated with MCDM techniques. The fuzzy analytic hierarchy process (FAHP) is used to obtain a fuzzy weight of factors and sub-factors. MCDM tools fuzzy technique for order of preference by similarity to ideal solution (FTOPSIS) and fuzzy complex proportional assessment (FCOPRAS) have been used to rank the selected sites. A centroid-based method for defuzzification and distance measure between two hexagonal fuzzy numbers (HFN) has been developed for this paper. A practical example in Howrah, India, is considered to show the applicability and usefulness of the model. The results depict the suitability of the proposed research. Comparative and sensitivity analyses have been demonstrated to check the reliability, robustness and effectiveness of the proposed method.

Keywords: site selection; FAHP; FTOPSIS; FCOPRAS; hexagonal fuzzy number

Citation: Ghosh, A.; Ghorui, N.; Mondal, S.P.; Kumari, S.; Mondal, B.K.; Das, A.; Gupta, M.S. Application of Hexagonal Fuzzy MCDM Methodology for Site Selection of Electric Vehicle Charging Station. *Mathematics* **2021**, *9*, 393. https://doi.org/10.3390/math9040393

Academic Editor: Michael Voskoglou

Received: 28 December 2020
Accepted: 9 February 2021
Published: 16 February 2021

1. Introduction

Electric vehicles play a very momentous role in addressing fossil fuel pollution and they are capable of making a paradigm shift in the entire transportation sector. Transportation is a significant contributor to urban air pollution and the reduction of urban city emission is the need of the hour. Electric vehicles make the world more liveable and provide a pollution-free mode of transportation in urban areas. The high level of pollution is degrading the environment and it has made the concept of sustainable development a fairy-tale phenomenon [1,2]. Sustainable consumption needs to be adopted by conducting timely environmental and sustainability assessments to prevent any large-scale ecological disaster from happening. This is where electrical vehicles come into the picture, with sustainability, environment-protecting and pocket-friendly being a few of their rewards. Electric vehicles use a minimum of one electric motor or traction motor for propulsion. They maybe self-contained with a generator or battery for converting the fuel into electricity or they may be power-drivenvia a collector scheme by using electricity from off-vehicle sources [3]. The problem of the energy crisis in the world can be tackled in the future by using this option. Ever-rising gas prices force people to look for alternative modes of energy supply, or to even switch over to walking or availing public transportation. The other reason why electrical vehicles are preferred over fossil-fueled vehicles is due to their eco-friendly nature and the cheaper cost of driving them [4,5]. Filling up an electric car costs less than filling up a full tank. Internal combustion engines emit more CO_2 emissions than electric vehicles [6–8].

Moreover, for the promotion of charging stations, we need to focus on two facets: the power of the government and the role of the market. The government influences taxation policies and subsidy policies directly on the stations, thus affecting the income of the charging station. On the other hand, from the market viewpoint, the key is the demand and supply relationship, i.e., the consumers who demand the electric vehicles and the supply of electric vehicle charging and the electric vehicle charging stations [9].Electric vehicles can usually be charged in three ways i.e., by using inductive charging, conductive charging and the battery replacement method [7].The inductive method of charging works through an electromagnetic transmission keeping no contact between the charging station and the electric vehicle. The conductive method has a battery connected by a cable which is directly plugged into an electricity provider, whereas the battery replacement method replaces the discharged batteries with new batteries in a charging station, keeping in mind the internal connections and the dimensions of the batteries, resulting in it being the least used method. The charging station operators mostly use conductive charging since it is more efficient and cheaper [10].The benefits of electric vehicles outweigh the problems. They are energy efficient since they convert a higher rate of electrical energy from the grid to power, compared to gasoline vehicles. They do not make noise or pollute the air since the electric motors have a quiet and smooth operation, requiring low maintenance. Their use of electricity reduces the need for fossil fuels [11]. The major problem with electric vehicles is their shorter driving range. The time taken to recharge the vehicle is also usually long and the battery packs are heavy and expensive [12]. Establishing an economic, efficient and convenient electric vehicle charging station will speed up the purchase of electric vehicles for consumers, thereby developing the sector and it vice-versa will hinder electric vehicle adoption leading to lesser incentives for investing in its development [13]. Selecting a suitable site for electric vehicle charging stations is of utmost importance since it has a direct linkage to the operation efficiency and the service quality of the charging stations throughout its entire life cycle [14–16]. Electric vehicle charging station (EVCS) site selection requires a comprehensive analysis of social, economic, environmental, operations and urban planning for which several prospective and potential alternative locations were considered with respect to a range of criteria. Hence, location selection is seen as a multiple-criteria decision-making (MCDM) problem. The technique for order of preference by similarity to ideal solution (TOPSIS) is an aggregation of the MCDM method, having the benefits of computational efficiency and the ability to measure in simple mathematical form the relative performance of each alternative criterion [17–23]. However, due to information deficiency and vagueness along with human qualitative judgements, some criteria fail to be measured by a crisp value and can only be a fuzzy value [24] such as the development planning of road networks and petrol stations. Fuzzy theory effectively tackles this issue [25].In today's times, finding a suitable EVCS site requires a multi-criteria approach as well as accuracy and reliability in the maps [26]. The efficacy of the final decision depends on the quality of the data which are used to produce criteria maps. The geographic information system-based multi-criteria decision analysis (GIS-based MCDA) method converts spatial as well as non-spatial data into information with respect to the decision-maker's own judgement [27]. To promote e-vehicle usage, long-term infrastructure installation is the need of the hour. For long-term infrastructure creation, detailed study of the criteria and in-depth analysis is required. This paper addresses the problem of charging infrastructure creation for e-vehicles in a city environment. This model can be replicated in different cities for a pollution-free mode of transport. Widespread usage of e-vehicles will make the environment greener.

1.1. Literature Review

The construction of an electric vehicle charging station is the turning point in its life cycle. Selecting an appropriate place for setting up the charging station and determining its capacity will benefit all the stakeholders and endorse sustainable development of the entire industry. Some studies have been shown this with respect to economy and technology. Lee et al. [28] examined the price competition among EVCSs using renewable power generators by employing game theory with relevant physical constraints. Rivera et al. [29] put forward a novel architecture for plug-in electric vehicles: a DC charging station using a grid-tied neutral point clamped converter. Wang et al. [30] utilized three-phase uncontrolled rectification chargers to study the harmonic amplification of EVCSs. Ding, Hu and Song [31] studied the energy storage system as a potential supplement of an electric bus fast charging station by employing mixed integer nonlinear programming for valuing the energy storage system. Fan, Sainbayar and Ren [32] calculated the effect of limiting electric vehicles' full state of charge to total charged energy and the revenue of EVCS. Capasso and Veneri [33] built a DC fast charging architecture for plug-in hybrid vehicles as well as fully electric vehicles by integrating the fleets of hybrid/road electric vehicles with renewable energy sources. Li et al. [34] studied the control of EVCSs and the management of energy and fit them in a dynamic price framework by developing a real-time simulation system for evaluating how the EVCS meets the charging and discharging requirements for grid-to-vehicle, vehicle-to-grid and vehicle-to-vehicle. Nansai et al. [35] conducted the life-cycle analysis on EVCSs in three phases of installation, transportation and production of electric vehicle charging equipment and then compared the carbon dioxide, carbon monoxide, sulphur oxides and nitrogen oxide emissions of electric vehicles and gasoline vehicles.

Research also focuses on the electric vehicle's size, its placement and the fields in which it is used. Khalkhali et al. [36] determined the optimal location and size of plug-in hybrid EVCSs using data envelopment analysis by maximizing the benefit of the distribution system management. Frade et al. [15] studied the location of EVCSs in Portugal by employing a maximal covering model for defining the capacity and number of EVCSs. Liu [37] investigated the nascent electric vehicle market in Beijing and then formulated an assignment model for different charging infrastructure assignment strategies. Wirges, Linder and Kessler [38] predicted several scenarios for charging infrastructure development until 2020 in Germany and also formulated a dynamic spatial model for the allocation of the EVCSs in the country. Wang et al. [39] developed an EVCS location model considering the electricity consumed along the roads and the oil sales, after which the EVCS quantity and layout were calculated. He et al. [40] developed an equilibrium modeling framework for deciding the optimal allocation of the charging stations in a metropolitan area. Liu et al. [41] proposed a two-step screening method for determining the optimal EVCS site, taking environmental factors and service radius into consideration.

There have been a large number of researches which have employed a type of decision-making method. Pashajavid and Golkar [42] put forward a scenario optimization algorithm by using multivariate stochastic modeling methodology for load demand by allocating the charging station of plug-in electric vehicles and also used a particle swarm optimization algorithm to minimize voltage deviation and energy loss in the distribution system. Chen, Kockelman and Khan [43] determined the optimal charging station location assignment in Seattle by developing a mixed-integer programming model which minimized the station access cost of the electric vehicle users and took trip attributes, parking demand, population density and local job as constraints. Sathaye and Kelley [44] used a continuous facility location model considering the demand uncertainty for finding the optimal location of the public-funded electric vehicle infrastructure on the highway corridors. Wang et al. [45] determined the optimal size and location of EVCSs with respect to smart grids by proposing a multi-objective EVCS planning method, maximizing the electric vehicle traffic flow under the constraint of battery capacity and the final optimal solution data-envelopment analysis was obtained and employed. Dong, Liu and Lin [46] formulated an activity-based assessment method for the evaluation of the feasibility of electric vehicles taking the heterogeneous travelling population and subsequently applying a genetic algorithm for determining the sub optimal location. Xu et al. [47] identified the candidate centralized charging station using a proposed mathematical model with the minimum total transportation distance. You and Hsieh [48] used the mixed-integer programming model to determinine the best location of the EVCS which would maximize the number of people who could complete round-trip itineraries, along with developing an efficient hybrid genetic algorithm which would obtain a compromised solution in a reasonable time. Lee et al. [49] first collected users' charging and traveling behaviors along with the batteries' state of charge and then proposed a location model of the rapid EVCS by using a probabilistic distribution function for the remaining fuel range. Baouche et al. [50] formulated an integer linear optimization model, taking the electric vehicle's input consumption as the optimal model. Sadeghi-Barzani [16] postulated a mixed-integer non-linear optimization approach for the determination of the optimal size and place of fast charging stations. Yao et al. [51] devised a multi-objective collaborative planning strategy to tackle the planning issue in electric vehicle charging systems and integrated power distributions by using a decomposition-based multi-objective evolutionary algorithm model and equilibrium-based traffic assignment model. Ma et al. [52] used an agent-based model to optimize the sketch of the initial EVCS. Chung and Kwon [53] devised a multi-period optimization model to perform the EVCS planning, basing it on the real traffic flow data of the Korean Expressway network in 2011. Lam, Leung and Chu [54] used four solution methods i.e., the greedy approach, chemical reaction optimization, iterative mixed-integer linear programming and effective mixed-integer linear programming for the EVCS place problem by checking them against practical and artificial cases. Cai et al. [55] found the relation between public charging infrastructure development and the travel patterns mined from big-data taken using large-scale trajectory data in Beijing.

Multi-criteria decision-making was used in transhipment site selection [56], shopping mall site selection [57], railway station site selection [58], waste management site selection [59] and macro-site selection of solar/wind hybrid power station [60].

The major drawback of the aforementioned studies which use multi-criteria decision-making approaches is the lack of incorporation between evaluation criteria and spatial data, since selecting a suitable site for an EVCS is a spatial decision predicament which involves conflicting, incommensurate and multiple evaluation criteria and a large set of evaluation criteria. Two unique parts of research i.e., multi-criteria decision-making and GIS can aid each other to overcome the intersections between evaluation criteria and spatial data. GIS integrates the spatially-referenced data in a problem-solving environment whereas multi-criteria decision-making refers to procedures and techniques used in structural decision problems by evaluating, prioritizing and designing the alternative decisions [61]. GIS-based multi-criteria decision-making techniques are usually used for spatial decision problems such as freight village site selection [62] wind power plant site selection [63] and refugee camp site selection [64]. Even though there have been multiple studies on the GIS-based multi-criteria decision-making approach on spatial decision problems as mentioned above, still there is a gap in the selection of EVCS sites using GIS.

1.2. Objectives of the Study

The present research has the following objectives:

(a) Identification of the most preferred site for the construction of an e-vehicle charging station.
(b) Application of hexagonal fuzzy numbers (HFN) in AHP-TOPSIS and AHP-COPRAS to obtain the ranking of the selected sites.

1.3. Novelties of the Study

Several researchers have explored fuzzy numbers with MCDM techniques AHP, TOPSIS, and COPRAS. Hardly any research has been done using hexagonal fuzzy numbers under MCDM methodology. The HFN defuzzification formulae has been developed and utilized. Distance measured between two HFN is also defined. The formulae has been developed to calculate the hexagonal fuzzy weight of factors and sub-factors. A technique has been developed to incorporate more than one decision maker's opinion into a single comprehensive value in terms of HFN. Two different ranking methods, fuzzy AHP-TOPSIS and fuzzy AHP-COPRAS has been used in this research. GIS software has been used for distance measurement and graphical presentation of the selected sites.

1.4. Structure of the Paper

The remainder of the paper is organized in the following way: Section 2 depicts the concept of fuzzy numbers, HFN, its properties, distance measure, and defuzzification formulae. It also includes the MCDM technique AHP, fuzzy TOPSIS and fuzzy COPRAS. Section 3 contains the numerical application and description of factors and sub-factors. Section 4 represents the comparison and sensitivity analysis, respectively. The results and discussion covered in Sections 5 and 6 describes the future scope and conclusion.

The structured framework of the study represented in Figure 1 describes the sequential steps followed in this paper. After the initial selection of factors and sub-factors, their weights are calculated by AHP and FAHP. Subsequently, FTOPSIS and FCOPRAS MCDM tools have been applied for the selection of preferred locations followed by comparative and sensitivity analysis, overall results and discussions.

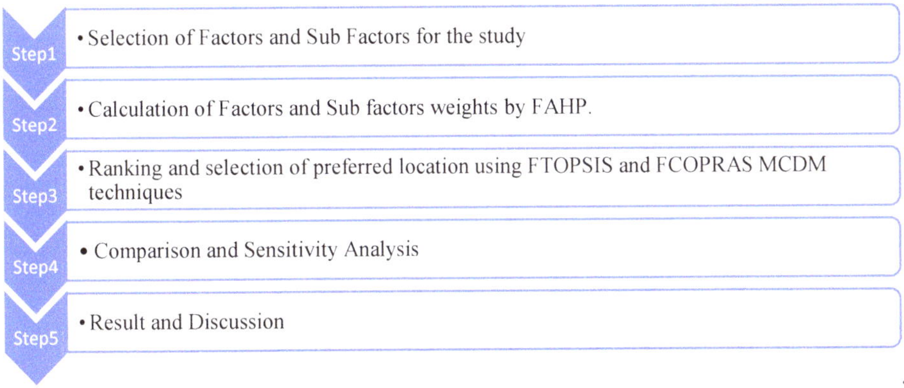

Figure 1. Structural framework of the study. FTOPSIS: fuzzy technique for order of preference by similarity to ideal solution; FCOPRAS: fuzzy complex proportional assessment; MCDM: multiple-criteria decision-making.

2. Preliminaries

2.1. Fuzzy Set

The fuzzy set theory was developed by the author [25] to deal with the impreciseness of real-life issues [65–73].

Definition 1. *A set $\hat{T}$, defined as $\hat{T} = \{(\tau, \mu_{\hat{T}}(\tau) : \tau \in \hat{T}, \mu_{\hat{T}}(\tau) \in (0,1)\}$, where $\mu_{\hat{T}}(\tau)$ represents the membership function of $\hat{T}$ which takes value from zero to one. In real life situations, where the information is vague and uncertain, fuzzy logic can be efficiently used to deal with these problems.*

Definition 2. *Hexagonal fuzzy number (HFN) A number $\alpha_{HFN} = \{(h_1, h_2, h_3, h_4, h_5, h_6), \mu_\alpha(x)\}$ is defined as HFN if it satisfies the following properties:*

a. $\mu_\alpha(x)$ is a continuous function [0, 1].
b. $\mu_\alpha(x)$ is strictly increasing continuous function in $[h_1, h_2]$ and $[h_2, h_3]$.

c. $\mu_{\alpha}(x)$ attains maximum value 1 in $[h_3, h_4]$.
d. $\mu_{\alpha}(x)$ is strictly decreasing continuous function in $[h_4, h_5]$ and $[h_4, h_5]$.

The following Figure 2 represents the membership function of symmetric HFN.

$$\mu_{\alpha}(x) = \begin{cases} 0, & for|x \leq h_1 \\ \frac{0.5(x-h_1)}{h_2-h_1}, & for|h_1 \leq x \leq h_2 \\ 0.5 + \frac{0.5(x-h_2)}{h_3-h_2}, & for\ h_2 \leq x \leq h_3 \\ 1, & for\ h_3 \leq x \leq h_4 \\ 1 - \frac{0.5(x-h_4)}{h_5-h_4}, & for\ h_4 \leq x \leq h_5 \\ 0.5 - \frac{0.5(x-h_5)}{h_6-h_5}, & for\ h_5 \leq x \leq h_6 \\ 0, & for\ x \geq h_6 \end{cases} \quad (1)$$

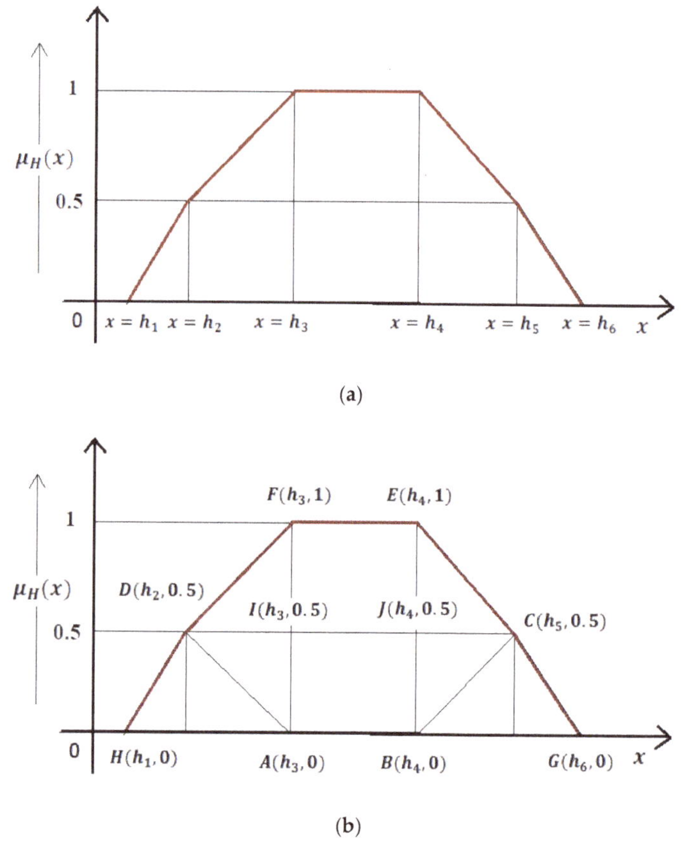

Figure 2. (**a**) Representation of the membership function of linear hexagonal fuzzy number (HFN). (**b**) Hexagonal fuzzy number as union of different regions.

Where h_1, h_2, h_3, h_4, h_5 *and* h_6 are real numbers such that $h_1 \leq h_2 \leq h_3 \leq h_4 \leq h_5 \leq h_6$.

Note 2.1: Hexagonal fuzzy numbers (HFN) capture the hesitancy and uncertainty in broader aspects compared to triangular fuzzy numbers (TFN), trapezoidal fuzzy numbers (TrFN), and pentagonal fuzzy numbers (PFN) as the latter undertakes three, four, and five numbers, respectively, to represent the impreciseness and ambiguity of the decision maker (DM). If we consider the linguistic term corresponding to TFN then it is represented numerically as (low, medium, high) where the medium value corresponds to the best possible chances of the quantity. For TrFN, the numerical presentation is expressed as (very low, low, high, very high) and the maximum possibility lies in a range of (low, high). Further, if we take PFN, it considers the numerical behavior as (very low, low, medium, high, very high), here the middle value denotes the maximum possibility, which is 1. HFN undertakes the highest level of distribution in the numerical form as (very very low, very low, low, high, very high, very very high), i.e., the maximum possible spread can be accommodated in HFN. With respect to a developing country in an unplanned city, many of the attributes under consideration have a wider range of linguistic representation which cannot be captured using TFN, TrFN and PFN. For example, the consumption level in a city with a heterogeneous earning pattern can be better captured using HFN with a wider range. The hesitancy of decision makers in such an environment requires a broader range of values to depict uncertainty; HFN enables this.

The hesitancy of decision makers (DMs) can be better captured in HFN. Different levels of hesitancy in linguistic terms corresponding to different HFN. It is easy to understand and analyze that the linguistic terms "weakly important" represented in HFN as (1.1, 1.2, 1.3, 1.4, 1.5, 1.6), "absolutely important" represented in HFN as (4.6, 4.8, 5, 5.2, 5.4, 5.7) and so on. These range from lower values of HFN to higher values. Similarly, the inverse of these HFN, specifically, the opposites of the linguistic terms can be obtained by using Equation (7).

2.2. Arithmetic Operations of Linear Symmetric HFN

Let $U = (u_1, u_2, u_3, u_4, u_5, u_6)$ and $V = (v_1, v_2, v_3, v_4, v_5, v_6)$ be two HFN, then their general arithmetic operations can be defined in the following way:

1. Addition:

$$(U + V) = (u_1 + v_1,\ u_2 + v_2,\ u_3 + v_3,\ u_4 + v_4,\ u_5 + v_5,\ u_6 + v_6) \tag{2}$$

2. Subtraction:

$$(U - V) = (u_1 - v_6,\ u_2 - v_5,\ u_3 - v_4,\ u_4 - v_3,\ u_5 - v_2,\ u_6 - v_1) \tag{3}$$

3. Multiplication:

$$(U \times V) = (u_1v_1,\ u_2v_2,\ u_3v_3,\ u_4v_4,\ u_5v_5,\ u_6v_6) \tag{4}$$

4. Scalar Multiplication:

$$\alpha U = (\alpha u_1,\ \alpha u_2,\ \alpha u_3,\ \alpha u_4,\ \alpha u_5,\ \alpha u_6) \tag{5}$$

5. Division:

$$\left(\frac{U}{V}\right) = \left(\frac{u_1}{v_6}, \frac{u_2}{v_5}, \frac{u_3}{v_4}, \frac{u_4}{v_3}, \frac{u_5}{v_2}, \frac{u_6}{v_1}\right) \tag{6}$$

6. Inverse:

$$U^{-} = \left(\frac{1}{u_6}, \frac{1}{u_5}, \frac{1}{u_4}, \frac{1}{u_3}, \frac{1}{u_2}, \frac{1}{u_1}\right) \tag{7}$$

2.3. Distance Measure of Two HFN

Definition 3. *Let* $\widetilde{A_d} = (a_1, a_2, a_3, a_4, a_5, a_6)$ *and* $\widetilde{B_d} = (b_1, b_2, b_3, b_4, b_5, b_6)$ *be two HFNs, then the distance between the two HFNs can be determined as:*

$$d\left(\widetilde{A_d}, \widetilde{B_d}\right) = \sqrt{1/6\left[(a_1 - b_1)^2 + (a_2 - b_2)^2 + (a_3 - b_3)^2 + (a_4 - b_4)^2 + (a_5 - b_5)^2 + (a_6 - b_6)^2\right]} \tag{8}$$

Example 1. Let $U = (1.2,\ 1.3,\ 1.5,\ 1.6,\ 1.8,\ 2)$ and $V = (1,\ 1.3,\ 1.7,\ 2,\ 2.1,\ 2.3)$ be two HFNs then, their distance

$$d(U,\ V) = \sqrt{1/6\left[(1.2-1)^2 + (1.3-1.3)^2 + (1.5-1.7)^2 + (1.6-2)^2 + (1.8-2.1)^2 + (2-2.3)^2\right]} = 0.26$$

2.4. Centroid-Based Method for the Defuzzification of Hexagonal Fuzzy Numbers

A HFN can be considered as the union of two triangles and two trapeziums, e.g., ΔADH, ΔBCG, ABCD and CDEF together form a HFN.

Further, the trapezium is a union of two triangles and one rectangle. Applying a centroid-based method to triangles and rectangles, and finally summing them, we obtain the centroid of the HFN. Since, the defuzzified value should remain within the range of a HFN, the formulae given below provides the required defuzzified value. Derivation of the HFN is executed in the following way:

(i). Centroid of

$$\Delta BCG = \left(\frac{h_4 + h_5 + h_6}{3}, \frac{0.5}{3}\right) \tag{9}$$

(ii). Centroid of

$$\Delta ADH = \left(\frac{h_1 + h_2 + h_3}{3}, \frac{0.5}{3}\right) \tag{10}$$

(iii). Centroid of Trapezium $ABCD$:

(a) Centroid of

$$\Delta ADI = \left(\frac{h_2 + 2h_3}{3}, \frac{1}{3}\right) \tag{11}$$

(b) Centroid of

$$\Delta BCD = \left(\frac{2h_4 + h_5}{3}, \frac{1}{3}\right) \tag{12}$$

(c) Centroid of Rectangle

$$ABIJ = (\frac{h_3 + h_4}{2}, \frac{0.5}{2}\right) \tag{13}$$

$$C_{ABCD}\left(\frac{2h_2 + 7h_3 + 7h_4 + 2h_5}{18}, \frac{11}{2}\right) \tag{14}$$

(iv). Centroid of Trapezium CDEF is calculated in the similar order and we obtain:

$$C_{CDEF} = \left(\frac{2h_2 + 7h_3 + 7h_4 + 2h_5}{18}, \frac{25}{12}\right) \tag{15}$$

The defuzzified value is determined by summing Equations (9), (10), (14) and (15) and dividing the denominator by the sum of the quantities of the numerator.

$$C_{HGCEFD} = \left(\frac{3h_1 + 3h_2 + 10h_3 + 10h_4 + 5h_5 + 3h_6}{34}, \frac{10}{3}\right) \tag{16}$$

Example 2. Let $U = (1.5,\ 1.6,\ 1.7,\ 1.9,\ 2,\ 2.1)$, then the defuzzified value of

$$U = \frac{4.5 + 4.8 + 17 + 19 + 10 + 6.3}{34} = 1.8 \tag{17}$$

2.5. Determination of Hexagonal Fuzzy Weights of Factors and Sub-Factors

We have extended the methodology developed by Buckley [74] for TFNs in the context of determining hexagonal fuzzy weight.

Step 1. The geometric mean value of the HFN is obtained using:

$$k_c = \left(\prod_{d=1}^{j} y_{cd}\right),\ c = 1, 2, \ldots, i$$

Step 2. Summation of each k_c
Step 3. To calculate the inverse of each k_c and arrange it in increasing order.
Step 4. To find the hexagonal fuzzy weight of factors and sub-factors using the following equation:

$$w_c = k_c * (k_1 + k_2 + \ldots + k_i)^{-1} \tag{18}$$

Step 5. The global hexagonal fuzzy weight of sub-factors are computed by the product of factor weight with the respective sub-factor fuzzy weight.

2.6. Fuzzy Analytic Hierarchy Process (FAHP)

The AHP was introduced by Satty [75]. It is used widely for the evaluation of factor and sub-factor weights. AHP helps in solving the real-life situations with a scientific approach. The comparison of factors and sub-factors, thereby giving preference in linguistic terms can be considered as a hesitant task for DMs, thus HFN with AHP methodology captures the vagueness of the problem. The determination of factors' and sub-factors' weights are important for ranking the electric vehicle charging station. AHP works with a problem hierarchy where a comparison matrix is constructed to represent subjective judgments regarding criteria and sub-criteria. In this work, FAHP is taken instead of AHP, keeping in mind the fuzzy setting represents the uncertainties of the decision experts. The FAHP concept with fuzzy logic allows the DMs in the evaluation of reliable results. The steps of FAHP are given below.

Step 1. Construction of a comparison matrix in terms of HFN by a group of decision experts.

Let a group of 'H' decision-makers assigned for the comparison of factors and sub-factors. Let each DM express their preference in the pairwise comparison of factors and sub-factors. Thus, 'h' set of matrices are obtained, $T_h = \{t_{cdh}\}$.

Where $t_{cdh} = (\widetilde{m}_{cdh}, \widetilde{n}_{cdh}, \widetilde{o}_{cdh}, \widetilde{p}_{cdh}, \widetilde{q}_{cdh})$ denotes the HFN of c factor to d factor as expressed by the 'h' DM.

$$\begin{cases} \widetilde{m}_{cd} = \min\limits_{h=1,2,\ldots,H} \widetilde{m}_{cdh} \\ \widetilde{n}_{cd} = \min\limits_{h=1,2,\ldots,H} \widetilde{n}_{cdh} \\ \widetilde{o}_{cd} = \sqrt[H]{\prod\limits_{h=1}^{H} \widetilde{o}_{cdh}} \\ \widetilde{p}_{cd} = \sqrt[H]{\prod\limits_{h=1}^{H} \widetilde{p}_{cdh}} \\ \widetilde{q}_{cd} = \max\limits_{h=1,2,\ldots,H} \widetilde{q}_{cdh} \\ \widetilde{r}_{cd} = \max\limits_{h=1,2,\ldots,H} \widetilde{r}_{cdh} \end{cases} \tag{19}$$

Step 2. Defuzzification of HFN:

A HFN can be defuzzified by using the centroid-based method used in this paper. Thus using Equation (16), the HFN is transformed to a crisp value.

Step 3. Normalization of the defuzzified matrix:

$$N_{cd} = \frac{M_{cd}}{\sum_{c=1}^{i} M_{cd}}, \text{ where } c = 1,2,\ldots,i;\ d = 1,2,\ldots,j; \tag{20}$$

Step 4. Estimation of factors' and sub-factors' weights:

$$E = \frac{N^{th} rootvalue}{\sum N^{th} root} \tag{21}$$

Step 5. To test the Consistence Index $(C.I)$ of the matrix:

$$(C.I) = \frac{\alpha_{max} - j}{j - 1} \tag{22}$$

where j denotes the size of the matrix.
Step 6. Determination of Consistence Ratio $(C.R)$:

$$C.R = \frac{C.I}{R.I} \tag{23}$$

where $R.I$ is stand for Random Index, and its value differs with the size of the matrix "n".

The assessment of $C.R \leq 0.1$ is acceptable and indicates that the weights obtained are justified. Thus further evaluation is not essential.

2.7. Technique for Order Preference by Similarity to Ideal Solution (TOPSIS) and (FTOPSIS)

The TOPSIS MCDM tool is an extensively used technique, developed by Hwang and Yoon [76] to rank the alternatives, thus giving an idea as to which choice is most preferred. The TOPSIS method is considered to be a distance measure method in which the optimal alternative obtained is farthest away from the negative ideal solution (NIS) and nearest to the positive ideal solution (PIS). The linguistic human decisions can be reflected suitably with Fuzzy TOPSIS (FTOPSIS). The approach is useful in handling the complexity of the situation involving several factors and their sub-factors. In this research, for the selection of the best site to construct an electric vehicle charging station, it is dependent on multiple conflicting factors and sub-factors, thus the MCDM method FTOPSIS introduced by Sodhi and Prabhakar [77] is one of the most helpful and reliable methods. The fuzzy logic extends our goal to obtain more sensitive results in this regard. The steps of FTOPSIS are described below.

Step 1: Construction of the decision matrix by the help of decision experts in terms of linguistic terms. The linguistic terms are then converted to a HFN.
Step 2: To evaluate the normalized HFN fuzzy decision matrix:

$$\begin{aligned} \check{N}D &= [n_{gh}]_{st}, g = 1,2,\ldots,s;\ h = 1,2,\ldots\ldots,t \\ N_{gh} &= \left(\frac{a_{1gh}}{a_6^*}, \frac{a_{2gh}}{a_6^*}, \frac{a_{3gh}}{a_6^*}, \frac{a_{4gh}}{a_6^*}, \frac{a_{5gh}}{a_6^*}, \frac{a_{6gh}}{a_6^*}\right) d \in B.A,\ a_6^* = \max a_{gh} \\ N_{gh} &= \left(\frac{a_h^*}{a_{6gh}}, \frac{a_h^*}{a_{5gh}}, \frac{a_h^*}{a_{4gh}}, \frac{a_h^*}{a_{3gh}}, \frac{a_h^*}{a_{2gh}}, \frac{a_h^*}{a_{1gh}}\right) d \in N.B.A, a_h^* = \min a_{1gh} \end{aligned} \tag{24}$$

where B.A and N.B.A signifies the benefit attributes and non-benefit attributes, respectively.
Step 3: To evaluate the weighted fuzzy normalized matrix, the sub-factors' fuzzy weights are multiplied with the normalized fuzzy value:

$$WN = [P_{gh}]_{st} g = 1,2,\ldots,s\ ; h = 1,2,\ldots,t \tag{25}$$

where

$$P_{gh} = N_{gh} \times \hat{W}_h,\ g = 1, 2, \ldots, s; h = 1, 2, \ldots, t \quad (26)$$

Step 4: Calculate the fuzzy positive ideal solution (FPIS) (PIS^+) and fuzzy negative ideal solution (FNIS) (NIS^-), where h_g^+ denotes the maximum value of h_{gh} and h_g^- denotes the minimum value of h_{gh}:

$$\left.\begin{array}{l} PIS^+ = \{a_1^+, a_2^+, \ldots, a_t^+\} = \left\{\left(\max a_{gh} \middle| h \in M_B\right), \left(\min a_{gh} \middle| h \in M_{NB}\right)\right\} \\ NIS^- = \{a_1^-, a_2^-, \ldots, a_t^-\} = \left\{\left(\min a_{gh} \middle| h \in M_B\right), \left(max h_{gh} \middle| h \in M_{NB}\right)\right\} \end{array}\right\} \quad (27)$$

where M_B denotes the benefit attributes and M_{NB} denotes the non-benefit attributes.
Step 5: Calculation of the distance measure of all alternatives from the PIS and NIS. The two Euclidean distances for individual alternatives can be calculated as follows:

$$\left.\begin{array}{l} L_g^+ = \sum_{h=1}^{t} d(P_{gh}, h_g^+),\ g = 1, 2, \ldots\ldots s. \\ L_g^- = \sum_{h=1}^{t} d(P_{gh}, h_g^-),\ g = 1, 2, \ldots\ldots, s. \end{array}\right\} \quad (28)$$

where d(., .) denotes the Euclidean distance between two fuzzy numbers.
Step 6: Determination of the relative closeness to the ideal alternatives:

$$R_g = \frac{L_g^-}{L_g^- + L_g^+},\ g = 1, 2, \ldots, s \quad (29)$$

Step 7: Rank the alternatives:

The alternatives are ranked based on the score obtained by R_g. The larger value of R_g signifies the better alternatives.

2.8. Fuzzy COPRAS Methodology

The complex proportional assessment (COPRAS) method was first introduced by Zavadskas, Kaklauskas and Sarka [78]. Fuzzy COPRAS is an extended approach of the COPRAS technique, widely used for decision-making problems. It uses the stepwise ranking and evaluation procedure for the alternatives with reference to significance and utility degree. A few applications of the COPRAS method are in economics, construction and property management. An extension of the COPRAS method is Fuzzy COPRAS, which is frequently used in decision-making problems. Ghose et al. [79] used a hybrid fuzzy COPRAS method for selecting the optimal material to be used for a solar car. They took into consideration 19 materials which had 14 different properties. Using a sensitivity analysis, the robustness of the model was checked. The reason for using the fuzzy-based MCDM technique was that it helps decision makers to get over the problems of ambiguous data. Tolga and Durak [80] used the fuzzy COPRAS technique in the air cargo sector for evaluating the potential capability. 18 criteria were chosen for selecting the best out of the present six alternatives. The steps of the COPRAS method are illustrated below:

Step 1. Decision matrix is constructed in terms of HFN, the alternatives are given linguistic terms by the decision experts with respect to the criteria.
Step 2. Normalized decision matrix is formulated using Equation (1), in the similar way, we constructed for TOPSIS normalized matrix.
Step 3. Weighted normalized matrix is constructed by multiplying the criteria weights with fuzzy normalized matrix using Equation (19).
Step 4. Aggregation of beneficial B_g^+ and non-beneficial indices NB_g^- for each alternative are evaluated.

$$B_g^+ = \{\textstyle\sum_{h=1}^{m} NW^{a_1},\ \sum_{h=1}^{m} NW^{a_2},\ \sum_{h=1}^{m} NW^{a_3},\ \sum_{h=1}^{m} NW^{a_4},\ \sum_{h=1}^{m} NW^{a_5},\ \sum_{h=1}^{m} NW^{a_6}\} \quad (30)$$

$$NB_g^- = \left\{ \sum_{h=m+1}^{t} NW^{a_1}, \sum_{h=m+1}^{t} NW^{a_2}, \sum_{h=m+1}^{t} NW^{a_3}, \sum_{h=m+1}^{t} NW^{a_4}, \sum_{h=m+1}^{t} NW^{a_5}, \sum_{h=m+1}^{t} NW^{a_6} \right\} \quad (31)$$

where $h = 1, 2, \ldots,$m represents the benefit attribute of the alternatives and $h = m + 1, m + 2, \ldots, t$ represents the non-benefit attributes of the alternative.
Step 5. Finally, the aggregated beneficial and non-beneficial indices are defuzzified using the Equation (9) and R_{+g} and R_{-g} are determined.
Step 6. Calculation of R_g using the following formulae:

$$R_g = R_{+g} + \frac{R_{-min} \sum_{g=1}^{l} R_{-g}}{R_{-g} \sum_{g=1}^{l} \frac{R_{-min}}{R_{-g}}} \quad (32)$$

Step 7. Ranking of the alternatives are done using the formulae:

$$R = \frac{R_g}{R_{max}} * 100\% \quad (33)$$

where R_g represents the g-thdefuzzified value and R_{max} represents the maximum defuzzified value of individual alternative.

3. Hexagonal Fuzzy MCDM Methodology for Site Selection of Electric Vehicle Charging Station (Numerical Application)

3.1. The Factors and Sub Factors Taken in This Research Have Been Explained in the Following Way

3.1.1. Economic Factors (C_1)

The prosperity of a nation is comprehended by the state of its economy. The assessment of the economic factors reveals the feasibility of the undertaken study. The factors considered are:

- Land cost (C_{11}): It is the crux of the entire planning for the optimum charging station location. Land costs are based on their use, i.e., non-agricultural urban land is more costly than agricultural land. Since the purpose is to build a charging station, we can minimize the land cost by utilizing an already existing utility station. If money is saved on the cost of procuring the land, then it can be utilized for setting up the station.
- Operating and management cost (OMC)(C_{12}): Yao, Bai and Xu [81] stated that a significant part of the budgeted amount is cut out for dealing with operating and management costs which arise in the day to day working of the project. The minimum management cost is essential since it helps in the smooth flow of the information from one department to the other. The charging station should be executed sothat the operations can be systematically planned, which will reduce the in-between costs. Electric vehicles will reduce the air pollution, hence initial operating costs are understandable since the long-term implications outweigh the costs.
- Consumption level (C_{13}): Modrak and Soltysova [82] studied the operational complexity measure. The measure of consumption level denotes how affluent the people in a particular locality are. In the case that consumption level in an area is high, it can be expected that people will be more willing to go to further distances in search of more options. A charging station can be built in a high consumption area since the throng of people will have more ways of traveling and procuring their wants.
- Construction cost (C_{14}): Manerba, Mansini and Perboli [83] researched the capacitated supplier selection problem considering total quantity discount policy and activation costs under uncertainty. Construction cost varies with the location, and to make the charging station a success, the initial fixed cost should be minimized as much as possible. If the location is well-connected by various transportation facilities, then the cost of transferring the construction materials will decrease, which will decrease the construction cost and the overall profitability of the charging station will increase initially.

- Public facilities (C_{15}): Kinay et al. [84] studied multi-criteria chance-constrained capacitated single source discrete facility location problems. Public facilities refer to schools, colleges, grocery stores, shopping malls, bus stops and the other everyday amenities which are used by commoners on a mass scale. In the case of a charging station being built near a location with a large density of public facilities, it will act as a boon since money will frequently change hands and thereby develop the area.

3.1.2. Environmental Factors (C_2)

This refers to those factors which will influence the immediate surroundings of the charging station. A clean and green environment helps in resonating the theme of electric vehicles and thereby makes a charging station built in such a location a success.

- Generation of noise and air pollution (GNAP)(C_{21}): In the current scenario, noise and air pollution are considered bigger hindrances than other forms of pollution. This is because they are experienced daily, which results in greater damage due to them. Electric vehicles will help in reducing both, since the batteries of the vehicles will not cause air pollution or make noise while being on the road.
- Petrol stations (C_{22}): The availability of petrol stations nearby signifies a greater number of vehicles in the area. Building a charging station near a petrol station will cause an increase in the number of vehicles and people visiting the area and thereby turn out to be more profitable since the cost will be less.
- Transportation stations (C_{23}): These refer to the various bus stops, bicycle charging stations, railway stations, car stops, and tram stops. The greater the frequency of the transportation vehicles, the more will be its impact on the environment since every form of transportation will bring more people, eventually harming the environment. Electric vehicles, on the other hand, help in the easy transportation of people without harming the environment.

3.1.3. Traffic Factors (C_3)

This refers to those sets of factors which are only noticeable when there is a huge population in the area. A charging station which takes into consideration the traffic factors is one which will be able to have huge implications on the state of travel in the area.

- Number of roads (C_{31}): Yao, Bai and Xu [81] studied the impact of the number of roads on China's thermal power industry. The number of roads stands for the various options which the vehicles can take in the case of congestion or for availing a shorter reach time. If the charging station is strategically built near an intersection of heavy-traffic roads, then it will help the drivers be more at ease while driving since they will have a backup nearby.
- Road potency (C_{32}): Hosseini and Sarder [85] studied the road potency for optimal site selection using a Bayesian network model. The higher the number of vehicles in the region, the greater the footfall will be, which will eventually increase the success rate of the charging station.
- Parking areas (C_{33}): The increase in parking areas will lead to an increase in the use of electric vehicles, since one of the major thoughts which arises in a prospective buyer's mind is where one will park the vehicle. If the charging station is built near a parking area, then the vehicle owners can directly charge and park it.

3.1.4. Societal Factors (C_4)

With the advancement in science and technology, the factors promoting wellness of human life have taken a back seat. The societal factors point out the major problems which may be looked at to improve the quality of the society at large.

- Adverse impact of noise and electromagnetic fields (AI)(C_{41}): (Due to the construction of the electric vehicle charging station). An electric vehicle charging station has a constant aura of noise and an electromagnetic field surrounding it during the construction phase which might cause a certain category of people to develop problems. If proper measures can be taken in the initial stage, then this impact may be minimized since public health is of utmost importance.

- Population density (C_{42}): This stands for the number of people living in each unit of area. When the population density in a locality is high, it shows that the area is overcrowded. There will be more consumption in such an area and the construction cost will also be high, but the quality of life will usually be low. The need of transportation in such an area is usually very high and an electric vehicle charging station constructed in such an area may just be what the people need.

Tables 1 and 2 describes the factors, sub-factors and alternatives considered for this research.

Table 1. Factors and sub-Factors considered in the present study.

Factors	Sub-Factors
Economic Factors (C_1)	Land cost (C_{11}) Operating and management cost (OMC)(C_{12}) Consumption level (C_{13}) Construction cost (C_{14}) Public facilities (C_{15})
Environmental Factors (C_2)	GNAP (C_{21}) Petrol stations (C_{22}) Transportation stations (C_{23})
Traffic Factors (C_3)	Number of roads (C_{31}) Road potency (C_{32}) Parking areas (C_{33})
Scheme 4	Adverse impact of noise and electromagnetic fields (AI)(C_{41}) Population density (C_{42})

Table 2. Alternatives selected in the study and their corresponding nearby location, latitude and longitude.

Alternative	Nearby Location	Latitude and Longitude
1. Dasnagar	HP petrol pump (Debi Service station)	22.599152, 88.307854
2. Santragachi	HPCL petrol pump	22.586515, 88.276026
3. Belgachia	Petrol pump	22.603168, 88.323001
4. Howrah Maidan	Near Kabra stores	22.581972, 88.332230
5. Liluah	Sur petrol pump	22.625105, 88.350044
6. Kadamtala	HP petrol pump	22.587778, 88.320151
7. Shibpur	Chowrabasti, Shibpur	22.562826, 88.326159
8. Salkia	Malipanchghara	22.600887, 88.349325
9. Bakultala, Shibpur	Botanical Garden west end	22.564016, 88.288795
10. Belur	SSBPCL petrol pump	22.639311, 88.350857

Figure 3 represents the hierarchical framework of the problem in the study.

The rural, urban and total population for the census year starting from 2001 and 2011 has been represented in a graph (Figure 4). In this figure, the projected rural, urban and total population also been forecasted for the years 2021, 2031, 2041 and 2051 in a chronological way. The graph clearly indicates a consistent increase in the population from 4.23 million in 2001 to 6.60 million in 2051 (projected) with a sharp increase in the urban population, and the urban population growth is much higher than the rural, which indicates higher increasing infrastructural demand. The projected population made us think about the future transport services in the area.

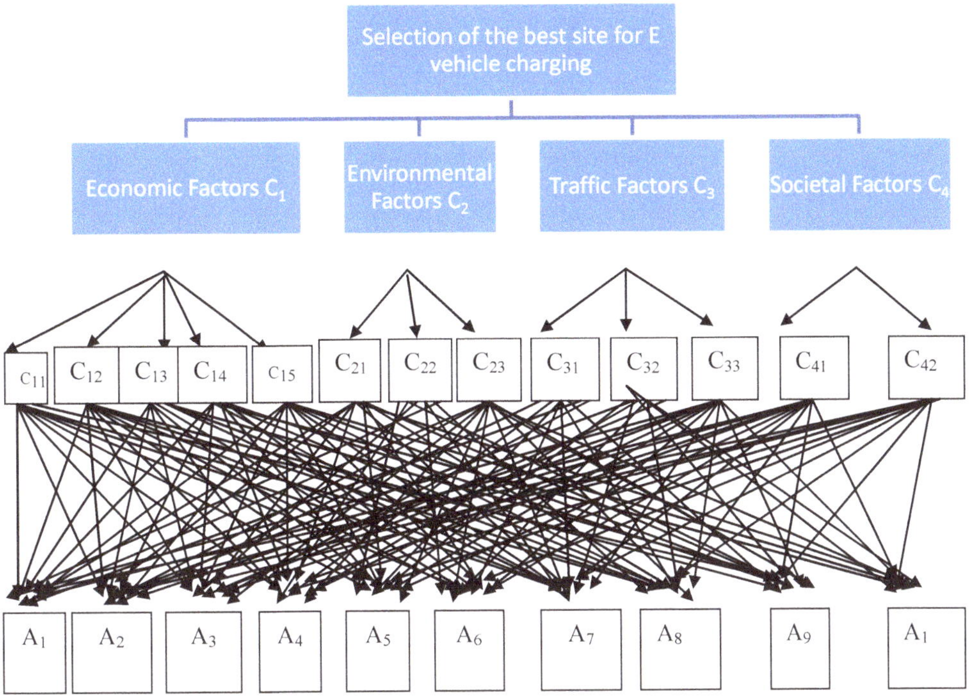

Figure 3. Hierarchical structure of the problem.

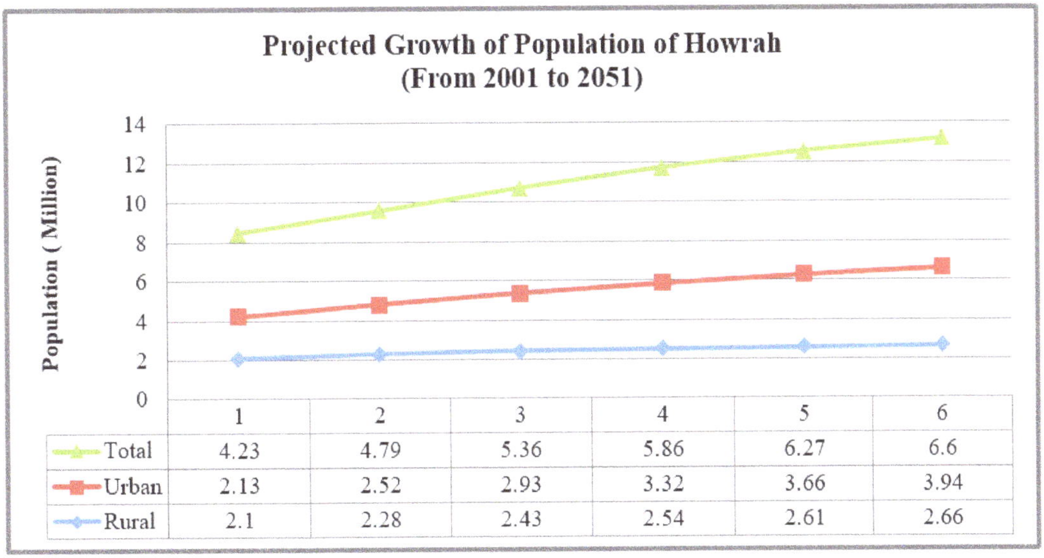

Figure 4. Projected growth of the population of rural and urban areas of the Howrah district.

Urbanization is considered one of the most noteworthy anthropogenic inputs of the environmental framework, and thus the present study considered the spatio-temporal characteristics of urban growth and its inference in the transport of Howrah. The built-up land (Figure 5) has been generated using the NDBI (normalized difference built-up index) with the following equation NDBI = MIR − (NIR/MIR) + NIR (Zha, 2003) [86]. Here, NIR is anear-infrared band such as ETM+ and TM, and LISSIII is a band no.4; MIR is a middle infrared band such as ETM+ and TM and LISSIII is a band no.5.The index is based on the unique spectral response of built-up lands that have a higher reflectance in the MIR wavelength range as compared to that in the NIR wave length range. Thereafter, the NDBI mapping has been prepared to understand the level of urbanization from 2000 (Figure 5a) to 2010 (Figure 5b) in the study area. It helps to correlate the changes in land use patterns and its consequences to the water storage of the study area. The NDBI values range from −1 to +1. Very low values of the NDBI (0.1 and below) correspond to non-urban features, while higher values indicate the covering of areas of impervious surfaces such as asphalt and concrete. To understand the levels of urbanization, the NDBI values have categorized into five zones, which range from −0.96 to 0.45 in 2000 and from −0.95 to 0.71 in 2010. The result with a negative value of NDBI represents the water bodies and vegetation covers, whereas the higher and positive values represent the build-up areas. The map of the year 2000 indicates a huge range of water bodies spread all over the district along with the two major rivers; the Rupnarayan River in west and south west and the Bhagirathi-Hooghly River in the east and south-east side. The major built-up areas are mostly concentrated over the north-east and north-west corners of the district covering the Bally-Jagacha and Udaynarayanpur blocks, respectively.

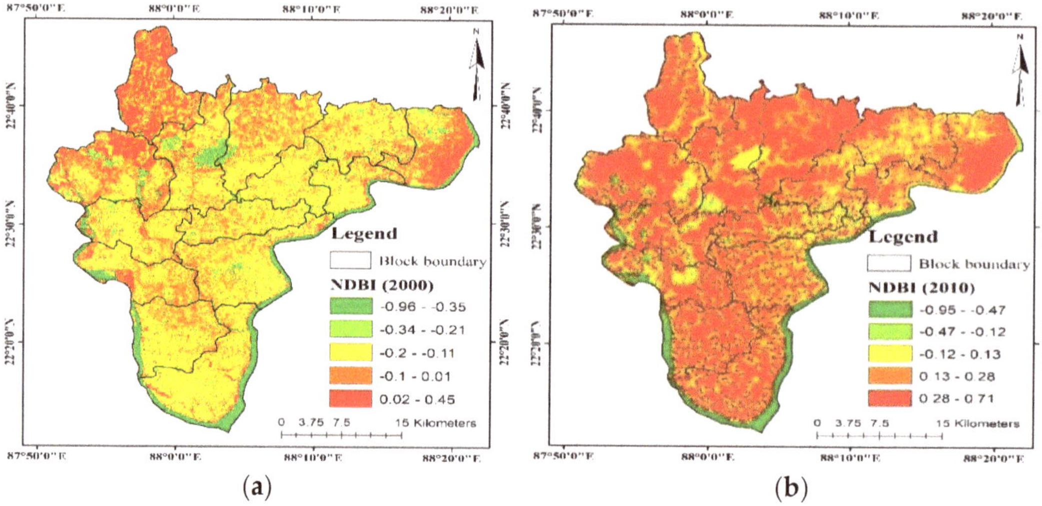

(a) (b)

Figure 5. Normalized difference built-up index mapping for level of urbanization during (**a**) 2000 (**b**) 2010.

The transport network of the Howrah district is mapped (Figure 6) and the closest census towns (Figure 7) to the Howrah Municipal Corporation (HMC) were also plotted to understand the importance of daily communication with the Howrah station or surrounding areas. The railway and the road transport connectivity with the Howrah station, which is located in the HMC, is very good. A large number of daily commuters are traveling to the area, mostly for economic and educational purposes. It indicates the concentration of traffic in the area, and in turn increases the public transport connectivity as well as the requirement for improving the local transport system.

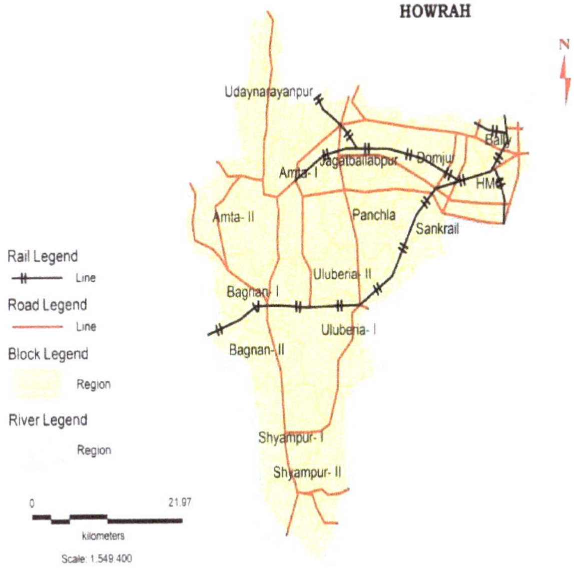

Figure 6. Transport Network in and around Howrah.

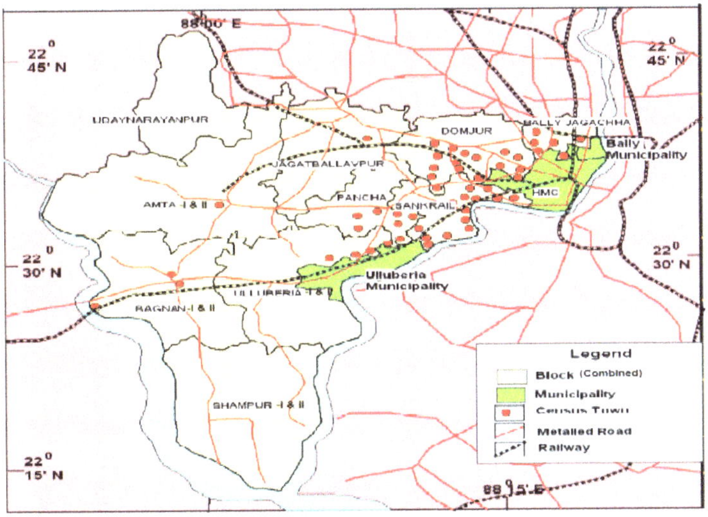

Figure 7. Transport Network of Howrah district. Along with the Census Town.

The ten selected points/locations (Figure 8) were mapped in the HMC to understand the spatial coverage and important transport nodes in the area of the current study. The population distribution of each of the 66 wards was also mapped (Figure 9) to observe the population pressure, which is also able to justify the present selection of the ten locations for the study. Most of the selected locations are in densely populated areas, where the public and local transport services are becoming very essential. Furthermore, the Howrah Maidan (S_4) and Salkia (S_8) are highly densely populated (Figure 9) and thus very important in terms of transport services, whereas, Liluah (S_5), Belur (S_{10}) and Shibpur (S_7) belong to highly densely populated and Kadamtala (S_6), Bakultala (S_9), Belgachia (S_3) belong to moderately populated and Dasnagar (S_1) and Santragachi (S_2) belong to comparatively less populated among the ten points. But all these points are almost equally important and essential in terms of either population or transport or both. Sub criteria scores of each location has been represented in Table 3.

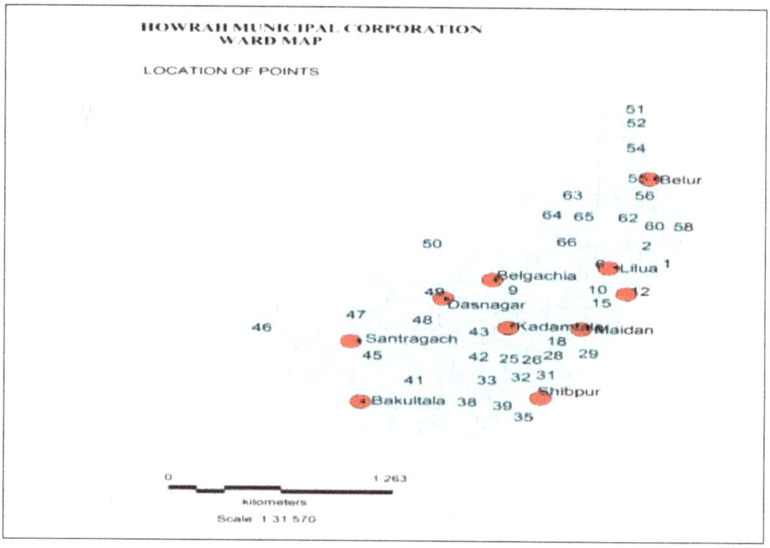

Figure 8. Location of points in HMC.

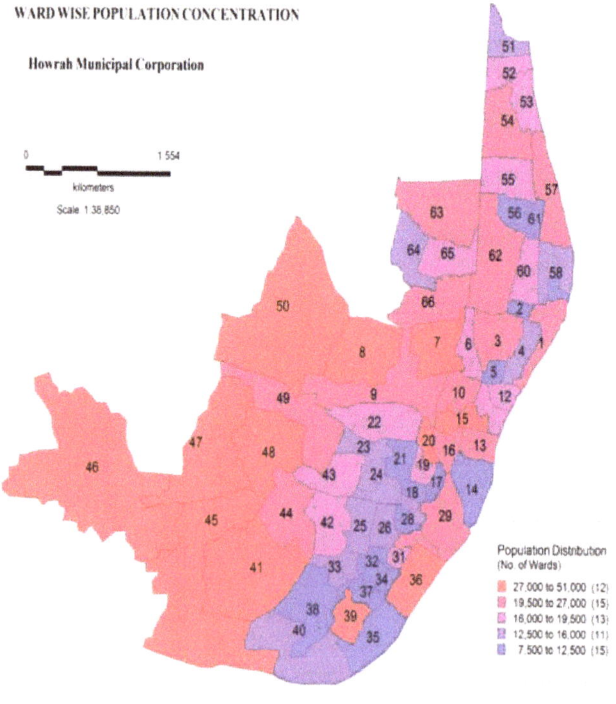

Figure 9. Ward wise population distribution of HMC.

Table 3. Sub-factors taken for the study.

Criteria		Sub-Criteria and Score				
Land cost	Cost in Million/720 ft^2)	<1.50	1.50–2.50	2.50–3.50	3.50–4.50	>4.50
	Score	1	3	5	7	9
Operating and Management cost		In a scale of 1,3,5,7,9				
Consumption Level		In a scale of 1,3,5,7,9				
Construction cost		In a scale of 1,3,5,7,9				
Public facilities		In a scale of 1,3,5,7,9				
Emission of Greenhouse gases		In a scale of 1,3,5,7,9				
Petrol stations	(Distance in mt.) (Using GIS)	<200	200–400	400–600	600–800	>800
	Score	9	7	5	3	1
Transportation stations	(Distance in mt.) (Using GIS)	<250	250–500	500–750	750–1000	>1000
	Score	9	7	5	3	1
Population density	(persons/km^2)	<10,000	10,000–13,000	13,000–16,000	16,000–19,000	>19,000
	Score	1	3	5	7	9
Number of Roads		Crisp value location wise				
Road Potency		In a scale of 1,3,5,7,9				
Parking Areas		In a scale of 1,3,5,7,9				
Adverse impact of noise and electromagnetic field due to construction of electric vehicle charging station		In a scale of 1,3,5,7,9				

Linguistic variables in HFN required for the comparison of factors and sub-factors are shown in Table 4.

Table 4. Linguistic term in HFN 1–9 scale.

Linguistic Terms	1–9 Scale	Hexagonal Fuzzy Number (HFN)
Equally Important (EI)	1	1
Weakly Important (WI)	2	(1.1,1.2,1.3,1.4,1.5,1.6)
Moderately Important (MI)	3	(1.8,2,2.2,2.5,2.7,3)
Strongly Important (SI)	5	(2.9,3,3.2,3.3,3.5, 3.9)
Very Strongly Important (VSI)	7	(3.6,4,4.1,4.4,4.5, 4.8)
Absolutely Important (AI)	9	(4.6,4.8,5,5.2,5.4, 5.7)
Absolutely Unimportant (AUI)	1/9	(0.17, 0.18, 0.19, 0.2, 0.21, 0.22)
Very Strongly Unimportant	1/7	(0.21, 0.22, 0.23, 0.24, 0.25, 0.28)
Strongly Unimportant	1/5	(0.26, 0.28, 0.3, 0.31, 0.33, 0.34)
Moderately Unimportant	1/3	(0.33, 0.37, 0.4, 0.45, 0.5, 0.55)

Comparison between factors in linguistic variables given by two DMs is presented in Table 5.

The above Table 6 represents preference of factors in defuzzified form using Equation (18).Normalized matrix is obtained using Equation (20); priority weight of factors are calculated using Equation (21). Societal factors obtain the maximum weight of 0.430, followed by environmental factor 0.37, followed by traffic factor 0.139 and economic factor 0.065.

Table 5. Representation of comparison between factors.

Factors	Economic Factors (C_1)		Environmental Factors (C_2)		Traffic Factors (C_3)		Societal Factors (C_4)	
Decision Makers (DMs)	DM1	DM2	DM1	DM2	DM1	DM2	DM1	DM2
Economic Factors (C_1)	EI	EI	AUI	VSUI	SUI	VSUI	SUI	AUI
Environmental Factors (C_2)	AI	VSI	EI	EI	SI	VSI	EI	MUI
Traffic Factors(C_3)	SI	VSI	SUI	VSUI	EI	EI	SUI	VSUI
Societal Factors (C_4)	SI	AI	EI	MI	SI	VSI	EI	EI

Note. Two DMs are considered here in the study, their opinions are combined using Equation (19).

Table 6. Representation of preference of factors in defuzzified form using Equation (18).

FACTORS	ECONOMIC	ENVIRONMENTAL	TRAFFIC	SOCIETAL
ECONOMIC	1	0.22	0.27	0.22
ENVIRONMENTAL	4.7	1	3.8	0.74
TRAFFIC	3.8	0.27	1	0.27
SOCIETAL	3.8	1.74	3.8	1

The weight of factors obtained is represented in Table 7.

Table 7. Representation of the priority weight of factors.

FACTORS	ECONOMIC	ENVIRONMENTAL	TRAFFIC	SOCIETAL	SUM	E/Sum
ECONOMIC	0.065	0.081	0.038	0.095	0.278	4.276795
ENVIRONMENTAL	0.305	0.366	0.529	0.318	1.518	4.148243
TRAFFIC	0.247	0.099	0.139	0.116	0.601	4.313789
SOCIETAL	0.247	0.637	0.529	0.430	1.843	4.287043

Using Equation (22), C.I is calculated to be 0.08. For R.I, as n = 4, the value is 0.09. Thus using Equation (23), $C.R = \frac{0.08}{0.9} = 0.09 < 0.1$, hence the matrix is consistent.

In the similar way, the comparison analysis of sub-factors with the help of two DMs has been calculated. The fuzzy weight of factors, sub-factors and global fuzzy weight are represented in Table 8.

Table 8. Hexagonal fuzzy weights of factors, sub-factors and global weight.

Factors Fuzzy Weight	Sub-Factor Fuzzy Weight	Global Weight
$C_1 = (0.04, 0.05, 0.07, 0.07, 0.10, 0.11)$	$C_{11} = (0.06, 0.07, 0.1, 0.11, 0.18, 0.20)$ $C_{12} = (0.09, 0.11, 0.16, 0.17, 0.29, 0.32)$ $C_{13} = (0.13, 0.15, 0.24, 0.25, 0.41, 0.46)$ $C_{14} = (0.03, 0.04, 0.06, 0.06, 0.09, 0.10)$ $C_{15} = (0.23, 0.27, 0.41, 0.43, 0.68, 0.76)$	$C_{11} = (0.003, 0.003, 0.007, 0.008, 0.018, 0.021)$ $C_{12} = (0.004, 0.005, 0.011, 0.012, 0.028, 0.035)$ $C_{13} = (0.006, 0.007, 0.016, 0.018, 0.040, 0.050)$ $C_{14} = (0.001, 0.002, 0.004, 0.004, 0.009, 0.011)$ $C_{15} = (0.01, 0.012, 0.027, 0.030, 0.067, 0.082)$
$C_2 = (0.21, 0.23, 0.35, 0.38, 0.54, 0.58)$	$C_{21} = (0.084, 0.09, 0.114, 0.125, 0.155, 0.177)$ $C_{22} = (0.17, 0.19, 0.24, 0.26, 0.34, 0.39)$ $C_{23} = (0.41, 0.46, 0.59, 0.65, 0.83, 0.93)$	$C_{21} = (0.02, 0.02, 0.04, 0.05, 0.08, 0.10)$ $C_{22} = (0.04, 0.04, 0.08, 0.10, 0.19, 0.23)$ $C_{23} = (0.08, 0.11, 0.21, 0.24, 0.45, 0.54)$
$C_3 = 0.09, 0.10, 0.14, 0.15, 0.20, 0.22)$	$C_{31} = (0.07, 0.08, 0.1.0.11, 0.14, 0.16)$ $C_{32} = (0.493, 0.55, 0.66, 0.71, 0.86, 0.96)$ $C_{33} = (0.142, 0.16, 0.197, 0.214, 0.27, 0.31)$	$C_{31} = (0.007, 0.008, 0.014, 0.016, 0.03, 0.035)$ $C_{32} = (0.045, 0.054, 0.09, 0.104, 0.18, 0.21)$ $C_{33} = (0.013, 0.015, 0.027, 0.031, 0.055, 0.07)$
$C_4 = (0.26, 0.27, 0.42, 0.45, 0.69, 0.76)$	$C_{41} = (0.3, 0.3, 0.49, 0.50, 0.82, 0.87)$ $C_{42} = (0.3, 0.3, 0.43, 0.43, 0.61, 0.63)$	$C_{41} = (0.08, 0.082, 0.202, 0.226, 0.57, 0.66)$ $C_{42} = (0.08, 0.082, 0.18, 0.195, 0.42, 0.48)$

The linguistic terms used in this study for the rating of alternatives with respect to sub-factors. Preference of alternatives with respect to sub-factors are expressed in linguistic terms, scores and crisp values are depicted in Table 9. Note 2. The sub-factors (C_{22}) and (C_{23}) are assigned score using Table 3; later, it is converted to HFN. A score of 9 implies "very high" and so on.

Table 9. Comparison analysis in linguistic variables for preference of alternatives with respect to sub-factors.

Locations	Sub-Factors												
	C_{11}	C_{12}	C_{13}	C_{14}	C_{15}	C_{21}	C_{22}	C_{23}	C_{31}	C_{32}	C_{33}	C_{41}	C_{42}
Dasnagar (S_1)	M	L	L	M	M	VL	9	9	3	H	M	VL	L
Santragachi(S_2)	H	L	L	M	L	VL	9	7	1	VH	H	VL	L
Belgachia (S_3)	M	M	VL	M	VL	H	9	3	2	H	L	H	M
Howrah Maidan (S_4)	VH	VH	VH	VH	VH	VH	7	7	4	VH	L	VH	VH
Liluah (S_5)	H	H	M	VH	VH	H	9	5	1	M	L	H	H
Kadamtala (S_6)	H	H	H	H	H	H	9	1	2	M	L	VH	M
Shibpur (S_7)	H	H	VH	H	H	M	9	1	1	M	L	M	H
Salkia (S_8)	H	VH	VH	VH	H	VH	5	1	2	M	L	VH	VH
Bakultala (S_9)	M	L	H	H	L	L	9	1	1	L	L	L	M
Belur (S_{10})	M	M	M	M	L	H	9	3	1	M	L	H	H

Note: For C_{22}, C_{23} and C_{31} are represented with crisp numbers.

3.2. Ranking of Alternatives Using Fuzzy AHP-TOPSIS Method

Following the steps represented in Section 2.7, distance measure, relative closeness and the ranking of sites has been computed as depicted in Table 10.

Table 10. Representation of distance measure, relative closeness and the ranking of sites.

Alternatives	L_g^+	L_g^-	$R_g = \frac{L_g^-}{L_g^- + L_g^+}$	Ranking
Dasnagar (S_1)	0.441	0.264	0.375	8
Santragachi (S_2)	0.468	0.238	0.337	9
Belgachia (S_3)	0.271	0.435	0.616	4
Howrah Maidan (S_4)	0.058	0.647	0.918	1
Liluah (S_5)	0.184	0.522	0.739	2
Kadamtala (S_6)	0.302	0.404	0.572	6
Shibpur (S_7)	0.285	0.422	0.596	5
Salkia (S_8)	0.259	0.446	0.633	3
Bakultala (S_9)	0.379	0.328	0.464	7
Belur (S_{10})	0.239	0.468	0.662	3

3.3. Ranking of Alternatives UsingFuzzy AHP-COPRAS

Following the steps represented in Section 2.8, R_{+g}, R_{-g}, R_g, R and ranking are computed as depicted in Table 11.

Table 11. Values of R_{+g}, R_{-g}, R_g, RR and ranking are represented.

Alternatives	R_{+g}	R_{-g}	R_g	R	Ranking
Dasnagar (S_1)	0.68	0.19	0.691	77.98	7
Santragachi (S_2)	0.66	0.19	0.703	79.35	6
Belgachia (S_3)	0.59	0.08	0.707	79.82	5
Howrah Maidan (S_4)	0.75	0.06	0.886	100.00	1
Liluah (S_5)	0.65	0.07	0.778	87.82	2
Kadamtala (S_6)	0.55	0.07	0.680	76.81	9
Shibpur (S_7)	0.58	0.08	0.685	77.33	8
Salkia (S_8)	0.58	0.06	0.716	80.82	4
Bakultala (S_9)	0.52	0.10	0.614	69.34	10
Belur (S_{10})	0.62	0.08	0.732	82.66	3

4. Comparison Analysis and Sensitivity Analysis

Two different MCDM techniques, fuzzy AHP-TOPSIS and fuzzy AHP-COPRAS were employed for the selection of the optimal site for aelectric vehicle charging station in and around the city of Howrah, West Bengal, India. Figure 10 denotes the comparative ranking obtained under the two methodologies, fuzzy AHP-TOPSIS and fuzzy AHP-COPRAS. We also tried to compare the alternatives ranking using the two different said proposed methods.

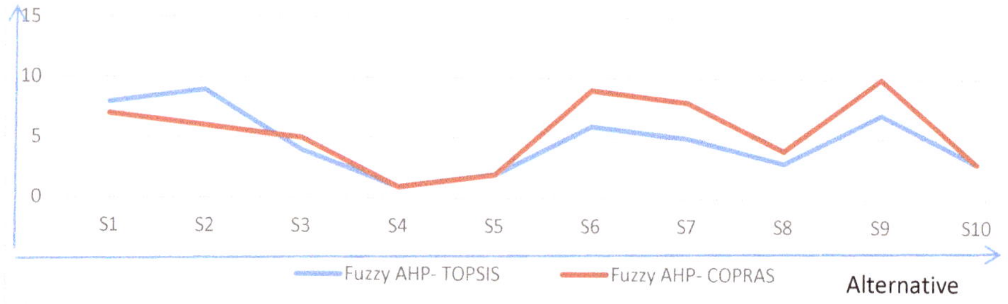

Figure 10. Representation of the ranking obtained under the two MCDM techniques.

A sensitivity analysis was conducted to see the ranking obtained under different changing conditions. Figures 11 and 12 represents the clustered column chart to compare the ranking with the interchange of the sub-factors' weight. Two different cases are taken. In the first case, the sub-factors parking facilities 'C_{15}' and population density 'C_{42}' weights were interchanged. In the second case, land cost 'C_{11}' and generation of noise and air pollution 'C_{21}' weights were interchanged. For both these cases, two different methodologies were used in this study, i.e., fuzzy AHP-TOPSIS and fuzzy AHP-COPRAS.

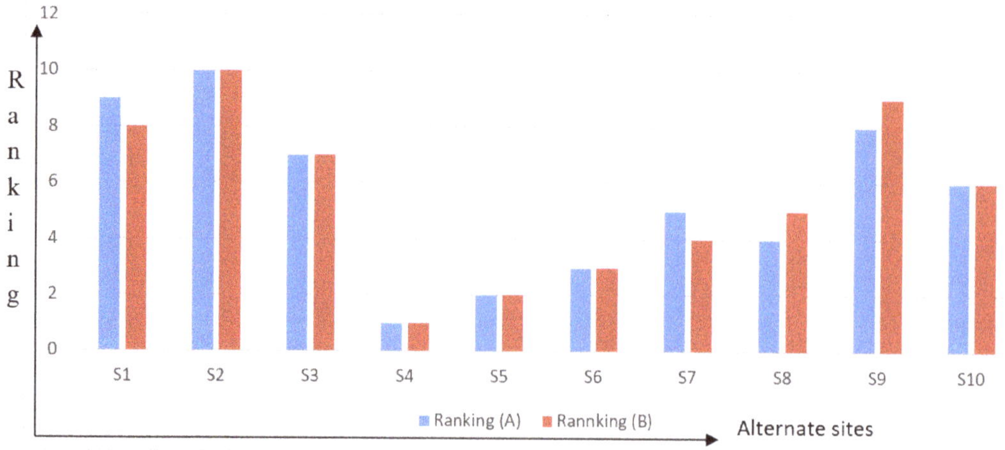

Figure 11. Sensitivity analysis ranking obtained under fuzzy AHP-TOPSIS.

Figure 11, i.e., ranking obtained by fuzzy AHP-TOPSIS under sensitivity analysis shows that the alternative (S_4), (S_5), (S_6), (S_{10}), (S_3) and (S_2) are consistent with first, second, third, fourth, fifth and sixth position, respectively, under the considered two cases, whereas following Figure 12, i.e., ranking yield by fuzzy AHP-COPRAS under sensitivity analysis shows that the sites (S_4), (S_5), (S_2), (S_{10}), (S_3) and (S_9) scores the rank of first, second, fifth, seventh, ninth and tenth position, respectively. The other sites' variation of rank is noticed under the sensitivity analysis and depicted in the mentioned figures.

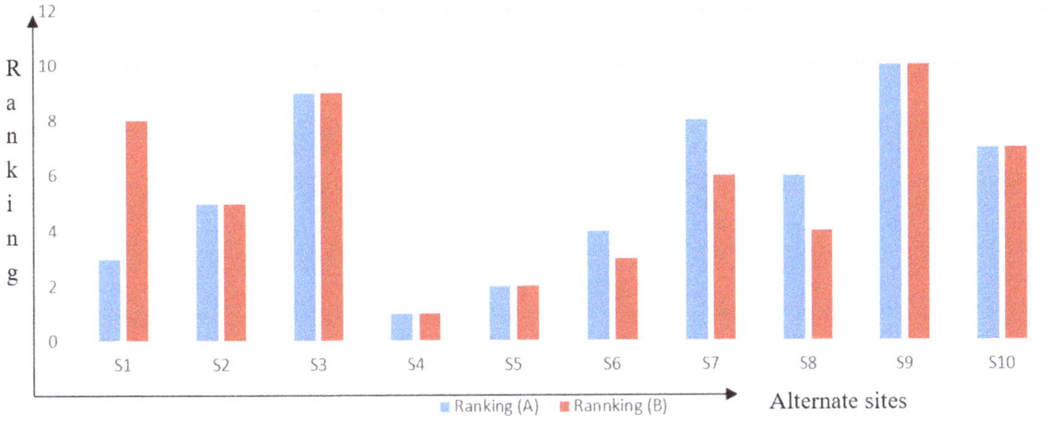

Figure 12. Sensitivity analysis ranking obtained under fuzzy AHP-COPRAS.

5. Results and Discussion

This section discusses the results obtained by the methodology FAHP-TOPSIS, FAHP-COPRAS and the sensitivity analysis. The ranking obtained under the two MCDM techniques yield the site "Howrah Maidan" (S_4) as the best alternative for e-vehicle site selection followed by "Liluah" (S_5) and "Belur" (S_{10}). The FAHP-TOPSIS ranked the alternative "Belur" (S_{10}) and "Salkia" (S_8) equally at the third position. Ranking obtained for all the sites are presented in Tables 10 and 11. In the sensitivity analysis, where the sub-factors' weight are interchanged as discussed in Section 5, it is seen that the site selected "Howrah Maidan" (S_4) consistently remains in the first position. The rankings obtained under the sensitivity analysis by the two methods are depicted in Figures 11 and 12.

6. Conclusionsand Future Scope

Ease of commutation, a pollution-free mode of transport, as well as employment generation are direct and indirect benefits of the e-vehicle. For developing countries where the pollution level is quite high and proportion of the roads is lower, e-vehicles can be a game changer. According to our study, considering ten locations across the city of Howrah Maidan due to its proximity to India's largest railway station Howrah and various other attributes ranked it as number one, followed by Liluah and Belur. These three locations are consistent with rank one, two and three, respectively, irrespective of the two MCDM methodologies applied in this research. High population density, enhanced level of consumerism, and the presence of various public facilities with higher footfalls led to this higher ranking.

This paper used the GIS and MCDM tools FAHP, F-TOPSIS and F-COPRAS to obtain the optimal site selection for the e-vehicle charging station. HFN has been used to give a preferential rating of factors, sub-factors and alternatives. The ranking obtained using MCDM tools are logical and scientific. The present findings provide important references for future potential work and problem solving.

For the site selection, factors such as environmental, economic, traffic, societal are incorporated with their respective sub-factors. Through this research, it is observed that the societal factor is the most significant and out of the sub-factors, population density is the most important.

HFN is used here as it captures the hesitancy and vagueness in an efficient way. To practice the qualitative criteria evaluation for imprecise information, FAHP, F-TOPSIS and F-COPRAS are used. Comparative analysis which uses F-TOPSIS and F-COPRAS in our example has depicted consistent results. The reliability, robustness and efficiency of this methodology has also been tested through sensitivity analysis. HFN captures a wider range of linguistic terms but usage of HFN makes computation a bit longer.

In Howrah, India, a very old and unplanned city, creating e-vehicle charging infrastructure across the city is important. A site like Santragachi (S_2) in this research which ranks lower at present could acquire a significant position due to huge infrastructural investment by the government in that area. A big railway terminus is coming up, which has the potential to change the demography of the area. Thus futuristic sub-factors can be incorporated into future research. The absence of futuristic sub-factors is a limitation at present. A larger number of decision makers based on the administrative point of view can be explored in future research. The proposed methodology used in this research can be applied in different fields such as new vendor selection, and treatment selection for new diseases where more ambiguity and uncertainty is prevalent. The other MCDM tools such as PROMETHEE, VIKOR, and WASPAS can be used in the future with intuitionistic, neutrosophic, and hesitant fuzzy numbers to yield improved, robust and practical solutions.

Author Contributions: Conceptualization, A.G., S.P.M.; methodology, S.P.M. and N.G.; software, B.K.M.; validation, A.D. and M.S.G.; formal analysis, A.G., N.G.; investigation, S.K. and A.G.; resources, B.K.M.; data curation, B.K.M. and S.P.M.; writing—original draft preparation, S.K. and A.G.; writing—review and editing, S.P.M. and N.G.; visualization, B.K.M.; supervision, A.G., S.P.M.; project administration, A.G. and S.P.M.; funding acquisition. All authors have read and agreed to the published version of the manuscript.

Funding: This research received no external funding.

Institutional Review Board Statement: Not applicable.

Informed Consent Statement: Not applicable.

Data Availability Statement: We mention the source of used data in the work.

Conflicts of Interest: The authors declare no conflict of interest.

References

1. Bilgen, S. Structure and environmental impact of global energy consumption. *Renew. Sustain. Energy Rev.* **2014**, *38*, 890–902. [CrossRef]
2. Zhao, H.R.; Guo, S.; Fu, L.W. Review on the costs and benefits of renewable energy power subsidy in China. *Renew. Sustain. Energy Rev.* **2014**, *37*, 538–549. [CrossRef]
3. Hernández, J.C.; Sutil, S.F.; Vidal, P.G.; Casas, R.C. Primary frequency control and dynamic grid support for vehicle-to-grid in transmission systems. *Int. J. Electr. Power Energy Syst.* **2018**, *100*, 152–166. [CrossRef]
4. Schoettle, B.; Sivak, M. The relative merits of battery-electric vehicles and fuel-cell vehicles. *Transp. Res. Inst. Mich.* **2016**. Available online: http://www.umich.edu/~{}umtriswt/PDF/UMTRI-2016-5.pdf (accessed on 15 February 2021).
5. Singh, M.; Kumar, P.; Kar, I. A multi charging station for electric vehicles and its utilization for load management and the grid support. *IEEE Trans. Smart Grid* **2013**, *4*, 1026–1037. [CrossRef]
6. Rowe, G.E.; Gardner, B.; Abraham, C.; Skippon, S.; Dittmar, H.; Hutchins, R.; Stannard, J. Mainstream consumers driving plug-in battery-electric and plug-in hybrid electric cars: A qualitative analysis of responses and evaluations. *Transp. Res. Part A Policy Pract.* **2012**, *46*, 140–153. [CrossRef]
7. Hernández, J.C.; Rodriguez, R.F.J.; Jurado, F. Modelling and assessment of the combined technical impact of electric vehicles and photovoltaic generation in radial distribution systems. *Energy* **2017**, *141*, 316–332. [CrossRef]
8. Mak, H.Y.; Rong, Y.; Shen, Z.J.M. Infrastructure planning for electric vehicles with battery swapping. *Manag. Sci.* **2013**, *59*, 1557–1575. [CrossRef]
9. Fang, Y.; Wei, W.; Mei, S.; Chen, L.; Zhang, X.; Huang, S. Promoting electric vehicle charging infrastructure considering policy incentives and user preferences: An evolutionary game model in a small-world network. *J. Clean. Prod.* **2020**, *258*, 753. [CrossRef]
10. Dericioglu, C.; Yirik, E.; Unal, E.; Cuma, M.U.; Onur, B.; Tumay, M. A review of charging technologies for commercial electric vehicles. *Int. J. Adv. Automot. Technol.* **2018**, *2*, 61–70.
11. Zivin, J.S.G.; Kotchen, M.J.; Mansur, E.T. Spatial and temporal heterogeneity of marginal emissions: Implications for electric cars and other electricity-shifting policies. *J. Econ. Behav. Organ.* **2014**, *107*, 248–268. [CrossRef]

12. Millo, F.; Rolando, L.; Fuso, R.; Mallamo, F. Real CO_2 emissions benefits and end user's operating costs of a plug-in hybrid electric vehicle. *Appl. Energy* **2014**, *114*, 563–571. [CrossRef]
13. Brown, S.; Pyke, D.; Steenhof, P. Electric vehicles: The role and importance of standards in an emerging market. *Energy Policy* **2010**, *38*, 3797–3806. [CrossRef]
14. Zhao, H.; Li, N. Optimal siting of charging stations for electric vehicles based on fuzzy Delphi and hybrid multi-criteria decision making approaches from an extended sustainability perspective. *Energies* **2016**, *9*, 270. [CrossRef]
15. Frade, I.; Ribeiro, A.; Gonçalves, G.; Antunes, A.P. Optimal location of charging stations for electric vehicles in a neighborhood in Lisbon, Portugal. *Transp. Res. Rec.* **2011**, *2252*, 91–98. [CrossRef]
16. Barzani, S.P.; Ghahnavieh, R.A.; Karegar, K.H. Optimal fast charging station placing and sizing. *Appl. Energy* **2014**, *125*, 289–299. [CrossRef]
17. Hwang, C.L.; Yoon, K. Methods for multiple attribute decision making. In *Multiple Attribute Decision Making*; Springer: Berlin/Heidelberg, Germany, 1981; pp. 58–191.
18. Hwang, C.L.; Lai, Y.J.; Liu, T.Y. A new approach for multiple objective decision making. *Comput. Oper. Res.* **1993**, *20*, 889–899. [CrossRef]
19. Tang, H.; Shi, Y.; Dong, P. Public blockchain evaluation using entropy and TOPSIS. *Expert Syst. Appl.* **2019**, *117*, 204–210. [CrossRef]
20. Sennaroglu, B.; Celebi, G.V. A military airport location selection by AHP integrated PROMETHEE and VIKOR methods. *Transp. Res. Part D Transp. Environ.* **2018**, *59*, 160–173. [CrossRef]
21. Gupta, P.; Mehlawat, M.K.; Grover, N. Intuitionistic fuzzy multi-attribute group decision-making with an application to plant location selection based on a new extended VIKOR method. *Inf. Sci.* **2016**, *370*, 184–203. [CrossRef]
22. Tian, Z.P.; Wang, J.Q.; Wang, J.; Zhang, H.Y. A multi-phase QFD-based hybrid fuzzy MCDM approach for performance evaluation: A case of smart bike-sharing programs in Changsha. *J. Clean. Prod.* **2018**, *171*, 1068–1083. [CrossRef]
23. Tian, Z.P.; Wang, J.Q.; Zhang, H.Y. An integrated approach for failure mode and effects analysis based on fuzzy best-worst, relative entropy, and VIKOR methods. *Appl. Soft Comput.* **2018**, *72*, 636–646. [CrossRef]
24. Zhao, H.; Guo, S. Selecting green supplier of thermal power equipment by using a hybrid MCDM method for sustainability. *Sustainability* **2014**, *6*, 217–235. [CrossRef]
25. Zadeh, L.A. Fuzzy sets. In *Fuzzy Sets, Fuzzy Logic, and Fuzzy Systems: Selected Papers by LotfiAZadeh*; World Scientific: Hackensack, NJ, USA, 1996; pp. 394–432.
26. Tang, Z.; Guo, C.; Hou, P.; Fan, Y. Optimal siting of electric vehicle charging stations based on voronoi diagram and fahp method. *Energy Power Eng.* **2013**, *5*, 1404–1409. [CrossRef]
27. Feizizadeh, B.; Roodposhti, M.S.; Jankowski, P.; Blaschke, T. A GIS-based extended fuzzy multi-criteria evaluation for landslide susceptibility mapping. *Comput. Geosci.* **2014**, *73*, 208–221. [CrossRef]
28. Lee, W.; Xiang, L.; Schober, R.; Wong, V.W. Electric vehicle charging stations with renewable power generators: A game theoretical analysis. *IEEE Trans. Smart Grid* **2014**, *6*, 608–617. [CrossRef]
29. Rivera, S.; Wu, B.; Kouro, S.; Yaramasu, V.; Wang, J. Electric vehicle charging station using a neutral point clamped converter with bipolar DC bus. *IEEE Trans. Ind. Electron.* **2014**, *62*, 1999–2009. [CrossRef]
30. Wang, Q.; Zhou, N.; Wang, J.; Wei, N. Harmonic amplification investigation and calculation of electric vehicle charging stations using three-phase uncontrolled rectification chargers. *Electr. Power Syst. Res.* **2015**, *123*, 174–184. [CrossRef]
31. Ding, H.; Hu, Z.; Song, Y. Value of the energy storage system in an electric bus fast charging station. *Appl. Energy* **2015**, *157*, 630–639. [CrossRef]
32. Fan, P.; Sainbayar, B.; Ren, S. Operation analysis of fast charging stations with energy demand control of electric vehicles. *IEEE Trans. Smart Grid* **2015**, *6*, 1819–1826. [CrossRef]
33. Capasso, C.; Veneri, O. Experimental study of a DC charging station for full electric and plug in hybrid vehicles. *Appl. Energy* **2015**, *152*, 131–142. [CrossRef]
34. Li, S.; Bao, K.; Fu, X.; Zheng, H. Energy management and control of electric vehicle charging stations. *Electr. Power Compon. Syst.* **2014**, *42*, 339–347. [CrossRef]
35. Nansai, K.; Tohno, S.; Kono, M.; Kasahara, M.; Moriguchi, Y. Life-cycle analysis of charging infrastructure for electric vehicles. *Appl. Energy* **2001**, *70*, 251–265. [CrossRef]
36. Khalkhali, K.; Abapour, S.; Tafreshi, M.S.M.; Abapour, M. Application of data envelopment analysis theorem in plug-in hybrid electric vehicle charging station planning. *IET Gener. Transm. Distrib.* **2015**, *9*, 666–676. [CrossRef]
37. Liu, J. Electric vehicle charging infrastructure assignment and power grid impacts assessment in Beijing. *Energy Policy* **2012**, *51*, 544–557. [CrossRef]
38. Wirges, J.; Linder, S.; Kessler, A. Modelling the development of a regional charging infrastructure for electric vehicles in time and space. *Eur. J. Transp. Infrastruct. Res.* **2012**, *12*. [CrossRef]
39. Wang, Z.; Liu, P.; Cui, J.; Xi, Y.; Zhang, L. Research on quantitative models of electric vehicle charging stations based on principle of energy equivalence. *Math. Probl. Eng.* **2013**, *2013*. [CrossRef]
40. He, F.; Wu, D.; Yin, Y.; Guan, Y. Optimal deployment of public charging stations for plug-in hybrid electric vehicles. *Transp. Res. Part B Methodol.* **2013**, *47*, 87–101. [CrossRef]

41. Liu, Z.; Wen, F.; Ledwich, G. Optimal planning of electric-vehicle charging stations in distribution systems. *IEEE Trans. Power Deliv.* **2012**, *28*, 102–110. [CrossRef]
42. Pashajavid, E.; Golkar, M.A. Optimal placement and sizing of plug in electric vehicles charging stations within distribution networks with high penetration of photovoltaic panels. *J. Renew. Sustain. Energy* **2013**, *5*, 3126. [CrossRef]
43. Chen, T.D.; Kockelman, K.M.; Khan, M. Locating electric vehicle charging stations: Parking-based assignment method for Seattle, Washington. *Transp. Res. Rec.* **2013**, *2385*, 28–36. [CrossRef]
44. Sathaye, N.; Kelley, S. An approach for the optimal planning of electric vehicle infrastructure for highway corridors. *Transp. Res. Part E Logist. Transp. Rev.* **2013**, *59*, 15–33. [CrossRef]
45. Wang, G.; Xu, Z.; Wen, F.; Wong, K.P. Traffic-constrained multiobjective planning of electric-vehicle charging stations. *IEEE Trans. Power Deliv.* **2013**, *28*, 2363–2372. [CrossRef]
46. Dong, J.; Liu, C.; Lin, Z. Charging infrastructure planning for promoting battery electric vehicles: An activity-based approach using multiday travel data. *Transp. Res. Part C Emerg. Technol.* **2014**, *38*, 44–55. [CrossRef]
47. Xu, H.; Miao, S.; Zhang, C.; Shi, D. Optimal placement of charging infrastructures for large-scale integration of pure electric vehicles into grid. *Int. J. Electr. Power Energy Syst.* **2013**, *53*, 159–165. [CrossRef]
48. You, P.S.; Hsieh, Y.C. A hybrid heuristic approach to the problem of the location of vehicle charging stations. *Comput. Ind. Eng.* **2014**, *70*, 195–204. [CrossRef]
49. Lee, Y.G.; Kim, H.S.; Kho, S.Y.; Lee, C. User Equilibrium–Based Location Model of Rapid Charging Stations for Electric Vehicles with Batteries That Have Different States of Charge. *Transp. Res. Rec.* **2014**, *2454*, 97–106. [CrossRef]
50. Baouche, F.; Billot, R.; Trigui, R.; El Faouzi, N.E. Efficient allocation of electric vehicles charging stations: Optimization model and application to a dense urban network. *IEEE Intell. Transp. Syst. Mag.* **2014**, *6*, 33–43. [CrossRef]
51. Yao, W.; Zhao, J.; Wen, F.; Dong, Z.; Xue, Y.; Xu, Y.; Meng, K. A multi-objective collaborative planning strategy for integrated power distribution and electric vehicle charging systems. *IEEE Trans. Power Syst.* **2014**, *29*, 1811–1821. [CrossRef]
52. Ma, T.; Zhao, J.; Xiang, S.; Zhu, Y.; Liu, P. An agent-based training system for optimizing the layout of AFVs' initial filling stations. *J. Artif. Soc. Soc. Simul.* **2014**, *17*, 6. [CrossRef]
53. Chung, S.H.; Kwon, C. Multi-period planning for electric car charging station locations: A case of Korean Expressways. *Eur. J. Oper. Res.* **2015**, *242*, 677–687. [CrossRef]
54. Lam, A.Y.; Leung, Y.W.; Chu, X. Electric vehicle charging station placement: Formulation, complexity, and solutions. *IEEE Trans. Smart Grid* **2014**, *5*, 2846–2856. [CrossRef]
55. Cai, H.; Jia, X.; Chiu, A.S.; Hu, X.; Xu, M. Siting public electric vehicle charging stations in Beijing using big-data informed travel patterns of the taxi fleet. *Transp. Res. Part D Transp. Environ.* **2014**, *33*, 39–46. [CrossRef]
56. Önüt, S.; Soner, S. Transshipment site selection using the AHP and TOPSIS approaches under fuzzy environment. *Waste Manag.* **2008**, *28*, 1552–1559. [CrossRef]
57. Ghorui, N.; Ghosh, A.; Algehyne, E.A.; Mondal, S.P.; Saha, A.K. AHP-TOPSIS Inspired Shopping Mall Site Selection Problem with Fuzzy Data. *Mathematics* **2020**, *8*, 1380. [CrossRef]
58. Mohajeri, N.; Amin, G.R. Railway station site selection using analytical hierarchy process and data envelopment analysis. *Comput. Ind. Eng.* **2010**, *59*, 107–114. [CrossRef]
59. Liu, H.C.; You, J.X.; Fan, X.J.; Chen, Y.Z. Site selection in waste management by the VIKOR method using linguistic assessment. *Appl. Soft Comput.* **2014**, *21*, 453–461. [CrossRef]
60. Jun, D.; Tian, T.F.; Sheng, Y.Y.; Yu, M. Macro-site selection of wind/solar hybrid power station based on ELECTRE-II. *Renew. Sustain. Energy Rev.* **2014**, *35*, 194–204. [CrossRef]
61. Malczewski, J. GIS-based multicriteria decision analysis: A survey of the literature. *Int. J. Geogr. Inf. Sci.* **2006**, *20*, 703–726. [CrossRef]
62. Özceylan, E.; Erbaş, M.; Tolon, M.; Kabak, M.; Durğut, T. Evaluation of freight villages: A GIS-based multi-criteria decision analysis. *Comput. Ind.* **2016**, *76*, 38–52. [CrossRef]
63. Atici, K.B.; Simsek, A.B.; Ulucan, A.; Tosun, M.U. A GIS-based Multiple Criteria Decision Analysis approach for wind power plant site selection. *Util. Policy* **2015**, *37*, 86–96. [CrossRef]
64. Çetinkaya, C.; Özceylan, E.; Erbaş, M.; Kabak, M. GIS-based fuzzy MCDA approach for siting refugee camp: A case study for southeastern Turkey. *Int. J. Disaster Risk Reduct.* **2016**, *18*, 218–231. [CrossRef]
65. Garg, H. A new generalized improved score function of interval-valued intuitionistic fuzzy sets and applications in expert systems. *Appl. Soft Comput.* **2016**, *38*, 988–999. [CrossRef]
66. Kumar, K.; Garg, H. TOPSIS method based on the connection number of set pair analysis under interval-valued intuitionistic fuzzy set environment. *Comput. Appl. Math.* **2018**, *37*, 1319–1329. [CrossRef]
67. Maity, S.; Chakraborty, A.; De, S.K.; Mondal, S.P.; Alam, S. A comprehensive study of a backlogging EOQ model with nonlinear heptagonal dense fuzzy environment. *RAIRO Oper. Res.* **2020**, *54*. [CrossRef]
68. Soni, H.; Sarkar, B.; Josh, M. Demand uncertainty and learning in fuzziness in a continuous review inventory model. *J. Intell. FuzzySyst.* **2017**, *33*, 2595–2608. [CrossRef]
69. Sarkar, B.; Mahapatra, A.S. Periodic review fuzzy inventory models with variable lead time and fuzzy demand. *Int. Trans. Oper. Res.* **2017**, *24*, 1197–1227. [CrossRef]

70. Rahaman, M.; Mondal, S.P.; Shaikh, A.A.; Ahmadian, A.; Norazak, S.; Salahshour, S. Arbitrary-order economic production quantity model with and without deterioration: Generalized point of view. *Adv. Differ. Equ.* **2020**, *16*, 1–30. [CrossRef]
71. Rahaman, M.; Mondal, S.P.; Alam, S.; Khan, N.A.; Biswas, A. Interpretation of exact solution for fuzzy fractional non-homogeneous differential equation under the Riemann-Liouville sense and its application on the inventory management control problem. *Granul. Comput.* **2020**, 1–24. [CrossRef]
72. Rahaman, M.; Mondal, S.P.; Shaikh, A.; Pramanik, A.P.; Roy, S.; Maity, M.K.; Mondal, R.; De, D. Artificial bee colony optimization-inspired synergetic study of fractional-order economic production quantity model. *Soft Comput.* **2020**, *24*, 15341–15359. [CrossRef]
73. Chakraborty, A.; Maity, S.; Jain, S.; Mondal, S.P.; Alam, S. Hexagonal fuzzy number and its distinctive representation, ranking, defuzzification technique and application in production inventory management problem. *Granul. Comput.* **2020**, 1–15. [CrossRef]
74. Buckley, J.J. Fuzzy Hierarchical Analysis. In *Fuzzy Sets Systems*; World Scientific: Hackensack, NJ, USA, 1985.
75. Satty, T.L. *The Analytic Hierarchy Process*; McGraw-Hill: New York, NY, USA, 1980.
76. Yoon, K.P.; Hwang, C.L. *Multiple Attribute Decision Making: An Introduction*; Sage Publications: Thousand Oaks, CA, USA, 1995.
77. Sodhi, B.; Prabhakar, T.V. A simplified description of Fuzzy TOPSIS. *arXiv* **2012**, arXiv:1205.5098. Available online: https://arxiv.org/abs/1205.5098 (accessed on 15 February 2021).
78. Zavadskas, E.K.; Kaklauskas, A.; Sarka, V. The new method of multicriteria complex proportional assessment of projects. *Technol. Econ. Dev. Econ.* **1994**, *1*, 131–139.
79. Ghose, D.; Pradhan, S.; Tamuli, P.; Shabbiruddin. Optimal material for solar electric vehicle application using an integrated Fuzzy-COPRAS model. *Energy Sources Part A Recovery Util. Environ. Eff.* **2019**, 1–20. [CrossRef]
80. Tolga, A.C.; Durak, G. Evaluating Innovation Projects in Air Cargo Sector with Fuzzy COPRAS. In *International Conference on Intelligent and Fuzzy Systems*; Springer: Cham, Switzerland, 2019; pp. 702–710.
81. Yao, R.; Bai, H.; Xu, H. Where should China's thermal power industry prioritize its B&R investment? A study based on an environmental site selection analysis. *J. Clean. Prod.* **2019**, *215*, 669–679.
82. Modrak, V.; Soltysova, Z. Development of operational complexity measure for selection of optimal layout design alternative. *Int. J. Prod. Res.* **2018**, *56*, 7280–7295. [CrossRef]
83. Manerba, D.; Mansini, R.; Perboli, G. The capacitated supplier selection problem with total quantity discount policy and activation costs under uncertainty. *Int. J. Prod. Econ.* **2018**, *198*, 119–132. [CrossRef]
84. Kınay, Ö.B.; da Gama, S.F.; Kara, B.Y. On multi-criteria chance-constrained capacitated single-source discrete facility location problems. *Omega* **2019**, *83*, 107–122. [CrossRef]
85. Hosseini, S.; Sarder, M.D. Development of a Bayesian network model for optimal site selection of electric vehicle charging station. *Int. J. Electr. Power Energy Syst.* **2019**, *105*, 110–122. [CrossRef]
86. Zha, Y.; Gao, J.; Ni, S. Use of normalized difference built-up index in automatically mapping urban areas from TM imagery. *Int. J. Remote Sens.* **2003**, *24*, 583–594. [CrossRef]

Σ mathematics

Article
Fuzzy Governance Model

Enriqueta Mancilla-Rendón [1,*], Carmen Lozano [1] and Enrique Torres-Esteva [2]

1 Dirección de Investigación, Universidad La Salle Mexico, Ciudad de México 25298, Mexico; carmen.lozano@lasalle.mx
2 Facultad de Negocios, Universidad La Salle Mexico, Ciudad de México 25298, Mexico; enrique.torres.esteva@gmail.com
* Correspondence: maenriqueta.mancilla@lasalle.mx; Tel.: +52-55-5278-9500

Abstract: This article aims to analyze the functions of corporate governance agents as a key part of the study and evaluation of the internal control by the independent auditor to propose a governance fuzzy model based on legality. This is a descriptive–hermeneutical study based on mercantile-securities law, the code of best practice of corporate governance, and auditing standards. The research design is cross-sectional and uses fuzzy logic theory as an alternative tool in contrast to classical mathematical models. The results suggest that corporate governance agents strongly influence the application of a management system. Evidence is given regarding the positive relationship between the functions of corporate governance agents as a management system. Additionally, the importance of an internal control management system as an inherent mechanism for governance is proven. The scientific value of this work lies in showing how the interaction between the application of mathematical models based on fuzzy set theory and the qualitative attributes of internal control policies and practices. It is a tool to evaluate governance as a management system for decision making. This work emphasizes that a model based on fuzzy sets is useful to evaluate a management system of internal control policies and procedures necessary to improve corporate governance.

Keywords: governance; fuzzy logic; management system

Citation: Mancilla-Rendón, E.; Lozano, C.; Torres-Esteva, E. Fuzzy Governance Model. *Mathematics* **2021**, 9, 481. https://doi.org/10.3390/math9050481

Academic Editor: Michael Voskoglou

Received: 6 January 2021
Accepted: 3 February 2021
Published: 26 February 2021

1. Introduction

Corporate collapses and scandals, recognized as the financial catastrophes of the late twentieth and the early twenty-first century, were models of business failures [1]. They were caused by a deficient or absent supervision of the internal control policies and procedures. This mainly affected the notes and amounts reported on the financial statements approved by their corporate governance. Several companies make up the amounts of the trial balance and tend to mislead investors, customers, and suppliers [2]. They include fictitious sales to obtain incentives or credits and omit commitments to show low indebtedness. In addition, they overvalue some assets through the capitalization of operating expenses, which changes the profits at the end of the year and, therefore, the stockholders' equity. They also register non-existent assets to place shares and convertible bonds in preferential fiscal territories and manage all kinds of inappropriate registers to decrease the tax base. The corporate structure of transnational companies is that of a corporate government. It comprehends senior management, the internal audit committee, an external auditor, and policies and procedures of the internal control system that define how corporations and their functions should operate. Corporate government extends to all areas in an organization to achieve the economic goals of the entity and maintain its assets. The breach and application of an internal control system leads corporations to issue unreliable financial reports and make wrong decisions that could trigger financial disasters, technical bankruptcy, and eventually, the dissolution of a company. However, this denotes the absence of internal controls and raises questions as: Is corporate governance a management system for internal control? Do corporate governance agents exercise a sound control system management? Does internal control compliance guarantee reliable accounting reports? Is fuzzy logic an evaluation system for internal control policies and procedures as an agent of corporate governance?

The aim of this work is to analyze the functions of the agents that make up the corporate governance, the shareholders' assembly, the board of directors, and the audit functions. This is carried out from the study and evaluation of the internal control conducted by the independent auditor. We seek to propose a governance model as a management system based on the mechanisms of legality, the law of the stock market, and the code of corporate best practices. The scientific value of this work lies in showing that the interaction between the application of fuzzy models based on fuzzy set theory and the qualitative attributes of internal control policies and processes is a tool to evaluate governance as a management system in decision making.

This study has a descriptive character under the critical hermeneutic approach [3]. We consider the use of the hermeneutical principles and the fact that "the explanation of the real (social and economic) limitations that act on the interpreter should be used". The work is based on the International Auditing Standards ([4], NIA 260, NIA 265, NIA 700, NIA 705), ([5], B3050, B5030, Guide 6060) related to the independent auditor's report ([6], p. 58), the securities market law, and the code of good corporate practice. This document is a proposal based on the fact that conventional quantitative methods and techniques are not appropriate for treating social systems [7]. Fuzzy logic emerges as the ideal tool for modeling complex situations since it allows for making decisions considering qualitative attributes of a dynamic complex system. It is an excellent alternative for the study of governance due to its complexity and non-linear behavior based on subjective estimates of available information and the expertise and experience of those who manage it.

The first section reviews the theoretical and normative literature related to governance while the second one discusses the actors of governance. The methodological proposal is presented in the third section and a proposal for a fuzzy Governance Model as a management system is explained in the fourth section. Finally, the discussion of this work are presented.

2. Literature Review

2.1. Board of Directors

Corporate governance is a model of administrative management and long-term leadership through which corporations must be controlled and operated with disclosure and transparency. It suggests a relevant influence on the top management leadership and is responsible for the implementation of procedures to verify and safeguard the integrity of the financial reporting in the company.

Governance is performed through management and control [8], which involves the relationship between the top management of the company, its board of directors, and the interested third parties [9]. The roles and responsibilities of the board of directors must be separated from those of the top management to monitor and hold shareholders accountable. Corporate governance is the way to follow organizations and defines their managerial style [10]. It is the internal control environment of an entity [11] established by the board of directors, the committee of best corporate practices, and the audit committee [12]. The shareholders' meeting is responsible for appointing the people that will become members of the board of directors, responsible for appointing the people that will become members of the board of directors, responsible for supporting the represented company ([13], p. 379), its institutions, creditors, and third parties involved ([14], p. 147).

Corporate governance guarantees decision making [15]. The management of the corporate government protects the assets, avoids conflicts of interest, and separates the property, functions, and activities of internal control [16].

The board of directors is expected to play a key role in corporate governance and must guarantee the successful perpetuation of the corporation. The Principles of Corporate Governance by the OECD [9] describe all their responsibilities. The most important are summarized below:

1. To ensure that board members are informed and act ethically and in good faith, with due diligence and care, in the best interest of the company and the shareholders.
2. To review and guide corporate strategy, objective setting, major plans of action, risk policy, capital plans, and annual budgets.
3. To ensure the integrity of the accounting and financial reporting systems of the corporations, including their independent audit.
4. To ensure that the board members are nominated and elected through a formal and transparent process.
5. To select, compensate, monitor, and replace key executives and oversee succession planning.
6. To ensure the set-up of an appropriate internal control system.
7. To oversee disclosure and communications processes.

Public companies and their related parties are managed under the supervision of the board of directors and the top management [13], which ensure that operations related to a corporate purpose are carried out [17]. The board of directors acts through the committee of best corporate practices and the audit committee. Together, not only do they approve the engagement letters regarding audit services and the financial statements but they also present the financial statements and the auditor's opinion to the general shareholders' meeting.

2.2. Audit Committee

In 1992, the final report of the Cardbury Committee issued the Code of Best Corporate Practices that recommends the audit committee include at least three non-executive members, independent from the board, the Chief Financial Officer (CFO), and the external auditors. Since the board of directors is responsible for governing the organization [18], the audit committee performs the oversight functions. This committee supports the board of directors to ensure that the registrations are performed in accordance with accounting principles and criteria. It also evaluates the external auditors' performance and independence, discusses the audited financial statements, and reports the strengths and weaknesses in regard to the internal control system and the audit department of the company to the board of directors.

Its principal duties and responsibilities include overseeing the hiring, performance, and independence of the external auditor and the financial reporting and disclosure process as well as monitoring the selection of accounting principles and the internal control policies and procedures. It oversees the regulatory compliance, ethics, and if necessary, report malpractice, unlawful or unethical behavior within the workplace.

In addition, the audit committee ensures that the processed information is relevant and reliable for internal and external users. The Committee on Corporate Governance (1998) establishes that the audit committee is the safeguard of investments and protects the assets of the organization. Nicolăescu [19] states that the quality of corporate governance influences the perception that investors have of the quality of financial information and the auditors' reputation. Then, Johnstone, Li, and Rupley [20] confirm that there is a positive association between corporate governance, the audit committee, and the top management.

2.3. Best Corporate Practices

The performance of the board of directors is evaluated through a tool called best corporate practices. Mexico was one of the first countries in Latin America to raise the issue of governance [8]. Calderón states [21] that the Code of Best Practice of Corporate Governance was issued at the same time as the principles of the OECD, before the Sarbanex–Oxley Act and the Spanish Aldama report [22].

The importance of these practices is that they are binding for public companies [23]. It is suggested to [24] adopt this code because it contains recommendations that companies must consider to infer good corporate governance. Best practices are useful for any company seeking to improve the proper functioning of the board of directors. Several authors suggest following principles and practices, such as the daily reporting of the agenda of the shareholders' meeting and the communication between them, along with their functions, integration, structure, operation, and responsibility. In addition, it is advised to consider the function of audit, the selection of the auditors and the financial information, the management system of internal control, and risk management. The code contains the guidelines to manage the compliance of the accounting standards and prepare the financial information.

2.4. Internal Control Policies and Procedures

Internal control policies and procedures are a management system that defines the structure of organizations to ensure the integrity of the financial information, promote accountability, and prevent fraud. On the other hand, they are defined as a set of rules or guidelines created by any organization to achieve a long-term set of goals. They are designed to influence and determine decisions and actions and take place within the boundaries set by them. Their main purpose is to protect assets and ensure that all the transactions are registered in the accounting system.

An internal control system is designed, implemented, and maintained by those in charge of corporate governance or administrative management. They provide reasonable assurance that the objectives of the entity are achieved in terms of financial information, effectiveness and efficiency of operations, and compliance with laws and regulations ([4], p. 25).

The clarity of the organization structure is another component that shows internal control works properly. The Report of the Committee of Sponsoring Organizations of the Treadway Commission (COSO) emphasizes the importance of operational manuals, processes, organization, and procedures that provide a reasonable degree of security to conduct business [25]. Management control is fundamental to obtain reliable financial information.

The audit committee together with the board of directors and the committee of best corporate practices selects the independent auditor. The auditor is responsible for planning and performing the audit to obtain reasonable assurance that the financial statements are free of material misstatements due to error or fraud [3].

2.5. Test of Compliance in the Independent Auditor's Examination

The audit committee together with the board of directors and the committee of best corporate practices selects the independent auditor. The auditor is responsible for planning and performing the audit to obtain reasonable assurance about whether the financial statements are free of material misstatement caused by error or fraud ([4], NIA 700).

Before issuing an opinion on the financial statements of the company, the auditor should design internal control tests to obtain reasonable assurance from the material weaknesses and their effectiveness. Similarly, the internal control management system should be able to prevent, detect, and correct potential misstatements that would cause the financial statements to become material errors ([5], B3050, paragraph 36). Internal control testing must be useful to obtain appropriate evidence that supports the financial reporting as of year-end and the auditor's responsibility on identifying its reliability. Probably there are internal control policies and procedures, but it does not necessarily mean that they are efficiently operating the management system ([5], B3050, paragraph 51). In some cases, the auditor may conclude that there is no confidence in the internal control, eliminating the compliance tests and determining new substantive ones. To perform this evaluation, the auditor conducts a statistical or non-statistical sampling. This decision requires professional judgment, skepticism, and experience, trying in all cases to gather sufficient and competent evidence. The auditor will seek to detect and correct possible deviations in the management system and determine an appropriate scope in the revision.

When the auditor uses statistical sampling, they rely on the guides focused on determining the size and set up the level of confidence regarding the universe. Statistical sampling aims to estimate the frequency of certain characteristics that appear in the universe and is represented as a percentage ([5], Guide 6060).

When the auditor uses non-statistical sampling, they inevitably draw on their professional judgment, skepticism, and experience to set up their own parameters. They consider their knowledge regarding the principal economic environment factors in which an economic entity operates. Among those factors are organizational culture, size, structure, internal control conditions, income level, number of transactions, invoicing, number of workers, industry leadership, market conditions, infrastructure, inflation, and government regulation [26].

During the audit, the auditor establishes communication with corporate governance ([4], NIA 260) and informs them of situations of strangeness, unavoidable risk, and errors or irregularities that may remain undiscovered and cause some distortion. The auditor provides information on the efficiency of the internal control management system ([4], NIA 265), and the detection or the possibility of fraud, relevant risks, deviations from the rules expected by the stock market, and limitations found during their revision ([4], NIA 260, paragraph 16 e), always delivering written notification to the management (paragraph 11 b). Because of the revision, the auditor has to issue a report including their opinion and must state that the audit was carried out with standards that comply with ethical requirements ([4], NIA 260; Sections 21, 22 and 23). They must also describe the responsibility of the management on the preparation and fair presentation of the financial statements.

2.6. *Corporate Governance as a Management System*

Transnational companies and their related parties converge in the stock market. These companies buy and sell financial assets, stocks, bonds, and long-term debt securities. Therefore, they are under strict regulations and required to provide their financial statements issued by an external auditor in an annual report. This report must include the corporate governance questionnaire and an assessment of the compliance level, including the relevant principles and the code [27] of best practices of corporate governance.

Garcia's assertion is taken to characterize corporate governance as a dynamic management system because the functions of its agents (the shareholders' meeting, the board of directors, the committee of best corporate practices, and the audit committee) are interdependent, yet coherently linked and organized at the same time ([28], p. 117). They agree on Mekler's opinion ([29], pp. 4–5) which emphasizes that a management system shares attributes that interact between their components, giving rise to phenomena with both collective and cooperative behaviors.

Corporate governance, as a management system, works together with the external auditor who, through the study and evaluation of internal controls, determines the scope of their revision. It is a management dynamic and multidimensional system because its agents contain qualitative attributes that define modalities of governance functions, which are linked at different hierarchical levels of management (Figure 1).

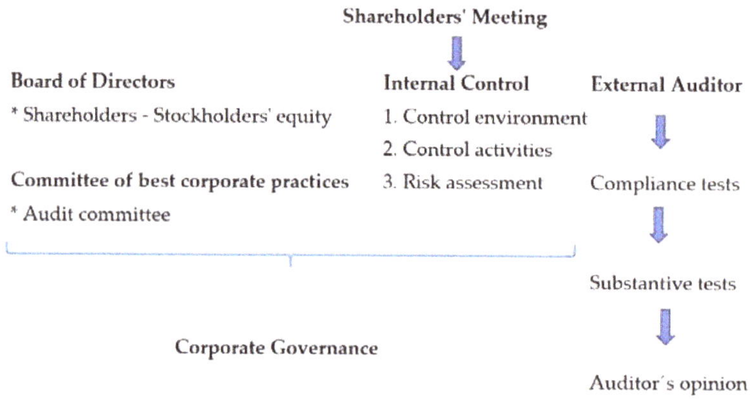

Figure 1. Corporate Governance structure.

3. Materials and Methods

3.1. Fuzzy Logic

In 1965, the Iranian–American mathematician, electrical engineer, computer scientist, and artificial intelligence researcher Lotfi Asker Zadeh published an article called Fuzzy Sets [30]. In it, he introduced the theory of fuzzy sets that are sets whose elements have degrees of membership. In 1973, he published an article called "Outline of a New Approach to the Analysis of Complex Systems and Decision Processes" [7]. The article points out that conventional quantitative techniques are inadequate to treat social systems.

Fuzzy logic emerges as the ideal tool for modeling complex situations since it allows for making decisions considering qualitative attributes of a dynamic complex system. It is an excellent alternative for the study of governance due to its complexity and non-linear behavior based on subjective estimates of available information and the expertise and experience of those who manage it. In addition, it provides a simple conclusion attributed to ambiguous, inaccurate, or incomplete information. Its main advantage is that it uses linguistic variables, not necessarily numerical, as human thinking. Unlike other statistical methods, fuzzy logic eliminates complex contents of mathematics and offers easy solutions to problems combining linguistic expressions with numerical data. On the other hand, it is worth highlighting that fuzzy logic models are based on mathematical ones. The latter allow labeling intermediate values to define estimates between true and false, black and white, hot and cold, little and much, small and large, short and tall, and close and far, among others.

Lozano and Fuentes [31] confirm that fuzzy logic is suitable for procedures based on intuitive rules that are hard to express in mathematical terms and its great potential comes from the probability of expressing operations and controlling rules in everyday language. Fuzzy logic was first applied in engineering and, although it is commonly handled by expert systems of artificial intelligence, there exist applications to several fields of science. For example, fuzzy logic was used to evaluate the sensitivity of the hydroelectric system, taking the human failure into consideration using the fuzzy analytic hierarchy process (AHP) approach [32]. On the other hand, [33] propose the first multicriteria model in a fuzzy environment to assist in decision making related to the renewal of healthcare equipment, this model tries to objectively analyze the different factors that must be taken into account in decision-making to renew healthcare technologies. Besides, we can find the application of fuzzy logic on financial risk indicators [34] and on accounting [31,35]. Research by Ji, Yu and Fu [36] proposes an assessment of personal default risk in peer to peer online lending platform, which reduces uncertainty while taking into account the psychological characteristics of lenders to avoid risk.

Most of the research is performed with statistical tools; however, sometimes there are no accurate data and the forecasts of statistical methods are far away from reality. Martínez [37] cited in Cruz [38] states that fuzzy logic is one of the mathematical disciplines that nowadays has the most followers. This is not only because of its easy understanding and flexibility but also because it is tolerant to inaccurate data and allows for conclusions based on uncertain assumptions.

In the following paragraphs, basic concepts regarding fuzzy logic are defined and fuzzy sets are explained. A fundamental concept for understanding classical and fuzzy logic theory is the membership function.

Bojadziev and Bojadziev [39] state that the fundamental notion in fuzzy sets is the one of membership. The concept of membership is the relationship that links each element with a set.

Let $X \subset \mathbb{R}$ a universal set and $A \subset X$. The membership function $\mu_A : X \to [0,1]$ denotes the membership grade of an element x to the set A, i.e.,

$$\mu_A(x) = \begin{cases} 0, & \text{if } x \notin A \\ r, & \text{if } r \text{ is the degree to which } x \text{ belongs to } A. \\ 1, & \text{if } x \in A. \end{cases}$$

Note that the range of the membership function is the interval $[0,1]$. A fuzzy set has the following representation

$$\widetilde{A} = \{(x, \mu_A(x)) : x \in X\}.$$

If X is a discrete universal set, i.e., $X = \{x_1, x_2 \ldots, x_n\}$, we have that:

$$\widetilde{A} = \{(x_1, \mu_A(x_1), (x_2, \mu_A(x_2)), \ldots, (x_n, \mu_A(x_n))\}. \tag{1}$$

The choice of the membership function depends on multiple parameters, such as context and application. Triangular functions, trapezoidal functions, and normal distribution functions, among others, are examples of membership functions. Still, simple functions, as the triangular ones, are usually chosen as a membership function (see Figure 2).

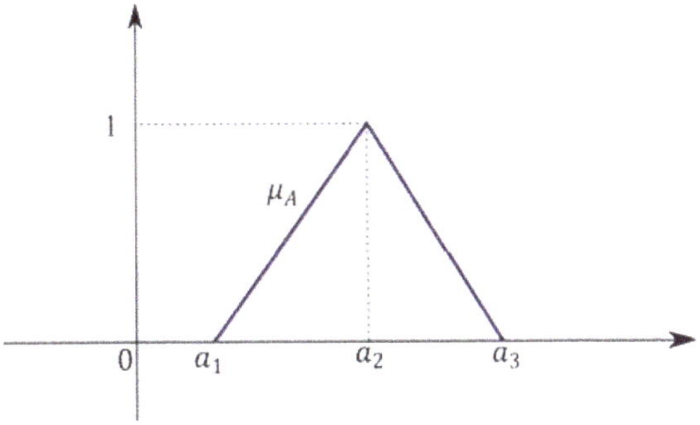

Figure 2. Membership function μ_A.

The operations between fuzzy sets are union, intersection, and idempotent. Let A and B be two fuzzy sets; then, the membership function of the union $A \cup B$ is represented by:

$$\mu_{A \cup B}(x) = \max\{\mu_A(x), \mu_B(x)\}.$$

The intersection $A \cap B$ has membership function given by:

$$\mu_{A \cap B}(x) = \min\{\mu_A(x), \mu_B(x)\},$$

and $A \cap A = A$, $A \cup A = A$.

A fuzzy number is an example of a fuzzy set. A triangular fuzzy number is determined by three parameters (a_1, a_2, a_3), where $a_1 < a_2 < a_3$ and $x = a_2$ is the peak of the triangle. See Figure 2.

When the range of the membership function μ_A is determined by experts, the membership function can be generalized by an interval valued function [40]. Thus, consider the set $\mathcal{L}([0,1])$ consisting of the whole closed subintervals in $[0,1]$. The φ fuzzy set, which is denoted by $\widetilde{A}^{\varphi}$, is the set with elements $(x, \mu^{\varphi}(x))$, where $\mu^{\varphi} : X \to \mathcal{L}([0,1])$ has the form $\mu^{\varphi}(x) = [a_x^1, a_x^2]$ and $x \in X$.

A fuzzy triangular number (a_1, a_2, a_3) can be written as follows:

$$\widetilde{A}^{\varphi} = \{(x, [\mu_A(x), \nu_A(x)]) : x \in X\},$$

where the functions $\mu_A, \nu_A X \to [0,1]$ are given by the following formulas:

$$\mu_A(x) = \begin{cases} 0 & \text{if } x < a_1 \\ \frac{x-a_1}{a_2-a_1} & \text{if } a_1 \leq x \leq a_2 \end{cases}$$
$$\nu_A(x) = \begin{cases} \frac{a_3-x}{a_3-a_2} & \text{if } a_2 \leq x \leq a_3. \\ 0 & \text{if } x > a_3 \end{cases}$$

The binary operations on a fuzzy triangular number are usually defined as follows. If $A = (a_1, a_2, a_3)$ and $B = (b_1, b_2, b_3)$ are two fuzzy triangular numbers and k is a real number, then

$$\begin{aligned} A + B &= (a_1, a_2, a_3) + (b_1, b_2, b_3) = (a_1 + b_1, a_2 + b_2, a_3 + b_3) \\ k \cdot A &= (ka_1, ka_2, ka_3). \end{aligned}$$

Decision making is key to an organization and its corporate governance. The management evaluates potential alternatives considering different tools to construct business objectives properly and timely before making a decision. For this reason, Mullor, Sansalvador, and Trigueros [41] state that fuzzy logic is the ideal tool to manage economic and administrative-accounting issues, such as personnel recruitment, supplier selection and evaluation, actuarial analysis, stock market prediction, inventory management, and stock level control.

Reducing the uncertainty and predicting the reality with inaccurate data is one of the greatest challenges for companies. Therefore, fuzzy sets constitute a novel tool that offers a strategic measurement. In this context, the use of fuzzy sets given in Equation (1) to back up information related to companies, people or processes allows to contrast and evaluate objects of the same nature. Fuzzy logic will be used to assess the degree of governance of corporate governance agents as a dynamic complex system. For each agent, a fuzzy set will be built along with the ideal agent, according to the professional judgment and experience of the independent auditor.

The similarity between agents will be measured using an addition competency index, which will indicate a higher level in compliance within internal controls. Consequently, there will be greater corporate governance when the intersection between the evaluated agents and the ideal is closer. The index between fuzzy sets

$$\widetilde{A}_1^{\varphi}, \widetilde{A}_2^{\varphi}, \ldots, \widetilde{A}_n^{\varphi}$$

is shown below:

$$\mu_{\widetilde{I}_1^{\varphi}}^{x_i}(\widetilde{A}_j^{\varphi}) = \frac{1}{n} \sum_{i=1}^{n} \mu_{\widetilde{I}_1^{\varphi}}^{x_i}(\widetilde{A}_j^{\varphi}), \quad (2)$$

where

$$\mu^{x_i}_{\widetilde{I}^{\varphi}_1}(\widetilde{A}^{\varphi}_j) = \frac{long\left([b^1_{x_i}, b^2_{x_i}] \cap [a_{x^1_i}, a^2_{x_i}]\right)}{long\left([b^1_{x_i}, b^2_{x_i}] \cup [a_{x^1_i}, a^2_{x_i}]\right)}.$$

Figure 3 illustrates the addition competency index. This graph orders the subjects and allows choosing those that have a higher adequacy index.

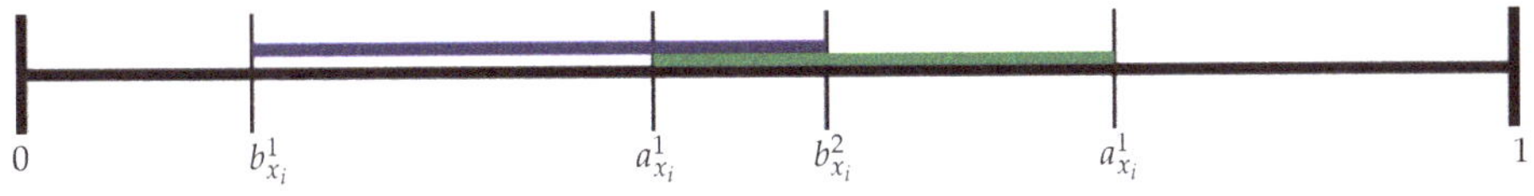

Figure 3. Similarity between agents.

3.2. Linguistic Variables

A linguistic variable is composed of words structured in an artificial language [30]. It has the form (T, Ω, G, M), where T is the linguistic variable, ω is the universe set, G represents the values of T and M is a map of T to fuzzy subsets of X.

In this work, T is the name of the linguistic variable while the characteristics of the enterprise are given by the set $G = \{$Very Good, Good, Regular, Bad, Very Bad$\}$ and the numerical values allocated to each linguistic label are within the interval $[0, 100]$. Then, the universe set is $\Omega = [0, 100]$ and M is the set consisting of whole the membership functions. See Figure 2.

Subsequently, each attribute will be qualified using the labels shown in the first column of Table 1. Afterwards, a fuzzy triangular set will be constructed with the gathered information as shown below:

Table 1. Fuzzy triangular numbers.

Linguistic Variables	Fuzzy Triangular Numbers	Fuzzy φ-Set: $[a^1_{x_j}, a^2_{x_j}]$
Very good	(66.4, 83, 100)	[66.4, 100]
Good	(49.8, 66.4, 83)	[49.8, 83]
Regular	(33.2, 49.8, 66.4)	[33.2, 66.4]
Bad	(16.6, 33.2, 49.8)	[16.6, 49.8]
Very bad	(0, 16.6, 33.2)	[0, 33.2]

The chart on each membership function of the fuzzy triangular set is given by Figure 4.

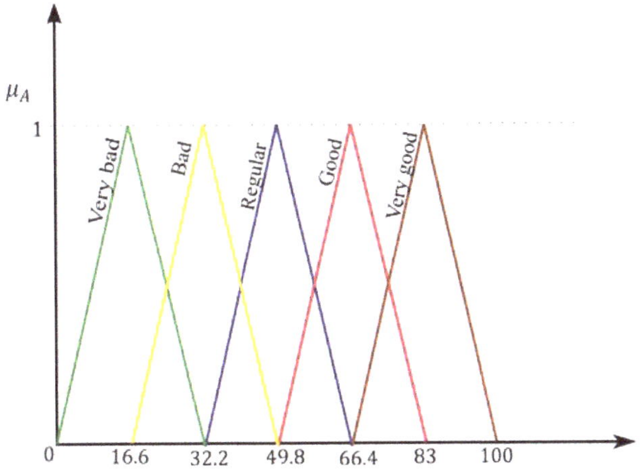

Figure 4. Graph of fuzzy triangular numbers.

4. Results

4.1. A Governance Fuzzy Model

As previously stated, the auditor should design an internal control testing to obtain reasonable assurance of the material weaknesses and effectiveness. This decision requires professional judgment, skepticism, and experience. The model proposed below is made of fuzzy sets and allows the auditor to perform their revision using both statistical and non-statistical sampling.

It will be considered a set of R categories evaluated to determine company size, income level, number of transactions, number of workers, industry leadership, and macroeconomic context, among others. These categories will be shown by the following set:

$$X = \{x_1, x_2, x_3, \ldots, x_R\}. \tag{3}$$

In addition, the external auditor will evaluate n agents:

$$A = \{A_1, A_2, \ldots, A_n\}.$$

Each of these agents must comply with a number of internal control policies and procedures. It can be understood that a model based on fuzzy logic is useful to assess the level of compliance in the application and supervision of policies and internal control procedures emerging as a need for improving corporate governance. The ideal agent I is that which best meets each category and is built according to the auditor's professional judgment and experience. Therefore, the fuzzy set is:

$$I = \{(x_1, [a^1_{x_1}, a_{x_1})^2]), (x_2, [a^1_{x_2}, a^2_{x_2}]), \ldots, (x_R, [a^1_{x_R}, a^2_{x_R}])\}.$$

Set I is built with the information in Table 1. Then, to determine the similarity between the agents in evaluation and the ideal, the categories of each one will be measured in relation to the ideal, using the addition competency index (Equation (2)). Some important remarks are listed below:

1. Some qualitative categories only allow for binary evaluations and other attributes, so the linguistic labels in Table 1 will be used for the analysis.
2. Categories such as company leadership can consider the use of information from some reputational evaluation instrument.
3. It is suggested to take the highest value obtained in categories corresponding to numbers of board members, reports of external auditors, and periodicity, among others.
4. Financial categories should be given greater attention since they deal with indicators of the economic position of the company; for example, profitability, liquidity, and leverage, among others. In order to evaluate each financial aspect, the highest value among all the obtained data is proposed.
5. Categories such as audit functions, conflict of interest, and code of ethics, among others, are dichotomous values; therefore, only Very good and Very bad will be considered in the evaluation.
6. This tool has the following advantages:
   (a) Economic: This is the greatest advantage since its low cost allows for its application in different contexts and situations and leads to low-cost audits.
   (b) Meticulous: It is a tool based on the auditor's professional judgment and experience to select non-random samples and reduce the number of categories to obtain meticulous information with satisfactory results.

The structure of the proposed model is shown in Figure 5.

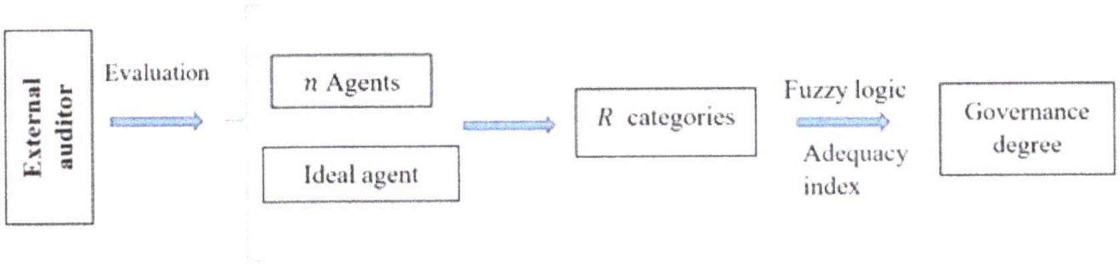

Figure 5. Governance structure.

4.2. Application

Most of the companies plan production at the lowest cost and offer products of the highest quality to the market. In this example, the Fuzzy Model for Governance presented seeks to meet three categories:

$$X = \{\text{Cost of raw material, quality and delivery time}\}.$$

Following the notation of Equation (3) we have a set of $R = 3$ categories that will be evaluated in its processes $X = \{x_1, x_2, x_3\}$.

The company has a purchasing policy for the acquisition of raw material that establishes purchase quotations with five suppliers. They correspond to agents A_1, A_2, A_3, A_4 and A_5. These five suppliers define the universe as follows:

$$\Omega = \{\text{Supplier 1, Supplier 2, Supplier 3, Supplier 4 and Supplier 5}\}.$$

According to the experience and the expertise of the acquisition, logistic, and quality areas, the three attributes are evaluated according to Table 1. The evaluations are shown in Table 2.

Table 2. Supplier evaluation (agents).

Attribute	A_1	A_2	A_3	A_4	A_5
Cost	Regular	Regular	Very bad	Bad	Good
Quality	Regular	Very good	Regular	Good	Regular
Delivery time	Very bad	Good	Very bad	Bad	Very good

Following the proposed model (Figure 5) and considering Table 1, fuzzy sets with interval-valued membership functions for each agent are built (see Table 3).

Table 3. Fuzzy sets for suppliers (agents).

Attribute	$\tilde{A}_1^{\varphi}$	$\tilde{A}_2^{\varphi}$	$\tilde{A}_3^{\varphi}$	$\tilde{A}_4^{\varphi}$	$\tilde{A}_5^{\varphi}$
x_1: Cost	[33.2, 66.4]	[33.2, 66.4]	[0, 33.2]	[16.6, 49.8]	[49.8, 83]
x_2: Quality	[33.2, 66.4]	[66.4, 100]	[33.2, 66.4]	[49.8, 83]	[33.2, 66.4]
x_3: Delivery time	[0, 33.2]	[49.8, 83]	[0, 33.2]	[16.6, 49.8]	[66.4, 100]

It must be considered that the ideal agent is built according to the independent auditor's professional judgement, skepticism, and experience. It complies with each category in the best way; therefore, the ideal agent (provider) complies with the attributes in Table 4, according to the acquisitions, quality, and logistics areas. To do so, the ideal provider must have a price degree not higher than 83% and a quality of at least 49.8%. It must also meet the delivery time at least 49.8% of the times.

Table 4. Fuzzy sets for suppliers (agents).

Attribute	Ideal I	$\widetilde{I}^{\varphi}$
x_1	Good	[49.8, 83]
x_2	Good	[49.8, 83]
x_3	Good	[49.8, 83]

Then, we use the addition competency index, Equation (2), between fuzzy sets $\widetilde{A}_1^{\varphi}, \widetilde{A}_2^{\varphi}, \widetilde{A}_3^{\varphi}, \widetilde{A}_4^{\varphi}, \widetilde{A}_5^{\varphi}$, and the ideal set I. To calculate $\mu_{\widetilde{I}^{\varphi}}(\widetilde{A}_1^{\varphi})$, the data in Table 3 and Equation (2). Then,

$$\mu_{\widetilde{I}^{\varphi}}(\widetilde{A}_1^{\varphi}) = \frac{1}{3}\sum_{i=1}^{n} \mu_{\widetilde{I}^{\varphi}}^{x_i}(\widetilde{A}_1^{\varphi}),$$

where

$$\mu_{\widetilde{I}^{\varphi}}^{x_1}(\widetilde{A}_1^{\varphi}) = \frac{long([33.2, 66.4] \cap [49.8, 83])}{long([33.2, 66.4] \cup [49.8, 83])} = \frac{long([49.8, 66.4])}{long([33.2, 83])} = 0.33,$$
$$\mu_{\widetilde{I}^{\varphi}}^{x_2}(\widetilde{A}_1^{\varphi}) = \frac{long([33.2, 66.4] \cap [49.8, 83])}{long([33.2, 66.4] \cup [49.8, 83])} = \frac{long([49.8, 66.4])}{long([33.2, 83])} = 0.33,$$
$$\mu_{\widetilde{I}^{\varphi}}^{3}(\widetilde{A}_1^{\varphi}) = \frac{long([0, 33.2] \cap [49.8, 83])}{long([0, 33.2] \cup [49.8, 83])} = \frac{long(\phi)}{long([33.2, 83])} = 0.$$

Figure 6 illustrates the intersection and union of sets for membership functions $\mu_{\widetilde{I}^{\varphi}}^{x_1}, \mu_{\widetilde{I}^{\varphi}}^{x_2}$.

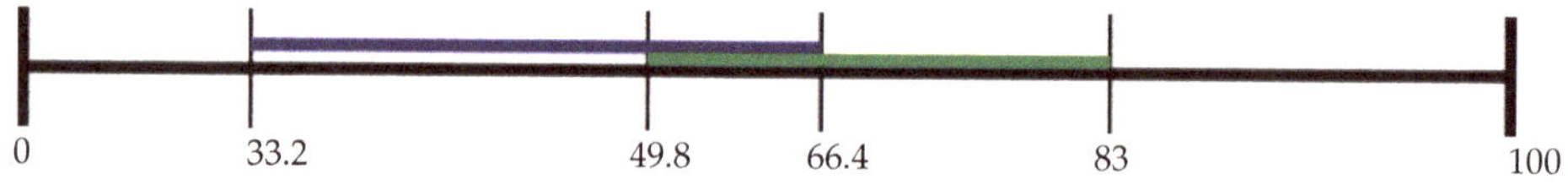

Figure 6. The unions and intersections of the supplier evaluation.

The model seeks to apply an addition competency index to measure the similarity of the agents under evaluation. Thus,

$$\mu_{\widetilde{I}^{\varphi}}(\widetilde{A}_1^{\varphi}) = \frac{1}{3}(0.33 + 0.33 + 0) = 0.22.$$

In an analogous way, the adequacy index for the rest of the agents is calculated. The results are summarized in Table 5.

Table 5. Adequacy index.

Agent	$\mu_{\widetilde{I}^{\varphi}}(\widetilde{A}_j^{\varphi})$
A_1	0.22
A_2	0.55
A_3	0.11
A_4	0.33
A_5	0.55

Clearly, agents A_2 and A_5 show a higher adequacy index and, in consequence, they (providers) comply with the internal control process for raw material in the best way, according to the acquisitions policies of the company. That is, both agents exert a good control environment, a governance aligned with the study, and an evaluation of internal control by the independent auditor.

Adequate segregation of the functions regarding purchase authorization and price quote when acquiring goods is key to evaluate internal control compliance. Then, product requirements must be followed according to the stock level indicated by the management. In addition, product reception at the warehouse must be observed.

The area manager must check that the process complies with the product quality attributes (weight, shape, smell, color, and packing, among others) for storage at the warehouse. The agent desirably implements an efficient control of the purchase orders and the pre-numbered reception notes related to every acquisition. They should also take frequent physical inventory counts. Both processes include the verification of legal-fiscal documentation, invoice review, prices and calculations, and a review against internal documents before accounting record and the creation of a liability (payment). Then, the manager's authorization degree and hierarchical level to contract and guarantee liabilities are reviewed.

Finally, logistic internal controls are fundamental to performing transactions, product sales by areas or online or any others determined by the management. In the end, the product is delivered to its destination efficiently. The authorization and sales documentation, along with lists of clients, prices, discounts, returns, and bonuses are also necessary to the right logistics.

To comply with the processes, the staff member in charge must authorize the appropriate segregation of clients' orders, payments (cash or credit), shipment, invoicing, credit notes, delivery dates and routes, physical custody, insurance, finance systems, and collaterals or pledges. Then the records are entered in the accounts.

The auditor evaluates the policies and internal control processes of the entity based on their professional assessment and experience, skepticism, and fuzzy governance model.

5. Discussion

The internal control system and the application of policies and procedures have long been studied by external auditors since they are enforceable in the professional-legal practice [21] within the framework of international audit standards [4]. Furthermore, the presentation of a financial notice is preceded by the study and evaluation of internal control [5]. This obligation is supported by the legislation [11] of international governments aiming at protecting investors and reducing financial disasters [1,2]. After the compliance with the code of best corporate practices [8], the governance system evaluates [9,22] its internal control system involved in accounting laws and standards [11] only from a qualitative perspective. The evaluation lacks mathematical analyses [39] without having a scientific confront.

The authors discuss the relevance of internal control as a cornerstone of governance [16], and good corporate practices [23] without alluding to a mathematical model. In our analysis, we state that mathematical models are binding to social sciences. It is, therefore, a novel way to relate laws [13,14] and audit (accounting) standards with fuzzy sets (fuzzy logic). Results show that control policies and procedures for the acquisition of products (pricing, quality, and delivery time) indicate a good governance management [12] since the company guarantees product supply to its clients. In addition, the fact of comparing a group of providers opens the door to different possibilities to acquire the product; that is, under better economic, resource, and speed conditions. These attributes promote the decision making of the management [25], in turn considering the rest of the providers as second or third sources of raw material.

Fuzzy logic as an agent of the governance system promotes decision making for product acquisition, leading to constant inventory rotation. The latter, in turn, facilitates corporate finances by opening market opportunities and thus charging clients due to sales increase. Liquidity increases and liabilities decrease, so investing in other assets, as financial instruments or companies, is possible. This is one of the benefits of fuzzy logic as a governance system.

The limitation of this study is its focus on the warehouse and purchasing areas of the corporation. Although we know that it is substantive, inventory control should be studied from a corporate system standpoint to relate it with internal control policies and procedures linked to other areas as sales and finance. This would complete the operation flow and allow a thorough understanding as to how supervising the application of policies and processes, or the lack thereof, increases corporate risk [38]. These research lines remain open for an integral study.

The fuzzy logic method is used since it is adequate for the study of attributes that can hardly be mathematically expressed. Its great potential lies in the possibility to express operations through everyday-use words. It is a tool used in administrative areas, as accounting and finance, of companies [7,38]. The fuzzy governance model is presented [30] as a management system for internal control policies and procedures, focusing on a legal framework to fulfill economic activities, necessary for decision making in corporate governance.

6. Conclusions

This work shows the inductive-deductive interaction between the implementation of fuzzy models and the qualitative attributes defined in the internal law of organizations. This document proposes a tool to evaluate the governance degree of agents belonging to corporate governance as a dynamic complex system.

Processes, including those non-written, are frequently absent and there is no documentary information to carry out the study and compliance evaluation of policies and internal control procedures within organizations. Fuzzy logic is a novel and efficient tool to study the compliance level of processes, especially that of those non-written.

The auditor observes process compliance and thus moves forward in the analysis of the study and evaluation of the internal control. All of this is based on the auditor's professional experience in the stock market, technical control, professional judgement, discretion, and skepticism exhibited during the examination of the financial information. The auditor uses a fuzzy model to identify the abilities and skills of other professionals, and materializes the technical control, professional judgement, skepticism, and professional experience. These qualities are hardly quantifiable using the known techniques.

Statutory commercial and securities law of accounting and auditing standards and agents of corporate governance concatenate into a complex system. Then, governance is exerted through inter- and transdisciplinary interactions involved in an established dynamic between areas and hierarchical levels of the organization.

Author Contributions: E.M.-R. and C.L. conceived the presented idea and they show how mathematics contains methodologies, such as fuzzy logic, which are useful tools for accounting and administrative sciences to evaluate qualitative attributes that are difficult to measure. E.M.-R. contributed to the acquisition, analysis, and interpretation of data for the work. C.L. developed the theory and performed the computations of fuzzy set theory. E.T.-E. searched the existing literature and data collection. E.M.-R., C.L. and E.T.-E. contributed to the analysis of the results and to the writing of the manuscript. All authors discussed the results and contributed to the final manuscript. All authors have read and agreed to the published version of the manuscript.

Funding: The APC was funded by the Universidad la Salle Mexico, project SAD-30/20.

Institutional Review Board Statement: Not applicable.

Informed Consent Statement: Not applicable.

Acknowledgments: This research was funded by Universidad La Salle México and is part of a research line and application of accounting and auditing of companies.

Conflicts of Interest: The authors declare no conflicts of interest.

References

1. García, M.; Vico, A. Los escándalos financieros y la auditoría: Pérdida y recuperación de la confianza en una profesión en crisis. *Rev. Valencia. Econ. Hacienda* **2003**, *7*, 25–48.
2. Ibarra Palafox, F. Enron o érase una vez en los Estados Unidos en El Poder de la Transparencia. In *Nueve Derrotas a la Opacidad*; Salazar, P., Ed.; UNAM-IFAI: Ciudad de Mexico, Mexico, 2007; Volume 251, pp. 1–44.
3. Álvarez Gayou, J. *Cómo Hacer Investigación Cualitativa, Fundamentos y Metodología*; Paidós Educador: Mexico, D.F., Mexico, 2003; p. 222.
4. IMCP. *Normas Internacionales de Auditoría*; Instituto Mexicano de Contadores Públicos: Ciudad de Mexico, Mexico, 2020.

5. IMCP. *Normas y Procedimientos de Auditoría y Normas Para Atestiguar*; Instituto Mexicano de Contadores Públicos: Ciudad de Mexico, Mexico, 2006.
6. Vieytes, R. *Campos de Aplicación y Decisiones de Diseño en la Investigación*; Merlino, A., Ed.; Investigación Cualitativa en Ciencias Sociales, Cengaje Learning: Buenos Aires, Argentina, 2009; pp. 43–84.
7. Zadeh, L.A. Outline of a New Approach to the Analysis of Complex Systems and Decision Processes. *IEEE Trans. Syst. Man Cybern.* **1973**, *3*, 28–44. [CrossRef]
8. CCE. *Código de Mejores Prácticas Corporativas*; Consejo Coordinador Emprearial: Miguel Hidalgo, Mexico, 2010.
9. OCDE. *Principios de Gobierno Corporativo de la OCDE*; Organización Para la Cooperación y el Desarrollo Económico: París, France, 2004.
10. ASF. *Fundamentos Conceptuales Sobre la Gobernanza*; Auditoría Superior de la Federación, OLACEFS, Organización Latinoamericana y del Caribe de Entidades Fiscalizadoras Superiores: Ciudad de Mexico, Mexico, 2015; pp. 12–42.
11. Visoso, F. *La Sociedad Anónima en la ley General de Sociedades Mercantiles y en la ley del Mercado de Valores*; Porrúa: México D.F., Mexico, 2007.
12. Canals, J. *Pautas de Buen Gobierno en los Consejos de Administración*; Universia Business Review, 2004. Available online: http://www.redalyc.org/articulo.oa?id=43300102 (accessed on 18 December 2020).
13. Paredes, L.; Meade, O. *Derecho Mercantil. Parte General y Sociedades*; Grupo Editorial Patria: México D.F., Mexico, 2008.
14. Macedo, J.; Macedo J. *Ley General de Sociedades Mercantiles. Anotada, Comentada, Concordada con Jurisprudencia y Tesis*; Cárdenas Editor y Distribuidor, Mexico, D.F., Mexico, 1993.
15. Ruiz, A.; Steinwascher, W. Gobierno corporativo, diversificación estratégica y desempeño empresarial en México. *MPRA* **2007**, *3819*, 1–16.
16. Tamborino, G. *Control Interno el Pilar del Gobierno Corporativo. Caso Portugal/España*; Cuaderno de Trabajo; Departamento de Contabilidad y Auditoría do ISCAC-IInstituto Superior de Contabilidade e Administração de Coimbra -IPC-Coimbra-Portugal: Coimbra, Portugal, 2011.
17. Mancilla-Rendón, E. ¿Si no está en el objeto social, no es deducible? In *Mitos Fiscales*; Burgóa, Ed.; Thompson Reuters: Mexico City, Mexico, 2017; pp. 37–48.
18. Ganga, F.; Vera, J. El gobierno corporativo consideraciones y cimientos teóricos. *Cuad. Adm.* **2008**, *21*, 93–126
19. Nicolăescu, E. Developments in corporate governance and regulatory interest in protecting audit quality. *Econ. Manag. Financ. Mark.* **2013**, *8*, 198–203.
20. Johnstone, K.; Li, C.; Rupley, K. Changes in corporate governance associated with the revelation of internal control material weaknesses and their subsequent remediation. *Contemp. Account. Res.* **2011**, *28*, 331–383. [CrossRef]
21. Calderón, M. Mejora de la Práctica Legal para Mitigar Riesgos en las Empresas Privadas en México. In *Derecho Mercantil*; Balino, P., Pablo, J., Páez, M., Alexandro, M., Eds.; Editorial Porrúa: Mexico City, Mexico, 2014; pp. 323–389.
22. OCDE. *Directrices de la OCDE Sobre el Gobierno Corporativo de las Empresas Públicas*; Organización para la Cooperación y el Desarrollo Económico: París, France, 2011.
23. Alonso Almeida, M.; Da Silva, J. Códigos de buen gobierno corporativo en Iberoamérica: Análisis comparativo entre Brasil y México. *Rev. Base (Adm. Contab.) UNISINOS* **2010**, *7*, 55–68. [CrossRef]
24. González, G.; Guzmán, A.; Prada, F.; Trujillo, M. Prácticas de gobierno corporativo en las asambleas generales de accionistas de empresas listadas en Colombia. *Cuad. Adm.* **2014**, *27*, 37–64. [CrossRef]
25. Martín Granados, V.; Mancilla-Rendón, E. Control en la administraciń para una información financiera confiable. *Contab. Negocios* **2010**, *5*, 68–75.
26. Chen, Y.-S.; Mardjono, E.S.; Yang, Y.-F. Competition and Sustainability: Evidence from Professional Service Organization. *Sustainability* **2020**, *12*, 7266. [CrossRef]
27. Moreno-Albarracín, A.L.; Licerán-Gutierrez, A.; Ortega-Rodríguez, C.; Labella, Á.; Rodríguez, R.M. Measuring What Is Not Seen—Transparency and Good Governance Nonprofit Indicators to Overcome the Limitations of Accounting Models. *Sustainability* **2020**, *12*, 7275. [CrossRef]
28. Carmona, D.; Sánchez, L. Teoría de sistemas complejos dinámicos. Una nueva reflexión sobre mercados financieros. In *La obra Desarrollo, Estructuras Económicas, Políticas Públicas y Gestión*; y Sánchez, V., Ed.; Reflexión Interdisciplinar: Ciudad de Mexico, Mexico, 2018; pp. 101–129.
29. Mekler, M.G. Sistemas Complejos. *Revista Digital Universitaria* **2012**, *13*, 1-xx–8-xx. Available online: http://www.revista.unam.mx/vol.13/num4/art44/art44.pdf (accessed on 15 December 2020).
30. Zadeh, L.A. Fuzzy sets. *Inf. Control* **1965**, *8*, 338–353. [CrossRef]
31. Lozano Gutiérrez, C.; Fuentes Martín, F. *Tratamiento Borroso del Intangible en la Valoración de Empresas en Internet*; Universidad Politécnica de Cartagena: Cartagena, España, 2003.
32. Ram, M.; Chandna, R. Sensitivity analysis of a hydroelectric production power plant under reworking scheme using fuzzy AHP approach. *J. Ind. Prod. Eng.* **2018**, *35*, 481–485. [CrossRef]
33. Domínguez, S.; Carnero, M.C. Fuzzy Multicriteria Modelling of Decision Making in the Renewal of Healthcare Technologies. *Mathematics* **2020**, *8*, 944. [CrossRef]
34. Córdova, J.F.D.; Molina, E.C.; López, P.N. Fuzzy logic and financial risk. A proposed classification of financial risk to the cooperative sector. *Contaduría y Adm.* **2017**, *62*, 1687–1703. [CrossRef]

35. Kwak, W.; Shi, Y.; Lee, C.F. The Fuzzy Set and Data Mining Applications in Accounting and Finance. In *Handbook of Quantitative Finance and Risk Management*; Lee, C.F., Lee, A.C., Lee, J., Eds.; Springer: Boston, MA, USA, 2010. [CrossRef]
36. Ji, X.; Yu, L.; Fu, J. Evaluating Personal Default Risk in P2P Lending Platform: Based on Dual Hesitant Pythagorean Fuzzy TODIM Approach. *Mathematics* **2020**, *8*, 8. [CrossRef]
37. Martínez, C. Uso de las Técnicas de Preprocesamiento de Datos e Inteligencia Artificial (*Lógica difusa*) en la Clasificación-Predicción del Riesgo Bancario. Bachelor's Thesis, Universidad de Los Andes, Mérida, Venezuela, 2007.
38. Cruz Martínez, A.; Alarcón Armenteros, A. La lógica difusa en la modelización del riesgo operacional. Una solución desde la inteligencia artificial en la banca cubana. *Cofín Habana* **2017**, *11*, 122–135.
39. Bojadziev, G.; Bojadziev, M. *Fuzzy Logic for Business, Finance and Management*, 2nd ed.; World Scientific Publishing Co.: London, UK, 2007.
40. Saad, R.; Ahmad, M.Z.; Abu, M.S.; Jusoh, M.S. Hamming Distance Method with Subjective and Objective Weights for Personnel Selection. *Sci. World J.* **2014**, *2014*, 865495. [CrossRef] [PubMed]
41. Mullor, J.R.; Sansalvador, S.M.E.; Trigueros, P.J.A. Lógica borrosa y su aplicación a la contabilidad. *Rev. Española Financ. Contab.* **2000**, *103*, 83–106.

mathematics

Article

Complex Uncertainty of Surface Data Modeling via the Type-2 Fuzzy B-Spline Model

Rozaimi Zakaria [1,*], Abd. Fatah Wahab [2], Isfarita Ismail [1] and Mohammad Izat Emir Zulkifly [3]

[1] Faculty of Science and Natural Resources, Universiti Malaysia Sabah, Jalan UMS, Sabah 88400, Malaysia; isfarita@yahoo.com
[2] Faculty of Ocean Engineering Technology and Informatics, Universiti Malaysia Terengganu, Kuala Nerus, Terengganu 21030, Malaysia; fatah@umt.edu.my
[3] Faculty of Science, Universiti Teknologi Malaysia, Skudai, Johor Bahru, Johor 81310, Malaysia; izatemir@utm.my
* Correspondence: rozaimi@ums.edu.my

Abstract: This paper discusses the construction of a type-2 fuzzy B-spline model to model complex uncertainty of surface data. To construct this model, the type-2 fuzzy set theory, which includes type-2 fuzzy number concepts and type-2 fuzzy relation, is used to define the complex uncertainty of surface data in type-2 fuzzy data/control points. These type-2 fuzzy data/control points are blended with the B-spline surface function to produce the proposed model, which can be visualized and analyzed further. Various processes, namely fuzzification, type-reduction and defuzzification are defined to achieve a crisp, type-2 fuzzy B-spline surface, representing uncertainty complex surface data. This paper ends with a numerical example of terrain modeling, which shows the effectiveness of handling the uncertainty complex data.

Keywords: type-2 fuzzy set; fuzzification; type-reduction; defuzzification; B-spline surface model function

Citation: Zakaria, R.; Wahab, A..F.; Ismail, I.; Zulkifly, M.I.E. Complex Uncertainty of Surface Data Modeling via the Type-2 Fuzzy B-Spline Model. *Mathematics* **2021**, *9*, 1054. https://doi.org/10.3390/math9091054

Academic Editor: Michael Voskoglou

Received: 18 February 2021
Accepted: 6 April 2021
Published: 7 May 2021

1. Introduction

Data points are the representation of the visible object in a digital system. The shape modification of items is carried out by tweaking the data points as the designer desires. However, the data points representing an object are not necessarily precise due to their characteristics, which are uncertain, leading to complexity. The complexities that are found in the data are due to various possible reasons, e.g., inappropriate analysis, human perception and logical assumption [1,2], limitations of tool accuracy, and the nature of data collection itself [3,4] with different types of uncertainty being based on BIPM (*Bureau International des Poids et Mesures)* concepts. The specific meaning of 'complex' in complex uncertainty data is the stack of the uncertainty of two arguments of the collected data points. It is impossible to accurately model the complex uncertainty data using an appropriate standard curve or surface function, such as the B-spline function, unless we formulate a new definition of the B-spline function with the complex uncertainty meaning.

To make the complex uncertainty data useful in modeling, we define complex uncertainty data in this paper. The complex uncertainty data can be explained by using type-2 fuzzy number (T2FN) concepts [5–7], which are defined by type-2 fuzzy set theory (T2FST) [8–10], especially interval type-2 fuzzy number (IT2FN) concepts [5,11]. The IT2FN concept is implemented to define the complex uncertainty data of real numbers subtly transformed into type-2 fuzzy data points (T2FDP), in line with the definition of type-2 fuzzy relation (T2FR). This is then followed by representing T2FDP as a type-2 fuzzy control point (T2FCP), which can be used for modeling the B-spline surface function [12–14]. Hence, we end up with a standard definition denoted as the type-2 fuzzy B-spline surface.

The interval type-2 fuzzy set (IT2FS) concept has been widely used to model higher-order uncertainty, which has been proven to be more suitable compared to the interval type-

1 fuzzy set. This concept is used when higher-order uncertainty exists in the measurement, and has been used for various applications [5,15–20]. This study employs the concept of the interval type-2 fuzzy set (IT2FN) to deal with the real data problem of complex uncertainty. The advantage of IT2FN is that it can define both complex uncertainty and data uncertainty for modeling. For the uncertainty case, it will first be reduced to become an interval type-1 fuzzy number. For the interval type-1 fuzzy number, this concept only deals with data uncertainty, but is not applicable when dealing with the complex uncertainty or a higher degree of uncertainty. Then, the IT2FN can define the complex uncertainty data or higher-level uncertainty data than the type-1 fuzzy number.

Therefore, the proposed approach in modeling the complex uncertainty data becomes important to obtain a better result instead of just modeling the perfect data, which does do not have complex uncertainty properties. Then, by excluding the complex uncertainty data, the main focus of generating a surface model based on all data is unreasonable and inaccurate. This is why we need an appropriate theory, such as T2FST, or more specifically, IT2FN, to define the complex uncertainty data and model those data to obtain a perfect model for better analysis, predictions and conclusions.

The interpolation method is usually used to produce the desired design. In Computer-Aided Geometric Design (CAGD), the interpolation method performs a pivotal role in modeling various techniques. Therefore, this method is practiced along with the surface data to find the control points as a reference data point. These control points are then represented with a surface function to produce the specific object's interpolation surface. With the complex uncertainty, the procedure in modeling the perfect interpolation surface using the B-spline surface function can be carried out using the type-2 fuzzy interpolation of the B-spline surface.

The remainder of this paper is organized as follows. Section 2 discusses the previous research about uncertainty data modeling via B-spline surface functions. Section 3 discusses the definition of T2FST, T2FN and T2FR. In this section, the definition of T2FDPs is developed along with the definitions of fuzzification, type-reduction and defuzzification processes. Section 4 discusses defining the type-2 fuzzy B-spline model, which uses the interpolation method for both the curve and surface. This section also elaborates on the construction of the type-2 fuzzy interpolation of the B-spline surface. Section 5 discusses the implementation of the type-2 fuzzy interpolation of the B-spline surface in seabed modeling. The alpha-cut operation of the fuzzification process with various alpha values is also discussed to indicate the influence of other alpha values concerning fuzzification, type-reduction, and the defuzzification surface. Section 5 discusses the accuracy of this developed model to sufficiently show the effectiveness of the type-2 fuzzy interpolation B-spline surface in modeling the uncertainty complex data.

2. Previous Work

The T2FST is useful in defining the uncertainty complex data to create a type-2 fuzzy curve and surface using the B-spline curve and surface functions. Regarding the designers' requirements in modeling the complex uncertainty data, the designers cannot choose and decide which data points are essential among the collectives of complex uncertainty data in modeling curves and surfaces. Therefore, we need the T2FST to define the complex uncertainty data and then model them through the B-spline curve and surface functions. This approach will make the complex uncertainty data become T2FDP/T2FCP, which can be modeled after integrating with the B-spline curve and surface functions.

Many types of research are carried out using the type-1 fuzzy set theory to model surface in dealing with uncertainty issues. Examples include the surface model proposed by Gallo et al. [21,22] of Mount Etna, Zakaria et al. [4] of lLakebed's modeling of Kenyir Lake, and Sarwar and Akram [23] proposed the fuzzy tensor product of Bezier surface. Note that the uncertainty level of these models is at level one. Suppose that the level of uncertainty increases due to specific errors, in particular the uncertainty at level two (complex uncertainty data). In that case, the current proposed approach for level one uncertainty data is not appropriate. The difference between uncertainty and error is the uncertainty defined as a range or interval where the actual value lies in this interval. An error is most likely the actual value obtained, which can be corrected by adding or subtracting the correction factor. Therefore, T2FST is proposed to treat the complex uncertainty data.

The expansion of modeling surfaces through T2FST is relatively new and it is at an early stage. Du proposed the example of T2FST in surface modeling, and Du and Zhu [24] had discussed about the modeling of spatial vagueness based on type-2 fuzzy set and was implemented in the Geographic Information System (GIS). However, this new method has been successfully developed, the surface results as a type-2 fuzzy condition which does not represent a type-2 fuzzy crisp surface. Furthermore, it does not deal with the fuzzification, type-reduction and defuzzification processes of the spatial vagueness modeling through the type-2 fuzzy set. In this article, the processes of fuzzification, type-reduction and defuzzification are discussed in detail based on [25]. The practicality of the proposed model is illustrated by the seabed modeling of Mengabang Telipot Beach.

3. Method: Type-2 Fuzzy Data Points

This section defines the complex uncertainty data with the definition of T2FST, IT2FN and T2FR [25].

Definition 1 ([25,26]). *A type-2 fuzzy set (T2FS), denoted as $\tilde{A}$ is characterized by a type-2 membership function $\mu_{\overleftrightarrow{A}}(x,u)$, where $x \in X$ and $\forall u \in [0,1]$, that is,*

$$\tilde{A} = \left\{ \left((x,u), \mu_{\tilde{A}}(x,u)\right) \middle| \forall x \in X, \forall u \in [0,1] \right\} \tag{1}$$

in which $0 \leq \mu_{\tilde{A}}(x,u) \leq 1$.

Definition 2 ([5]). *A T2FN is broadly defined as a type-2 fuzzy set (T2FS) that has a numerical domain. An interval of T2FS is defined using the following four constraints, where $\tilde{A}_\alpha = \{[a^\alpha, b^\alpha], [c^\alpha, d^\alpha]\}, \forall \alpha \in [0,1], \forall a^\alpha, b^\alpha, c^\alpha, d^\alpha \in \mathbb{R}$ (Figure 1):*

1. $a^\alpha \leq b^\alpha \leq c^\alpha \leq d^\alpha$.
2. *$[a^\alpha, d^\alpha]$ and $[b^\alpha, c^\alpha]$ generate a function that is convex and $[a^\alpha, d^\alpha]$ generate a normal function.*
3. $\forall \alpha_1, \alpha_2 \in [0,1] : (\alpha_2 > \alpha_1) \Rightarrow ([a^{\alpha_1}, c^{\alpha_1}] \supset [a^{\alpha_2}, c^{\alpha_2}], [b^{\alpha_1}, d^{\alpha_1}] \supset [b^{\alpha_2}, d^{\alpha_2}])$, *for* $c^{\alpha_2} \geq b^{\alpha_2}$.
4. *If the maximum of the membership function generated by $[b^\alpha, c^\alpha]$ is the level α_m, that is $[b^{\alpha_m}, c^{\alpha_m}]$, then $[b^{\alpha_m}, c^{\alpha_m}] \subset [a^{\alpha=1}, d^{\alpha=1}]$.*

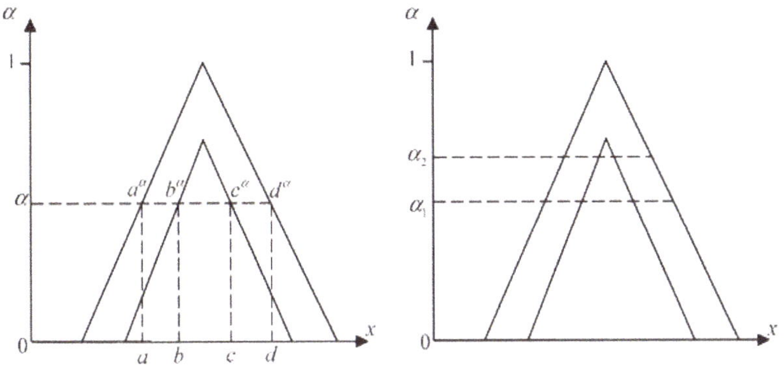

Figure 1. Definition of an IT2FN.

Definition 3 ([25]). *Let* $X, Y \subseteq R$, $\widetilde{A} = \{((x,u), \mu_{\widetilde{A}}(x,u)) | \forall x \in X, \forall u \in [0,1]\}$ *and* $\widetilde{B} = \{((y,v), \mu_{\widetilde{B}}(y,v)) | \forall y \in Y, \forall v \in [0,1]\}$ *are two T2FSs. Then,* $\widetilde{R} = \{(((x,u),(y,v)), \mu_{\widetilde{R}}(\mu_{\widetilde{A}}(x,u), \mu_{\widetilde{B}}(y,v))) | (\forall x \in X, \forall u \in [0,1]) \times (\forall y \in Y, \forall u \in [0,1])\}$ *is a type-2 fuzzy relation (T2FR) on* $\widetilde{A}$ *and* $\widetilde{B}$ *if* $\mu_{\widetilde{R}}(\mu_{\widetilde{A}}(x,u), \mu_{\widetilde{B}}(y,v)) \leq \mu_{\widetilde{A}}(x,u)$, $\forall ((x,u),(y,v)) \in (\forall x \in X, \forall u \in [0,1]) \times (\forall y \in Y, \forall v \in [0,1])$.

The following definition is T2FDP which is defined using the previous definitions as stated before [25].

Definition 4. *Let* $P = \{x | x \text{ type} - 2 \text{ fuzzy point}\}$ *and* $\overset{\leftrightarrow}{\overset{\leftrightarrow}{P}} = \{P_i | P_i \text{ data point}\}$ *be the set of the type-2 fuzzy data point with* $P_i \in P \subset X$, *where X is a universal set and* $\mu_P(P_i) : P \rightarrow [0,1]$ *is the membership function defined as* $\mu_P(P_i) = 1$ *and formulated as* $\overset{\leftrightarrow}{\overset{\leftrightarrow}{P}} = \{(P_i, \mu_P(P_i)) | P_i \in \mathbb{R}, \}$ $\{ i = 0, 1, 2, \ldots, n\}$. *Therefore,*

$$\mu_P(P_i) = \begin{cases} 0 & \text{if } P_i \notin X \\ c \in (0,1) & \text{if } P_i \overset{\leftrightarrow}{\in} X \\ 1 & \text{if } P_i \in X \end{cases} \tag{2}$$

with $\mu_P(P_i) = \left\langle \mu_P(\overset{\leftrightarrow}{\overset{\leftrightarrow}{P}}{}_i^{\leftarrow}), \mu_P(P_i), \mu_P(\overset{\leftrightarrow}{\overset{\leftrightarrow}{P}}{}_i^{\rightarrow}) \right\rangle$ *which* $\mu_P(\overset{\leftrightarrow}{\overset{\leftrightarrow}{P}}{}_i^{\leftarrow})$ *and* $\mu_P(\overset{\leftrightarrow}{\overset{\leftrightarrow}{P}}{}_i^{\rightarrow})$ *are left and right footprint of membership values with* $\mu_P(\overset{\leftrightarrow}{\overset{\leftrightarrow}{P}}{}_i^{\leftarrow}) = \langle \mu_P(a_i), \mu_P(e_i), \mu_P(b_i) \rangle$ *where* $\mu_P(a_i)$, $\mu_P(e_i)$ *and* $\mu_P(b_i)$ *are left–left, left, right–left membership grade values and* $\mu_P(\overset{\leftrightarrow}{\overset{\leftrightarrow}{P}}{}_i^{\rightarrow}) = \langle \mu_P(c_i), \mu_P(f_i), \mu_P(d_i) \rangle$ $\mu_P(d_i)$, $\mu_P(f_i)$ *and* $\mu_P(c_i)$ *are right–right, right, left–right membership grade values, which can be written as follows:*

$$\overset{\leftrightarrow}{\overset{\leftrightarrow}{P}} = \left\{ \overset{\leftrightarrow}{\overset{\leftrightarrow}{P}}_i : i = 0, 1, 2, \ldots, n \right\} \tag{3}$$

For every i, $\overset{\leftrightarrow}{\overset{\leftrightarrow}{P}}_i = \left\langle \overset{\leftrightarrow}{\overset{\leftrightarrow}{P}}{}_i^{\leftarrow}, P_i, \overset{\leftrightarrow}{\overset{\leftrightarrow}{P}}{}_i^{\rightarrow} \right\rangle$ *with* $\overset{\leftrightarrow}{\overset{\leftrightarrow}{P}}{}_i^{\leftarrow} = \langle a_i, e_i, b_i \rangle$ *where* a_i, e_i *and* b_i *are left–left, left and right–left T2FDPs and* $\overset{\leftrightarrow}{\overset{\leftrightarrow}{P}}{}_i^{\rightarrow} = \langle c_i, f_i, d_i \rangle$ *where* c_i, f_i *and* d_i *are left–right, right and right–right T2FDPs, respectively. This can be illustrated as shown in Figure 2.*

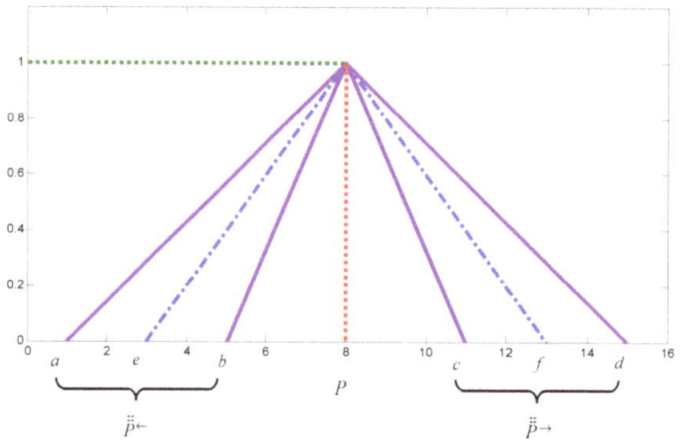

Figure 2. T2FDP around eight.

From Figure 2, T1FDP becomes the primary membership function bounded by upper bound, $[a,d]$ and lower bound, $[b,c]$ respectively.

After T2FDP has been defined, the next procedure is the fuzzification process that applies the alpha-cut of IT2FN [25]. This definition is determined based on the fuzzification process of type-1 fuzzy data points discussed in [4]. Therefore, the description of the fuzzification process against T2FDP can be given through Definition 5, as follows.

Definition 5. *Let $\overset{\leftrightarrow}{\overset{\leftrightarrow}{P}}$ be the set of T2FDPs with $\overset{\leftrightarrow}{\overset{\leftrightarrow}{P}}_i \in \overset{\leftrightarrow}{\overset{\leftrightarrow}{P}}$ where $i = 0, 1, \ldots, n-1$. Then $\overset{\leftrightarrow}{\overset{\leftrightarrow}{P}}_{i_\alpha}$ is the alpha-cut operation of T2FDPs with $i = 0, 1, 2, \ldots, n$ which is given as follows.*

$$\begin{aligned} \overset{\leftrightarrow}{\overset{\leftrightarrow}{P}}_{i_\alpha} &= \left\langle \overset{\leftrightarrow\leftarrow}{\overset{\leftrightarrow}{P}}_{i_\alpha}, P_i, \overset{\leftrightarrow\rightarrow}{\overset{\leftrightarrow}{P}}_{i_\alpha} \right\rangle \\ &= \langle \langle a_i^\alpha ; e_i^\alpha ; b_i^\alpha \rangle, P_i, \langle c_i^\alpha ; f_i^\alpha ; d_i^\alpha \rangle \rangle \\ &= \langle [(P_i - \langle a_i; e_i; b_i \rangle)\alpha + \langle a_i; e_i; b_i \rangle], P_i, \\ &\quad [-(\langle c_i; f_i; d_i \rangle - P_i)\alpha + \langle c_i; f_i; d_i \rangle] \rangle \end{aligned} \tag{4}$$

This definition is illustrated in Figure 3.

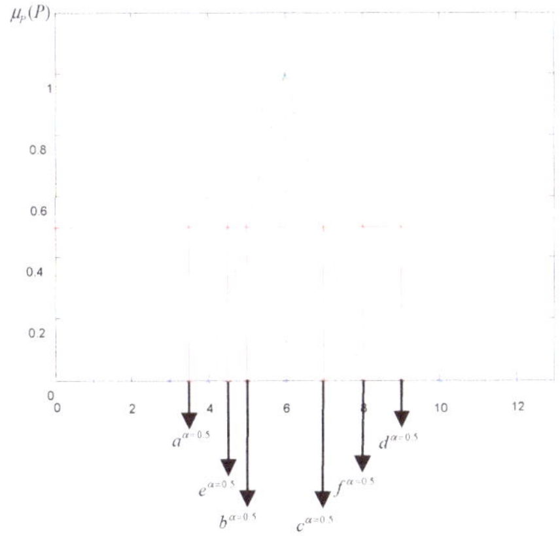

Figure 3. The alpha-cut operation toward T2FDP.

Figure 3 shows the implication of the alpha-cut operation against T2FDP, which is the fuzzification process with the specific value of alpha (membership value). The alpha value of this operation is 0.5. If the alpha value increases to one, then the crisp data point is obtained. This can be illustrated by Figure 4 as follows.

After performing the fuzzification process, then the following process is type-reduction. Type-reduction is defined, then used against T2FDP to allow the defuzzification of the type-1 fuzzy set. The type-reduction is an approach to simplify type-2 defuzzification. The first type-reduction had been proposed by Nie and Tan [27]. On the other hand, type-reduction on discretized interval type-2 fuzzy sets have been discussed in [28]. The consistent linear and quadratic type-reduction methods have been introduced by [29].

The type-2 defuzzification has been already discussed, such as the Karnik–Mendel algorithm [30]. Subsequently, type-2 defuzzification with explicit models of the uncertainty was proposed in [31]. Moreover, the various methods of type-2 defuzzification have different mathematical properties discussed by [31].

In this paper, the proposed methods for type-reduction and type-2 defuzzification against T2FDP are based on the centroid method based on Definition 6 and Definition 7, respectively.

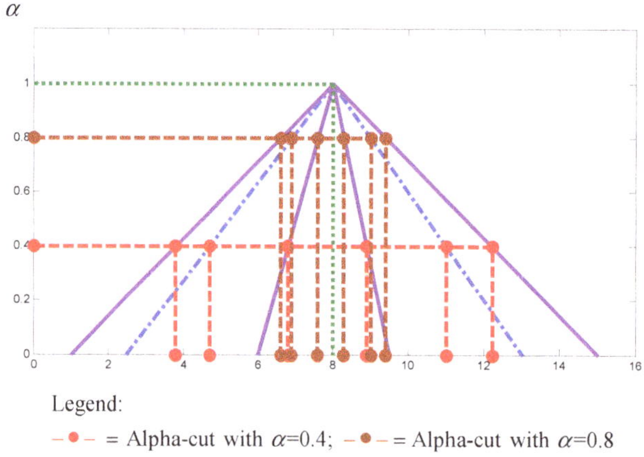

Figure 4. The illustration of the correlation between alpha values and T2FDPs.

Definition 6. *Let $\overleftrightarrow{P}_i$ be a set T2FDP and $\overleftrightarrow{P}_{i_\alpha}$ are the set of T2FDP after the fuzzification process for $i = 0, 1, 2, \ldots, n$, then the type-reduction of $\overleftrightarrow{P}_{i_\alpha}$ which is represented as $\overline{\overleftrightarrow{P}}_{\alpha}$ can be defined as follows:*

$$\overline{\overleftrightarrow{P}}_{\alpha} = \left\{ \overline{\overleftrightarrow{P}}_{i_\alpha} = \left\langle \overline{\overleftrightarrow{P}}^{\leftarrow}_{i_\alpha}, P_i, \overline{\overleftrightarrow{P}}^{\rightarrow}_{i_\alpha} \right\rangle ; i = 0, 1, 2, \ldots, n \right\} \quad (5)$$

where P_i is crisp data points and $\overline{\overleftrightarrow{P}}^{\leftarrow}_{i_\alpha}$ and $\overline{\overleftrightarrow{P}}^{\rightarrow}_{i_\alpha}$ are left and right fuzzified type-reduction T2FDP respectively with their formulation given by the following:

$$\begin{aligned} \overline{\overleftrightarrow{P}}^{\leftarrow}_{i_\alpha} &= \tfrac{1}{3} \sum_{i=0,1,\ldots,n} \left\langle a_i^\alpha + e_i^\alpha + b_i^\alpha \right\rangle \\ \overline{\overleftrightarrow{P}}^{\rightarrow}_{i_\alpha} &= \tfrac{1}{3} \sum_{i=0,1,\ldots,n} \left\langle c_i^\alpha + f_i^\alpha + d_i^\alpha \right\rangle \end{aligned} \quad (6)$$

Through the implementation of Definition 6, the fuzzified type-reduction T2FDP is obtained. Then, the next procedure to get the crisp T2FDP is the defuzzification process of type-1. The defuzzification process has been defined in [4,32,33] and described as follows.

Definition 7. *Let $\overline{\overleftrightarrow{P}}_{i_\alpha}$ be the fuzzified type-reduction T2FDP with $i = 0, 1, 2, \ldots, n$. Then, $\overline{\overline{P}}_{i_\alpha}$ is the defuzzification process of $\overline{\overleftrightarrow{P}}_{i_\alpha}$ if for every $\overline{\overleftrightarrow{P}}_{i_\alpha} \in \overline{\overleftrightarrow{P}}_{\alpha}$,*

$$\overline{\overline{P}}_{\alpha} = \left\{ \overline{\overline{P}}_{i_\alpha} \middle| i = 0, 1, 2, \ldots, n \right\} \quad (7)$$

where each $\overline{\overline{P}}_{i_\alpha}$ can be formalized as:

$$\overline{\overline{P}}_{i_\alpha} = \frac{1}{3} \sum_{i=0,1,\ldots,n} \left\langle \overline{\overleftrightarrow{P}}^{\leftarrow}_{i_\alpha}, P_i, \overline{\overleftrightarrow{P}}^{\rightarrow}_{i_\alpha} \right\rangle \quad (8)$$

The process of obtaining T2FDP defuzzified from defining T2FDP, including the fuzzification, type-reduction and defuzzification processes can be summarized and illustrated in the following Figure 5.

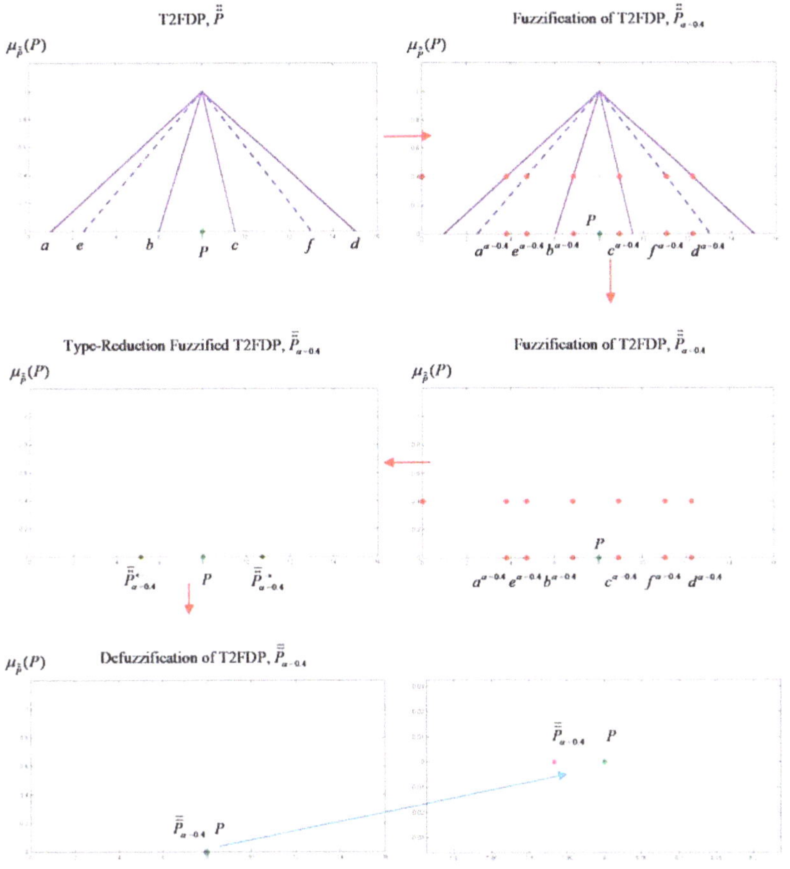

Figure 5. The processes of defining, fuzzification, type-reduction and defuzzification towards T2FDP.

4. Results: Type-2 Fuzzy B-Spline Model

This section will discussing type-2 fuzzy B-spline model which specifically used the interpolation method. When creating a type-2 fuzzy curve and surface, the T2FDP are integrated into the B-spline curve and surface function where the end result is known as a type-2 fuzzy B-spline curve and surface. This type-2 fuzzy model meets the processes of fuzzification, type-reduction and defuzzification to obtain the crisp type-2 fuzzy curve and surface solution (single curve and surface solution).

The construction of the type-2 fuzzy B-spline model is based on the studies carried out by Zakaria et al. [4], Zakaria and Wahab [32], Wahab et al. [33], and Karim et al. [34]. These studies discussed the construction of the type-1 fuzzy B-spline model. The type-2 fuzzy interpolation B-spline model [25] is defined in Definition 8 and illustrated by Figure 6.

Definition 8. *Let $\overset{\leftrightarrow}{\overset{\leftrightarrow}{D}}_i \in R$ be a list of T2FDP with $i = 0, 1, 2, \ldots, n$, then the type-2 fuzzy interpolation B-spline curve (T2FIBsC) can be defined as follows:*

$$\overset{\leftrightarrow}{\overset{\leftrightarrow}{BsC}}(t) = \sum_{i=0}^{k+h-1} \overset{\leftrightarrow}{\overset{\leftrightarrow}{P}}_i N_{i,h}(t) \text{ which } \overset{\leftrightarrow}{\overset{\leftrightarrow}{BsC}}(t_i) = \overset{\leftrightarrow}{\overset{\leftrightarrow}{D}}_i \tag{9}$$

where $\overset{\leftrightarrow}{\overset{\leftrightarrow}{P}}_i$ *are T2FCP,* $N_{i,h}(t)$ *is a basic function of B-spline and* t *is crisp knot sequences* $t_1, t_2, \ldots, t_{m=d+n+1}$ *in which* d *represents the degree of B-spline function and* n *represents the numbers of control points.*

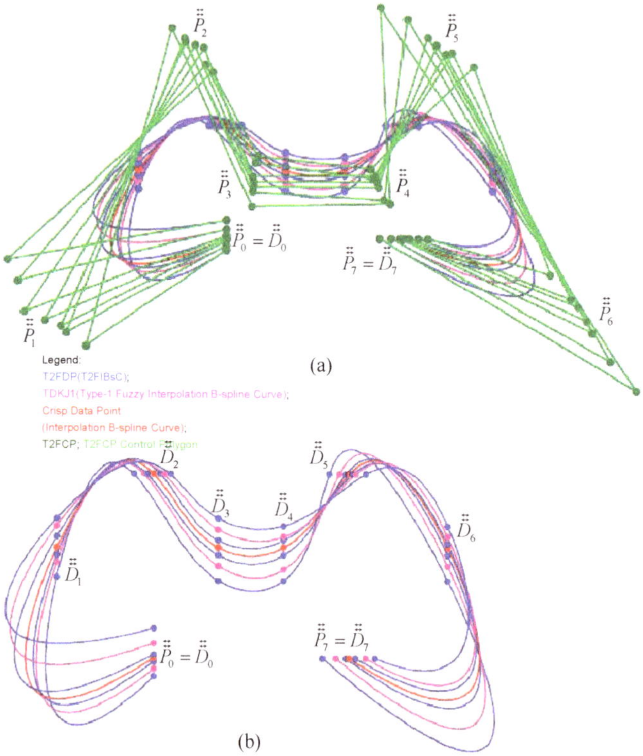

Figure 6. The example of T2FIBsC model: (**a**) with T2FCP; (**b**) without T2FCP.

Definition 9. *Let* $\overset{\leftrightarrow}{\overset{\leftrightarrow}{D}}_{a,b}$ *be given as a set of T2FDP with* $a = 0, 1, \ldots, m$ *and* $b = 0, 1, \ldots, n$ *which has degree p and q. Then, the type-2 fuzzy interpolation B-spline surface (T2FIBsS) model with degree p and q can be given as follows:*

$$\overset{\leftrightarrow}{\overset{\leftrightarrow}{BsSi}}(s_a, t_b) = \sum_{l=0}^{n} \sum_{k=0}^{m} N_{k,p}(s_a) N_{l,q}(t_b) \overset{\leftrightarrow}{\overset{\leftrightarrow}{P}}_{k,l} = \overset{\leftrightarrow}{\overset{\leftrightarrow}{D}}_{a,b} \tag{10}$$

where $\overset{\leftrightarrow}{\overset{\leftrightarrow}{P}}_{k,l}$ *is the set of T2FCP which is the unknown value.*

Equation (10) can be rewritten as

$$\overset{\leftrightarrow}{\overset{\leftrightarrow}{D}}_{a,b} = \sum_{l=0}^{n} N_{k,p}(s_a) \left[\sum_{k=0}^{m} N_{l,q}(t_b) \overset{\leftrightarrow}{\overset{\leftrightarrow}{P}}_{k,l} \right] = \sum_{l=0}^{n} N_{k,p}(s_a) \overset{\leftrightarrow}{\overset{\leftrightarrow}{R}}_{k,b} \tag{11}$$

with

$$\overset{\leftrightarrow}{\overset{\leftrightarrow}{R}}_{k,b} = \sum_{k=0}^{m} N_{l,q}(t_b) \overset{\leftrightarrow}{\overset{\leftrightarrow}{P}}_{k,l} \tag{12}$$

Therefore, Equation (12) is resolved first which used $\overset{\leftrightarrow}{\overset{\leftrightarrow}{D}}_{a,b}$ before $N_{k,p}(s_a)$. It is followed by solving $\overset{\leftrightarrow}{\overset{\leftrightarrow}{P}}_{k,l}$ using Equation (10). Therefore, we obtain T2FCP values that allow the surface to interpolate the T2FDP. Thus, the illustration of the T2FIBsS model can be illustrated in Figure 7 as follows.

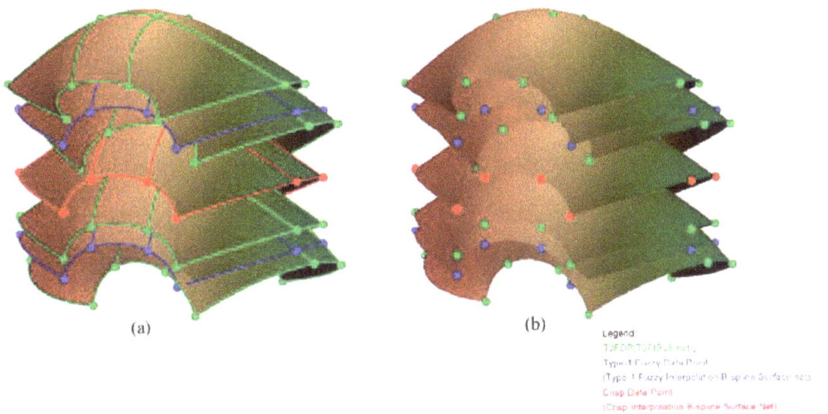

Figure 7. The example T2FIBsS model: with (**a**) and; without (**b**) type-2 fuzzy data net.

Figure 7 shows that the T2FIBsS model in bicubic shapes was constructed based on the T2FCP net obtained using Equation (10). The T2FDP net of T2FIBsS of Figure 7a illustrates the first impression of the against the type-2 fuzzy surface.

The next process is the fuzzification process. This fuzzification is performed against the T2FDPs before being integrated with the basic B-spline surface function. Therefore, the fuzzification of T2FIBsS can be defined as the following equation

$$\overset{\leftrightarrow}{\overset{\leftrightarrow}{BsSi}}_{\alpha_j}(s_a, t_b) = \sum_{l=0}^{n}\sum_{k=0}^{m} N_{k,p}(s_a)N_{l.q}(t_b)\overset{\leftrightarrow}{\overset{\leftrightarrow}{P}}_{(k,l)_{\alpha_j}} = \overset{\leftrightarrow}{\overset{\leftrightarrow}{D}}_{(a,b)_{\alpha_j}} \tag{13}$$

where α_j is the value of alpha-cut operation of the type-2 triangular fuzzy number represents the fuzzification process with $\alpha \in (0,1]$, $j = 1,2,\ldots,y$. Therefore, based on Figure 7, the illustration after fuzzification process can be shown in Figure 8.

After the fuzzification process has been applied, the next step is the type-reduction process. Therefore, the reduction of fuzzified T2FIBsS can be formalized as follows, based on Definition 6.

$$\overline{\overset{\leftrightarrow}{BsSi}}_{\alpha_j}(s_a, t_b) = \sum_{l=0}^{n}\sum_{k=0}^{m} N_{k,p}(s_a)N_{l.q}(t_b)\overline{\overset{\leftrightarrow}{P}}_{(k,l)_{\alpha_j}} = \overline{\overset{\leftrightarrow}{D}}_{(a,b)_{\alpha_j}} \tag{14}$$

The defuzzification of type-1 fuzzy then follows it to obtain the final result as the crisp T2FIBsS. Therefore, the defuzzification of type-reduced fuzzified T2FIBsS is given by the following equation.

$$\overline{\overline{BsSi}}_{\alpha_j}(s_a, t_b) = \sum_{l=0}^{n}\sum_{k=0}^{m} N_{k,p}(s_a)N_{l.q}(t_b)\overline{\overline{P}}_{(k,l)_{\alpha_j}} = \overline{\overline{D}}_{(a,b)_{\alpha_j}} \tag{15}$$

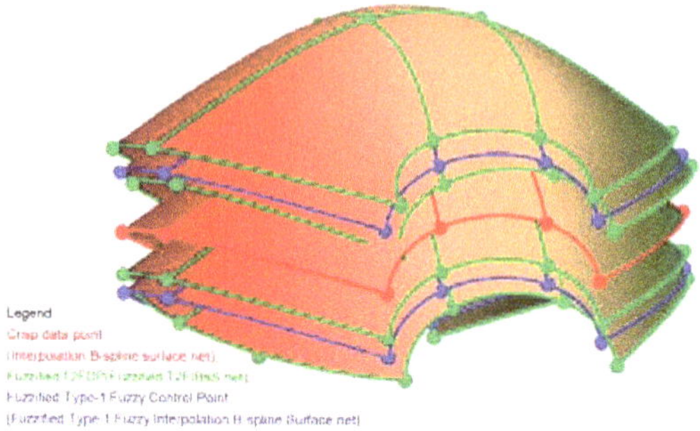

Figure 8. The example fuzzified T2FIBsS model with fuzzified type-2 fuzzy data net.

Then, the illustration of this type-reduction process can be illustrated in Figure 9 as follows.

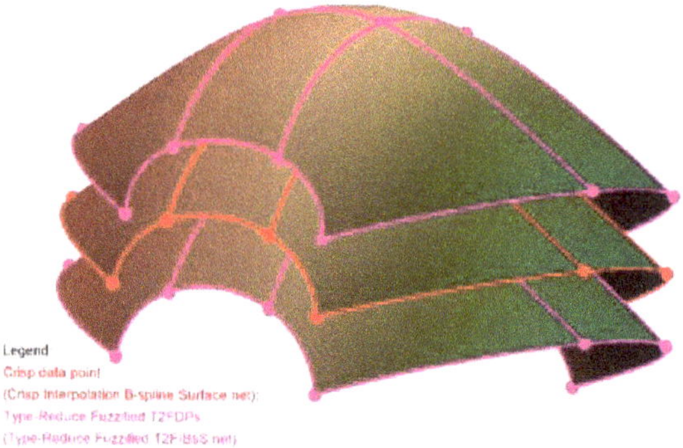

Figure 9. The example of type-reduced fuzzified T2FIBsS together with type-reduced fuzzified T2FDPs net.

The illustration of the defuzzification process against the type-reduced fuzzified T2FIBsS is shown in Figure 10.

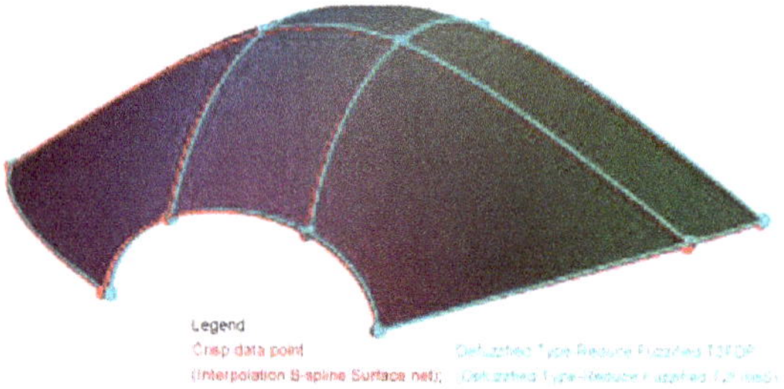

Figure 10. The example of defuzzification-reduced T2FIBsS with defuzzification-reduced T2FDPs.

Figure 10 shows that the defuzzification-reduced T2FIBsS model along with crisp interpolation B-spline surface. Both surfaces were built by finding the control points that interpolates data. The crisp data points were marked by red and the defuzzfication-reduced T2FDPs were marked by cyan.

5. Application: Seabed Modeling

This section discusses the practical application of the T2FIBsS model of seabed modeling. In the seabed modeling, multiple uncertainties or errors occur due to the nature of collecting data points, i.e., the wavy water surface and the collector's uncertain perception and truth level. Thus, the errors of the data collected are bounded by two uncertainties. This scenario is illustrated in Figure 11, which is also the extension of uncertainty data for lakebed modeling [4].

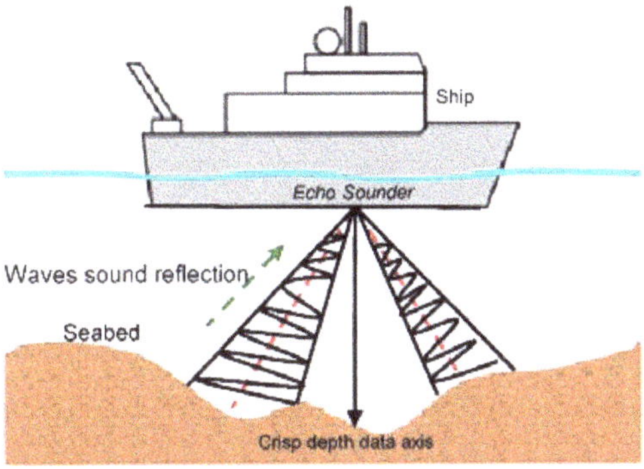

Figure 11. Illustration of procedure in taking depth data point (in meter) of the seabed, which consists of uncertainty complex data.

Figure 11 shows that obtaining the depth sea by means of an echo sounder where the data points obtained have the complex uncertainty properties. Therefore, it necessary to use the T2FIBsS method to model those complex uncertainty data after the data points are defined through the T2FST.

The following algorithm shows a step-by-step process of defining uncertainty complex data until the final stage of defuzzification-reduced T2FDP B-spline surface interpolation function.

The result of Algorithm 1 is illustrated in Figures 12–14 as follows.

Figure 12 shows the implementation of T2FIBsS against the seabed data modeling. In getting the final output named as crisp T2FIBsS, the fuzzifying process involves the output of T2FIBsS for seabed surface from the complex uncertainty data. Then, the fuzzification process is applied to shorten the T2FIBsS interval with the alpha value of 0.2 by using the alpha-cut of a triangular type-2 fuzzy number. After the fuzzification process, the following process applied is the type-reduction process, which reduces the type-2 fuzzy to become type-1 fuzzy to allow the defuzzfication process. Later, the defuzzification process is applied to obtain the crisp type-2 fuzzy solution that shows the distinct surface.

For Figures 13 and 14, getting the crisp T2FIBsS is the same as the processes in Figure 12. The only difference between all these three figures are the alpha values of the fuzzification process. The main focus is on showing the relation between alpha values crisp T2FIBsS output, and crisp B-spline surface seabed modeling, where the crisp T2FIBsS also tends to crisp B-spline surface seabed modeling when the alpha values are used to one side. On the other hand, percentage errors of all three figures have also demonstrated that the errors are smaller as the alpha-values increases.

Therefore, the different fuzzification processes involving alpha-cut, 0.2, 0.5 and 0.8 as shown in the figures. The average error percentage for each surface is 0.0018403 m, 0.0011502 m and 0.00046008 m, respectively. This average error percentage shows T2FDPs tends to crisp data points if their alpha values increase and tends to towards 1.

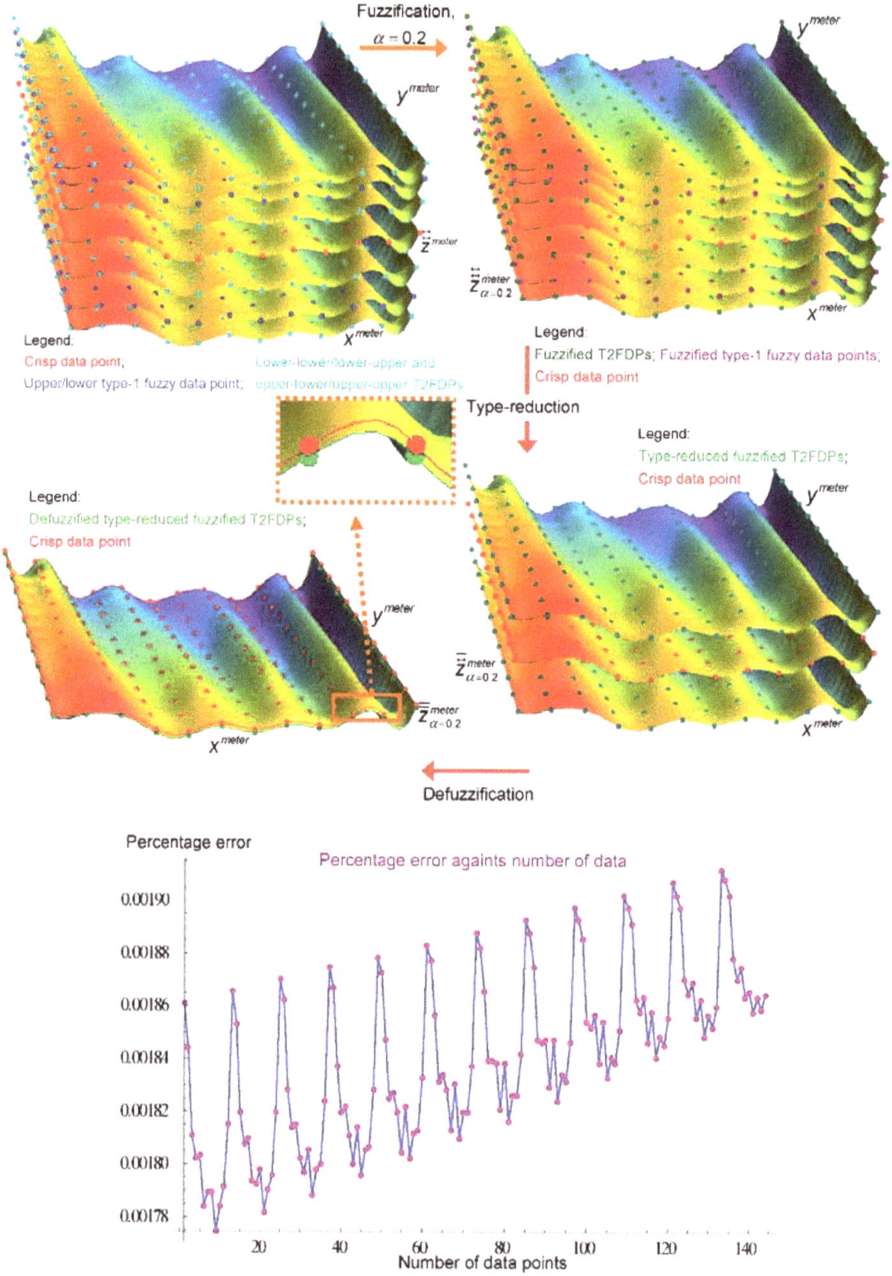

Figure 12. The T2FIBsS modeling through fuzzification until defuzzification processes along with the error plot.

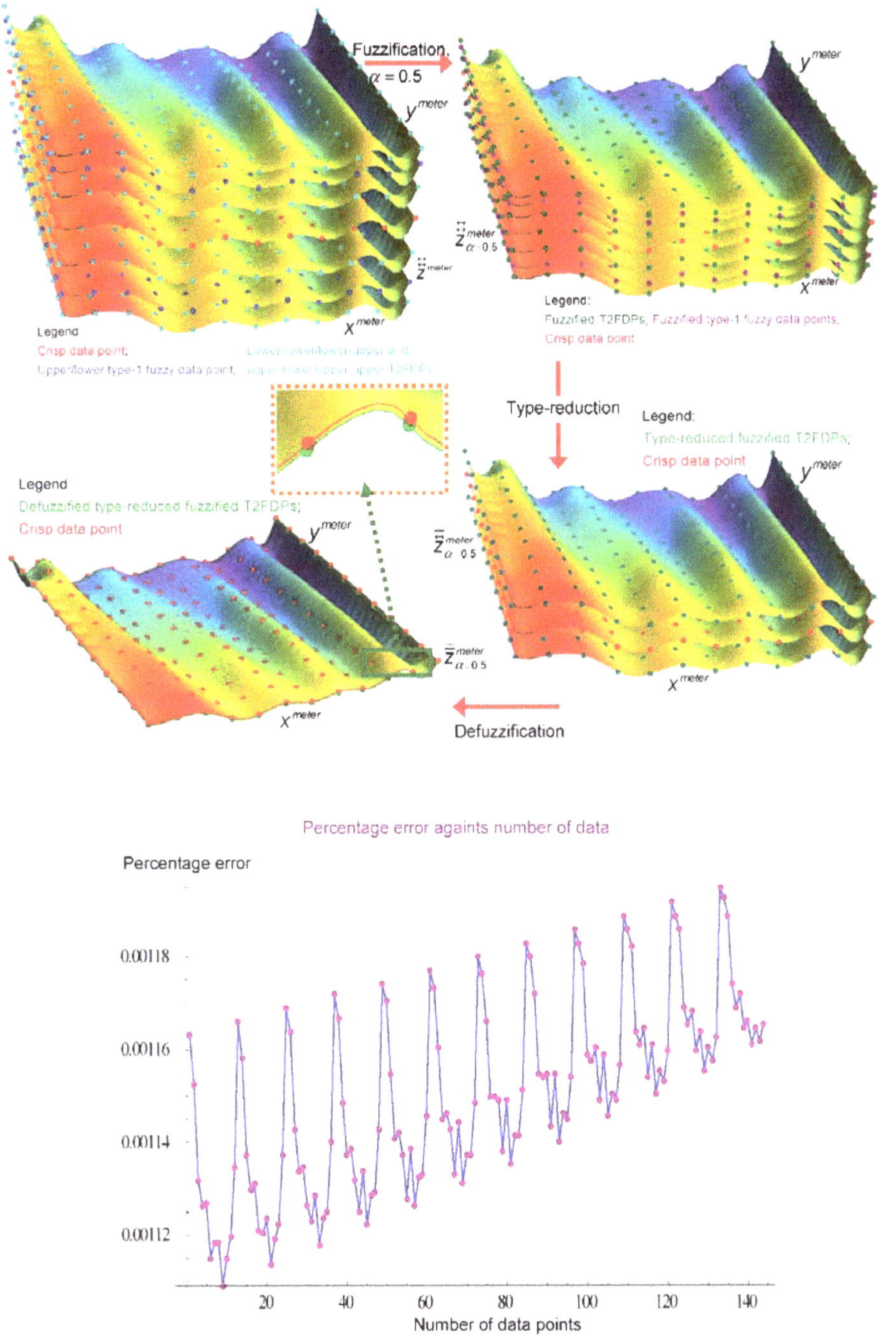

Figure 13. The T2FIBsS modeling through fuzzification until defuzzification processes with $\alpha = 0.5$ along the error plot.

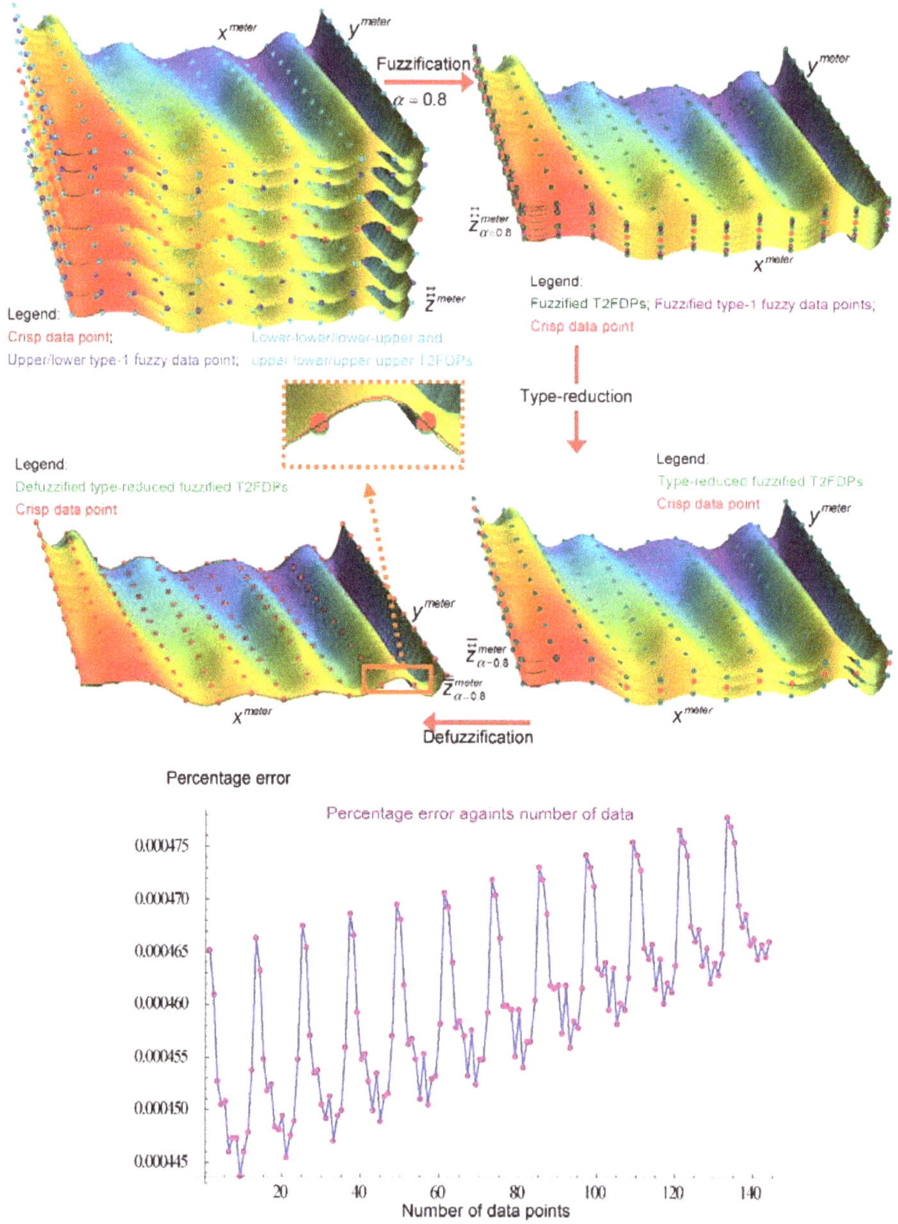

Figure 14. The T2FIBsS modeling through fuzzification until defuzzification processes with $\alpha = 0.8$ along the error plot.

Algorithm 1. Modeling T2FDP using interpolation type-2 B-spline surface.

Step 1: *Define the uncertainty complex data of seabed by using Definition 4.*
Step 2: *Use T2FDP ($\overset{\leftrightarrow}{\overset{\leftrightarrow}{D}}_{a,b}$) of seabed to solve Equation (12).*
Step 3: *Find the values of $N_{k,p}(s_a)$ after Equation (12) has been solved.*
Step 4: *Find the points of $\overset{\leftrightarrow}{\overset{\leftrightarrow}{P}}_{k,l}$ through Equation (10) by using Equation (12).*
Step 5: *Plot the $\overset{\leftrightarrow}{\overset{\leftrightarrow}{D}}_{(a,b)}$ and then model $\overset{\leftrightarrow}{\overset{\leftrightarrow}{P}}_{k,l}$ via Equation (10).*
Step 6: *Fuzzification process: Apply Equation (4), then apply step 2 until step 4 and model $\overset{\leftrightarrow}{\overset{\leftrightarrow}{P}}_{(k,l)_{\alpha_j}}$ as in Equation (13).*
Step 7: *Type-reduction process: Apply Equation (6), then apply step 2 until step 4 and model $\overline{\overset{\leftrightarrow}{P}}_{(k,l)_{\alpha_j}}$ as in Equation (14).*
Step 8: *Defuzzification process: Apply Equation (8), then apply step 2 until step 4 and model $\overline{\overline{P}}_{(k,l)_{\alpha_j}}$ by Equation (15).*

Step 9: *Find and plot the error between defuzzification-reduced T2FDPs and crisp data points of seabed depth data using the following equation:*

$$\frac{\sum\limits_{k=0,\ldots,n}^{n} {}^{L}D_{k_e}}{\sum\limits_{k=0,\ldots,n} n({}^{L}D_{k_e})} \text{ where } {}^{L}D_{k_e} = \frac{{}^{L}\overline{\overline{D}}_k - {}^{L}D_k}{{}^{L}D_k} \text{ with } k = 0, 1, 2, \ldots, n. \quad (16)$$

6. Discussion and Conclusions

This paper proposed a new method for defining complex uncertainty data, and modeled it on the hybrid model constructed called T2FIBsS. The complex uncertainty data was defined through IT2FN concepts obtaining T2FDPs and integrated with the B-spline surface function to produce the T2FIBsS model. This model has an advantage in modeling complex uncertainty data, as shown in seabed modeling. Meanwhile, the error and the percentage error have been calculated between defuzzification-reduced T2FDPs and crisp data points demonstrating the feasibility of the proposed model.

This developed model as T2FIBsS also deals with uncertainty modeling, which type-1 fuzzy modeling can do when the T2FIBsS first is reduced to the type-1 fuzzy model. However, the type-1 fuzzy model, which has been discussed in the literature, cannot be used in defining and modeling the complex uncertainty data. Therefore, T2FIBsS can be used to define and model the uncertainty and complex uncertainty compared to the type-1 fuzzy model, which only can be used to define and model the uncertainty data, but not applicable for complex uncertainty data.

The limitation of this study is the properties of the data that we want to model. The data that we obtained is secondary data, which had been filtered from raw data. Next is the surface function, which generates a desired surface based on the properties of the data. This function generates the surface entirety, rather than as a suitable surface patch. This surface that is generated entirety will make the surface sometimes unreasonable due to the data point position.

This research can be improved in the future by taking into account the complex uncertainty data, which has more uncertainty based on decision making or perceptions. This research can be expanded by using a complicated surface function, such as the Non-Uniform Rational B-spline (NURBS) surface, for modeling parts. The NURBS surface function has the advantage of local control on the surface other than the control points and knots, but the added value as weight in the NURBS formulation can modify the surface locally.

Author Contributions: R.Z., A.F.W. and M.I.E.Z. conceived of the presented the idea, conceptualization and methodology. I.I., gives the seabed data based on GIS from ArcGIS. All authors have read and agreed to the published version of the manuscript.

Funding: This research received no external funding and the APC was funded by Universiti Malaysia Sabah.

Institutional Review Board Statement: Not applicable.

Informed Consent Statement: Not applicable.

Data Availability Statement: Not applicable.

Acknowledgments: The authors acknowledge the Research and Innovation Management Centre (RIMC) of Universiti Malaysia Sabah (UMS) for funding this article and providing the facilities to conduct this research. They further acknowledge anonymous referees for their constructive comments which improved the readability of this article.

Conflicts of Interest: The authors declare no conflict of interest.

Notations Index

Notations	**Explanation**
$\widetilde{A}$	Type-2 fuzzy set (T2FS)
$\widetilde{A}_\alpha$	Interval T2FS
$\widetilde{R}$	Type-2 fuzzy relation
P	Data point
$\overleftrightarrow{\overleftrightarrow{P}}$	Type-2 fuzzy data point (T2FDP)
$\mu_P(P_i)$	Membership function of T2FDP
$\mu_P(\overleftarrow{\overleftrightarrow{\overleftrightarrow{P}}_i})$	Left footprint of membership values
$\mu_P(\overrightarrow{\overleftrightarrow{\overleftrightarrow{P}}_i})$	Right footprint of membership values
$\mu_P(a_i)$	Left-left membership grade value
$\mu_P(e_i)$	Left membership grade value
$\mu_P(b_i)$	Right-left membership grade value
$\mu_P(c_i)$	Left-right membership grade value
$\mu_P(f_i)$	Right membership grade value
$\mu_P(d_i)$	Right-right membership grade value
$\overleftrightarrow{\overleftrightarrow{P}}_{i_\alpha}$	Fuzzification process (alpha-cut process) against *i*th T2FDP
$\overleftarrow{\overleftrightarrow{\overleftrightarrow{P}}}_{i_\alpha}$	Left interval fuzzification process (alpha-cut process) against *i*th T2FDP
$\overrightarrow{\overleftrightarrow{\overleftrightarrow{P}}}_{i_\alpha}$	Right interval fuzzification process (alpha-cut process) against *i*th T2FDP
a_i^α	Left-left fuzzification process (alpha-cut process) against *i*th T2FDP
e_i^α	Left fuzzification process (alpha-cut process) against *i*th T2FDP
b_i^α	Right-left fuzzification process (alpha-cut process) against *i*th T2FDP
c_i^α	Left-right fuzzification process (alpha-cut process) against *i*th T2FDP
f_i^α	Right fuzzification process (alpha-cut process) against *i*th T2FDP
d_i^α	Right-right fuzzification process (alpha-cut process) against *i*th T2FDP
$\overline{\overleftrightarrow{P}}_\alpha$	Type-reduction of T2FDP after fuzzification process
$\overleftarrow{\overline{\overleftrightarrow{P}}}_{i_\alpha}$	Left type-reduction of *i*th T2FDP after fuzzification process
$\overrightarrow{\overline{\overleftrightarrow{P}}}_{i_\alpha}$	Right type-reduction of *i*th T2FDP after fuzzification process
$\overline{\overline{P}}_{i_\alpha}$	Defuzzification of *i*th T2FDP after type-reduction process
$\overleftrightarrow{\overleftrightarrow{D}}_i$	*i*th T2FDP of type-2 fuzzy interpolation B-spline curve (T2FIBsC)

Notations	Explanation
$\overset{\leftrightarrow}{\overset{\leftrightarrow}{BsC}}(t)$	T2FIBsC model
$\overset{\leftrightarrow}{\overset{\leftrightarrow}{P}}_i$	ith type-2 fuzzy control point (T2FCP) of T2FIBsS
$\overset{\leftrightarrow}{\overset{\leftrightarrow}{D}}_{a,b}$	T2FDP of Type -2 fuzzy interpolation B-spline surface (T2FIBsS)
$\overset{\leftrightarrow}{\overset{\leftrightarrow}{BsSi}}_{\alpha_j}(s_a, t_b)$	T2FIBsS function modeling after fuzzification process
$\overset{\leftrightarrow}{\overset{\leftrightarrow}{D}}_{(a,b)_{\alpha_j}}$	Fuzzification (alpha-cut operation) against T2FDP of T2FIBsS with $\alpha \in (0,1]$
$\overset{\leftrightarrow}{\overset{\leftrightarrow}{P}}_{(k,l)_{\alpha_j}}$	Fuzzification (alpha-cut operation) against T2FCP of T2FIBsS with $\alpha \in (0,1]$
$\overline{\overset{\leftrightarrow}{BsSi}}_{\alpha_j}(s_a, t_b)$	T2FIBsS function modeling after type-reduction process
$\overline{\overset{\leftrightarrow}{D}}_{(a,b)_{\alpha_j}}$	Type reduction process against T2FDP of T2FIBsS after fuzzification process
$\overline{\overset{\leftrightarrow}{P}}_{(k,l)_{\alpha_j}}$	Type reduction process against T2FCP of T2FIBsS after fuzzification process
$\overline{\overline{BsSi}}_{\alpha_j}(s_a, t_b)$	Defuzzify modeling of T2FIBsS after type-reduction process
$\overline{\overline{D}}_{(a,b)_{\alpha_j}}$	T2FDP defuzzification of T2FIBsS
$\overline{\overline{P}}_{(k,l)_{\alpha_j}}$	T2FCP defuzzification of T2FIBsS
$\overset{\leftrightarrow}{\overset{\leftrightarrow}{z}}^{meter}$	T2FIBsS model of Seabed data collection (depth) in meter
$\overset{\leftrightarrow}{\overset{\leftrightarrow}{z}}^{meter}_{\alpha=0.2}$	T2FIBsS fuzzification model with the alpha value is 0.2
$\overline{\overset{\leftrightarrow}{z}}^{meter}_{\alpha=0.2}$	T2FIBsS type-reduction model after fuzzification process
$\overline{\overline{z}}^{meter}_{\alpha=0.2}$	T2FIBsS defuzzification model after type-reduction process

References

1. Zulkifly, M.I.E.; Wahab, A.F.; Zakaria, R. B-Spline Curve Interpolation Model by using Intuitionistic Fuzzy Approach. *IAENG Int. J. Appl. Math.* **2020**, *50*, 1–7.
2. Bidin, M.S.; Wahab, A.F.; Zulkifly, M.I.E.; Zakaria, R. Generalized Fuzzy Linguistic Cubic B-spline Curve Model for Uncertainty Fuzzy Linguistic Data. *Adv. Appl. Discret. Math.* **2020**, *25*, 285–302. [CrossRef]
3. Adesah, R.S.; Zakaria, R. The Definition of Complex Uncertainties in B-spline Surface by using Normal Type-2 Triangular Fuzzy Number. *ASM Sci. J.* **2020**, *13*, 1–8. [CrossRef]
4. Zakaria, R.; Wahab, A.F.; Gobithaasan, R.U. Fuzzy B-Spline Surface Modeling. *J. Appl. Math.* **2014**, *2014*, 8. [CrossRef]
5. Aguero, J.R.; Vargas, A. Calculating Functions of Interval Type-2 Fuzzy Numbers for Fault Current Analysis. *IEEE Trans. Fuzzy Syst.* **2007**, *15*, 31–40. [CrossRef]
6. Coupland, S.; John, R. An Approach to Type-2 Fuzzy Arithmetic. In Proceedings of the UK Workshop on Computational Intelligence, Bristol, UK, 1–3 September 2003; pp. 107–114.
7. Dinagar, D.S.; Anbalagan, A. A New Type-2 Fuzzy Number Arithmetic Using Extension Principle. In Proceedings of the International Conference on Advances in Engineering, Science and Management (ICAESM), Tamil Nadu, India, 30–31 March 2012; pp. 113–118.
8. Zadeh, L.A. The concept of a linguistic variable and its application to approximate reasoning-Part I. *Inf. Sci.* **1975**, *8*, 199–249. [CrossRef]
9. Zadeh, L.A. The concept of a linguistic variable and its application to approximate reasoning-Part II. *Inf. Sci.* **1975**, *8*, 301–357. [CrossRef]
10. Zadeh, L.A. The concept of a linguistic variable and its application to approximate reasoning-Part III. *Inf. Sci.* **1975**, *9*, 43–80. [CrossRef]
11. Hu, J.; Chen, P.; Yang, Y. An Interval Type-2 Fuzzy Similarity-Based MABAC Approach for Patient-Centered Care. *Mathematics* **2019**, *7*, 140. [CrossRef]
12. Rogers, D.F. *An Introduction to NURBS: With Historical Perspective*; Academic Press: San Diego, CA, USA, 2001. [CrossRef]
13. Farin, G. *Curves and Surfaces for CAGD: A Practical Guide*, 5th ed.; Academic Press: San Diego, CA, USA, 2002.
14. Salomon, D. *Curves and Surfaces for Computer Graphics*; Springer: New York, NY, USA, 2006.
15. Aminifar, S.; Marzuki, A. Uncertainty in Interval Type-2 Fuzzy System. *Math. Probl. Eng.* **2013**, *2013*, 16. [CrossRef]
16. Nie, M.; Tan, W.W. Modeling Capability of Type-1 Fuzzy Set and Interval Type-2 Fuzzy Set. In Proceedings of the IEEE International Conference on Fuzzy Systems (FUZZ-IEEE), Brisbane, QLD, Australia, 10–15 June 2012; pp. 1–8. [CrossRef]

17. Shen, W.; Mahfouf, M. Multi-Objective Optimisation for Fuzzy Modelling Using Interval Type-2 Fuzzy Sets. In Proceedings of the IEEE International Conference on Fuzzy Systems (FUZZ-IEEE), Brisbane, QLD, Australia, 10–15 June 2012; pp. 1–8. [CrossRef]
18. Liang, Q.; Mendel, J.M. Interval type-2 fuzzy logic systems: Theory and design. *IEEE Trans. Fuzzy Syst.* **2000**, *8*, 535–550. [CrossRef]
19. Türk, S.; Deveci, M.; Özcan, E.; Canıtez, F.; John, R. Interval type-2 fuzzy sets improved by Simulated Annealing for locating the electric charging stations. *Inf. Sci.* **2021**, *547*, 641–666. [CrossRef]
20. Karagöz, S.; Deveci, M.; Simic, V.; Aydin, N. Interval type-2 Fuzzy ARAS method for recycling facility location problems. *Appl. Soft Comput.* **2021**, *102*, 107107. [CrossRef]
21. Gallo, G.; Spagnuolo, M.; Spinello, S. Fuzzy B-Splines: A Surface Model Encapsulating Uncertainty. *Graph. Models* **2000**, *62*, 40–55. [CrossRef]
22. Gallo, G.; Spagnuolo, M.; Spinello, S. Rainfall Estimation from Sparse Data with Fuzzy B-Splines. *J. Geogr. Inf. Decis. Anal.* **1998**, *2*, 194–203.
23. Sarwar, M.; Akram, M. Certain Algorithms for Modeling Uncertain Data Using Fuzzy Tensor Product Bézier Surfaces. *Mathematics* **2018**, *6*, 42. [CrossRef]
24. Du, G.N.; Zhu, Z.Y. Modelling spatial vagueness based on type-2 fuzzy sets. *J. Zhejiang Univ. Sci. A* **2006**, *7*, 250–256. [CrossRef]
25. Zakaria, R.; Wahab, A.F.; Gobithaasan, R.U. Perfectly Normal Type-2 Fuzzy Interpolation B-spline Curve. *Appl. Math. Sci.* **2013**, *7*, 1043–1055. [CrossRef]
26. Mendel, J.M. *Uncertain Rule-Based Fuzzy Logic Systems: Introduction and New Directions*; Prentice Hall PTR: Upper Saddle River, NJ, USA, 2001.
27. Nie, M.; Tan, W.W. Towards an Efficient Type-Reduction Method for Interval Type-2 Fuzzy Logic Systems. In Proceedings of the IEEE International Conference on Fuzzy Systems (FUZZ-IEEE), IEEE World Congress on Computational Intelligence, Hong Kong, China, 1–6 June 2008; pp. 1425–1432. [CrossRef]
28. Greenfield, S.; Chiclana, F.; John, R. Type-Reduction of the Discretised Interval Type-2 Fuzzy Set. In Proceedings of the IEEE International Conference on Fuzzy Systems (FUZZ-IEEE), Jeju, Korea, 20–24 August 2009; pp. 738–743. [CrossRef]
29. Runkler, T.A.; Chen, C.; John, R. Type reduction operators for interval type–2 defuzzification. *Inf. Sci.* **2018**, *467*, 464–476. [CrossRef]
30. Karnik, N.N.; Mendel, J.M. Centroid of a type-2 fuzzy set. *Inf. Sci.* **2001**, *132*, 195–220. [CrossRef]
31. Runkler, T.A.; Coupland, S.; John, R. Properties of Interval Type-2 Defuzzification Operators. In Proceedings of the 2015 IEEE International Conference on Fuzzy Systems (FUZZ-IEEE), Istanbul, Turkey, 2–5 August 2015; pp. 1–7. [CrossRef]
32. Zakaria, R.; Wahab, A.F. Pemodelan Titik Data Kabur Teritlak. *Sains Malays.* **2014**, *43*, 799–805.
33. Wahab, A.F.; Ali, J.M.; Majid, A.A.; Tap, A.O.M. Fuzzy Set in Geometric Modeling. In Proceedings of the International Conference on Computer Graphics, Imaging and Visualization (CGIV 2004), Penang, Malaysia, 2–2 July 2004; pp. 227–232. [CrossRef]
34. Karim, N.A.A.; Wahab, A.F.; Gobithaasan, R.U.; Zakaria, R. Model of Fuzzy B-Spline Interpolation for Fuzzy Data. *Far East J. Math. Sci. (FJMS)* **2013**, *72*, 269–280.

MDPI
St. Alban-Anlage 66
4052 Basel
Switzerland
Tel. +41 61 683 77 34
Fax +41 61 302 89 18
www.mdpi.com

Mathematics Editorial Office
E-mail: mathematics@mdpi.com
www.mdpi.com/journal/mathematics

www.ingramcontent.com/pod-product-compliance
Lightning Source LLC
LaVergne TN
LVHW070126100526
838202LV00016B/2235